NATO ASI Series

Advanced Science Institutes Series

A Series presenting the results of activities sponsored by the NATO Science Committee, which aims at the dissemination of advanced scientific and technological knowledge, with a view to strengthening links between scientific communities.

The Series is published by an international board of publishers in conjunction with the NATO Scientific Affairs Division

A Life Sciences B Physics	Plenum Publishing Corporation London and New York
C Mathematical and Physical Sciences D Behavioural and Social Sciences E Applied Sciences	Kluwer Academic Publishers Dordrecht, Boston and London
F Computer and Systems Sciences G Ecological Sciences H Cell Biology	Springer-Verlag Berlin, Heidelberg, New York, London, Paris and Tokyo

Series E: Applied Sciences Vol. 140

Mathematical Modeling in Combustion and Related Topics

edited by

Claude-Michel Brauner and Claudine Schmidt-Lainé

Département de Mathématiques – Informatique – Systèmes et Unité Associée au CNRS 740, Ecole Centrale de Lyon, Ecully Cedex, France

1988 **Martinus Nijhoff Publishers**
Dordrecht / Boston / Lancaster
Published in cooperation with NATO Scientific Affairs Division

Proceedings of the NATO Advanced Research Workshop on
"Mathematical Modeling in Combustion and Related Topics",
Lyon, France,
April 27–30, 1987

NATO Advanced Research Workshop on "Mathematical Modeling in Combustion and Related Topics" (1987 : Lyon, France)
Mathematical modeling in combustion and related topics / editors, Claude-Michel Brauner and Claudine Schmidt-Lainé.
p. cm. -- (NATO ASI series. Series E, Applied sciences ; no. 140)
"Proceedings of the NATO Advanced Research Workshop on 'Mathematical Modeling in Combustion and Related Topics,' Lyon, France, April 27-30, 1987"--T.p. verso.
Includes bibliographies and index.
ISBN 9024736560
1. Combustion--Mathematical models--Congresses. 2. Flame--Mathematical models--Congresses. 3. Fluid dynamics--Mathematical models--Congresses. 4. Ludford, G. S. S.--Congresses. I. Brauner, Claude-Michel, 1947- II. Schmidt-Lainé, Claudine. III. Title. IV. Series.
QD516.N317 1987
541.3'61'0724--dc19 87-34798
CIP

ISBN 90-247-3656-0
ISBN 90-247-2689-1 (series)

Distributors for the United States and Canada: Kluwer Academic Publishers, 101 Philip Drive, Norwell, MA 02061, USA

Distributors for the UK and Ireland: Kluwer Academic Publishers, MTP Press Ltd, Falcon House, Queen Square, Lancaster LA1 1RN, UK

Distributors for all other countries: Kluwer Academic Publishers Group, Distribution Center, P.O. Box 322, 3300 AH Dordrecht, The Netherlands

Printed in The Netherlands

This book is dedicated to the memory of Geoffrey S. S. Ludford

Geoffrey S. S. Ludford
(1928–1986)

PREFACE

This volume contains invited lectures and contributed papers presented at the NATO Advanced Research Workshop on Mathematical Modeling in Combustion and related topics, held in Lyon (France), April 27 - 30, 1987. This conference was planned to fit in with the two-month visit of Professor G.S.S. Ludford to the Ecole Centrale de Lyon. He kindly agreed to chair the Scientific and Organizing Committee and actively helped to initiate the meeting. His death in December 1986 is an enormous loss to the scientific community in general, and in particular, to the people involved in the present enterprise.

The subject of mathematical modeling in combustion is too large for a single conference, and the selection of topics reflects both areas of recent research activity and areas of interest to Professor G.S.S. Ludford, to whose memory the Advanced Workshop and this present volume are dedicated.

The meeting was divided into seven specialized sessions : detonation theory, mathematical analysis, numerical treatment of combustion problems, flame theory, experimental and industrial aspects, complex chemistry, and turbulent combustion. It brought together researchers and engineers from University and Industry (see below the closing remarks of the workshop by Prof. N. Peters). The articles in this volume have been judged and accepted on their scientific quality, and language corrections may have been sacrificed in order to allow quick dissemination of knowledge to prevail.

Thanks are due to many people for the role that they played in the success of the meeting. These include the speakers ; the chairs (C. Bardos, P. Clavin, G. Comte-Bellot, G. Duvaut, P. Fife, A. Linan, A. Samarskii, D. Serre); the contributors to the panel on the modeling of industrial problems (P.A. Baudart, G. Bayle-Labouré, G. Fratacci, X. Lys, and A. Moiroux) ; those who contributed to the organization, especially Francoise Chatelin, Florence Garcia, Ms. Gouteyron, Michel Teissieux, and Sylvie Thiollier ; and all participants for their enthusiasm.

In addition to the financial support of the NATO Office of Scientific Affairs, this meeting received the sponsorship of several organizations and industries. Their support is gratefully acknowledged.

Finally we thank our spouses Danielle and Pierre for their understanding and patience during the preparation of the conference.

Claude-Michel Brauner
ARW-Director

Claudine Schmidt-Lainé
ARW-Director

Closing remarks by Professor Norbert Peters

In resuming the scientific contents of this conference, I first want to point out that it was the merit of C.M. Brauner and Claudine Schmidt-Lainé to bring mathematicians and combustion scientists together in this interdisciplinary workshop. Very often mathematicians do excellent but not very applied work, while engineers, in developing very useful approximate methods, sometimes lack the necessary mathematical rigour. In view of the development of high speed computers I believe that there is a decade of rapid development of numerical methods in engineering science ahead of us. Therefore the collaboration between applied mathematicians and engineers is extremely necessary. In developing this interaction this workshop has achieved a great deal. I also want to mention that in serving the development of science, the results of this interaction will be appreciated by the industry and also by those engineers who work for the defence of our countries.

Scientific and Organizing Committee

Ph.Arques (ECL), G.Bayle-Labouré (SNECMA), R.Borghi (Université Rouen), C.M.Brauner (ECL), M.Brun (ECL), S.Candel (ECP/ONERA), G.Duvaut (Université Paris VI/ONERA), D.Escudié (ECL), P.Fife (University Arizona), D.Jeandel (ECL), P.Lallemand (DRET), B.Larrouturou (INRIA), G.S.S.Ludford (Chairman of Committee), X.Lys (Renault), B.Nicolaenko (Los Alamos), R.Pellat (CNES), J.Périaux(AMD-BA), Cl.Schmidt-Lainé (CNRS,ECL), D.Serre (ENS Lyon), R.Temam (Université Paris XI)

Sponsors

Aderly Lyon-Technopolys, Commissariat à l'Energie Atomique (CEA), Centre National d'Etudes Spatiales (CNES), Centre National de la Recherche Scientifique (CNRS), Conseil Général du Rhône, Crédit du Nord, Direction des Recherches, Etudes et Techniques (DRET), Electricité de France (EDF), Groupe pour l'Avancement des Méthodes Numériques de l'Ingénieur (GAMNI), Office National d'Etudes et de Recherches Aéronautiques (ONERA), Société de Mathématiques Appliquées et Industrielles (SMAI), SNECMA, Technology Resources, Turboméca, U.S.Army.

TABLE OF CONTENTS

REMINISCENCES ON THE LIFE AND WORK OF G.S.S. LUDFORD[1]

J. Buckmaster

University of Illinois

As many of you know, Geoffrey Ludford had strong professional ties with France, and was chairman of the organizing committee for this meeting, so it is fitting that we should start by recalling something of the man and his life. This task has fallen to me since I probably knew him for as long and as well as anyone here. In fact, I knew him for nearly 22 years. We first met in the spring of 1965. I was working in Buffalo, New York, at Cornell Aeronautical Laboratory, owned at that time by Cornell University. Geoffrey was a professor at Cornell and that spring he periodically flew up from Ithaca to teach a graduate class on tensor analysis and the calculus of variations. It was a remarkably well-taught class in which the material was presented in an attractively pragmatic, no-nonsense style. It was not surprising then that when I went to Cornell University that fall to start my graduate studies, I asked him to be my advisor.

My relationship with him as a student was conventional. When I heard the familiar "slap-slap" of his sandles in the corridor outside my office, I would make sure I looked busy, hastily discarding such things as newspapers. He didn't frighten or intimidate, but he created a serious atmosphere which encouraged dedication.

After I got my degree, I returned to the world and my contact with Geoffrey was fairly limited until 1974, when I faced a minor crisis. I was at Yale, preparing to move to Illinois, and because of the different fiscal calendars of these institutions, found myself facing a three-month period with no salary and with two small children to feed. Somehow Geoffrey found this out, or guessed it, because he telephoned--and I was first introduced to an essential part of his character. If you were his student, or friend, Geoffrey regarded the relationship as a commitment for life, and if he could do anything to help when needed, he would. So I spent that summer at Cornell, supported by his ARO grant, and my children didn't starve.

In the obituary that was printed in the March 1987 SIAM News, I wrote, "We have lost a remarkable colleague, friend, father, and husband, a man whose most striking attribute was an unswerving and selfless concern for his family, his friends, his students, his science, and his profession." These words came straight from the heart. One little known example of this side of his nature was his efforts on behalf of Russian refuseniks--Jewish scientists denied exit visas. He was active in generating political support for these unfortunate people; he once attended the famous weekly Moscow Seminar; and he wrote to them, often sending gifts--a book or calculator--paid for from his own pocket.

1 A -- more-or-less -- verbatim record of the oral presentation, recalled from memory.

C.-M. Brauner and C. Schmidt-Lainé (eds.), Mathematical Modeling in Combustion and Related Topics, 1–7.

Let me now say something about his early life. He was born in 1928 on February 2--Ground Hog's Day in the United States, but of no special significance in England so far as I know--and grew up in a middle-class home in London. Early on he was exposed to a singular example of the value of mathematics. His father's job required the execution of a large number of numerical multiplications, and this was done by longhand until his father introduced logarithm tables to his employers. The latter were so impressed by the increase in efficiency, they offered to make him a director! One can imagine Geoffrey, as a young man, thinking that if logarithms could be so useful, what might be possible with calculus!

I don't know much about his schooling, but presumably it was typical for a very bright boy from his class. He went to a grammar school at the age of 10 or 11, a school designed for university preparation, and in addition to the usual subjects would have taken Latin, since in those days that was required for Oxford and Cambridge entrance in both the sciences as well as the humanities. As an outstanding student, he would have been coached for the Oxbridge entrance examinations, a successful endeavor, since he went to Cambridge to read mathematics, obtaining his degree in 1948 at the age of 20.

Whilst at Cambridge he was a rower, and since he never did anything by halves, a very successful one. In 1947 he rowed for the winning crew in the Thames Challenge Cup race of the Henley Regatta. This was very high-class rowing but, in later years, if one remarked on this accomplishment he would shrug it off with the suggestion that in the immediate post-war years nobody was interested in rowing so the competition was weak. In 1949, as a first-year graduate student at Cambridge, he rowed (again victoriously) in the annual Oxford-Cambridge boat race.

Rowing, like cross-country skiing, or professional-bicycle-road-racing is an activity that requires the ability to sustain enormous power output, and to endure the pain that goes along with this. There is a natural tendency for such athletes to "let things go" when they retire. Many a one-time Tour de France rider has a substantial belly in middle-age. This was not Geoffrey's style however. He sustained a remarkable level of fitness throughout his life. During the years that I knew him, I occasionally was approached by someone asking why Geoffrey looked ill. He looked ill for the same reason that a marathon runner looks ill--he didn't have an extra ounce of fat on his body. I shall always remember the time that he spent in hospital some weeks before his death. Everyday he was required to blow into a device that tested his lung capacity. This included a small ball in a tube. The tube was graduated and there was a table of expected performance as a function of age, sex, and height. The topmost mark was for a 21 year old male, 6-$^1/_2$ feet tall, or thereabouts. Everyday, with little discernable effort, Geoffrey blew the ball past that mark to the top of the tube. We were not to be impressed by this, he argued, since there was a "trick" to such things.

But to return to his early years: Geoffrey commenced his graduate studies at Cambridge with Howarth but found no mutual interests and so moved to Cambridge, Massachusetts to work with Richard von Mises at Harvard. (For the record, although von Mises supervised his thesis research, all three of his degrees were from Cambridge.) At some time between the mathematics and the rowing he found time for romance and marriage, and his wife accompanied him to the United States.

It is clear that the association with von Mises was in some ways special. In the early papers of the '50's von Mises is often mentioned, and after the latter's death Geoffrey collaborated with Hilda Geiringer, von Mises's wife, and herself an accomplished mathematician, in the completion of von Mises's Mathematical Theory of Compressible Flow.

With a Ph.D. in hand at age 23, Geoffrey took up his first academic appointment at the University of Maryland. Most of his early work was in compressible flow, a subject that I am not expert in, so rather than try to identify the most important work from that era, I have just extracted a number of examples which looked interesting and representative from the large pile of reprints (160 or so) that I have in my office.

Structure of Shock Waves Using the Navier-Stokes Equations

The one-dimensional, steady, Navier-Stokes equations can be written in the form

$$\frac{dt}{dv} = \frac{4P}{3\gamma} \cdot \frac{t - t_0}{t - t_\infty}$$

where P is the Prandtl number, γ the specific heat ratio, t is essentially the temperature, v essentially the square of the velocity, and

$$t_0 = (\gamma - 1)(v - \sqrt{2v} + c) \quad , \quad t_\infty = \sqrt{2v} - 2v$$

where c is a constant. A complete discussion of the integral curves is given.

Subsonic Compressible Joukowsky Theorem

For incompressible flow past a body, it was well known that the complex velocity potential has the far-field representation

$$w(z) \sim A_0 z + B_0 \ln z + C_0/z + 0\ (1/z^2)$$

where B_0 and C_0 are, respectively, proportional to the lift and the moment acting on the body. For subsonic compressible flow with no shocks, the analogous result for the potential is

$$\Phi/U = r\cos\theta + A \ln r + \sum_{m=0}^{\infty} \sum_{n=0}^{\infty} f_{mn}(\theta)\left(\frac{\ln r}{r}\right)^m \frac{1}{r^n} \ .$$

The logarithmic terms are necessary to ensure that the coefficient functions are periodic in θ .

Interior Sphere Problem

Suppose $\phi_0(r) \equiv \phi_0(r,\theta,\phi)$ is the potential due to a distribution of total mass zero inside a sphere $r = a$. Then

$$\phi_1 = \frac{a}{r}\,\phi_0(a^2/r) + \frac{2}{ar}\int_a^\infty \lambda\phi_0(\lambda^2/r)\,d\lambda$$

is harmonic in $r \leq a$, and $\frac{\partial}{\partial n}(\phi_0 + \phi_1) = 0$ on $r = a$.

Extension of Riemann's Method

Several of the early papers are concerned with extensions of classical results for the inital value problem defined by

$$u_{xt} + a(x,t)u_x + b(x,t)u_t + c(x,t)u = 0 ,$$

$$u_x, u_t \text{ prescribed on a curve } C .$$

These extensions include:

(i) C is tangential to a characteristic at some point;

(ii) C and the data have certain singularities;

(iii) Some characteristic meets C in more than one point.

A discussion of the nonuniqueness is the essential ingredient of (i) and (iii) (these cases don't arise in the physical plane of course).

These and other contributions were well received by the scientific and university community. By 1955 Geoffrey had tenure, and in 1957, at the age of 29, he won a Guggenheim fellowship.

In 1960 he moved to Cornell, by which time magneohydrodynamics engaged most of his scientific attentions. Here I shall describe what I think was one of his most important contributions.

During the time that MHD was a fashionable subject, a lot of work was concerned with the limit of infinite magnetic Reyonlds number $(Rm \to \infty)$, corresponding to perfect conductivity $(\sigma \to \infty)$. A lot of elegant mathematics is possible in this limit and the results are striking, but unfortunately they are of little physical interest on anything but astrophysical scales--liquid conductivities, even of Hg and Na , are not particularly large. Geoffrey did his share of work in this area, but he also considered a much more important limit, namely $\sigma \to 0$, $\sigma H_0^2 \to \infty$ so that there is no induced magnetic field H but there is a significant ponderomotive force. As an example, consider the nondimensional equations governing inviscid steady MHD in the form

$$\vec{\nabla} \cdot \vec{v} = 0 , \qquad \vec{\nabla} \times \vec{E} = 0 , \qquad \vec{\nabla} \cdot \vec{H} = 0 ,$$

$$\vec{\nabla} \times \vec{H} = Rm(\vec{E} + \vec{v} \times \vec{H})$$

$$(\vec{v} \cdot \vec{\nabla})\vec{v} = - \vec{\nabla}p + N(\vec{E} + \vec{v} \times \vec{H}) \times \vec{H} .$$

The limit corresponds to $Rm \to 0$, N (the interaction parameter) $\to \infty$. In the case of unbounded two-dimensional flow over a body with a transverse magnetic field, these equations are (after the limit $Rm \to 0$)

$$\vec{H} = \vec{i} \quad , \quad \vec{v} - \vec{i} = (u,v)$$

$$\frac{\partial u}{\partial x} + u\frac{\partial u}{\partial x} + v\frac{\partial u}{\partial y} = - \frac{\partial p}{\partial x} - Nu ,$$

$$\frac{\partial v}{\partial x} + u\frac{\partial v}{\partial x} + v\frac{\partial v}{\partial y} = - \frac{\partial p}{\partial y} \quad , \quad \frac{\partial u}{\partial x} + \frac{\partial v}{\partial y} = 0 .$$

Then when $N \to \infty$ we have

$$P = Nf(x) , \qquad u = -f'(x) , \qquad v = yf''(x) + g(x)$$

where f,g are arbitrary functions. The tangency condition on the body is not sufficient to uniquely determine these functions and the solution can only be completed by considering a region of nonuniformity in the neighborhood of $y = \infty$. This region is characterized by the scalings

$$y = \sqrt{N}\, Y \; , \qquad u = U/\sqrt{N} \; , \qquad P = \sqrt{N}\, P \; , \qquad v = v$$

yielding the limit equation

$$\frac{\partial^3 v}{\partial x^3} + \frac{\partial^2 v}{\partial Y^2} = 0 \; .$$

This describes an inertial-Lorentz force balance. The streamlines eventually predicted by this analysis are sketched below.

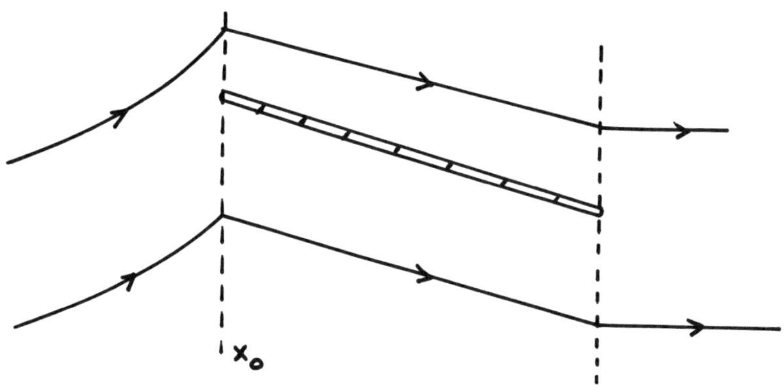

The discontinuity at $x = x_0$ can be resolved using new variables

$$x - x_0 = N^{-1/3} X \; , \qquad u = N^{-1/3} U \; , \qquad P = N^{1/3} P$$

which again leads to the equation

$$\frac{\partial^3 v}{\partial x^3} + \frac{\partial^2 v}{\partial y^2} = 0 \; ;$$

this is known as a Ludford layer.

The key idea here plays a central role in a large number of papers written in the '60's in collaboration with Julian Hunt and John Walker on MHD duct flows, work of enduring value and practical importance.

In 1974 Geoffrey started working in combustion. It is not necessary for me to describe his contributions in that subject to this audience, but I would like to pay tribute to the vision that led to our collaboration on Theory of Laminar Flames. In 1976, when neither of us knew much about combustion, Geoffrey suggested that we should write such a book. He argued that, as the use of asymptotics in combustion grew, we would have an increasing amount of material to draw upon, and eventually there would be enough for a complete book. So he started to work on Chapter 1, the Introduction, whilst I started on Chapter 6, drawing on the pioneering work of Liñan on diffusion flames, all that could be done at that time. He was right of course, and eventually there was more than enough material.

It would be grossly misleading to describe Geoffrey's contributions to science without saying something of his role as a teacher. He created and taught a remarkable sequence of 'methods' courses at Cornell, courses which educated and influenced literally hundreds of science and engineering students. In addition, he supervised nearly 30 Ph.D. students, more than a third of whom are now on the faculties of leading American universities. It is fair to say that his influence on applied mathematics through his students is at least the equal of his influence through his scientific work.

Let me finish by reading part of a eulogy that I delivered at a memorial service on the Cornell campus shortly after his death. Attending that service were members of his family and close friends--people who loved him--and these words were addressed directly to them. I know that this audience also includes close friends.

"These past few months have been a difficult and sad time for all who cared for Geof, especially for Pamela, Susan, Jennifer and the rest of the family. As we now mourn his loss, there are two things that give me some comfort, make the pain a little less than it might otherwise be, and perhaps they can do the same for you.

Up to the time that Geof became ill, 1986 was a remarkable and wonderful year for him professionally. The Mathematical Sciences Institute was won by Cornell, largely through his efforts, and he was appointed its first director. He was extremely proud of that. When, during the early phase of the illness, we were discussing with the surgeon the different courses of action that might be taken, Geoffrey made a point of telling him that he was the director of the finest center for applied mathematics in North America, and possibly the world. This was not boasting--in 21 years I never heard Geoffrey brag. But he wanted the doctor to know the nature of the man he was treating, and he saw this as the crowning achievement of his career, a role that was the natural expression of his scientific talents and values. It gives me some comfort to know that he experienced that success, even though it was so tragically cut short.

One of the concerns that Geoffrey had when he first learned of his illness was that he would spend the last few weeks of his life, in hospital, hooked up by tubes and wires to various machines. He had a horror of that. He had a "living will" prepared and when the idea for that first came to him it had to be dealt with immediately, of course. Someone had to call his lawyer from the hopital right away. I think that if she had been in her office at that time, he would have insisted that she deliver the document that very hour. As it turned out, the will wasn't necessary. And his last few weeks were spent, not in hospital, but in a bed set up in his living room where he could see the birds at the feeder outside the window, which gave him a great deal of pleasure. And there, in familiar surrounding, Pam, helped by Sue and Jenny during their frequent visits, by friends, and by people from HospiCare, cared for him, comforted him, and surrounded him with love. Surely this was one of the finest gifts that human beings can give to another."

Bibliography

G.S.S. Ludford, "The behavior at infinity of the potential function of a two-dimensional subsonic compressible flow", J. Math. & Phys. XXX, No. 3., p. 117, 1951.

G.S.S. Ludford, "The classification of one-dimensional flows and the general shock problem of a compressible, viscous, heat-conducting fluid", J. Aero. Sci. 18, No. 12, p. 830, 1951.

G.S.S. Ludford, "The boundary layer nature of shock transition in a real fluid", Quart. Appl. Math. X, No. 1, p. 1, 1952.

G.S.S. Ludford, "On an extension of Riemann's method of integration, with applications to one-dimensional gas dynamics", Proc. Cambr. Phil. Soc. 48, p. 499, 1952.

G.S.S. Ludford, "Extensions on the applicability of Riemann's formula", J. Rat. Mech. and Anal. 3, No. 1, p. 77, 1954.

G.S.S. Ludford, J. Martinek, and G.C.K. Yeh, "The sphere theorem in potential theory", Proc. Cambr. Phil. Soc. 51, p. 389, 1955.

G.C.K. Yeh, J. Martinek, and G.S.S. Ludford, "A general sphere theorem for hydrodynamics, heat, magnetism, and electrostatics", ZAMM, 36, p. 111, 1956.

G.S.S. Ludford, "The effect of a very strong magnetic cross-field on steady motion through a slightly conducting fluid", J. Fluid Mech. 40, p. 141, 1961.

G.S.S. Ludford and D.W. Yannitell, "The super-aefvénic nature of Sears Resfer flow", J. de Mécanique 6, p. 11, 1967.

J.S. Walker and G.S.S. Ludford, "Three dimensional MHD duct flows with strong transonic magnetic fields. Part 4. Fully insulated, variable-area rectangular ducts with small divergences", J. Fluid Mech. 56, p. 481, 1972.

J. Buckmaster and G.S.S. Ludford, Theory of Laminar Flames, Cambridge University Press, NY, 1982.

PART I
INVITED LECTURES

A NONLINEAR ELLIPTIC PROBLEM DESCRIBING THE PROPAGATION OF A CURVED PREMIXED FLAME

H. BERESTYCKI[(1,2)], B. LARROUTUROU[(2)], L. NIRENBERG[(3)]

(1) Université Paris Nord, C.S.P., Département de Mathématiques, 93430 Villetaneuse, FRANCE.

(2) INRIA, Sophia-Antipolis, 06560 Valbonne, FRANCE.

(3) Courant Institute of Mathematical Sciences, NYU, 251 Mercer Street, New-York NY 10012, USA.

1. INTRODUCTION

We shall report here on some results and methods from [5], [8], [9] concerning some nonlinear elliptic equations arising from a two-dimensional combustion model. More precisely, we consider the following problem, set in the strip $S = \{(x,y) \in \mathbb{R}^2, 0 < y < L\}$:

$$\begin{aligned} &\Delta T - c\alpha(y) T_x + g(T) = 0 \ \text{ in } S , \\ &T(-\infty,y) = 0 , \ T(+\infty,y) = 1 \ \text{ for } 0 \le y \le L , \\ &T_y(x,0) = T_y(x,L) = 0 \ \text{ for } x \in \mathbb{R} , \end{aligned}$$

where α is a given function and g is some nonlinear source term satisfying certain conditions to be specified below. The unknowns are the positive parameter c and the function T which represents a "travelling front". This type of problem is studied in several contexts (see e.g. [2], [11]); in particular, it arises in the "thermo-diffusive" model of combustion for the propagation of a steady curved premixed flame; the unknown T then represents the temperature distribution. The derivation of this model is described in the next section.

Our main results concern:

* The existence of a solution [5];
* The monotonicity of the profile [8];
* The uniqueness of c and T (up to shifts of the origin) [9].

There is a vast literature devoted to problems of this type, but in a *one-dimensional* setting, that is for ordinary differential equations, dating back to the celebrated paper of Kolmogorov, Petrovskii and Piskunov [17] (we also mention for instance the works of Aronson and Weinberger [2], Berestycki, Nicolaenko and Scheurer [7], Johnson and Nachbar [13], Kanel [14], [15] and Marion [19]).

The results we present here are in some sense the first extensions to higher dimensions and partial differential equations of those previous works. It is hoped that we have

C.-M. Brauner and C. Schmidt-Lainé (eds.), Mathematical Modeling in Combustion and Related Topics, 11-28.

now come to a point when the introduction of these new methods from nonlinear partial differential equations might prove useful in furthering the understanding of the mathematical models of combustion theory.

REMARK 1: The classical singular limit of the above problem corresponding to an activation energy becoming infinite leads to a free boundary problem: the free boundary is actually the flame front. The existence of a singular limit is derived in [9]. The uniqueness of the solution to the free boundary problem (even in a weak sense), as well as its regularity, are established in [4]. •

REMARK 2: The results of this paper hold if the strip S is replaced by an infinite cylindrical domain $\mathbb{R} \times \omega$ in any dimension n ($\omega \subset \mathbb{R}^{n-1}$) with Neumann boundary conditions on $\mathbb{R} \times \partial\omega$. For the sake of simplicity in the presentation, we only state the results for the case $n = 2$ and $\omega = (0,L)$. •

2. FORMULATION OF THE PROBLEM

In this section, we briefly present the classical models used in combustion theory to describe the two-dimensional propagation of a curved premixed flame, and we formulate the mathematical problem which is considered in this paper.

Consider a wrinkled flame propagating in a gaseous mixture confined in an infinite rectangular channel $S = \{(x,y) \ , \ x \in \mathbb{R} \ , \ 0 < y < L\}$ of width L. The same model would arise in the case of a three-dimensional tube (see Remark 2 above). If one assumes for simplicity that a single one-step chemical reaction $A \rightarrow B$ takes place in the gas, the phenomenon is described by the following system of governing equations (see [10], [22]):

$$\begin{aligned}
&\rho_t + \vec{\nabla}.(\rho\vec{V}) = 0 \ , \\
&(\rho u)_t + \vec{\nabla}.(\rho u\vec{V}) + p_x = 0 \ , \\
&(\rho v)_t + \vec{\nabla}.(\rho v\vec{V}) + p_y = 0 \ , \\
&\rho c_p \Theta_t + \rho c_p \vec{V}.\vec{\nabla}\Theta = \vec{\nabla}.(\lambda\vec{\nabla}\Theta) + p_t + \vec{V}.\vec{\nabla}p + Q\omega(\Theta,Y) \ , \\
&\rho Y_t + \rho\vec{V}.\vec{\nabla}Y = \vec{\nabla}.(\rho D\vec{\nabla}Y) - m\omega(\Theta,Y) \ , \\
&mp = \rho R\Theta \ .
\end{aligned} \tag{1}$$

The notations are the following: ρ is the density, $\vec{V} = (u,v)$ the velocity, p the pressure, Θ the temperature of the mixture, and Y represents the mass fraction of the reactant A. Furthermore, c_p is the specific heat at constant pressure of the mixture, λ the heat conductivity, D the diffusion coefficient of the reactant A, m its molecular mass, and R is the universal gas constant. Lastly $Q > 0$ is the energy released by the exothermic chemical reaction per unit mass of the reactant, and $\omega(\Theta, Y)$ represents the rate at which this reaction proceeds.

Because of the complexity of system (1), some approximation is often used in order to obtain a set of governing equations which might be tractable from an analytical or a mathematical point of view. In addition to the classical approximation which consists in considering the coefficients c_p , λ, D or ρD, and Q as constant, there exist in combustion theory two basic ways of simplifying the system (1). The first one is known as the ***isobaric approximation:*** it follows from the highly subsonic character of the flame propagation and consists in neglecting the spatial variations of the pressure in the gaseous mixture (see [10] for the derivation of the equations and [18] for a rigorous mathematical analysis in a one-dimensional context). The second usual simplification is more drastic and is the well-known ***constant-density approximation,*** in which the density fluctuations of the gas are neglected. It is essentially valid for situations where the hydrodynamical effects play a secondary role with respect to the reactive and diffusive effects, in other words when the gas flow is not too far from being uniform (see [21]). This approximation retains many features of the phenomenon, including cellular flame instabilities, and leads to the well-known Kuramoto-Sivashinsky equation for the evolution of the flame front (see [21]).

In the framework of the constant-density approximation, the equations (1) become:

$$\begin{aligned} \vec{\nabla}.\vec{V} &= 0 \ , \\ u_t + \vec{V}.\vec{\nabla} u &= - p_x \ , \\ v_t + \vec{V}.\vec{\nabla} v &= - p_y \ ; \end{aligned} \tag{2}$$

$$\begin{aligned} T_t + \vec{V}.\vec{\nabla} T &= \Delta T + \Omega(T,Z) \ , \\ Z_t + \vec{V}.\vec{\nabla} Z &= \frac{1}{Le}\Delta Z - \Omega(T,Z) \ . \end{aligned} \tag{3}$$

We have used normalized temperature and mass fraction variables T, Z instead of Θ and Y ; Le is a constant positive parameter, the Lewis number of the reactant ($Le = \dfrac{\lambda}{\rho c_p D}$), and Ω is the normalized reaction rate.

A major effect of the constant-density approximation is therefore to decouple the equations (2) for the velocity and pressure from system (3) for the "combustion variables" T, Z. If one wants to study the so-called thermo-diffusive system (3), one may consider the velocity field as given, satisfying (2). For the case of a flame propagating in the infinite rectangular channel S, it is natural to choose a velocity field parallel to the tube walls: $v \equiv 0$. Then, (2) yields $\vec{V} = (u(y),0)$, and the model can be rewritten as:

$$\begin{aligned} T_t + u(y) T_x &= \Delta T + \Omega(T,Z) \ , \\ Z_t + u(y) Z_x &= \frac{1}{Le}\Delta Z - \Omega(T,Z) \ . \end{aligned} \tag{4}$$

These equations are associated to initial data, and to boundary conditions of the form:

$$T(-\infty,y,t) = 0 \ , \ \ Z(-\infty,y,t) = 1 \ \ \text{(fresh mixture)} \ , \tag{5.a}$$

$$T(+\infty,y,t) = 1 \ , \ \ Z(+\infty,y,t) = 0 \ \ \text{(burnt gases)} \ , \tag{5.b}$$

$$\frac{\partial T}{\partial y}(x,0,t) = \frac{\partial T}{\partial y}(x,L,t) = 0 \ , \tag{5.c}$$

$$\frac{\partial Z}{\partial y}(x,0,t) = \frac{\partial Z}{\partial y}(x,L,t) = 0 \ . \tag{5.d}$$

The conditions (5.c) express that the tube walls are adiabatic. We refer the reader to [3] for a mathematical analysis of this mixed initial-boundary value problem.

We are interested in this paper in the investigation of some steady solutions of the system (4)-(5) when Le = 1. As always in combustion problems, the assumption of a unit Lewis number leads to a simpler model, since it yields the identity $T + Z = 1$. Setting $g(T) = \Omega(T,1-T)$, we are then left with the following problem:

$$\begin{aligned} & u(y)T_x - \Delta T = g(T) \ , \\ & T(-\infty,y) = 0 \ , \ \ T(+\infty,y) = 1 \ \ \text{for } y \in [0,L] \ , \\ & T_y(x,0) = T_y(x,L) = 0 \ \ \text{for } x \in \mathbb{R} \ . \end{aligned} \tag{6}$$

Our study is therefore a first generalization to a higher dimensional context of the many works about the corresponding one-dimensional problem describing a steady planar flame:

$$\begin{aligned} & -T'' + cT' = g(T) \ , \ \ T = T(x) \ \ \text{defined on } \mathbb{R} \ , \\ & T(-\infty) = 0 \ , \ \ T(+\infty) = 1 \ . \end{aligned} \tag{7}$$

(see e.g. [2], [7], [13], [14], [15], [17], [19]).

Following what is well-known for problem (7), in which the function T *and* the real c are unknown, we will investigate *a family of solutions of (6)* given by the following system, where $\alpha(y)$ is a given function of $y \in [0,L]$, and where $T \in C^2(S)$ and $c \in \mathbb{R}$ are the unknowns:

$$c\alpha(y)T_x - \Delta T = g(T) \ \ \text{in } S \ , \tag{8.a}$$

$$T(-\infty,y) = 0 \ , \ \ T(+\infty,y) = 1 \ \ \text{for } 0 \le y \le L \ , \tag{8.b}$$

$$T_y(x,0) = T_y(x,L) = 0 \ \ \text{for } x \in \mathbb{R} \ . \tag{8.c}$$

The hypotheses used to study this problem and our main results about the solutions of (8) are stated in the next section.

REMARK 3: Some of the results presented below can be extended to a similar problem in which $c\alpha(y)$ is replaced by $c + \alpha(y)$ (see [5]). The latter problem describes travelling-wave solutions $T(x - ct)$, $Z(x - ct)$ of (4)-(5) when the velocity field $u(y)$ is a given function $\alpha(y)$ and when $Le = 1$, and is also the subject of some current work done in collaboration with P. L. Lions [6]. •

3. ASSUMPTIONS AND MAIN RESULTS

From now on, we will assume that α and g are two given functions satisfying the hypotheses:

$$\alpha \in C([0,L],\mathbb{R}) \ , \ \min_{0 \le y \le L} \alpha(y) = \alpha_0 > 0 \ ; \tag{9}$$

$$g \in C([0,1],\mathbb{R}_+) \ , \ g(1) = 0 \ , \tag{10.a}$$

$$\exists\, \theta \in (0,1) \ , \ g \equiv 0 \text{ on } [0,\theta] \ , \ g > 0 \text{ on } (\theta,1) \ , \tag{10.b}$$

where θ denotes an ignition temperature (see [7], [19]).

Our main result concerning problem (8) is the following:

THEOREM 4:

Under the assumptions (9) and (10), there exists a solution (T,c) of problem (8) which satisfies $T \in C^2(\bar{S})$, $0 < T < 1$ in $\bar{S}$ and $c > 0$. This solution is monotone increasing in the x-direction: $T_x > 0$ in $\bar{S}$.

Moreover, this solution is unique in the following sense: if (T_1,c_1) and (T_2,c_2) are two solutions of (8), then $c_1 = c_2$ and there exists $X \in \mathbb{R}$ such that:

$$T_1(x,y) = T_2(x+X,y). \ \bullet$$

4. EXISTENCE

In this section we briefly sketch the proof of the existence of a solution (T,c) to problem (8). The reader is referred to [5] for the details.

In a first step, we consider an analogous problem posed on a bounded domain. For $a > 0$, we set $S_a = \{(x,y) \in S, -a < x < ka\}$, where k is a positive parameter (its precise value is to be specified later). We consider the problem:

$$-\Delta T + c\ \alpha(y) T_x = g(T) \ \text{ in } S_a \ , \tag{11.a}$$

$$T(-a,y) = 0 \ , \ T(ka,y) = 1 \ \text{ for } 0 \le y \le L \ , \tag{11.b}$$

$$T_y\,(x,0) = T_y(x,L) = 0 \ \text{ for } -a \le x \le ka \ , \tag{11.c}$$

$$\max_{\substack{-a \le x \le 0 \\ 0 \le y \le L}} T(x,y) = \theta , \tag{11.d}$$

where the unknowns are again T and c. The condition (11.d) is related to the translation invariance of the solutions of (8), and will also be shown to determine the right value of c. Indeed, for any $c > 0$, there exists a solution T to equations (11.a)-(11.c); this is easily obtained by using Schauder's fixed point theorem. However, if c is not chosen adequately, the solution on S_a does not converge, when $a \to +\infty$, towards the solution of (8). Actually, it will be seen that the value of c is uniquely determined for the problem (8) in the whole strip.

We are first going to prove the existence of a solution of (11) in S_a, and then solve problem (8) by investigating the passage to the limit as $a \to +\infty$.

4.1. A priori estimates for problem (11)

We begin with some a priori estimates for a solution (T,c) of problem (11). It follows obviously from the maximum principle that a solution (T,c) of (11) satisfies:

$$0 < T < 1 \text{ in } S_a \ , \quad \max_{0 \le y \le L} T(0,y) = \theta \ .$$

The crucial point is to obtain estimates on c. Indeed, it then follows from the elliptic estimates that T is bounded in $W^{2,p}(S_a)$. We start with the upper estimate:

LEMMA 5:

Let (T,c) be a solution of (11) in S_a. There exists a constant K (independant of a) such that for any $a \ge 1$, the following estimate holds:

$$c < K \ . \bullet$$

PROOF: Let $M = \max_{0 \le s \le 1} g(s)$, and let $H(x)$ be the Heavyside function. We consider the function w satisfying:

$$- w'' + c\alpha_0 w' = MH(x) \text{ for } x \in (-a,ka) \ ,$$
$$w(-a) = 0 \ , \quad w(ka) = 1 \ .$$

From the maximum principle, we easily derive that $T(x,y) \le w(x)$ for $(x,y) \in S_a$. Hence, $w(0) \ge \theta$. Using the explicit value of $w(0)$, this implies the upper bound on c. $\bullet$

A similar argument based on the maximum principle shows that, if $c \ge 0$:

$$T(x,y) \leq \theta \, \frac{e^{c\alpha_0 x} - e^{-c\alpha_0 a}}{1 - e^{-c\alpha_0 a}} \quad \text{for } (x,y) \in S_a \text{ and } x \leq 0 \,. \tag{12}$$

We can then prove the following lower estimate on c:

LEMMA 6:

Assume that k satisfies $(1 + k)^{-1} > \theta$. Let (T,c) be a solution of (11) in S_a. There exist $a_0 \geq 1$ and a constant $\delta > 0$ (independant of a) such that, if $a \geq a_0$, then the following estimate holds:

$$c > \delta > 0 \,. \bullet$$

PROOF: Again from the maximum principle, one can show that $(1 + k)^{-1} > \theta$ implies $c > 0$. We then combine two integral identities obtained by multiplying equation (11.a) successively by $(1-T)$ and T_x. These yield:

$$L \int_0^1 g(s)\,ds \leq \frac{1}{2} \int_0^L T_x^2(-a,y)\,dy + c\beta \int_0^L T_x(-a,y)\,dy + \beta c^2 \,,$$

where β is some positive constant. Since (12) implies that:

$$T_x(-a,y) \leq \frac{\theta}{a} \quad \text{for all } y \in [0,L] \,,$$

we easily conclude. •

4.2. Solution in S_a

We can then prove the next result about problem (11):

PROPOSITION 7:

Let $a > 0$ and $k > 0$ be such that $(1 + k)^{-1} > \theta$. Under the hypotheses (9)-(10), there exists a solution of problem (11). •

PROOF: In order to solve problem (11), we write it as a fixed point problem $X = F(X)$. The mapping F is defined in the Banach space $E = C^1(\overline{S_a}) \times \mathbb{R}$ by:

$$F(T,c) = \Big(\, U(T,c), \max_{\substack{-a \leq x \leq 0 \\ 0 \leq y \leq L}} T(x,y) - \theta + c \, \Big) \,,$$

where $U = U(T,c)$ satisfies the boundary conditions (11.b)-(11.c) and the equation:

$$-\Delta U + c\,\alpha(y) U_x = g(T) \text{ in } S_a \tag{13}$$

[we extend the domain of definition of g by setting $g \equiv 0$ in $\mathbb{R}_- \bigcup [1,+\infty)$].

The existence of a fixed point for F is obtained using the Leray-Schauder topological degree. For $\tau \in [0,1]$, let F_τ be the mapping obtained by substituting g by τg and $\alpha(y)$ by $\tau\alpha(y) + (1-\tau)\alpha_0$ in (13) (note that $F_1 = F$). Then F_τ is a compact mapping from E into itself. Furthermore, using a priori estimates (similar to the above ones) for the solutions of the equation $F_\tau(X) = X$, we can construct an open set Ω of the form

$$\Omega = \{(T,c) \in E,\ ||T||_{C^1} < \text{K},\ c_1 < c < c_2\},$$

such that $Id - F_\tau \neq 0$ on $\partial\Omega$ (Id denotes the identity in E). The degree $d(Id - F_\tau, \Omega, 0)$ is therefore defined for any $\tau \in [0,1]$, and does not depend on τ. Its value can be evaluated using F_0 : indeed, the mapping F_0 is of the form:

$$F_0(T,c) = (U_c\ ,\ \max_{\substack{-a \leq x \leq 0 \\ 0 \leq y \leq L}} T(x,y) - \theta + c\ ),$$

where U_c does not depend on T any longer. Using another homotopy argument, one can then show that the degree $d(Id - F_0, \Omega, 0)$ simply equals $d[h, (c_1,c_2), 0]$, where the function h is defined by $h(c) = \max\limits_{-a \leq x \leq 0} U_c - \theta + c$. Showing that h is a decreasing function, we can explicitly evaluate the last degree and finally obtain:

$$d(Id - F_0, \Omega, 0) = d(Id - F, \Omega, 0) = -1,$$

which proves the existence of a fixed point for F, that is the existence of a solution of (11). •

4.3. Passage to the limit as $a \to +\infty$

Henceforth we assume that $(1+k)^{-1} > \theta$. Let (a_n) be a sequence of positive reals such that $\lim a_n = +\infty$, and let (T_n,c_n) be a solution of (11) in S_{a_n}. The a priori estimates on the sequence (T_n,c_n) allow us to extract a subsequence (which we still denote by (T_n,c_n)) such that $T_n \to T$ locally and $c_n \to c$, with T and c satisfying (8.a), (8.c), and the properties $c > 0$, $0 \leq T \leq 1$, $\max\limits_{0 \leq y \leq L} T(0,y) = \theta$. In this section, the notation (T,c) will always be used for the limit of the sequence (T_n,c_n).

Taking the limit as $a \to +\infty$ in (12), we easily get:

$$T(x,y) \leq \theta e^{c\alpha_0 x} \quad \text{for } x \leq 0,$$

which implies $T(-\infty,y) = 0$. In order to prove that (T,c) actually is a solution of (8), it remains to investigate the limit $T(+\infty,y)$. We begin with the following lemma:

LEMMA 8:

The following integrals are finite:

$$\int_S g(T) < +\infty \ , \ \int_S |\nabla T|^2 < +\infty \ . \bullet \tag{14}$$

PROOF: On $\mathbb{R}_- \times [0,L]$, we have $g(T) \equiv 0$ since $T \leq \theta$. On the other hand, for $z > 0$, we set $Q_z = (0,z) \times (0,L)$ and $H(z) = \int_0^L T(z,y)dy$. Integrating the equation (8.a) in Q_z, we get $\int_{Q_z} g(T) = A(z) - A(0)$, with $A(z) = c\int_0^L \alpha(y)T(z,y)dy - H'(z)$. If $\int_{Q_z} g(T) \to +\infty$ as $z \to +\infty$, then $H'(+\infty) = -\infty$, which is impossible since H is bounded. For the second integral in (14), we multiply the equation (8.a) by T and then integrate in Q_z, and we conclude in the same manner. •

For $z \in \mathbb{R}$, let R_z be the rectangle $(z,z+1) \times (0,L)$. In order to study the limit $T(+\infty,y)$, we are going to consider the sequence of functions T_m defined on R_0 by:

$$T_m(x,y) = T(x + m,y) \ \text{ for } (x,y) \in R_0 \text{ and } m \in \mathbb{N}. \tag{15}$$

Since T_m satisfies the equation (8.a) on R_0, we can use classical elliptic estimates to extract a subsequence (T_{m_k}) which converges to a limit T_0 in $C^1(\overline{R_0})$. If we choose the subsequence (m_k) such that the rectangles R_{m_k} are pairwise disjoint, Lemma 8 implies that the series with general term $\int_{R_0} g(T_{m_k})$ and $\int_{R_0} |\nabla T_{m_k}|^2$ converge. Thus $g(T_0) = \nabla T_0 = 0$: T_0 is a constant l, and $l \in [0,\theta] \cup \{1\}$ from (10).

It remains to prove that $l = 1$. Arguing as above, we can show that the subsequence (m_k) can be chosen so that the sequence (T_{-m_k}) converges to 0 in $C^1(\overline{R_0})$ ((T_{-m_k}) is defined by extending (15) to negative integers). Integrating on $(-m_k,m_k) \times (0,L)$ the equation (8.a) multiplied by T_x, we obtain a relation whose limit as $m_k \to +\infty$ gives:

$$c\int_S \alpha(y)T_x^2 = L\int_0^l g(s)ds \ .$$

The left-hand side of this equality is positive since $c > 0$, $\alpha(y) \geq \alpha_0 > 0$ and $\max_{0 \leq y \leq L} T(0,y) = \theta \neq T(-\infty,y)$. We then get $l = 1$ from (10.b). A classical argument by contradiction then shows that the whole sequence (T_m) converges to 1, which ends the proof: $T(+\infty,y) = 1$, and (T,c) satifies (8). •

5. UNIQUENESS AND MONOTONICITY

We report in this section on the following result from [8], [9]:

THEOREM 9:

Suppose the assumptions (10)-(11) hold. Then, the solution (T,c) of problem (8) is unique. That is, c is uniquely determined and T is uniquely determined up to translations of the origin: if (T_1,c_1) and (T_2,c_2) are two solutions of (8), then $c_1 = c_2$ and there exists $X \in \mathbb{R}$ such that:

$$T_1(x,y) = T_2(x+X,y). \bullet$$

We just outline here the proof and refer the reader to [8], [9] for the detailed arguments.

5.1. Monotonicity of the profile

Since the results hold for more general equations than (8), we first introduce some notation and describe the more general setting. Henceforth we assume that $L = 1$ for simplicity. Let γ: $[0,1] \to \mathbb{R}$ and h: $[0,1] \times \mathbb{R} \to \mathbb{R}$ be two continuous given functions; we will consider in this section the following problem:

$$-\Delta T + \gamma(y) T_x = h(y,T) \text{ in } \mathbb{R} \times [0,1] , \tag{16.a}$$

$$T_y(x,0) = T_y(x,1) = 0 \text{ for } x \in \mathbb{R} , \tag{16.b}$$

$$T(-\infty,y) = 0 , \quad T(+\infty,y) = 1 \text{ for } 0 \leq y \leq 1 . \tag{16.c}$$

The following conditions will be assumed on γ and h:

$$\gamma(y) > 0 \text{ for } y \in [0,1] ; \tag{17.a}$$

$$h(y,0) = h(y,1) = 0 \text{ for } y \in [0,1] ; \tag{17.b}$$

$$\begin{array}{l} h = h(y,s) : [0,1] \times \mathbb{R} \to \mathbb{R} \text{ has continuous derivatives} \\ \text{in the neighborhood of } s = 0 \text{ and } s = 1 ; \end{array} \tag{17.c}$$

$$h_s(y,0) \le 0 \ , \ \ h_s(y,1) < 0 \ \text{ for } y \in [0,1] \ . \tag{17.d}$$

Hence, the term $h(y,s)$ is of one of the following types (for fixed $y \in [0,1]$):

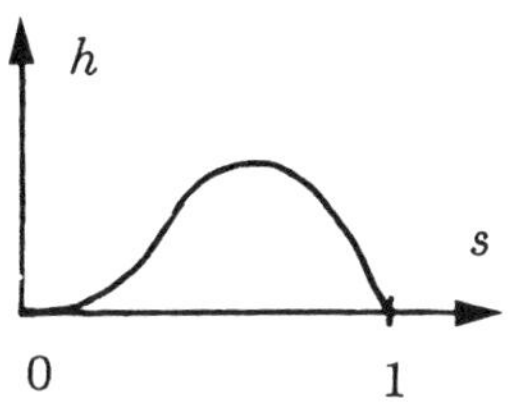

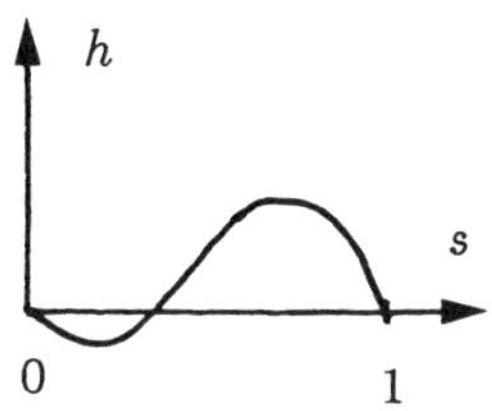

The first case corresponds to problems in combustion, while the second case arises in some models in biology (see [2], [11]).

The basic fact to start with is given in the next result:

PROPOSITION 10:

Suppose γ and h satisfy the assumptions (17) and let T be a solution of problem (16). If $0 < T < 1$ in $S = \mathbb{R} \times (0,1)$, then $T_x > 0$ in S. •

This result is established in [8] as one of several qualitative properties that hold in general for solutions of nonlinear elliptic equations in cylindrical domain. Inspired by the work of Gidas, Ni and Nirenberg [12], it relies on the maximum principle and a method of reflection accross a family of parallel hyperplanes and moving these up to a critical position.

For each $\lambda \in \mathbb{R}$, we let $S_\lambda = \{(x,y) \in S,\ x < \lambda\}$. We consider the function w_λ defined in S_λ by $w_\lambda(x,y) = T(2\lambda - x,y) - T(x,y)$ (notice that $(2\lambda - x,y)$ is the reflection of the point (x,y) accross the line $x = \lambda$). Owing to the expansion of $T(x,y)$ as $x \to -\infty$ which is discussed below, we know that for λ close to $-\infty$, $T_x > 0$ and $w_\lambda > 0$ in S_λ. Thus, we may define $\Lambda = Sup\{\lambda \in \mathbb{R},\ T_x > 0 \text{ and } w_\lambda > 0 \text{ in } S_\lambda\}$. Using a strong form of the maximum principle it is shown in [8] that Λ cannot be finite, i.e. $\Lambda = +\infty$, whence the result. •

5.2. Determining asymptotic exponential solutions

We denote: $b_-(y) = -\ h_s(y,0)$, $b_+(y) = -\ h_s(y,1)$. Hence, $b_- \ge 0$ and $b_+ > 0$ in $[0,1]$. The linearized equations as $x \to -\infty$ and $x \to +\infty$ are respectively of the form:

$$-\ \Delta T + \gamma(y) T_x + b_-(y) T = 0 \ ,$$

$$-\ \Delta T + \gamma(y) T_x - b_+(y)(1-T) = 0 \ .$$

Hence, looking for exponential solutions $T = e^{\lambda x}\varphi(y)$ and $1-T = e^{\mu x}\psi(y)$ respectively, one is led to the following problems:

$$\begin{aligned} &-\varphi'' + b_-(y)\varphi = [\lambda^2 - \lambda\gamma(y)]\varphi \text{ in } (0,1) , \\ &\varphi'(0) = \varphi'(1) = 0 ; \end{aligned} \tag{18}$$

$$\begin{aligned} &-\psi'' + b_+(y)\psi = [\mu^2 - \mu\gamma(y)]\psi \text{ in } (0,1) , \\ &\psi'(0) = \psi'(1) = 0 . \end{aligned} \tag{19}$$

Although linear in φ and ψ, these are "nonlinear" eigenvalue problems because of the fashion that λ and μ appear. One can prove the following facts (see [8] for the proof):

PROPOSITION 11:

Assume that $b_- \geq 0$. Then, there exists a sequence of *positive* "eigenvalues" $0 < \lambda_1 < \lambda_2 < \lambda_3 \cdots < \lambda_n < \cdots$ with associated eigenfunctions $(\varphi_k)_{k=1}^{\infty}$ for problem (18) such that $\lim_{n\to\infty} \lambda_n = +\infty$.

Moreover, λ_1 is "simple", and the unique eigenfunction φ of (18) which does not change sign in (0,1) is φ_1 (up to multiplication by a constant) and corresponds to λ_1. •

PROPOSITION 12:

Assume that b_+ is a continuous function in (0,1) such that $b_+ \geq 0$ and $b_+ \equiv 0$ in (0,1). Then, there exists a sequence of *negative* "eigenvalues" $0 > \mu_1 > \mu_2 > \mu_3 \cdots > \mu_n > \cdots$ with associated eigenfunctions $(\psi_k)_{k=1}^{\infty}$ for problem (19) such that $\lim_{n\to\infty} \mu_n = -\infty$.

Moreover, μ_1 is "simple", and the unique eigenfunction ψ of (19) which does not change sign in (0,1) is ψ_1 (up to multiplication by a constant) and corresponds to μ_1. •

REMARK 13: One can further characterize the eigenfunctions (φ_k) and (ψ_k) by their number of oscillations. Thus there is a complete analogy with the usual Sturm-Liouville theory. •

5.3. Comparison of principal eigenvalues

Let us now consider two functions γ and γ^* and the corresponding eigenvalue problems (18) and (19). We denote by λ_1, λ_1^* and μ_1, μ_1^* (with obvious notations) the corresponding "principal eigenvalues". It will be important to compare these eigenvalues. We just mention here the result (see [9] for the proof):

PROPOSITION 14:

Assume that γ and γ^* are continuous and that $0 < \gamma \leq \gamma^*$ in (0,1). Then:

$$0 < \lambda_1 \leq \lambda_1^* \text{ and } \mu_1 \leq \mu_1^* < 0 \ . \bullet$$

5.4. Asymptotic behaviour near infinity

A crucial tool in the study of elliptic equations in a cylindrical domain is the possibility to expand the solution asymptotically as $x \to \pm\infty$ in terms of the exponential solutions we have just discussed above. This follows from a general theory developped by Agmon and Nirenberg [1] for general linear elliptic or parabolic partial differential equations. Here, actually, we require an extension of these results for problems with a linearization at infinity which is due to Pazy [20].

Consider a solution T of (16) and assume that $0 < T < 1$ in S. Comparing T with some exponential functions, it can be shown that T cannot vanish faster than any exponential as $x \to -\infty$. Using the notations introduced above, we can write:

$$T(x,y) = a_{j_1} e^{\lambda_{j_1} x} \varphi_{j_1}(y) + o(e^{\lambda_{j_1} x}) \text{ as } x \to -\infty \ , \tag{20}$$

$$T_x(x,y) = \lambda_{j_1} a_{j_1} e^{\lambda_{j_1} x} \varphi_{j_1}(y) + o(e^{\lambda_{j_1} x}) \text{ as } x \to -\infty \ , \tag{21}$$

$$T(x,y) = 1 + b_{j_2} e^{\mu_{j_2} x} \psi_{j_2}(y) + o(e^{\mu_{j_2} x}) \text{ as } x \to +\infty \ , \tag{22}$$

for some integers j_1 and j_2, with $a_{j_1} \neq 0$, $b_{j_2} \neq 0$. Similar expansions hold for T_x and T_y as $x \to -\infty$ and $x \to +\infty$.

Since the leading term in these "expansions" determines the behaviour near infinity, it must be the case that $j_1 = 1$ and $a_1 > 0$ (we recall that we have assumed $T > 0$ and that all φ_k with $k \geq 2$ change sign). Similarly, since $T < 1$, $j_2 = 1$ and $b_1 < 0$.

Writing $\lambda = \lambda_1$, $\mu = \mu_1$, $\varphi = \varphi_1$ and $\psi = \psi_1$, we therefore have:

PROPOSITION 15:

Suppose T is a solution of problem (16) with $0 < T < 1$ in S. Then, there are positive constants a and b such that:

$$T(x,y) = a \ e^{\lambda x} \varphi(y) + o(e^{\lambda x}) \text{ as } x \to -\infty \ , \tag{23.a}$$

$$T(x,y) = 1 - b \ e^{\mu x} \psi(y) + o(e^{\mu x}) \text{ as } x \to +\infty \ . \bullet \tag{23.b}$$

5.5. A change of variable

Since $T_x > 0$ in S, we can invert T with respect to the x variable. Therefore, for fixed $y \in (0,1)$, we set:

$$v(.,y) = T^{-1}(.,y) ,$$

that is, for all $s \in (0,1)$:

$$T(v(s,y),y) = s . \tag{24}$$

Thus v is defined in the square $Q = (0,1) \times (0,1)$. Translating problem (16) into this new setting, we find that v satisfies:

$$L[v] = \gamma(y) \text{ in } Q , \tag{25}$$

$$v_y(s,0) = v_y(s,1) = 0 \text{ for } 0 \leq s \leq 1 , \tag{26}$$

where:

$$L[v] = -\frac{v_{ss}}{v_s^2}(1 + v_y^2) - v_{yy} + \frac{2}{v_s} v_y v_{ys} + h(y,s) v_s .$$

The essential property that we use here is the fact that the nonlinear differential operator $-L$ is *elliptic* (we recall that an operator $-L[v] = F(\underline{x},v,\partial_i v,\partial_{ij} v)$ is called elliptic if the Hessian matrix $F''_{\partial_{ij} v}$ is definite positive). The fact that $-L$ is an elliptic operator was already used in the work of Kinderlehrer and Nirenberg [16] to which we refer for more details. Notice also that there is no zero order terms in $-L[v]$ (i.e. $-L[v] = F(\underline{x},\partial_i v,\partial_{ij} v)$ is a function of the derivatives of v, not of v itself). A consequence of these properties is that the maximum principle applies:

LEMMA 16:

Assume that v_1 and v_2 are two functions defined on Q, both satisfying (26), and such that $L[v_1] \leq L[v_2]$ in Q. Then:

$$\operatorname*{Sup}_{Q} (v_1 - v_2) \leq \operatorname*{Sup}_{\substack{0 \leq y \leq 1 \\ s=0 \text{ or } s=1}} (v_1 - v_2) . \bullet$$

5.6. Boundary conditions for v

Thus, v defined as above verifies (25) and (26). In order to determine the conditions on the remaining part of the voundary ∂Q, that is on the segments

$\{s = 0, 0 \leq y \leq 1\}$ and $\{s = 1, 0 \leq y \leq 1\}$, we observe that:

$$v(y,s) \to -\infty \quad \text{as } s \to 0^+ ,$$

and:

$$v(y,s) \to +\infty \quad \text{as } s \to 1^- ,$$

for $y \in [0,1]$. More precisely, one can derive from Proposition 15 the following behavior of v:

$$v(y,s) \sim \frac{1}{\lambda} Log(s) \quad \text{as } s \to 0^+ , \tag{27.a}$$

$$v(y,s) \sim \frac{1}{\mu} Log(1-s) \quad \text{as } s \to 1^- . \tag{27.b}$$

5.7. Uniqueness of c

Let us now examine the problem (16) with:

$$\gamma(y) = c\ \alpha(y) \quad \text{and } h(y,s) = g(s) , \tag{28}$$

where α and g are given functions satisfying (9) and (10), and where c is an unknown positive parameter. We will now show that c is uniquely determined.

Since $h \geq 0$, any solution T of (16) (with (28)) satisfies $0 < T < 1$ in S from the maximum principle. Thus, Propositions 10 and 15 apply, and the "inverse function" v satisfying (25)-(27) can be introduced for any solution of (16).

Suppose there are two values of c, say $0 < c_1 < c_2$ for which the system (16) has solutions T_1 and T_2 respectively. Let v_1 and v_2 be the corresponding inverse functions defined by (24). Thus, v_1 and v_2 satisfy (26), and (25) and (28) imply:

$$L[v_1] < L[v_2] \text{ in } Q .$$

Using (27) and the comparison of "eigenvalues" stated in Proposition 14, we are able to compare v_1 and v_2 near $s = 0$ and $s = 1$. We obtain:

$$v_1 - v_2 \to -\infty \quad \text{as } s \to 0^+ ,$$

$$v_1 - v_2 \to -\infty \quad \text{as } s \to 1^- .$$

Lemma 16 then implies:

$$v_1 - v_2 \equiv -\infty \text{ in } Q ,$$

which is impossible. Hence, c is uniquely determined.

5.8. Uniqueness of T

A shift of the origin modifies the leading coefficient in the expansion (23.a) near $x \to -\infty$. Indeed, if:

$$T(x,y) = a\ e^{\lambda x}\ \varphi(y) + o(e^{\lambda x}) \quad \text{as } x \to -\infty\ ,$$

then:

$$T(x+X,y) = (a\ e^{\lambda X})\ e^{\lambda x}\ \varphi(y) + o(e^{\lambda x}) \quad \text{as } x \to -\infty\ .$$

We now assume that for $c = c_1 = c_2$, we still have two solutions T_1 and T_2. After the appropriate shift for T_2, that is, setting $\hat{T}_2(x,y) = T_2(x+X,y)$, we may assume that T_1 and $\hat{T}_2$ have the same leading coefficient in the expansion near $x \to -\infty$:

$$T_1(x,y) = a\ e^{\lambda x}\ \varphi(y) + o(e^{\lambda x}) \quad \text{as } x \to -\infty\ , \tag{29.a}$$

$$\hat{T}_2(x,y) = a\ e^{\lambda x}\ \varphi(y) + o(e^{\lambda x}) \quad \text{as } x \to -\infty\ . \tag{29.b}$$

Using the functions v_1 and v_2 associated with T_1 and $\hat{T}_2$ as before, and the expansions (27), one can show using the maximum principle that lest $T_1 \equiv \hat{T}_2$, $v_1 - v_2$ and therefore $T_1 - \hat{T}_2$ have a sign.

Say $T_1 < \hat{T}_2$. We know that $\hat{T}_2 - T_1$ satisfies a linear equation in the half strip $\{(x,y),\ x < x_0\}$ for some x_0 (i.e. $g(T_1) = g(\hat{T}_2) \equiv 0$ for $x < x_0$). Furthermore, one can show that $\hat{T}_2 - T_1$ cannot decay faster than any exponential. Hence, we have for some $k > 0$:

$$(\hat{T}_2 - T_1)(x,y) = a_k\ e^{\lambda_k x}\ \varphi_k(y) + o(e^{\lambda_k x}) \quad \text{as } x \to -\infty\ ,$$

with $a_k = 0$. Since (29) yields that $a_1 = 0$ in the previous expansion, we have $k \geq 2$. But then, φ_k changes sign, whence a contradiction. Therefore:

$$T_1(x,y) = T_2(x+X,y)\ .$$

REMARK 17: There is a somewhat more direct proof of uniqueness by working directly with the function T rather than with its inverse v. It again relies on the maximum principle and is more general; it is presented in [9]. •

ACKNOWLEDGEMENTS: The first and third authors were partly supported under grant ARO, DAA G 29-84-K-0150

REFERENCES:

[1] S. AGMON & L. NIRENBERG, "Properties of solutions of ordinary differential equations in Banach spaces", Comm. Pure Appl. Math., **16,** (2), pp. 121-239, (1963).

[2] D. G. ARONSON & H. F. WEINBERGER, "Nonlinear diffusion in population genetics, combustion and nerve propagation", in "Partial differential equations and related topics", Lect. Notes in Math., **446,** Springer Verlag, New York, pp. 5-49, (1975).

[3] F. BENKHALDOUN & B. LARROUTUROU, "Numerical analysis of the two-dimensional thermo-diffusive model for flame propagation", *to appear.*

[4] H. BERESTYCKI, L. CAFFARELLI & L. NIRENBERG, *in preparation.*

[5] H. BERESTYCKI & B. LARROUTUROU, "A semilinear elliptic equation in a strip arising in a two-dimensional flame propagation model", Prepublication Univ. Paris-Nord 73, (1987).

[6] H. BERESTYCKI, B. LARROUTUROU & P. L. LIONS, *in preparation.*

[7] H. BERESTYCKI, B. NICOLAENKO & B. SCHEURER, "Traveling wave solutions to combustion models and their singular limits", SIAM J. Math. Anal., **16,** (6), pp. 1207-1242, (1985).

[8] H. BERESTYCKI & L. NIRENBERG, "Some qualitative properties of positive solutions to semilinear elliptic equations in cylindrical domains", *in preparation.*

[9] H. BERESTYCKI & L. NIRENBERG, "Uniqueness and singular limit of the solution of a two-dimensional elliptic model for flame propagation", *in preparation.*

[10] J. D. BUCKMASTER & G. S. S. LUDFORD, "Theory of laminar flames", Cambridge Univ. Press, Cambridge, (1982).

[11] P. C. FIFE, "Mathematical aspects of reacting and diffusing systems", Lect. Notes in Biomathematics, **28,** Springer Verlag, New York, (1979).

[12] B. GIDAS, W. M. NI & L. NIRENBERG, "Symmetry and related properties via the maximum principle", Comm. Math. Phys., **68,** pp. 209-243, (1979).

[13] W. E. JOHNSON & W. NACHBAR, "Laminar flame theory and the steady linear burming of a monopropellant", Arch. Rat. Mech. Anal., **12,** pp. 58-91, (1963).

[14] Ja. I. KANEL', "Stabilization of solutions of the Cauchy problem for equations encountered in combustion theory", Mat. Sbornik, **59,** pp. 245-288, (1962).

[15] Ja. I. KANEL', "On steady state solutions to systems of equations arising in combustion theory", Dokl. Akad. Nauk USSR, **149,** (2), pp. 367-369, (1963).

[16] D. KINDERLEHRER & L. NIRENBERG, "Regularity in free boundary value problems", Ann. Scuola Norm. Sup. Pisa, Ser. IV, **4**, pp. 373-391, (1977).

[17] A. N. KOLMOGOROV, I. G. PETROVSKII & N. S. PISKUNOV, "A study of the equation of diffusion with increase in the quantity of matter and its application to a biological problem", Bjul. Moskowskogo Gos. Univ., **17**, pp. 1-26, (1937).

[18] B. LARROUTUROU, "The equations of one-dimensional unsteady flame propagation: existence and uniqueness", *to appear in* SIAM J. Math. Anal..

[19] M. MARION, "Etude mathématique d'un modèle de flamme laminaire sans température d'ignition: I - Cas scalaire", Ann. Fac. Sci. Toulouse, **6**, pp. 215-255, (1984).

[20] A. PAZY, "Asymptotic expansions of solutions of ordinary differential equations in Hilbert spaces", Arch. Rational Mech. Anal., **24**, (3), pp. 193-218, (1967).

[21] G. I. SIVASHINSKY, "Instabilities, pattern formation and turbulence in flames", Ann. Rev. Fluid Mech., **15**, pp. 179-199, (1983).

[22] F. A. WILLIAMS, "Combustion theory", (second edition), Benjamin Cummings, Menlo Park, (1985).

MATHEMATICAL MODELING IN THE AGE OF COMPUTING: IS IT REDUNDANT?[1]

J. Buckmaster
University of Washington[2]
Seattle, Washington 98195
U.S.A.

INTRODUCTION

The remarkable growth in computing power that we have witnessed in recent years provides a formidable tool for unraveling scientific mysteries in many fields, not least that of combustion. At the same time, the euphoria that such power can engender has led some practitioners of the silicon arts--otherwise sensible men and women--to suggest, in some cases explicitly, that this new age of computing will write the epitaph for mathematical modeling. I know of no such claims in print, and so risk the charge of raising a straw man, but I am content to let the reader's own experience provide the answer to whether or not such a charge is accurate.

---

1 This differs from the title of the oral presentation so that an explanation is in order. Of the three combustion problems described in these pages, only detonation stability was described at the Lyon Meeting--a review of previously published work on the linear problem [1], and a discussion of the non-linear problem, the details of which will be published elsewhere [2]. Rather than just reproduce an extended abstract of this work here--the most that would be proper--I have taken the opportunity to put on paper a talk recently given at the SIAM Conference on Numerical Combustion, held in San Francisco in March 1987. This includes a brief summary of the Lyon material, enables me to respond to several requests for the San Francisco material, and will perhaps be of interest to those who share my belief that funding agencies need to be reminded how important it is to reserve some of their largesse for mathematical modeling and analysis, despite the insatiable demand for computing resources that is an inevitable characteristic of this 'super-computer' age.

2 Permanent address: University of Illinois, Urbana.

C.-M. Brauner and C. Schmidt-Lainé (eds.), Mathematical Modeling in Combustion and Related Topics, 29–42.

In contemplating a particularly vigorous claim recently heard from a prominent figure in the computational community, I decided that it might be of interest to examine the theoretical combustion problems that have, in some sense, engaged my recent attentions, to see to what extent they refute such claims. Thus the three problems to be discussed have not been chosen as the best available examples with which to make the point.

Example 1. Lifted Turbulent Diffusion Flames

The first problem is a familiar one, but one of the finest examples in combustion in which relatively simple mathematical modeling provides remarkable insight into a problem that, on its face, is intimidatingly complex. The key ideas are contained in work of Liñán, Bilger, and, most importantly, Peters, and my connection with the problem is peripheral but adequate to support the claim that it 'engaged my recent attentions'.

Recently, D.S. Stewart and I, helped by A. Ignatiadis and M. Williams, looked at what happens when a diffusion flame is extinguished [3]. As the temperature drops, the density rises, so that the fluid is drawn in from the far-field, a straightforward conclusion. However, when reaction is confined to a thin, large-activation-energy flame-sheet, the initial temperature drop is small but occurs very rapidly so that the inflow is large--the flame inhales sharply as it expires. Subsequently we were concerned with applications, and a good example of diffusion flame quenching is afforded by the lifted flames observed on tube burners at large Reynolds Numbers.

The notion that lifting is controlled by the quenching of the flame by turbulent eddies is due to Peters [4], and the key ideas are easily described. Before doing so, however, it should be noted that a direct numerical attack on this problem would be extremely difficult, perhaps impossible in the near future, and unlikely to provide information and insights distinct from those that can be obtained from careful experiment. Real understanding can only come with the help of good modeling.

Consider diffusion flame quenching. If some variable characterizing the state of the flame, e.g., the maximum temperature, is plotted against the Damköhler Number, the familiar S-shaped response is obtained. The transition from the top branch to the middle branch is associated with quenching, which therefore can be achieved by reducing the Damköhler Number a sufficient amount.

There are various ways in which this can be done, depending on the specific physical context. One way, relevant to the problem at hand, is to place the flame in a straining flow generated by the counter flow of jets of fuel and oxidizer. Then the Damköhler Number is inversely proportional to the jet speed, so that quenching will occur when the latter is large [5]. This is the quenching mechanism in the lifted flame: in some neighborhood of the burner rim, a large straining flow is generated by turbulent eddies.

To put this in mathematical terms, consider a large-activation-energy combustion model. More precisely, consider a one-step irreversible reaction that occurs at a rate

$$\sim XYe^{-E/RT} \tag{1}$$

where X,Y are the oxidizer and fuel mass fractions, T the temperature, R the gas constant, and E the (large) activation energy. Reaction is then confined to a thin flame-sheet whose structure is defined by Liñán's equation [6], which has the form

$$\underbrace{\frac{d^2\phi}{d\xi^2}}_{d^2T/dn^2} = \underbrace{\tilde{D}}_{D}\ \underbrace{(\phi + A\xi)}_{X}\ \underbrace{(\phi + B\xi)}_{Y}\ \underbrace{e^{-\phi}}_{e^{-E/RT}} \tag{2}$$

Here D is the Damköhler Number, n is the distance normal to the sheet, and we have indicated the connection between the scaled variables and primitive ones. Time derivatives can be neglected because the sheet is thin.

Appropriate boundary conditions for (2) come from matching with the temperature gradient in the exterior, non-reactive flow, i.e.,

$$\begin{aligned} &\underline{\zeta \to -\infty} \qquad \underbrace{\frac{d\phi}{d\xi}}_{dT/dn} \to -A \\ &\underline{\xi \to +\infty} \qquad \frac{d\phi}{d\xi} \to -B \quad . \end{aligned} \tag{3}$$

If A and B are fixed, solution of this two-point boundary problem yields a graph of $\phi_{maximum}$ vs. $\tilde{D}$ that is C-shaped, corresponding to the upper half of the S-response.

This model predicts quenching but the result cannot be directly applied to the lifted flame problem since there the temperature gradients A and B are stochastic variables that are difficult to characterize. Further progress requires additional assumptions. Consistent with (1), suppose that non-dimensional equations for X, Y and T are

$$\begin{aligned} L(T) &= -L(X) = -L(Y) = \Omega \\ L &\equiv \rho \frac{D}{Dt} - \nabla^2 \end{aligned} \tag{4}$$

where Ω is the reaction rate. The key assumption here is that the two Lewis Numbers are equal to 1. Then the linear combinations $(T + X)$, $(T + Y)$, and $(X - Y)$ are solutions of the homogeneous equation

$$L(\psi) = 0 \tag{5}$$

and so are linearly related if they assume prescribed constant values in the supply volumes. If now, following Bilger, $Z \equiv X - Y$ is used as the independent variable to describe the combustion field, a simple universal picture is obtained in which T , X and Y are piecewise linear functions of Z determined solely by the supply parameters, Fig. 1. In particular, the temperature gradients in Z are known constants whether the combustion field is steady or highly turbulent.

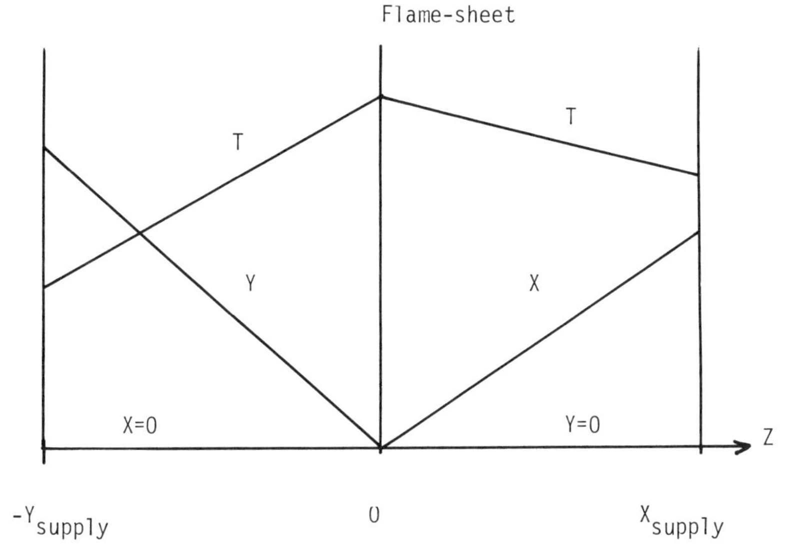

Fig. 1. All diffusion flame fields in the limit of infinite activation energy, when the Lewis Numbers equal 1, and when there are well defined supplies in which T , X and Y assume constant values.

Returning to the structure equation, but still using Z as the independent variable,it has the same form as (2) except for the introduction of the multiplicative factor $|\vec{\nabla} z|^{-2}_{z=0}$ into the right-hand side, where Z , recall, satisfies (5). Thus the transformation identifies a Damköhler Number which controls the quenching process. When $|\vec{\nabla} z|$ evaluated at the flame-sheet (Z = 0) is large, quenching will occur [4].

The final step in the explanation of lifting is to argue that if the mean value of $|\vec{\nabla} z|_{z=0}$ is large enough, quenching occurs over such a large portion of the flame-sheet and for such a long time that the entire sheet in some neighborhood is destroyed. For the lifted flame $\overline{|\vec{\nabla} z|}_{z=0}$ is bigger than some critical value near the rim, but is smaller at greater distances.

Once modeling and analysis have provided crucial insights into the phenomenon, it is possible to define useful numerical experiments. A fine example, part of a program underway at Flow Research in Renton, Washington, is described in [7]. An initial value problem is solved in which the diffusion flame is located in a parallel shear flow. The Kelvin-Helmholtz instability makes this shear layer roll up, stretching and distorting the flame. In regions where the stretch is large enough, quenching occurs, and, by also calculating Z, Peters' theory can be tested. Moreover it is possible--although this has not yet been done--to explore the effects of Lewis Numbers different from 1, something difficult to do analytically.

Example 2. Unstable Detonation Waves[3]

A characteristic of combustion problems that can lead to serious numerical difficulties is the presence of regions (spatial or temporal) in which gradients are extremely large. Different computations of identical problems will agree qualitatively if done by careful workers, but sometimes differ quantitatively because of these difficulties. Analysis can provide valuable insight into the structure of rapidly changing regions, and analytical results can be incorporated into a computational framework to minimize errors.

An example of a combustion problem which severely tests present day computing resources, and is characterized by extreme transients, is the unstable detonation wave. There is a great deal of interest in this problem, much of it concerned with predicting the spacing of the transverse waves generated by the instability. The nature of the transients can be elucidated by quite simple analysis.

Detonation waves can be modeled using Euler's equations for a compressible gas, together with one-step irreversible

[3] This is a brief summary of the material presented at the Lyon Meeting.

kinetics proceeding at a rate $\sim Ye^{-E/RT}$. Many numerical calculations have used this or similar models [8], [9]. The steady solution, the point of departure for a stability analysis, is then easily described when $\theta \equiv E/RT_s$ is large (T_s is the shock temperature). Behind the shock there is an induction zone of length L (say) in which the temperature is perturbed by just an $O(1/\theta)$ amount from the constant value T_s. This perturbation (T_1) has a logarithmic singularity at $x = L$, i.e.,

$$T_1 \propto \ell n(1 - x/L) \quad , \quad x \uparrow L \quad , \tag{6}$$

indicating the transition to the 'fire', a thin region in which the temperature increases from T_s to the burnt value T_b. This steady structure defines a length scale (L) and a time (L/c_s) where c is the speed of sound.

For a wide range of parameter values the steady solution is unstable, as shown, for example, by one-dimensional numerical calculations [8]. These one-dimensional solutions are characterized by very sharp periodic increases in shock pressure. An initial value analysis, starting with small perturbations from the steady structure, and valid for large θ, can also describe sharp pressure increases [1], [2].

With appropriate initial data, the early evolution occurs quite slowly on a time scale $O(\frac{\theta L}{c_s})$, and is characterized by the following equation for the shock displacement $x = h(t)$,

$$\nu h_t + h_{tt} \exp(\frac{\theta\beta}{T_s} h_t) = 0 \quad , \tag{7}$$

(ν, β constants).

As some critical time t_c is approached (determined by the initial data), the solution has a logarithmic singularity (cf. (16) but note that here the entire perturbation structure 'blows-up' in time), marking the transition to an extremely rapid evolution that occurs for $O(1)$ values of τ where

$$t_c - t = t_{ref.} e^{-\theta\tau} \quad , \quad t < t_c \tag{8}$$

($t_{ref.}$ is a constant). Such exponentially small time scales often occur when the activation energy is large. During this phase the temperature in the induction zone is

$$T/T_s \sim (1 - \tau)^{-1} = \left[1 + \frac{1}{\theta} \ln\left(\frac{t_c - t}{t_{ref.}}\right)\right]^{-1} \tag{9}$$

and there is a corresponding increase in the pressure on the same time scale. Clearly, the resolution of transients of this kind gets more difficult as θ is increased.

Associated with these transients are large spatial gradients, revealed by the description of the burnt gas behind the fire. There are two distinct regions, both exponentially close to the fire, but having quite different thicknesses. In the inner layer, for which

$$x - x_{fire} = O(Le^{-\theta\tau}) \tag{10}$$

we have

$$T \sim T_b(\tau) \tag{11}$$

where T_b can be calculated from a simple system of non-linear ordinary differential equations. In the outer layer, where

$$x - x_{fire} = Le^{-\theta\chi} \quad , \quad \chi = O(1) \ , \ \chi < \tau \quad , \tag{12}$$

we have

$$T \sim T_b(\chi) \tag{13}$$

the same function but with a different argument.

The situation is more complicated for the two-dimensional problem. Now the initial 'blow-up' occurs at discrete points defined by the initial data, rather than along the entire front. If $y = 0$ is such a point, then in its neighborhood [8] and [9] are still valid but now t_c is a function of y,

$$t_c = c_1 + c_2 \left(\frac{y}{L\theta}\right)^2 \quad , \quad t < c_1 \leq t_c \tag{14}$$

(c_1, c_2 constants)

so that

$$[c_1 - t] + c_2 \left[\frac{y}{L\theta}\right]^2 = t_{ref.} e^{-\theta\tau} \quad . \tag{15}$$

Each of the terms in brackets is positive, and so each must be separately very small. As $t \uparrow c_1$, $y \neq 0$,

$$T/T_s \sim \left[1 + \frac{1}{\theta} \ell n \left(\frac{c_2}{t_{ref.}}\right) \frac{y}{L\theta}^{\ 2} \right]^{-1} \tag{16}$$

with a corresponding (although much more complicated) expression for the pressure. Intense pressure spots form as a consequence of the instability, and these will generate blast-waves from which will emerge the familiar transverse waves.

It seems likely that the final spacing of these waves is determined by nonlinear interactions. If the spacing at any time is larger than the equilibrium value, new waves can be generated by the mechanism described here. Thus accurate prediction of the spacing by computation is only possible if pressure spot formation is not suppressed by inadequate resolution.

Example 3. The Flow-Field and Stability of Flame-Bubbles

The third and final problem to be discussed is that of 'flame-bubbles'. If a very lean hydrogen-air mixture, confined to a vertical tube, is ignited at the bottom, a number of small 'bubbles' of flame are generated, rather than a single tent-shaped flame [10]. Each bubble is approximately 4mm in diameter and its luminous portion forms approximately a hemispherical cap, but the extent of the region in which reaction is significant is not known. The bubbles appear to have stationary structure and rise at a steady speed.

Recently S. Weeratunga, a student of mine, set out to learn something about the flow-field associated with a single bubble [11]. To do this exactly requires solution of the Navier-Stokes equations for a variable density (but incompressible) fluid coupled with an energy (temperature) equation containing a reaction controlled heat-source term. This is a formidable problem so that the computations of [11] are done for a simplified model which eliminates the complication of reaction. This model is characterized by a constant temperature spherical sheet of prescribed radius moving at a prescribed speed U relative to the uniform far field. The ability to predict the rise speed is lost, but the calculated flow-field will have much in common with the exact solution, and what has been learned from this work will be of great value in any more realistic attempt.

At first sight it might be thought that analysis has no role to play in this problem, but that is not the case. The heat carried away from the sheet is swept into the wake so that the fluid there has an elevated temperature and a reduced density. The associated buoyancy force generates a momentum deficit which persists to an infinite distance behind the bubble where a similarity solution can be constructed [12]. If x is measured axially, σ radially, the independent similarity variable is $\eta \propto \sigma^2/x$ and the axial velocity-u is a function of η that can be determined from the numerical solution of ordinary differential equations. When the Prandtl Number is equal to 1, the center-line velocity is

$$\frac{u}{U}(\eta = 0) = [1 - \dot{q}/Fr^2]^{1/2} \tag{17}$$

where $\dot{q}$ is the non-dimensional heat shed by the bubble and Fr is the Froude Number. This result shows that for a given heat release there is a minimum allowable Froude Number, i.e., a minimum rise speed. This is also true when the Prandtl Number is different from 1, but no simple formula like (17) can then be found.

This wake has a displacement effect on the exterior flow like a paraboloidal body of revolution. Thus as $x \to \infty$, $\sigma/\sqrt{x} \to \infty$, the non-dimensional stream function is

$$\psi \sim \frac{1}{2}\sigma^2 - \frac{c}{Re}[x^2 + \sigma^2)^{1/2} + x] + \ldots \qquad (18)$$

where the first term is the uniform flow, c is a constant defined by the wake similarity solution, and Re is the bubble Reynolds Number (typically ~ 5).

The significance of these results insofar as the numerical computations are concerned is that they must be used to define the far-field boundary conditions if meaningful, accurate results are to be obtained. Ignoring them and simply applying uniform far-field conditions leads to significant error.

Analysis can also provide useful insights into the stability of these bubbles. There is good reason to believe that the very strong curvature characteristic of the physical flames is de-stabilizing. The evidence comes from analysis of a model for stationary flame-bubbles first proposed by Zeldovich, in which reaction is modeled by a Dirac delta-function whose strength is an Arrhenius function of temperature. The location of this delta-function defines a flame-sheet. For the stationary bubble, a spherically symmetric solution can be constructed in which the gas velocity is everywhere zero, and the mixture supplied at infinity reaches the flame-sheet by diffusion alone. However, this solution is unstable for all Lewis Numbers [12]: a small one-dimension displacement of the flame-sheet from its equilibrium position will lead either to collapse or unbounded growth.

It is likely that this instability can persist in the presence of convection. Evidence comes from unpublished work of Weeratunga, Buckmaster, and R.E. Johnson, in which the flame-sheet is assumed to be a spherical cap (rather than a complete ball) and the temperature along the axis is described by the model equation

$$\rho u C_p \frac{dT}{dx} = \lambda\left[\frac{d^2T}{dx^2} + \frac{2}{x}\frac{dT}{dx}\right] + \Omega \qquad , \qquad x < 0 \qquad (19)$$

where ρu is constant, and the flame-sheet, of radius a, cuts the axis at $x = -a$. When the Reynolds Number is zero, this is equivalent to Zeldovich's model, and as it approaches infinity the description is that of a plane flame, so that (19) can be thought of as an iterpolation between these well-defined large and small curvature limits.

Despite the fact that the plane flame is stable to one-dimensional disturbances for a wide range of Lewis Numbers, the solution described by (19) (and the corresponding equation for the mixture concentration) is unstable for all Lewis Numbers when the Reynolds Number is large but finite, and undoubtedly this instability persists to smaller Reynolds Numbers.

A possible stabilizing mechanism is suggested by the flow-field solutions (loc. sit.). Far upstream on the x-axis ρu is defined by the uniform incoming flow. Downstream on the axis it is defined by the wake solution and is relatively small, perhaps zero. Thus ρu on the axis has a negative gradient in the neighborhood of the flame-sheet. Collapse of the flame is associated with a downstream displacement into a region of decreased mass flux, so that the tendency for the flame to propagate at a fixed speed relative to the mixture provides a restoring mechanism. A similar argument can be applied to a growing flame cap moving into the region of increasing mass flux. Analysis of the model (19) with ρu different from a constant should clarify this possibility, and this is presently being explored.

Concluding Remarks

The three problems described here make the point. Analysis can explain numerical results; it can lead to the definition of useful numerical problems; it can aid in preventing errors due to inadequate resolution; and it can ensure the proper definition of a numerical problem. Those who suggest that computers are a panacea are wrong, and are likely to remain so in the forseeable future.

Acknowledgement

This work was supported by the Air Force Office of Scientific Research. A NATO research grant facilitating collaboration with Dr. Claudine Schmidt-Lainé of the Ecole Centrale de Lyon is gratefully acknowledged.

References

[1] J.D. Buckmaster and G.S.S. Ludford, "The effect of structure on the stability of detonations, I. Role of the induction zone," Proceedings of the 21st International Symposium on Combustion, The Combustion Institute, Pittsburgh, 1986-7, to appear.

[2] J. Buckmaster, "Pressure transients and the genesis of transverse shocks in unstable detonations," in review.

[3] J. Buckmaster, D.S. Stewart, A. Ignatiadis and M. Williams, "On the wind generated by a collapsing diffusion flame," Combustion Science and Technology, 46, 145, 1986.

[4] N. Peters, Prog. Energy Combustion Sci., 10, 319-339, 1984.

[5] J. Buckmaster and G.S.S. Ludford, Lectures on Mathematical Combustion, SIAM Press, Philadelphia, 1983.

[6] A. Liñán, "The asymptotic structure of counterflow diffusion flames for large activation energies," Acta Astronautica, 1, 1007-39, 1974.

[7] P. Givi, W.-H. Jou and R.W. Metcalfe, "Flame extinction in a temporally developing mixing layer." Proceeding of the 21st International Symposium on Combustion, The Combustion Institute, Pittsburgh, 1986-7, to appear.

[8] G.E. Abousief and T.Y. Toong, "Theory of unstable one-dimensional detonations," Combustion and Flame, 45, 67-94, 1982.

[9] K. Kailasanath, E.S. Oran, J.P. Boris, and T.R. Young, "Determination of detonation cell size and the role of transverse waves in two-dimensional detonations," Combustion and Flame, 61, 199-209, 1985.

[10] G.H. Markstein, "Instability phenomena in combustion," Fourth International Symposium on Combustion, The Williams and Wilkins Co., Baltimore, 44-59, 1953.

[11] S. Weeratunga, "Numerical solution of buoyancy induced motion of a hydrogen flame bubble," Ph.D. thesis, U. of Illinois, Urbana-Champaign, 1986.

[12] J. Buckmaster and S. Weeratunga, "The stability and structure of flame bubbles," Combustion Science and Technology, 35, 287-296, 1984.

COMBUSTION AND COMPRESSIBILITY IN GASES

J F CLARKE

AERODYNAMICS, CRANFIELD INSTITUTE OF TECHNOLOGY, BEDFORD, ENGLAND

1. INTRODUCTION

This paper first examines the local character of a general flow within which chemical reactions and compressibility effects coexist. Local Mach numbers must not be small in such circumstances, and it is deduced that the "inviscid" Euler equations are adequate to describe all behaviour outside the interior of shock waves (which are therefore treated as discontinuities) when the reactions are of the large-activation energy type.

Second, possible steady or quasi-steady-state plane-flow solutions, in both half-space and unbounded spaces, are examined. As a result, essentially unsteady-states are defined, and identified with conditions that arise during the development or initiation of a detonation wave.

Finally, the association of dilatational-wave propagation and combustion-reactions in unsteady induction domain behaviour is outlined, and linked with the existence of deflagration waves within which reaction goes quickly to completion, but in a quasi-steady manner. It is shown how such quasi-steady deflagrations can sustain a shock wave in a non-steady environment. The model affords some description of events that precede the birth of fully-established detonations.

A number of strands of the whole pattern have been through some individual development in recent times, and the opportunity has been taken to draw them together into a more coherent picture of some aspects of plane combustion-wave propagation through a compressible medium.

2. LOCAL PARAMETERS

2.1 The species equation

It is possible to learn a lot about broad features of any flow by inspecting the relative magnitudes of the terms that appear in the conservation equations. Each term describes the role of a particular physical mechanism and as such is contrasted with other mechanisms through a local dimensionless number or parameter. The matter will become clear as the present analysis proceeds, first, to a discussion of the reactant species conservation equation.

Assume that the combustion reaction is a simple irreversible uni-molecular decomposition

$$F \rightarrow P, \tag{2.1}$$

so that the conservation equation for reactant species F, whose mass fraction is c, can be written in the form

$$(\rho c)_t = -\underline{\nabla}.(\rho c\underline{u} - \rho\tilde{D}\underline{\nabla}c) - \rho W\, c/\tau_{ch} \tag{2.2}$$

C.-M. Brauner and C. Schmidt-Lainé (eds.), Mathematical Modeling in Combustion and Related Topics, 43–63.

The mixture density is ρ, $\underline{u}$ is the flow velocity and W is the (dimensionless) molecular weight of F. It has been assumed that the diffusion of mass can be described by Fick's law, so that $-\rho\tilde{D}\nabla c$ is equal to the diffusive mass flux of F per unit area per unit time and $\tilde{D}$ is the local diffusion coefficient.

2.2 Reaction rates

The chemical time τ_{ch} must first recognize that a chemical reaction occurs as a result of an interaction between (at least) a pair of molecules, even in the present case of a unimolecular reaction (cf the Lindemann mechanism described by Williams (1985, §B.2.2.)). Thus τ_{ch} must have as a factor the average time between molecular collisions τ_{coll}. Elementary kinetic theory (e.g. §1.10, Clarke & McChesney, 1976) shows that τ_{coll} is given by the kinematic viscosity ν divided by the square of the mean thermal molecular speed. It is very appropriate to choose the latter to be given by the frozen sound speed a_f (§2.1, Clarke & McChesney, 1976), so that

$$\tau_{coll} = \nu/a_f^2. \tag{2.3}$$

Arrhenius hypothesised that only a fraction of all molecular collisions results in a reaction, since a certain minimum (activation) energy is required in any encounter before such an event can take place. Thus τ_{ch} is larger than τ_{coll} by an exponential factor

$$\exp(E_e/\tilde{R}T),$$

where E_e is the activation energy of the reaction per mole of reactant, $\tilde{R}$ is the Universal Gas Constant and T is the absolute temperature. Evidently

$$E_e/\tilde{R} \equiv T_e \tag{2.4}$$

defines an activation temperature T_e, so that one can write

$$\tau_{ch} \propto \tau_{coll} \exp(T_e/T). \tag{2.5}$$

The ideas leading to this sort of expression for τ_{ch}, though naive, are broadly correct in physical terms. However it would usually not be acceptable, numerically, to simply make τ_{ch} equal to the right-hand side of expression (2.5). A corrective factor, that helps to bring matters into a more acceptable line, often associated with notions of the participation of internal molecular energy and molecular orientation in a successful encounter, can be applied to (2.5) so that it now reads

$$\tau_{ch} = (\nu/a_f^2)(T/T_e)^s \exp(T_e/T), \tag{2.6}$$

where s an index whose size is modest and roughly in the range ± 1.

2.3 Local length and time scales and their relation to derivatives

It will be assumed that convection always has a part to play in the combustion processes discussed in the present paper. Whether it is always a major part or not will depend upon circumstances, as will be seen. Since $\nabla.(\rho c\underline{u})$ is the term that summarises the influence of convection in (2.2) it is clear that convection implies two important local features of any flow system; (i) there is a local flow direction defined by the vector velocity $\underline{u}$ and (ii) there is a local length-scale l_s such that $\nabla.\rho c\underline{u}$ is of order $\rho cu/l_s$ where

$$u = |\underline{u}| \tag{2.7}$$

In fact $\underline{\nabla}$ can be treated locally as a vector of the general form

$$\underline{s}/l_s + \underline{n}/l_n + \underline{b}/l_b, \tag{2.8}$$

where $\underline{s}$ is the unit vector $\underline{u}/|\underline{u}|$ in the streamwise direction; likewise $\underline{n}$ and $\underline{b}$ are unit vectors in the directions of the principal normal and binormal at the same point on the particle path. The lengths l_s, l_n, l_b are indicative of the local scales for significant rates of local spatial change, or, in other words, they normalise the appropriate element of the gradient operator $\underline{\nabla}$.

In general (unsteady) conditions there will be a typical time scale for local time variations at a fixed point in space, so that, calling this quantity τ_u, it can be seen that the right-hand side of (2.2) is equal, locally, to $\rho c/\tau_u$.

Therefore (2.2) can be re-expressed as follows:

$$\underset{U}{\rho c/\tau_u} = \underset{C}{-\rho c u/l_s} + c\rho\tilde{D}\{\underset{D_s}{l_s^{-2}} + \underset{D_c}{\underbrace{l_n^{-2} + l_b^{-2}}}\} \quad \underset{R}{-W\rho c/\tau_{ch}} \tag{2.9}$$

The terms identified as U, C, D_s, D_c and R represent, respectively, unsteadiness, convection, diffusion along the particle path, diffusion across the particle path, and reaction. With C as the standard, their relative sizes are evidently in the ratios

$$\tau_{fl}/\tau_u : 1 : \tilde{D}/ul_s : (\tilde{D}/ul_s)(l_s/l_c)^2 : W\,\tau_{fl}/\tau_{ch}, \tag{2.10}$$

where

$$\tau_{fl} \equiv l_s/u_s \tag{2.11}$$

is indicative of local flow times, and l_c is either l_n or l_b.

In term R the molecular weight W is often a number of order 20 or 30, and must obviously not be neglected in any assessment of relative magnitudes. It is even more important to observe that, in a bi-molecular reaction (say), it will not be simply W but W times the mass fraction of the other reactant that appears as a multiplier in R, this quantity may be very small and clearly therefore very influential when it comes to assessing the relative size of R. Whether one incorporates quantities like W, or W times another reactant concentration, into the definition of τ_{ch} is a matter for choice.

Significant local parameters can now be identified as follows:

$\nu/\tilde{D}$ = Sc; local Schmidt number (≈ 1);
$Re_s \equiv ul_s/\nu$; local streamwise Reynolds number; 2.12
$Re_c \equiv ul_c/\nu$; local cross-flow Reynolds number; 2.13
$\Delta \equiv \tau_{fl}/\tau_{ch}$; local Damkohler number; 2.14
$\omega \equiv \tau_{fl}/\tau_u$; local frequency parameter. 2.15

2.4 Local Damkohler number

Combining (2.6, 11 and 15) reveals the structure of Δ in terms of local parameters that are descriptive of both flowfield and chemical kinetics, as follows;

$$\Delta = (Re_s/\mathcal{M}^2)\{(T_E/T)^S \exp(-T_e/T)\} \equiv (Re_s/\mathcal{M}^2)\mathcal{A} \tag{2.16}$$

where

$$u/a_f = \mathcal{M} \qquad 2.17$$

is the local frozen Mach number and $\mathcal{A}$ defines a local "activation-energy" number.

It is central to any future discussion of flows that contain chemical reactions to take note, not only of the appearance of the local Mach number $\mathcal{M}$, but particularly of its intimate association with the kinetics of any chemical activity. It is particularly important to note that $\mathcal{A}$ is very sensitive to changes in the absolute temperature T, since T_e is usually large for typical combustion reactions. In fact, any increment $d\mathcal{A}$ in $\mathcal{A}$ is related to the increment dT in T via

$$d\mathcal{A}/\mathcal{A} = [(T_e/T) - s](dT/T), \qquad 2.18$$

so that a local Damkohler number Δ will change very rapidly with changing temperature (such as might occur as a result of the combustion reaction itself) since T_e/T is usually in excess of ten and $|s|$ is of order unity. Note, too, that

$$\mathcal{A} \ll 1 . \qquad 2.19$$

In passing it is interesting to observe that any Reynolds number can be expressed as a quotient of a Mach number and a Knudsen number Kn, which is the ratio of a molecular mean free path, such as, l_m (equal to ν/a_f), and a flow length, such as l_s; thus, in the present case,

$$Re_s = \mathcal{M}/Kn_s \; ; \; Kn_s = l_m/l_s. \qquad 2.20$$

The local Damkohler number (2.16) can therefore be written in the form

$$\Delta = \{(T_e/T)^s \exp(-T_e/T)\}/(\mathcal{M}Kn_s) = \mathcal{A}/(\mathcal{M}Kn_s). \qquad 2.21$$

Since we are only concerned with combustion in continuum flows, for which Kn is essentially very small, the role of Kn_s as a divisor in the local Damkohler number Δ is evidently important.

3. COMBUSTION WAVES AND COMPRESSIBILITY

3.1 Role of diffusion in compressible flows

A key result from §2 is contained in (2.16), which is reiterated here;

$$\Delta = (Re_s/\mathcal{M}^2)\mathcal{A}. \qquad 3.1$$

With activation energy large, T_e/T is usually of order 10 or more and $\mathcal{A}$ is always a very small number, even in the hottest parts of the flow.

The case for which the flow is extremely hot as a consequence of some outside, combustion-unrelated, phenomena, such as shock-heating in a hypersonic field is excluded from the present discussion.

When the local Mach number is itself of order unity it follows that Δ can only be of that same order when $Re_s \gg 1$. As a consequence diffusive terms D_s will be negligible compared with the convective C terms and the field will be governed by convective-reactive (or CR) balances, at least in steady or quasi-steady situations. It can be seen from (2.10) that quasi-steady means $\tau_{fl} \ll \tau_u$; this point must be taken up in more detail below.

It is perfectly possible to find local cross-stream scales small enough to make the D_c terms of order one; in fact $l_c = O(l_s Re_s^{-\frac{1}{2}})$ is all that is required. However, this streamwise, boundary-layer-like, situation will not be considered here and attention will be confined to broadly one-dimensional geometries for which the combustion activity takes place between surfaces that lie across rather than along the flow. Such phenomena can usefully be referred to as a combustion waves.

Since the main theme of this article is the relationship between combustion and compressibility it is important to make the following (summary) observations:

Values of $\mathcal{M}$ of order unity are the proper indication that compressibility is of central importance or, in other words, that thermal and flow-kinetic energies are comparable. In that case those parts of the field that involve a combustion reaction are, in essence, free of diffusive effects, and combustion waves in these conditions will not be thermal flames.

3.2 Deflagration

It would be helpful to have another name for those parts of a field in which combustion and compressibility effects co-exist and it is proposed to use the word deflagration for that purpose here and, hopefully, elsewhere.

3.3 Shock Waves

One must anticipate the presence of shock waves in any supersonic compressible flow. These are essentially thin regions within which convection and viscous "diffusion" are in a steady or quasi-steady state of balance; l_s must be small enough to make Re_s of order unity in such regions, and so ensure the necessary CD_s balance. From (3.1), the implication is that Δ is of order $\mathcal{A}$ in magnitude and reaction has, in all essentials, no role to play within the shock-wave, which can therefore be treated as a chemically-frozen phenomenon.

In the context of a "diffusionless", or Euler-equation, model of combustion in compressible flow, shocks will be treated as quasi-steady discontinuities that obey the Rankine-Hugoniot relations for a gas mixture of fixed or frozen chemical composition.

Cases for which a shock is so strong that the translational temperature T of the immediate post-shock gas is high enough to make $\mathcal{A}$ not very much less than unity will not be considered here but are none-the-less interesting topics for study.

3.4 Dilatational Waves

It is now opportune to review some recent studies of deflagrations, their relationship to low-speed flames, and other matters, and to attempt to put such matters into their proper context with special reference to the initiation of, or transition to, gaseous detonation waves.

It is at this point that one meets that distinctive property of all compressible media, namely their capacity to sustain longitudinal dilatational waves, of which the shock is but one example. Indeed it is mostly in association with the matter of wave propagation that one must make statements about terms like U (in (2.9)), and the time-scale τ_u. Although §2 deals in detail exclusively with the species-conservation equation (because of its direct involvement with the reaction term R) it is clear that similar local parameterisations apply to all of the other necessary conservation equations.

4. EQUATIONS FOR ONE-DIMENSIONAL UNSTEADY MOTION

The work that follows will deal almost exclusively with one-dimensional unsteady flow and therefore with the following set of conservation equations, for mass, momentum and energy:

$$\rho_t = -(\rho u)_x, \tag{4.1}$$

$$(\rho u)_t = -(p + \rho u^2 - \tfrac{4}{3}\rho\nu(\partial u/\partial x))_x, \tag{4.2}$$

$$(\rho h_{st})_t = p_t - (\rho u h_{st} - \lambda T_x - \rho\tilde{D}Qc_x - \tfrac{4}{3}\rho\nu(\tfrac{1}{2}u^2)_x)_x . \tag{4.3}$$

Apart from those quantities already defined in §2, p is the pressure and h_{st} is the stagnation enthalpy, namely

$$h_{st} = C_p T + \tfrac{1}{2}u^2 + cQ. \tag{4.4}$$

The specific heat at constant pressure, C_p, has been assumed to be independent of both c and T, in line with the broadly heuristic character of the present exercise. The foregoing equations must be augmented by the thermal equation of state

$$p = \rho RT, \tag{4.5}$$

where R here is the Universal Gas Constant divided by molecular weight W, and the particular version of (2.2) namely

$$(\rho c)_t = -(\rho uc - \rho\tilde{D}c_x)_x - \rho W(a_f^2/\nu)\, c. \tag{4.6}$$

The frozen sound speed is a_f, where

$$a_f^2 = \gamma RT = [C_p/(C_p - R)]RT \tag{4.7}$$

and γ is the ratio of frozen specific heats (assumed constant).

Evidently h_{st} can be re-written in the form

$$h_{st} = C_p T\{1 + \tfrac{1}{2}(\gamma - 1)\mathcal{M}^2\} + cQ \tag{4.8}$$

where $\mathcal{M}$ is the local Mach number, defined in (2.17). Present concern is with the situation for which $\frac{1}{2}(\gamma - 1)\mathcal{M}^2$ is <u>not negligible</u> compared with unity.

5. ONE DIMENSIONAL STEADY OR QUASI-STEADY FLOW

5.1 Introduction

Analysis of the effects of energy addition to inviscid, non-heat-conducting, compressible flow (it will be helpful to refer to such flows in a general way as "inviscid" or Euler flows from now on, meaning that all terms like D_S, D_C are negligible compared with C, R and, where necessary, U (cf equation (2.9))) is a relatively simple matter and wholly suitable for undergraduate texts on gas dynamics. At least this is true when the energy addition is a <u>specified</u> quantity, <u>not</u> linked to local gas-dynamical conditions, or when one is only interested in the final equilibrium output conditions for a given net heat addition to a given input flow. The classical Hugoniot-curve and Rayleigh-line treatments are usually all that is required in this latter case. It is clearly necessary to have a good reason for any re-examination of such a problem in the present context, and, simply stated, such a reason lies in the elucidation of the <u>structure</u> of the flow between input and output states or, in other words, in the detailed connections between local chemical rates and gas-dynamical states. Re-examination also provides an opportunity to put subsequent studies of <u>un</u>steady flows in context, and also to make some useful observations of a general kind about the steady-state.

5.2 Simplified equations

Using the arguments of §3, equations (4.1, 2, 3 and 4) can be simplified and integrated to give

$$\rho u = m, \quad p + mu = mf, \quad C_p T + \tfrac{1}{2}u^2 + cQ = H, \qquad 5.1$$

where m, f and H are all constants, in a steady flow. The species equation (4.6) becomes

$$mc_x = - \{W(a_f^2/\nu)(T_e/T)^S\}\exp(-T_e/T)\rho c, \qquad 5.2$$

so that c must be a monotonic decreasing function of distance x perpendicular to the combustion wave. Together with (4.5) and (4.7) these equations form a closed system, given that ν, T_e and W are known for any given system.

5.3 Temperature-velocity diagrams

By eliminating p and ρ from the momentum relation (5.1b) in favour of temperature T the following temperature-velocity relationship can be derived

$$\tilde{L}: T' = (1 + \gamma M^2)u' - \gamma M^2 u'^2 : \quad M^2 = u_i^2/a_{fi}^2, \qquad 5.3$$

where T' is T measured in units of T_i and u' is u measured in units of u_i; T_i and u_i are values of T and u in the "inlet" stream and M is the (frozen) Mach number of this stream. $\tilde{L}$ is in fact a version of the Rayleigh line relation. Similarly (5.1c) can be written as

$$\tilde{E}: \quad T' + \tfrac{1}{2}(\gamma - 1)M^2u'^2 + c'Q' = 1 + \tfrac{1}{2}(\gamma - 1)M^2 + Q'. \qquad 5.4$$

where

$$c = c_i c', \quad c_i Q = C_p T_i Q'. \qquad 5.5$$

Thus $\tilde{E}$ is a second temperature-velocity relationship parameterised by c', which lies in the range 0<c'<1; c' = 1 marks the initial (unburnt) condition and c' = 0 the final (burnt) state.

The following relationships for T' > 0, u' > 0 are readily established for any value of M; they are exmplified in Fig. 1 for $M^2 > 1$:

$\tilde{L}$ has a maximum value at P, where

$$T'_{max} = (1 + \gamma M^2)^2/4\gamma M^2 = \gamma M^2 u'^2_{max}; \qquad 5.6$$

$\tilde{E}$: c' = 0 is tangent to $\tilde{L}$ at what is called the Chapman-Jouguet, or CJ, point, where

$$T'_{CJ} = (1 + \gamma M^2)^2/(\gamma + 1)^2 M^2 = M^2 u'^2_{CJ} = 4\gamma T'_{max}/(\gamma + 1)^2, \qquad 5.7$$

provided that

$$Q' = Q'_{max} = (M^2 - 1)^2/2(\gamma + 1)M^2. \qquad 5.8$$

Translating (5.7) back into dimensional terms shows that u_{CJ}^2 is equal to a_{fCJ}^2 and flow out of a combustion wave at the CJ point is sonic. When $Q' < Q'_{max}$ $\tilde{E}$: c' = 0 intersects $\tilde{L}$ at points like S_1, S_2.

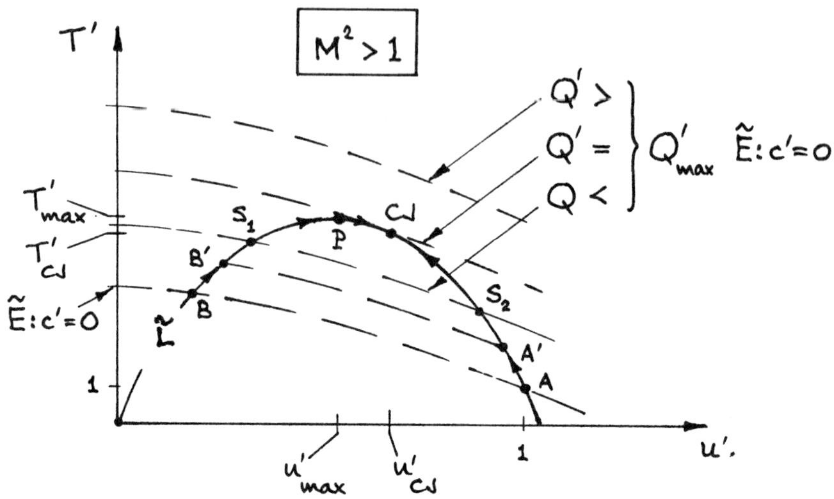

FIGURE 1. The Rayleigh line $\tilde{L}$ and energy relationship $\tilde{E}$ on the T', u' plane (see (5.3) and (5.4)). T'_{max}, u'_{max}, T'_{CJ} and u'_{CJ} are defined in (5.6) and (5.7) while Q'_{max} is defined in (5.8). The diagram is drawn for $M^2 > 1$ (i.e. a supersonic flow at "inlet"). The arrows on $\tilde{L}$ show the direction of x - increasing and these loci represent the continuous parts of the "inviscid" flow-field.

5.4 Solutions for $M^2 > 1$

The possible types of steady-state solution for a supersonic "inlet" stream can now be described as follows:

(i) The "inlet" stream at conditions relating to point A flows into a frozen Rankine-Hugoniot shock-wave and jumps from conditions A to conditions B, where T' is equal to

$$T'_B = [2 + (\gamma - 1)M^2][2\gamma M^2 - \gamma + 1]/(\gamma + 1)^2 M^2 \qquad 5.9$$

and u' is equal to

$$u'_B = [2 + (\gamma - 1)M^2]/(\gamma + 1)M^2. \qquad 5.10$$

Thereafter the flow proceeds in continuous fashion to a final all-burnt (c' = 0) solution-point, such as S_1, on that part of $\tilde{L}$ between B and CJ, provided that $Q' \leq Q'_{max}$. The flow between B and CJ is subsonic:

(ii) The inlet stream undergoes a continuous change from A to point A' on $\tilde{L}$ as a result of energy release into the supersonic flow. At A', with the fuel partially burnt ($0 < c' < 1$), the flow passes through a frozen Rankine-Hugoniot shock to point B' on the subsonic branch of $\tilde{L}$; reaction proceeds to completion at points like S_1 on this branch:

(iii) The flow proceeds from inlet to all-burnt, at S_2 for example, on the supersonic branch of the Rayleigh-line $\tilde{L}$.

5.5 Remarks

Several comments on these results are in order. First, it should be observed that all of the three types of solution just outlined have been found in a study of solutions of the full equations of §4 in a half-space $x > 0$ for the case of Prandtl number equalt to $\frac{3}{4}$ and Lewis number equal to 1 (Clarke, 1983). Of course the discontinuities described in that paper have proper continuous CD_S structure (of continuum Navier-Stokes type) as a solution of the full equations, and this must be joined, or matched, into the CR "inviscid" domains that lie between B and A on $\tilde{L}$ by transitional segments within which C, D_S and R all have a part to play. The discussion in §3 makes it clear that the R term will usually be smallest in such situations, since any non-zero value of Q' will usually ensure that final temperatures are greater than T_B and $\mathcal{A}(T_B)$ will therefore be smaller than its final all-burnt value.

This fact is important when one remembers that, for any solution that is not a CJ (or $Q' = Q'_{max}$) solution, the flow behind the shock is subsonic and disturbances can overtake the shock and its downstream CD_SR domain. As a consequence the delicate CD_SR balance that links the shock and its associated CR domains, within which most of the chemical energy is released, can be perturbed and the CD_S shock itself will move in response to the disturbances; the nature of such responses and their stability is an important topic on which much work remains to be done.

For flow in a half-space, the exact momentum equation for steady flow integrates to give (cf (4.2.))

$$p + mu - \tfrac{4}{3}\rho\nu(du/dx) = p_i + mu_i - \tfrac{4}{3}\rho_i\nu_i(du/dx)_i, \qquad 5.11$$

where $(\;)_i$ indicates an inlet value at $x = 0$. The present "inviscid" analysis is based on the idea that the term in du/dx is negligible (of order $1/Re_S$; cf §2) compared with p + mu, but this need not be true of the term in $(du/dx)_i$ relative to $p_i + mu_i$. It can be verified that $(du/dx)_i$ must be negative and that its effect on $\tilde{L}$ in (5.3) is to add a positive term to the right-hand side of that relation,proportional to $(du/dx)_i$ in magnitude; as a result, $\tilde{L}$ in Figure 1 is moved upwards so that solutions then exist for Q' values greater than Q'_{max}. However, the flow then must start at $x = 0$ with a portion of a frozen CD_S shock adjacent to the $x = 0$ plane and all energy release must then take place under subsonic conditions.

5.6 Waves in unbounded space

It is perfectly possible to deal with solutions like ABS_1 in Figure 1 as if they occupied a complete x-space ($-\infty < x < \infty$), provided that one is able to ignore changes of the state ahead of the shock with time. Since state A on Figure 1 must involve a non-vanishing reaction rate, as can be seen from (5.2), it may not always be possible to do this (a problem often referred to as the "cold-boundary difficulty", e.g. Williams (1985) §2.1.2). If the temperature at A is low enough (e.g. normal room temperature the ambient atmosphere, into which the shock is travelling, changes so slowly that it is well within the accuracy implied by the use of the Euler equations (5.1) and (5.2) to ignore such changes during the brief period of travel of the whole shock and combustion wave system across any finite part of the domain that is of practical significance. In passing it is worth noting two formulations of the problem of large combustion-driven disturbances to a spatially uniform atmosphere, whose temperature T_O varies with time alone as a result of the ambient (slow) combustion activity; one, due to Aldushin, Khudyaev and Zeldovich (1981) examines T/T_O and another (Clarke, 1985a) uses the difference $T - T_O$. Although both papers deal explicitly with $M \ll 1$ their results are instructive in the present context and illustrate something of the character of slowly-varying or quasi-steady combustion-wave propagation.

5.6.1 Lack of quasi-steady solutions. For an unbounded space $(-\infty < x < \infty)$ it is clear that no mechanism exists to sustain values of $(du/dx)_i$ other than zero, since "inlet" here must mean a uniform stream emanating from $x \to -\infty$. Thus any wave system that travels at Mach number $M > 1$ with $Q' > Q'_{max}$ cannot have a steady or quasi-steady solution (cf Fig. 1). This case is important for the question of detonation initiation and will be raised again below in Section 6.

5.7 Solutions for $M^2 < 1$

As M^2 diminishes the point (1,1) on Figure 1 moves to the left along $\tilde{L}$ on Figure 1. When $M^2 = 1$ it coincides with the CJ point and no heat at all can be added to a steady flow since $Q'_{max} = 0$ (cf (5.8)). When $M^2 = 1/\gamma$ the initial point (1,1) coincides with the maximum point P on $\tilde{L}$ and moves to the left of this point as M^2 decreases further. Quasi-steady solutions are possible, with energy addition taking place entirely into a subsonic flow, culminating in a just-sonic flow at CJ when $Q' = Q'_{max}$ for the given wave speed.

Such behaviour is exactly what takes place behind a shock like AB on Figure 1, except that $T'_B > 1$ behind the shock and so reaction-rates are invariably considerably faster in the shock-led flows.

5.8 Deflagration structures for $M^2 > 1$

It is clear that when $M^2 > 1$ and $Q' < Q'_{max}$ (see (5.8)) a quasi-steady wave exists, with a deflagration following at a fixed distance behind a frozen Rankine-Hugoniot shock wave. For the reasons given earlier it will be assumed that the deflagration starts at point B on $\tilde{L}$ in Figure 1, where the flow Mach number is $M_B < 1$. Note from (5.9), (5.10) and the definition of T' and u' that

$$M_B^2 = M^2 u_B'^2/T_B' = (2 + (\gamma - 1)M^2)/(2\gamma M^2 - \gamma + 1) < 1, \; M > 1. \qquad 5.12$$

In order to simplify (5.2) for c it will be assumed that the quantity in brackets { } in that equation has a constant value Ω. With dimensions of frequency, or reciprocal time. The approximation is a mild (additional) one, especially since it is the exponential $\exp(-T_e/T)$ that dominates behaviour of the reaction rate, apart from c itself, of course, and simplifies subsequent analysis without losing any of the essential physics. Thus (5.2) is considered in the form

$$u\, c_x = -\Omega \exp(-T_e/T)c \qquad 5.13$$

which must be solved together with (5.1) in order to uncover the structure of the quasi-steady deflagration.

Defining the set of dimensionless variables

$$f'' = f/f_B; \quad f = T,\, u,\, c,\, p \text{ and } \rho; \; \{NBc_B \equiv c_i\}, \qquad 5.14$$

and the parameters

$$\varepsilon = T_B/T_e, \quad Q'' = c_i Q/C_p T_B, \qquad 5.15$$

equations (5.1) and (5.13) can be re-expressed in the form

$$\rho'' u'' = 1; \; p'' + \gamma M_B^2 u'' = 1 + \gamma M_B^2;$$

$$T'' + \tfrac{1}{2}(\gamma - 1)M_B^2 u''^2 + c''Q'' = 1 + \tfrac{1}{2}(\gamma - 1)M_B^2 + Q''; \qquad 5.16$$

$$u'' c''_x = -(\Omega/u_B)c'' \exp(-1/\varepsilon T'').$$

In addition (4.5) gives

$$p'' = \rho''T'' = T''/u'' \tag{5.17}$$

where the last result follows from (5.16a).

Downstream of the shock, quantities vary rather slowly, especially when $\varepsilon \ll 1$. In such circumstances the induction domain immediately behind the shock has the following form:

$$T'' - 1 \sim (u'' - 1)(1 - \gamma M_B^2) \sim -(p'' - 1)(1 - \gamma M_B^2)/\gamma M_B^2 \sim -(c'' - 1)(1 - \gamma M_B^2)/(1 - M_B^2) \tag{5.18}$$

where

$$T'' - 1 \sim - \varepsilon \ln(1 - \chi/\chi_{ign}). \tag{5.19}$$

The local coordinate χ in the induction domain is given by

$$\chi = (x - x_{sh})(\Omega \exp(-1/\varepsilon)/\varepsilon u_B), \tag{5.20}$$

with the shock-wave located at $x = x_{sh}$. The "distance to ignition" behind the shock is given by χ_{ign}, where

$$\chi_{ign} = (1 - M_B^2)/Q''(1 - \gamma M_B^2). \tag{5.21}$$

The induction domain, within which changes of gas properties are comparatively slow, terminates when χ is close to the ignition distance χ_{ign}. It is important to note the direct role of compressibility, as exemplified by the post-shock flow Mach number M_B, in determining the latter; if $M_B^2 > 1/\gamma$, χ_{ign} is negative and ignition, in the sense of a localised precipitate rise in T does not occur; the chemical energy is liberated preferentially into kinetic-energy of the flow at the expense of thermal energy and it is true to say that compressibility suppresses ignition. The critical M_B^2 value of $1/\gamma$ occurs when the Mach number of the leading-shock is given (see (5.12)) by

$$M^2 = M_{min}^2 = (3\gamma - 1)/\gamma(3 - \gamma). \tag{5.22}$$

Then $M > M_{min} > 1$ is necessary for ignition within the deflagration attached to a quasi-steady shock. Note that M_{min} is $(10/7)^{\frac{1}{2}} \simeq 1.2$ for $\gamma = 7/5$. It is a point of some interest that a succession of steady weak shocks of this kind will not each individually give rise to a quasi-steady deflagration but will, by their cumulative influence, heat the gas and so prepare it to respond more readily to the ignition-provoking event when it does arrive. It is interesting to re-examine schlieren pictures (such as those of Urtiew & Oppenheim, 1966) of ultimately-detonating mixtures in this light.

Assuming that the Mach number M satisfies (5.22), so that χ_{ign} is positive, behaviour in the neighbourhood of this point is best elucidated by using Kassoy's (1975) non-linear transformation from thermal-explosion theory, namely

$$\chi = \chi_{ign} - H(T'':\varepsilon)\exp\{-(1 - 1/T)/\varepsilon\}, \tag{5.23}$$

where H is a function to be found. This evaluation has been carried out by Kassoy & Clarke (1985) in exactly the present context; since the details of the analysis can be found in the paper just cited only the briefest resume of some of the results will be given here.

Differentiating (5.16) with respect to χ, eliminating $dc''/d\chi$ by using (5.16d) and the relation (5.20), and finally introducing (5.23) gives the (exact) result

$$u''\{1+(\gamma-1)M_B^2\,u''\,(du''/dT'')\} = \{HT''^{-2}-\varepsilon(dH/dT'')\}c''Q'' \tag{5.24}$$

Combining (5.16b) and (5.17) gives the local Rayleigh line relation

$$T'' = (1+\gamma M_B^2)u'' - \gamma M_B^2 u''^2, \tag{5.25}$$

while (5.16c) with (5.25) gives

$$c''Q'' = Q'' + 1 + \tfrac{1}{2}(\gamma-1)M_B^2 - (1+\gamma M_B^2)u'' + \tfrac{1}{2}(\gamma+1)M_B^2 u''^2. \tag{5.26}$$

In view of $\varepsilon \ll 1$ (5.24) for H can be simplified by using the asymptotic expression

$$H \sim H^{(1)} + \varepsilon H^{(2)} \tag{5.27}$$

provided that one is not too near to either one of a pair of critical values. As can be seen from (5.24) these occur as $c'' \to 0$, and one reaches the all-burnt condition, or as $dT''/du'' \to 0$. The latter only occurs when T'' is near to T''_c and u'' is near to u''_c, where

$$T''_c \equiv (1+\gamma M_B^2)^2/4\gamma M_B^2 = \gamma M_B^2 u''^2_c. \tag{5.28}$$

Of course T''_c and u'' are just T'_{max} and u'_{max} from (5.6) but written in the new system (5.14) of dimensionless variables. The singular character of (5.24) at $c'' = 0$ and $dT''/du'' = 0$ is evident from (5.24); special treatments are then necessary to evaluate H and Kassoy and Clarke (1985) show that H is $O(1/\varepsilon)$ near $c'' = 0$ and $O(1/\varepsilon^{\frac{1}{2}})$ near T''_c. It is otherwise easy to see that H is O(1) as $\varepsilon \to 0$, so that $H^{(1)}$ relates simply to u'' (and hence T'') via

$$H^{(1)} = \{1+(\gamma-1)M_B^2 u''(du''/dT'')\}(T''^2u''/Q''c'') \tag{5.29}$$

(NB (5.25) and (5.26)).

The crucial point that emerges from these solutions is that T'' makes the transition from the order unity values that exist in the induction domain, of length χ_{ign}, to its final all-burnt value (obtained by putting $c'' = 0$ in (5.26) to find the final value of u'' and then using (5.25) to find the corresponding value of T'') within distances of $\chi_{ign} - \chi$ that are exponentially small compared with χ_{ign} itself. Thus, although there is no question of a thin flame sheet in the usual sense implicit in thermal flame thoery, there is most certainly a very thin region on the scale of χ_{ign} itself within which reaction goes to completion with no assistance from diffusion.

Some qualitative pictures of the way in which temperature, gas velocity (relative to a stationary shock) and pressure vary behind the shock are displayed in Fig. 2; it has been assumed that Q'' is near enough to Q''_{max}, where

$$Q''_{max} = (1-M_B^2)^2/2(\gamma+1)M_B^2 \tag{5.30}$$

Q''_{max} is exactly Q'_{max} from (5.8) but written in the present dimensionless system of variables.

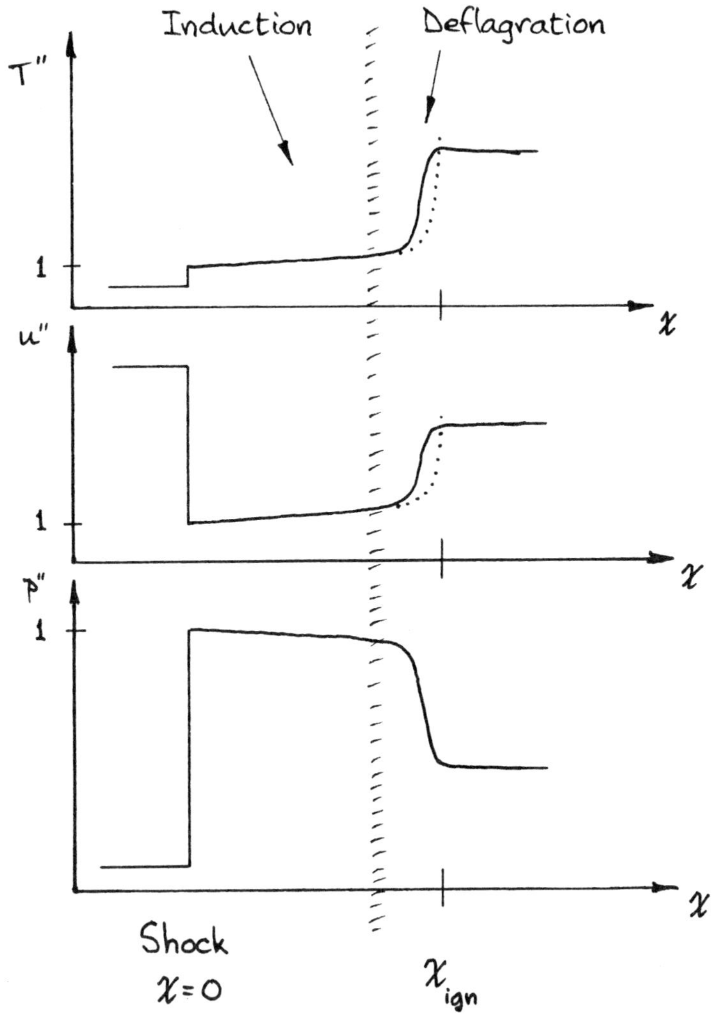

FIGURE 2. Variations of T", u" and p" with χ (see (5.20)) behind a shock in a steady-state condition. The dotted lines are continuations of the induction-domain solutions (5.18). (Not to scale.)

5.8.1 Evolving states. To conclude this section on quasi-steady combustion waves reference is made back to the remark in §5.6.1. about the absence of any quasi-steady-state solutions for situations in which $Q' > Q'_{max}$ (or $Q'' > Q''_{max}$, of course). An alternative way to describe such a situation is to say that the propagation Mach number of the leading shock-wave M is less than the CJ value, obtained from (5.8) with Q'_{max} given. The association of such conditions with evolution towards quasi-steady detonation will be taken up in the next Section.

6. EVOLUTION IN CASES FOR WHICH THE INITIAL SHOCK IS TOO WEAK

6.1 Switch-on shocks

When no steady or quasi-steady state is possible (cf §§5.6.1 and 5.8.1) the term U in (2.9) or (2.10) must play a part in developments from whatever initial states exist or are brought about by some outside agency. Possibilities here are many and varied,but there is obviously prime interest in circumstances *for which* it is *the chemical activity* itself that *drives events*. Then (2.10) makes it clear that the time-scale of local temporal rates of change τ_u must be comparable with the local chemical time τ_{ch}, at least in some parts of the flow.

In particular, situations like the one sketched in Figure 3 can be hypothesised to exist at some initial instant of time. Some confirmation of this possibility comes from numerical computations by Clarke, Kassoy and Riley (1986). Taking lessons from the previous Section (5) and dealing only with those cases for which the atmosphere ahead of the shock is spatially uniform, and reacting so slowly that changes in the atmosphere are of no significance, it is assumed that the shock is strong enough to switch on chemical activity downstream that *is* significant during the time of passage of the wave through the apparatus. It is convenient to be able to refer to shocks in these situations as *switch-on shocks*.

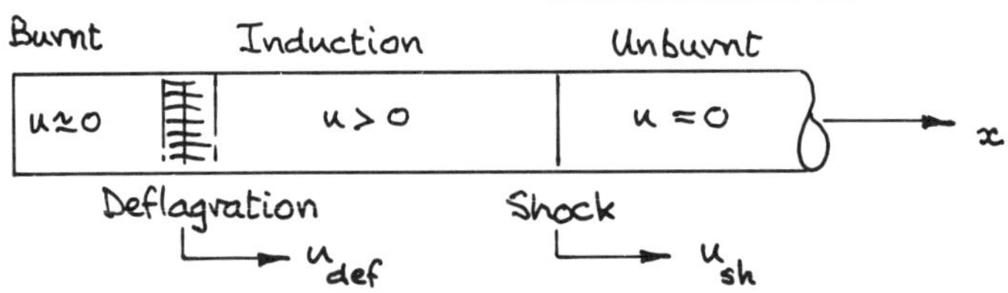

FIGURE 3. General configuration of a shock, induction-domain and deflagration in a tube. Energy released by the deflagration is responsible for driving the shock in this case. Flow velocity behind the shock is $u_{sh} - u_B$ relative to the tube walls.

6.2 The unsteady state

Interest will centre on the situation for which a deflagration follows the switch-on shock. As a consequence of the fact that Q exceeds Q_{max} (cf (5.30) and (5.15) which define Q_{max} in terms of the subsonic Mach number M_B behind the leading shock wave) this deflagration will be seen to fall behind the shock as time progresses. An essentially unsteady induction domain must therefore lie between shock and deflagration (Fig. 3). The chemical time will be greatest in this unsteady induction domain; adapting the notation of §5.8, τ_{ch} in the induction domain has the value

$$\varepsilon e^{1/\varepsilon}/\Omega, \tag{6.1}$$

where ε is now the *initial value of* ε from (5.15a) when T_B has its *initial* value. With (5.14) modified to read

$$f'' = f/f_B \; ; \; f_B \equiv \text{initial post-shock value}, \tag{6.2}$$

the full set of equations from §4 without transport terms can be re-expressed as follows:

$$(\partial/\partial\tau)\rho'' + (\partial/\partial\chi)(\rho''\upsilon'') = 0 = (D\rho''/D\tau) + \rho''(\partial\upsilon''/\partial\chi),$$

$$\gamma M_B^2\,\rho''(D\upsilon''/D\tau) + (\partial p''/\partial\chi) + \gamma M_B^2\dot{u}''_{sh}\rho'' = 0,$$

$$\rho''(DT''/D\tau) - (\gamma - 1/\gamma)(Dp''/D\tau) = Q''\varepsilon\exp[(1 - 1/T'')/\varepsilon]\rho''c'', \qquad 6.3$$

$$(Dc''/D\tau) = -\varepsilon\exp[(1 - 1/T'')/\varepsilon]\,c'',$$

$$p'' = \rho''T''. \quad (D()/D\tau) \equiv (\partial()/\partial\tau) + \upsilon''(\partial()/\partial\chi).$$

The independent variables τ and χ are defined as follows:

$$\tau = t\Omega/\varepsilon\exp(1/\varepsilon);\ \chi = (x - x_{sh}(t))\Omega/u_B\varepsilon\exp(1/\varepsilon) \qquad 6.4$$

where $x_{sh}(t)$ is the time-dependent position of the shock wave. The velocity component υ'' is given by

$$\upsilon'' = u'' - u''_{xh}(\tau) \qquad 6.5$$

where u''_{sh} is $\dot{x}_{sh}/u_B$ and u'' is the dimensionless flow velocity relative to the fixed ground; $\dot{u}''_{sh}$ is the dimensionless acceleration of the shock.

A useful alternative form of the foregoing equations can be written in terms of the Lagrangian variable ψ (NB χ_{sh} is χ at $x = 0$);

$$\psi = \int_{\chi_{sh}}^{\chi} \rho''(\tau,\tilde{\chi})d\tilde{\chi}. \qquad 6.6$$

Thus

$$\rho''_\tau + \rho''^2 u''_\psi = 0;\quad \gamma M_B^2\,u''_\tau + p''_\psi = 0;$$

$$\gamma\rho''T''_\tau - (\gamma - 1)p''_\tau = \gamma Q''\varepsilon\exp[(1 - 1/T'')/\varepsilon]\rho''c''; \qquad 6.7$$

$$c''_\tau = -\varepsilon\exp[(1 - 1/T'')/\varepsilon]c''.$$

6.3 Evolving induction domain

Immediately downstream of the shock, perturbations from initial post-shock values can be written, by analogy with the steady or quasi-steady-state expressions in §5.8, as follows;

$$f'' \sim 1 + \varepsilon f^{(1)}(\tau,\psi)\ ;\quad f = T,\ u,\ p,\ \rho,\ c. \qquad 6.8$$

Since time and distance derivatives are of comparable order in present situations, the perturbation functions $f^{(1)}$ now satisfy the set of partial differential equations

$$\rho_\tau^{(1)} + u_\psi^{(1)} = 0;\ \gamma M_B^2\,u_\tau^{(1)} + p_\psi^{(1)} = 0;$$

$$\gamma T_\tau^{(1)} - (\gamma - 1)p_\tau^{(1)} = \gamma Q''\exp(T^{(1)});\ c_\tau^{(1)} = -\exp(T^{(1)}); \qquad 6.9$$

$$p^{(1)} = \rho^{(1)} + T^{(1)},$$

which can be manipulated to give a single equation for $T^{(1)}$, namely

$$M_B^2(T_\tau^{(1)} - \gamma Q''\exp(T^{(1)}))_{\tau\tau} - (T_\tau^{(1)} - Q''\exp(T^{(1)}))_{\psi\psi} = 0 \qquad 6.10$$

This equation encapsulates the essence of the wave/reaction interactions. Clarke & Cant (1984) and Jackson & Kapila (1985) have described the evolution of an induction domain, situated between a solid piston and the shock wave that is driven by a mixture of heat-release and piston motion. Cant (1984) gives examples for a similar situation which has a contact-surface in place of the solid piston, and recently Clarke & Wang (1986) have described what happens in an induction domain that is terminated by a thermal-flame, with a piston behind the flame to maintain a satisfactorily strong switch-on shock. The general status of (6.10) and its appearance in a number of situations has been described by Clarke (1985b).

It will be shown below how satisfactorily-strong switch-on shocks can be maintained by a deflagration, but meanwhile it is important to observe that all of the previous studies show that temperature and pressure behave, broadly, as depicted in Fig.4. Between the switch-on shock and the leading "driving" mechanism (piston, contact-surface or thermal flame) it is notable that not only temperature but also pressure rises as the location of the "driver" at $x_d(t)$ is approached in the manner sketched in Figure 4, and the switch-on shock itself accelerates, albeit only gently in the time scales of the foregoing calculations (roughly $0 \leq \tau < 1$). The induction domain must behave similarly when the "drive" is from a deflagration.

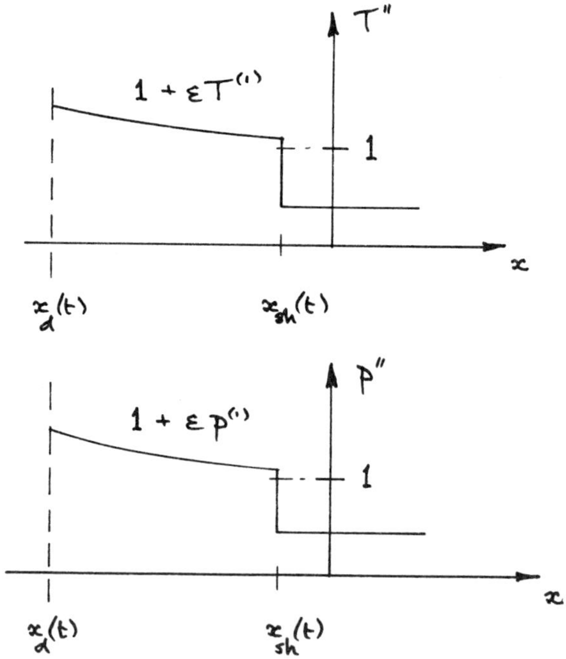

FIGURE 4. Temperature and pressure versus x for some time $t > 0$ in the evolving induction domain behind a switch-on shock at $x_{sh}(t)$ and ahead of the "driving-mechanism" at $x_d(t)$. Compare the rise of pressure p" in the present unsteady induction domain with its fall in the steady-state case, as depicted in Figure 2.

Note, particularly, the contrast between the present unsteady-state pressure field and the steady-state induction behaviour sketched in Figure 2.

6.4 The deflagration

It is hypothesised that the temperature T" rises from its level in the induction domain to its final all-burnt value in the neighbourhood of location $x_d(t)$ (Figure 4). In order to analyse conditions in this region one can adapt the non-linear transformation given in (5.23) so that it now reads

$$X = X_{ign}(\tau) - H(T'', \tau; \varepsilon)\exp\{-(1 - 1/T'')/\varepsilon\}. \qquad 6.11$$

Transformation of equations (6.3) into a new set which has time τ and temperature T" as the independent variables is straightforward but lengthy and only the principal results will be described here.

In essence, it is found that flow structure in the neighbourhood of X_{ign} (as defined in (6.11) for $T'' > 1 + O(\varepsilon)$) is quasi-steady. On the assumption that H and its derivatives are all $O(1)$ as $\varepsilon \to 0$ equations (6.3) reduce to the following set, satisfied by the leading-order estimates for the flow variables:

$$\rho''V'' \sim m(\tau); \quad \gamma M_B^2 m(\tau)V'' + p'' \sim m(\tau)f(\tau);$$
$$Q''c'' \sim -\{T'' + \tfrac{1}{2}(\gamma - 1)M_B^2 V''^2\} + g(\tau); \qquad 6.12$$
$$V''T''^2\{1 + (\gamma - 1)M_B^2 V''V''_{T''}\} \sim -c''Q''H,$$

where $m(\tau)$, $f(\tau)$ and $g(\tau)$ are all functions of integration to be evaluated by matching the quasi-steady deflagration structure defined by (6.12) with the surrounding flow particularly, of course, with the induction domain (cf §6.3 and Figure 4). The velocity V" is defined to be

$$V'' \equiv \dot{X}_{ign} - v'' = \dot{X}_{ign} + u''_{sh} - u'', \qquad 6.13$$

where the last result follows from (6.5), and V" is evidently the velocity of the deflagration relative to the local fluid. It is convenient to define u''_{def}, the dimensionless velocity of the deflagration relative to the tube walls (Figure 3), so that

$$u''_{def} = \dot{X}_{ign} + u''_{sh} \qquad 6.14$$

and (6.13) is the same as

$$V'' = u''_{def} - u''. \qquad 6.15$$

6.5 Evaluation of m, f and g

To leading order ρ'', T" and c" are equal to one in the induction domain, certainly for early times, and u" has a value $u''_{sh} - 1 \equiv u''_I$, say. Therefore

$$m \simeq u''_{def} - u''_I \equiv V''_I; \quad mf = 1 + \gamma M_B^2 V''^2_I;$$
$$g \simeq Q'' + 1 + \tfrac{1}{2}(\gamma - 1)M_B^2 V''^2_I. \qquad 6.16$$

But $u''_{def} - u''_I$ is the speed of flow of gas from the induction domain into the deflagration and $M_B(u''_{def} - u''_I)$ is the Mach number of this flow; to be consistent with §5.3 this is written here as M. Note that

$$M_B V''_I = M = (u_{def} - u_I)/a_{FB} \tag{6.17}$$

and that the first and second equations from (6.12) can be written as

$$\begin{aligned} p'' + \gamma M_B M V'' &\simeq 1 + \gamma M^2, \\ Q''c'' + T'' + \tfrac{1}{2}(\gamma - 1)M_B^2 V''^2 &\simeq Q'' + 1 + \tfrac{1}{2}(\gamma - 1)M^2 \end{aligned} \tag{6.18}$$

6.6 Deflagration alone driving the switch-on shock

Equations (6.18) can now be used to find the conditions described in Figure 3, which implies that the deflagration alone drives the shock. Behind the deflagration where c" vanishes u" is zero and V" is equal to u''_{def}; ahead of the deflagration u" is $u''_{sh} - 1$ and V" is V''_I.

The equation of state in (6.3e) can be written with the aid of (6.12a) and (6.16) in the form

$$p'' = V''_I T''/V'' \tag{6.19}$$

whence (6.18a,b) can be used to find u''_{def} as a function of M, γ and Q". Calculating the difference $u''_{def} - V''_I = u''_{def} - (u''_{def} - u''_I)$ it can be shown after a little algebra that the following relation holds

$$\begin{aligned} u''_{def} - V''_I &= \{(1 - M^2) - [(1 - M^2)^2 - 2(\gamma + 1)M^2 Q'']^{\frac{1}{2}}\}V''_I/(\gamma + 1)M^2 \\ &= u''_I = 2(1 - M_B^2)/(\gamma + 1)M_B^2, \end{aligned} \tag{6.20}$$

and it must be remarked that $V''_I = M/M_B$.

To reiterate, M is the Mach number of flow into the deflagration which releases energy $c_i Q = C_p T_B Q''$; M_B is the Mach number of the post-shock stream relative to the shock; if the situation is steady M and M_B must be equal and this case has been described in Section 5.

For the sake of illustration, assume that the deflagration releases the maximum amount of energy so that Q" is Q'_{max} from (5.8); or, in other words, it is a CJ deflagration. Then (6.19) shows that

$$(1/M) - M = 2((1/M_B) - M_B), \tag{6.21}$$

and it is evident that $M < M_B$ in all such circumstances, because $M_B < 1$ is essential.

Figure 5 illustrates the temperature and pressure profiles that are implied by the model situation discussed in the present §. With pressure in the (stagnant) all-burnt domain written as p_o it can be shown from (6.18), etc, that

$$p''_o \equiv (p_o/p_B) = 1 - 2\gamma M(M_B^{-1} - M_B)/(\gamma + 1) \tag{6.22}$$

and that

$$\begin{aligned} T''_o &\equiv (T_o/T_B) = Q'' + 1 + \tfrac{1}{2}(\gamma - 1)M^2(1 - (u''_{def}/V''_I)^2), \\ (u''_{def}/V''_I) &= 1 + (1 - p''_o)/\gamma M^2. \end{aligned} \tag{6.23}$$

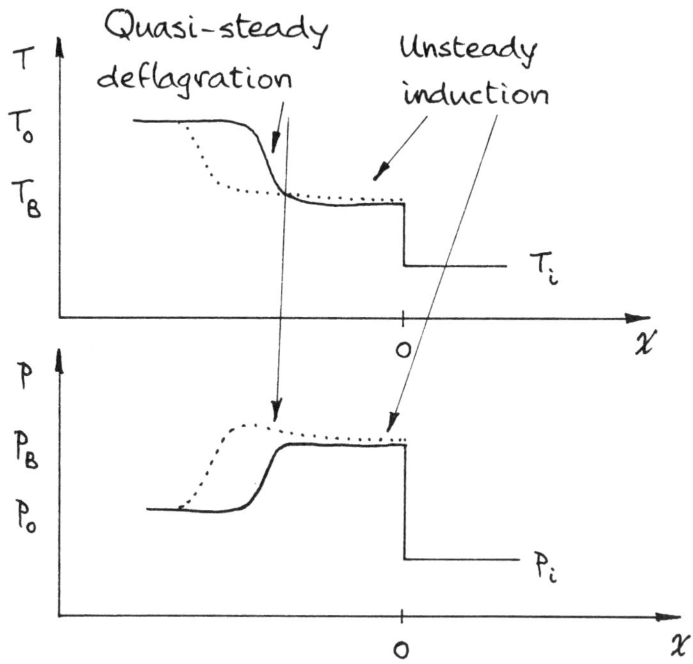

FIGURE 5. Pressure and temperature versus distance from the switch-on shock (diagrammatic). The full line is a proposed initial state (t = 0) and the dotted lines indicate how matters appear to develop, prior to formation of a detonation, according to the numerical work of Clarke, Kassoy & Riley (1986). These dotted lines are also consistent with the amalgamation of the quasi-steady deflagration of the present § with the unsteady induction zone of §6.3 and Figure 4.

Consider the following numerical example which has $Q'' = 2.63$, $M_B \simeq 0.5$ and $T_B \simeq 2.28T_i$, where T_i is the temperature ahead of the shock; (6.20) shows that, with $\gamma = 1.4$, $M = 0.259$ [thus Q''_{max} at 2.7 is not much larger than Q''] and (6.22) and (6.23) make $T_o \simeq 7.27\ T_i$ and $p_o \simeq 4.4\ p_i$. The numbers accord reasonably well with some early-time results computed by Clarke, Kassoy & Riley (1986) for initiation of a detonation by energy deposition through the end-wall of a tube, like the one illustrated in Figure 3; p_o here is rather low compared with the value 5.2 in the numerical experiments,which also give 0.276 for M. These experiments do show that mass flux through the rapid-reaction region is very nearly constant at any given time, and there is evidently some support for the present model from the numerical computations (which, incidentally, solve the Navier-Stokes equations).

The early stages of a detonation-initiation therefore appear to consist of relatively slow evolutions from an essentially non-steady combination of switch-on shock and deflagration. The wave/reaction interaction in the induction domain has a central role in the developments, as outlined in §6.3.

7. CONCLUSIONS

The association of combustion activity with a relatively high speed, compressible, flow has been investigated in general terms. For reactions that follow the Arrhenius hypothesis it has been shown that local Damkohler numbers must be such that diffusion, in its general sense (for mass, momentum and energy), is not associated with any regions of significant chemical reaction, but must be confined to the interiors of shock waves.

For any steady one-dimensional flow there is a maximum amount of energy that can be released into a stream of given Mach number. This result is usually quoted in the form of the statement that there is a minimum steady-state supersonic flow speed for a given energy addition (the CJ speed). When steady-states are possible the reaction zone behind the shock consists, in essence, of a weakly-reacting induction domain followed by a thin region, or deflagration, within which reaction proceeds rapidly to its conclusion. This basic structure is revealed rather simply by asymptotic methods based on large activation energy. If the leading shock is too weak this rapidly-reacting region does not appear and chemical energy is deposited preferentially into kinetic energy of the flow whilst temperature actually falls. Successions of such steady weak waves may have a part to play in more violent events by processing the gas towards hotter, more reactive, states and making the gas more receptive to ignition by subsequent stronger waves.

Any attempt to add more energy than the maximum to a flow of a given Mach number must result in an unsteady configuration in a one-dimensional unbounded space. This state of affairs must arise when a shock that is initially weaker than the CJ shock propagates into a combustible gas. The natural time scale in such a case is the induction time behind the initial shock, and a strong coupling between dilatational waves and the combustion reaction is set up behind the shock in such circumstances. Disturbance amplitudes in such an evolving induction zone are small when activation energy is large. However, a large disturbance in the form of a deflagration can be found at the end of the induction region, where it can exist in a quasi-steady condition. This deflagration releases all of the available combustion energy, which is too large to allow propagation of a steady-state combustion wave; in order to do this the deflagration must itself propagate at a Mach number smaller than the Mach number of the flow out of the shock. Deflagration and shock will therefore start by moving further and further apart as time increases.

It is shown that, in this combination, a deflagration can drive a switch-on shock wave on its own, without any assistance from other shock-driving mechanisms. As such, it provides an explanation of phenomena observed in the early stages of detonation initiation in some recent numerical experiments. In association with previously-studied behaviour in the evolving induction region, it is now possible to see how the wave/reaction interactions in the induction zone are instrumental in bringing the early-time configurations through to full ZND-type waves. Large disturbances are ultimately involved, and detailed late histories will almost certainly need to be studied by numerical methods.

The early-time configuration of switch-on shock, wave/reaction-dominated induction domain, and quasi-steady deflagration provides a fruitful background for the sustenance of disturbances, including dilatational waves, in one, two or three dimensions. It will undoubtedly be very profitable to make theoretical studies of disturbances in this otherwise slowly-evolving system.

REFERENCES

Aldushin, A.P., Khudyaev, S.I. & Zeldovich, Ya. B. (1981). Archivum Combustionis, 1, 9-21.

Cant, R.S. (1984) Ph.D. Thesis, Cranfield Institute of Technology, Cranfield, England.

Clarke, J.F. (1983) J. Fluid Mech., 136, 139-166.

Clarke, J.F. (1985a) J. Engg. Math., 19, 105-119.

Clarke, J.F. (1985b) Chapter V in "The Mathematics of Combustion", Ed. J.D. Buckmaster. SIAM Pa, USA.

Clarke, J.F. & Cant, R.S. (1984) Prog. in Astro. & Aeronautics, 95, 142-163.

Clarke, J.F., Kassoy, D.R. & Riley, N. (1986). Proc. Roy. Soc. Lond. A, 408, 129-148.

Clarke, J.F. & McChesney, M. (1976) "Dynamics of Relaxing Gases", Butterworths, London.

Clarke, J.F. & Wang, Z.W. (1986) College of Aeronautics Report No. 8616, Cranfield Institute of Technology, Cranfield, England.

Jackson, T.L. & Kapila, A.K. (1985) SIAM J. Appl. Math., 45, 130-137.

Kassoy, D.R. (1975) Q. J. Mech. Appl. Math., 28, 63-78.

Kassoy, D.R. & Clarke, J.F. (1985) J. Fluid Mech., 150, 253-280.

Urtiew, P.A. & Oppenheim, A.K. (1966) Proc. Roy. Soc. Lond., A 295, 13-28.

Williams, F.A. (1985) "Combustion Theory" 2nd ed. Benjamin Cummings Pub. Co. Inc., Menlo Park, California.

COOL FLAME PROPAGATION

P. Clavin
Laboratoire de Recherche en Combustion, CNRS/Université de Provence, case 252, Centre St Jérôme, 13397 Marseille cedex 13. France.

1. INTRODUCTION

The structure of planar and steady premixed flames is controlled by a system of nonlinear reaction diffusion equations. The stiff character of some of the nonlinear reaction terms makes flames different from most of the other reaction-diffusion waves. Two well known examples which were carried out in 1938 illustrate clearly the above mentioned difference : the Kolmogorov Petrovskii Piskunov /1/ (KPP) solution and the Zeldovich Frank Kamenetskii /2/ (ZFK) solution of the simplest model for the propagation of a reaction diffusion wave. The model consists of a single equation for a scalar θ representative of either the reduced temperature or the reduced mass fraction of the products of the reaction :

$$M\theta'_{\xi} - \theta''_{\xi\xi} = w(\theta) , \quad (1a)$$

with the following boundary condition(ξ is the nondimensional space variable)

$$\xi = -\infty \ : \quad \theta = 0 , \quad (1b)$$

$$\xi = +\infty \ : \quad \theta = 1 , \quad (1c)$$

The initial condition (θ=0) corresponds to an unstable state, $w(\theta=0)=0$ and $w'_0 \equiv |dw/d\theta|_{\theta=0} >0$. The final condition ($\theta$=1) corresponds to a stable equilibrium state, $w(\theta=1)=0$ and $|dw/d\theta|_{\theta=1} <0$. The flame speed M is an eigenvalue of the problem. The only difference between the models used by KPP and ZFK lies in the form of the reaction rate $w(\theta)$. The one used by KPP in a biological context is a weak nonlinearity of quadratic type:

$$w(\theta) = \theta(1-\theta) \quad (2a)$$

and represents an autocatalytic reaction. The one used by ZFK for premixed flames is representative of an Arrhenius law with a high value of the reduced activation energy β (Zeldovich number) (stiff nonlinearity) :

$$w(\theta) = \frac{\beta^2}{2} \ (1-\theta)\left[e^{-\beta(1-\theta)} - e^{-\beta}\right] \quad (2b)$$

The factor $\beta^2/2$ is used here for convenience in order to make the reduced flame speed M of order unity in the asymptotic limit $\beta\to\infty$. The term $e^{-\beta}$ in the r.h.s. of Eq.(2b) is introduced to make the problem (1a) well posed from a mathematical point of view. This term which avoids the famous cold boundary difficulty does not play any role from a physical point of view as soon as the Zeldovich number β is large enough. Such an artifice is not necessary when the time dependent version of Eq.(1a) is considered. In this case, a multitime scales analysis providesus with the correct solution /3/.

The KPP result shows that there is a continuous set of solution for M, bounded from below, with a minimum value M_{KPP} which is proportional to the square root of w'_0. The minimum solution has been proved to be the one

C.-M. Brauner and C. Schmidt-Lainé (eds.), Mathematical Modeling in Combustion and Related Topics, 65–75.

selected asymptotically in the long time limit by the time dependent version of Eq.(1a) from a large class of initial conditions /4/ including the physical situations.

The asymptotic solution for $\beta \to \infty$ obtained by ZFK with the reaction rate (2b) consists in a single solution with a flame speed M_{ZFK} proportional to the square root of $\int_0^1 w(\theta)d\theta$. Notice that, contrary to the KPP case, the ZFK solution cannot be selected by a linearization of the problem around the initial condition $\theta=0$ /5/.

By noticing that for small β, the reaction rate (2b) reduces to a similar solution to (2a), a transition must occur in the solutions of Eqs.(1a) and (2b) when β varies from 0 to ∞. Such a transition is described in Refs./3/ and /5/. It turns out that for $0<\beta<3.04$ the minimum speed is effectively given by M_{KPP} but for $\beta>3.04$ the minimum speed M^* is larger than M_{KPP} and increases with β to tend to M_{ZFK} in the limit $\beta \to \infty$. Moreover the $\theta(\xi)$ profiles of all the solutions corresponding to $M>M^*$ in the continuous spectrum, present for large values of β a tail toward the initial condition ($\theta=0$, $\xi \to -\infty$) which is very long at the scale of the ordinary flame thickness taken as unity of length in Eq.(1a). In this non physical long tail ahead of the place where the major part of the reaction takes place, the diffusion process are negligible and the reaction rate is balanced by the convection associated with the front propagation : $M\theta_\xi' \sim \theta w_0'$. Thus the reduced thickness of this long tail region is of order M/w_0' ($=2Me^\beta/\beta^3$ for the Arrhenius law represented by Eq.(2b)). It is also worthwhile to consider the limit $w_0' \to 0$. In this limit, each of the long tailed solutions corresponding to $M>M^*$ can no more satisfy the upstream boundary condition (1b) (see Figs. I).

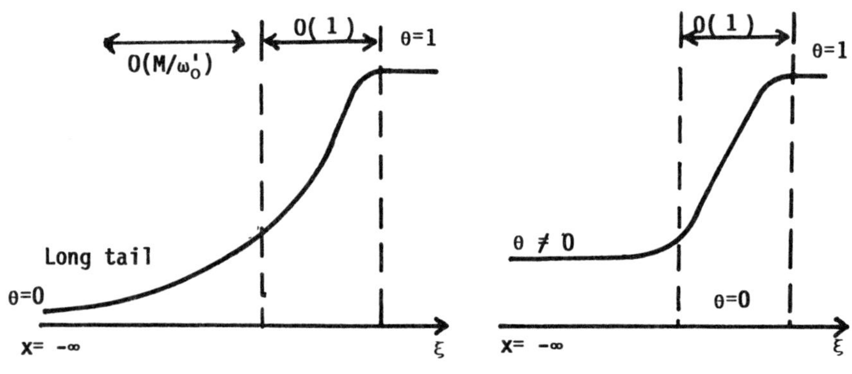

FIGURE 1a - θ-profile for $M>M^*$

FIGURE 1b - Limit of the θ profile of Fig.1a when $w_0' \to 0$.

When $\beta \gg 1$, the solution corresponding to the minimum flame speed in the only one for which the θ-profile toward the fresh is controlled by a balance between diffusion and advection (short tail : $\xi=O(1)$).

The only solution of Eqs.(1a) with the boundary conditions (1b-c) which remains in the limit $w_0' \to 0$ is the one corresponding to the minimum speed $M=M^*$. Moreover, M^* tends to M_{ZFK} when the major part of the reaction rate becomes localized around the final equilibrium state $\theta=1$ $(\beta \to \infty)$.

To summarize, when the reaction rate of the initial unstable state is negligible comparedto the one associated with conditions encountered inside the reaction diffusion wave, the propagation of the travelling wave is entirely controlled by the fast reaction rate. The fact that the chemical reaction around the initial state can be completly neglected modifies the nature of the travelling wave solutions.

In a recent theoretical study of a model for the propagation of cool flames, we have shown in collaboration with Amable Linan and Colette Nicoli /6/ that these particular flames exhibit a transition from KPP to ZFK travelling wave solutions when the initial temperature of the fresh mixture is modified. This work will be published elsewhere /6/. Some of the results are summarized in this paper.

Other types of complex chemistry effects upon flame propagation have also been recently investigated such as the ones concerning competing reactions /7/ or chain branching reactions in halogen-hydrogen combustion /8/. Modellisations of hydrocarbons "hot flames" are presented in the lectures of F.A. Williams /9/ and N. Peters /10/.

2. GRAY AND YANG MODEL

Cool flames are phenomena which are observed during the incomplete combustion at low temperature $500°K<T<800°K$ of most hydrocarbons /11/. Weak exothermicity, oscillations and a negative temperature coefficient of reaction rate leading to self quenching are general characteristics of cool flames observed in closed vessel or in well stirred reactors /12/. For many years cool flame kinetics was an intriguing question. It is now well established that chain branching reactions and thermal feedback are essential ingredient of cool flames. A simple kinetic model proposed by B.F. Gray and C.H. Yang /13/ was successfully used to describe the main characteristics of cool flame /14/. The backbone of the skeleton of this model which has been shown to partially represent the acetaldehyde oxidation, is a weakly exothermic chain branching reaction ($Q_b>0$) with a moderate activation energy (activation temperature $T_{ba} \sim 3500$ K). This chain branching reaction competes with two chain breaking reactions. The first one which is very exothermic ($Q_1>Q_b$) presents a high action energy ($T_{1a} \sim 8000°K > T_{ba}$) and is responsible for the quenching of the overall reaction at high temperature ($T>T^*$) above a cross over temperature $T^* \sim 710°K$ (see Fig.2). The second chain breaking reaction which is a athermic ($Q_2 \sim 0$) with a zero activation energy ($E_2= 0$), quenches the reaction at low temperature under an ignition temperature $T_i \sim 535°K$ (see Fig.2).

Chain branching : $X \xrightarrow{(b)} 2X + Q_b \nearrow, \quad T_{ba} \sim 3500°K.$ (3a)

Chain breaking :
$$\begin{cases} X \xrightarrow{(1)} P_1 + Q_1 \nearrow, & T_{1a} \sim 8000°K. & (3b) \\ X \xrightarrow{(2)} P_2, \quad Q_2=0, & T_{2a} = 0. & (3c) \end{cases}$$

X is a radical and P_1, P_2 are products considered as inert species. An exothermic chain breaking reaction with an activation energy larger than the one of the chain branching reaction is the typical feature of this model leading to self quenching at high temperature. The very slow initiation reaction which produces the radical X by decomposition of the reactants R can be neglected in the propagating flame regime and is not written in Eqs.(3). The rest of this paper is precisely devoted to the study of travelling waves in the cool flame regime described by Eqs.(3).

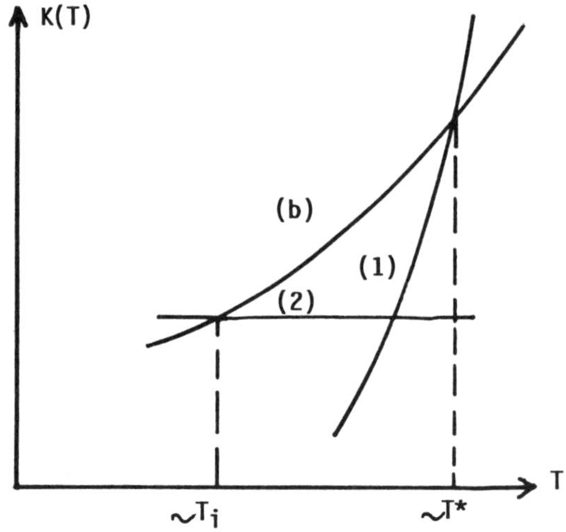

FIGURE 2 - Temperature dependence of the constant of reaction of the Gray and Yang model.

3. TRAVELLING WAVES

Because of the irreversible approach to equilibrium described by (3), travelling waves are expected to exist for propagating an equilibrium stable state ($T=T_f>T^*$ and zero mass fraction of the radical, Y=0) into an initial state ($T=T_o<T^*$, Y=0) which can be either unstable ($T_i<T_o<T^*$) or metastable ($T_o<T_i$) depending on the initial condition T_o. Such steady and 1-D travelling waves are governed by a system of two coupled reaction-diffusion equations which can be written in an adimensional form as :

$$m\,\phi'_x - \phi''_{xx} = (\alpha e^{b\phi} + e^{\phi})X\ , \quad \alpha<1, \quad b<1 \tag{4a}$$

$$m\,X'_x - (1/L_e)X''_{xx} = \omega(\phi)X\ , \tag{4b}$$

with $$\omega(\phi) = (e^{b\phi} - e^{\phi} - k) \tag{4c}$$

and with the following boundary conditions :

$$x = -\infty\ ,\ \text{fresh mixture}\ ,\ X = 0\ ,\ \phi = \phi_o<0\ , \tag{4d}$$

$$x = +\infty\ ,\ \text{burnt gases}\ ,\ X = 0\ ,\ \phi = \phi_f>0\ , \tag{4e}$$

where the reduced flame speed m and the reduced flame temperature ϕ_f (final state) are unknown. $\phi \equiv \beta(T-T^*)(C_p/Q_1)$ is the reduced temperature, $\beta \equiv (T_{1a}/T^{*2})(Q_1/C_p)$ is the reduced activation energy and C_p is the specific heat of the gaseous mixture which is assumed to be constant. $X = Y/\beta$ and Y are the reduced mass fraction and the mass fraction of the radical respectively. The reduced flame speed m is defined by $\beta m = U/(D_T/\tau_r)^{\frac{1}{2}}$ where $D_T = \lambda/\rho C_p$ is the heat diffusion coefficient and τ_r the characteristic reaction time defined at the cross over temperature T* :

$$\tau_r \equiv B_b^{-1} \exp(T_{ba}/T^*) \equiv B_1^{-1} \exp(T_{1a}/T^*) \ . \qquad (5a)$$

B_b and B_1 are the prefactors of the chain branching reaction (3a) and of the first chain breaking reaction (3b) respectively,

$$K_b(T) = B_b \exp(-T_{ba}/T), \ K_1(T) = B_1 \exp(-T_{1a}/T) \qquad (5b)$$

and K denote the constants of reactions. The Lewis number Le is defined as the ratio of the thermal diffusion coefficient of the mixture by the molecular diffusion coefficient of the radical. The ratio of the activation temperature and of the heat release of (3a) and (3b) are $b \equiv (T_{ba}/T_{1a})$ and $\alpha \equiv (Q_b/Q_1)$ respectively. The relative rate of the chain breaking reaction at low temperature is defined by $k \equiv K_2\tau_r$ where K_2 is the reaction rate of (3c). The reaction rate $\omega(\phi)$ of the overall radical production or consumption is given by Eq.(4c) and is sketched in Fig.3. The reaction rates in the r.h.s. of Eqs.(4a-c) are obtained from Arrhenius laws (5b) in the limit of large values of $\beta(\beta >> 1)$ and for $(T-T^*)/T^* = o(1)$. The unit of length used to adimensionalized Eqs.(4) is the thickness of the reaction layer $\sqrt{D_T\tau_r}/\beta$, $(x=\beta\xi)$.

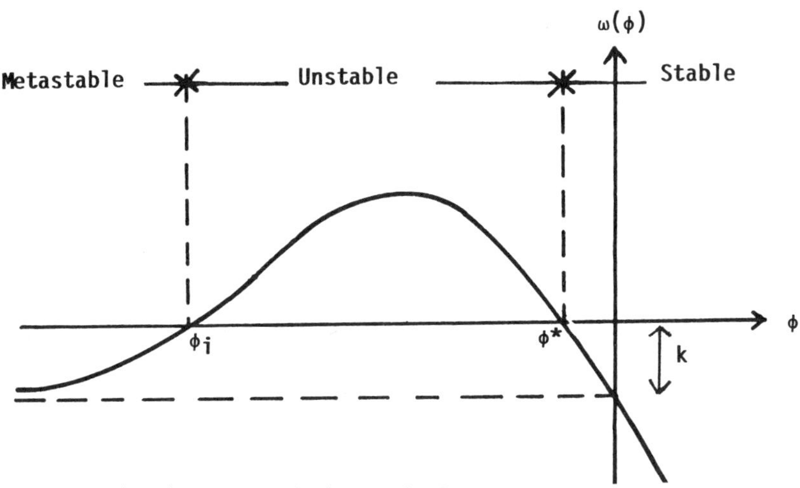

FIGURE 3 - Production rate of the radical
$\phi_i < \phi < \phi^* \Rightarrow \omega(\phi) > 0$ corresponds to an unstable medium.

The two zeros of $\omega(\phi)$ are denoted ϕ_i and ϕ^* and they correspond to the ignition temperature T_i and to the cross over temperature T^*.

Typical profiles of the reduced temperature and concentration associated with travelling waves propagating from right to left into an initial state at $x=-\infty$ and with a final state at $x=+\infty$, are sketched in Fig.4.

By linearization, the final state $(\phi=\phi_f>\phi^*, X=0)$ appears to be a saddle point in the phase space. Here, the concentration profile is found to be of the form :

$$x \to +\infty \quad : \quad X = e^{+x\mu_f^-} \ , \qquad (6a)$$

with $\mu_f^- = (+m - \sqrt{m^2 - 4\,\omega_f/Le})L_e/2$ and $\omega_f \equiv \omega(\phi_f) < 0$. (6b)

By taking advantage of the translation invariance in x, a factor unity has been arbitrarily used in Eq.(6a). Such a behavior in the final state is in principle well adapted to integrate the reaction diffusion equations (4)

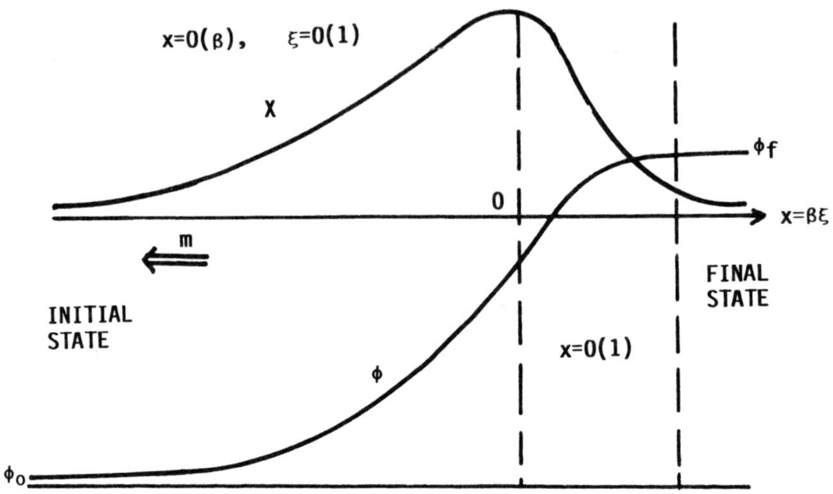

FIGURE 4 - Profiles of the reduced temperature ϕ and concentration X.

by using a shooting method starting from the final state. But the presence of two unknowns m and ϕ_f makes the method very difficult to be used for the initial conditions ($\phi=\phi_o$, X=0) for which a single travelling wave solution exist.

The nature of the initial conditions depends on the initial temperature ϕ_o compared to the ignition temperature ϕ_i. For $\phi_o<\phi_i$, one has $\omega_o \equiv \omega(\phi_o)<0$ and the initial state corresponds to a saddle point with a single root μ_o^+. In such a case, by comparison with the case of one scalar (1a), one may expect zero, one solution or at most a discrete spectrum. For $\phi_i<\phi_o<\phi^*$, one has $\omega_o>0$ and the initial state corresponds to a node with two possible roots μ_o^- and μ_o^+. The corresponding concentration profiles present a tail in the direction of the initial state which has the following form :

$$x\to-\infty : \quad X(x) \simeq A_+ \exp(x\mu_o^+) + A_- \exp(x\mu_o^-) , \tag{7a}$$

$A_\pm$ are constant and the roots $\mu_o^\pm$ are given by :

$$2\mu_o^\pm = Le(m \pm \sqrt{m^2 - 4\omega_o/Le}) , \quad (\omega_o>0 \Rightarrow \mu_o^\pm>0). \tag{7b}$$

In such a case one may expect a continuous spectrum of solutions bounded from below by the solution corresponding to the minimum speed m_{KPP}

$$m_{KPP} = 2\sqrt{\omega_o/Le} , \tag{7c}$$

which is defined by :

$$\mu_o^- = \mu_o^+ . \tag{7d}$$

4. ASYMPTOTIC ANALYSIS FOR A NEGLIGIBLE CHAIN BREAKING REACTION RATE AT LOW TEMPERATURE

When k=0 in Eqs(4) ($\phi_i = -\infty$, $\phi^* = 0$ in Fig.3) all the possible initial states are unstable. In this case an asymptotic analysis of the travelling waves solution can be carried out for large negative values of the initial reduced temperature :

$$|\phi_o| = O(\beta) \quad \text{with} \quad \beta \to \infty . \tag{8}$$

The results obtained by this analysis are also valid for $|\phi_i| = O(\beta)$ which corresponds to a transcendentally small value of k. As explained in section I, the details of a transcendentally small reaction rate in the initial condition do not play any role from a physical point of view. In the limit described by Eq.(8), the profiles have a similar shape as in Fig.4 but the reaction zone for the chain breaking and chain branching reactions collapses to a relatively thin layer following a relatively large preheated zone. The ratio of the thickness of these two layers is of order $O(1/\beta)$ and one has $|\phi_o| = O(\beta)$ and $\phi_f = O(1)$. In the preheated zone the profiles are

$$\phi(x) = (\phi_f - \phi_o)e^{mx} , \quad m = O(\tfrac{1}{\beta}) , \quad x = O(\beta), \quad |\phi| = O(\beta), \tag{9a}$$

$$X = X_f \, e^{mx} \quad \text{with} \quad X_f = O(1) \tag{9b}$$

where the origin of the x axis is taken at the reaction layer, in such a way that the diffusive fluxes going into the preheated zone and coming from the reaction layer are given by

$$(d\phi/dx)_{0_-} \sim (-\phi_o)m \equiv \Lambda = O(1) , \tag{9c}$$

in such a way that m must be of order $O(1/\beta)$.

$$(dX/dx)_{0_-} = mX_f = O(1/\beta) . \tag{9d}$$

In the reaction layer, the leading order of Eqs.(4) yields :

$$- \phi''_{xx} = (\alpha e^{b\phi} + e^{\phi})X, \quad x = O(1), \quad X = O(1), \quad \phi = O(1), \tag{10a}$$

$$-(1/Le)X''_{xx} = (e^{b\phi} - e^{\phi})X, \quad b < 1 . \tag{10b}$$

The variable x of Eqs.(4) and (10) is in fact of order unity in the reaction layer and of order β in the preheated zone for which Eqs.(9a-b) hold. Notice that only one unknown (i.e. ϕ_f) appears in Eq (10), (m has disappeared). Then Eqs.(10) can be solved by a shooting method starting from the final state $\phi = \phi_f$, X=0. One has to determine the value of ϕ_f which corresponds to (dX/dx)= 0 in the upstream direction ($x \to -\infty$) for satisfying the leading order of the matching condition with (9d). The corresponding value of the asymptote $(d\phi/dx)_{x \to -\infty}$ gives the unknown eigenvalue of order unity Λ defined by Eq.(9c) which leads to the flame speed m. Typical results for Λ in terms of b and Le are plotted in Fig.5 for a value of $\alpha = 0.2$. The results of the asymptotic analysis can be summarized as follows. For given values of Le, α and b, the flame temperature ϕ_f and the scalar $\Lambda \equiv (-\phi_o)m$ are fixed quantities independent from the initial condition ϕ_o. Thus, the flame speed is found to vary as $1/(-\phi_o)$. According to this analysis, only one solution is obtained for the wave travelling in the initial mixture at a given temperature ϕ_o. This solution corresponds to the ZFK solution of the one scalar problem of section 1. The reason is that both the chain branching and chain breaking reactions behave as having a large reduced activation energy whenever Eq.(8) is satisfied. No singularity occurs for $\phi_o = 0$ because the present result is valid only for $|\phi_o| = O(\beta)$ and a transition appears in the solution for $\phi_o = O(1)$ as it will be shown in the next section.

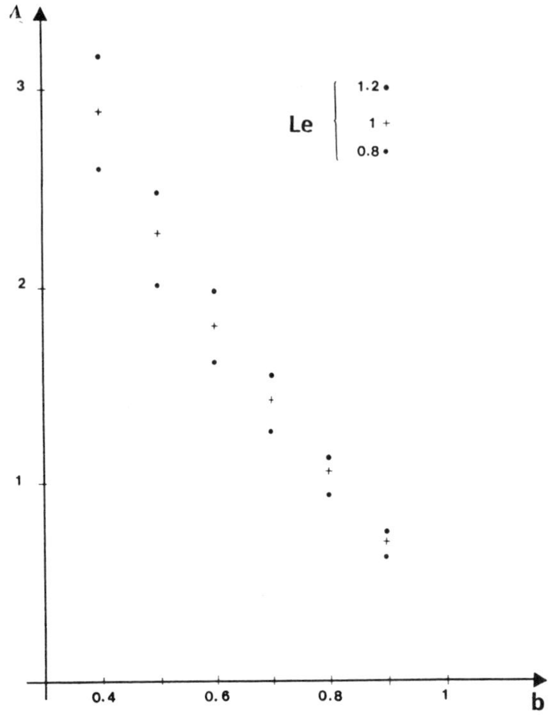

FIGURE 5 -

Leading order of the asymptotic expansion ($\beta\to\infty$) for the eigenvalue Λ expressed in terms of b for three different values of the Lewis number Le.
(Courtesy of Colette Nicoli)

Another instructive case corresponding to a zero activation energy of the chain branching reaction ($T_{ba}=0$) can be easily solved analytically when $K_b= B_b= C^t$, $K_2= 0$ and $\beta\to\infty$. For $Q_b=0$, the governing equations of the corresponding, travelling waves can be written in an adimensional form as :

$$m\phi'_{\xi} - \phi''_{\xi\xi} = \frac{\beta^2}{2Le} e^{\phi} X, \qquad (11a)$$

$$mX'_{\xi} -(1/Le)X''_{\xi\xi} - X = - \frac{\beta^2}{2Le} e^{\phi} X, \qquad (11b)$$

where the cross over temperature T* is here defined by

$$B_b^{-1} = \frac{\beta}{2Le} B_1^{-1} \exp(T_{1a}/T^*) \ . \qquad (11c)$$

Such a definition of T* is slighly different from the one introduced in Eq.(5a). It has been choosen arbitrarly to make the reduced flame temperature ϕ_f of order unity, $\phi_f=O(1)$. Notice that the space variable ξ which has been used in Eqs.(11a-b) is such that $\xi=O(1)$ in the preheated zone. The radical concentration is produced in this zone by the chain branching reaction (b). The chain breaking reaction (1) is localized in a thin layer $\xi=O(1/\beta)$. The structure of this thin reaction layer is similar to the ordinary one step flame model in the limit of high reduced activation energy. The only difference comes from the fact that the species which are consumed have to be produced in the preheated zone by the autocatalytic reaction (b). The asymptotic solution of this problem can be easily obtained for $|\phi_o|=O(\beta)$ ($\beta\to\infty$) to give $\phi_f=O(1)$ and $m=O(1/\beta)$. One obtains in this case a continuous set of solutions for m bounded from below by $m_{KPP}=2/Le^{\frac{1}{2}}$ and the flame temperature ϕ_f is given by :

$$m(-\phi_o) = \exp(\phi_f/2) \ . \tag{12}$$

The fundamental role which is played by the autocatalytic reaction in the tailed toward the initial mixture, explains the KPP nature of the travelling waves solutions of (11a-b). It is only when all the reactions are negligible in this tail that the ZFK nature of the solutions are valid as in Eq.(4) for b=0(1) (0<b<1). Furthermore notice that when a cut off temperature ϕ_i is introduced in the chain branching reaction rate (b) of Eq(11c) in the following form :

$$B_b = 0 \quad \text{for} \quad \phi<\phi_i \quad , \tag{13a}$$

$$B_b = C^t \quad \text{for} \ \phi_i<\phi<\phi^* \ , \ \text{with} \ |\phi_i| = O(\beta) \ , \tag{13b}$$

the travelling waves solutions are not modified for $\phi_i<\phi_o<\phi^*$ but there is no more solution for $\phi_o<\phi_i$.

Coming back to the original cool flame model described by Eqs.(3) and (4), one can expect that a transition similar to the one described in Section 1 must occur when ϕ_o varies to $-\infty$ to zero. Such a transition is necessary to go continuously from the ZFK situation described at the beginning of Section 4 ($|\phi_o| = O(\beta)$) to KPP types of solutions which must appear in Eqs.(4) when ϕ_o=0(1). This problem has been studied numerically /6/ with a particular attention devoted to the case $\phi_o=\phi_i$. The corresponding results are briefly presented in the next section.

5. NUMERICAL ANALYSIS

Numerical analysis of Eqs.(4) have been carried out by Colette Nicoli /6/. Typical results corresponding to α=0.2, b=0.7 and k=0.01 (ϕ_i= -6.34, ϕ^*= -0.034) are presented in Fig.6 in which the flame speed m is plotted versus the initial condition ϕ_o. Two different numerical methods have been used. A direct numerical simulation of the solutions of the time dependent version of Eqs.(4) have been carried out to describe the waves travelling into a given initial state (ϕ_o given) by using a adaptative gridding method developed by B. Larrouturou /15/. A shooting method has also been used in the phase space. Starting from the saddle point of a stable final equilibrium state corresponding to a given set of m and ϕ_f, the Eqs.(4) are integrated up to the initial condition to obtain the corresponding value of ϕ_o . This method is particularly well adapted when there is a continuous set of solutions corresponding to a given value of ϕ_o. As anticipated at the end of paragraph 3, this is found to be the case for $\phi_i<\phi_o<\phi^*$. Minimum flame speed are obtained by decreasing m for a given value of ϕ_f. Under a critical value for m, the concentration profiles X(x) become negative in the tail toward the initial condition. This critical value is found to be either the KPP solution (7c) for ϕ_o>-4 or the ZFK solution described in Paragraph 4 ($-\Phi_o m=\Lambda$) for ϕ_o<-4. The theoretical results for KPP and ZFK are also plotted in Fig.6. No difference between those theoretical values and the numerical results for the minimum flame speed can be noticed in Fig.6. Moreover the flame speed which is obtained asymptotically in time by the direct numerical simulation are found to correspond precisely to the critical minimum value which has been obtained by the shooting method. The transition from KPP to ZFK solution for the minimum flame speed is found to be rather sharp at about $\phi_o \sim -4$. The fact that the ZFK solution is so well accurated can be explained by the small value of k(=0.01) which has been used in Fig.6 in such a way that the corresponding critical value of $|\phi_o|$($\sim$4) is sufficiently large.

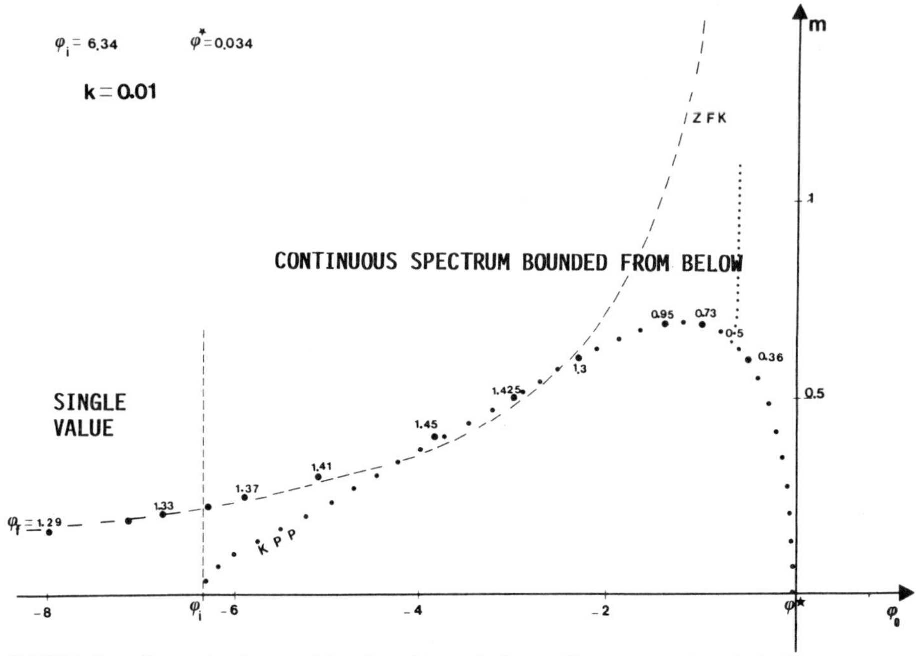

FIGURE 6 - Numerical results for the minimum flame speed : • • •
KPP solution : ....
ZFK solution : ---
Typical curve $\phi_f = C^t$: ....

(Courtesy of Colette Nicoli)

Another transition is found to appear for $\phi_o=\phi_i$. The initial state which is unstable for $\phi_o>\phi_i$ becomes stable for $\phi_o<\phi_i$. All the solutions except the one corresponding to the minimum flame speed are found to disappear at $\phi_o=\phi_i$. The corresponding temperature profiles X(x) obtained by the shooting method blow up in the tail for large negative values of the space coordinate ($x\to -\infty$) and the upstream boundary condition (X=0) can no more be verified. The only solution crossing $\phi=\phi_i$ is the one corresponding to the minimum flame speed and the steady solution for the travelling wave problem become single valued for $\phi_o<\phi_i$.

6. CONCLUDING REMARKS

The 1-D and steady solutions for travelling waves sustained by the Gray and Yang kinetic model (3) have been showed to experience two transitions when the initial condition ϕ_o is varied. A first transition occurs at the inflamation temperature $\phi_o=\phi_i$ which is found to separate a domain of a single valued solution ($\phi_o<\phi_i$ corresponding to an initial state which is metastable) from a domain of a continuous set of solutions ($\phi_i<\phi_o<\phi^*$). In this last domain in which the initial state is unstable, another transition occurs between a KPP and a ZFK scenario.

The theoretical results presented in this paper predict that travelling waves associated with cool flame chemistry can propagate into an initial mixture at an ordinary temperature well below the ignition temperature. Such waves could be initiated in a region where the mixture is locally warmed up to an intermediate temperature corresponding to an unstable state

$T_i<T<T^*$. Observations of such cool flame propagations can be make difficult by the occurence of ordinary hot flames which, propagating at higher speeds, are thus susceptible to overcome the front of the cool flame.

Moreover when volumetric heat losses representative of cooling at the walls of tubes or vessels are taken into account, a limit cycle may appear around T*. In such a case, one dimensional travelling wave train as well as spirals or targets in two dimensional geometry are expected to occur.

ACKNOWLEDGEMENTS

The results presented in this paper have been obtained in collaboration with Professeur A. Linan and C. Nicoli. This work was supported by the ECC contract n° STI-070-J-C (CD)

REFERENCES

1. Kolmogorov, A.N., Petrovskii, I.G., Piskunov, N.S. : Bull. Univ. Moscow, Ser. Internat. Sec. A 1, 1-25 (1937)
2. Zeldovich, Ya.B., Frank-Kamenetskii, D.A. : Zhur. Fiz. Khim. 12, 100 (1938)
3. Aldushin, A.P., Khudyaev, S.I., Zeldovich, YA.B. : Archivum Combustionis, 1 n°½, 9-21 (1981)
4. Aronson, D.G., Weinberger, H.F. : Advances in Mathematics 30, 33-76 (1978)
5. Clavin, P., Liñan, A. : in NATO ASI Series B vol.116, 291-346 Plenum Press (1984)
6. Clavin, P., Liñan, A., Nicoli, C. : "Travelling waves in the cool flame regime" submitted to Combust. Flame (1987).
7. Clavin, P., Fife, P., Nicolaenko, B. : SIAM J. Appl. Math. 47 (2), 87 (1987)
8. Clavin, P., Liñan, A. : "Premixed flames with nonbranching chain reaction". To appear in Combust. Flame (1987)
9. Williams, F.A. : "Influence of detailed chemistry on asymptotic approximations for flame structure". These Proceedings (1987)
10. Peters, N. : "Grid requirements due to the inner structure of premixed hydrocarbon flames". These proceedings (1987)
11. Lignola, P.G., Reverchon, E. : Prog. Energy Combust. Sci. 13, 75-96 (1987)
12. Gray, P. : Sixteenth Symp. Int. on Combustion, p.919 (1977)
13. Gray, B.F., Yang, C.H. : Trans. Faraday Soc. 65, 1614 (1969)
14. Griffiths, J.F. : in Oscillations and Traveling waves in Chemical Systemps, p.529, John Wiley & Sons (1985)
15. Larrouturou, B. : Lectures in Applied Mathematics, AMS, vol.24, 415 (1986)

MODELING THE CHEMISTRY IN FLAMES

PAUL C. FIFE

MATHEMATICS DEPARTMENT, UNIVERSITY OF ARIZONA,
TUCSON, AZ, 85721, USA

1. INTRODUCTION

Flames typically involve a large number of reactions and species. This complexity makes it very difficult to comprehend on an intuitive level the essential character of the combustion events. Moreover, computations are expensive and sometimes not immediately enlightening as to the reasons for the phenomena they display. It is therefore extremely important to be able, if possible, to systematically reduce the complex problem to understandable terms. For example, one would like to know (i) which of the reactions are really important, and why; (ii) what the typical orders of magnitude of the various species, especially the radicals, are during the combustion process; and (iii) whether the essential features of the flame can be deduced from a simpler skeleton model with many fewer reactions and species.

There have been several approaches used in the past to move toward this goal. Sensitivity analysis, for example, is a very popular avenue to singling out the important reaction paths (for a recent example of this, see [16]). Other kinds of information have been used to reduce complex chemical kinetics to simpler proportions, as in [2], [9 - 13]. Finally, under the assumption that at least one reaction has high activation energy and that for all temperatures the various reaction constants k(T), except for 2 or 3 of them, are sufficiently far apart from one another, then the scheme proposed by Fife and Nicolaenko [6 - 8] is applicable, and provides a way to gauge the importance of the various reactions, along with the spatial structure of the flame and the order of magnitude of its velocity.

The last of the three types of approaches mentioned is the most systematic; however there are important cases when the assumptions on which it is based are not warranted. An example is the combustion of hydrogen and oxygen. In fact, it was the desire to understand H_2 - O_2 flames that motivated the recent development of the approach which will be described here. It is designed to give qualitative answers to the above questions in some cases when the assumptions of the Fife - Nicolaenko scheme are not fulfilled.

This approach is based, firstly, on the assumption that there is a thin flame layer at a characteristic temperature T in which either all or part of the reaction network under consideration comes to chemical equilibrium. Such a layer would be occasioned by some reaction(s) with high activation energy which trigger the other reactions taking place in the layer. If only part of the network does this, then the rest of the network may come to equilibrium in another (not necessarily localized) site on the flame profile at a slower rate.

Secondly, it is assumed that each reactant has a characteristic average level of concentration in the layer.

Thirdly, it is based on often exaggerating the differences in magnitudes of rates and concentrations. Exaggeration is the essence of asymptotic models such as this. Consequently the results obtained are expected to be only qualitatively accurate in many cases. Nevertheless, they provide a clear picture of the main chemical processes at work to produce the flame.

One form of this model was applied in [4] to H_2 - O_2 flames using a 19 - reaction network. The analysis there was done by hand and encouraging results were obtained. Later in [15], [14] the procedure was implemented on a computer using the language of MACSYMA. At the same time, an interpretation of the procedure in terms of vertices of polytopes was given and the concept of critical temperatures developed. The latter are temperatures at which the geometry of crucial vertices changes.

The purpose of the present paper is to give a more general formulation of the model and to explain it in the context of suggestive simple examples.

C.-M. Brauner and C. Schmidt-Lainé (eds.), Mathematical Modeling in Combustion and Related Topics, 77–87.

2. THE FRAMEWORK

For convenience we operate within the framework of a thermo-diffusive model, although a more general context will lead to the same traveling wave equation (2) below. First, we shall develop the equations for steady plane flames governed by a general network $\mathcal{M}$ of m possibly reversible reactions,

$$\mathcal{M} = \{R1^+, R1^-, R2^+, ..., Rm^-\}. \tag{1}$$

Those equations take the form (as in [6])

$$DU'' - MU' + K\cdot(\omega^+(U) - \omega^-(U)) = 0. \tag{2}$$

The meaning of the symbols is as follows. The independent variable is the traveling wave coordinate z. The (n+1) vector U is defined as

$$U = \begin{bmatrix} T \\ Y_1 \\ \dots \\ Y_n \end{bmatrix} \tag{3}$$

where T is temperature and the Y_i are the concentrations. We nondimensionalize T so that one unit of T is the difference ΔT between a typical burned temperature T_+ in the flame and the unburned temperature T_-. The concentration Y_i will be defined as the mass fraction of species A_i in the gas at any position and time divided by the molecular weight of that species.

The constant M is the unknown rescaled mass flux. D is a diffusion matrix, assumed to be constant. The wave moves to the left, so that the unburned conditions far upstream are attained at $z = -\infty$: $U(-\infty) = U_-$ (say).

The (n+1)x(m) stoichiometric matrix K is constructed directly from the sets of reactions Rj and their heat releases. First, a matrix K^+ is defined by setting each element κ^+_{ij} equal to the stoichiometric coefficient of A_i in the left side of reaction Rj. These coefficients are those appearing in the representation of Rj as

$$\Sigma\kappa^+_{ij}A_i \rightleftarrows \Sigma\kappa^-_{ij}A_i.$$

Thus for example if R4 is $A_2 + A_3 \rightarrow 2A_1$, then $\kappa^+_{24} = \kappa^+_{34} = 1$ and all the other elements in column 4 with $i \geq 1$ are zero. There is also a 0th row in the matrix, with $\kappa^+_{04} = -Q_4$, a suitable measure of the heat released by reaction R4. This completely defines the matrix K^+. The matrix K^- is defined in the same way by looking at the right sides of the reactions. However, the 0th row of K^- is set equal to zero. Finally,

$$K \equiv K^- - K^+. \tag{4}$$

The m-vectors ω^+ and ω^- give the rates of the forward and reverse reactions respectively. The rates are assumed to be of mass action type:

$$\dot{\omega}^\pm_j(U) \equiv k^\pm_j(T) \prod_i Y_i^{\kappa^\pm_{ij}}. \tag{5}$$

The products are mass-action monomials, and the k's are reaction constants, generally determined (if indeed possible) by experimental measurements.

This defines all the symbols in the basic equation (2).

3. THE MODEL PROBLEM

It is assumed that part (at least) of the original network $\mathcal{M}$ (call it $\mathcal{M}'$) comes to chemical equilibrium in a thin layer (primary reactive zone) at a characteristic temperature T. The rest of the network may also come to equilibrium, but this is assumed to happen at a slower rate and elsewhere in the flame profile. This other part of the profile will have a larger characteristic length scale. The profile will therefore be relatively flat downstream of the primary combustion layer.

In view of that, we may integrate (2) with respect to z from $-\infty$ to the relatively flat region immediately downstream of the layer. The integral of U″ vanishes approximately because of the flatness assumption. Recall that U_- is the unburned state far to the left (where the integration begins). Let $U_+ = (T_+, Y_+)$ be the state attained by the chemical processes in the layer, so that $U = U_+$ at the end of the integration region behind the layer. Then we obtain

$$Y_+ = Y_- + \hat{K}\cdot\boldsymbol{\alpha}, \tag{6}$$

where the "allocation" vector $\boldsymbol{\alpha}$ is defined by

$$\boldsymbol{\alpha} \equiv \frac{1}{M}\int(\omega^+(U(z)) - \omega^-(U(z)))dz, \tag{7}$$

and $\hat{K}$ is the stoichiometric matrix with first row deleted. In the same way, the integrated 0th component of (2) yields

$$T_+ = T_- + Q\cdot\boldsymbol{\alpha}, \tag{8}$$

where Q is the vector of heat releases.

In accordance with the second point in §1 above, we associate with each concentration Y_i a characteristic average order of magnitude in the reaction layer

$$Y_i \simeq e^{-\nu_i} \leq 1, \tag{9}$$

where the $\nu_i \geq 0$. (Actually, we also want to ensure that at least one of the Y's be an O(1) quantity, so we may normalize them to achieve that purpose, as was done in [4], [15].)

We are interested in examining the relative magnitudes of the ω's and do this by taking the negative natural logs of the expressions (5), and subtracting an (unknown) parameter s which is intended to be chosen so that the minimum of the resulting expressions is equal to zero. (This is not quite true when "partial equilibrium" conditions exist, as we shall see later.) We thus define

$$\beta_j^{\pm}(\boldsymbol{\nu},s,T) \equiv \ell_j^{\pm} + \Sigma\kappa_{ij}^{\pm}\,\nu_i - s. \tag{10}$$

In this expression, the ℓ's are negative logarithms of the k's, and are assumed to be given non-increasing functions of T. The summation is over i. Of course $\boldsymbol{\nu} \equiv (\nu_1,\nu_2,..,\nu_n)$.

For a given Y_-, we attempt to judge the relative importance of the reactions at a given temperature T (in the flame layer) and at the same time the magnitudes of the concentrations (measured by $\boldsymbol{\nu}$) in the following way. We choose nonnegative "figures of merit" or "resolution parameters" a and b, and seek values of $\boldsymbol{\nu}$, s, and $\boldsymbol{\alpha}$ such that

i) for each j, $\beta_j^{\pm} \geq 0$ or $\beta_j^{+} = \beta_j^{-} < 0$, (11a)

ii) $\nu \geq 0$, (11b)
iii) $\beta^+_j > a \Rightarrow \alpha_j \leq 0$; $\beta^-_j > a \Rightarrow \alpha_j \geq 0$, (11c)
iv) with Y_+ defined in terms of $\boldsymbol{\alpha}$ by (6), $Y_+ \geq 0$, (11d)
v) $\nu_j > b \Rightarrow Y_{+j} = 0$, (11e)
vi) Y_+ is in equilibrium for the subnetwork $\mathcal{M}' = \{Rj: \beta_j \leq a\}$. (11f)

The first part of i) is the role of the normalization parameter s. The second part will be explained later in §6. The requirement ii) was mentioned before. iii) expresses the fact that large β's mean small rates, hence by (7) the corresponding α's (which, by the way, are all $\leq$ O(1) because of (6)) are approximated by zero. If the rate of a reaction is small in only one direction, then again by (7) the corresponding $\boldsymbol{\alpha}$ will be of predetermined sign. Requirement iv) is obvious; v) simply expresses the approximation that when ν_j is large, Y_j is small and we approximate it in the final state by 0. Finally vi) says that excluding those reactions whose rates are so small they can be neglected anyway, the network goes to equilibrium at the end of the primary burning process. If it did not, the burning would not be complete. A later combustion process in a larger region behind the primary layer may take care of those remaining reactions which were first neglected. Reaching equilibrium here means reaching a state U_+ such that the reaction term in (2) vanishes:

$$\tilde{K}\cdot(\omega^+(U_+) - \omega^-(U_+)) = 0, \tag{12}$$

the matrix $\tilde{K}$ denoting the stoichiometic matrix of the reduced network $\mathcal{M}'$.

In seeking solutions of (11), the strategy will be to concentrate on solutions for which as many as possible of the inequalities in (11a), (11b) are equalities. The reason is most easily seen in the case we choose a = b = 0 (this bold choice turns out to give meaningful results and was often used in [14], [15]). Then because of (11c), the inequalities β^+_j, $\beta^-_j > 0$ forces $\alpha_j = 0$, which effectively eliminates Rj from $\mathcal{M}'$. Likewise by (11c), $\nu_j > 0$ imposes an additional constraint on the α's. Our strategy therefore leaves as few as possible constraints on the α's, hence gives a wider range of possibilities for the participating network. On the one hand, this is the most secure strategy in that it better provides that viable possibilities are not inadvertently missed. On the other hand, it may well (and in fact often does) result in the existence of a nontrivial solution of (11), which might not exist away from the vertices; freedom in the choice of α's will better provide for the existence of allocations satisfying (11d-f).

Notice that the temperature T_+ of the layer where we are doing our analysis itself depends on $\boldsymbol{\alpha}$. We shall see later that this coupling generates some interesting phenomena.

There is no guarantee that, for a given T, a solution of (11) exists, and likewise there is no guarantee that it is unique if it does exist. The consequences and possibilities of nonuniqueness are under current investigation. We feel that nonexistence generally implies a kinetic extinction phenomenon.

Finally, the obvious crudeness of the model can be alleviated by passing to a second approximation, as will be explained in §8 below.

We illustrate the process by two simple examples.

4. EXAMPLE: The sequential reaction $A_1 \rightarrow A_2 \rightarrow A_3 \rightarrow 0$.

We denote the three reactions, proceeding from left to right, by R1, R2, and R3. We suppose that k_2 and k_3 do not depend on temperature, but that k_1 does very strongly (has high activation energy). The initial state is $Y_- = \begin{bmatrix} 1 \\ 0 \\ 0 \end{bmatrix}$.

The following rough picture is fairly clear. The high activation energy of the first reaction causes a primary layer like the classical one for a one-step reaction, at a specific temperature T.

a) If, at the given temperature T, $k_1(T)$ is much smaller than k_2 and k_3, then R2 rapidly consumes any amount of A_2 produced by R1, and its product A_3 is again rapidly consumed by R3.

b) If, on the other hand, $k_1(T)$ is much larger than k_2, then all A_1 is consumed in a primary flame layer in which R1 goes to completion (i.e. the subnetwork consisting only of R1 comes to equilibrium). The remaining reactions R2 and R3 are not yet in equilibrium, but approach it at a much slower rate behind the primary layer (notice this is true even if k_3 is very large; no assumption is being made about its magnitude). Those reactions are spread out over a much larger domain.

c) If k_1 is much smaller than k_2 but much larger than k_3, then R1 and R2 quickly come to completion in the primary layer, and R3 does so later in a trailing zone.

These results are trivial to grasp and we do not need any complex machinery to deduce them. However, the systematic procedure described in the previous section is necessary for more involved networks in which the results are not so intuitive, and so it will be good to illustrate that procedure in the context of this simple example. Moreover, this analysis will clarify some of the points in the above categorization, including some of the assumptions made. We therefore proceed as outlined in the previous section.

Clearly $K^- = 0$ and

$$K = \begin{bmatrix} Q_1 & Q_2 & Q_3 \\ -1 & 0 & 0 \\ 1 & -1 & 0 \\ 0 & 1 & -1 \end{bmatrix}. \tag{13}$$

We shall need to estimate the width of the flame layer. According to the standard theory of a one-step high activation energy reaction [1], the ratio of the width of the preheat zone to that of the flame layer is roughly equal to the Zel'dovich number

$$\theta \equiv \frac{d}{dT}\ln k_1(T),$$

this expresssion being evaluated at the layer's temperature T. Moreover, relative to the width of the preheat zone the slope of the concentration profile $Y_1(z)$ will be an O(1) quantity. Given that $Y_1(z)$ also vanishes at the back of the layer, it follows that its average value in the layer itself will be $O(\theta^{-1})$. Therefore a priori from (9) we may set

$$\nu_1 = \ln \theta.$$

The other concentration parameters ν_2 and ν_3, as well as the parameter s, are not known beforehand, however.

Recalling $\ell_j = -\ln k_j$, we have by (10)

$$\beta_1 = \ell_1 + \nu_1 - s, \qquad \beta_2 = \ell_2 + \nu_2 - s, \qquad \beta_3 = \ell_3 + \nu_3 - s. \tag{14}$$

From (6) we also have

$$Y_+ = \begin{bmatrix} 1 - \alpha_1 \\ \alpha_1 - \alpha_2 \\ \alpha_2 - \alpha_3 \end{bmatrix}. \tag{15}$$

Since there are no reverse reactions, the second part of (11a) can be overlooked, and (11a) says

$$s \le \ell_1 + \nu_1, \quad \nu_2 \ge s - \ell_2, \quad \nu_3 \ge s - \ell_3. \tag{16a}$$

Condition (11d), with (15), says that

$$\alpha_1 \le 1; \quad \alpha_2 \le \alpha_1; \quad \alpha_3 \le \alpha_2. \tag{16b}$$

Therefore in order for any combustion to occur at all, we have to have $\alpha_1 > 0$, so from (11f), reaction R1 must go to completion, i.e. $Y_{1+} = 0$, hence from (15), $\alpha_1 = 1$. Let us take the resolution constants a and b to be zero, a very conservative stance. Then since $\alpha_1 > 0$, we must have $\beta_1 = 0$ from (11c). Hence from (14) and (16a),

$$s = \ell_1 + \nu_1; \quad \nu_2 \ge \ell_1 + \nu_1 - \ell_2; \quad \nu_3 \ge \ell_1 + \nu_1 - \ell_3. \tag{17}$$

We consider three cases:

a) $\ell_1 + \nu_1 > \ell_2$, $\ell_1 + \nu_1 > \ell_3$. Then the region allowed by (17) in the positive quadrant of the (ν_2, ν_3) plane is as shown in the shaded part of Fig. 1a. It has one vertex V, defined by making the inequalities in (17) equalities. We test the vertex of the region to see if it satisfies the rest of (11). [We do this in general, working only at the vertices, in accordance with the strategy given at the end of the last section.] The vertex V has all β's equal to zero, so all reactions participate, and all α's are positive. Hence by (11f), the entire network must come to equilibrium, which means that $Y_+ = 0$. From (15) and equalities in (17), we get

$$\boldsymbol{\alpha} = (1, 1, 1), \quad \boldsymbol{\nu} = (0, \ell_1 + \nu_1 - \ell_2, \ell_1 + \nu_1 - \ell_3), \quad s = \ell_1 + \nu_1.$$

This is the only solution of (11) in this case.

b) $\ell_1 + \nu_1 > \ell_2$, $\ell_1 + \nu_1 < \ell_3$. The region defined by (17) and (11b) in this case is as shown in Fig. 1b. The only vertex is at $(\ell_1 + \nu_1 - \ell_2, 0)$. Now $\beta_3 > 0$, so $\alpha_3 = 0$ by (11c). But $\beta_2 = 0$, so by (11f) in fact R1 and R2 must pass to equilibrium, meaning $Y_{1+} = Y_{2+} = 0$, or $\alpha_1 = \alpha_2 = 1$. Thus

$$\boldsymbol{\alpha} = (1, 1, 0), \quad \boldsymbol{\nu} = (0, \ell_1 + \nu_1 - \ell_2, 0), \quad s = \ell_1 + \nu_1.$$

This is the only case satisfying (11). There is a residual amount of A_3 left over after the primary layer.

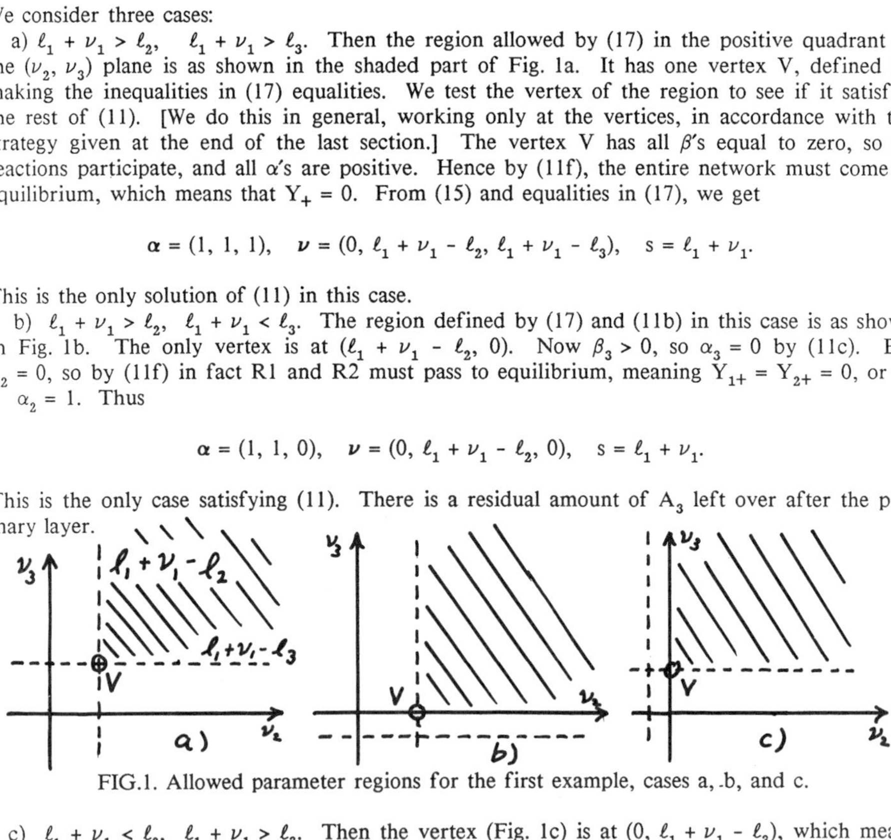

FIG.1. Allowed parameter regions for the first example, cases a, b, and c.

c) $\ell_1 + \nu_1 < \ell_2$, $\ell_1 + \nu_1 > \ell_3$. Then the vertex (Fig. 1c) is at $(0, \ell_1 + \nu_1 - \ell_3)$, which means that $\beta_3 = 0$ but $\beta_2 > 0$. The latter says that $\alpha_2 = 0$, hence by (16b), $\alpha_3 = 0$ as well. Thus $\boldsymbol{\alpha} = (1,0,0)$, $\boldsymbol{\nu} = (\nu_1, 0, \ell_1 + \nu_1 - \ell_3)$, $s = \ell_1 + \nu_1$. This is the only solution of (11). One could then argue that in fact $\alpha_2 = 0$ implies there is no production of A_3, so we should have $\nu_3 = \infty = \beta_3$. This shifts the above named vertex to one at ∞, namely $(0,\infty)$. But in view of our conservative choice a = b = 0, this does not change the original conclusion that A_3 exists in a negligible amount, since $\nu_3 > 0$.

d) $\ell_1 + \nu_1 < \ell_2$, $\ell_1 + \nu_1 < \ell_3$. The analysis in this case is essentially the same as that in c).

These conclusions reproduce the original easy intuituve conclusions if inequalities such as "$\ell_1 + \nu_1 < \ell_2$" are interpreted to mean that "k_1 is <u>much</u> greater than k_2". Of course it does not really imply the latter, but this only emphasizes the fact that our qualitative model exaggerates differences.

5. EXAMPLE: The chain branching - chain terminating network

$$A_1 + A_2 \rightleftarrows 2A_2, \quad A_2 \to 0.$$

Take the initial concentration to be $Y_- = \begin{bmatrix} 1 \\ 0 \end{bmatrix}$, and the first forward reaction to have high activation energy, so that when the initial supply of A_1 is consumed, it will be in a thin layer with characteristic temperature T. For simplicity, let us assume the other reactions have zero activation energy.

As in the previous example we first give an intuitive account. If the reaction rate $k_2(T)$ of the second reaction is much larger than $k_1^+(T)$, then R2 outcompetes R1$^+$ for any available supply of A_2, and since there is none to begin with, no combustion can be sustained, and there is no flame.

Now suppose k_2 is much smaller than k_1^+. Then R1 comes to equilibrium on a time scale which is relatively short, compared to that associated with R2. This first equilibration may be the only thing which occurs in the primary layer. The equilibrium values of Y_1 and Y_2 are related by

$$k_1^+ Y_1 = k_1^- Y_2.$$

If k_1^-/k_1^+ is large, then Y_2 is small, meaning that not much A_2 has been produced; we therefore have α_1 small, and we approximate it by 0. Again, the only possibility for any primary flame would be $\boldsymbol{\alpha} = 0$, and we conclude that there in fact is no primary layer. On the other hand, there will be a small amount of Y_2 (of order k_1^+/k_1^-) present, and so R2 will gradually operate to use up all the A_2 produced. It is essentially produced from A_1 through R1$^+$, and so the initial supply of A_1 will be slowly consumed. This will not occur in a layer. In fact, the heat release from R2 will in general cause the temperature to rise gradually during the process, and there will be no one characteristic temperature.

On the other hand if k_1^-/k_1^+ is small, then the first equilibration event occurs in a layer and converts most of the initial supply of A_1 into A_2. This is the primary layer; the new supply of A_2 is then slowly consumed by R2.

Now let us corroborate these quite easy conclusions by use of the formalism of (11). As before, we argue that if combustion really occurs, then R1$^+$ must indeed go to completion in a layer, and therefore that we may set $\nu_1 = \ln \theta$ on an a priori basis. Then from (10) and (6)

$$\beta_1^+ = \ell_1^+ + \nu_1 + \nu_2 - s, \qquad \beta_1^- = \ell_1^- + 2\nu_2 - s, \qquad \beta_2 = \ell_2 + \nu_2 - s, \tag{18}$$

$$Y_{1+} = 1 - \alpha_1, \qquad Y_{2+} = \alpha_1 - \alpha_2. \tag{19}$$

<u>Case I.</u> $\ell_2 < \ell_1^+ + \nu_1$. Then from (18) and (11a) we have that $\beta_1^+ > 0$ always, so by (11b), $\alpha_1 \le 0$. But (19) and (11c) imply $\alpha_2 \ge 0$ and $\alpha_1 - \alpha_2 \ge 0$, so in all, $\boldsymbol{\alpha} = 0$. No combustion is possible.

<u>Case II.</u> $\ell_2 > \ell_1^+ + \nu_1$. In the (ν_2, s) plane the region defined by the inequalities stating that all β's and all ν's are ≥ 0 appears as either (a) or (b) in Fig. 2.

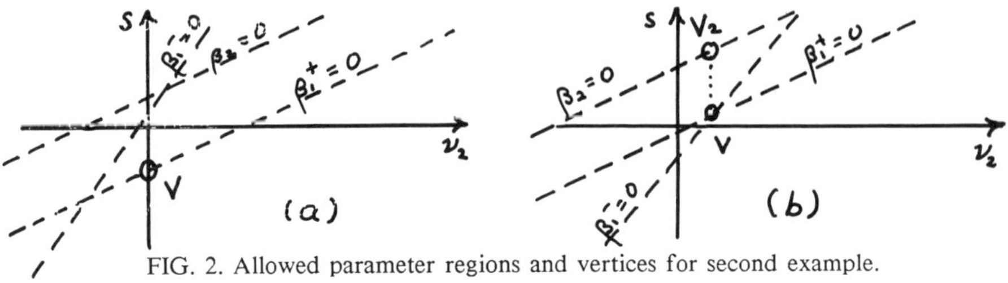

FIG. 2. Allowed parameter regions and vertices for second example.

In case (a), we have a vertex V at $\nu_2 = 0$, $s = \ell_1^+ + \nu_1$. At V, $\beta_2 > 0$ so by (11c) we set $\alpha_2 = 0$. We also have $\beta_1^- > 0$ so we neglect $R1^-$. Thus $R1^+$ must come to equilibrium ($\mathcal{M}' = \{R1^+\}$), hence $Y_{1+} = 0$ and by (19) $\alpha_1 = 1$. In short, the primary layer has

$$\alpha = (1,0), \quad \nu_2 = 0, \quad s = \ell_1^+ + \nu_1.$$

This, of course, will be followed by the slow consumption of A_2 outside the layer.

In case (b), the analogous vertex V is at $\nu_2 = \ell_1^+ + \nu_1 - \ell_1^-$, $s = 2\ell_1^+ + 2\nu_1' - \ell_1^-$. Again, $\alpha_2 = 0$ for the same reason, but it can no longer be said that $\alpha_1 = 1$, because now $\beta_1^- = 0$. Also $\nu_2 > 0$, so by (11e) and (19) we must have $\alpha_1 = \alpha_2 = 0$, and there is no primary layer.

There is another vertex in case (b), however, which we have so far overlooked. According to (11a), we must allow the possibility that $\beta_1^+ = \beta_1^- < 0$. This is shown as vertex V_2 on the above diagram. At that vertex, $\beta_2 = 0$, so by (11f), R2 must come to equilibrium, meaning $Y_{2+} = 0$. By (19) then, $\alpha_1 = \alpha_2$. On the other hand, by (11f) the entire network must come to equilibrium. Equilibrium for R1 means that Y_{1+} and Y_{2+} are linearly related. Since the latter was just shown to be zero, the former must be also. Therefore by (19), $\alpha_1 = 1$. We conclude that the unique <u>nontrivial</u> solution of (11) is

$$\alpha = (1,1), \quad \nu_2 = (0, \ell_1^+ + \nu_1 - \ell_1^-), \quad s = \ell_2 + \ell_1^+ + \nu_1 - \ell_1^-.$$

We may now see that the systematic treatment mirrors the intuitive treatment. In fact, Case I corresponds to the case when k_2 is much larger than k_1^+, Case IIa to the case when it is much smaller than k_1^+, as is k_1^-, and Case IIb corresponds to the remaining case when k_1^+ is much larger than k_2, but much smaller than k_1^-.

All of this leaves open the question of what happens in the intermediate cases, when the inequalities are not so extreme. Then of course the character of the chemistry is intermediate between those of the extreme cases. It is best and easiest first to delineate the extreme cases, and then proceed from there by modifying and moderating the results.

6. THE CONCEPT OF PARTIAL EQUILIBRIUM

In case (b) of the example in §5, we have $\alpha_1 = \alpha_2 = 1$, even though the rate of reaction $R1^+$, as measured by $\beta_1^+ < 0$, is much larger than that of R2, since $\beta_2 = 0$. This is no paradox, however; it simply means that the reverse reaction $R1^-$ also has large rate, and that rate partially balances the forward rate, so that the net rate of R1 is the same order of magnitude as that of R2. When this happens the reaction R1 is said to be in partial equilibrium; it is a very common phenomenon in reaction kinetics. We could define the degree of partial equilibrium by the difference between $\beta_2 = 0$ and $\beta_1^+ = \ell_1^+ + \nu_1 - \ell_2$. When temperature varies by a small amount, causing ℓ_1^+ to vary, this degree of partial equilibrium will also change so that $R1^+$ and $R1^-$ balance to a different extent, but the essential character of the flame will not change.

The rationale for the second part of stipulation (11a) is now clear. It is designed to account for reactions being in partial equilibrium.

7. PARAMETER VARIATION

It is interesting to study the changes in the solutions of (11) when one of the parameters in the problem (the most meaningful probably being T) changes. It is natural to characterize the vertex generating the solution by specifying which of the inequalities in (11a) and (11b) are actually equalities at the vertex. Even though the position of a vertex will move around in parameter space, let us agree to say that the vertex retains its identity if and only if that characterization does not change. Then as the one variable parameter changes, the corresponding vertex will only change at discrete values of the parameter, and then the change will be abrupt. Usually (as in these examples) $\boldsymbol{\alpha}$ will also remain constant, changing only when the vertex changes in the above sense. On the other hand, $\boldsymbol{\nu}$ will of course change continuously.

In Example 1, let us choose ℓ_1 to be the variable parameter. We begin with large enough ℓ_1 so that case a) holds, and $\boldsymbol{\alpha} = (1,1,1)$. As ℓ_1 is decreased, at a critical point case a) changes into case b) or case c), depending on which of ℓ_2 and ℓ_3 is the larger. In the first case, α suddenly shifts to (1,1,0), the third reaction essentially being quenched in the primary layer; and in the second, to (1,0,0).

If all three reactions are exothermic and T_- is fixed, then the temperature T in the layer depends on $\boldsymbol{\alpha}$ according to (8), and so of course will change at the critical parameter point. Considering that k_1 and ℓ_1 themselves depend on T, we may find ourselves in a situation that increasing T_- (say) will cause T to increase, hence ℓ_1 to decrease, until it reaches the critical point. At that point the allocation changes and because of that change, T suddenly decreases. This in turn raises ℓ_1 again, which turns the old allocation back on, etc. What will finally result is a compromise in which the flame layer is host to a mixture of the two allocations in such a proportion that the temperature there is exactly the critical temperature. This critical temperature then will be insensitive to changes in T_- within a certain range. This is an example of a temperature plateau, as pointed out in [3] and [5].

It should be born in mind, however, that this temperature plateau is only with regard to the primary layer. The final burned temperature of the gas will always be given by (8) with $\alpha = (1,1,1)$, but it may be attained partly through combustion in the zone behind the layer.

A similar critical transition occurs in Example 2: in this case the change involves the onset of partial equilibrium in the primary layer.

Finally, the changes in the solution of (11) are abrupt, but they correspond to more gradual changes in any real flame being modeled. It has to be understood that the discontinuities in the model will be smoothed out in reality.

8. FURTHER APPROXIMATIONS; DISCUSSION

The procedure as described here is fraught with exaggerations and so may result in a crude approximation, but in a clear qualitative picture. At the same time, the result may well serve as a basis for better approximations, as was shown in [4]. Any viable solution of (11), in fact, corresponds to a vertex in $(\boldsymbol{\nu},s)$ parameter space. At that vertex, some of the β's are $\leq$ a and result in nonzero allocation components. But all of the β's, whether eliminated on the first round or not, can be evaluated at that value of $(\boldsymbol{\nu},s)$, and the extent of participation of the various reactions estimated. This way, it will in general be found that some of the reactions which were eliminated in the first approximation actually do participate to some (perhaps minor) extent. A further effect of this second approximation will be to "smooth out" the sharp changes which occurred at the critical temperatures mentioned in §7.

Solving the model problem as explained here gives a reduced reaction network consisting of only those reactions which have nonzero α's. This is expected to be a skeleton network giving the essential chemistry for the flame. However, the analysis is all temperature-dependent, and the skeleton network so obtained may very well change at critical temperatures. It may therefore be true that there is no single simplified network which will be valid for the entire temperature range of interest.

The reduced network obtained by this approach may of course imply overall effective chemistry through a further reduced network in which those radicals which do not persist behind the primary combustion zone are eliminated. However, this overall network will no longer satisfy mass action kinetics, and so will be of limited use. The rates of its reactions can be obtained from the β's in our calculations, but they may be rather complicated functions of temperature and concentrations.

As mentioned in §3, the allocation α obtained in the primary layer itself determines the temperature of that layer through (8). On the other hand, the vertex of interest, hence α, was determined as a function of temperature in the layer. This poses an obvious compatibility problem: will the temperature computed by (8) be the same as the one with which one started? This question of compatibility produces interesting consequences. First of all, it should be realized that although the vertex does change continuously with temperature, α does not. In fact the α's satisfy certain linear constraints (11d,f) to which additional constraints (11e) have been added; the latter are the only constraints on the α's which depend on the ν's, hence on the temperature. They change only at critical temperature values. It therefore follows that α will be constant between those critical temperature values. This makes the compatibility problem less serious than it may appear at first glance, because the temperature computed from (8) now need be only in a certain interval.

On the other hand, examples in [15] show that it may not be possible to get a compatible temperature this way, and the actual resulting flame temperature may be a compromise such as was indicated in §7. Then a temperature plateau results. Such plateaus were found in [15]. Another possibility is nonuniqueness due to this temperature coupling effect, as explored in [3], [5]. Examples of this kind of nonuniqueness were not found in the H_2 - O_2 system.

REFERENCES

1. Buckmaster J and Ludford GSS: Theory of Laminar Flames. New York: Cambridge Univ. Press, 1982.
2. Clavin P: Dynamic behavior of premixed flame fronts in laminar and turbulent flows. Prog. Energy Combust. Sci. 11, 1 - 59 (1985).
3. Clavin P, Fife PC and Nicolaenko B: Multiplicity and related phenomena for flames with competitive networks. SIAM J. Appl. Math., in press.
4. Fife PC: Modeling and analysis of hydrogen-oxygen flames, preprint (1986).
5. Fife PC and Nicolaenko B: The singular perturbation approach for flame theory with chain and competing reactions. Ordinary and Partial Differential Equations, W. N. Everitt and B. D. Sleeman, eds., Lecture Notes in Mathematics No. 964, pp 232-250. Berlin: Springer-Verlag, 1982.
6. Fife PC and Nicolaenko B: Asymptotic flame theory with complex chemistry. Nonlinear Partial Differential Equations, J. Smoller, ed. Contemporary Mathematics 17, pp. 235-256. Providence: Amer. Math. Soc., 1983.
7. Fife PC and Nicolaenko B: How chemical structure determines spatial structure in flame profiles. Modelling of Patterns in Space and Time, W. Jager and J. D. Murray, eds., Lecture Notes in Biomathematics, No. 55, pp. 73-86. Berlin: Springer-Verlag, 1984.
8. Fife PC and Nicolaenko B: Flame fronts with complex chemical networks. Physica 12D, 182-197 (1984).
9. Paczko G, Lefdal PM, and Peters N: Reduced reaction schemes for methane, methanol and propane flames. Twenty-first Intern. Symp. on Combustion, The Combustion Institute, 1986.
10. Peters N: Numerical and asymptotic analysis of systematically reduced reaction schemes for hydrocarbon flames. pp. 90-109 in: Numerical Simulation of Combustion Phenomena, R. Glowinski, B. Larrouturou and R. Temam, eds. Lecture Notes in Physics 241. Berlin: Springer-Verlag, 1985.
11. Peters N and Kee RJ: The computation of stretched laminar methane-air diffusion flames using a reduced four-step mechanism, preprint.
12. Peters N and Smooke MD: Fluid dynamic-chemical interactions at the lean flammability limit. Combustion and Flame 60, 171-182 (1985).
13. Peters N and Williams FA: The asymptotic structure of stoichiometric methane-air flames, preprint (1986).
14. Roten CD: An algorithm for allocation and temperature, and its consequences for the chemistry of H_2 - O_2 combustion. These proceedings.

15. Roten CD and Fife PC: Modelling and analysis of H_2 - O_2 combustion between 1100^0 and 1600^0, preprint.
16. Smooke MD, Rabitz H, Reuven Y, and Dryer FL: Application of sensitivity analysis to premixed hydrogen-air flames, preprint.

COMPUTER SIMULATION OF 2D/3D REACTING FLOWS IN COMPLICATELY SHAPED REGIONS FOR ENGINEERING

B.P. GERASIMOV

KELDYSH INSTITUTE OF APPLIED MATHEMATICS, USSR ACADEMY OF SCIENCES, MOSCOW

The computer simulation becomes a common research tool not only in science but in industry too and this tool should be available to the wide community of engineers and technologists, who do not possess deep knowledge of numerical mathematics. The highly efficient CAD or diagnostic systems cannot be created without proper involvement of computer simulation. So a major problem arises concerning the development of special computational methods and appropriate multipurpose software designed for applied science and industry.

Heat and mass transfer as well as fluid flows are the most common transport mechanisms in technical devices and technological processes. Because of the complicated nature of basic differential equations the customary methods and software, e.g. in computational aerodynamics, are divided in two categories. The first is designed for experienced "computators" and allows them to simulate various complex problems ; the second is designed for ordinary users, any particular code providing routine solution of a fixed usually narrow class of problems. Both of them are inadequate for industrial applications, which demand the combination of complexity and variety of problems with simplicity and unification of computations. The respective multipurpose applied code packages (ACP) involve sufficiently comprehensive multi-dimensional models adapting actual geometric configurations and may be easily modified by a user due to a standardized modular structure as well as simple, reliable and versatile algorithms.

In this report the numerical methods and the software developed for the engineering computational experiment in hydrodynamics are discussed, and their capabilities are demonstrated on a number of problems. Three basic models were chosen from a great variety of fluid flow models - the Navier-Stokes equations for incompressible fluid, the Euler equations for compressible gas, and large systems of stiff ordinary differential equations, which cover nearly the whole area of applications. The respective packages were named NEPTUNE, AEOL, and STIFSP. Their standard structure was based on the well known OLYMPUS system [10]. A common Fortran subset provides the package portability. The main advantage of the first two packages, which made them particularly suitable for engineering, is conservative scheme on a rectangular grid, which allows computing flows in regions of complex varying shape.

The NEPTUNE computer code package is designed for computing the nonstationary two-or quasi three dimensional viscous incompressible fluid flows in a plane or cylindrical (with three velocity components) geometry [1]. It involves convective heat and mass transfer (with kinetics of chemical reactions and phase transitions) ; the k - ε turbulence model ; a multi-component, porous or dispersive nature of a medium ; the radiation heat transfer, etc.

C.-M. Brauner and C. Schmidt-Lainé (eds.), Mathematical Modeling in Combustion and Related Topics, 89–101.

Examples of computer simulation of mixed convection in high temperature chemical reactors [5,9] of various configurations (in a cylindrical geometry) are shown in Fig. 1 the isotherms being in the light and streamlines being in the right. The deteriorated intensity distribution of the initially circular Gaussian laser beam after a considerable travel in a weakly absorbing medium with free convection is shown in Fig. 2a and the corresponding temperature plume in Fig. 2b.

The AEOL computer code package [2,3] is designed for computing 2D and 3D nonstationary compressible gas flows (including detachment effects) in regions of complex and arbitrarily varying configurations. The Eulerian variables in a plane or cylindrical geometry are used. The isobar fields at the 3 moments of shock diffraction on a deformable cone and a hinged wedge are shown, respectively, in Figs. 3a and 3b. The velocity and isobar fields for a breaked flow in the initially straight cylindrical tube with elastic walls (pressure surge) [8] are plotted in Fig. 4. The aircraft-like body is shown in Fig. 5 by the lines of intersection between grid planes and body surface, and respective isobars in symmetry plane of 3D flow are shown in Fig. 6. The flow patterns in super - and transonic nozzles are shown in Fig. 7 [4].

The STIFSP computer code package [6] is designed for solving large systems of algebraic and ordinary differential first order equations that are resolved or not with respect to derivatives. A special matrix storage and inversion algorithm is provided for large sparse systems. This code involves three groups of methods : the Adams methods for nonstiff problems, the Gear methods for stiff problems, and the optimized methods that need no stiffness analysis. The code contains an algorithm for the automated control of integration step and method order to minimize computational time at given accuracy.

REFERENCES

1. B.P. Gerasimov et al. The NEPTUNE computer code for the computer simulation of incompressible viscous fluid flows. Preprint, Keldysh Inst. Appl. Math., N 65, 1985.

2. B.P. Gerasimov et al. The AEOL-2 code package for calculating two-dimensional gasdynamic problems. Preprint, Keldysh Inst. Appl. Math., N 28, 1986.

3. B.P. Gerasimov et al. Conservative boundary conditions for three-dimensional gasdynamic problems. Preprint, Keldysh Inst. Appl. Math., N 66, 1986.

4. V.I. Gavrilyuk, B.P. Gerasimov, S.A. Semushin. On one formulation of boundary conditions for the compressible gas flow calculations. Preprint, Keldysh Inst. Appl. Math., N 31, 1987.

5. B.P. Gerasimov, A.I. Lesunovsky. Thermodiffusion phenomenon in the convective heat and mass transfer in a epitaxial gas reactor. preprint, Keldysh Inst. Appl. Meth., N 126. 1986.

6. B.P. Gerasimov, I.A. Kul'chitskaya. STIFFSP - the computer code package

for integrating differential-algebraic systems of large dimension. Preprint, Keldysh Inst. Appl. Math., N 103, 1984.

7. B.P. Gerasimov et al. The temperature dependence of nonstationary radiative conductivity of polymers. Preprint, Keldysh Inst. Appl. Math., N 146, 1984.

8. B.P. Gerasimov, S.A. Semushin. Calculations of flows around bodies with a varying geometry on a fixed Eulerian grid. Differential Equations, v. 17, N 7, p. 1214, 1981.

9. B.P. Gerasimov et al. Gasdynamics investigations of chambers in installations for growing the silicon monocrystals. In : Mathematical Modeling, Moscow, Nauka, 1986, p. 133.

10. J.P. Shristiansen, K.V. Roberts. OLYMPUS. A standard control and utility package for initial-value Fortran programs. Comp. phys. comm., v.7, p. 245, 1974.

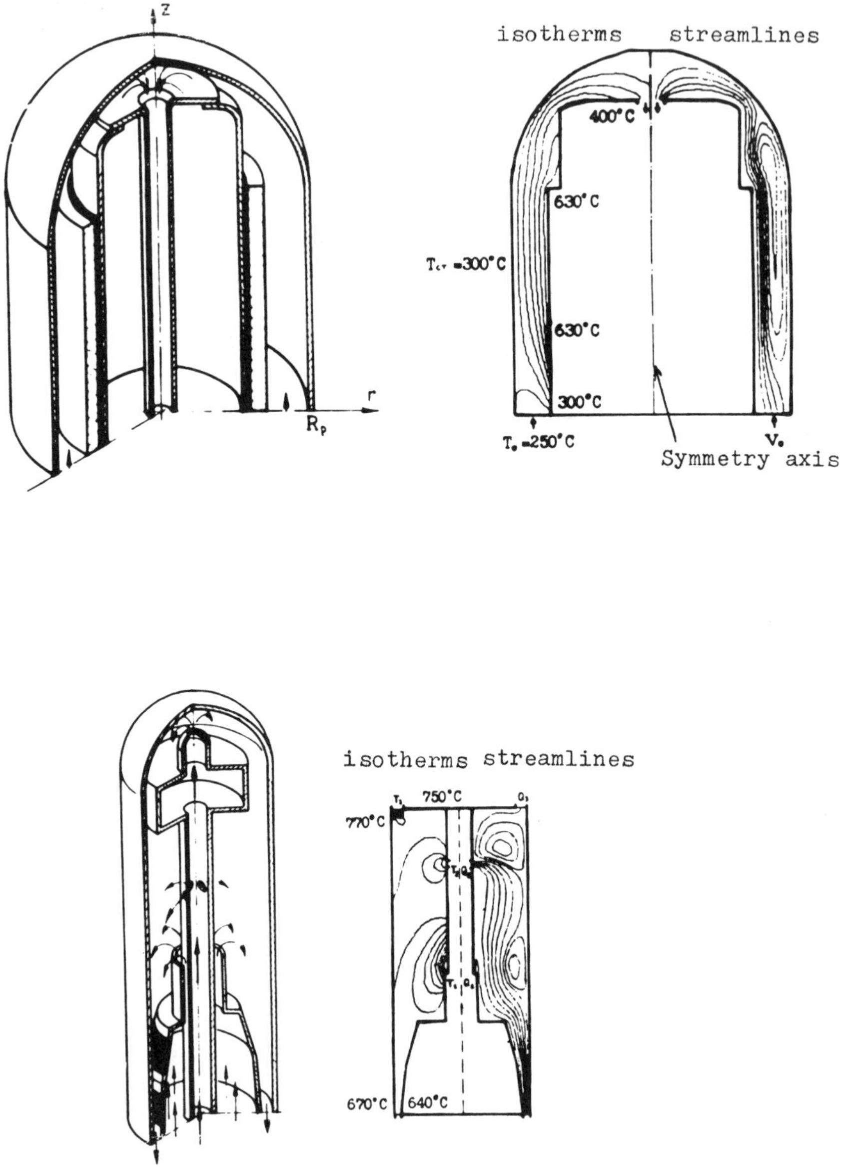

z
r
R_p
isotherms
streamlines
400°C
630°C
T_{cr} =300°C
630°C
300°C
T_0 =250°C
V_0
Symmetry axis
isotherms streamlines
750°C
770°C
670°C
640°C

Fig : 1

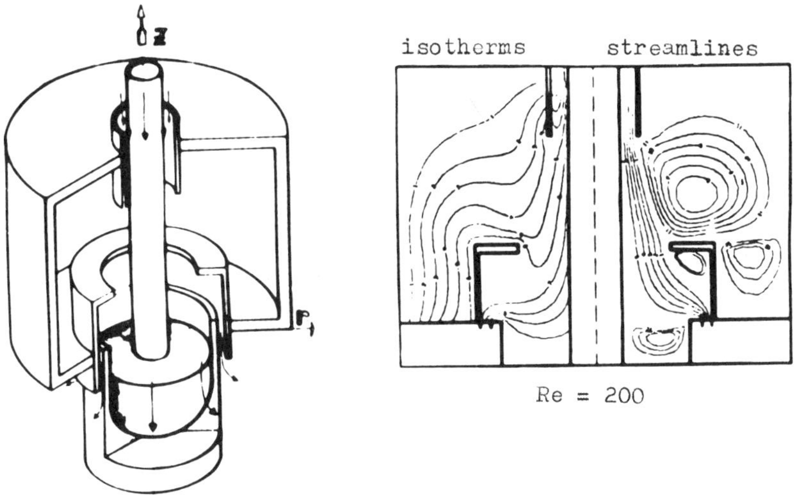

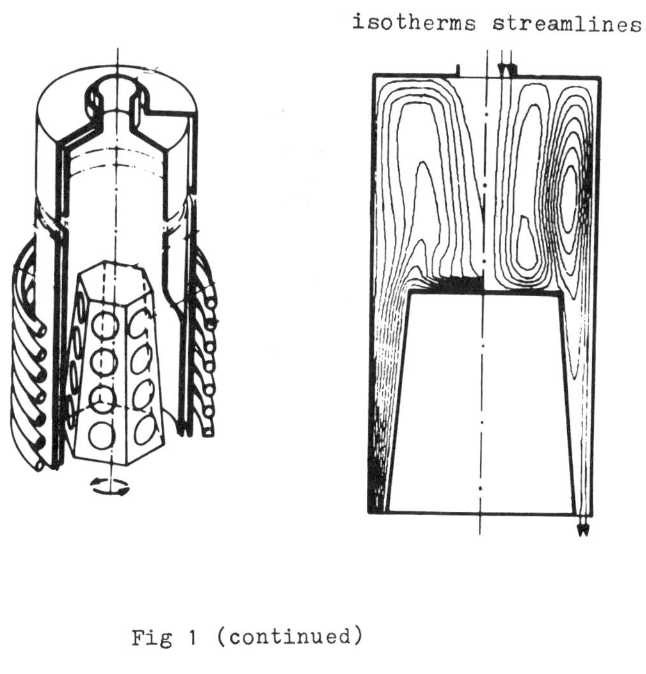

Fig 1 (continued)

a) intensity

b) isotherms

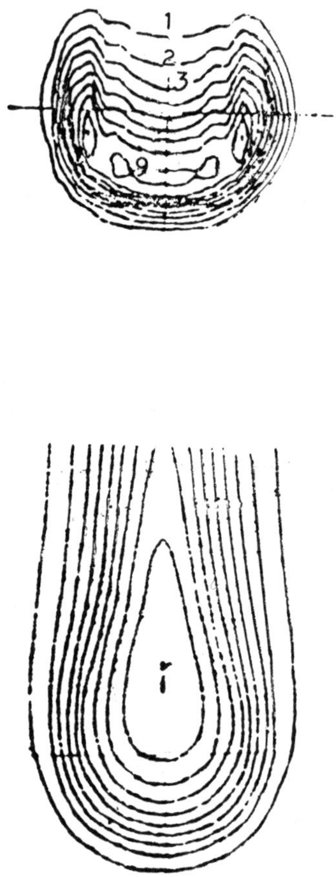

Fig : 2

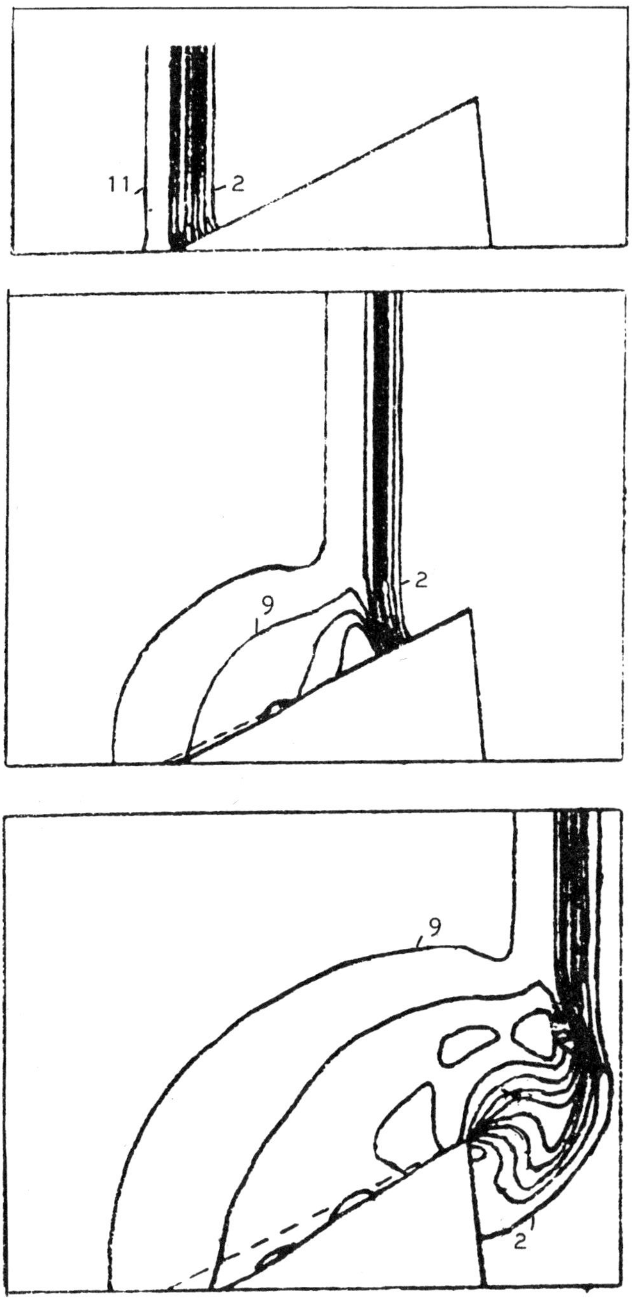

Fig : 3 a) cone

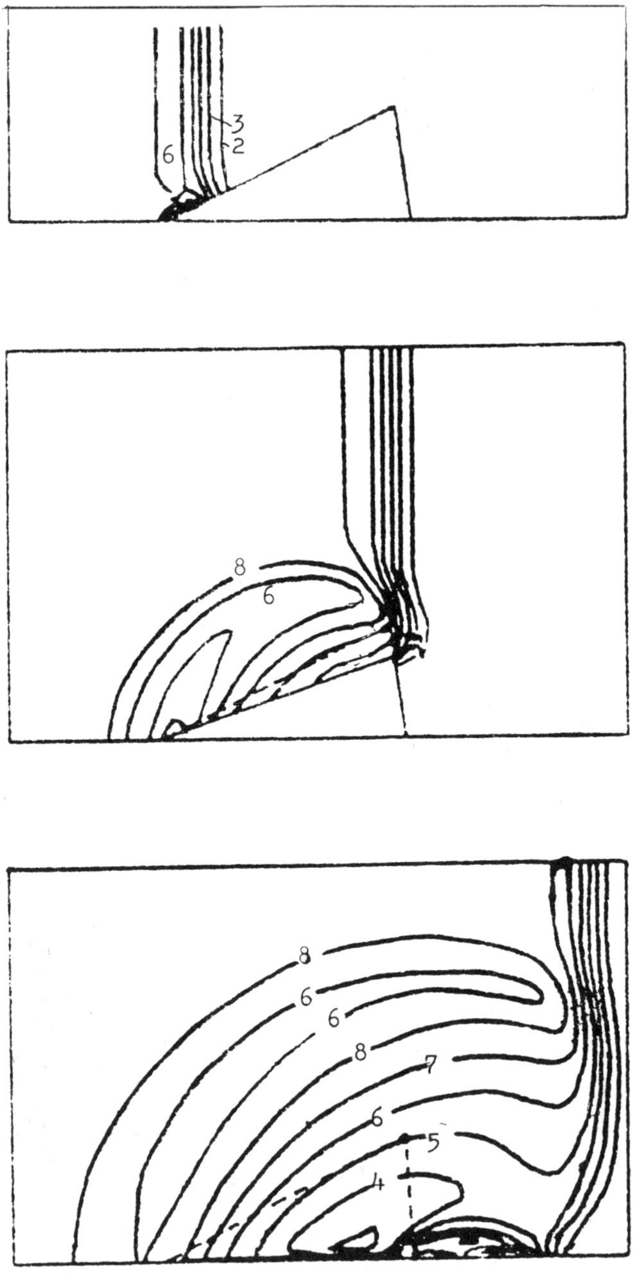

Fig : 3 b) wedge

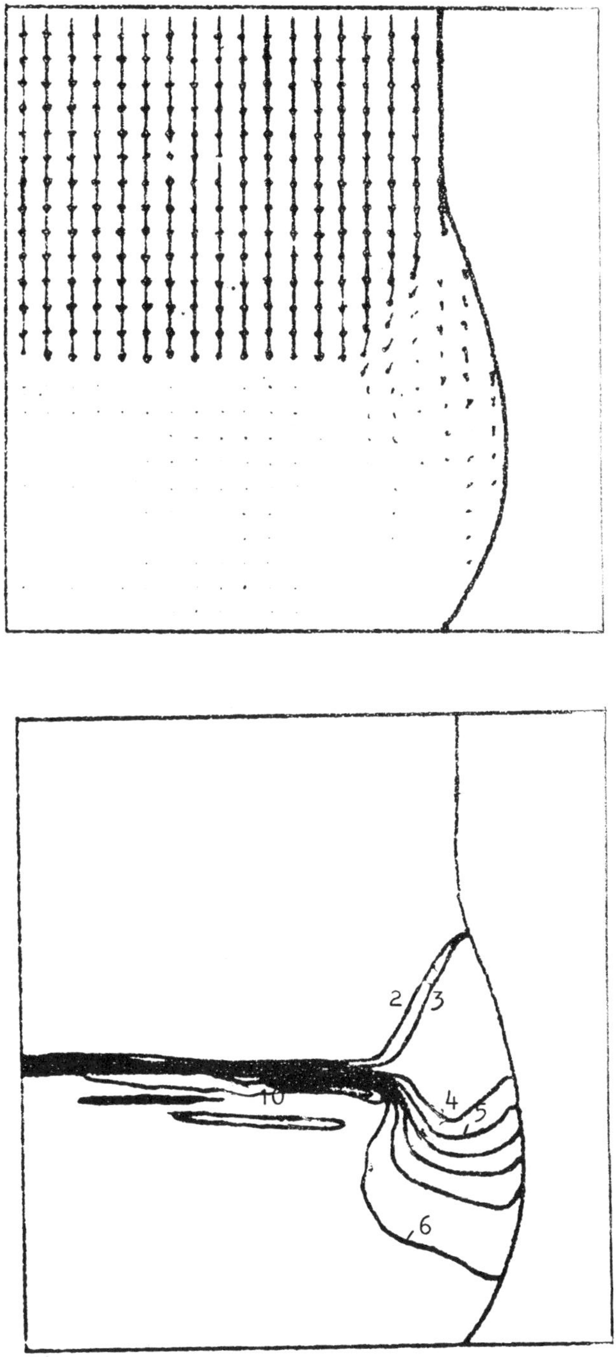

Fig : 4

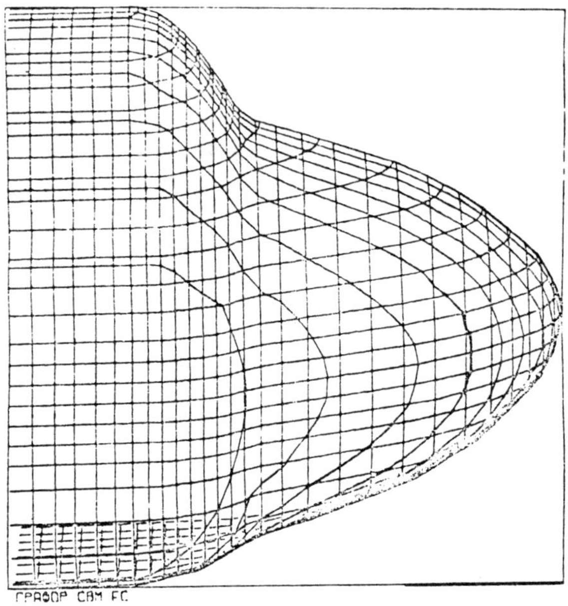

Fig : 5 a

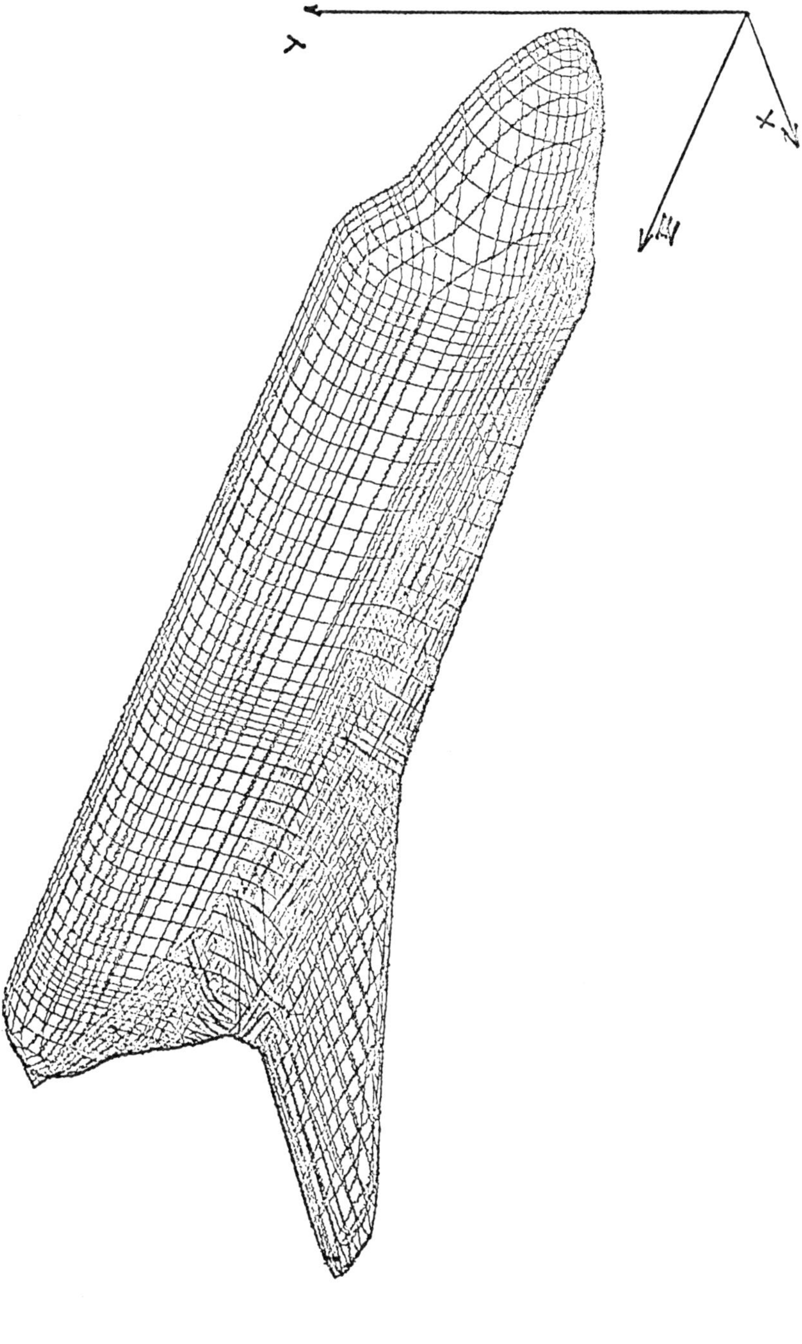

Fig. 5b

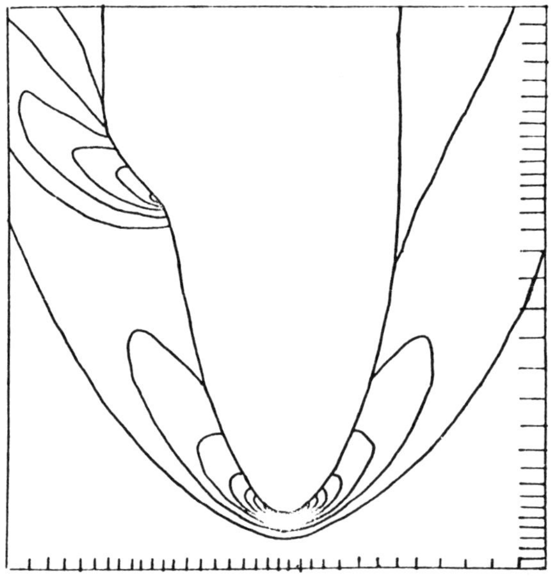

Fig. 6

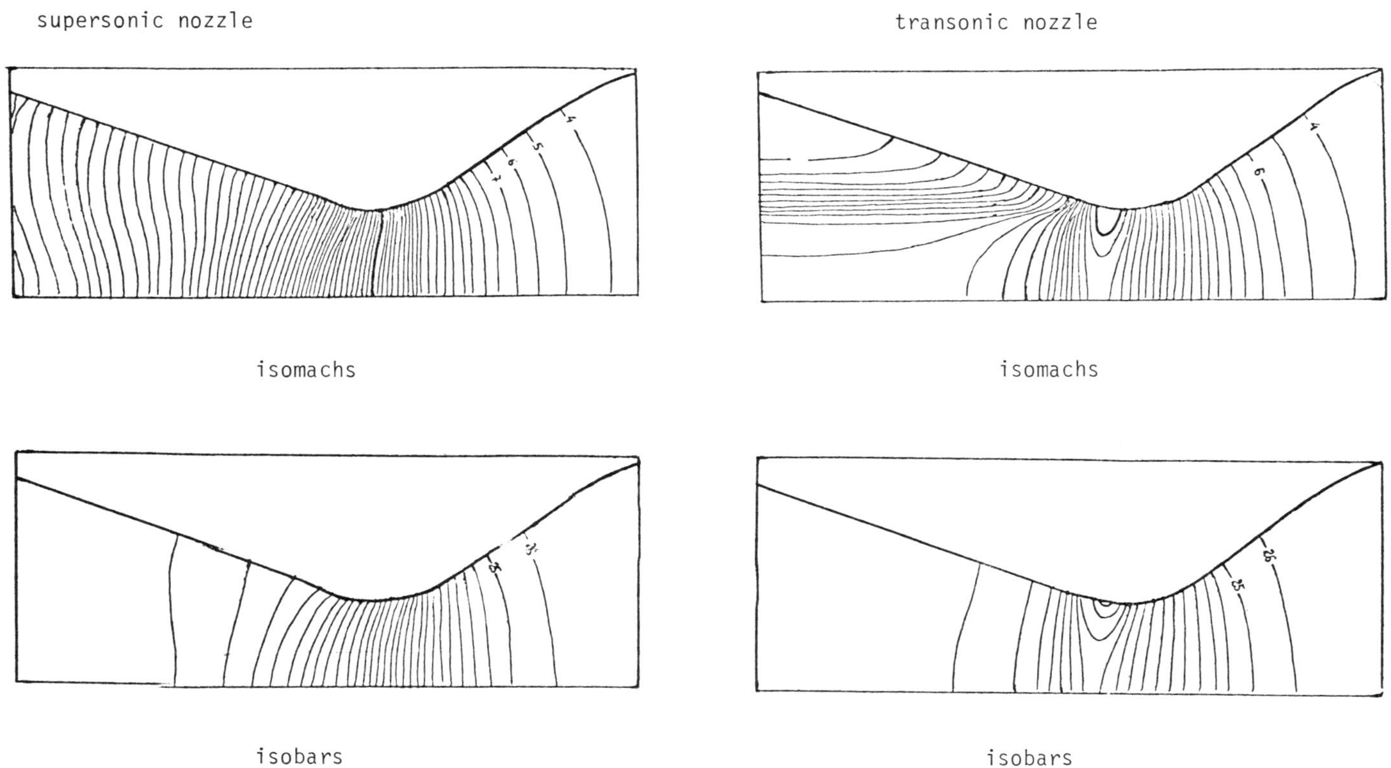
supersonic nozzle
transonic nozzle
4
5
6
7
isomachs
isomachs
isobars
isobars
25
26

Fig. 7

RADIATIVE TRANSFER IN UNSTEADY, WEAKLY CURVED, PARTICLE-LADEN FLAMES

Guy JOULIN

U.A. 193 C.N.R.S., Rue Guillaume le Troubadour, 86034 Poitiers (FRANCE)

0. ABSTRACT

Recent theoretical results on flames propagating in gaseous premixtures seeded with inert particles are presented. The attention is focused on the influence of the radiative exchanges, which are incorporated as a singular perturbation in the limits of a small Boltzmann number (radiant to convective flux ratio) of a large Zel'dovich number (activation to maximum temperature ratio) and of optically-thin flame fronts. The effects of mixture loading by the suspension, and of temperature-lags between the phases are also taken into account.

We describe planar flames, first steady ones, for which cases of regime multiplicity are exhibited. Then we obtain evolution equations for unsteady plane flames, the analysis of which allows one to study unsteady transitions between stable regimes and to display a mechanism leading to relaxation oscillations in burning speed. The case of expanding spherical flames is also briefly evoked.

In a second part of the talk we consider a model of weakly curved, steady flame, for which the radiative transfer leads to non-local curvature effects on the burning velocity, hence on the flame shape. Angled front shapes are shown to be permitted, in some instances. We finally investigate unsteady, weakly curved flames, and we obtain a non-linear evolution equation to compute their shapes, which is non-local in space and in time.

1. INTRODUCTION

During the last fifteen years, many problems concerning the existence, the structure and the dynamics of gaseous premixed flames, including some turbulent fronts, have been elucidated by first using well chosen, steady one-dimensional configurations as starting points, then by employing stability analyses, perturbations and/or bifurcation techniques successively to takle more involved situations : appearance of cells, of self-turbulence, of vibrations..., response to flow gradients... /1-5/.

It is tempting to try applying the same kind of methodology to gas particle premixtures, about which the theory is far less developed than for gaseous flames, even though they are often involved in the industry (e.g. coal-dust-fired burners) and/or in hazardous explosions /6,7/.

Given the huge number of independent parameters that are needed to characterize a gas particle premixture, and the complexity of the corresponding conservation equations, trying to solve the problem in its full generality represents a rather hopeless challenge, however, even in 1-D situations. A classification, even tentative and with somewhat fuzzy boundaries, according to the dominant processes, is thus welcomed. For example, one may imagine the various reactive two-phase mixtures classified according to the reactivity of the particles, their volatile yields, and of the carrier gas(es). One can then identify a continuous spectrum of media,

C.-M. Brauner and C. Schmidt-Lainé (eds.), Mathematical Modeling in Combustion and Related Topics, 103–130.

ranging from fuel particles (e.g. : coal dust) with no volatile yield dispersed in a gaseous oxidizer, to slowly reacting or even inert particles dispersed in reactive gaseous premixtures ; even in the latter case, the amount of particles and their distribution in size are still degrees of freedom. Similarly, the reaction waves occurring in these multiphase media can be also ordered from radiation-affected to radiation-dominated, and from optically-thick to a optically-thin, waves, according to the intensity and the range of a new mechanism introduced by the existence of solids, namely the radiative transfer of energy.

In the same way as for gaseous flames, it may prove useful to first focus one's attention on the most extreme configurations, with the ultimate goal of progressively filling the gaps between them and/or identifying crossover behaviors. For example, a problem of radiation-dominated combustion wave in mixtures of a gaseous oxidizer and gasless fuel particles has been recently solved analytically /8/, and we currently use the results to study the influence of a small amount of particle-generated reactive volatile. In the present article, we review recent theoretical results concerning another "boundary" of the spectrum evoked above, namely the radiation-affected propagation of flames in reactive mixtures seeded with inert particles. In addition to being a first step towards a better understanding of flame propagations in reactive gases seeded with slowly reacting suspensions (e.g. graphite in CH_4-air mixtures) the study is also of interest for explosions in mines or above dusty grounds, or to better predict the efficiency of certain dusts used as fire-suppressant (e.g. sand curtains). In addition, this system leads to mathematically interesting and new phenomena, related to the peculiarities of radiative transfer, which is a non-local mechanism.

The starting point of our analyses is the statement that the overall chemical rate of heat release - hence the flame burning velocity - corresponding to the combustion of most premixed gases is a strongly increasing function of temperature, since the activation to maximum thermal energy ratio (Zel'dovich number) is usually large. As a consequence, in the physically realistic, and mathematically convenient, limit of large Zel'dovitch numbers, one may first content oneself with assuming that the inert particles only weakly (but not negligibly-) perturb the structure of a gaseous flame ; it is not the whole story, but a convenient way to start with.

The influence of the particles is threefold. First of all, loading the reactive mixture by some inert material increases the heat capacity. By contrast to what happens with gaseous diluents, however, the inert suspension and the reactive gas may be locally out of thermal equilibrium ; then, temperature-lags appear between the phases. Last, but not least, the radiative transfer between the particles which provides the burned, hot medium with a means to preheat (at a distance) the incoming, yet unreacted, one. Those three effects influence any particle-laden flame, even the steady, planar ones. In unsteady and/or wrinkled flames, this influence may be quite peculiar, however, owing to the non-local characteristics of radiation. Showing this is the purpose of the present review paper.

2. THE BASIC GASEOUS FLAME

When the reactive medium does not contain any particle, the structure of a steady, plane flame propagating through it can be obtained by solving the energy equation (1) and the mass balance (2) for the deficient reactant :

$$\rho u\ C_p \frac{dT}{dn} = \frac{d}{dn}\left(\lambda \frac{dT}{dn}\right) + Qw \qquad (1)$$

$$\rho u \frac{dy}{dn} = \frac{d}{dn}\left(\rho D \frac{dy}{dn}\right) - w \qquad (2)$$

provided the burning process is modelled by a one-step, irreversible overall reaction "reactant → products + heat". Two reactants, e.g. fuel and oxidizer, could be used explicitely ; the final results we are to show would not be changed, however, since they are of thermal origin. In (1) (2), T, ρ, λ and C_p are the gase temperature, its density, its heat conductivity and its specific heat at constant pressure, respectively. ρu is the burning rate, an eigenvalue of the problem. y is the reactant mass fraction in the gas, and D is its diffusion coefficient. Q is a heat of reaction. We further assume that the reaction rate w follows the Arrhenius law :

$$w = Z \ .\ y\ .\ \exp(- E/RT) \qquad (3)$$

where the pre-exponential factor Z only changes mildly with T in comparison to its exponential companion. It is now widely known /1-5/ how the method of matched asymptotic expansions allows one to solve (1) - (3) for T, y and the adiabatic burning speed $U_{ad} \equiv (\rho u)/\rho_u$ (the subscript "u" refers to the far-upstream, unreacted mixture), when the Zel'dovitch number, $\beta = E(T_{ad}-T_a)/RT_{ad}^2$, is large, even only moderately. In the formal limit $\beta \to \infty$, the flame exhibits three zones (Figure 1), namely :

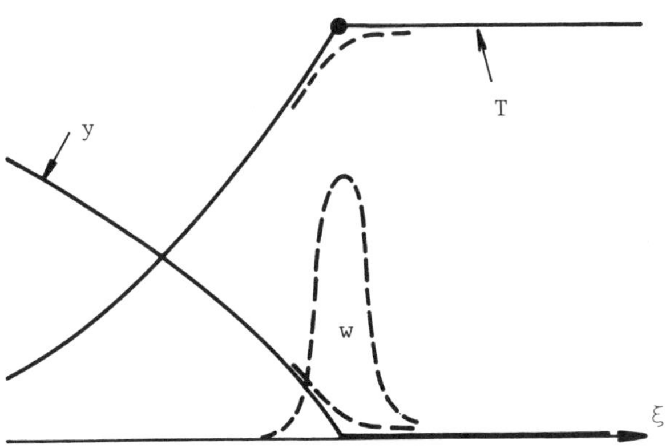

Figure 1 : Schematic structure of a particle-free, planar flame, when $\beta \gg 1$. (——) : outer profiles, (----) inner profiles.

i) the burned gas region, where $T(n) = T_{ad} \equiv T_u + Q\, y_u / C_p$.
ii) the preheat zone, the width of which is comparable to the conduction / convection length $\ell = \lambda_u / \rho u\ C_p$, where the chemistry is frozen, hence negligible ; the corresponding outer profiles, for $T(n < 0)$ and $y\ (n < 0)$, are :

$$T = T_u + (T_{ad} - T_u)e^{\xi} \quad , \quad y/y_u = 1 - e^{Le\ \xi} \tag{4a}$$

where $Le = \lambda/\rho DC_p$ is the Lewis number, and the reduced coordinate ξ is defined by :

$$\xi = \rho_u U_{ad} C_p \int_o^n dn/\lambda \tag{4b}$$

iii) a reaction zone, which is thinner than the previous one by an $O(1/\beta)$ factor, where convection is weak compared to conduction, diffusion and chemistry, at least at the leading order in the limit $\beta \to \infty$. Solving this convection-free problem gives :

$$\frac{dT_{inner}}{dn}\Big|_{-\infty} = \left(\frac{2Le\ Z_{ad}\ e^{-E/RT_{ad}} R^2 T_{ad}^4 C_p}{\lambda_{ad}\ E^2}\right)^{1/2} + \ldots \tag{5}$$

for the leading order temperature slope at the upstream side (defined by $\eta \equiv \beta n \to -\infty$) of the reaction zone. Identifying it with the outer slope dT_{outer}/dx deduced from (4a.b), via a standard matching, gives :

$$U_{ad}^2 = \frac{2\ Le\ Z_{ad}\ e^{-E/RT_{ad}}\ R^2 T_{ad}^4 \lambda_{ad}}{\rho_u^{\ 2}\ C_p\ E^2 (T_{ad} - T_u)^2} + \ldots \tag{6}$$

A very important conclusion is drawn from (6) : U_{ad} is very sensitive to variations in the reaction temperature T_R (here $T_R = T_{ad}$) owing to the large value of E/RT_{ad}, since $O(RT_{ad}^2/E)$ changes in T_R affect U_{ad} by an $O(1)$ factor. In addition, the reaction zone is seen to be asymptotically thinner than the preheat zone and the burned gas region (both spatially and thermally), when $\beta \to \infty$, so that the reaction rate may be viewed as concentrated along a reaction sheet. This can be shown to be still true even if the flame is non-adiabatic /9/, curved and/or nonsteady, provided the flame evolution time and radii of curvature are not smaller than $O(\ell/U_{ad})$ and $O(\ell)$, respectively, and the temperature gradients in the burned gas region are $O((T_{ad} - T_u)/\beta\ell)$ at most. Then, the outer profiles for T and y satisfy the following jump relations across the reaction sheet:

$$\begin{aligned} &[T]_-^+ = o((T_{ad} - T_u)/\beta) \\ &y \quad = 0 \\ &\left(\lambda \frac{dT}{dn} + Q\rho D \frac{dy}{dn}\right)_-^+ = O(\rho_u U_{ad} C_p (T_{ad} - T_u)/\beta^2) \end{aligned} \tag{7}$$

where $(\ )_-^+$ denotes a jump, and n is the normal coordinate to the reaction sheet. Equations (7), and the boundary conditions, upstream and downstream, allow one to compute (in principle) the reaction temperature T_R in terms of what happens (heat losses, unsteadiness...) on both sides of the reaction sheet, as well as $dT/dn|_o$ on the fresh side of it. Once this is done, the flame shape, speed and/or evolution is determined (again in principle) by imposing the extra condition :

$$\lambda_{ad} \frac{dT_{outer}}{dn}\Big|_{o^-} \underset{\beta\to\infty}{\to} (T_{ad} - T_u)\, \rho_u\, U_{ad}\, C_p \exp\left(\frac{E}{RT_{ad}} \frac{T_R - T_{ad}}{2\, T_{ad}}\right) \tag{8}$$

in which $(T_R - T_{ad})/T_{ad}$ has to be $O(1/\beta)$ at most.

3. FLAMES IN TWO-PHASE MIXTURES

We now assume that inert particles are present in the mixture. For the sake of simplicity, we only consider a monodisperse suspension of spherical solid particles, each one having a radius r_s, a mass $4\pi r_s^3\, \rho_s/3$ and a heat capacity $4\pi r_s^3 \rho_s C_s/3$. Provided the two-phase mixture is viewed as a one-velocity medium, and the total volume of the particles is neglected in comparison to that of the gas, the energy equations for the two phases are:

$$\rho C_p \frac{DT}{Dt} = \operatorname{div}(\lambda \operatorname{grad} T) + Qw - N\, 4\pi r_s \lambda (T - T_s) \tag{9}$$

$$N \frac{4}{3} \pi r_s^3 \rho_s C_s \frac{DT_s}{Dt} = N\, 4\pi\, r_s\, \lambda (T - T_s) - \operatorname{div} q \tag{10}$$

where $D/Dt \equiv \partial/\partial t + \underline{u}.\mathrm{grad}$; T_s is the particle temperature, $\underline{u}$ is the velocity vector, N is the local particle number density ($N \sim \rho$ is a one-velocity medium), and $\underline{q}$ is the radiant net flux of energy. To write (9)-(10) we assumed that the gas is transparent to radiation, and that the temperature is uniform in each particle owing to the large heat conductivity of most solids ; we also considered that the two phases exchange energy by conduction through the particles surface.

4. DISTINGUISHED LIMITS

Studying two-phase systems governed by (9)-(10) in their full generality still represents a very difficult problem. We simplify the picture further by considering special - but physically realistic-situations, in which the parameters at our disposal (U_{ad}, amount and size of the particles...) are chosen in such a way that the presence of inert particles only gives rise to small, but by no means negligible, perturbations of the particle-free, radiation-free flame. In other words, we consider particular distinguished limits, in addition to the basic one $\beta \to \infty$.

4.1 Loading

Adding some inert material to a reactive mixture increases its heat capacity, thereby tending to decrease the reaction temperature... and the flame speed. Since $O(RT_{ad}^2/E)$ changes in reaction temperature are enough to modify the reaction rate, we require that the fractional change in heat capacity due to the particles be $O(1/\beta)$. More precisely, we assume that the loading parameter C, defined by :

$$2C/\beta = (N_u\, 4\pi \rho_s C_s r_s^3)/(3\rho_u C_p) \tag{11}$$

is O(1) in the limit $\beta \to \infty$. Therefore the final temperature T_{eq} no longer equals T_{ad}, but is still close to it : $T_{eq} = T_{ad} - 2(T_{ad} - T_u) \quad C/\beta + \ldots$

4.2 Temperature-lags

For the loading effet to have its maximum influence on the (gas-) temperature, the gas and the inert material must be in local thermal equilibrium throughout the whole flame. This is true for inert gases, but may cease to be so in the case of inert particles, because the latter have a noticeable thermal inertia. Owing to the finite rate of conductive exchanges between the phases, reaching the thermal equilibrium needs some time, whose magnitude t_c is estimated from (10) to be :

$$t_c = r_s^2 \rho_s C_s / 2\lambda_u \tag{12}$$

If $t_c \ll t_T$, where $t_T = \ell/U_{ad}$ is the typical time of transit of a particle across a flame, and if the flame speed is $O(U_{ad})$, then the local thermal equilibrium prevails even within the flame front ; otherwise, temperature-lags appear. A convenient parameter controlling these lags is the thermal response parameter K, defined as :

$$K \equiv t_T/t_c = 3(\ell/r_s)^2 (\rho_u C_p/\rho_s C_s) \tag{13}$$

Depending on U_{ad} (from 1 cm/s to 1 m/s) and the particle radius r_s (from 0.1 µm to 0.1 mm), K ranges from $K \ll 1$ to $K \gg 1$ for usual flames at the atmospheric pressure.

We therefore assume $K = O(1)$, to cover the widest spectrum. Then the temperature-lags between the phases appearing in the flame fronts will depend on the actual flame speed (and conversely). Moreover, the two-phase medium will be in local thermal equilibrium wherever the temperatures T and T_s only vary over well thicker zones than the flame front.

4.3. Intensity of radiative exchanges

Before comparing the radiative exchanges to other mechanisms, one has to give a means to compute the radiant net flux $\underline{q}$ featuring in (11). The theory could takle the general equation of radiative transfer, at the expense of "notably more lengthy" computations, but we shall content ourselves with using the Eddington differential approximation /19/, which is comparatively easy to handle ; we further specialize it to an isotropically scattering, grey medium, in which case it reads as :

$$L\,\underline{\text{grad}}\,(L\,\text{div}\,\underline{q}) = 3\underline{q} + L\,\underline{\text{grad}}\,(4\sigma\sqrt{1-\omega}\,T_s^4) \tag{14}$$

where $L = \left[(a + s)a\right]^{-1/2}$ is the local attenuation length of radiation, σ is the Stefan-Boltzmann constant, and $\omega = s/(a + s)$ is the single-scattering albedo. The definitions of L and ω involve the volumetric coefficients of absorption (a) and scattering (s), which are both proportional to $N\pi r_s^2$, hence to ρ in a one-velocity medium.

From (14), and the anticipated result that $T_s - T_u = O(T - T_u)$, one can estimate $\underline{q}$ to be $O(\sigma(T_{ad}^4 - T_u^4)\sqrt{1-\omega})$. If alone this radiative flux would produce an $O(q/\rho_u U_{ad} C_p)$ change in the temperature of the medium, provided $\underline{u} = O(U_{ad})$ and the conductive exchanges are fast enough (a point to be checked later on). Requiring that this change in temperature be also $O(R\,T_{ad}^2/E)$ leads one to assume that the following scaled Boltzmann number B :

$$B \equiv \frac{\sigma(T_{ad}^4 - T_u^4)\sqrt{1-\omega}\;E}{\rho_u\,U_{ad}\,C_p\,T_{ad}\,\sqrt{3}\,RT_{ad}} \tag{15}$$

is O(1) (the factor $1/\sqrt{3}$ is introduced for future convenience). In addition to being mathematically convenient, considering B as O(1) in the limit $\beta \to \infty$ is also a physically realistic assumption for a class of two-phase flames: with $C_p = 10^3$ J/kg.K, $E/RT_{ad} = 10$, $T_{ad} = 1.5\ 10^3$ K, $T_u = 300$ K, $\rho_u = 1$ kg/m³, $U_{ad} = 1$ m/s, and considering moderately scattering particles such that $\sqrt{1-\omega}/\sqrt{3} = 0.5$, one obtains $B \cong 1$.

4.4. Range of radiative exchanges

The range of the radiative exchanges, anticipated to be O(L), can be compared to the conduction/convection length $\ell = \lambda_u/\rho_u U_{ad} C_p$ characterizing the thickness of a usual flame. It is readily seen that the optical flame thickness ε :

$$\varepsilon = \ell/L_u \tag{16}$$

is a very small number in most actual situations involving moderately fast flames. Using $L \cong 1/(N\pi r_s^2)$, $C_p \cong C_s$, $\rho_s = 2\ 10^3$ kg/m³, $r_s = 1$ µm, and an initial concentration of particles such that $N\ 4\pi r_s^3 \rho_s/3 = \rho/10$, one obtains $L_u \cong 2.5$ cm ; this is quite large a value, compared to the conduction / convection length $\ell \cong 25$ µm of a flame propagating at $U_{ad} = 1$ m/s. ε is inversely proportional to r_s, to U_{ad}, to pressure, and increases linearly with the particle number density. We shall therefore consider ε as a very small quantity, i.e. smaller than all the β - related relevent small numbers appearing in the analysis.

5. TWO COMPLEMENTARY PROBLEMS

The facts that C, K, B are O(1) in the limit $\beta \to \infty$, and that ε may be considered as infinitely small, have many important consequences, provided one further assume that the typical lengths characterizing the front shape are also $O(L_u)$, a very reasonable assumption if one wants the non-local properties of the radiative exchanges to be retained.

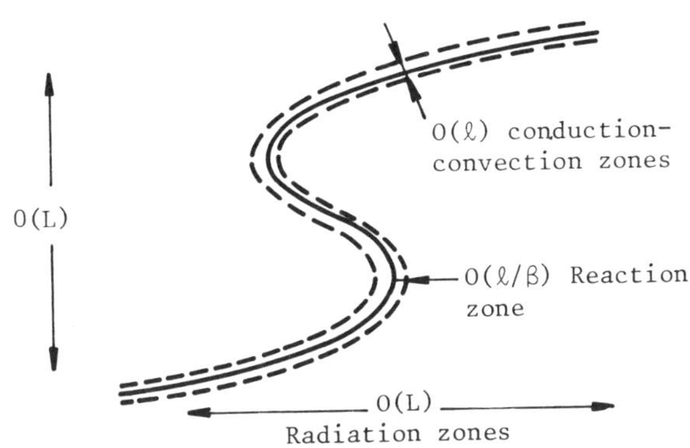

Figure 2 : The three main scales.

These consequences may be summarized as follows. The class of radiation-affected, particle-laden flames, which we consider presently, consists of :

- Optically-transparent ($\varepsilon \cong 0$), hence "radiation-free", locally-planar fronts, the width of which is $O(\ell) = O(\varepsilon L_u)$; there are embedded in :

- much thicker radiation-zones, of $O(L_u)$ dimensions, in which conduction may be neglected, the two-phase mixture behaves like a one-temperature medium, and where the temperature is piecewise constant at the leading order.

5.1. The flame fronts

Let us first examine the structure of the aforementionned fronts in a coordinate system attached to the reaction sheet (it is anticipated to subsist, consistently with i)). The flux divergence div $\underline{q}$ featuring in (10) can be estimated as $O(q/L_u)$ since $\varepsilon \cong 0$ and $T_{eq} \cong T_{ad}$, whereas the convective term of (10) is $O(C\rho_u U_{ad} C_p (T_{ad} - T_u)/\beta\ell)$. The corresponding radiant-to-convective flux ratio is thus $O(\varepsilon B/C)$, hence negligible in any case (ε/C does not depend on the amount of particles). Accordingly, the gas and particle temperatures rises across the flame front are the same as if the radiant exchanges were absent, apart from the radiation - induced shift in temperature at the front entrance and the corresponding change in local flame speed : in the limit $\varepsilon \to 0$, the front is too thin to allow for any significant local absorption, scattering and emission of radiant energy.

Furthermore, the structure of these thin layers of steep gradients is locally the same as that of a planar front, because the front radii of curvature have been assumed $O(L_u)$: the normal derivatives of T, T_s, $\underline{q}$ are there in excess over the tangential ones by $O(1/\varepsilon)$ factors, and are thus the only ones that matter in the limit $\varepsilon \to 0$.

We now make the following further assumption : the typical evolution time of these fronts is $O(L_u/U_{ad}) = O(\varepsilon\ell/U_{ad})$. This is not always true ; we shall see later when difficulties may arise, and how to get rid of them in some cases. If L_u/U_{ad} does represent the relevent, typical evolution time, the flame fronts are quasi-steady, and their structure at $O(\ell)$ distances from the reaction sheet is described by :

$$m\, \rho_u U_{ad} C_p \frac{\partial T}{\partial n} = \frac{\partial}{\partial n} (\lambda \frac{\partial T}{\partial n}) - 4\pi\, N\, r_s \lambda (T - T_s) \tag{17}$$

$$m\, \rho_u U_{ad} \frac{\partial y}{\partial n} = \frac{\partial}{\partial n} (\rho D \frac{\partial y}{\partial n}) \tag{18}$$

$$m\, \rho_u U_{ad} \frac{4}{3} \pi\, r_s^3 \rho_s C_s \frac{\partial T_s}{\partial n} = 4\pi\, r_s \rho\, \lambda (T - T_s) \tag{19}$$

where $\rho_u U_{ad}$ m represents the local burning speed. Recall that $N \sim \rho \sim 1/T$; we further assume that $\lambda^2 \sim T$, an assumption which is not too unrealistic for gases, to simplify the algebra.

The boundary conditions associated with (17) - (19) are (Figure 3) :

$$T(-\infty) = T_s(-\infty) = T_* \tag{20}$$

$$T(+\infty) = T_s(+\infty) = T_* + T_{eq} - T_u \tag{21}$$

$$y(-\infty) = y_u \quad , \qquad y(\infty) = 0 \tag{22}$$

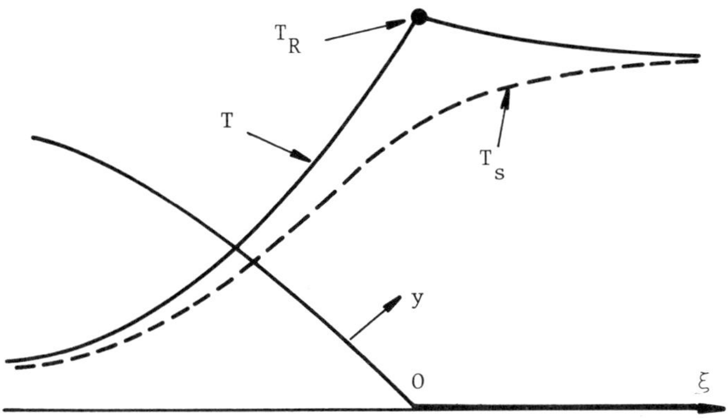

Figure 3 : Schematic structure of the flame front, along the normal coordinate n.

In (20), T_* represents the mixture temperature at the entrance of the flame front (Figure 3), and may vary along it. T_* is not known, and can only be computed by examining the radiation zones ; at the present stage, we merely know that T_* is close to T_u, since the radiative preheating is weak ($B = O(1)$, $\beta \to \infty$). That T and T_s reach the same values for $n \to \pm \infty$ follows from (19), account taken of the assumption that the particles have O(1) thermal response parameters (K).

The solution y(n) is almost the same as in (4a), viz :

$$y/y_u = 1 - e^{m\xi} \tag{23}$$

for $\xi < 0$, and $y \equiv 0$ for $\xi > 0$; ξ is still defined by Eq. (4b).

Owing to the distinguished limits introduced in section 4 (especially $C = O(1)$ and $K = O(1)$), the last term in the R.H.S. of (17) is only a small contribution to the energy budjet in the gas phase. Accordingly, the leading order solution $T°(n)$ to (17) is given by :

$$T° = \begin{cases} T_u + (T_{ad} - T_u) \exp(m\xi) & \xi < 0 \\ T_{ad} & \xi > 0 \end{cases} \tag{24}$$

to be compared to (4a). $T°(\xi)$ is used in (19) to compute T_s at the leading order :

$$T_s° = \begin{cases} T_u + (T_{ad} - T_u) e^{m\xi} K/(K + m^2) & \xi < 0 \\ T_{ad} - (T_{ad} - T_u)\, e^{-K\xi/m}\, m^2/(K + m^2) & \xi > 0 \end{cases} \tag{25}$$

We note the temperature-lag, which of course vanishes if $K \to \infty$. Once known, $T_s°$ is used back in (17) to find :

$$\left(\lambda \frac{\partial T}{\partial n}\right)_{0^-} = m\, \rho_u U_{ad} C_p (T_R - T_*) + 4\pi r_s \int_{-\infty}^{0} N\, \lambda (T° - T_s°) dn + \ldots \tag{26a}$$

$$(\lambda \frac{\partial T}{\partial n})_{0^+} = - \rho_u U_{ad} C_p T_{ad} \frac{4N \pi r_s^3 \rho_s C_s}{3\rho C_p} \frac{K m^3}{(K + m^2)^2} \tag{26b}$$

Using (23) - (25) in (7), then by employing (8), one obtains a relationship between m, C, K and $\Theta_* \equiv (T_* - T_u)E/2RT^2_{ad}$, viz :

$$H(m, C, K) = \Theta_* \tag{27a}$$

where H(m, C, K) is defined by :

$$H(m, C, K) \equiv \text{Log } m + CK(K + 2m^2)/(K + m^2)^2 \tag{27b}$$

Besides (27), another set of useful relationship is deduced from the transfer equation (14), which degenerates locally (i.e. within the front) into :

$$L \frac{\partial}{\partial n} (L \text{ div } q) \cong L \frac{\partial}{\partial n} (4\sigma\sqrt{1-\omega}\, T_s^4) \tag{28}$$

in the limit $\varepsilon \cong 0$. From (28), and from the use of standard techniques of asymptotic matching /10/, one can show that the profile of L div $\underline{q}$ corresponding to the radiation zones undergoes a jump across the flame front, which is given by :

$$\{L \text{ div } \underline{q}\}^+_- = 4\sigma\sqrt{1-\omega}\, \{T_s^4\}^+_- \tag{29}$$

The outer profile of $\underline{q}$ (i.e. outside the front) is continuous, and therefore satisfies (again when $\varepsilon \cong 0$) :

$$\{\underline{q} \,.\, \underline{n}\}^+_- \cong 0 \tag{30}$$

5.2. Radiation-zones

On the range of the radiative exchanges, namely at $O(L_u)$ distances from the front, the (radiation-induced) temperature variations are too slow for any lag to appear, and the local thermal equilibrium prevails.

In addition, since the mixture is weakly loaded (C = O(1)) and the radiative exchanges are weak (B = O(1)), the temperature variations are $O((T_{ad} - T_u)/\beta)$ there. The leading order temperature profile is piecewise constant (viz : T_u (upstream) or T_{ad} (downstream)) ; so are the density ($\rho \sim 1/T$) and attenuation length ($L \sim 1/\rho$) profiles. More precisely, we write T in the upstream radiation zone as :

$$T = T_u + 2\beta^{-1}(T_{ad} - T_u)\Theta + \ldots \tag{31}$$

$\Theta = O(1)$ accounts for the radiative preheating ; it is such that $\Theta = 0$ far upstream of the front, and Θ_* is its value at the entrance (Figure 4).

To compute Θ, hence Θ_*, one "only" has to solve a convection/radiation balance, because temperature variations on the $O(L_u)$ space-scale only induce negligibly small conduction effects. This entails finding the radiation and convection fields.

Before coming to this, we define the non-dimensional radiant flux $\underline{Q}$ and velocity vector $\underline{U}$ by $\underline{q} = \rho_u U_{ad} C_p (T_{ad} - T_u) \underline{Q}/\beta$ and $\underline{u} = \underline{U}\, U_{ad}$, respect-

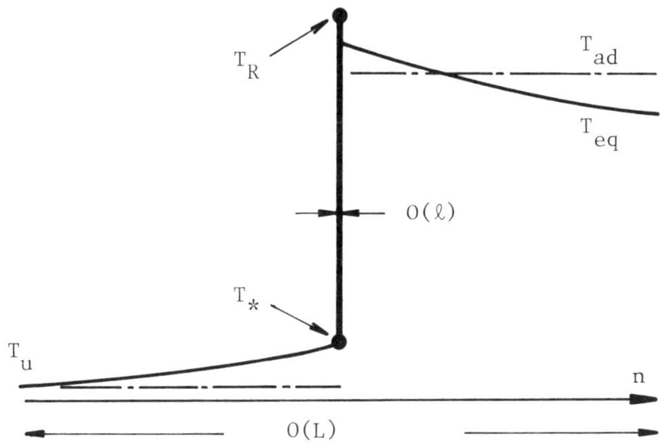

Figure 4 : Schematic temperature profile in the radiation zones, along a normal to the front.

ively, and the operators Grad and Div refering to spatial coordinates measured in units of $L_u/\sqrt{3}$. Then, the aforementionned convection radiation balance reads as follows, at the leading order :

$$\frac{\partial \Theta}{\partial \tau} + \underline{U} \cdot \underline{\text{Grad}}\ \Theta = - G/2 \tag{32}$$

where $\tau = \sqrt{3}\ t\ U_{ad}/L_u$. $G \equiv \text{Div}\ \underline{Q}$ and $\underline{Q}$ are given by the leading order radiative-transfer equation :

$$\underline{\text{Grad}}\ G = \begin{cases} \underline{Q} & \text{upstream} \\ \underline{Q}\ \Lambda^{-2} & \text{downstream} \end{cases} \tag{33}$$

where $\Lambda = L_{ad}/L_u = T_{ad}/T_u$. We note that (33) only involves constant coefficients (but a moving boundary, the front) and is source-free (outside the front) because both L and T are piecewise constant functions at the leading order. Equation (33) has to be supplemented with boundary conditions at the flame front (now considered as discontinuous), in addition to $\underline{Q}(\pm\infty) = 0$, namely :

$$\{\underline{Q} \cdot \underline{n}\}_-^+ = 0 \tag{34}$$

as a consequence of (30), and :

$$\Lambda G|_+ = G|_- + 4B \tag{35}$$

which follows from (29). In principle, (33) - (35) are enough to compute G; upon substitution in (32), G allows one to have Θ_* along the flame front hence, owing to (27), the local burning velocity and the flame shape.

5.3. Flow-field

Unfortunately, m and the flame front shape are both needed to compute the velocity vector $\underline{U}$ appearing in (32). $\underline{U}$ has to meet the usual Hugoniot jump relations, and :

$$\underline{n} \cdot \underline{U}|_- = V_n + m \tag{36}$$

where $\underline{n}\, V_n$ is the local front normal velocity. In addition, $\underline{U}$ has to be solution to the (piecewise-) incompressible Euler's equations !

6. STEADY PLANAR FLAMES

When the flame is planar and steady, the upstream velocity vector $\underline{U}$ may be written as $\underline{U} = (\bar{m},0,0)$, where $\bar{m}$ is an eigenvalue (the unknown value of m), and $\underline{Q} = (\bar{Q}(Z),0,0)$, where Z is a streamwise coordinate. Equation (33) gives :

$$\bar{Q} = \begin{cases} -2B\, e^{Z} & Z < 0 \\ -2B\, e^{-Z/\Lambda} & Z > 0 \end{cases} \tag{37}$$

Thus :

$$G = \begin{cases} -2B\, e^{Z} & Z < 0 \\ +2B\, e^{-Z/\Lambda}/\Lambda & Z > 0 \end{cases} \tag{38}$$

and $\bar{\Theta}(Z) = B\, e^{Z}/\bar{m}$ follows from (32). Therefore $\bar{\Theta}_* = B/\bar{m}$, and $\bar{m}$ is found from (27) as a root of :

$$H(\bar{m}, C, K) - B/\bar{m} = 0 \tag{39}$$

Had we employed the exact equation of transfer, the resulting equation for $\bar{m}$ would have had the same form as (39), provided the definition of B is slightly modified /11/ : in (15), $\sqrt{1-\omega}/\sqrt{3}$ ought to be replaced by $j(\omega)$, where $2j(\omega) \cong \sqrt{1-\omega}\,(1 + \omega(2/\sqrt{3} - 1))$. Spectral effects and anisotropic scattering could also be included in the theory.

The response curves $H - B/m = 0$ are plotted in Figure 5, in the (C, m) plane, for different values of C and two values of B, namely $B = 0$ and $B = 1.3 \text{ Log } 1.3 \cong 0.34$. For $B = 0$, the response curves may be monotonic or S-shaped, depending on whether K is larger or less than a critical value $K_c \cong 0.08$. Note that dm/dC is positive along the middle branch of solutions (when any), thereby indicating that the corresponding regimes are unstable (increasing the amount of particles would make the flame propagate faster !). We also note that $m \sim \exp(-C)$ when $K \gg 1$, i.e. when the local thermal equilibrium prevails throughout the flame, whereas $K \ll 1$ yields $m^2 \text{ Log } m + 2CK \cong 0$, i.e. because the particles do not have time enough to be heated when crossing the flame. Including the radiative transfer does not change the picture qualitatively. Notice, however, that a comparatively small value of B (here : 0.34) makes the flame much some resistant to "quenching" by addition of inert particles, in a more and more pronounced way as B increases. This is due to the fact that preheating the incoming gas has two

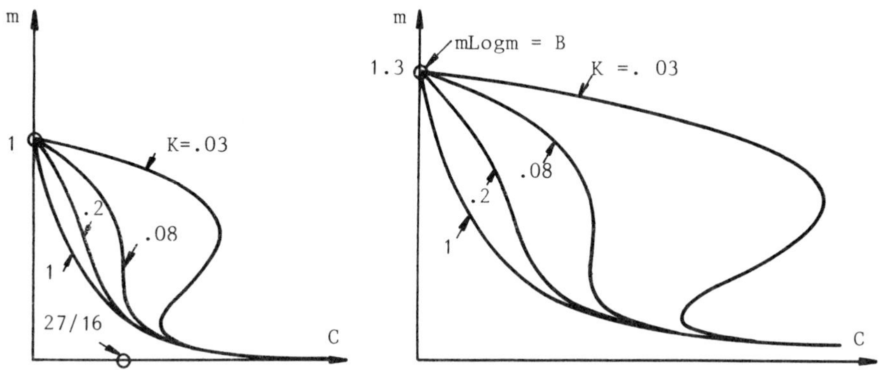

Figure 5 : The response curves $H - B/\bar{m} = 0$, in the $(\bar{m}, C)$ plane for different values of K. 5a : B = 0 ; 5b : B = 1.3 Log 1.3 ≅ 0.34.

consequences : i) increasing the reaction temperature, like in ordinary gaseous flames and ii) shortening the transit time of a particle across the front, thereby diminishing the influence of conductive losses (Figure 5a). Comparing Figures 5a and 5b shows that omitting the radiant exchanges may lead to the erroneous conclusion that adding inert particles always slows down the propagation : in some instances, the opposite trend is observed.

7. UNSTEADY PLANE FLAMES

7.1 Slow motions

If the flame is still planar, but unsteady (m is then a function of time), the analysis leading to (37) is unchanged, because $\underline{Q}$ is computed from the leading over temperature profile in the radiation zones, which is known a priori ; therefore (38) still applies. The energy balance (32) now reads :

$$\frac{\partial\Theta}{\partial\tau} + m(\tau)\,\frac{\partial\Theta}{\partial Z} = B\,e^{Z} \tag{40}$$

provided Z = 0 still denotes the flame front location. Given an initial condition $\Theta_o(Z)$, (40) is solved to give :

$$\Theta_* = \Theta_o(-\zeta) + B\int_o^{\zeta} e^{\zeta'-\zeta}\,d\zeta'/m(\zeta') \tag{41a}$$

where :

$$\zeta = \int_o^{\tau} m(\sigma)\,d\sigma \tag{41b}$$

Using (27) and (41), the evolution equation for $m(\zeta)$ is then found to be :

$$H(m,\ C,\ K) = \Theta_*(\zeta) \tag{42a}$$

$$d\Theta_*/d\zeta + \Theta_* = B/m(\zeta) + p(\zeta) \tag{42b}$$

with :

$$p(\zeta) \equiv (\Theta_o(Z) - d\Theta_o/dZ)_{Z=-\zeta} \tag{42c}$$

Once this algebraic-differential system (42) is solved to give $m(\zeta)$, the evolution of m with τ is recovered by integrating

$$\frac{d\tau}{dm} = 1/m(\zeta) \tag{43}$$

a posteriori.

7.2. Fast motions

Equation (42) needs to be examined carefully before any integration is performed. For (42) to hold, the evolution of the flame front structure must follow that of the radiation zones, i.e. the former must adjust very quickly to changes in the latter. In other words, the flame front structure must be stable on the $O(\ell/U_{ad}) = o(L_u/U_{ad})$ time-scale. To see whether it is the case, or not, we first note that (42a), with Θ_* assigned a known, constant value $\bar{\Theta}$, gives the burning speed of a radiation-free, particle-laden flame when the far-upstream medium is $T_u + 2(T_{ad} - T_u)\Theta/\beta$ instead of T_u. It has been demonstrated /12/ that such radiation-free regimes are unstable on the "fast" time-scale $t = O(\ell/U_{ad})$, if the root of $H(m,C,K)=\bar{\Theta}$ is such that $dm/dC > 0$ (burning speed which increases with C !), i.e., since $\partial H/\partial C > 0$, if $\partial H/\partial m < 0$. The condition $\partial H/\partial m > 0$ is not sufficient - in general - to guarantee the stability of a radiation-free, particle-laden flame on the $t = O(\ell/U_{ad})$, fast scale : other instabilities, related to a differential diffusion of heat and reactant, may appear /21/ even if $\partial H/\partial m > 0$. We shall assume that the Lewis number Le and the Zel'dovich number β have been chosen so as to insure that such instabilities do not show up. Then,

$$\partial H/\partial m > 0 \tag{44}$$

is a necessary and sufficient condition for radiation-free, particle-laden flames to be stable on the $t = O(\ell/U_{ad})$ scale. It is important to stress that (44) is also a necessary and sufficient condition for a radiation-affected, particle-laden front to follow the slow evolutions in $\Theta_*(\tau)$, thus for the "slow equations" (42) to hold : when (44) is met, the flame front structure does have time enough to adjust itself to the "slow" evolution of the radiation zone structure. If, owing to the variations of $\Theta_*(\tau)$, the condition $\partial H/\partial m = 0$ is satisfied, for some value Θ_J of Θ_*, then m evolves quickly (i.e. on the $t = O(\ell/U_{ad})$ scale) until it reaches the "closest stable regime". We are not able to describe the transient exactly since the unsteady version of (17) - (19) cannot be solved analytically. However, the following remarks are enough for our present purpose. Even during a fast transient, (20) - (22) still hold, even though $T(n)$ and $T_s(n)$ vary on the $O(\ell/U_{ad})$ scale ; so do (28) - (30), hence (33) - (35). Then (42) indicate that Θ_* is continuous (but $d\Theta_*/d\tau$ and $d\Theta_*/d\zeta$ are not) even if m changes jumpwise. Therefore, if $\varepsilon \cong 0$, Θ_* takes the same value Θ_J during a transient, and the "closest stable regime" corresponds to the closest root of $H(m, C, K) = \Theta_J$ satisfying $\partial H/\partial m > 0$. Then m evolves according to (42), until a stable steady state is reached, or a new transient occurs.

7.3. Relaxation oscillations

With H(m, C, K) given by (27b), $\partial H/\partial m$ reads :

$$\frac{\partial H}{\partial m} = \frac{1}{m}\left(1 - \frac{4C\,K\,m^4}{(K + m^2)^3}\right) \tag{45}$$

and two cases have to be distinguished.

7.3.1. $C < 27/16$. If $C < 27/16$ (eg. $C \ll 1$), H is a monotonically increasing function of m (Figure 6a) and the fast scale instability cannot appear. Equations (42) are then meaningful. Since $\partial H/\partial m > 0$ implies $\partial H/\partial m + B/m^2 > 0$, the roots $\bar{m}$ of (39) are also stable on the slow scale ($\tau = O(1)$), as it is readily checked upon linearization of (42) for m close to $\bar{m}$.

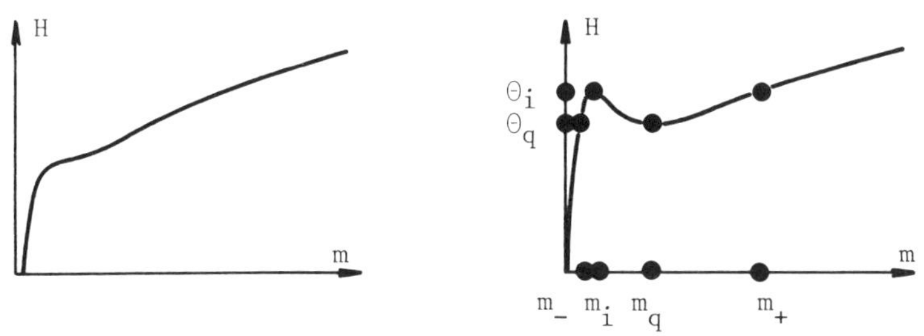

Figure 6 : H(m, C, K) v.s. m, for a fixed value of K. 6a : $C < 27/16$; 6b : $C > 27/16$.

7.3.2. $C > 27/16$. The function H(m, C, K) exhibits two maxima (Figure 6b). Of special interest are the points of coordinates (Θ_i, m_i) ("i" for "ignition") and (Θ_q, m_q) ("q" for "quenching") which are the starting points of jumps, and the values m_+ and m_- at which H takes the same values as at its extrema. If $\Theta_*(\tau)$ reaches Θ_i (resp. Θ_q) from below (resp. from above), m jumps from m_i to m_+ (resp. from m_q to m_-), then both Θ_* and m evolve according to (42).

The procedure used to select the stable states therefore consists of plotting the J-line $\partial H/\partial m = 0$ ("J" for "Jumps") in the m, C plane of Figure 5b, along with the curve $H(m, C, K) - B/m = 0$. All the steady states corresponding to the right of the J-line are unstable on the fast scale. This is the case for the regimes for which $dm/dC > 0$ along the response curve $H(m, C, K) - B/m = 0$ (a very reasonable result, since $dm/dC > 0$ implies that m increases with the amount of inert material !). If $dm/dC < 0$, no simple rule exists ; depending on the relative positions of the J-line ($\partial H/\partial m = 0$) and the response curves $H - B/m = 0$, the number of stable steady states may be one, two or zero (Figures 7).

Of special interest are the situations where no stable steady state exists. For the sake of argument, we consider the case in which the unstable steady state $m = \bar{m}$ is unique.

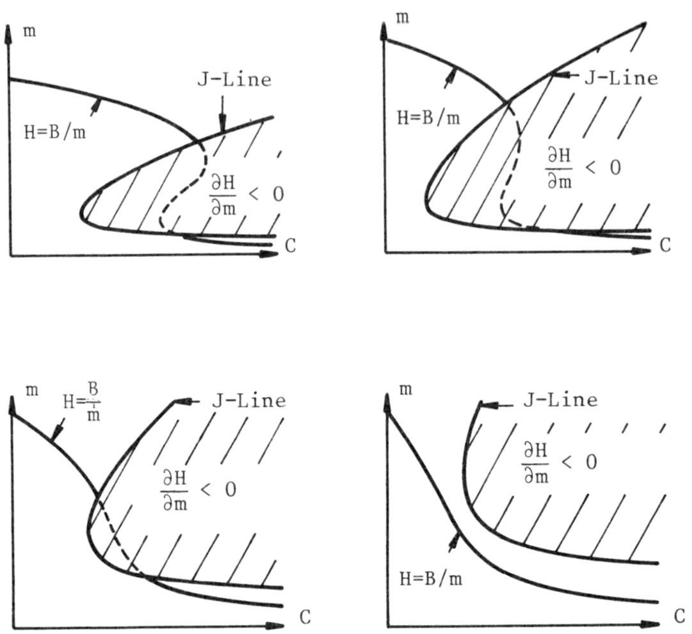

Figure 7 : The response curve (H - B/m = 0) and the J-line (∂H/∂m = 0) in four different situations. The dotted lines correspond to unstable steady states.

To determine the behavior of m for τ (or ζ) tending to infinity, it is enough to consider (42) with $p(Z) = 0$ (since $p(\infty) = 0$). We also envisage an initial condition such that m and Θ_* satisfy (42), with m(o) less than m_i (e.g. : $m(o) = m_-$). Equations (42a) (42b) can be combined to give :

$$\frac{\partial H}{\partial m}\frac{dm}{d\zeta} = B/m - H$$

For $m < m_i$, $\partial H/\partial m$ and $B/m - H$ are positive, so that $m(\zeta)$ increases with ζ, until m_i is reached from below, then m jumps to m_+. Since $m_+ > m_q$, $\partial H/\partial m$ is still positive, but $B/m - H$ is now negative ($m_i < \bar{m} < m_+$). Therefore m decreases as ζ increases, until m_q is reached from above ; then m jumps to m_-. We are then back to the initial situation, and the process can start all over again. Consequently, when the unique steady state is unstable, the reduced flame velocity evolves to a discontinuous periodic function (Figure

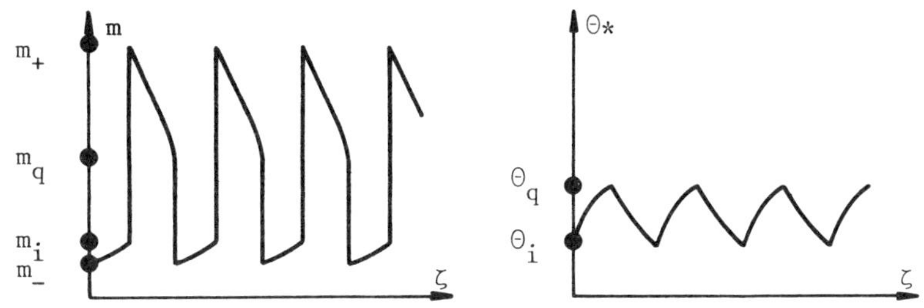

Figure 8 : $m(\zeta)$ and $\Theta_*(\zeta)$ histories, when the steady states lie between m_q and m_i.

8a). Meanwhile, $\Theta_*(\zeta)$ is continuous, but $d\Theta_*/d\zeta$ is not (Figure 8b) /13/. The physical origin of this oscillation is the fact that decreasing m has two antagonist effects, which operate on different time scale, namely increasing both the radiative preheating and the conduction losses to the particles. Similar behaviours are often encountered in chemical systems /14/.

On the other hand, when two stable state coexists, one may trigger transitions between them /13/, for example by using adequate initial conditions $\Theta_o(Z)$, or because of flame / flame interactions.

8. EXPANDING SPHERICAL FLAMES

Besides the planar fronts, there is only a very limited number of problems of radiation affected flames which one can solve exactly, even though (32) and (33) are well simplified forms of the energy balance and the radiative transfer equation. Among these special cases, a situation of great interest, namely : the spherically expanding flame.

Indeed the Euler equations (and the associated hydrodynamical jump conditions) are readily solved to yield :

$$u = \begin{cases} 0 & r < R \\ (1 - 1/\Lambda)R_\tau (R/r)^2 & r > R \end{cases} \qquad (46)$$

for the radial velocity component ; in (46), $R(\tau)$ and r are the flame radius and the radial coordinate (both measured in units of $L_u/\sqrt{3}$). Recall that $R_\tau = \Lambda.m$, owing to the continuity equation integrated over the flame front thickness. On the other hand, (33) is also easily integrated to give

$$G(r > R(\tau)) = - 2g(R(\tau), r)B \qquad (47)$$

where $g(R(\tau), r)$ can be factorized as $r^{-1}e^{-r}\,N(R, \Lambda)$, N being a complicated function of its arguments which we do not give here /15/. (For $R \to \infty$, $N(R) \sim R \exp(R)$, thereby matching (38)).

Then, the energy equation (32) reads as :

$$\frac{\partial\Theta}{\partial\tau} + (1 - 1/\Lambda)\, R_\tau \left(\frac{R}{r}\right)^2 \frac{\partial\Theta}{\partial r} = Bg(R(\tau), r) \qquad (48)$$

$$\Theta(0, r) \equiv 0$$

Equation (48) can be solved by quadratures (by using it in characteristic form), giving :

$$\Theta_* = B \int_0^R g(\alpha\zeta^3 + (1-\alpha)R^3)^{1/3}, \zeta) \frac{d\zeta}{R_\tau} \quad ; \quad \alpha = 1 - 1/\Lambda \tag{49}$$

for the scaled temperature increase due to radiative preheating. Then, owing to (27), the evolution of m with $\zeta = \int^\tau m \, d\tau = R/\Lambda$ is found by the solving a non-linear integral equation of the following type :

$$H(m, C, K) = B \int_0^\zeta \kappa(\zeta, \zeta', \Lambda) \, d\zeta'/m(\zeta') \tag{50}$$

where H(m, C, K) is still defined by (27b), and $\kappa(\zeta, \zeta')$ is an involved kernel which is no longer of the convolution type, except for $\zeta, \zeta' \to \infty$ and $\zeta - \zeta' = O(1)$, to match (41a), by contrast to what happened for planar fronts (the system is no longer invariant against shifts of the front position). For $R \to 0$, $\Theta_* = o(R) \to 0$; therefore (50) describes the transition between a radiation-free (but already modified by the conductive losses) propagation, to a radiation-affected one. Its numerical integration is in progress /15/, the task being complicated by the relaxation phenomena evidenced in the preceding section.

9. WEAKLY CURVED, STEADY FLAMES

The now fashionable (and often efficient !) way of going further would be to move directly to numerics, to solve (32) (33) and the Euler equations especially when multidimensional fields are involved. Before undertaking such a task, however, it is not unreasonable to try to grasp informations on the way the radiative transfer affects wrinkled flames. Indeed, it has to be recalled that radiation is a non-local process, which therefore introduces a reference length (viz : L_u) by contrast to conduction or diffusion. To do this, however, one has to restrict ones goal, since no analytical tool exists to solve the Euler equations and the deceiptively simple equation (33) when a free boundary is involved.

We first note that the influence of stronger and stronger intensities of radiant energy exchanges are very likely to be simpler to introduce via perturbation methods than the non-local characteristics of radiation. This leads to the idea of considering the Boltzmann number B (defined in (15)) as numerically small, and using series in B. In principle, series expansions in B should apply only to very fast flames and/or highly scattering particles, but are believed to give qualitatively correct results in many cases.

For $B \ll 1$ to yield qualitatively new phenomena, it is expected that one has to study situations in which the radiative exchanges compete with other mechanisms which are themselves weak ; this leads to the other idea of considering weakly distorbed flames. For these weakly distorted flames to exist, despite the Landau-Darrieus instability /2/, one could envisage only very special boundary conditions which would lead to nearly uniform incoming flows, such as upstream of Bunsen flames when the burner is well designed /16/. Another possibility could be to content oneself with a model in which the Landau-Darrieus instability does not show up, i.e. a constant-density model ; despite its a priori restricted range of applicability, the second strategy allows one to study the response of a flame front to prescribed incoming flowfields and to see what it looks like when radiation is the only mechanism of transverse energy transfer. Because our primary goal is to show the difference between the influences of local and non-local transports on the local flame speed, we adopt the second possibility, i.e. we consider Λ as unity. Once the radiation-induced curvature effects are fully understood, one may think of introducing small density changes ($\Lambda \gtrsim 1$) and the corresponding "weak" Landau-Darrieus instability a posteriori, since this methodology gave very valuable informations on the dynamics of particle-free flames /2/ /3/.

9.1 Steady Bunsen flames

To begin with, we consider the simplest situation, in which the incoming flow is uniform, and has a velocity component $M > \bar{m}$ along the Z axis (Figure 9).

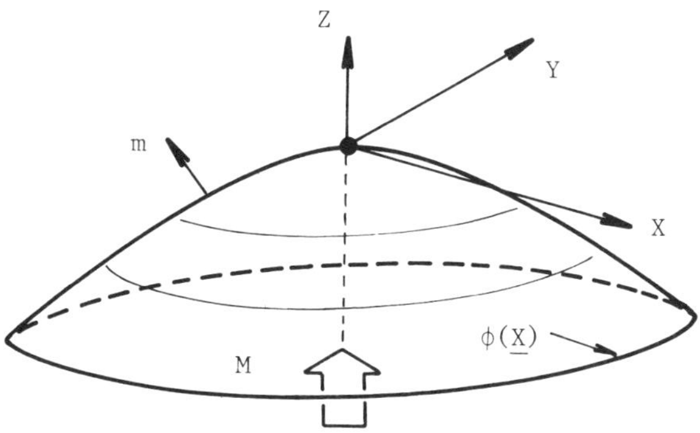

Figure 9 : Geometry of a Bunsen-like flame, and the coordinate system.

We shall restrict ourselves to steady flames and use $\phi(\underline{X})$ to denote the shape of this radiative analog of a Bunsen flame ($\underline{X} = (X, Y)$). As indicated above, we assume $B \ll 1$ and $M \cong \bar{m}$ to have a radiation-affected, weakly distorted front.

By analogy with the theory of radiation-free, particle-free flames, one expects the following "result" to hold for the local flame speed m :

$$m = \bar{m} + \text{"curvature effect"} \tag{51}$$

with :

$$m = M/(1 + |\nabla\phi|^2)^{1/2} \tag{52}$$

and $\nabla = (\partial_X, \partial_Y)$. The invoked above "curvature effect" results from radiation and curvature, and vanishes when both B and $\Delta\phi$ vanish ($\Delta \equiv \partial_{X^2} + \partial_{Y^2}$). In a situation where B and $\Delta\phi$ are both small, $m - \bar{m}$ should be $O(B\Delta\phi)$. Accordingly, when $M - \bar{m}$ is small, it is reasonable to assume :

$$M - \bar{m} \sim |\nabla\phi|^2 \sim B\Delta\phi \tag{53}$$

This results in :

$$M - \bar{m} = O(B^2) \;;\; \phi = O(B) \tag{54}$$

because X and Y must be viewed as O(1) if one wants the non-local properties of radiation to be retained, and the geometrical non-linearity is known to be an important ingredient of flame dynamics. With $\phi \ll 1$, Eqs (33) - (35) can be solved to give :

$$G(Z < \phi) + 2Be^{Z} = B\pi^{-1} \int e^{i\underline{K}.\underline{X} + Z\sqrt{1+k^2}} f(\underline{K})(1 + k^2)^{-1/2} \, d\underline{K} + \ldots \tag{55}$$

where $k^2 = \underline{K}.\underline{K}$, and $f(\underline{K})$ is the Fourier transform of $\phi(X)$. Upon substitution in (32), and a subsequent integration (with $\underline{U} = (0, 0, M)$, one obtains the following equations to compute the flame shape :

$$H(m, C, K) = \Theta_*(\underline{X}) \tag{56a}$$

$$M\Theta_*(\underline{X}) = B + B \,.\, (2\pi)^{-1} \int e^{i\underline{K}.\underline{X}} k^2(1 + k^2)^{-1} f(\underline{K}) \, d\underline{K} + \ldots \tag{56b}$$

It must be stressed that the argument leading to (53) (54) may cease to be valid close to the turning points of the response curves even though (56) still holds. These special configurations are not studied here. It is then enough to consider the case $C \ll 1$ only ; considering arbitrary values of C, K would not affect the results qualitatively, except close the turning points. Then (56) yields :

$$M - \bar{m} - |\nabla\phi|^2/2 = + B(2\hat{\pi})^{-1} \int e^{i\underline{K}.\underline{X}} k^2 f(\underline{K})(1 + k^2)^{-1} \, d\underline{K} + \ldots \tag{57}$$

since $M \sim \bar{m} \sim 1$ when $C \ll 1$ and B is small.

When the flame is wedge-shaped, (57) is equivalent to :

$$M - \bar{m} - \phi_X{}^2/2 = - B \int_{-\infty}^{+\infty} \phi_{XX} (X - \xi) \, e^{-|\xi|} \, d\xi + \ldots \tag{58}$$

and to :

$$\psi_X{}^2 = 1 + \nu \int_{-\infty}^{+\infty} \psi_{XX} (X - \xi) e^{-|\xi|} \, d\xi \tag{59}$$

once $\psi = \phi/(2(M - \bar{m}))^{1/2}$ and

$$\nu = B/(2(M - \bar{m}))^{1/2} \tag{60}$$

are introduced /17/. Together with the boundary conditions :

$$\psi_X(\pm \infty) = \mp 1 \tag{61}$$

(59) allows one to compute the flame shape. Clearly, (59) means that the radiative exchanges induce <u>non-local curvature effects</u>, which may differ qualitatively from those due to local transports, such as conduction. To see whether this is the case, we first consider (59) when $\nu \gg 1$. Anticipating that the corresponding $\psi(X)$ varies on the $X = O(\sqrt{\nu})$ scale, one may pull ψ_{XX} out of the integral to obtain :

$$\psi_X^2 = 1 + 2\nu\psi_{XX} + O(\nu\psi_{XXXX}) \tag{62}$$

thereby finding that ψ does vary on the anticipated scale.

According to (62), when the radiant exchanges are comparatively strong ($B^2 \gg M - \bar{m}$) the result looks like what is obtained with local transports. In this optically-thick limit, the flame shape belongs to the C^∞-class ; in particular, ψ_X vanishes at the flame tip. On the other hand, when $\nu = 0$, $\psi_X^2 \equiv 1$ (if the RHS of (59) remains bounded). One deduces that $\psi_{XX} = -2\delta(X)$ when $\nu = 0$. Therefore, when $\nu \ll 1$, one obtains :

$$\psi_X^2 = 1 - 2\nu \exp(-|X|) + \ldots \tag{63}$$

Hence the result : when the radiative exchanges are comparatively weak ($B^2 \ll M - \bar{m}$) ψ is continuous at the tip, but its slope is not ; in other words, angles exist even though the incoming flow is "smooth". Of course, comparing (62) and (63) leads to the question : what happens in between, for $\nu = O(1)$? Two plausible answers offer themselves namely :

i) $\psi_X(0^\pm) \to 0$ faster than a power of $1/\nu$, when $\nu \to \infty$

ii) $\psi_X(0^\pm) = 0$ if and only if $\nu \geq \nu_2$, where ν_2 is a critical value of ν.

To know which possibility is the correct one, we first note that assuming $\psi_X = 0 = (\psi_X \psi_{XXX})$ at $X = 0$ allows one to rewrite (59) as :

$$2(\psi_X\psi_{XXX} + \psi_{XX}^2 + \nu\psi_{XX}) + 1 - \psi_X^2 = 0 \tag{64}$$

for any X, because $\exp(-|X|)$ is the Green function of d^2/dX^2-1 on the real line. From (64), one may compute $\psi_{XX}(0)$ as the root of :

$$2\psi_{XX}^2(0) + 2\nu\, \psi_{XX}(0) + 1 = 0 \tag{65}$$

Equation (65) has real roots only if $\nu \geq \sqrt{2}$, we therefore conjecture that ii) is the correct answer, with $\nu_2 = \sqrt{2}$. To go further, we note that the restriction Ψ of ψ to $X \pm 0$ satisfies :

$$2(\Psi_{XX}^2 + \Psi_X\Psi_{XXX} + \nu\Psi_{XX}) + 1 - \Psi_X^2 = 0 \tag{66a}$$

$$\Psi_X(-\infty) = 1 \tag{66b}$$

$$\Psi_X^2(0^-) - 1 + 2\nu\Psi_X(0^-) = 2\nu\int_{-\infty}^{0} e^\xi\, \Psi_{XX}\, d\xi \tag{66c}$$

The last equation is obtained from (59) by realizing that ψ_{XX} contains the contribution $-2\Psi_X(0-)\delta(X)$, and that Ψ_{XX} may be viewed as an <u>even, continuous function</u>. We integrated (66a-b) numerically with $\Psi_X(X \to \infty) = 1 - C\exp(\gamma X) + \ldots$ (γ : positive root of $a^2 + \nu a = 1$) as starting value. The shooting parameter $C > 0$ was changed iteratively until (66c) is met, then $\Psi_X(0-)$ was available.

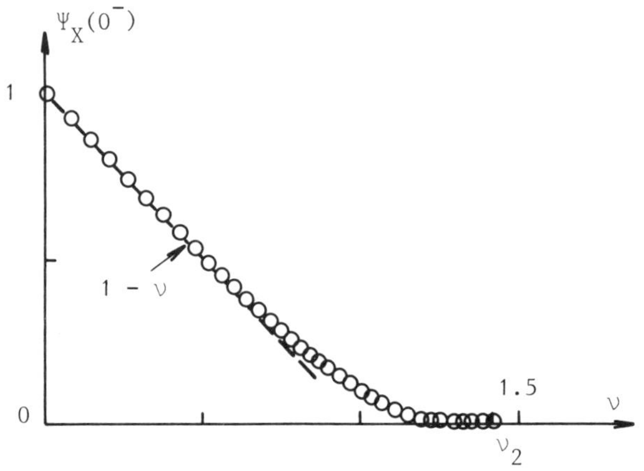

Figure 10 : $\psi_X(0^-)$ v.s. ν, as obtained numerically from (66).

Our findings /17/, summarized in Figure 10, are fully consistent with the conjecture. As a further check, we noted that $\psi_X(X \lessgtr 0) = \pm 1 \mp \exp(\pm X/2)$ is an exact solution to (66) for $\nu = \nu_3 = 3/2 > \sqrt{2}$; it is such that $\psi_X(0\pm) = 0$.

We also noted that peculiar events occur when ν crosses special values larger than $\nu_2 = \sqrt{2}$. For example, if one tries to compute $\psi_{XXX}(0)$ under the assumption that $\psi_X(0) = 0$ and ψ_{XX} exists, one gets $\psi_{XXX}(0)(1 \mp 2\psi_{XX}(0)) = 0$; hence $\psi_{XXX}(0) = 0$ or $1 + 2\psi_{XX}(0) = 0$, i.e., by using (64), $\nu = \nu_3 = 3/2$. That ψ_{XXX} is discontinuous at $X = 0$ when $\nu = 3/2$ can be checked directly on the exact solution found in this case. Similar events occur for $\nu^2 = \nu_n^2 = n^2/2(n - 1)$; apparently, ψ gains one more continuous derivative at the flame tip when ν crosses ν_n from below... and for $\nu = \infty$, ψ belongs to the C^∞ class. Equation (57) is separable in cartesian coordinates, and similar events happen for 3-dimensional flames, but the cross over values of ν are smaller than for wedge-shaped fronts (in 3.D radiative preheating is more efficient). For 3-D flames, the generalization of (59) is :

$$|\underline{\nabla}\psi|^2 = 1 - 2\nu(2\pi)^{-1} \int e^{i\underline{K}.\underline{X}}\, k^2\, F(\underline{K})\, (1 + k^2)^{-1}\, d\underline{K} + \ldots \qquad (67)$$

where F is the Fourier transform of ψ ; (67) is equivalent to :

$$\Delta(|\nabla\psi|^2) + 2\nu\Delta\psi + 1 - |\nabla\psi|^2 = 0 \qquad (68)$$

almost everywhere ; axisymmetric flames can be studied with (68).

9.2 Steady response to a shear flow

If the incoming velocity vector is not constant any longer, but reads $\underline{U} = (0, 0, M + V(X))$, the preceding analysis is readily generalized, if $\bar{V} = O(B^2)$. Indeed, (55) is still valid, because the radiation field given by (33) only depends on the flame shape. Integrating the energy does not pose major problems either, since assuming $V = O(M - \bar{m}) = O(B^2)$ does not change it at the two first orders ; then (56b) still holds. The only change is of kinematic origin, because the expression of m now accounts for the modulations in the incoming flow.

One finally obtains :

$$\psi_X^2 = 1 + \nu \int_{-\infty}^{+\infty} \psi_{XX}(X - \xi)e^{-|\xi|}\, d\xi + S(X) \tag{69}$$

as generalized form of (59), in which $S \sim V(X)/(M - \bar{m})$ may be assumed to have a zero transverse average. The above equation can be written as :

$$2\psi_{XX}^2 + 2\psi_X\psi_{XXX} + 2\nu\psi_{XX} + 1 - \psi_X^2 = S_{XX} - S \tag{70}$$

as least when ψ is smooth enough. To fix the ideas, we use $S = S_o \cos k_o X$, with $S_o > 0$, as example of spatial modulation of the incoming flow. Using this function S(X) in (70), one may try to compute ψ_{XX} at the extrema of $\psi(X)$, when we assume $\psi_X = 0$ (Figure 11).

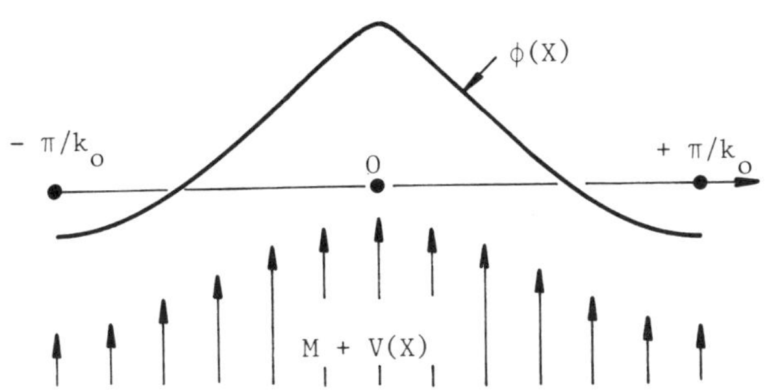

Figure 11 : Flame shape corresponding to an incoming shear flow $\underline{U} = (0, 0, M + V(X))$.

At the crests, which correspond to $X = 0 \pm 2\pi/k_o$, ..., the resulting equation for $\psi_{XX}(0)$ ceases to have real roots if :

$$(1 + k_o^2)S_o > (\nu^2/2 - 1) \tag{71}$$

When S_o and k_o meet (71) (for example if $\nu^2 < 2$, since $S_o > 0$) angles appear at the crests. In other words, if the incoming flowfield tends to make the front slope have a too large variation, radiation alone is unable to smooth the front shape ; for the Bunsen - like flames, ν was the only parameter controlling the appearance of angles, whereas (71) also involves the strength of the flow inhomogeneity and its wavelength. By analogy with what we found previously, one may expect that ψ gains continuous higher derivatives when S_o, k_o and ν fulfill well defined relationships. When S is no longer a periodic function, but is more representative of a random, smooth field, the appearance of angles or singularities in the higher derivatives of ψ should have consequences on the spectra associated with ψ; this question is currently under study /18/.

9.3 Back to local effects

The equation giving ψ_{XX} at the troughs, i.e. at $X = \pm \pi/k_o$, $\pm 3\pi/2\, k_o$..., may also fail to have real roots, if :

$$S_o(1 + k_o^2) < (1 - \nu^2/2) \tag{72}$$

We not yet know whether (72) may be fulfilled. That equations such as (70) may lead to angles pointing towards the fresh mixture can nevertheless be inferred from (66). When $\nu = 3/2$, $\psi_X(X \gtrless 0) = \pm 1 = C \exp(\pm X/2)$ fits (66a, b) for any value of C. Inserting this in (66c) yields $(C - 1)(C - 3) = 0$; $C = 1$ has already been considered, giving $\psi_X(0\pm) = 0$, whereas $C = 3$ yields $\psi_X(0\pm) = \pm 2$, i.e. an angle pointing towards the fresh mixture.

The following arguments, however, must be taken into account. Close to angle vertices, the <u>local</u> curvature effects can no longer be neglected, because the local variations of ψ_X cease to take place on the $X = O(1)$ scale. Equation (59), for example, has to be modified into :

$$\psi_X^2 = 1 + \nu \int_{-\infty}^{+\infty} \psi_{XX}(X - \xi)\, e^{-|\xi|}\, d\xi + \mu\, \psi_{XX} \tag{73}$$

where μ is proportional to ℓ/L and is thus <u>very</u> small ; (4) is then replaced by :

$$(\psi_X^2 - \mu\, \psi_{XX})_{XX} + 2\nu\, \psi_{XX} + 1 + \mu\, \psi_{XX} - \psi_X^2 = 0 \tag{74}$$

Close to a vertex of a solution to (64), ψ_X reads $\psi_X = F_\sigma + \ldots$, $\sigma = X/\mu$ being an O(1) stretched variable, and (74) gives :

$$F_\sigma^2 = F_{\sigma\sigma} + \psi_X^2(0^-) + \ldots \tag{75}$$

upon use of standard techniques of asymptotic matching in the limit $\mu \to 0$. According to (75) $F_\sigma = -\,\mathrm{Log}\cosh(\sigma|\psi_X(0^-)|)$ so that $\psi_X(0^-) \equiv F_\sigma(-\infty) = |\psi_X(0^-)|$; therefore, $\psi_X(0^-)$ must be <u>positive</u>, and the solutions of (64) exhibiting angles which point towards the fresh mixture are spurious. This shows how a <u>very small local</u> effect can select large-scale patterns (for a more famous example, see /20/). According to our previous discussion (section 9.2), very small local effects could also be of importance about the spectra associated to $\psi(X)$, even in the wave-number range $1/L \ll k \ll 1/\ell$; for $k \gtrsim O(1/\ell)$, the local effects are of course needed, to provide a cut-off.

10. STRONGLY CURVED FRONTS

When $B = O(1)$ and the front is highly curved, and in the framework of a constant density model, the equations (33 - 35) determining the radiation field can be summarized as :

$$\begin{aligned} &\Delta G = G + 4B\Delta(H(Z - \phi(X))) \\ &G(Z = \pm\infty) = 0 \end{aligned} \tag{76}$$

where H(.) is the Heaviside function. Since $\Delta H(Z - \phi(X)) = \delta'(Z - \phi)(1+\phi_X^2) - \delta(Z - \phi)\phi_{XX}$, (at least for wedged shape flames) the solution to (76) can be written as $-2Bg(Z, X ; \phi)$, where g reads :

$$g \equiv -2 \int_{-\infty}^{+\infty} \{(1 + \phi_X^2(\xi))\, \frac{\partial\gamma}{\partial Z}(X-\xi, Z-\phi(\xi)) - \phi_{XX}(\xi)\gamma(X-\xi, Z-\phi(\xi))\}d\xi \tag{77}$$

with $2\pi\gamma(u, v) = K_o(\sqrt{u^2 + v^2})$, $K_o(.)$ being the modified Bessel function of zeroth order and 2nd kind (a mess !). When the incoming flowfield corresponds to $\underline{U} = (0, 0, M + V(X))$, with $V(X) = O(1)$ and $\langle V(X) \rangle = 0$, the equations giving $\phi(X)$ and M in terms of B, C, K have the following structure :

$$H(m, C, K) = BI(\phi)/(M + V(X)) \tag{78a}$$

where H us still given by (27a),

$$m = (M + V(X))/(1 + \phi_X{}^2)^{1/2} \tag{78b}$$

and ! :

$$I(\phi) = \int_{-\infty}^{+\infty} d\xi (1 + \phi_X{}^2(\xi))\gamma(X - \xi, \phi(X) - \phi(\xi))$$
$$- \int_{-\infty}^{+\infty} d\xi\, \phi_{XX}(\xi) \int_{-\infty}^{0} \gamma(X - \xi, \phi(X) - \phi(\xi) + u)du \tag{78c}$$

In principle (78) allows one to compute the steady response of a particle-laden flame to a shear-flow. Unfortunately, this integro-differential equation involves little-tabulated kernels, and may be the most expedient way of getting the answer would be to treat (33) - (31), (32) and (27) numerically ; one has to remember that angles may appear along the flame front (e.g. : for B = 0), however !

11. UNSTEADY FLAMES IN SHEAR FLOWS

Although (77) still applies (if $\Lambda = 1$) when the flame is unsteady, the picture is even darker than for steady fronts. Indeed the energy equation (33) has now the form :

$$\Theta_\tau + W\,\Theta_Z = Bg(Z, X ; \phi(X, \tau)) \tag{79}$$

where $W = M(\tau) + V(X, \tau)$ is now allowed to depend on X $\underline{\text{and}}$ τ. Equation (72) can be integrated formally, once an initial condition $\Theta_o(Z, X)$ is given, to yield :

$$\begin{aligned} \Theta_* = B\int_o^\tau g\{\phi(X,\tau) + \int_\tau^{\tau'} W(X,\sigma)d\sigma, X ; \phi(X,\tau')\} + \\ + \Theta_o(\phi(X,\tau) - \int_o^\tau W(X,\sigma)d\sigma, X) \end{aligned} \tag{80}$$

and $\Theta_*(X,\tau)$ defined in (80) can be used in (27) with

$$m(\tau,X) = (W(X,\tau) - \phi_\tau)/(1 + \phi_X{}^2)^{1/2} \tag{81}$$

to compute the time evolution of the mean flame speed $M(\tau)$ ($\langle V \rangle = 0$) and of the flame shape $\phi(X,\tau)$ ($\langle \phi \rangle = 0$). However, (80) (81) and (27) yield an highly non-linear integro-differential equation which is now non-local both in $\underline{\text{space}}$ and in $\underline{\text{time}}$. That (80)-(81) have to involve the histories of M, ϕ, V is easily understood : $\tau = O(1)$ corresponds to the transit time of a fluid element across the radiative preheat zone. These non-local characteristics are beleived to be important ingredients of the dynamics of particle-laden, radiation-affected, flames propagating in time-dependent shear-flows, but they are very difficult to take into account in an analytical theory, simultaneously : the form of (80) - (81) when g is given by (77) should convince the reader ! To open up this subject matter, we suggest to study an approximate model equation, built as follows. Assume that $|\phi|$ is small (recall that $\langle\phi\rangle = 0$) ; then g is given by (55), whatever the value of B and the flame evolution. In addition we assume that $M - \bar{m}$, B, V are small. Then (80) takes the form :

$$\Theta_* = B/\bar{m} + \Theta_1 + \Theta_2 \tag{82}$$

to be used in (27), where :

$$\frac{\partial \Theta_1}{\partial \tau} + \bar{m}\,\Theta_1 \cong B\,\phi - B\,(M - \bar{m} + V - \phi_\tau)/\bar{m} + \ldots \tag{83}$$

and the Fourrier transform of Θ_2, viz $T_2(k,\tau)$, is given by :

$$\frac{\partial T_2}{\partial \tau} + \bar{m}\,\sqrt{1 + k^2}\,T_2 = -\,B\,f(k,\tau)/\sqrt{1 + k^2} + \ldots \tag{84}$$

Clearly, once used in (27) Eqs (81) - (84) allow one to study the unsteady response of a weakly curved, unsteady particle-laden flame, with the non-local characteristics of radiative preheating taken into account, with respect to both time and space. The study of (27), (81) - (84) is the subject of current works, which will be reported elsewhere.

12. CONCLUDING REMARKS

Constructing non-linear equations to describe the steady or unsteady response of weakly-curved, radiation-affected flames to given incoming flowfields is a step towards a better understanding of the dynamics of reaction fronts propagating in particle-laden mixtures. The work presented here is not complete, however. The next step is of course to integrate these equations, numerically. About the last point, it has to be stressed again that the radiation-induced curvature effect - if alone -, allows for angled flame shapes, even spurious ones, or tolerate singularities in the higher order derivatives of the function describing the shape of the flame sheet. This may lead to difficulties in a numerical treatment (e.g. : aliasing) ; knowing that small local curvature effects can have important consequences (e.g. : ruling out spurious solutions) is certainly a usefull information.

In this paper, we have focused our attention on constant-density models of weakly curved fronts. Incorporating the change in medium opacity, related to the jump is density, should not pose a major problem. In addition, a qualitative account of the hydrodynamical effects on a nearly planar flame can also be included in the theory, by considering the density jump $((\rho_u - \rho_{ad})/\rho_u)$ as weak when it is involved in the fluid mechanics, with the result that a weak form of the Landau-Darrieus instability is incorporated in the theory : it is enough to use the relevent form of the corresponding weakly perturbed incoming flow, as given in /3/ or /22/. Such a model would help us in understanding the couplings between an hydrodynamical instability and the radiation-induced curvature effects (and local ones, if needed locally).

Apart from special cases (spherically expanding fronts, in a stagnation flow...), it is almost certainly impossible to study the influence of large density changes on a curved flame with analytical methods ; numerics should then be employed, and we hope that the simplified set of equations (27) and (32) - (35) we have given in the paper will help people when studying highly curved, unsteady, radiation-affected flames.

Finally, one has to recall that the models of a radiation-affected propagation may cease to be valid if the flame velocity is too low, such as close to kinetics-induced flammability limits, in which case the propagation may well be radiation-dominated. This is, however, an other story..., which is in the process of being written.

13. ACKNOWLEDGEMENTS

In addition to C.M. Brauner and C. Schmidt-Lainé, who organized the NATO workshop, I thank P. Cambray and B. Deshaies, who co-authored papers on the present subject matter. Last, but not least, I would like to dedicate this review paper to the memory of G.S.S. Ludford, a fine man and scientist.

14. REFERENCES

/1/ J. Buckmaster, G.S.S. Ludford (1982) "Theory of Laminar flames", Cambridge University Press, Cambridge.

/2/ P. Clavin (1985) "Dynamical behavior of premixed flames in Laminar and Turbulent flows", Prog. En. and Comb. Sci., 11, pp. 1-59.

/3/ G.I. Sivashinsky (1983) "Instabilities, pattern formation and turbulence in flames" ; Ann. Rev. Fluid. Mech., 15, pp. 179-199.

/4/ F.A. Williams (1985) "Combustion theory", 2nd Edition, Benjamin - Cummings, N.Y..

/5/ Ya. B. Zeldovich et al (1980) "Mathematical theory of combustion and explosions", Nauka Moscow. English Translation National Bureau, N.Y. (1985).

/6/ W.E. Baker et al (1983) "Explosion Hazard and Evaluation", Elsevier Publishing Company.

/7/ K.N. Palmer (1973) "Dust-explosions and fires", Chapman and Hall Ltd, London.

/8/ B. Deshaies, G. Joulin (1986) "Radiative transfer as a propagation mechanism for combustion waves in rich suspensions", SIAM Jal Appl. Math., 46.

/9/ G. Joulin, P. Clavin (1979) "Linear Stability analysis of non-adiabatic flames", Comb. and Flame, 35, 139.

/10/ J. Kevorkian, J.D. Cole (1981) "Perturbation methods in applied mathematics", Appl. Math. Sci., n°34, Springer Verlag, N.Y..

/11/ G. Joulin, B. Deshaies (1986) "Radiation-affected steady flame propagation in reactive mixtures seeded with inert particles", Comb. Sci. and Tech., 47, pp. 299-315.

/12/ G. Joulin (1986) "Temperature-lags and radiative transfer in particle laden gaseous flames. Part I : Steady, planar fronts", Comb. Sci. and Tech., 52, 377.

/13/ G. Joulin, P. Cambray (1986) "Temperature-lags and radiative transfer in particle-laden gaseous flames. Part II : unsteady propagations", Comb. Sci. and Tech., 52, 397.

/14/ E. Köros et al (1974) in "Physical chemistry of oscillatory phenomena, Faraday Symposium of the Chemical Society n°9, University Press, Aberdeen (and references therein).

/15/ G. Joulin, P. Cambray "Radiation-affected dynamics of unconfined spherical particle-laden flames", in preparation.

/16/ B. Lewis, G. Von Elbe (1961) "Combustion flames and explosion of gases", Academic Press, N.Y..

/17/ G. Joulin (1987) "On the radiation-induced curvature effects on particle-laden flames", to appear in Comb. Sci. and Tech.

/18/ G. Joulin (1987) "On the shape of a particle-laden flame in a steady shear flow" ; in preparation.

/19/ M.N. Ozisic (1972) "Radiative transfer and interaction with conduction and convection", Wiley and Sons, N.Y..

/20/ R. Combescot et al (1986), Phys. Rev. Lett., 56, 2036.

/21/ G. Joulin (1980) "Asymptotic analysis of non-adiabatic flames : heat losses towards small inert particles", 18th Symposium (Int.) on Combustion, pp. 1395-1404.

/22/ G.I. Sivashinsky, P. Clavin (1986) "On the non linear hydrodynamic flame stability", submitted, private communication.

NUMERICAL SIMULATION OF COHERENT STRUCTURES IN FREE SHEAR FLOWS

M. LESIEUR, P. COMTE, J.P. CHOLLET and P. LE ROY

Institut de Mécanique de Grenoble*

Institut National Polytechnique and Université Scientifique et Médicale de Grenoble
BP 68, 38402 Saint-Martin d'Hères, France

ABSTRACT

Direct numerical simulations (or large-eddy simulations) of two-dimensional turbulence developing from a basic inflexional velocity profile are presented, with applications to mixing-layers, plane jets or wakes. The temperature is considered within the passive scalar approximation.

The flows are periodic in the mean flow direction. Turbulence is generated with the aid of a random perturbation of small amplitude superposed upon the basic flow. The numerical code is a finite differences $\psi - \omega$ code with a resolution up to 256×512 grid points. The dissipative operator is either of the harmonic type (direct simulation, Reynolds number 1 000), or biharmonic (large-eddy simulation, Reynolds number 10 000), with approximately the same results. Vorticity, stream function and temperature fields are visualized.

In the mixing-layer case, the appearance and successive pairings of the coherent structures (Kelvin-Helmholtz eddies) are displayed. These structures are shown to be unpredictable. A spanwise two-mode truncation of the three-dimensional mixing-layer allows the three-dimensional kinetic energy of a particular spanwise perturbation to be associated with the error energy in the two-dimensional predictability problem. Thus the predictability calculations imply that the spreading of the layer is accompanied with an exponential growth of three-dimensionality. Such an instability might lead to a complete destruction of a coherent structure, once formed. It is proposed that the instability of the mean inflexional shear will then recreate a new two-dimensional coherent structure.

In the case of the Bickley jet, it is shown that the flow evolves towards a turbulent Karman street, made of coherent structures which can either pair (if they are both of the same sign) or form dipole-like structures (if of opposite sign). Moreover, and if the jet is also forced at its fundamental frequency by a perturbation of small amplitude, one obtains in certain cases a very stable Karman street.

Finally the temperature field is shown to be a much more efficient indicator of the coherent structures than the vorticity field.

1 INTRODUCTION

The understanding of the mixing due to turbulence is of prior importance in most of the combustion studies. It is a very difficult problem if posed within its whole complexity, since turbulence and chemical reactions are generally coupled in a complicated manner.

* The Institut de Mécanique de Grenoble is sponsored by the CNRS (U.A. 6).

C.-M. Brauner and C. Schmidt-Lainé (eds.), Mathematical Modeling in Combustion and Related Topics, 131–150.

Fortunately, it may happen that the modification of the structure of turbulence due to a flame and to the associated thermal heating can be ignored: this is the case for instance in the study of diffusion flames under certain conditions (see e.g. Linan, 1987; Laverdant and Candel, 1987). This paper will then consider turbulence alone, in the simplest case when the temperature is only a passive scalar. We believe however that the following results will be of interest for the understanding of turbulence combustion.

The turbulent mixing is generally due both to the action of the small scale three-dimensional structures, and to the large scale spatially organized structures. The latter, usually called *coherent*, can be either three-dimensional (like in the turbulent boundary-layers for instance, see e.g. Kline et al. 1967) or quasi-two-dimensional (like the Kelvin-Helmholtz eddies in the mixing layers, see e.g. Brown and Roshko 1974).

The present study will consider only the coherent structures of the free shear flows like mixing layers, plane jets and wakes. These flows are shown on Figure 1. They will be studied here by means of unstationary numerical simulations of the two-dimensional Navier-Stokes equation (with constant density). These numerical simulations will show the *turbulent* character of the coherent structures. For the mixing layer, this character will be investigated through predictability studies. For the plane jet in an approximation derived from the work of Bickley (1937), a turbulent Karman street with pairing of same sign vortices will occur when the jet is forced by a white noise perturbation only. The study of the diffusion of a passive scalar will be done in both situations, showing that a passive scalar (such as a dye or a passive temperature) is a much better indicator of the coherent structures than the vorticity or the stream function. Such a study neglects all the three-dimensional effects. We will however investigate the stability of a Kelvin-Helmholtz billow under certain three-dimensional perturbations.

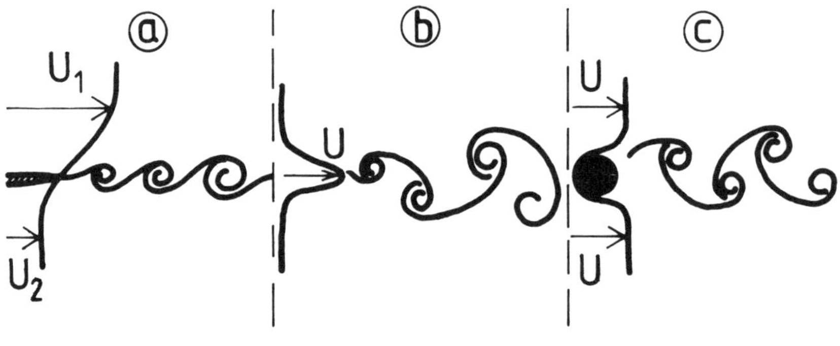

Fig. 1: schematic view of a) a mixing layer, b) a plane jet c) a wake.

2 PRESENTATION OF THE TWO-DIMENSIONAL SIMULATIONS

2.1 Numerical methods

We consider the two-dimensional Navier-Stokes equation for a fluid of density having small fluctuations with respect to a constant basic value ρ_0. Then, to the lowest order, the velocity field $\underline{u}(x,y,t)$ of components $u(x,y,t)$ and $v(x,y,t)$ parallel to the $\underline{x},\underline{y}$ plane satisfies a two-dimensional Navier-Stokes equation with the constant density ρ_0: introducing the stream function $\psi(x,y,t)$ such as

$$u(x,y,t) = \frac{\partial \psi}{\partial y}, \quad v(x,y,t) = -\frac{\partial \psi}{\partial x}, \tag{2-1}$$

the vorticity vector $\nabla \times \underline{u} = \omega \, \underline{z}$ is given by

$$\omega = -\nabla^2 \psi \tag{2-2}$$

where the ∇ operator corresponds to the variables x and y. The two-dimensional Navier-Stokes equation simply expresses that the vorticity ω is transported with the flow motion and diffused by molecular viscous effects, that is

$$\frac{D\omega}{Dt} = \nu \nabla^2 \omega \tag{2-3}$$

where

$$\frac{D}{Dt} = \frac{\partial}{\partial t} + u\frac{\partial}{\partial x} + v\frac{\partial}{\partial y} \tag{2-4}$$

ν being the molecular viscosity and u and v being given in terms of ψ by (2-1). Within the passive scalar approximation, the temperature $\theta(x,y,t)$ satisfies an equation analogous to (2-3)

$$\frac{D\theta}{Dt} = \kappa \nabla^2 \theta \tag{2-5}$$

where κ is the molecular conductivity. Like the vorticity, the temperature is transported by the flow and diffused by molecular exchanges. The important difference between eqs (2-3) and (2-5) lies in the fact that the temperature is *passive*, in the sense that an external perturbation brought to the temperature field will not affect the velocity field. This is, on the contrary, not true for the vorticity, because of (2-2). Notice however that, in the case of strictly identical initial and boundary conditions for the vorticity and the temperature (and for $\nu = \kappa$), the two fields will be identical for any time. This remark permits to characterize the vorticity as an *active scalar*. In the calculations which follow, the Prandtl number[1] ν/κ is taken equal to unity.

The calculations done on the basis of eqs (2-3) and (2-5) will be called direct numerical simulations. In certain cases, and in order to increase artificially the Reynolds and Peclet numbers, the harmonic dissipative terms $\nu\nabla^2\omega$ and $\kappa\nabla^2\theta$ will be replaced by biharmonic diffusion operators $-\nu_1(\nabla^2)^2\omega$ and $-\kappa_1(\nabla^2)^2\theta$. Such simulations are no more direct, but will be referred to as *large-eddy simulations*. In fact, two-dimensional Navier-Stokes equations can never be an accurate representation of physical flows, which are always three-dimensional in the small scales, at least at the level of the dissipative scales. Consequently, these biharmonic subgrid-scale[2] the operators have to be considered more as a way of modelling the action of the small three-dimensional subgrid-scales upon the large scales explicitely simulated. As

[1] or Schmidt number if the scalar is not a temperature, but a dye which marks the flow for instance

[2] The subgrid scales correspond to the motions of wave length smaller than Δx, the computational grid mesh.

a matter of fact, the large scales dynamics will reveal to be quite insensitive to the particular subgrid-scale procedure chosen. The equivalent Prandtl number ν_1/κ_1 will be also taken equal to one.

Calculations are done in a square or rectangular domain, using a finite differences $\psi - \omega$ method. One assumes periodicity at the boundaries in the x-direction. This corresponds to the so-called *temporal approximation*, by opposition to the spatially developing layers shown on Figure 1. In the mixing layer for instance, the temporal approximation corresponds to a study from a reference frame moving with the mean velocity

$$\frac{U_1 + U_2}{2},$$

and neglects locally the spatial spreading of the layer. In the temporal mixing layers, the boundary conditions are

$$u = \pm U \text{ for } y = \pm\infty, \text{ with } U = \frac{U_1 - U_2}{2}.$$

Thus if δ_i, defined below, is a characteristic scale of the coherent structures in the temporal problem, the characteristic associated time will be δ_i/U. In the spatial problem, it will correspond to a distance δ_i/α downstream of the splitter plate, where

$$\alpha = \frac{U_1 - U_2}{U_1 + U_2} \tag{2-6}$$

is called the velocity ratio. Let us stress finally that within the temporal approximation, the mean velocity profiles of the plane jet and of the wake give rise to the same flow pattern.

The Jacobian operator $J(.,\psi) = u\ \partial./\partial x + v\ \partial./\partial y$ in (2-4) is approximated using Arakawa's 1966 formulation. The $\partial/\partial t$ operator is a second order leapfrog operator. In the y-direction, boundary conditions are of the free slip type ($\psi = \text{constant}, \omega = 0$). Calculations are done at resolutions up to 256^2 grid points (in a square domain) and 256×512 grid points (in a rectangular domain). The initial condition for the velocity consists in a basic flow $\bar{u}(y)$ upon which is superposed a small perturbation defined below. The initial temperature field $\theta(y)$ is identical to the basic velocity $\bar{u}(y)$. The choice of the domain length L will be justified in the next subsection.

2.2 Linear instability and fundamental modes

In order to precise the initial conditions and to fix the length L of the computational domain, it is of interest to recall some basic results of the linear instability theory, for both the temporal mixing layer and the Bickley jet.

In this theory, it is assumed that the stream function of the velocity field is of the form

$$\psi(x,y,t) = \bar{\psi}(y) + \psi'(x,y,t), \ |\psi'| << |\bar{\psi}| \tag{2-7}$$

where $\bar{\psi}(y)$ is the stream function corresponding to a basic flow $\bar{u}(y)$, and ψ' is a small perturbation of the form

$$\psi'(x,y,t) \sim \exp i(kx - ct) \tag{2-8}$$

When $\Im(c)$, the imaginary part of c, is positive, the perturbation will amplify exponentially, and the flow will be linearly unstable. The longitudinal wavenumber k_a for which $\Im(c)$ is maximum will be called the most amplified wavenumber.

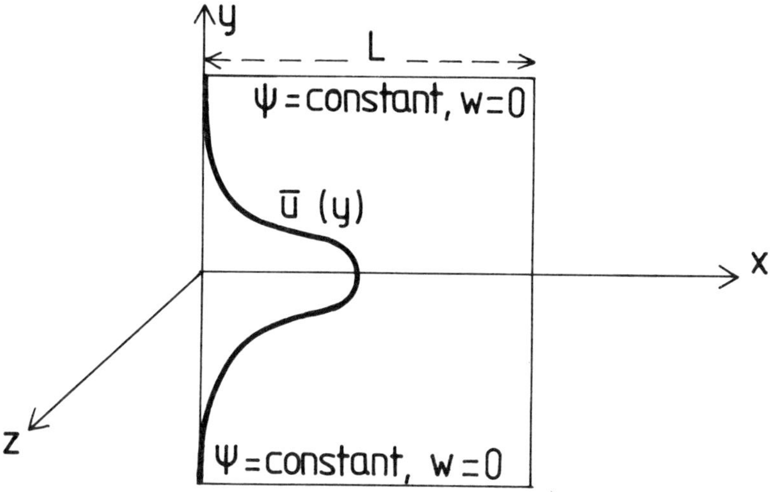

Fig. 2: computational domain, with the basic velocity profile of the plane jet.

2.2.1 *the mixing layer*

Let the basic velocity be given by

$$\bar{u}(y) = U \tanh(2y/\delta_i) \tag{2-9}$$

The schematic linear stability diagram, derived from the resolution of the Orr-Sommerfeld equation of the linear instability theory applied to this problem[3], is shown on Figure 3, after Betchov and Szewczyk (1963). For a given Reynolds number $R = U\delta_i/\nu$ of the basic flow, there exists a span of unstable longitudinal wavenumbers k. If the perturbation ψ' contains the whole range of unstable modes, as will be assumed below, it is the most amplified wavenumber which will appear the first. When R is greater or of the order of 20, the instability is *inviscid* in the sense that k_a does not depend anymore on R. The corresponding wave length λ_a is then equal to (Michalke, 1964)

$$\lambda_a \approx 7\delta_i \tag{2-10}$$

Thus the initial rotational zone corresponding to the inflexion point of the basic velocity profile will develop an undulation of longitudinal wave length λ_a: indeed, the particular form of the stream function perturbation (2-8) corresponds to a velocity perturbation parallel to y which convects the rotational stripe. Thus "low speed fluid" (that is of velocity $-U$) will be lifted into the high speed region (of velocity $+U$), while high speed fluid will dive into the low speed region. The result will be a tendency to a roll up, and hence the formation of a Kelvin-Helmholtz type vortex of wave length λ_a. This phenomenon is schematically represented on Figure 4.

[3] Details on the theory can be found in Drazin and Reid (1981).

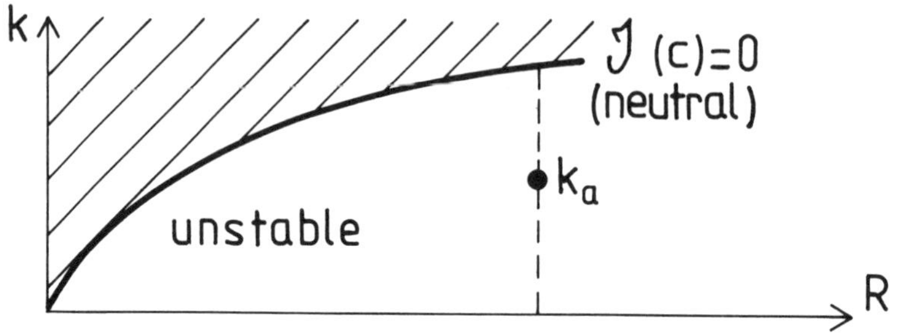

Fig. 3: schematic stability diagram of the temporal mixing layer.

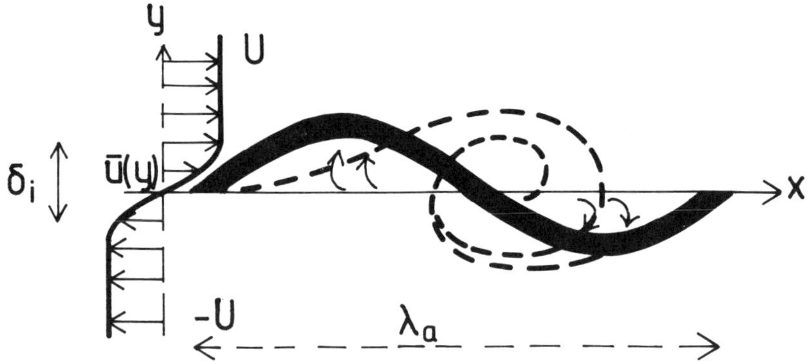

Fig. 4: formation of a Kelvin-Helmholtz vortex of wave length λ_a, due to the linear instability of the inflexional basic velocity profile.

2.2.2 *the Bickley jet*

The basic velocity profile of the Bickley jet is

$$\bar{u}(y) = U \operatorname{sech}^2(y/\delta_0) \tag{2-11}$$

corresponding to the stream function

$$\bar{\psi}(y) = U\delta_0 \tanh(y/\delta_0) \tag{2-12}$$

where $\delta_i = 1.3\ \delta_0$ is the distance between the two inflexion points, and U the velocity at the center of the jet (see Figure 5). The most amplified mode

$$\lambda_a \approx 6.83\ \delta_0 \tag{2-13}$$

(see Drazin and Reid, 1981) is a *sinuous* mode, corresponding approximately to the superposition in phase opposition of two mixing layers of vorticity thickness δ_i. This mode actually corresponds to a Karman street. The *varicous* mode, corresponding to a symmetry of the two mixing layers with respect to the axis $\underline{x}$, is less amplified than the sinuous mode, and we could not excite it in all our calculations. Notice however that in the case of the round jet (axisymmetric mixing layer), it is the varicous mode which is the most amplified.

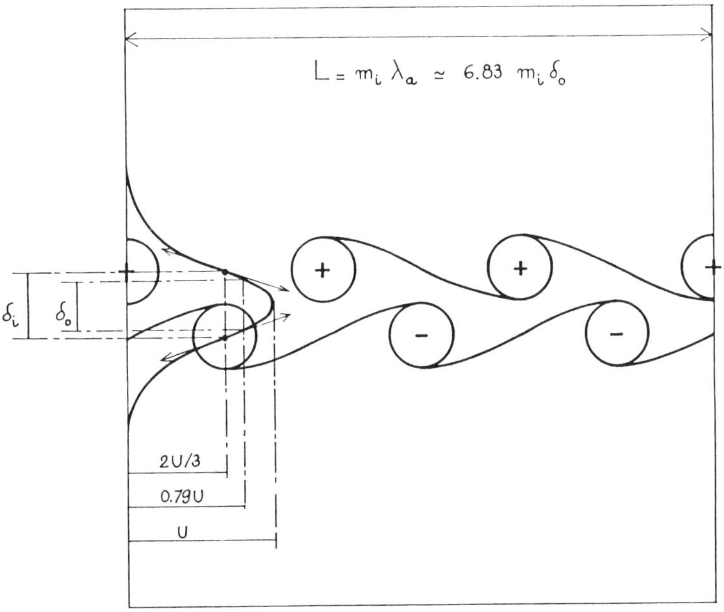

Fig. 5: schematic representation of the sinuous mode of the Bickley jet, corresponding to a Karman street (from Comte et al., 1987).

In all the following calculations, The length L of the computational domain will be chosen as a multiple of the fundamental wave length λ_a,

$$L = m_i \lambda_a \qquad (2-14)$$

where m_i is an integer.

2.3 The initial conditions

The Reynolds number is equal to

$$R = U\delta_i/\nu = 1\ 000$$

in the direct numerical simulations, and

$$R_1 = U\delta_i^3/\nu_1 = 10\ 000$$

in the large-eddy simulations. Initial conditions for the stream function are given in both cases (mixing-layer and Bickley jet) by the basic flows defined above, to which are superposed small perturbations defined as follows.

2.3.1 *the mixing layer*

The initial stream function perturbation consists in a white noise of amplitude 10^{-4} relative to the maximum of the basic flow stream function, injected at all the grid points of the calculation, and multiplied by $\exp(-y^2/\delta_i^2)$. This white noise will inject energy into all the unstable modes, and will thus be responsible for the appearance of the fundamental mode k_a. It represents (in a very rough way) the residual turbulence which triggers the Kelvin-Helmholtz instability in the experimental mixing-layers behind a splitter plate. In some of the calculations, a deterministic longitudinal sine perturbation of wave number k_a will be added to the white noise, in order to fix the phase of the Kelvin-Helmholtz eddies on the x-axis: we have checked that this does not change the turbulent and unpredictable character of the layer, contrary to the Bickley jet case.

Finally, the role of the initial subharmonic perturbation at the mode $k_a/2$ contained in the white noise is to trigger the subharmonic instability leading to the pairing, as pointed out by Kelly (1967).

2.3.2 *the Bickley jet*

Here, two cases will be considered:

• A white-noise perturbation of same type as in the mixing-layer case. This case will be called the *free jet*.

• The same white-noise perturbation, plus a deterministic longitudinal sine perturbation at the fundamental mode. This case will correspond to the *forced jet*. The amplitude of this perturbation with respect to the white noise will be equal to the parameter n_i.

2.4 The results

The vorticity and temperature fields resulting from the calculations have been processed in colour, and are here presented in black and white.

2.4.1 *the mixing layer*

The initial evolution corresponds exactly to the predictions of the linear stability theory, and presented schematically on Figure 4. The fundamental Kelvin-Helmholtz vortices appear at about $t = 15\ \delta_i/U$. Figure 6 shows these vortices at $t = 25\ \delta_i/U$ in a calculation involving 4 eddies ($m_i = 4$, Reynolds number 10 000).
At about $t = 35\ \delta_i/U$, the eddies have completed the first pairing, and there remain two eddies in the domain. Figure 7 shows these eddies at $t = 65\ \delta_i/U$. They are about to undergo the second pairing, which will be completed at about $t = 80\ \delta_i/U$ (see Figure 8).

Figure 9 shows the whole story of the vorticity field in a 4-eddy calculation with $R = 10\ 000$, from $t = 0$ to $t = 80\ \delta_i/U$.

Figure 10 shows the temperature field during the second pairing in a calculation involving initially 8 eddies, at a Reynolds number of 1 000 (direct simulation). The coherent structures of the temperature field which appear correspond to vortical coherent structures, as we have verified. The first remark which can be done is that the temperature is a much better visual indicator of the coherent structures than the vorticity[4]. This is encouraging for the interpretation of the experiments where the flow

[4] But the vorticity is itself a better indicator of the eddies shape than the stream

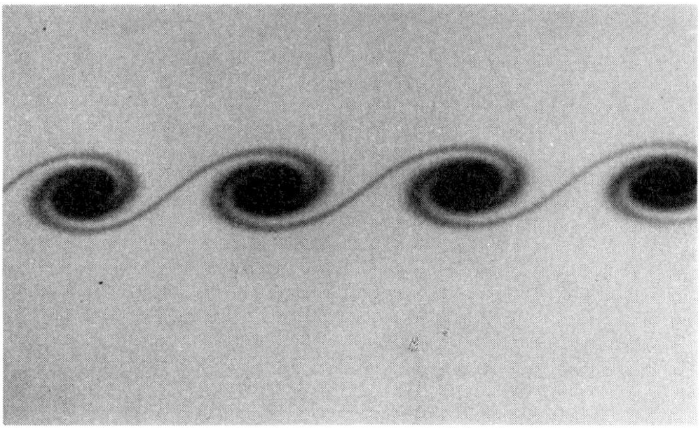

Fig. 6: vorticity field obtained at $t = 25$ characteristic times δ_i/U in a 4-eddy mixing-layer calculation with $R = 10\ 000$. Dark regions correspond to intense vorticity.

Fig. 7: same calculation as in Figure 6; vorticity field obtained at $t = 65\ \delta_i/U$, between the first and the second pairing.

---

function, as checked in our calculations. This is why wheather maps showing the vorticity field instead of the pressure field (proportional to the horizontal stream function within the geostrophic approximation), would visualize more properly the cyclonic or anticyclonic eddies in the atmosphere.

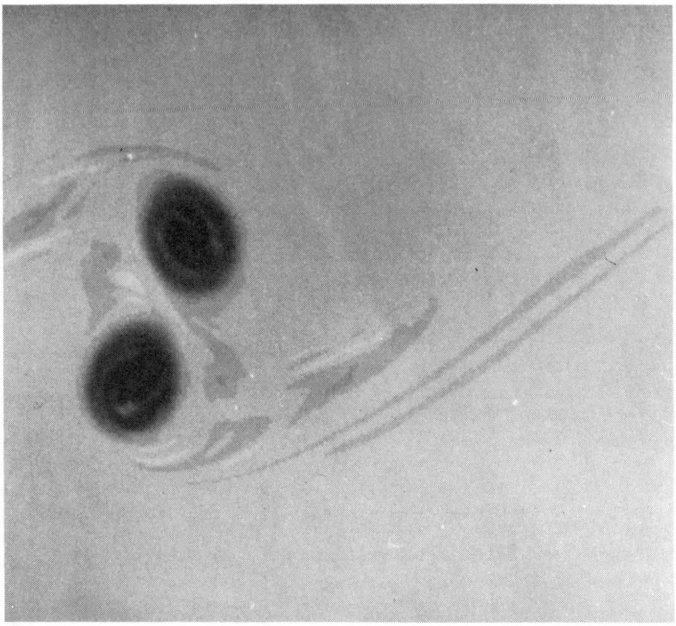

Fig. 8: same calculation as in Figure 6; vorticity field obtained during the second pairing.

is generally visualized with the aid of a passive temperature or a dye (which is also a passive scalar). The second remark concerns the way the mixing occurs, through thin tongues of warm fluid which winds around the vortices, with for consequence the development of localized patches of warm fluid within the cold fluid. Thus the mixing is in no way a homogeneous process.

Another argument justifying the turbulent character of the mixing-layer coherent structures lies in the fact that the longitudinal spatial kinetic energy spectrum is a broad band spectrum of slope intermediate between k_x^{-4} and k_x^{-3} extending above k_a, as shown by Lesieur et al. (1987). This spectrum develops after the first pairing. Its maximum corresponds to the fundamental wave number k_a, but without any marked peak. The two-dimensional coherent structures constitute thus a continuous hierarchy of scales, built by the shearing of small scale fluid particles by larger scale velocity gradients, which is particularly active during the pairing. Such a mechanism is reminiscent of the enstrophy cascade of two-dimensional isotropic turbulence (Kraichnan, 1967, see also Lesieur, 1987). In real mixing layer flows, three-dimensional turbulence, of spectrum following the Kolmogorov law $k_x^{-5/3}$, will superpose upon the coherent structures spectrum, and the resulting spectrum will be a $k_x^{-5/3}$ spectrum for $k_x > k_a$, as measured by Perry et al. (1982) and Browand and Ho (1983). The coherent structures kinetic energy spectrum is therefore hidden behind the three-dimensional turbulence spectrum. However, it is possible that, to the two-dimensional coherent structures spectrum, could be associated a temperature spectrum of slope k_x^{-1}, as for two-dimensional isotropic turbulence (see Lesieur and Herring, 1985, and Babiano

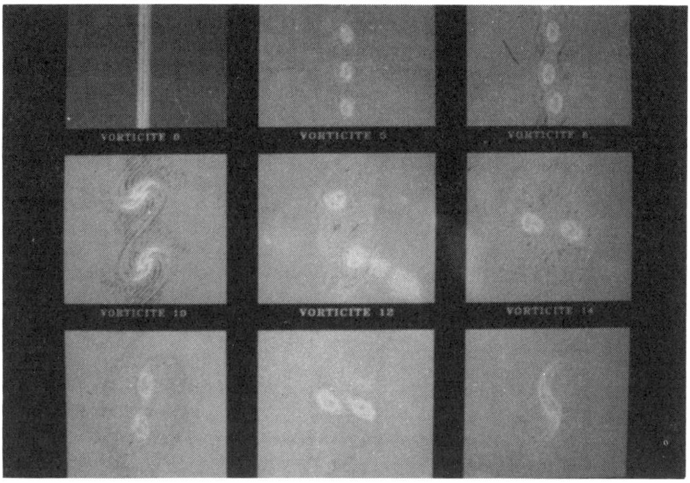

Fig. 9: time-evolution of the vorticity field in a 4-eddy mixing layer (R = 10 000); interval of time between slides: 10 δ_i/U.

et al., 1987). Hence, since the passive scalar in three-dimensional turbulence follows a $k_x^{-5/3}$ Corrsin-Oboukhov law (see e.g. Lesieur, 1987), the temperature spectrum resulting from the coherent structures should overwhelm the temperature spectrum of three-dimensional turbulence. Consequently, the temperature spectrum in a real mixing layer should possess a k_x^{-1} range for $k_x > k_a$.

2.4.2 *the Bickley jet*

It has been already stressed that the fundamental mode corresponds to a Karman street. It is possible to excite this mode in a very regular way in the forced case under certain conditions: Figure 11 shows the vorticity and temperature fields obtained at $t = 75\ \delta_i/U$ in a forced jet with 8 pairs of vortices ($m_i = 8$), when the deterministic perturbation is 200 times larger than the white-noise amplitude ($n_i = 200, R = 1\ 000$).

In the case of the free jet, a Karman street forms initially, but it is unstable and evolves towards a very irregular flow, where coherent structures of the same sign may pair: such pairings are displayed on Figure 12, corresponding to a Reynolds number of 1 000. Similar pairings are observed in the calculations at higher Reynolds number (10 000). In the free case, the jet can be approximately interpreted as a double mixing layer, since the fundamental eddies form at $t = 30\ \delta_i/U$, and the first pairings occur at about $t = 75\ \delta_i/U$. These times are very close to the comparable times obtained in the mixing layer, if we remark that the two mixing layers associated to a Bickley jet have a vorticity thickness of approximately δ_i and a velocity difference $U/2$. Other types of interaction between eddies are encountered, such as the formation of dipoles of opposite sign eddies, which are ejected from the vortex street. The same phenomena have been observed both experimentally and numerically by Couder et al. (1984). Eventually, the eddies will scatter throughout the whole domain, resembling then the coherent structures found by Mc Williams (1984) in the case of isotropic

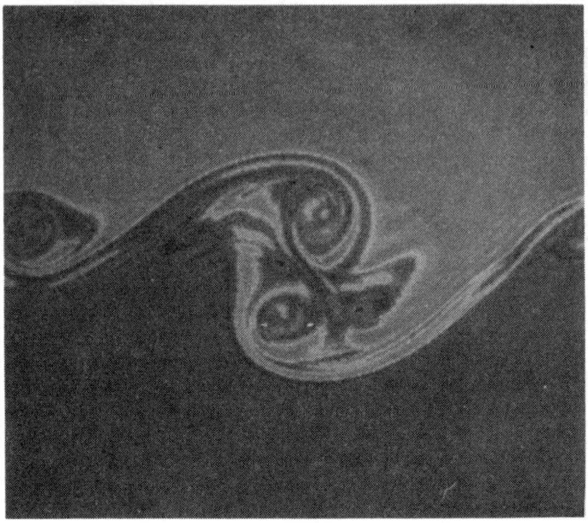

Fig. 10: temperature field in a mixing layer involving initially 8 eddies, during the second pairing ($R = 1\ 000$).

two-dimensional turbulence.

3 PREDICTABILITY OF THE MIXING-LAYER

For dynamical systems which possess only a limited number of degrees of freedom (and hence no developed spatial energy spectrum), the turbulent character is generally characterized by the unpredictability of the system. The latter consists in considering two figurative points in the phase space, initially very close: if the separation between the two points grows exponentially, the system will be unpredictable. It is for instance the case for the dynamical system introduced by Lorenz (1963) on the basis of a three-mode truncation of the Navier-Stokes equation within the Boussinesq approximation, yielding the famous Lorenz attractor (see e.g. Bergé et al., 1984). The concept of predictability can also be introduced for fluid turbulence (see the review by Thompson, 1984), where it has been shown, mainly on the basis of stochastic models, that a small difference in the small scales between two velocity fields would amplify, both in three and two dimensions, leading eventually to a complete decorrelation between the two fields (see e.g. Leith and Kraichnan, 1972, and Métais and Lesieur, 1986).

3.1 The predictability problem in the mixing layer

In the case of the mixing layer, the predictability problem will be posed in the following manner: let $\underline{u}^{(1)}(x,y,t)$ and $\underline{u}^{(2)}(x,y,t)$ be two velocity fields in the same mixing layer calculation, differing initially only by the white noise realization. Let $\mathcal{D}$ be a square

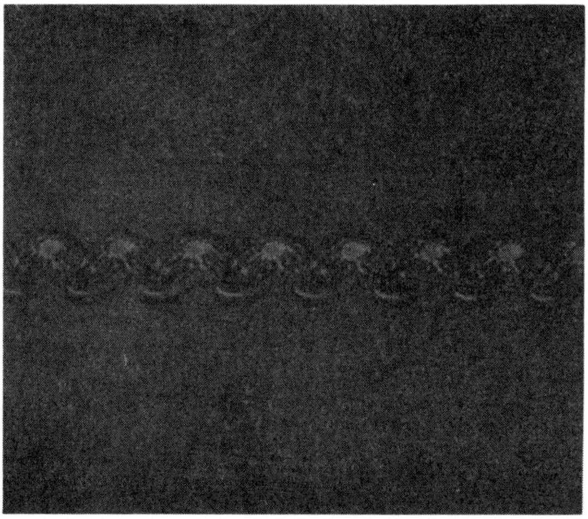

Fig. 11-a: vorticity field at $t = 75\ \delta_i/U$ in a forced Bickley jet ($m_i = 8, R = 1\ 000$).

domain included into the computational domain, and upon which spatial averages of the difference between the two fields will be taken. All the calculations considered will have the same basic velocity profile of width δ_i, and $\mathcal{D}$ will be chosen so as to contain 4 fundamental eddies, hence with a side equal to 28 δ_i. The kinetic energy of the error will be defined as

$$E_\Delta = \frac{1}{2}\int_{\mathcal{D}} [\underline{u}^{(1)}(x,y,t) - \underline{u}^{(2)}(x,y,t)]^2 dx\ dy \qquad (3-1)$$

and the error rate by

$$r(t) = E_\Delta / \int_{\mathcal{D}} \underline{u}^2 dx\ dy \qquad (3-2)$$

where $\underline{u}$ stands for one of the two fields $\underline{u}^{(1)}(x,y,t)$ or $\underline{u}^{(2)}(x,y,t)$. The computational square is $\mathcal{D}$ itself in a so-called *4-eddy* calculation. We have also performed *8-eddy* and *16-eddy* calculations, in computational squares of side 56 δ_i and 112 δ_i. In each of the computational domains, the evolution of the two fields $\underline{u}^{(1)}(x,y,t)$ and $\underline{u}^{(2)}(x,y,t)$ is considered. The 4 and 8-eddy calculations are schematically represented on Figure 13. Initially, the two fields will differ only by the white noise, and $r(0)$ will be very small, but the same for the 4, 8 and 16-eddy calculations. On the other hand, a complete decorrelation between the two fields will correspond to $r(t) = 1$.

3.2 Growth of the layer thickness

Before looking at the numerical results concerning the error rate, let us consider the

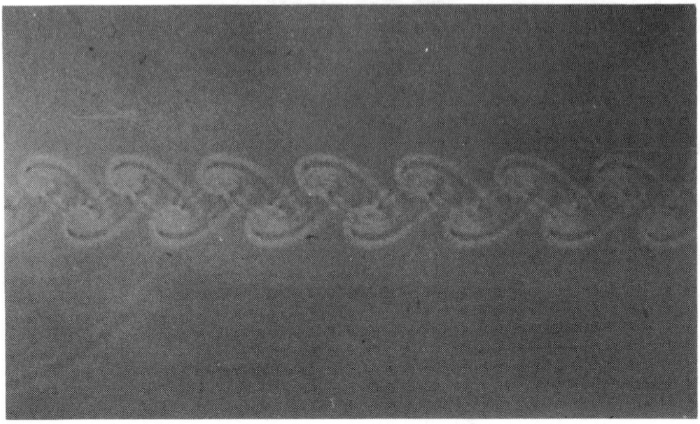

Fig. 11-b: temperature field in the calculation of Figure 11-a.

time evolution of the vorticity thickness $\delta(t) = 2U/(d\bar{u}/dy)_{max}$, the bar representing an average in the x-direction. These results are presented on Figure 14, for the 4, 8 and 16-eddy calculations and a Reynolds number of 10 000. In each case, the vorticity thickness is an average between $\underline{u}^{(1)}(x,y,t)$ and $\underline{u}^{(2)}(x,y,t)$. On the figure are indicated the times corresponding to the appearance of the fundamental eddies ($15\ \delta_i/U$), the end of the first pairing ($35\ \delta_i/U$), second pairing ($80\ \delta_i/U$) and third pairing ($155\ \delta_i/U$). The vorticity thickness starts growing linearly, with a slope [5] of about 0.10, then saturates. This saturation is clearly a domain size effect, since it is delayed when the computational square increases. Actually, the saturation is due to the longitudinal periodicity of period L, as we have checked by performing an 8-eddy calculation in a rectangular domain of height $2L$, which has not shown a sensible difference in the spreading of the layer. This saturation occurs when there remain two eddies in the computational domain. These eddies pair in the calculation, but without any vorticity thickness growth. Let us remark that in a spatial mixing layer, this periodicity assumption does not exist, which explains that the linear spreading can keep on.

---

[5] Experiments concerning spatial mixing layers indicate a faster growth, with a value of 0.17 for the associated spreading rate $\alpha^{-1}\ d\delta/dx$ (α being the velocity ratio defined by (2-6)), see Lesieur et al. 1987 for a discussion on this point.

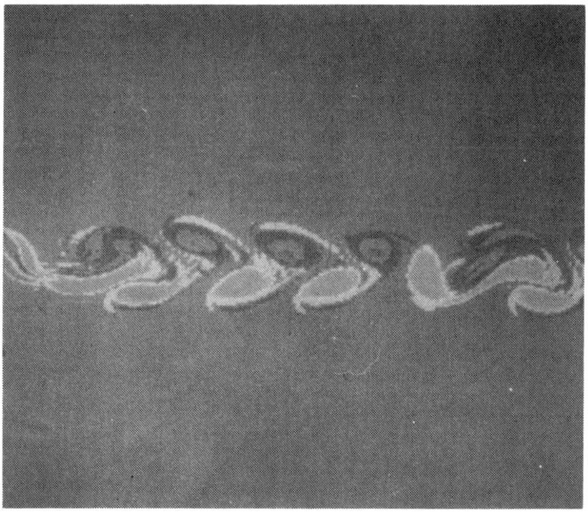

Fig. 12: vorticity field in a free jet, showing a pairing ($R = 1\ 000$).

3.3 Growth of the error rate

Figure 15 shows the error rate in the same calculations as in Figure 14. It presents strong fluctuations, due to the fact that only one pair of velocity fields is considered. A smoother curve would have required a statistical average on several pairs, which would have been too costly. Nevertheless the results indicate a growth of the error, which saturates together with the vorticity thickness. Thus the error saturation is clearly a domain effect, and a larger domain (within the temporal approximation) or the loss of longitudinal periodicity (in the spatial problem) would probably yield a growth of unpredictability accompanying the spreading of the layer. This error rate, as shown in Lesieur et al. (1987), exhibits two distinct exponential growths: the first one, up to $t = 25\ \delta_i/U$, corresponds to

$$r(t) = r(0) \exp \frac{Ut}{3\ \delta_i} \qquad (3-3)$$

Afterwards, the exponential growth is much gentler: for a layer of thickness δ at a given time, the characteristic time of the exponential is of $15\ \delta/U$.

4 GROWTH OF THREE-DIMENSIONALITY

The preceding results concern a strictly two-dimensional analysis. We intend now to show how this predictability analysis allows to study some three-dimensional processes developing in the mixing layer. More details can be found in Lesieur et al. (1987).

Let us first consider a three-dimensional velocity field of the form

$$\underline{u}(x, y, z, t) = \underline{u}_{2D}(x, y, t) + \sqrt{2}\underline{u}_{3D}(x, y, t) \sin k_z z \qquad (4-1)$$

where $\underline{u}_{2D}(x, y, t)$ and $\underline{u}_{3D}(x, y, t)$ are two non divergent fields parallel to the (x, y) plane. Considering for instance a quasi two-dimensional Kelvin-Helmholtz billow with

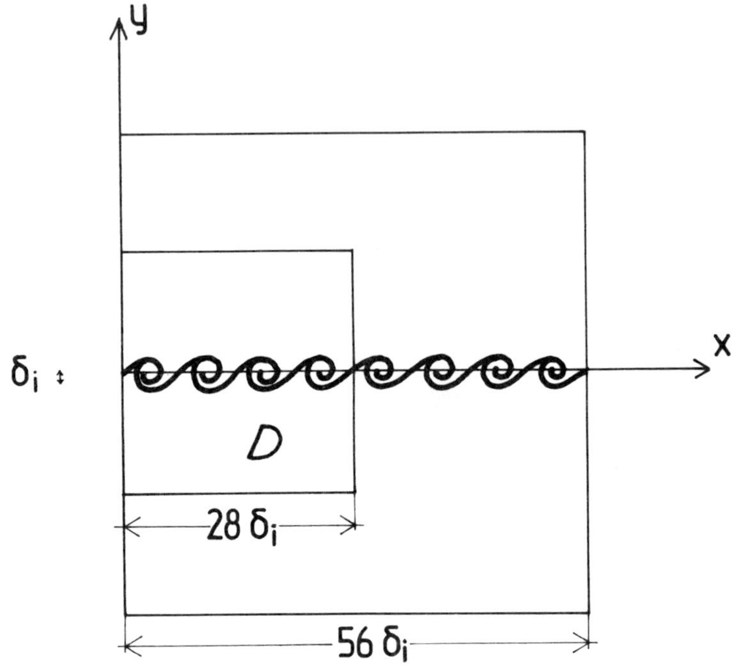

Fig. 13: calculations involving respectively 4 and 8 fundamental eddies.

a sine like oscillation in the spanwise direction z, the two fields correspond to a basic two-dimensional flow and the amplitude of the spanwise oscillation. If one considers a similar expansion for the pressure field, projects the three-dimensional Navier-Stokes equations on these two modes, and truncates by discarding the terms corresponding to the harmonic $2k_z$, it is found (Lesieur et al., 1987) that the two fields

$$\underline{u}^{(1)}(x,y,t) = \underline{u}_{2D} + \underline{u}_{3D}, \; \underline{u}^{(2)}(x,y,t) = \underline{u}_{2D} - \underline{u}_{3D} \qquad (4-2)$$

satisfy two *independant* Navier-Stokes equations. Thus, for a particular velocity field of the form (4-1), the growth of the error energy defined in (3-1) can be identified with the growth of the three-dimensional kinetic energy $\int_D \underline{u}_{3D}^2$, since $\underline{u}_{3D} = [\underline{u}^{(1)} - \underline{u}^{(2)}]/2$. It can then be concluded, from the results of the preceding section, that a spanwise sine perturbation of a Kelvin-Helmholtz billow will amplify as long as the vorticity thickness grows. It is possible that the development of this three-dimensional instability will first be responsible for the formation of the longitudinal streamwise vortices observed in several experiments[6] (see e.g. Bernal, 1981), then will cause a breakdown of the billow into three-dimensional turbulence. As envisaged in Lesieur et al. (1987), it is possible that the linear instability of the mean velocity profile will then recreate a two-dimensional Kelvin-Helmholtz billow: indeed, the role of

[6] But of course the two-mode expansion (4-1) is unable to take these vortices into account.

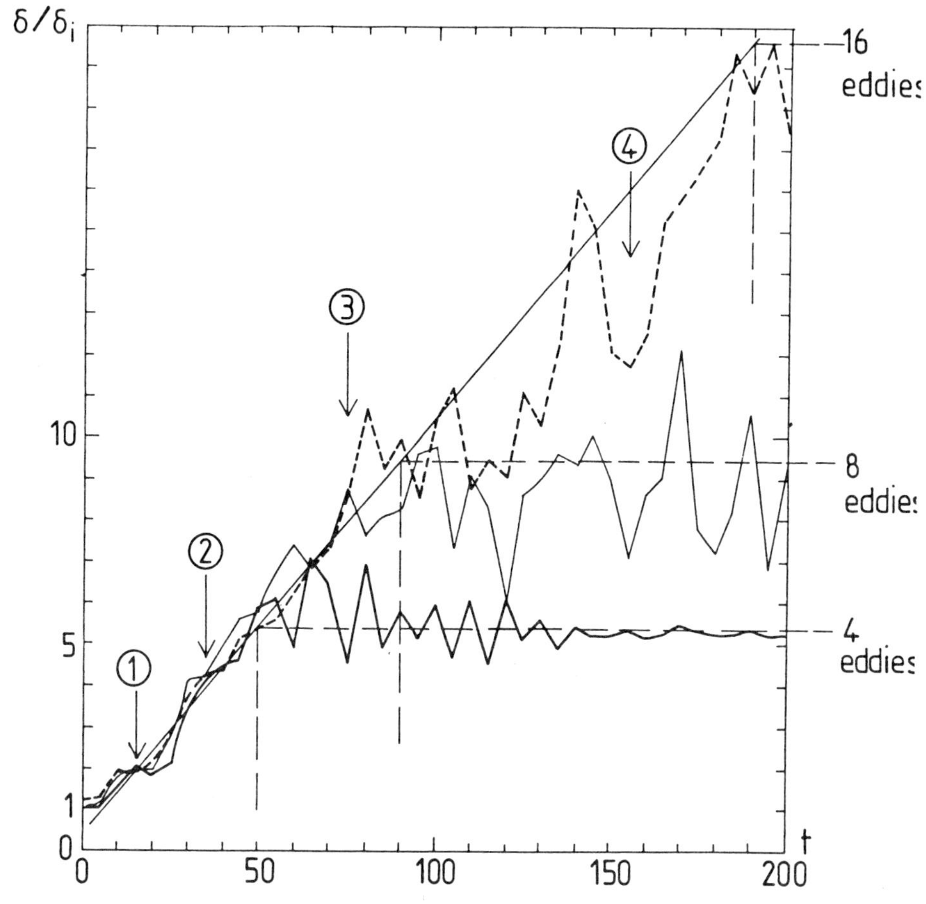

Fig. 14: vorticity thickness evolution in the 4, 8 and 16-eddy calculations (R = 10 000). Arrows indicate the formation of the fundamental vortices (1), and three successive pairings [(2), (3) and (4)] (from Lesieur et al., 1987).

three-dimensional turbulence upon the mean velocity shear (which is inflexional) can be described with the aid of an eddy-viscosity still allowing the linear instability to develop. Thus, for a layer of vorticity thickness δ, and because of the perturbation due to the three-dimensional turbulence itself, a Kelvin-Helmholtz eddy of wave length 7δ will be recreated. This could explain the persistence, extremely far downstream, of the coherent structures found in the experiments.

5 CONCLUSION AND DISCUSSION

We have developed direct or large-eddy simulations of large Reynolds number two-dimensional free shear flows such as mixing layers and plane jets (or wakes), which show that coherent structures similar to those observed experimentally may result, through linear instability mechanisms, from the development of tiny random perturbations superimposed upon the basic inflexional velocity profile. The initial randomness of the perturbation gives to the flow a disordered and chaotic character,

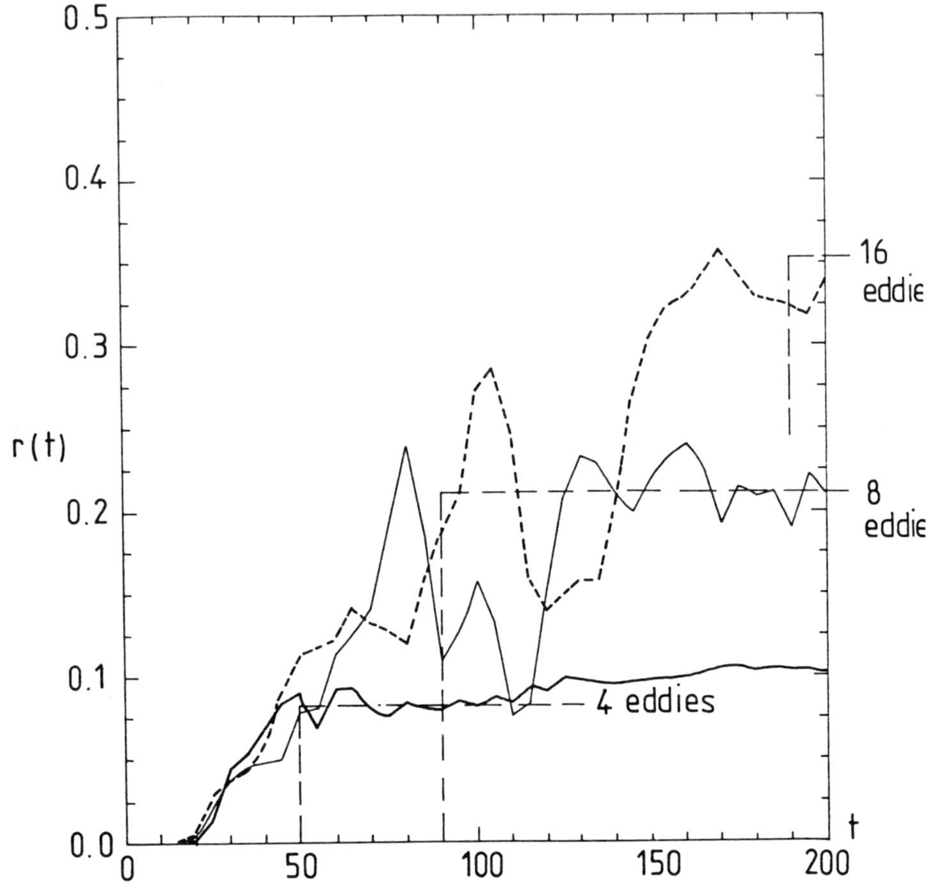

Fig. 15: error rate in the same calculation as in Figure 14. (from Lesieur et al., 1987).

characterized by an unpredictable behaviour with an exponential growth of the initial uncertainty. This permits to consider the coherent structures of these flows as a field of two-dimensional turbulence. The development of this turbulence can be inhibited (or at least delayed) in the case of the Bickley jet initially forced at the fundamental frequency.

We have also shown that the growth of two-dimensional unpredictability implies the development of a particular three-dimensional spanwise perturbation of a Kelvin-Helmholtz billow. In this sense, a quasi two-dimensional coherent structure will eventually return to three-dimensionality and breakdown into three-dimensional turbulence. Finally, a two-dimensional Kelvin-Helmholtz billow can be recreated by the instability of the mean velocity profile.

The result of these various processes is a superposition of two-dimensional and three-dimensional turbulence, which might mix the quantities transported by the flow in a very different way. It is certainly important, when considering turbulent combustion in these free shear flow, to keep in mind the existence of these two types of dynamics.

AKNOWLEDGEMENTS
This work was suported by the D.R.E.T. under contract 85 220, and by the C.N.E.S.-A.M.D. under contract 650 577.

REFERENCES

Arakawa, A., 1966, "Computational design for long term numerical integration of the equations of fluid motion: two-dimensional incompressible flow", J. Comp. Phys., **1**, pp 119-143.

Babiano, A., Basdevant, C., Legras, C. and Sadourny, R., 1987, "Vorticity and passive scalar dynamics in two-dimensional turbulence", J. Fluid Mech., in Press.

Bergé, P., Pomeau, Y. and Vidal, C., 1984, "L'ordre dans le chaos", Hermann, Paris.

Bernal, L., 1981, "The coherent structures of turbulent mixing-layers", PhD, Calif. Inst. for Technology, Pasadena.

Betchov, R. and Szewczyk, G., 1963, "Stability of a shear layer between parallel streams", Phys. Fluids, **6**, pp 1391-1396.

Bickley, W.G., 1937, "The plane jet", Phil. Mag., S7, Vol. 23, **156**, pp 727-731.

Browand, F.K. and Ho, C.M., 1983, "The mixing layer: an example of quasi two-dimensional turbulence", J. Mec. Theor. Appl., Special issue, "Two-dimensional turbulence", R. Moreau ed., pp 99-120.

Brown, G.L. and Roshko, A., 1974, "On density effects and large structure in turbulent mixing layers", J. Fluid Mech., **64**, pp 775-816.

Comte, P., Lesieur, M. and Chollet, J.P., 1987, "Simulation numérique d'un jet plan turbulent", preprint.

Couder, Y., Basdevant, C. and Thomé, H., 1984, "Sur l'apparition de couples solitaires de tourbillons dans les sillages bidimensionnels turbulents", C.R. Acad. Sci., Paris, ser II, **t 299**, pp 89-94.

Drazin, P.G. and Reid, W.H., 1981, "Hydrodynamic stability", Cambridge University Press.

Kelly, R.E., 1967, "On the stability of an inviscid shear layer which is periodic in space and time", J. Fluid Mech., **27**, pp 657-689.

Kline, S.J. , Reynolds, W.C., Schraub, F.A. and Runstadler, P.W., 1967, "The structure of turbulent boundary layers", J. Fluid Mech., **30**, pp 741-773.

Kraichnan, R.H., 1967, "Inertial ranges in two-dimensional turbulence", Phys. Fluids, **10**, pp 1417-1423.

Laverdant, A. and Candel, S., 1987, "Etude d'une flamme de diffusion dans un tourbillon à noyau visqueux", present proceedings.

Leith, C.E. and Kraichnan, 1972, "Predictability of turbulent flows", J. Atmos. Sci., **29**, pp 1041-1058.

Lesieur, M., 1987, "Turbulence in fluids", Nijhoff Publishers, Dordrecht.

Lesieur, M. and Herring, J.R., 1985, "Diffusion of a passive scalar in two- dimensional turbulence", J. Fluid Mech., **161**, pp 77-95.

Lesieur, M., Staquet, C., Le Roy, P. and Comte, P., 1987, "The mixing layer and its coherence examined from the point of view of two-dimensional turbulence", preprint.

Linan, A., 1987, "Some aspects of the hydrodynamics of diffusion flames", these proceedings.

Lorenz, E.N., 1963, "Deterministic non periodic flow", J. Atmos. Sci., **20**, pp 130-141.

Mc Williams, J.C., 1984, "The emergence of isolated coherent vortices in turbulent flows", J. Fluid Mech., **146**, pp 21-43.

Métais, O. and Lesieur, M., 1986, "Statistical predictability of decaying turbulence", J. Atmos. Sci., **43**, pp 857-870.

Michalke, A., 1964, "On the inviscid instability of the hyperbolic tangent velocity profile", J. Fluid Mech., **19**, pp 543-556.

Perry, A.E., Chong, M.S. and Lim, T.T., 1982, "Vortices in turbulence", in "Vortex motion", pp 106-121, Proceedings of the 75th anniversary of the Aerodynamische Versuchsanstalt, Goettingen, edited by H.G. Hornung and E.A. Muller.

Thompson, P.D., 1984, "A review of the predictability problem", in "Predictability of fluid motion", La Jolla Institute 1983, AIP Conference Proceedings **106**, edited by G. Holloway and B.J. West, pp 1-10.

DIFFUSION FLAME ATTACHMENT AND FLAME FRONT PROPAGATION ALONG MIXING LAYERS

A. Liñán

E.T.S.Ingenieros Aeronáuticos
Universidad Politécnica de Madrid
Pza. Cardenal Cisneros 3
28040-Madrid (SPAIN)

1. INTRODUCTION

In diffusion flames mixing of the reactants takes place simultaneously with the chemical reaction. If the Reynolds number is large mixing takes place only in thin layers, which in most cases of interest are also vorticity layers, and thereby subject to the Helmholte-Kelvin instabilities that lead to turbulence. When a diffusion flame is established in the mixing layer surrounding a fuel jet, the heat release due to the chemical reaction leads to changes in the flow structure and to a lowering of the effective Reynolds number around the flame sheet, so that it is often possible to observe very beautiful coherent structures. See, for example, Dumont and Borghi (1986).

The diffusion flame, under stationary conditions, is anchored close to the splitter plate, or injector lip, only for low values of the fuel jet velocity. If the splitter plate is cold the reaction is quenched close to the plate, and there is a small region ahead of the flame where the fuel and oxygen mix and are heated by upstream heat conduction from the flame. The size of the region of flame attachment is such that the Reynolds number, $U_N\delta_N/\nu_\infty$, based on the characteristic size δ_N, typical local flow velocity U_N, and initial kinematic viscosity ν_∞ of the air, is of order one. This is so as to allow for a balance of convection and heat conduction. For the typically large values of the Reynolds numbers, based on the fuel jet radius a, and maximum velocity U_1, which are encountered in practice the flame attachment region is very small compared with a, and it is embedded in the wake region of the splitter plate.

Due to the large sensitivity of the reaction rates with temperature, in order to establish a flame in the attachment region, the mixture must be ignited. This is so, because the spontaneous ignition length of the fuel air mixture is of the order of U_1 times the adiabatic explosion time, evaluated for the stoichiometric mixture of the fuel with air at their initial temperatures T_∞, and this is too long for the reaction to occur in the combustion chamber. Therefore, ignition must be accomplished artificially by local heating of the mixture; then propagation along the mixing layer must take place until the flame is anchored at the attachment region of the splitter plate, if the jet velocity is smaller than a critical value of flame detachment; for higher values the flame will be lifted to a lift-off distance, see Peters (1984).

2. CURVED FLAME FRONT PROPAGATION

In order to describe the process of flame propagation along mixing layers, Dold (1987) has presented an asymptotic analysis based on the large typical values of the activation energy of the reaction. Even though the effects of the thermal expansion are not accounted for in Dold's original analysis, it can be generalized to include these important effects.

C.-M. Brauner and C. Schmidt-Lainé (eds.), Mathematical Modeling in Combustion and Related Topics, 151–154.

When the flame in established in a mixing layer, where the mixture fraction Z varies from 0, on the air side, to 1, on the fuel side, it becomes curved because it moves faster along the stoichiommetric surface $Z=Z_s$ than on the air side, where it is lean, or on the fuel side, where it is rich. Notice that a diffusion flame, lying along the stoichiommetric surface $Z=Z_s$ is left behind the curved flame front. In this flame the temperature is the adiabatic flame temperature T_f for the stoichiommetric mixture fuel and air, if the Lewis numbers are equal to one.

The premixed flame temperature T_b drops rapidly from T_f, when moving away from the stoichiommetric surface. If the Zeldovich number $\beta=E/RT_f$ is large, the local flame velocity decreases by a factor of order unity when the variation of Z from Z_s is of order $1/\beta$, and the flame bends backwards rapidly in a distance δ_m/β. This is the characteristic size of the nose region of the curved flame front, small compared with the characteristic width δ_m of the mixing layer.

If we use a reference system moving with the curved flame front the process appears steady and quasi two-dimensional. Even when using the thermal diffusive model, the analysis becomes very complicated if one looks, as Dold does at the distinghished limit for which δ_m/β is of the order of the characteristic preahet zone δ_p, of the stoichiommetric flame (of velocity U_p such that $U_p\delta_p/\nu_\infty=1$). If we consider the flame front propagation process for values of $\delta_m/\beta\delta_p>>1$ viscous and diffusion effects can be neglected outside the thin curved flame, and flow deflection, due to the thermal expansion in the flames, can be accounted for in the analysis. The density can be considered constant, equal to ρ_u, upstream of the flame, and also downstream, where the changes in ρ from the adiabatic flame value ρ_f are of order ρ_f/β.

Similarly, we can also neglect the changes in the flow velocity, relation to the front, of the incoming stream from its value U associated with the stoichiommetric surface, because they are $1/\beta$ times the velocity difference accross the mixing layer. Hence the flow upstream of the flame front is irrotational with a stream function ψ that satisfies the equation

$$\Delta\psi = 0. \tag{1}$$

The mass fraction of fuel and oxidizer as well as the mixture fraction are conserved along the streamlines, and are therefore known linear functions of ψ determined by the incoming upstream state. The temperature and the stagnation pressure are also constant ahead of the flame front.

Downstream of the flame front there is vorticity ω generated at the flame, which is conserved together with the stagnation pressure along the streamlines, so that the stream function ψ satisfies the equation

$$\Delta\psi = -\omega(\psi). \tag{2}$$

The function $\omega(\psi)$ is obtained from the jump conditions accross the flame, in terms of the local normal flame propagation velocity u_n, a known function of the local upstream composition and therefore of ψ.

The resulting non-linear problem includes the flame front position and front velocity U as unknowns to be obtained as part of the solution.

As a consequence of the gas expansion associated with the moderately large value of ρ_u/ρ_f the streamlines diverge before approaching the flame, and the velocity along the stoichiommetric surface is reduced from U to u_p at the flame.

The ratio $U/u_p>1$ is only a function of Z_s and the density ratio ρ_u/ρ_f. If the flame front velocity U is larger than the local velocity at the stoichiommetric line the front will move upstream, otherwise it will be locally blown down. In the first case there is the possibility that it will move upstream until it reaches the splitter plate or lip of the injector.

3. STRUCTURE OF THE FLAME ATTACHMENT REGION

As indicated before, the size of this region δ_N is such that the local Reynolds number is of order unity, and the complete quasi-steady form of the Navier-Stokes equations, augmented by the conservation equations of energy and species, must be used for the description of the flow in the region, and therefore of the flame attachment process. However, for the typically large values of the Reynolds number $U_1 a/\nu_\infty$, this region is embeded in a much larger region where the flow can be described using the boundary layer approximation, and the Burke-Schumann limit of infinite reaction rates can be used to describe the diffusion controlled combustion. See Williams (1985).

The structure of the outer region is slightly different in the cases where there is a non-zero flow velocity U_0 on the air side, and the case where the air is originally stagnant, see Daniels (1977). In the first case the boundary layer approximation fails in a tripple deck region, close to the splitter plate end but external to the Navier-Stokes region, where the pressure decreases below the ambient. In the second case this lowering of the pressure in the near wake region is not so strong, and matching of the Navier-Stokes (N-S) region with the outer boundary layer region is possible. It must be pointed out that in this region, the viscous and diffusion effects are confined, just downstream of the plate, to a thin mixing sublayer placed along the deviding streamline that without diffusion would separate the fuel and air. The velocity along this line and the thickness of the mixing layer grow, in this Goldstein region, as the cubic root of the downstream distance.

The size δ_N of the Navier-Stokes region is related to the size δ_B of the fuel boundary layer when it reaches the splitter plate end. This is equal roughly to a if the flow is of the Poisseuille type upstream, otherwise $\delta_B \ll a$. When leaving the (N-S) region toward the fuel side the velocity should grow at the same linear rate shown by the fuel boundary layer close to the wall. On the air side of the (N-S) region the longitudinal velocity component should tend to zero, if the air is stagnant. These matching conditions allow us to obtain the scales for the velocity U_N and length δ_N of the Navier Stokes region. Namely,

$$U_N/U_1 = \delta_N/\delta_B \tag{3}$$

which together with the relation

$$U_N \delta_N/\nu_\infty = 1 \tag{4}$$

determines both U_N and δ_N.

When comparing the residence time, δ_N/U_N, in this region with the characteristic reaction time at the adiabatic flame temperature T_f, or, equivalently, with the residence time accross the stoichiommetric planar flame δ_p/U_p, we obtain a Damköhler number D. This, together with the Zeldovich number $\beta = E/RT_f$ and the temperature ratio T_f/T_∞, are the main parameters left, when the conservation equations and boundary conditions are written using U_N and δ_N as scales.

One can anticipate that the solution of the resulting problem is multivalued for $\beta \gg 1$, showing the existence of two solutions with attached flames for values of D larger than a critical value D_c. If the plate is not thermally insulated the flame will be quenched upstream of a flame standoft distance from the plate δ_q of the order of δ_N, which is a function of D of the form shown in the figure.

The branch corresponding to the larger values of δ_q may correspond to a flame front anchored at the Goldstein region, but it is unstable.

The details of the analysis sketched here will be presented in a paper under preparation to be published elsewhere.

ACKNOWLEDGMENT

This research has been partially sponsored by the Spanish CAICYT under Project no. 2291-83.

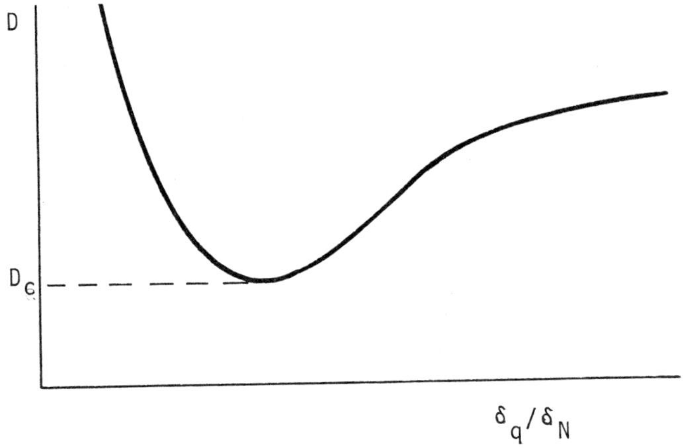

REFERENCES

1. Dumont, J.P. and Borghi, R., A Qualitative Study by Laser Tomography of the Structure of Turbulent Flames. Combust. Sci. and Techn., Vol. 48, p. 107 (1986).
2. Peters, N., Laminar Diffusion Flamelet Models in Non-premixed Turbulent Combustion. Prog. Energy Combust. Sci., Vol. 10, p. 319 (1984).
3. Dold, J.W., Flame Propagation in a Non-uniform Mixture: Analysis of a Slowly-Varying Tripple Flame. Bristol Univ. Report No. AM-87-02 (1987).
4. Williams, F.A., Combustion Theory, Benjamin Cunmings (1985).
5. Danields, P.G., Viscous Mixing at a Trailing Edge, Quart, Jl. Mech. Appl. Math., Vol. 30, p. 319 (1977).

SOME REMARKS ON TURBULENT COMBUSTION FROM THE ATTRACTOR POINT OF VIEW.

Martine MARION and Roger TEMAM

Laboratoire d'Analyse Numérique, Université Paris-Sud, Bâtiment 425, 91405 Orsay (France).

INTRODUCTION.

The constant-density model is derived from the basic equations of small Mach number combustion by neglecting the thermal expansion of the gas. This yields a fundamental simplification of the equations since the hydrodynamic equations may then be solved independently of the rest of the equations. The gas velocity turns out to be determined as the solution of the incompressible Navier-Stokes equations. This simplified model has allowed many interesting phenomena in flame propagation to be described (see for instance Sivashinsky [11], Joulin-Clavin [6]).

In this work, we shall consider a one step chemistry and a two dimensional geometry. Our aim is to study the long time behaviour of the solutions of the constant-density model on the point of view of the existence of an attractor and its dimension. This kind of question has been studied for various types of equations, among them the Navier-Stokes equations (Babin-Vishik [1], Constantin-Foias-Temam [4]) and the Kuramoto--Sivashinsky equations (Nicolaenko-Scheurer-Temam [10]). In particular, it was shown in [4] that the bound on the dimension of the Navier-Stokes attractor is the same as that of the number of degrees of freedom of a turbulent flow, according to the Kolmogorov theory of turbulence. Hence, mathematically sharp bounds on the dimension of the attractors describing turbulent flows can yield physically relevant information on the complexity of the flow. We apply here the same ideas to the constant-density model and this work may be viewed as a first step in the study of more complex situations in turbulent combustion.

The paper is organized as follows. In Section 1 we first assume that the mixture is at rest ; this leads to a system of reaction-diffusion equations known as the thermo-diffusive model. We prove for this problem the existence of an attractor with a finite dimension. We then recall in Section 2 some results on attractors describing turbulent flows governed by the incompressible Navier-Stokes equations. Finally, in Section 3, we consider the constant-density model ; in particular we derive there a bound on the dimension of the corresponding attractor which is the sum of that for the thermo-diffusive model and that for the Navier-Stokes equations. The final form of the estimate appears in (3.13) (see also (3.12)). Depending on the explicit values of the parameters it permits to compare the contribution (to the bounds) of the dynamical and reacting effects.

The authors wish to thank P. Clavin and O. Manley for their interest in this work and for helpful comments on the physics aspect.

C.-M. Brauner and C. Schmidt-Lainé (eds.), Mathematical Modeling in Combustion and Related Topics, 155–172.

CONTENTS.

1. THE THERMO-DIFFUSIVE MODEL.

1.1. The equations.

We consider a flame propagating through a gas *at rest* whose density is assumed to be constant. We consider a one-step chemical mechanism : $mR \rightarrow mP$. Such a flame is fully described by the following non-dimensional system of two reaction-diffusion equations (see Buckmaster-Ludford [2])

$$(1.1)\qquad \frac{\partial\theta}{\partial t} - \Delta\,\theta = \omega(\theta,\Psi),$$

$$(1.2)\qquad \frac{\partial\Psi}{\partial t} - \frac{1}{Le}\,\Delta\Psi = -\,\omega(\theta,\Psi),$$

where the unknowns are the reduced temperature θ and the reduced concentration of the reactant Ψ ; Le is the Lewis number ([1]) and the normalized reaction rate ω is given by the Arrhenius law

$$(1.3)\qquad \omega(\theta,\Psi) = \frac{\beta^{m+1}}{2m!Le^{m}}\,\Psi^{m}\exp\left(-\,\frac{\beta(1-\theta)}{1+\gamma(\theta-1)}\right),$$

where $m \in \mathbb{N}^*$ is the order of the reaction, $\beta > 0$ is the reduced activation energy and $\gamma = 1 - T_u/T_b$ ([2]) is a heat release parameter.

We consider the reactive flow in a finite tube represented by a bounded two dimensional domain $\Omega =]0,a[\times]0,b[$ with the following boundary conditions (see Figure 1)

$$(1.4)\qquad \theta = 0 \text{ and } \Psi = 1 \text{ at } x_1 = 0 \quad \text{(unburnt gas)}$$

$$(1.5)\qquad \theta = 1 \text{ and } \Psi = 0 \text{ at } x_1 = a \quad \text{(burnt gas)}$$

$$(1.6)\qquad \frac{\partial\theta}{\partial x_2} = \frac{\partial\Psi}{\partial x_2} = 0 \text{ at } x_2 = 0 \text{ and } x_2 = b \quad \text{(adiabatic wall).}$$

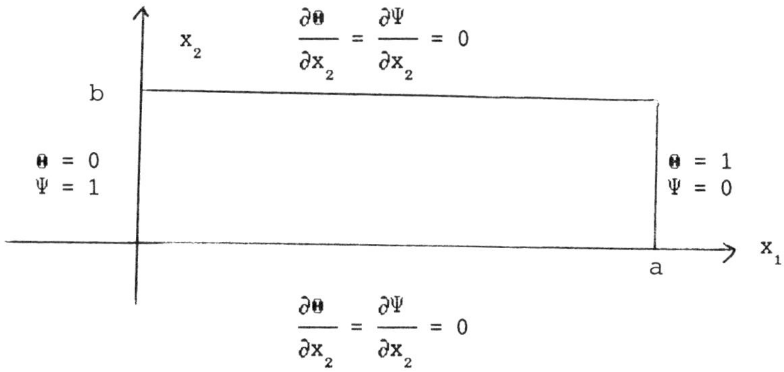

Figure 1

Finally, we supplement (1.1)-(1.2) (1.4)-(1.6) with the initial conditions

(1.7) $\quad \boldsymbol{\theta}(x,0) = \boldsymbol{\theta}_0(x), \quad \Psi(x,0) = \Psi_0(x).$

Before recalling the existence and uniqueness results for the initial-boundary value problem (1.1)-(1.2) (1.4)-(1.7), we introduce a slightly modified formulation of the problem. Let δ and δ_1 be two functions defined by

(1.8) $$\delta(x) = \frac{x_1}{a}, \quad \delta_1(x) = 1 - \frac{x_1}{a} = 1 - \delta(x).$$

Then the problem (1.1)-(1.2), (1.4)-(1.7) can be rewritten in terms of the new unknowns

$$\theta(x,t) = \boldsymbol{\theta}(x,t) - \delta(x), \quad \psi(x,t) = \Psi(x,t) - \delta_1(x),$$

as

(1.9) $$\begin{cases} \dfrac{\partial \theta}{\partial t} - \Delta\theta = \omega(\theta + \delta, \psi + \delta_1), \\ \dfrac{\partial \psi}{\partial t} - \dfrac{1}{Le}\Delta\psi = -\omega(\theta + \delta, \psi + \delta_1), \end{cases}$$

(1.10) $$\theta(x,0) = \theta_0(x) = \boldsymbol{\theta}_0(x) - \delta(x), \quad \psi(x,0) = \psi_0(x) = \Psi_0(x) - \delta_1(x),$$

(1.11) $$\begin{cases} \theta = \psi = 0 \quad \text{at } x_1 = 0 \text{ and } x_1 = a, \\ \dfrac{\partial \theta}{\partial x_2} = \dfrac{\partial \psi}{\partial x_2} = 0 \quad \text{at } x_2 = 0 \text{ and } x_2 = b. \end{cases}$$

Mathematical setting.

The existence and uniqueness results for this problem are classical (see [3,9]). There exists a positively invariant region

$$\mathcal{D} = \{(\theta,\Psi),\ \theta \geqslant 0,\ 0 \leqslant \Psi \leqslant 1\}.$$

This means that if $\theta_0(x) \geqslant 0$ and $0 \leqslant \Psi_0(x) \leqslant 1$, then these inequalities remain true for all later time t for which the solution of (1.1) exists. Thus, it follows that, given $(\theta_0,\psi_0) \in H_1$ where (3)

$$(1.12)\quad H_1 = \{(\theta,\psi) \in L^2(\Omega)^2,\ (\theta+\delta,\psi+\delta_1) \in \mathcal{D} \text{ for a.e. } x \in \Omega\},$$

then there exists a unique solution $(\theta,\psi) \in \mathcal{C}(\mathbb{R}^+,H_1)$ of Problem (1.9)-(1.11).

The dynamical system.

We introduce the family of nonlinear operators $S_1(t)$, $t \geqslant 0$, from H_1 into H_1 defined by

$$S_1(t) : (\theta_0,\psi_0) \longrightarrow (\theta(t),\psi(t)).$$

These operators are continuous and satisfy the usual semi-group properties

$$\begin{cases} S_1(0) = Id, \\ S_1(t+s) = S_1(t) \circ S_1(s), \quad \forall\, t,s \geqslant 0. \end{cases}$$

1.2. Existence of a maximal attractor.

We first recall some general definitions and results. Let H be a Hilbert space and S(t), $t \geqslant 0$, be a semi-group of operators from H into itself.

A set $\mathcal{X} \subset H$ is a *functional invariant set* for the semi-group S(t) if

$$S(t)\mathcal{X} = \mathcal{X}, \quad \forall\, t \geqslant 0 .$$

The *maximal attractor* $\mathcal{A}$, when it exists, is the unique compact invariant set which attracts every bounded set in H, i.e. for any bounded set $\mathcal{B}$

$$\lim_{t \to +\infty} \text{dist}\,(S(t)\mathcal{B},\mathcal{A}) = 0$$

where, for two bounded sets $\mathcal{B}_1$ and $\mathcal{B}_2$ of H,

$$\text{dist}\,(\mathcal{B}_1,\mathcal{B}_2) = \sup_{x \in \mathcal{B}_1} \inf_{y \in \mathcal{B}_2} d(x,y)$$

where d(.,.) denotes the distance on H.

A set $\mathcal{B} \subset H$ is said to be *absorbing in H* for the semi-groupe S(t) if, for every bounded set $\mathcal{B}_0$ of H, there exists $T = T(\mathcal{B}_0)$ such that $S(t)\mathcal{B}_0 \subset \mathcal{B}$, $\forall\, t \geqslant T(\mathcal{B}_0)$.

We can now state a very useful result ensuring the existence of an attractor (see for instance [13] :

THEOREM 1.1.

If the semi-group $S(t)$ is such that

(1.13) *there exists an absorbing set $\mathcal{B}$,*

(1.14) *there exists $t_0 > 0$ such that $S(t_0)$ is compact,*

then there exists a maximal attractor.

We now return to the semi-group $S_1(t)$ associated to the thermo--diffusive model.

THEOREM 1.2.

For the semi-group $S_1(t)$ associated to (1.9)-(1.11), there exists a maximal attractor $\mathcal{A}_1$ which is bounded in $H^1(\Omega)^2$ [4].

This result is proved by applying Theorem 1.1 to the semi-group $S_1(t)$. One checks that $S_1(t)$ satisfies (1.13) and (1.14) thanks to uniform in time a priori estimates on the solutions of (1.9)-(1.11).

1.3. *Estimate of the fractal dimension of the attractor.*

We first recall the definition of the fractal dimension of a compact set $\mathcal{C}$ in a metric space H. Denoting by $n_{\mathcal{C}}(\epsilon)$ the minimum number of balls of radii ϵ necessary to cover $\mathcal{C}$, the capacity or fractal dimension of $\mathcal{C}$ (Mandelbrot [8]) is defined by

$$(1.15)\quad \dim_M(\mathcal{C}) = \limsup_{\epsilon \to 0} \frac{\log n_{\mathcal{C}}(\epsilon)}{\log 1/\epsilon} .$$

This definition is a natural extension of the notion of euclidian dimension in the sense that for smooth manifolds (curves, surfaces, etc.) in euclidian spaces, (1.15) coïncides with the usual dimension.

The next theorem gives an upper bound on the fractal dimension of the thermo-diffusive attractor. It shows in particular that the long-time behaviour of the reactive flow is governed by a finite number of parameters. We set

$$(1.16)\quad \alpha_1 = \max\left(\sup_{\substack{\theta \geqslant 0 \\ 0 \leqslant \Psi \leqslant 1}} \left|\frac{\partial\omega}{\partial\theta}\right| , \sup_{\substack{\theta \geqslant 0 \\ 0 \leqslant \Psi \leqslant 1}} \left|\frac{\partial\omega}{\partial\Psi}\right| \right)$$

$$= \frac{\beta^{m+1}}{2m!Le^m} \exp\left(\frac{\beta}{\gamma}\right) \max\left(m , \frac{4\gamma^2}{\beta} \exp(-2)\right).$$

THEOREM 1.3.

The fractal dimension of the thermo-diffusive attractor $\mathcal{A}_1$ is finite. Moreover it is bounded by

$$(1.17)\quad c\,\{1 + ab\,\alpha_1 \max(1,Le)\},$$

where c denotes a constant depending only on the shape factor b/a and α_1 is given by (1.16).

The proof of this result (and of (1.19) below) can be found in Marion [9] where more general reaction-diffusion systems are considered. It relies on a general method of Constantin-Foias-Temam [4] that will be partially recalled in Section 3.3 below.

To get a better insight in the estimate (1.17), let us recall the order of magnitude of the different parameters. The Lewis number Le, the order of the reaction m and the volume of the rectangle Ω are of order unity while the activation energy β is large. Hence, (1.17) can be rewritten (see (1.16)) as

$$(1.18)\quad \dim_M \mathcal{A}_1 \leqslant c\,\beta^{m+1}\exp\left(\frac{\beta}{\gamma}\right).$$

Typical values such as $\beta = 20$, $\gamma = 0{,}8$ produce already a value of the right hand side of (1.18) which is fairly large.

Remark 1.4. A similar analysis can be carried out for a reactive flow in an open bounded set Ω of $\mathbb{R}^3$. Then the fractal dimension of the corresponding thermo-diffusive attractor is bounded by

$$(1.19)\quad c(1 + |\Omega|\,\alpha_1^{3/2} \max(1,Le)^{3/2}),$$

where c and α_1 are as in Theorem 1.3 and $|\Omega|$ denotes the volume of Ω.

On the contrary, the results of the following sections can not be generalized to the three dimensional case since the mathematical theory of the three-dimensional Navier-Stokes equations is not available.

2. SURVEY OF RESULTS ON THE 2D NAVIER-STOKES ATTRACTORS.

Assume that an homogeneous and incompressible Newtonian fluid fills a region $\Omega =]0,a[\times]0,b[$ of $\mathbb{R}^2$. Then the velocity $u = (u_1,u_2)$ and the pressure p are governed by the equations

$$(2.1)\quad \frac{\partial u}{\partial t} - \nu\,\Delta u + (u.\nabla)u + \nabla p = f,$$

$$(2.2)\quad \operatorname{div} u = 0,$$

where $\nu > 0$ is the kinematic viscosity and f represents a density of force per unit volume. We assume that f is time independent and satisfies $f \in L^2(\Omega)^2$.

We supplement (2.1)(2.2) with the initial condition

(2.3) $u(x,0) = u_0(x)$

and the boundary conditions (see Figure 2)

(2.4) $u = 0$ at $x_1 = 0$ and $x_1 = a$ (non slip)

(2.5) $\dfrac{\partial u_1}{\partial x_2} = u_2 = 0$ at $x_2 = 0$ and $x_2 = b$ (no penetration and no tangential stresses).

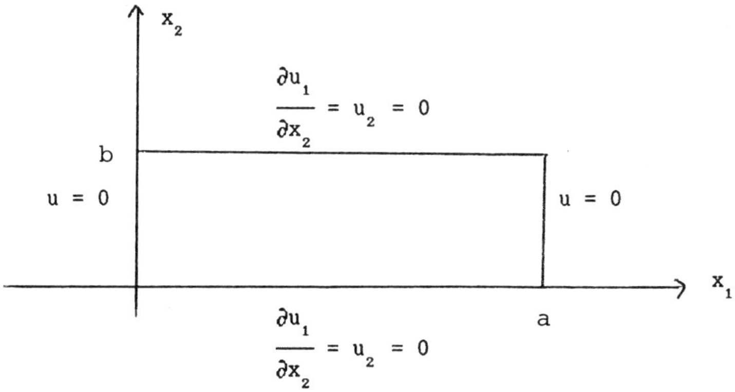

Figure 2.

The existence and uniqueness results for the 2D-Navier-Stokes equations are classical. One introduces

(2.6) $H_2 = \{u \in L^2(\Omega)^2 \ , \ \mathrm{div}\, u = 0 \ , \ u.\nu = 0\}$,

where ν = the unit vector normal to the boundary. Then, given $u_0 \in H_2$, there exists a unique solution $u \in \mathcal{C}(\mathbb{R}^+, H_2)$ of (2.1)-(2.5) (the pressure p being determined up to a constant). We refer to Temam [12] for more details.

The existence of a maximal attractor for Problem (2.1)-(2.5) was proved in Constantin-Foias-Temam [4] (see also Babin-Vishik [1]) and we now recall some results on the fractal dimension of this attractor.

Estimate of the fractal dimension in terms of the Reynolds number.

The Reynolds number is defined as follows

(2.7) $$\mathrm{Re} = \frac{1}{\nu\sqrt{\lambda_1}} \left(\int_\Omega |f(x)|^2 dx\right)^{1/4},$$

where λ_1 denotes the smallest eigenvalue of $-\Delta$ associated to the boundary conditions $u = 0$ at $x_1 = 0$ and $x_1 = a$ and then $\lambda_1^{-1/2}$ is a typical length-scale. The chosen typical velocity scale is

$$\left(\int_\Omega |f(x)|^2 dx\right)^{1/4} . \tag{2.8}$$

When the Reynolds number is small, Problem (2.1)-(2.5) possesses a unique stationary solution and all solutions of (2.1)-(2.5) converge exponentially as $t \longrightarrow +\infty$ towards this stationary solution (see Temam [12]). The attractor is trivial and reduces to the stationary solution. As the Reynolds number increases, it is observed experimentally that the stationary solution loses its stability and the behaviour of the flow may become more complicated, up to full chaos. Constantin-Foias-Temam [4] derived the following estimate of the fractal dimension of the attractor :

THEOREM 2.1.

The fractal dimension of the attractor associated to (2.1)-(2.5) is bounded by

$$c\, R_e^2, \tag{2.9}$$

where c denotes a constant depending only on the shape factor b/a.

Remark 2.3. In the definition of the Reynolds number (2.7) (and in (2.8)), one can replace f by Pf which is the projection in $L^2(\Omega)^2$ of f on the divergence free function space H_2 (see (2.6)).

Remark 2.4. In a particular case, Babin-Vishik [1] have obtained a *lower* estimate of the dimension of the attractor. These authors consider periodic boundary conditions on a rectangle $\Omega_\epsilon =]0,b/\epsilon[\times]0,b[$, with $\epsilon \ll 1$. Then

$$\dim_M \mathcal{A}_2 \geq c\, Re^{4/3}.$$

An estimate of the fractal dimension in connection with the theory of turbulence.

At each point $x \in \Omega$, at each time τ, we consider the rate of dissipation of energy $\epsilon(x,\tau) = \nu|\nabla u(x,\tau)|^2$ and the Kolmogorov dissipation length

$$\ell_d(x,\tau) = \left(\frac{\nu^3}{\epsilon(x,\tau)}\right)^{1/4} .$$

To these quantities we associate the number ϵ which is the time average of the supremum of the (local) dissipation rate of energy

$$\epsilon = \nu \limsup_{t \longrightarrow \infty} \left(\sup_{u_0 \in \mathcal{A}_2} \frac{1}{t} \int_0^t \sup_{x \in \Omega} \epsilon(x,\tau) d\tau \right),$$

and

$$\ell_d = \left(\frac{\nu^3}{\epsilon}\right)^{1/4}.$$

Let $\ell_0 = \lambda_1^{-1/2}$ be a typical macroscopic length. Then

THEOREM 2.2. (Constantin-Foias-Temam [4]).

The fractal dimension of the attractor associated to (2.1)-(2.5) is bounded by

$$c \left(\frac{\ell_0}{\ell_d}\right)^2, \tag{2.10}$$

where c denotes a constant depending only on the shape factor b/a.

This result establishes a connection between the mathematical theory of the Navier-Stokes equations and the conventional theory of turbulence where (2.10) appears as an heuristic estimate of the number of degrees of freedom of a turbulent flow ; cf. Landau-Lifschitz [7].

3. THE CONSTANT-DENSITY MODEL.

3.1. The equations.

For a one-step chemistry : mR $\longrightarrow$ mP, the low Mach number combustion equations reduce, when assuming the density constant, to the following non dimensional system (cf. Buckmaster-Ludford [2])

$$\frac{\partial u}{\partial t} - \Pr \Delta u + (u.\nabla)u + \nabla p = 0, \tag{3.1}$$

$$\operatorname{div} u = 0, \tag{3.2}$$

$$\frac{\partial \theta}{\partial t} - \Delta\theta + u.\nabla\theta = \omega(\theta,\Psi), \tag{3.3}$$

$$\frac{\partial \Psi}{\partial t} - \frac{1}{Le}\Delta\Psi + u.\nabla\Psi = -\omega(\theta,\Psi), \tag{3.4}$$

where the unknowns are the velocity u, the pressure p, the temperature θ and the concentration of the reactant Ψ. Here, the Prandtl number Pr is a positive constant (ratio of the kinematic viscosity to the thermal conductivity) ; the Lewis number Le and the Arrhenius term ω are as in Section 1.2 (see (1.3)).

The main feature of this system is that u and p are determined as the solutions of the Navier-Stokes equations (3.1)(3.2) independently of θ and Ψ. Hence, from the point of view of the combustion problem, they can be considered as given quantities and we shall study in the following the attractor of a turbulent flame whose turbulence is given.

Actually, we shall investigate a slightly different system where

(3.1) is replaced by

$$(3.1)' \quad \frac{\partial u}{\partial t} - \Pr \Delta u + (u.\nabla)u + \nabla p = f,$$

where f represents a density of force per unit volume. Indeed, we could have considered Problem (3.1)-(3.4) associated to a boundary condition $u = \Phi$ on $\partial\Omega$, but this leads because of mathematical technicalities to a high physically unrealistic estimate of the dimension of the attractor due to the lack of a complete mathematical understanding of the nonhomogeneous Navier-Stokes equations. Hence, we prefer to study here Problem (3.1)'-(3.4).

We consider the reactive flow in a bounded region $\Omega =]0,a[\times]0,b[$ and we associate to (3.1)'-(3.4) the boundary conditions introduced in Sections 1 and 2 (namely (1.4)(1.5)(1.6)(2.5)) except (2.4) which is replaced by

$$u = U = (U_1,0) \text{ at } x_1 = 0 \text{ and } x_1 = a,$$

where U_1 is some given constant ; thus $U = (U_1,0)$ represents the basic velocity of the flow of reactant entering the "combustion chamber" Ω.

Introducing the new unknowns

$$v = u - U, \ \theta = \boldsymbol{\theta} - \delta, \ \phi = \Psi - \delta_1,$$

where δ and δ_1 are given by (1.8), the problem can be rewritten : to find v,p,θ and ϕ satisfying

$$(3.5) \quad \begin{cases} \dfrac{\partial v}{\partial t} - \Pr \Delta v + (v.\nabla)v + (U.\nabla)v + \nabla p = f, \\ \\ \operatorname{div} v = 0, \end{cases}$$

$$(3.6) \quad \begin{cases} \dfrac{\partial \theta}{\partial t} - \Delta\theta + v.\nabla\theta + U.\nabla\theta + v.\nabla\delta = \omega(\theta+\delta,\phi+\delta_1) - U.\nabla\delta, \\ \\ \dfrac{\partial \phi}{\partial t} - \dfrac{1}{Le}\Delta\phi + v.\nabla\phi + U.\nabla\phi + v.\nabla\delta_1 = -\omega(\theta+\delta,\phi+\delta_1) - U.\nabla\delta_1, \end{cases}$$

together with the initial conditions

$$(3.7) \quad v(x,0) = v_0(x), \ \theta(x,0) = \theta_0(x), \ \phi(x,0) = \phi_0(x),$$

and the boundary conditions

$$(3.8)\qquad \begin{cases} v = \theta = \psi \text{ at } x_1 = 0 \text{ and } x_1 = a, \\ \dfrac{\partial v_1}{\partial x_2} = v_2 = \dfrac{\partial \theta}{\partial x_2} = \dfrac{\partial \psi}{\partial x_2} = 0 \text{ at } x_2 = 0 \text{ and } x_2 = b. \end{cases}$$

Combining the techniques of Sections 1 and 2, it is easy to derive for Problem (3.5)-(3.8) an existence and uniqueness result. More precisely, let H_3 be defined by

$$H_3 = H_2 \times H_1$$

where H_2 (resp. H_1) is given by (2.6) (resp. (1.12)). We assume that f is time-independent and $f \in L^2(\Omega)^2$. Then, given $(v_0, \theta_0, \psi_0) \in H_3$, there exists a unique solution $(v, \theta, \psi) \in \mathcal{C}(\mathbb{R}^+, H_3)$ of (3.5)-(3.8) (as usual the pressure p is determined up to a constant). Hence, one can associate to this problem a semi-group of continuous operators $S_3(t), t \geqslant 0$ defined by

$$S_3(t) : H_3 \longrightarrow H_3$$

$$(v_0, \theta_0, \psi_0) \longrightarrow (v(t), \theta(t), \psi(t)).$$

3.2. Estimate of the fractal dimension of the attractor.

The existence of a maximal attractor $\mathcal{A}_3$ for the semi-group $S_3(t)$ follows from Theorem 1.1 and a priori estimates on the solutions of (3.5)-(3.8) (details are omitted). We now give an estimate of its fractal dimension that will be proved in Section 3.3 below.

By analogy with (2.7), we define the number

$$(3.9)\qquad R = \frac{1}{Pr\sqrt{\lambda_1}} \left(\int_\Omega |f(x)|^2 dx \right)^{1/4},$$

where λ_1 denotes again the smallest eigenvalue of $-\Delta$ associated to the boundary conditions $u = 0$ at $x_1 = 0$ and $x_1 = a$. We also set

$$(3.10)\qquad \Lambda = \inf\left(Pr, 1, \frac{1}{Le}\right)$$

and

$$(3.11)\qquad \alpha_2 = \sup_{\substack{\boldsymbol{\theta} \geqslant 0 \\ 0 \leqslant \Psi \leqslant 1}} \omega(\boldsymbol{\theta}, \Psi) = \frac{\beta^{m+1}}{2m!Le^m} \exp\left(\frac{\beta}{\gamma}\right).$$

THEOREM 3.1.

The fractal dimension of the maximal attractor $\mathcal{A}_3$ describing the long time behaviour of the solutions of the constant-density model is bounded by

$$(3.12)\quad c\left[1+\frac{1}{\Lambda}\{\Pr R^2+\alpha_1 ab+\alpha_2(ab)^{3/2}+ab\max(1,Le)\}\right],$$

where c denotes a constant depending only on the shape factor b/a and Λ, R, α_1 and α_2 are respectively given by (3.10), (3.9), (1.16) and (3.11).

Remark 3.2. In order to simplify the estimate (3.12), we now recall the order of magnitude of the different parameters. The Prandtl number Pr is often of order unity and the number R is large while the orders of magnitude of the other parameters Le, m, ab, and β are as stated in Section 1.3 above. Hence, (3.12) yields

$$(3.13)\quad \dim_M \mathcal{A}_3 \leqslant c(R^2+\beta^{m+1}\exp(\frac{\beta}{\gamma})).$$

This bound of $\dim_M \mathcal{A}_3$ is the sum of the estimate (1.18) for the thermo-diffusive model and of the estimate (2.9) for the Navier-Stokes equations. For ordinary mixtures, β is about 20 and $\gamma \sim 0,8$. Thus, for an order one reaction, one has

$$\beta^{m+1}\exp(\frac{\beta}{\gamma}) \sim 10^{13}.$$

In conclusion, for this model, and unless the Reynolds number (= O(R)) exceeds $10^{6.5}$, the complexity of the flow seems to be predominantly due to the chemical effects. However for other values of the parameters the inverse can be true or the two factors can play competing roles.

3.3. Proof of Theorem 3.1.

In order to estimate the fractal dimension of the attractor $\mathcal{A}_3$, we use a general method of Constantin-Foias-Temam [4] which relies on estimates of the global Lyapunov exponents of the attractor. The Lyapunov exponents are directly related to the linearized flow around a solution (v,θ,ψ) of (3.5)-(3.8)

$$(3.14)\quad \frac{\partial w}{\partial t} - \Pr\Delta w + (w.\nabla)v + (v.\nabla)w + (U.\nabla)w + \nabla\Pi = 0,$$

$$(3.15)\quad \operatorname{div} w = 0,$$

$$(3.16)\quad \frac{\partial\xi}{\partial t} - \Delta\xi + v.\nabla\xi + w.\nabla\theta + U.\nabla\xi + w.\nabla\delta - \frac{\partial\omega}{\partial\theta}\xi - \frac{\partial\omega}{\delta\Psi}\eta = 0,$$

$$(3.17)\quad \frac{\partial\eta}{\partial t} - \frac{1}{Le}\Delta\eta + v.\nabla\eta + w.\nabla\psi + U.\nabla\eta + w.\nabla\delta_1 + \frac{\partial\omega}{\partial\theta}\xi + \frac{\partial\omega}{\partial\Psi}\eta = 0,$$

$$(3.18)\quad \begin{cases} w = \xi = \eta \text{ at } x_1 = 0 \text{ and } x_1 = a, \\ \dfrac{\partial w_1}{\partial x_2} = w_2 = \dfrac{\partial \xi}{\partial x_2} = \dfrac{\partial \eta}{\partial x_2} = 0 \text{ at } x_2 = 0 \text{ and } x_2 = b. \end{cases}$$

Let (v_0,θ_0,ψ_0) be a point on the maximal attractor and $(v(t),\theta(t),\psi(t))$ be the corresponding solution of (3.5)-(3.8). For $m \in \mathbb{N}^*$, we introduce the quantity (5)

$$(3.19)\quad T_m(v_0,\theta_0,\psi_0,\{w_j,\xi_j,\eta_j\}_{j=1}^m,t) =$$

$$= \sum_{j=1}^{m} \int_\Omega \{\Pr|\nabla w_j|^2 + |\nabla\xi_j|^2 + \frac{1}{Le}|\nabla\eta_j|^2 + (w_j.\nabla)v.w_j +$$

$$+ (w_j.\nabla\theta)\xi_j + (w_j.\nabla\delta)\xi_j + (w_j.\nabla\psi)\eta_j + (w_j.\nabla\delta_1)\eta_j -$$

$$- \frac{\partial\omega}{\partial\theta}\xi_j^2 - \frac{\partial\omega}{\partial\Psi}\xi_j\eta_j + \frac{\partial\omega}{\partial\theta}\xi_j\eta_j + \frac{\partial\omega}{\partial\Psi}\eta_j^2\}dx ,$$

where $\{(w_j,\xi_j,\eta_j)\}_{j=1}^m$ is a family of functions of $H^2(\Omega)^4$ satisfying (3.15) and (3.18) and which is orthonormal in $L^2(\Omega)^4$. We set

$$(3.20)\quad q_m =$$

$$\limsup_{t\to+\infty} \left\{ \inf_{(v_0,\theta_0,\psi_0)\in\mathcal{A}_3} \frac{1}{t}\int_0^t \inf_{(w_j,\xi_j,\eta_j)} T_m(v_0,\theta_0,\psi_0,\{w_j,\xi_j,\eta_j\},s)ds \right\}.$$

According to [4], we can now state the following result : if there exists some $m \in \mathbb{N}^*$ such that

$$(3.21)\quad q_m > 0,$$

then the fractal dimension of $\mathcal{A}_3$ is finite and is bounded by

$$(3.22)\quad m \max_{1\leqslant\ell\leqslant m-1} \left\{ 1 + \frac{(-q_\ell)_+}{q_m} \right\}.$$

Theorem 3.1 will be derived by applying this result and our aim now is to obtain (3.21) for some $m \in \mathbb{N}^*$. We start by estimating the different integrals in (3.19).

Thanks to the Schwarz inequality, we write the pointwise inequality

$$|(w_j.\nabla)v.w_j| \leqslant |\nabla v|\ |w_j|^2.$$

Integrating by part, we then have

$$\int_\Omega (w_j.\nabla\psi)\eta_j\,dx = -\int_\Omega \psi(w_j.\nabla\eta_j)\,dx.$$

The function $\psi = \Psi - \delta_1$ satisfies $-1 \leqslant \psi \leqslant 1$ a.e. in Ω. Hence, using also the Cauchy-Schwarz inequality,

$$\left|\sum_{j=1}^m \int_\Omega (w_j.\nabla\psi)\eta_j\,dx\right| \leqslant \sum_{j=1}^m \int_\Omega |w_j|\,|\nabla\eta_j|\,dx,$$

$$\leqslant \left(\int_\Omega \sum_{j=1}^m |w_j|^2 dx\right)^{1/2} \left(\int_\Omega |\nabla\eta_j|^2 dx\right)^{1/2},$$

$$\leqslant \frac{1}{8\ \mathrm{Le}} \sum_{j=1}^m \int_\Omega |\nabla\eta_j|^2 dx + 2\ \mathrm{Le} \sum_{j=1}^m \int_\Omega |w_j|^2 dx.$$

By similar arguments, we get

$$\left|\sum_{j=1}^m \int_\Omega (w_j.\nabla\delta_1)\eta_j\,dx\right| \leqslant \frac{1}{8\ \mathrm{Le}} \sum_{j=1}^m \int_\Omega |\nabla\eta_j|^2 dx + 2\ \mathrm{Le} \sum_{j=1}^m \int_\Omega |w_j|^2 dx,$$

$$\left|\sum_{j=1}^m \int_\Omega (w_j.\nabla\delta)\xi_j\,dx\right| \leqslant \frac{1}{8} \sum_{j=1}^m \int_\Omega |\nabla\xi_j|^2 dx + 2 \sum_{j=1}^m \int_\Omega |w_j|^2 dx.$$

Also

$$\left|\sum_{j=1}^m \int_\Omega (w_j.\nabla\theta)\xi_j\,dx\right| \leqslant \sum_{j=1}^m \int_\Omega |w_j|\ |\nabla\theta|\ |\xi_j|\,dx,$$

$$\leqslant \frac{1}{2} \int_\Omega |\nabla\theta| \left(\sum_{j=1}^{m} (|w_j|^2 + |\xi_j|^2) \right) dx \ .$$

Finally, using the constant α_1 given by (1.16),

$$\left| \sum_{j=1}^{m} \int_\Omega \left\{ - \frac{\partial\omega}{\partial\theta} \xi_j^2 - \frac{\partial\omega}{\partial\Psi} \xi_j \eta_j + \frac{\partial\omega}{\partial\theta} \xi_j \eta_j + \frac{\partial\omega}{\partial\Psi} \eta_j^2 \right\} dx \right|$$

$$\leqslant 2 \alpha_1 \sum_{j=1}^{m} \int_\Omega (|\xi_j|^2 + |\eta_j|^2) dx \ .$$

Thus, combining the above inequalities and (3.19) and using also the constant Λ defined in (3.10), we obtain

$$\text{(3.23)} \quad T_m \geqslant \frac{3}{4} \Lambda \sum_{j=1}^{m} \int_\Omega (|\nabla w_j|^2 + |\nabla\xi_j|^2 + |\nabla\eta_j|^2) dx -$$

$$- \int_\Omega \left(|\nabla v| + \frac{|\nabla\theta|}{2} + 6 \max(1, Le) + 2 \alpha_1 \right) \rho \, dx,$$

where we have set

$$\rho = \sum_{j=1}^{m} \{ |w_j|^2 + |\xi_j|^2 + |\eta_j|^2 \} \ .$$

Thanks to a generalization of the Sobolev-Lieb-Thirring inequalities proved in Ghidaglia-Marion-Temam [5], we know that there exist two constants κ and χ depending only on the shape factor b/a such that

$$\int_\Omega \rho^2 dx \leqslant \kappa \sum_{j=1}^{m} \int_\Omega (|\nabla w_j|^2 + |\nabla\xi_j|^2 + |\nabla\eta_j|^2) dx + \frac{\chi}{\delta(\Omega)^2} \int_\Omega \rho \, dx,$$

where $\delta(\Omega)$ = the diameter of Ω ; this yields in particular

$$(3.24)\qquad \int_\Omega \rho^2 dx \leqslant \kappa_2 \sum_{j=1}^{m} \int_\Omega (|\nabla w_j|^2 + |\nabla \xi_j|^2 + |\nabla \eta_j|^2) dx ,$$

where $\kappa_2 = \kappa + \dfrac{\chi}{\lambda_1 \delta(\Omega)^2}$ depends only on b/a.

The last integral in (3.23) can be majorized as follows (we set $g = |\nabla v| + |\nabla\theta|/2 + 6\max(1,Le) + 2\alpha_1$)

$$\left|\int_\Omega g\ \rho\ dx\right| \leqslant \left(\int_\Omega g^2 dx\right)^{1/2} \left(\int_\Omega \rho^2 dx\right)^{1/2} ,$$

$$\leqslant \text{(by (3.24))},$$

$$\leqslant \left(\int_\Omega g^2 dx\right)^{1/2} \kappa_2^{1/2} \left(\sum_{j=1}^{m} \int_\Omega (|\nabla w_j|^2 + |\nabla \xi_j|^2 + |\nabla \eta_j|^2) dx \right)^{1/2}$$

$$(3.25)\qquad \leqslant \frac{\Lambda}{4} \sum_{j=1}^{m} \int_\Omega (|\nabla w_j|^2 + |\nabla \xi_j|^2 + |\nabla \eta_j|^2) dx + \frac{\kappa_2}{\Lambda} \int_\Omega g^2\ dx .$$

Also, since the family (w_j, ξ_j, η_j) is orthonormal in $L^2(\Omega)^4$, we have

$$\int_\Omega \rho(x) dx = m .$$

Hence, (3.24) and Hölder inequality give

$$(3.26)\qquad \sum_{j=1}^{m} \int_\Omega (|\nabla w_j|^2 + |\nabla \xi_j|^2 + |\nabla \eta_j|^2) dx \geqslant \frac{m^2}{\kappa_2\ ab} .$$

Combining (3.23), (3.25) and (3.26), we obtain

$$T_m \geqslant \frac{\Lambda}{2\ \kappa_2\ ab} m^2 - \frac{\kappa_2}{\Lambda} \int_\Omega (|\nabla v| + \frac{|\nabla\theta|}{2} + 6\max(1,Le) + 2\alpha_1)^2 dx.$$

This inequality is independent from the family $\{(w_j, \xi_j, \eta_j)\}_{j=1}^{m}$. Thus, returning to the definition (3.21), we get the following lower bound

$$(3.27) \quad q_m \geq \frac{\Lambda}{2\,\kappa_2\, ab}\, m^2 - \frac{\kappa_2}{\Lambda}\,\alpha,$$

where

$$\alpha = \liminf_{t \to +\infty} \sup_{(v_0,\theta_0,\psi_0)\in\mathcal{A}_3} \frac{1}{t}\int_0^t\int_\Omega (|\nabla v| + \frac{|\nabla\theta|}{2} + 6\max(1,Le) + 2\alpha_1)^2 dxds.$$

It is now easy to conclude the proof of Theorem 3.1. Let $\bar{m}$ be the unique integer such that

$$(3.28) \quad (\bar{m} - 1)^2 \leq 4\,\frac{\kappa_2^2\, ab\, \alpha}{\Lambda^2} < \bar{m}^2 .$$

Then, $q_{\bar{m}} > 0$. The fractal dimension of $\mathcal{A}_3$ is finite ; moreover, using the bound (3.22) and (3.27), it is majorized by

$$\dim_M \mathcal{A}_3 \leq 2\,\bar{m} .$$

This last inequality yields the estimate (3.12) by expressing the bound on $\bar{m}$ given by (3.28). Details are omitted.

FOOTNOTES.

(1) Le is the ratio of the thermal diffusivity to the diffusion coefficient of the reactant R.

(2) T_u (resp. T_b) denotes the temperature of the unburnt (resp. burnt) gas.

(3) $L^2(\Omega)$ denotes the space of functions which are square integrable on Ω.

(4) $H^1(\Omega)$ denotes the subspace of $L^2(\Omega)$ of functions whose first order derivatives are square integrable.

(5) This quantity occurs naturally when one consider the mth trace of the linear operator that generates the linearized flow (3.14)-(3.18).

REFERENCES.

[1] Babin A.V. and Vishik M.I., Attractors of partial differential equations and estimates of their dimension, Russian Math. Surveys, 38:4, 1983, 151-213.
[2] Buckmaster J.D. and Ludford G.S.S., *Theory of Laminar Flames*, Cambridge

Univ. Press, Cambridge, 1982.
[3] Chueh K.N., Conley C.C. and Smoller J.A., Positively invariant regions for systems of nonlinear diffusion equations, Indiana Univ. Math. J., 26, 1977, 373-392.
[4] Constantin P., Foias C. and Temam R., Attractors representing turbulent flows, Memoirs of AMS, 53, n° 314, 1985.
[5] Ghidaglia J.M., Marion M. and Temam R., Sur quelques inégalités fonctionnelles, C.R. Acad. Sc. Paris, série I, 1987, 287-290, and Generalization of the Sobolev-Lieb-Thirring inequalities and applications, to appear.
[6] Joulin G. and Clavin P., Linear stability analysis of non-adiabatic flames : diffusional-thermal model, Combustion and Flame, 35, 1979, 139-153.
[7] Landau L. and Lifschitz I.M., *Fluid Dynamics,* Addison-Wesley, New-York, 1953.
[8] Mandelbrot B., *Fractals : Form, Chance and Dimension,* Freeman, San Francisco, 1977.
[9] Marion M., Attractors for reaction-diffusion equations : existence and estimate of their dimension, Applicable Analysis, to appear.
[10] Nicolaenko B., Scheurer B. and Temam R., Some global dynamical properties of the Kuramoto-Sivashinsky equations : nonlinear stability and attractor, Physica 16D, 1985, 155-183.
[11] Sivashinsky G.I., Diffusional-thermal theory of cellular flames, Combustion Science and Technology, 15, 1977, 137-145.
[12] Temam R., *Navier-Stokes Equations,* 3rd ed., North-Holland, Amsterdam, 1984.
[13] Temam R., *Infinite dimensional dynamical systems in mechanics and physics,* Springer-Verlag, to appear in 1988.

GRID REQUIREMENTS DUE TO THE INNER STRUCTURE OF PREMIXED HYDROCARBON FLAMES

Norbert Peters

Institut für Allgemeine Mechanik, RWTH Aachen
5100 Aachen, West-Germany

1. INTRODUCTION

Numerical computations of structures of planar, steady methane flames with full kinetics have become relatively straightforward in recent years. However, they are still somewhat expensive and time-consuming, and their complexity precludes their general use in multidimensional and time-dependent problems of engineering interest. Therefore, there is a need to investigate reductions of the full kinetic mechanisms to a smaller number of steps. Reductions that are systematic and that can be justified within reasonable accuracy contribute to our understanding of the flame structure as well as fulfilling the engineering need. Such reductions have been performed in a series of recent papers /1/, /2/, /3/ showing that for hydrocarbon flames the number reduces to four global reactions, if steady state assumptions for intermediates are introduced. In /1/ and /2/ it was shown that the kinetic information of only 12 elementary reactions was necessary to provide a satisfactory description of lean to stoichiometric premixed methane flames /1/ and methane-air diffusion flames /2/. In /3/ reduced four- and five-step mechanisms were derived for methane-, methanol-, and propane flames, covering the entire range of stoichiometry from lean to rich flames. Again only a much reduced number of elementary reactions entered into the final description of the mechanism.

On the basis of the reduced schemes, an asymptotic analysis of the flame structure of premixed methane-air flames has been performed in /4/. To illustrate the consequences of the flame structure on numerical calculations and grid size restrictions in particular a brief description of the procedures and results of /4/ is given here. The detailed development, as well as the references to background and related studies, may be found in the full-length publication /4/.

2. A FOUR-STEP MECHANISM FOR METHANE FLAMES

A four-step mechanism for methane oxidation has been identified and analysed previously /1/, /2/. This mechanism, which is the starting point of the present investigation, may be presented most simply by beginning with a suitable subset of reactions 1-22 of Table 1. Assumptions of steady states for O, OH, HO_2, CH_3, CH_2O, and CHO are introduced, and for algebraic convenience the additional but inessential postulates of partial equilibria for reactions 2 and 3 are made. Summing the fastest reactions of the C_1-chain,

11	$CH_4 + H \rightarrow CH_3 + H_2$
13	$CH_3 + O \rightarrow CH_2O + H$
17	$CH_2O + H \rightarrow CHO + H_2$
22	$CHO + M \rightarrow CO + H + M$

together with the partial-equilibrium reactions

C.-M. Brauner and C. Schmidt-Lainé (eds.), Mathematical Modeling in Combustion and Related Topics, 173–184.

TABLE 1. The elementary reaction mechanism and associated rate coefficients

No.	Reaction			B	α	E (= RT_a)
1	$O_2 + H$	→	$OH + O$	$2.00\ 10^{14}$	0.00	70.30
1b	$OH + O$	→	$O_2 + H$	$1.40\ 10^{13}$	0.00	3.20
2	$O + H_2$	→	$H + OH$	$1.50\ 10^{7}$	2.00	31.60
2b	$H + OH$	→	$O + H_2$	$6.73\ 10^{6}$	2.00	22.35
3	$OH + H_2$	→	$H + H_2O$	$1.00\ 10^{8}$	1.60	13.80
3b	$H + H_2O$	→	$OH + H_2$	$4.62\ 10^{8}$	1.60	77.50
4	$OH + OH$	→	$H_2O + O$	$1.50\ 10^{9}$	1.14	0.42
4b	$H_2O + O$	→	$OH + OH$	$1.49\ 10^{10}$	1.14	71.14
5	$H + O_2 + M$	→	$HO_2 + M$	$2.30\ 10^{18}$	- 0.80	0.00
6	$HO_2 + H$	→	$OH + OH$	$1.50\ 10^{14}$	0.00	4.20
7	$HO_2 + H$	→	$H_2 + O_2$	$2.50\ 10^{13}$	0.00	2.90
8	$HO_2 + H$	→	$H_2O + O$	$3.00\ 10^{13}$	0.00	7.20
9	$HO_2 + OH$	→	$H_2O + O_2$	$6.00\ 10^{13}$	0.00	0.00
10	$CO + OH$	→	$CO_2 + H$	$4.40\ 10^{6}$	1.50	- 3.10
10b	$CO_2 + H$	→	$CO + OH$	$4.96\ 10^{8}$	1.50	89.71
11	$CH_4 + H$	→	$H_2 + CH_3$	$2.20\ 10^{4}$	3.00	36.60
11b	$H_2 + CH_3$	→	$CH_4 + H$	$8.83\ 10^{2}$	3.00	33.53
12	$CH_4 + OH$	→	$H_2O + CH_3$	$1.60\ 10^{6}$	2.10	10.30
13	$CH_3 + O$	→	$CH_2O + H$	$7.00\ 10^{13}$	0.00	0.00
14	$CH_3 + OH$	→	$CH_2O + H + H$	$9.00\ 10^{14}$	0.00	64.80
15	$CH_3 + OH$	→	$CH_2O + H_2$	$8.00\ 10^{12}$	0.00	0.00
16	$CH_3 + H$	→	CH_4	$6.00\ 10^{16}$	- 1.00	0.00
17	$CH_2O + H$	→	$CHO + H_2$	$2.50\ 10^{13}$	0.00	16.70
18	$CH_2O + OH$	→	$CHO + H_2O$	$3.00\ 10^{13}$	0.00	5.00
19	$CHO + H$	→	$CO + H_2$	$2.00\ 10^{14}$	0.00	0.00
20	$CHO + OH$	→	$CO + H_2O$	$1.00\ 10^{14}$	0.00	0.00
21	$CHO + O_2$	→	$CO + HO_2$	$3.00\ 10^{12}$	0.00	0.00
22	$CHO + M$	→	$CO + H + M$	$7.10\ 10^{14}$	0.00	70.30
23	$CH_3 + H$	→	$CH_2 + H_2$	$1.80\ 10^{14}$	0.00	63.00
24	$CH_2 + O_2$	→	$CO_2 + H + H$	$6.50\ 10^{12}$	0.00	6.30
25	$CH_2 + O_2$	→	$CO + OH + H$	$6.50\ 10^{12}$	0.00	6.30
26	$CH_2 + H$	→	$CH + H_2$	$4.00\ 10^{13}$	0.00	0.00
26b	$CH + H_2$	→	$CH_2 + H$	$2.79\ 10^{13}$	0.00	12.61
27	$CH + O_2$	→	$CHO + O$	$3.00\ 10^{13}$	0.00	0.00
28	$CH_3 + OH$	→	$CH_2 + H_2O$	$1.50\ 10^{13}$	0.00	20.93

Continuation of Table 1

No.	Reaction			B	α	E (= RT_a)
29	$CH_2 + OH$	→	$CH_2O + H$	$2.50\ 10^{13}$	0.00	0.00
30	$CH_2 + OH$	→	$CH + H_2O$	$4.50\ 10^{13}$	0.00	12.56
31	$CH + OH$	→	$CHO + H$	$3.00\ 10^{13}$	0.00	0.00

The units are cm, mol, K and kJ.

2b $\quad H + OH \rightarrow O + H_2$

3b $\quad H + H_2O \rightarrow OH + H_2$

(where b identifies the backward direction of the reactions of Table 1) gives the overall reaction

I' $\quad CH_4 + 2\,H + H_2O \rightarrow CO + 4\,H_2 \quad .$

Similarly, the combination

10, 10b $\quad CO + OH \rightleftarrows CO_2 + H$

3b, 3 $\quad H + H_2O \rightleftarrows OH + H_2$

yields the overall water-gas shift

II' $\quad CO + H_2O \rightleftarrows CO_2 + H_2 \quad ,$

and the combination

5 $\quad O_2 + H + M \rightarrow HO_2 + M$

9 $\quad OH + HO_2 \rightarrow H_2O + O_2$

3b $\quad H + H_2O \rightarrow OH + H_2$

results in the overall recombination

III' $\quad 2\,H + M \rightarrow H_2 + M \quad .$

Finally, a sum of the steps

1, 1b $\quad O_2 + H \rightleftarrows OH + O$

2, 2b $\quad O + H_2 \rightleftarrows OH + H$

3, 3b $\quad OH + H_2 \rightleftarrows H_2O + H$

(with the third taken twice) produces the fourth overall reaction

IV' $\quad O_2 + 3\,H_2 \rightleftarrows 2\,H_2O + 2\,H \quad .$

Only ten elementary reactions have been introduced (four with reverse steps included) in obtaining this four-step mechanism.

In terms of the elementary rate constants k_i and equilibrium constants K_i, the rates of each of the four step may be written as

$$\begin{aligned} w_I &= k_{11}\,[CH_4]\,[H] \quad , \\ w_{II} &= (k_{10}/K_3)\,([H]/[H_2])\,([CO]\,[H_2O] - [CO_2]\,[H_2]\,/\,K_{II}) \quad , \\ w_{III} &= k_5\,[O_2]\,[H]\,[M] = k_{III}\,p\,[O_2]\,[H] \quad , \\ w_{IV} &= k_1\,[H]\,([O_2] - [H]^2\,[H_2O]^2\,/\,([H_2]^3\,K_{IV})) \quad , \end{aligned} \tag{1}$$

where the partial-equilibrium results,

$$\begin{aligned} [OH] &= [H]\,[H_2O]\,/\,([H_2]\,K_3) \quad , \\ [O] &= [H]\,[OH]\,/\,([H_2]\,K_2) \quad , \end{aligned} \tag{2}$$

have been introduced. With the temperature T in Kelvin, the equilibrium constants here are given by

$$\begin{aligned} K_2 &= 2.23\exp(-1112/T) \quad , \quad & K_3 &= 0.216\exp(7658/T) \quad , \\ K_{II} &= 0.035\exp(3652/T) \quad , \quad & K_{IV} &= 1.48\exp(6133/T) \quad . \end{aligned} \tag{3}$$

In terms of the catalytic efficiency η_i of molecule i (i = 1,...,N) and its molecular weight W_i and mass fraction Y_i, the concentration of the third body M is

$$[M] = \frac{p\,\overline{W}}{R\,T} \sum_{i=1}^{N} (\eta_i\,Y_i\,/\,W_i) \quad , \tag{4}$$

where $R = 82.05$ atm cm^3/mol K , p denotes pressure (atm) and $\overline{W}$ the average molecular weight. In the following development further use will be made of the ideal-gas equation of state, and the approximation of constant $\overline{W}$, good for methane flames, will be introduced. In eq. $(1)_3$, the third expression in eq. (1), eq. (4) has been used in the second equality to transfer the three-body rate constant k_5 into an effectively two-body form through

$$k_{III} = k_5\,\overline{W}\,/\,(R\,T) \sum_{i=1}^{N} (\eta_i\,Y_i\,/\,W_i) = 1.6\,k_5\,/\,(R\,T) \quad .$$

The last equality is obtained by use of the efficiencies of Warnatz /1/ (0.4, 1.5 and 6.5 for N_2, CO_2 and H_2O, respectively) and the respective downstream mole fractions (0.715, 0.095 and 0.190).

3. THE THREE-STEP MECHANISM

At sufficiently high pressures (typically above 1 atm) the pressure dependence of reaction III causes a reduction in the H-atom concentration to a point at which its steady-state balance

$$w_{IV} = w_I + w_{III} \tag{5}$$

becomes accurate. From eqs. (1) and (5) the expression

$$[H] = F^{1/2}\,K_{IV}^{1/2}\,[O_2]^{1/2}\,[H_2]^{3/2}\,/\,[H_2O] \tag{6}$$

is obtained, where

$$F = 1 - (k_{III}\, p / k_1) - (k_{11} [CH_4] / (k_1 [O_2])) \quad . \tag{7}$$

As a consequence of this additional steady-state approximation, the four-step mechanism reduces to the three-step mechanism

I $\quad CH_4 + O_2 \rightarrow CO + H_2 + H_2O \quad ,$

II $\quad CO + H_2O \rightleftharpoons CO_2 + H_2 \quad ,$

III $\quad O_2 + 2\,H_2 \rightarrow 2\,H_2O \quad ,$

which is the basis of the following analysis.

4. THE TWO-STEP MECHANISM

In the limit of infinite Damköhler number D_{II} for reaction II the water-gas equilibrium is obtained and can be written as

$$[CO] = \alpha\, [H_2] \quad , \tag{8}$$

where α depends on $[CO_2]$, $[H_2O]$ and temperature T. The kinetics then are reduced to two steps involving fuel consumption (at rate w_I)

$$CH_4 + O_2 \rightarrow \left(\frac{2}{1+\alpha}\right)(H_2 + \alpha\ CO) + \left(\frac{2\,\alpha}{1+\alpha}\right) H_2O + \left(\frac{1-\alpha}{1+\alpha}\right) CO_2$$

and oxidation (at rate w_{III})

$$\left(\frac{2}{1+\alpha}\right)(H_2 + \alpha\ CO) + O_2 \rightarrow \left(\frac{2}{1+\alpha}\right) H_2O + \left(\frac{2\alpha}{1+\alpha}\right) CO_2 \quad .$$

The key rates of elementary steps that remain are those of

$$H + O_2 \rightarrow OH + O \quad ,$$

which appears in w_{IV} in the four-step mechanism but only in F in the three-step and two-step mechanisms, of

$$H + O_2 + M \rightarrow HO_2 + M \quad ,$$

which appears in the oxidation rate w_{III}, of

$$H + CH_4 \rightarrow H_2 + CH_3 \quad ,$$

which appears in the fuel-consumption rate w_I, and of

$$CO + OH \rightarrow CO_2 + H \quad ,$$

which does not enter into the two-step mechanism but which occurs in the water-gas shift in the three-step and four-step mechanism.

Although the three-step mechanism with its nonequilibrium water-gas shift must be included to achieve reasonable accuracy in predictions, the general structure can be understood on the basis of the two-step mechanism.

5. FLAME STRUCTURE

The structure obtained is illustrated in Fig. 1. There is a thin fuel-consumption layer (of nondimensional thickness δ compared with the total flame thickness)

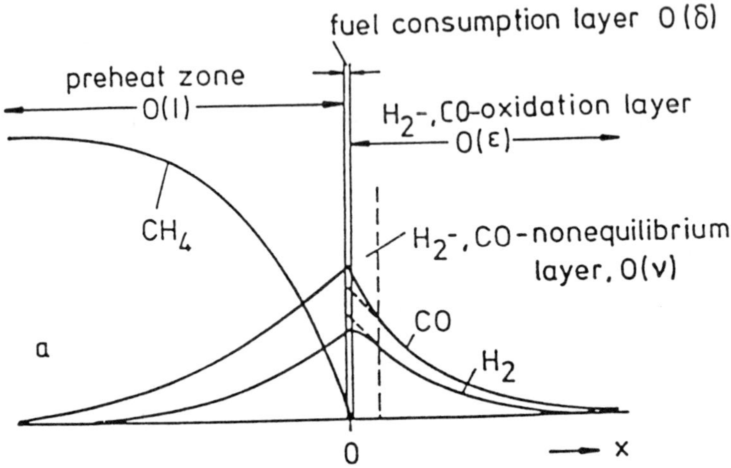

FIGURE 1. Asymptotic inner structure of premixed stoichiometric methane-air flames.

followed by an oxidation layer of corresponding thickness ε, large compared with δ but still small compared with the flame and preheat-zone thicknesses. In the two-step mechanism there is nothing more, but in the three-step mechanism at the upstream end of the oxidation layer there is a region of water-gas nonequilibrium of relative thickness ν, small compared with ε but large compared with δ. The theory involves separate analyses of these layers and appropriate matching. The thickness of the different layers determines the grid size required in computational schemes. Since the thinnest layer is the fuel consumption layer of thickness δ where all hydrocarbon kinetics take place, this layer must be resolved. The dependence of δ on pressure and preheat temperature must be considered in particular.

6. THE FUEL-CONSUMPTION LAYER

The small parameter appropriate to the fuel-consumption layer is

$$\delta = (k_1/k_{11})_o ([O_2]_o W_{CH_4}) / (L_{CH_4} \rho_o Y_{CH_4u}) \quad , \tag{9}$$

where the subscript o identifies conditions at this layer, W is molecular weight, L Lewis number, ρ density and Y mass fraction, with u denoting fresh-mixture conditions. The stretched variables are

$$\begin{aligned} \zeta &= x/\delta \quad , \\ y &= [CH_4] W_{CH_4} / (\delta L_{CH_4} \rho_o Y_{CH_4u}) = (k_{11}/k_1)_o [CH_4]/[O_2]_o \end{aligned} \tag{10}$$

and the appropriate Damköhler number

$$D_I = (1/v_u^2)(\lambda/c_p)_o (T_u/T_o)^2 L_{CH_4} ([H]_o/\rho_o) k_{11_o} \quad , \tag{11}$$

where v_u is the burning velocity. In the first approximation the differential equation obtained is

$$d^2y/d\zeta^2 = \delta^2 D_I y \sqrt{F} \quad , \tag{12}$$

with the boundary conditions $y \to 0$ as $\zeta \to \infty$ and $y \to -\zeta$ as $\zeta \to -1$ (the position of the radical-consumption layer). Here $F = 1-y$, but inclusion of additional chemistry is seen later to modify F. Integration yields

$$D_I = (15/8\ \delta^2)(1 + ...) \quad . \tag{13}$$

In eq. (12) $\sqrt{F}$ performs the function served by the activation energy in activation-energy asymptotics, cutting off the reaction rate at the upstream end of the fuel-consumption layer. This cutoff is seen to be due to radical depletion, not temperature decrease. The preheat zone thus exists here for kinetic reasons associated with the smallness of δ.

7. THE OXIDATION LAYER

Numerical simplification in the analysis of the oxidation layer is achieved by fortuitous energetics. If Q is the heat released in $CH_4+O_2 \rightarrow CO_2+2H_2O$, then that released in $H_2+\frac{1}{2}O_2 \rightarrow H_2O$ is about 0.3116 Q, and that released in $CO+\frac{1}{2}O_2 \rightarrow CO_2$ is about 0.3479 Q. The approximation $0.3116 \approx 0.33 \equiv q$ is then introduced, and appropriate variables for the oxidation layer become

$$\begin{aligned} \eta &= 2qx/\varepsilon \quad , \\ z &= (q(1+\alpha)/\varepsilon)([H_2]\,W_{CH_4}/(L_{H_2}\,\rho_o\,Y_{CH_4u})) \\ &= (2q/\varepsilon)([O_2]\,W_{CH_4}/(L_{O_2}\,\rho_o\,Y_{CH_4})) \\ &= (1/\varepsilon)(T_b - T)/(T_b - T_u) \quad , \end{aligned} \tag{14}$$

where T_b is the final adiabatic flame temperature, and the parameter ε is related to the Damköhler number D_{III} by

$$\varepsilon = 1/D_{III}^{1/4} \quad , \tag{15}$$

where

$$D_{III} = \frac{1}{v_u^2}\left(\frac{\lambda}{c_p}\right)_o \left(\frac{T_u}{T_o}\right)^2 \frac{(L_{O_2}L_{H_2})^{3/2}(\rho_o Y_{CH_4u}/W_{CH_4})^2\,[H]_o\,[M_o]\,k_{5_o}}{2^{5/2}\,q^4\,(1+\alpha_o)^{3/2}\,[O_2]_o^{1/2}\,[H_2]_o^{3/2}\,\rho_o} \tag{16}$$

In the first approximation for small ε the oxidation layer maintains a reactive-diffusive balance and is described by the differential equation

$$d^2z/d\eta^2 = z^3 \tag{17}$$

with the boundary conditions that $z \rightarrow 0$ as $\eta \rightarrow \infty$ and $dz/d\eta \rightarrow -1$ as $\eta \rightarrow 0$, the last obtained from integration across the fuel-consumption layer. In eq. (17) the z^3 may be traced to a z from $[O_2]$ and a z^2 from $[H]$. Equation (14) shows that $[H_2]$ and $[O_2]$ are related to temperature in this layer through the necessity of satisfying downstream boundary conditions. Integration of eq. (17) across the oxidation layer then gives conditions at the fuel-consumption layer, viz., $T_o = T_b-(T_b-T_u)\varepsilon z_o$, etc., with

$$z_o = 2^{1/4}(1 + ...) \quad . \tag{18}$$

8. WATER-GAS NONEQUILIBRIUM

The water-gas nonequilibrium is analysed by perturbation methods in the oxidation layer and leads to a definition of ν in terms of D_{II} and a modification to relationships among Damköhler numbers /4/. For simplicity this complication is not detailed here.

9. THE BURNING VELOCITY

Equation (11) can be solved for v_u, and eq. (19) can be solved for v_u. By equating these two expressions, and by using eqs. (9), (13), (15) and (18), it is found that the structure requires

$$(15/16)\, k_{11_o} k_{5_o} [M]_o = k_{1_o}^2 (L_{O_2}/L_{CH_4}) \tag{19}$$

in the first approximation. This equation determines T_o, the temperature of the fuel-consumption layer, as a function of the pressure p, mainly from the T-dependence of k_1. Inclusion of water-gas nonequilibrium introduces substantial numerical modifications to this simple formula, although the general qualitative character of the balance is unmodified. Equation (19) plays off the branching effects of k_1 against the breaking effects of k_{11} and k_5 to establish T_o. After T_o is determined, the burning velocity may be calculated, e.g. from

$$D_{III} = z_o^4 (T_b - T_u)^4 / (T_b - T_o)^4 \quad , \tag{20}$$

in which eq. (16) is employed and the result solved for v_u.

When the predictions for v_u are compared with experimental results and with results of full numerical integrations agreements within a factor of 2 are found for 1 atm < p < 10 atm , but the predicted burning velocities are much too low at higher pressures. The disagreement arises from the kinetic mechanism adopted above for the fuel-consumption layer. Therefore, the approach was extended to include more detailed fuel-consumption kinetics.

10. AUGMENTED KINETICS FOR THE FUEL-CONSUMPTION LAYER

A systematic approach was developed for estimating the importance of various elementary steps in the fuel-consumption layer /4/. The approach involves straightforward order-of-magnitude estimates and resulted in an augmented kinetic scheme for the hydrocarbon fuel consumption. Steady-state approximations are introduced for all new intermediates so that the number of unknown concentrations returns to the seven species of the four-step mechanism. These estimates can be completed for any desired mechanism of hydrocarbon chemistry because only fuel species are involved, so that changes in the analysis are restricted to the fuel-consumption layer, in which the temperature and major-species concentration are known from the analysis just outlined. It is found that eq. (12) is modified by replacing y by y-G, where G depends on ratios of rates of the new reactions introduced to w_1, and F is modified from eq. (7) by these same kinds of rate-ratio terms. The only change in the analysis then is to replace eq. (13) by

$$\delta^2 D_I = \left(2 \int_0^\infty (y - G) \sqrt{F}\, dy \right)^{-1} \tag{21}$$

with the modified functions F and G. Although the integral now must be evaluated numerically, the task is straightforward, and burning velocities are readily calculated quickly.

11. KINETIC EFFECTS INVESTIGATED

Consideration was given to augmenting the kinetic scheme by various steps that compete with

$$CH_3 + O \rightarrow CH_2O + H$$

for removal of the methyl radical. These competitions become stronger with increasing p and enhance branching by reducing the extent of radical removal associated with fuel consumption. The consequent higher radical concentrations are exactly what is needed to increase the predicted burning velocity at high p thus improving agreement with experiment.

The step

$$CH_3 + OH \rightarrow CH_2O + H + H \quad ,$$

possibly through the intermediate CH_3O, was added with Roth/Just rate constants and found to greatly improve the agreement of the prediction (solid line in Fig. 2) with full numerical integrations employing the same rate parameters (points in Fig. 2, helpfully supplied by Warnatz). This agreement certainly is within the accuracy of the asymptotics over the entire range of pressure.

As another possibility, step 14 was deleted and replaced by a CH_2-chain beginning with

$$CH_3 + OH \rightarrow CH_2 + H_2O \quad ,$$
$$CH_2 + O_2 \rightarrow CO + OH + H \quad ,$$

etc., suggested by Miller. The results, shown by the dashed line in Fig. 2, lead to some improvement but not as much as obtained by step 14.

Consideration also was given to C_2-chemistry through

$$CH_3 + CH_3 + M \rightarrow C_2H_6 + M$$
$$C_2H_6 + H \rightarrow C_2H_5 + H_2$$
$$C_2H_5 + M \rightarrow C_2H_4 + H + M$$

etc., and although this was of some help when steps 14 and 28 were neglected, it provided negligible change when either of these steps was included. Thus the C_2-chain seems possibly to be not of importance to the burning velocity.

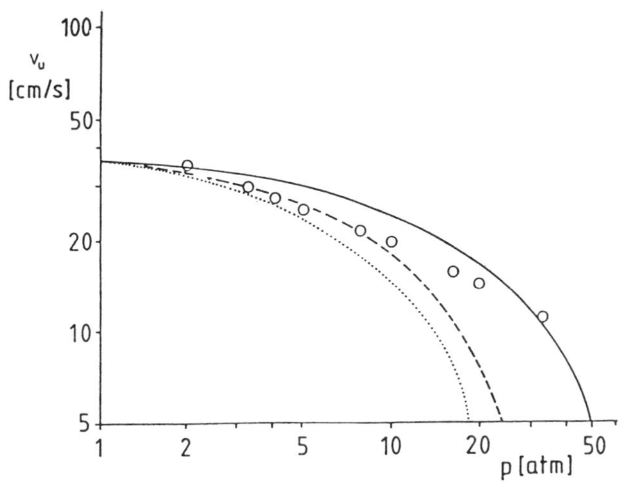

FIGURE 2. The dependence of the burning velocity on pressure at an initial temperature of 300 K according to numerical calculations of Warnatz (points), asymptotic expansions that include reaction 14 (solid line) and that include reaction 28 (dashed line).

3a 3b 3c

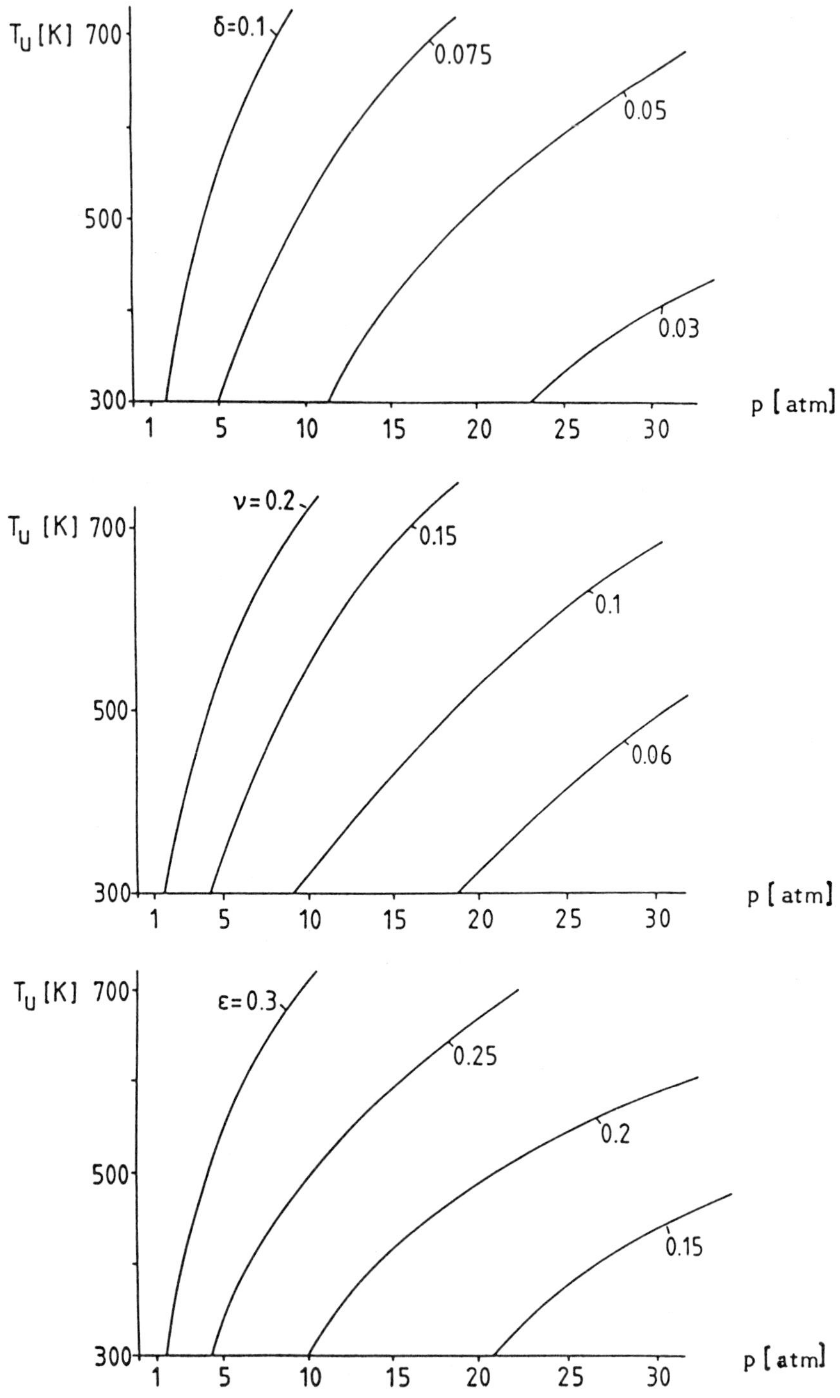

FIGURE 3. Lines of constant values of the expansion parameters δ, ν and ε, respectively, in a plane of the pressure and initial temperature.

12. THE THICKNESSES OF THE DIFFERENT LAYERS AND MESH REQUIREMENTS

In Fig. 3 the small expansion parameters δ, ν and ε have been plotted as isoparameter contours. They decrease in magnitude, leading to greater accuracy from the asymptotics, as the initial temperature decreases or as the pressure increases. Over the entire range covered in Fig. 3, the inequalities $\delta < \nu < \varepsilon < 1$ apply, the last of which suggests that the asymptotic approach remains appropriate throughout the diagram. Unlike other systems, such as the ozone flame /5/, crossovers in relative magnitudes of different expansion parameters do not occur here, so that the merging of different layers, leading to modifications in the basic structure, is not encountered. The structure illustrated in Fig. 1 applies throughout the range of conditions of Fig. 3. The boundaries between the fuel-consumption layer, the water-gas nonequilibrium layer and the oxidation layer are not very sharp because the three expansion parameters do not differ quantitatively very much. However, the results that have been obtained suggest that, beyond aiding in understanding the flame structure, separate consideration of the different layers leads to satisfactory prediction of burning velocities through asymptotics.

As a consequence of the fact that the asymptotic description yields satisfactory results for flame velocities, it follows that in a numerical calculation the thinnest layer that was necessary must be resolved accurately. This is the fuel-consumption layer. The grid size Δx across the flame structure must therefore be in physical coordinates

$$\Delta x \leq \delta \, \ell_F / n \tag{22}$$

where ℓ_F is the flame thickness

$$\ell_F = \lambda / (\rho \, c_p \, v_u) \quad , \tag{23}$$

and n of the order 10.

The flame thickness itself is pressure-dependent. Since $\rho \sim p^{-1}$, and λ / c_p being independent of pressure, its pressure-dependence is influenced by that of the burning velocity. Assuming approximately $v_u \sim p^{-1/2}$ it follows that

$$\ell_F = \ell_{F,o} \, p^{-1/2} \tag{24}$$

where $\ell_{F,o} = \ell_F \, (p = 1 \text{ atm}) \approx 0.1$ mm , such that Δx must decrease with pressure even more rapidly than Fig. 3a would indicate. Similar arguments, leading to opposite conclusions, apply for the influence of the preheat temperature on flame thickness.

13. CONCLUSIONS

Asymptotic analysis of the flame structure of hydrocarbon flames may provide a guidance for the minimum mesh size necessary to accurately resolve the flame structure of real flames. Unlike model flames based on one-step kinetics, the inner structure of hydrocarbon flames consists on several layers whose thicknesses depend strongly on pressure and preheat temperature. It is necessary to resolve the thinnest of these layers, the fuel-consumption layer, in order to obtain an accurate numerical prediction of the burning velocity.

REFERENCES

1. Peters, N.: Numerical and asymptotic analysis of systematically reduced reaction schemes for hydrocarbon flames, in: Numerical simulation of combustion phenomena, R. Glowinski, B. Larroutourou and R. Temam (Eds.). Lecture Notes in Physics 241, p. 90-109, Springer-Verlag, 1985.
2. Peters, N., Kee, R.J.: The computation of stretched laminar methane-air diffusion flames using a reduced four-step mechanism. Sandia Report SAND 86-8836, July 1986.
3. Paczko, G., Lefdal, P.M., Peters, N.: Reduced reaction schemes for methane, methanol and propane flames. Paper presented at the 21st Symposium (International) on Combustion, Munic, August 1986.
4. Peters, N., Williams, F.A.: The asymptotic structure of stoichiometric methane-air flames. Paper to appear in Combustion and Flame (1987).
5. Rogg, B., Liñán, A., Williams, F.A.: Combust. Flame 65, p. 79-101, 1986.

NONLINEAR STUDIES OF LOW-FREQUENCY COMBUSTION STABILITIES

Thierry Poinsot and Sebastien M. Candel

E.M2.C Laboratory, CNRS
Ecole Centrale des Arts et Manufactures, 92295 Châtenay-Malabry, France
and ONERA, 92322 Châtillon, France

Abstract
This paper presents a direct simulation of low frequency combustion instabilities in premixed turbulent combustors. The simulation is based on the thin flame sheet (TFS) model proposed by Marble and Candel in their investigation of ramjet and afterburner instabilities. This model is solved numerically to allow the prediction of the flame response to any excitation. The numerical calculations are performed with an original implicit extension of a two phase flow TRAC method. Results obtained for reactive flows show that stability is mainly determined by the steady flow configuration and by the interaction parameter introduced to represent the effects of pressure on the local normal flame speed. Quarter-wave excitations induce the strongest instabilities when convective perturbations travelling on the flame front couple with the acoustic oscillation. A criterion for quarter- wave instability is derived and compared with experimental results.

1. INTRODUCTION

Combustion instabilities occur in many practical systems such as ramjets, afterburners, liquid and solid rocket motors, industrial powerplants etc. The analysis and prediction of these instabilities has been the subject of many theoretical and experimental studies during the last forty years.

Important progress in the area of liquid rocket instability was achieved in the early fifties and sixties by Crocco and his group (see for example Crocco 1951, Crocco and Cheng 1956, Crocco 1965, Crocco 1969). Other important studies in this field are those of Tsien (1952), Marble and Cox (1953), Barrere and Williams (1968). An extensive report on this subject may be found in Harrje and Reardon (1972).

Studies of turbulent ducted flame instabilities have been less extensive and the current understanding of the mechanisms involved is also less profound. The mechanism of high-frequency (screech) instability of flames stabilized by bluff bodies is clarified in an early study of Rogers and Marble (1956). Under screeching combustion, a transverse acoustic mode is excited in the duct. This mode produces large slosking velocities near the stabilizer lips which trigger the alternate shedding of vortices. Fresh mixture entrained by these vortices into the hot wake of the bluff body reacts after a certain characteristic time and the heat pulse produced feeds energy into the transverse acoustic mode. The process becomes self sustained if certain phasing conditions are satisfied. Low frequency ducted flame oscillations are studied by Blackshear (1958) and an attempt is made to interpret observations in terms of a linear instability analysis of the flow in the channel. Blackshear assumes that the flow velocity in the fresh stream is uniform

C.-M. Brauner and C. Schmidt-Lainé (eds.), Mathematical Modeling in Combustion and Related Topics, 185–215.

in each section and that the velocity increases linearly in the hot stream. The most unstable wave numbers corresponding to this configuration are determined and compared to experimental observations of the flame front motion. Many different types of combustion driven oscillations are reviewed by Putnam (1971). Current interest in the problem of afterburner and ramjet instabilities has led to many recent experiments on turbulent ducted flames (Keller et al. 1981, Pitz and Daily 1981, Bray et al. 1983, Smith and Zukoski 1985, Poinsot et al. 1987). Spectacular visualisations obtained in these studies indicate that unstable oscillations are associated with large scale motions of the flame front. In certain cases the amplitude of the motion becomes so large that the flame front rolls up into periodic discrete vortices.

This new experimental evidence and some of the earlier observations indicate that theoretical models of ducted flame instabilities must (1) take into account the flame front dynamics (2) contain descriptions of the fresh and burned gas motion. These features are included in the thin flame sheet (TFS) model proposed by Marble and Candel (1978) to study the nonsteady behavior of large combustors. Further studies based on the TFS model are contained in Marble et al. (1979) and Subbaiah (1983). A recent extension of the model is given by Yang and Culick (1986) who include compressibility effects and a description of the recirculating flow behind the flameholder. All these contributions are limited to small amplitude perturbations because the dynamic equations are solved after linearization.

The present paper contains a nonlinear extension of these linear studies. The objective is to establish a method allowing the direct simulation of low frequency ducted flame oscillations based on the thin flame sheet model. This model is attractive because it includes a convenient description of the flame dynamics in a quasi-one dimensional form. Nonlinear oscillations may be investigated with two-dimensional direct simulations as shown for example by Kailasanath et al. (1985). This direct approach is certainly of interest but requires large amounts of super computer time and still raises many unsolved modeling problems.

The thin flame sheet model on the other hand makes a compromise between the details included in the description and the complexity of the numerical solution. Because the flame front position is explicitely used as a variable, the model may be qualified as "one and a half dimensional". Another interesting feature of the TFS model is that it allows inclusion of simple or more complex submodels.

The basic equations of the TFS model are presented in Section 2. Integral dynamic equations are established in Section 3. In the integral formulation a quasi one-dimensional description of the nonsteady flow is obtained. The unknown variables are the pressure, densities and velocities on both sides of the flame sheet and the instantaneous flame location. An important aspect in the derivation of integral relations is the choice of closure assumptions. This important point is also discussed in Section 3. The choice of a numerical method adapted to the mathematical nature of the problem is another important part of the method. The numerical method and the corresponding tests used to determine the code dispersion and dissipation are described in (Poinsot and Candel 1987). Section 4 describes the steady flame spreading calculations. Finally, studies of the flame response under initial spatial excitations are presented in Section 5. Stable and unstable configurations are examined and an instability mechanism is described in some detail.

2. THE THIN FLAME MODEL

Consider a turbulent two-dimensional flame in a duct. The duct has parallel walls, its width is , the blocage is designated by B and the duct length is L (Fig. 1). In this model the turbulent flame thickness is supposed to be small compared to the typical wavelength of acoustic perturbations in the combustor. As a consequence the flame front location may be described by its distance   (x,t) to the lower wall. The flame is viewed as a surface of discontinuity separating the fresh mixture, region 1, and the burned combustion products, region 2. All the flow variables such as the pressure, density, temperature velocity are indicated by subscript 1 in the fresh gases of region 1 and by subscript 2 in the hot products of region 2.

The previous description may be recognized as the compactness condition employed in a similar context by Marble and Candel 1978. Compactness assumptions are commonly used in the analysis of the nonsteady behavior of nozzles and blade rows (see for example Marble and Candel 1977 and Cumpsty and Marble 1977). The compactness condition limits the validity of the model to low frequency oscillations for which the ratio of the pressure perturbation wavelength to the turbulent flame thickness is large.

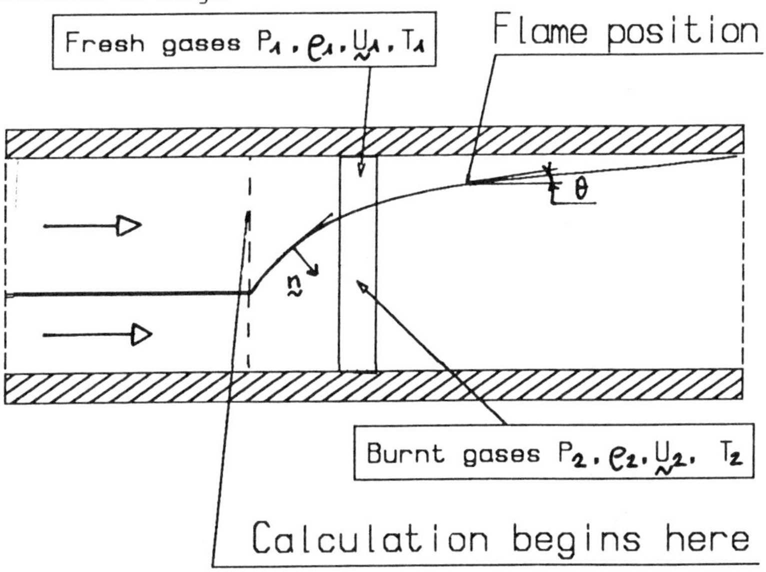

Fig. 1 The thin flame model for a premixed turbulent combustor

To be consistent with the representation of the flame as a thin sheet one also has to specify a normal flame speed w_1 which describes the time rate of displacement of the flame with respect to the fresh gases. In addition it will be convenient to define a normal flame speed relative to the downstream side of the flame which is designated by w_2.
The mathematical definitions of these normal speeds are given by

$$w_1 = (\mathbf{v}_1 - \mathbf{v}_i) \cdot \mathbf{n} \qquad (1)$$

and

$$w_2 = -(\mathbf{v}_2 - \mathbf{v}_i) \cdot \mathbf{n} \qquad (2)$$

where $\underset{\sim}{v}_1$ and $\underset{\sim}{v}_2$ are the flow velocities in the fresh and burned gases and the normal n is oriented in the fresh to burned direction. The absolute velocity of the flame front is $\underset{\sim}{v}_i$ and its projection on the normal n is given by

$$\underset{\sim}{v}_i \cdot \underset{\sim}{n} = - \frac{\partial \eta}{\partial t} / \left[1 + \left(\frac{\partial \eta}{\partial x}\right)^2\right]^{1/2} = - \frac{\partial \eta}{\partial t} \cos\theta \qquad (3)$$

where designates the flame angle with respect to the duct axis (see Fig. 1) and is such that

$$\tan\theta = \partial \eta / \partial x \qquad (4)$$

Clearly the normal flame speed concept summarizes many of the detailed mechanisms involved in turbulent combustion. Its definition and the closure rules specified to relate this quantity to the other flow variables constitute an important aspect of the model which will be discussed in Section 3.

Let us consider now the flame stabilization process. In most practical applications such as afterburners and ramjets, stabilization is achieved by recirculation of hot combustion products behind bluff bodies or backward facing steps (Fig. 2). Now, recirculation poses some important difficulties if one wishes to develop a one dimensional nonsteady model of the combustor. An integral (one-dimensional)

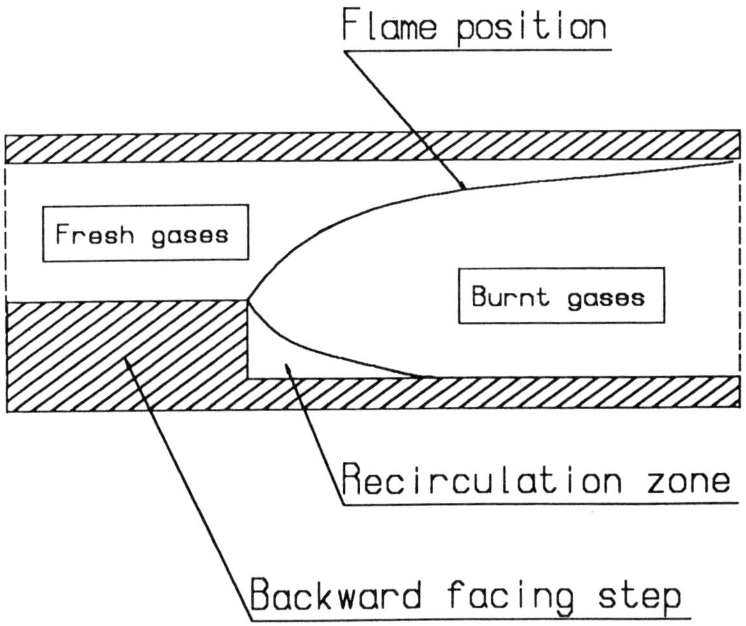

Fig. 2 Stabilization of the premixed flame by a backward facing step

description of the flow with recirculation may be carried for the steady state (see Yang and Culick 1986) but the dynamics of the recirculation region under nonsteady perturbations is not easily included. This difficulty may be overcome if one considers a flame stabilized by a parallel flow of hot gases acting as a pilot flame. The simplest situation is that where the hot gases are at the adiabatic flame temperature pertaining to the fresh mixture (Fig. 3). This geometry is well adapted to a one-dimensional (integral) treatment because no recirculation takes place. Furthermore it is found in some experimental devices (see for example Moreau et al. 1981). This configuration is adopted in this analysis.

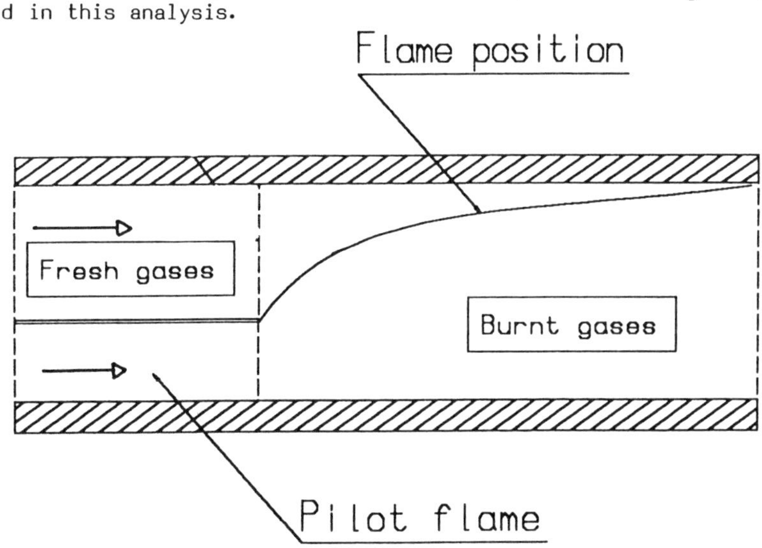

Fig. 3 Stabilization of the premixed flame by a pilot flame

Because the flow in the duct is turbulent and the mean velocity is large, gravitational and viscous effects may be neglected. In this situation the governing equations are those corresponding to each region and a set of matching conditions at the flame.

In regions 1 and 2 the conservation equation of mass and momentum are

$$\frac{\partial \rho_k}{\partial t} + \frac{\partial}{\partial x}(\rho_k u_k) + \frac{\partial}{\partial y}(\rho_k v_k) = 0 \tag{5}$$

$$\frac{\partial}{\partial t}(\rho_k u_k) + \frac{\partial}{\partial x}(\rho_k u_k^2) + \frac{\partial}{\partial y}(\rho_k u_k v_k) = -\frac{\partial p_k}{\partial x} \tag{6}$$

$$\frac{\partial}{\partial t}(\rho_k v_k) + \frac{\partial}{\partial x}(\rho_k u_k v_k) + \frac{\partial}{\partial y}(\rho_k v_k^2) = -\frac{\partial}{\partial y} p_k \tag{7}$$

where the subscript k designates the fresh (k = 1) or burned gases (k = 2). Variables ρ , u, v and p are respectively the density, velocity components and pressure (Fig. 1). Since the flame is a thin sheet, entropy production only takes place at the interface between the fresh and burned gases. Entropy s_1 is constant in the fresh mixture so that

$$p_1 = K_1 \rho_1^{\gamma} \qquad \text{where} \qquad K_1 = e^{s_1/c_v} = \text{const.} \tag{8}$$

In the burned gases entropy remains constant on a streamline but because of flame curvature its value changes across the streamlines. However since the integral model handles mean values it is consistent to assume that s_2 is also a constant and therefore

$$p_2 = K_2 \rho_2^{\gamma} \qquad \text{where} \qquad K_2 = e^{s_2/c_v} = \text{const.} \tag{9}$$

The matching relations at the interface are obtained by specifying kinematic and dynamic conditions.
The kinematic conditions are deduced from (1), (2) and (3) :

$$\frac{\partial \eta}{\partial t} + u_1(x,\eta,t)\frac{\partial \eta}{\partial x} - v_1(x,\eta,t) = w_1 \sec\theta \tag{10}$$

$$\frac{\partial \eta}{\partial t} + u_2(x,\eta,t)\frac{\partial \eta}{\partial x} - v_2(x,\eta,t) = w_2 \sec\theta \tag{11}$$

These conditions express the fact that the flame moves with respect to the upstream and downstream flows with normal velocities w_1 and w_2 respectively.

Mass and momentum balance relations applied across the flame front yield

$$\rho_1 w_1 = \rho_2 w_2 \tag{12}$$

$$p_1 + \rho_1 w_1^2 = p_2 + \rho_2 w_2^2 \tag{13}$$

$$u_1(x,\eta,t) - v_1(x,\eta,t)\tan\theta = u_2(x,\eta,t) - v_2(x,\eta,t)\tan\theta \tag{14}$$

Equations (13) and (14) are the projections of momentum conservation on the normal and tangential directions to the flame sheet. The standard pressure jump across the flame is obtained from (13) while (14) indicates that the tangential velocity is conserved.

The conservation of energy across the flame provides

$$h_1 + \frac{1}{2}(\underset{\sim}{v_1} - \underset{\sim}{v_i})^2 = h_2 + \frac{1}{2}(\underset{\sim}{v_2} - \underset{\sim}{v_i})^2 \tag{15}$$

where h designates the enthalpy per unit mass. Assuming that the fresh and burned mixtures follow the same perfect gas law, that their specific heat c_p is constant and designating the difference in formation enthalpy per unit mass by Δh_f° it is possible to write (15) as follows

$$\frac{T_2}{T_1} = 1 + \frac{(-\Delta h_f^\circ)}{c_p T_1} + \frac{1}{2 c_p T_1}\left[(\underset{\sim}{v_1} - \underset{\sim}{v_i})^2 - (\underset{\sim}{v_2} - \underset{\sim}{v_i})^2\right] \tag{16}$$

Now $(-\Delta h_f^\circ)$ is generally large in comparison with the change in kinetic energy and equation (16) is well approximated by

$$\frac{T_2}{T_1} = 1 + \frac{(-\Delta h_f^\circ)}{c_p T_1} \tag{17}$$

The boundary conditions for the mean flow are

$$
\begin{aligned}
\eta\,(x=0,t) &= B \\
u_1(0,y,t) &= u_{10} \\
v_1\,(0,y,t) &= 0 \\
\rho_1\,(0,y,t) &= \rho_{10} \\
p_1\,(0,y,t) &= p_{10}
\end{aligned}
\tag{18}
$$

The upstream mean flow variables are u_{10}, ρ_{10}, p_{10}.

For the unsteady flow, the boundary conditions at the combustor inlet and outlet are obtained by making use of the Rieman invariants on characteristic lines. At the outlet the pressure is imposed and equal to the ambient. At the inlet we specifically consider situations where the perturbations in axial velocity or pressure are vanishing. In the acoustic limit of small perturbations the vanishing pressure condition corresponds to a zero impedance and is a good approximation of a duct exhaust in the subsonic regime. In that same limit the vanishing axial velocity condition is associated with the existence of a pressure antinode at the flame origin. This situation is representative of 1/4 wave or 3/4 wave resonant modes which are often found in practice (Mugridge 1980, Yang and Culick 1986).

3. INTEGRAL FORMULATION OF THE THIN FLAME MODEL

To obtain a flowfield model in terms of integral variables the conservation equations (5) and (6) are integrated in the transverse direction. Mean values of the flow variables are defined in the fresh and burned gases respectively by

$$\bar{f}_1 = \frac{1}{\ell-\eta}\int_{\eta}^{\ell} f_1\,dy \tag{19}$$

where $f_1 = \rho_1, u_1, p_1$ or T_1 and

$$\bar{f}_2 = \frac{1}{\eta}\int_{0}^{\eta} f_2\,dy\,. \tag{20}$$

where $f_2 = \rho_2, u_2, p_2$ or T_2.

The mathematical derivation of integral equations follows standard lines and it is only carried in this paper for the conservation of mass in the fresh gases. Consider equation (5) for k = 1. Integrating this equation between $y=\eta(x,t)$ and $y=\ell$ one obtains

$$\int_{\eta}^{\ell}\frac{\partial\rho_1}{\partial t}\,dy + \int_{\eta}^{\ell}\frac{\partial}{\partial x}(\rho_1 u_1)\,dy + \int_{\eta}^{\ell}\frac{\partial}{\partial y}(\rho_1 v_1)\,dy = 0$$

When the definitions (19) are introduced in this relation one finds

$$\frac{\partial}{\partial t}\bar{\rho}_1(\ell-\eta) + \rho_1(x,\eta,t)\frac{\partial\eta}{\partial t} + \frac{\partial}{\partial x}\left(\int_{\eta}^{\ell}\rho_1 u_1\,dy\right) +$$

$$+ \rho_1(x,\eta,t)u_1(x,\eta,t)\frac{\partial \eta}{\partial x} + \rho_1(x,\ell,t)v_1(x,\ell,t)$$

$$- \rho_1(x,\eta,t)\, v_1(x,\eta,t) = 0 \tag{21}$$

From the momentum equation (7) projected in the y-direction it is possible to infer that the variation of pressure in each section is small so that the pressure p(x,y,t) may be assimilated to its mean value $\bar{p}_1(x,t)$. The temperature of the fresh gases is also nearly uniform and as a consequence the density of the fresh gases may be replaced by its mean value

$$\rho_1(x,y,t) = \bar{\rho}_1(x,t)$$

Equation (21) becomes

$$\frac{\partial}{\partial t}\bar{\rho}_1(\ell-\eta) + \frac{\partial}{\partial x}\bar{\rho}_1\bar{u}_1(\ell-\eta) + \bar{\rho}_1 u_1(x,\eta,t)\frac{\partial \eta}{\partial x}$$

$$+ \bar{\rho}_1 v_1(x,\ell,t) - \bar{\rho}_1 v_1(x,\eta,t) + \bar{\rho}_1\frac{\partial \eta}{\partial t} = 0 \tag{22}$$

Using the kinematic condition (10) and noting that the transverse velocity v_1 vanishes at the duct wall one obtains

$$\frac{\partial}{\partial t}\bar{\rho}_1(\ell-\eta) + \frac{\partial}{\partial x}\bar{\rho}_1\bar{u}_1(\ell-\eta) + \rho_1 w_1 \sec\theta = 0 \tag{23}$$

Similar procedures lead to integrated forms of the other balance equations

$$\frac{\partial}{\partial t}\bar{\rho}_1\bar{u}_1(\ell-\eta) + \frac{\partial}{\partial x}\bar{\rho}_1\bar{u}_1^2(\ell-\eta) + (\ell-\eta)\frac{\partial \bar{p}_1}{\partial x} =$$

$$- \bar{\rho}_1 u_1(x,\eta,t) w_1 \sec\theta \tag{24}$$

$$\frac{\partial}{\partial t}\bar{\rho}_2\eta + \frac{\partial}{\partial x}\bar{\rho}_2\bar{u}_2\eta - \bar{\rho}_2 w_2 \sec\theta = 0 \tag{25}$$

$$\frac{\partial}{\partial t}\bar{\rho}_2\bar{u}_2\eta + \frac{\partial}{\partial x}\bar{\rho}_2\bar{u}_2^2\eta + \eta\frac{\partial \bar{p}_2}{\partial x} =$$

$$+ \bar{\rho}_2 u_2(x,\eta,t) w_2 \sec\theta \tag{26}$$

To obtain these equations one has to use the conventional assumption that the mean value of u^2 may be approximated by the square of the mean value $(u)^2$. It is worth noting however that the velocities $u_1(x,\eta,t)$ and $u_2(x,\eta,t)$ at the flame front may not be simply replaced by the mean velocities $\bar{u}_1(x,t)$ and $\bar{u}_2(x,t)$ without violating the global conservation of momentum in the duct.

Additional relations are obtained from interfacial matching conditions

$$\bar{\rho}_1 w_1 = \bar{\rho}_2 w_2 \tag{27}$$

$$\bar{p}_1 + \bar{\rho}_1 w_1^2 = \bar{p}_2 + \bar{\rho}_2 w_2^2 \tag{28}$$

$$u_2(x,\eta,t) = u_1(x,\eta,t) + (\bar{\rho}_1/\bar{\rho}_2 - 1) w_1 \sin\theta \tag{29}$$

and from the geometrical expression (4). Entropy conservation expressions for the mean values are

$$p_1 = K_1 \rho_1^{\gamma} \quad \text{and} \quad p_2 = K_2 \rho_2^{\gamma} \tag{30}$$

The system of equations (23) to (26) together with the jump relations (27) to (29) and the isentropic expressions (30) form the basis of the thin flame integral model. Like all integral descriptions and two-phase flow formulations this model requires closure rules.

These rules also designated as phenomenological are suggested by intuition or experiments (Delhaye et al. 1984). Such rules have already been introduced by stating that the pressure and density profiles were uniform in each section and by replacing u^2 by $(u)^2$. These statements are reasonable in the case of highly turbulent ducted flames. Another closure rule needed by the model concerns the velocities at the flame. The difficulty is to provide an expression of $u_1(x, \eta, t)$ (the velocity of the fresh gases next to the front) in terms of the mean velocity of the fresh gases. Subbaiah (1983) gives an analysis of this problem by considering the two-dimensional flow in the fresh stream. He shows that after a short distance from the flame origin $u_1(x, \eta, t)$ may be taken equal to $\bar{u}_1(x,t)$. This closure rule is used in the present calculations

$$u_1(x,\eta,t) = \bar{u}_1(x,t) \tag{31}$$

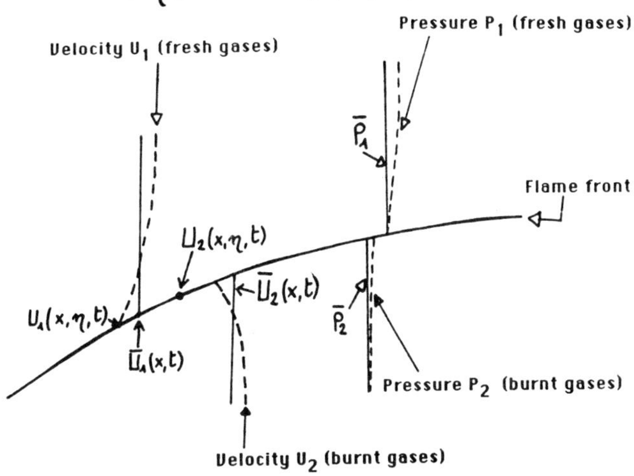

Fig. 4 Closure rules used in the TFS model for pressure, density and velocities

Combining equations (24) and (26) with the continuity equations (23) and (25) and introducing the closure (31) the following system is obtained

$$\frac{\partial}{\partial t}\bar{\rho}_1(\ell-\eta) + \frac{\partial}{\partial x}\bar{\rho}_1\bar{u}_1(\ell-\eta) = -\bar{\rho}_1 w_1 \sec\theta \tag{32}$$

$$\frac{\partial}{\partial t}\bar{\rho}_2\eta + \frac{\partial}{\partial x}\bar{\rho}_2\bar{u}_2\eta = \bar{\rho}_2 w_2 \sec\theta \tag{33}$$

$$\frac{\partial \bar{u}_1}{\partial t} + \bar{u}_1\frac{\partial \bar{u}_1}{\partial x} + \frac{1}{\bar{\rho}_1}\frac{\partial \bar{p}_1}{\partial x} = 0 \tag{34}$$

$$\frac{\partial \bar{u}_2}{\partial t} + \bar{u}_2\frac{\partial \bar{u}_2}{\partial x} + \frac{1}{\bar{\rho}_2}\frac{\partial \bar{p}_2}{\partial x} = \frac{u_{2\eta}-\bar{u}_2}{\eta} w_2 \sec\theta \tag{35}$$

where $u_{2\eta}$ stands for $u_2(x, \eta, t)$
Using the entropy relations the pressures may be directly expressed in terms of the densities.

It is worth indicating how the k_2 constant appearing in (9) is determined. This is done by first writing the interfacial temperature ratio at the flame origin x = 0

$$\left(\frac{\bar{T}_2}{\bar{T}_1}\right)_{x=0} = 1 + \frac{(-\Delta h_f^o)}{c_p\, T_1(x=0)} = R_T \tag{36}$$

Here R_T stands for the temperature ratio of burned to fresh gases and is a given parameter. Now, at x = 0 the following relations must be satisfied

$$\begin{aligned} &\bar{p}_1 + \bar{\rho}_1 w_1^2 = \bar{p}_2 + \rho_2 w_2^2 \\ &\bar{\rho}_1 w_1 = \bar{\rho}_2 w_2 \\ &\bar{p}_1 = \bar{\rho}_1 r \bar{T}_1 \quad , \quad \bar{p}_2 = \bar{\rho}_2 r \bar{T}_2 \end{aligned} \tag{37}$$

where r is the perfect gas constant. In this system all the values pertaining to the fresh gases are imposed as well as the temperature ratio R_T. The set (36)-(37) may be solved exactly and provides all thermodynamic variables for the burned gases at x = 0. For example the density ratio is given by

$$\left(\frac{\bar{\rho}_2}{\bar{\rho}_1}\right)_{x=0} = \frac{1}{2R_T}\left\{1 + \gamma\left(\frac{w_1}{c_{10}}\right)^2 + \left[\left(1+\gamma\frac{w_1^2}{c_{10}^2}\right)^2 - 4\gamma\frac{w_1^2}{c_{10}^2}R_T\right]^{1/2}\right\} \tag{38}$$

where c_{10} is the speed of sound in the fresh gases at x = 0 and the constant K_2 is then obtained from

$$K_2 = K_1 R_T\left[\left(\frac{\bar{\rho}_2}{\bar{\rho}_1}\right)_{x=0}\right]^{1-\gamma}$$

In the limit $w_1 = 0$ expression (38) yields $(\bar{\rho}_2/\bar{\rho}_1)_{x=0} = 1/R_T$ which is the approximation used by Subbaiah (1983). However if one is interested in nonsteady perturbations which affect the density, expression (38) is more precise.

A question of compatibility may be raised between this initial calculation based on the enthalpy equation and the isentropic laws used downstream. It is possible to show that a correct value of the enthalpy jump is obtained in each section (see Poinsot 1987). The temperature ratio $\bar{T}_2/\bar{T}_1$ may be obtained at any point in the combustor.
Its axial variation is of the order M_{10} where M_{10} is the inlet Mach number $M_{10} = u_{10}/c_{10}$. Usually M_{10} is less than 0.1 so that T_2/T_1 is essentially constant.

To complete this integral model one has to specify how the normal flame speed w_1 is obtained. One has to remember that w_1 represents the local turbulent flame speed. It is known that this uantity does not have the universal character of the laminar flame speed because it strongly depends on the flow configuration. However, many experiments performed on ducted turbulent flames indicate that the mean turbulent flame speed is mainly a function of the turbulence intensity as represented for example by the turbulent kinetic energy k

$$w_1 = f\,(k^{1/2})$$

In general the ratio $k^{1/2}/u_{10}$ is a constant so that a simple approximation of w_1 consists of taking

$$w_1 / u_{10} = \text{const.}$$

This approximation has been successfully used to predict steady flame spreading in ducts (see for example the early studies of Tsien 1951, Scurlock 1948 and the recent review of Zukoski 1978). It is also used in the linear analysis of Marble and Candel (1978) and Subbaiah (1983).

Now, many recent experiments (Keller et al. 1981, Bray et al. 1983, Poinsot et al. 1987) indicate that in nonsteady circumstances, w_1 is sensitive to the pressure and its gradient. Due to the pressure perturbations the flame accelerates or decelerates with respect to the incoming stream. The process is not well elucidated at this point but it may be described in terms of sensitivity parameters

$$\frac{\delta w_1}{w_1} = N\frac{\delta \bar{p}}{\bar{p}} + G\frac{\delta \bar{u}}{\bar{u}} \tag{39}$$

This expression indicates that the pressure and velocity fractional variations produce fractional changes of the flame speed. The corresponding pressure and velocity coupling are described by interaction indices N and G. These parameters are functions of the conditions prevailing locally.

In the absence of a more complete understanding of the flame acceleration mechanism we will only consider a pressure coupling. In addition N will be taken to be constant and used as an input parameter of the model.

Under these circumstances the nonlinear extension of (39) is

$$w_1(x,t)/u_{10} = W^+\left[\,p_1(x,t)/p_{10}(x)\,\right]^N$$

Furthermore W^+ will be allowed to vary along the flame to represent the thermal extinction mechanism when the flame reaches the wall

$$w_1(x,t)/u_{10} = W_1^+ f(\eta^+) \left[p_1(x,t)/p_{10}(x) \right]^N$$

where $f(\eta^+)$ is unity far from the wall ($\eta^+ \neq 1$) and goes to zero at the wall ($\eta^+ \longrightarrow 1$) and W_1^+ is a constant.

4. STEADY STATE FLAME SPREADING

The linear analysis of the thin flame model was carried by Marble and Candel (1978), Subbaiah (1983) and Yang and Cuclick (1986). Our scope is to perform a nonlinear study based on that model. One may wish to use analytical methods for that purpose but a numerical treatment is more appropriate.

To begin with, it is worth writing the dynamic equations (32) to (35) in terms of the following reduced coordinates and variables

$$\begin{aligned} &x^+ = x/\ell \;,\; \eta^+ = \eta/\ell \;,\; t^+ = u_{10} t/\ell \;,\; w_1^+ = w_1/u_{10} \\ &u_1^+ = \bar{u}_1/u_{10} \;,\; u_2^+ = \bar{u}_2/u_{20} \;,\; p_1^+ = \bar{p}_1/\rho_{10} u_{10}^2 \\ &p_2^+ = \bar{p}_2/\rho_{10} u_{10}^2 \;,\; \rho_1^+ = \bar{\rho}_1/\rho_{10} \;,\; \rho_2^+ = \bar{\rho}_2/\rho_0 \end{aligned} \tag{40}$$

With these notations the system (32) to (35) becomes

$$\begin{aligned} &\frac{\partial}{\partial t^+}\rho_1^+(1-\eta^+) + \frac{\partial}{\partial x^+}\rho_1^+ u_1^+(1-\eta^+) = -\rho_1^+ w_1^+ \sec\theta \\ &\frac{\partial}{\partial t^+}\rho_2^+ \eta^+ + \frac{\partial}{\partial x^+}\rho_2^+ u_2^+ \eta^+ = \rho_2^+ w_2^+ \sec\theta \\ &\frac{\partial}{\partial t^+} u_1^+ + u_1^+ \frac{\partial}{\partial x^+} u_1^+ + \frac{1}{\rho_1^+}\frac{\partial}{\partial x^+} p_1^+ = 0 \\ &\frac{\partial}{\partial t^+} u_2^+ + u_2^+ \frac{\partial}{\partial x^+} u_2^+ + \frac{1}{\rho_2^+}\frac{\partial}{\partial x^+} p_2^+ = \frac{u_{2p}^+ - u_2^+}{\eta^+} w_2^+ \sec\theta \end{aligned} \tag{41}$$

The same reduced variables may be introduced in the interfacial conditions (27) to (29), in the state relations (30) and in the boundary conditions (18).

To proceed one must solve two problems. The first is the computation of the steady state flame spreading. The second concerns the transient response of the system to a given excitation. These two problems are considered successively in Section 4 and 5.

The steady flame spreading in the duct is easily calculated with a standard marching algorithm applied to system (41) with the time dependent terms discarded. Numerical details may be found in Poinsot (1987). To fix the ideas it is worth giving some results for the steady flame spreading. Reduced parameters which determine this steady state are the duct blocage and length, the reference normal flame speed, the temperature ratio and the upstream Mach number $B^+ = B/\ell$, $L^+ = L/\ell$, $W_1^+ = W_1/u_{10}$, $\bar{T}_2/\bar{T}_1 = R_T$, $M_{10} = u_{10}/c_{10}$.
Figure 5 shows the evolution of the flame position and of the velocities u_1 and u_2 for a typical steady state corresponding to the following set of parameters : $B^+ = 0.85$, $L^+ = 2$, $W_1^+ = 0.1$, $R_T = 4.5$ $M_{10} = 0.06$.

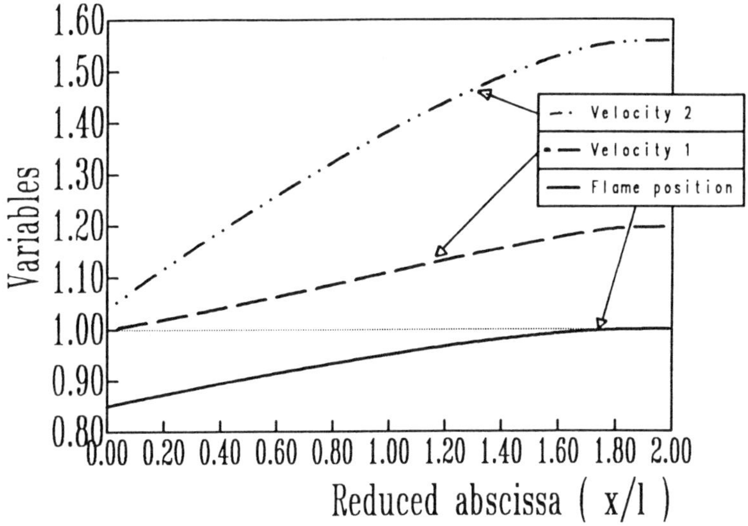

Fig. 5 Variation of the flame position η , the fresh gases velocity u_1 for a steady state flame spreading with $B^+ = 0.85$, $L^+ = 2$, $W_1^+ = 0.1$, $R_T = 4.5$, $M_{10} = 0.06$

This case corresponds to a short combustion chamber and the flame reaches the duct wall at $x/\ell = 1.7$ just before the duct end ($L/\ell = 2$). It must be pointed out that the whole computation is made in a single step. When the flame reaches the wall, the flame speed goes to zero and computation goes on with only burned gases. The normal flame speed variations used for this computation are shown in Figure 6. The great advantage of the model in this case is that it is able to describe large flapping motions of the flame in the upstream and downstream directions as well as acoustic waves inside and outside the flame zone in a single

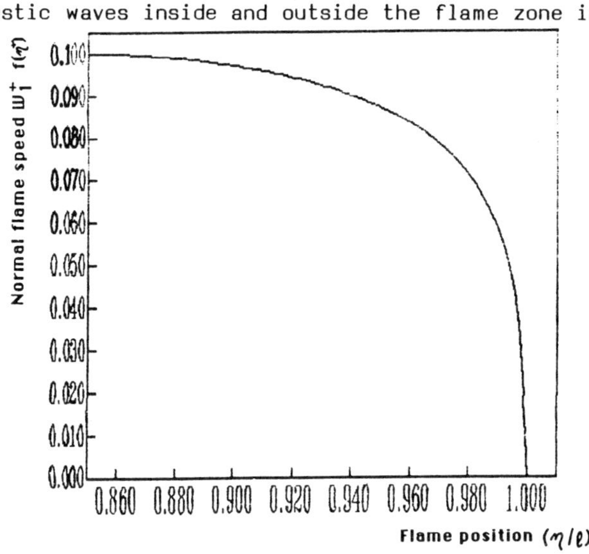

Fig. 6 Variations of the normal flame speed for a steady flame spreading with $B^+ = 0.85$, $W_1^+ = 0.1$

calculation. To verify the effects of these variable flame speed on flame spreading, the conditions used by Subbaiah (1983) and Yang and Culick (1986) : $B^+ = 0.1$, $L^+ = 10$, $W_1^+ = 0.15$, $R_T = 4.5$, $M_{10} = 0.06$ are treated as a test case.

Figure 7 shows that the agreement between our calculation and Subbaiah's results is quite good. One important difference, however, is that the model used by Subbaiah does not apply beyond the point where the flame reaches the wall while our method does not break down at that point. This is essential if one wishes to calculate the nonlinear response of the ducted flame.

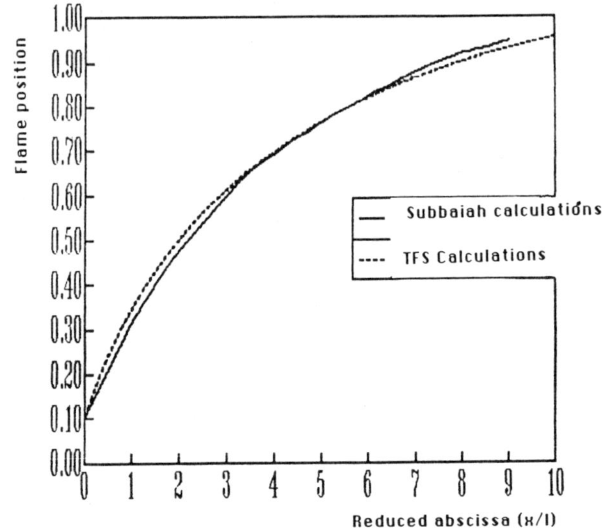

Fig. 7 Comparison of TFS results and Subbaiah calculations for a steady state flame spreading with $B^+ = 0.1$, $L^+ = 10$, $W_1^+ = 0.15$, $R_T = 4.5$, $M_{10} = 0.06$

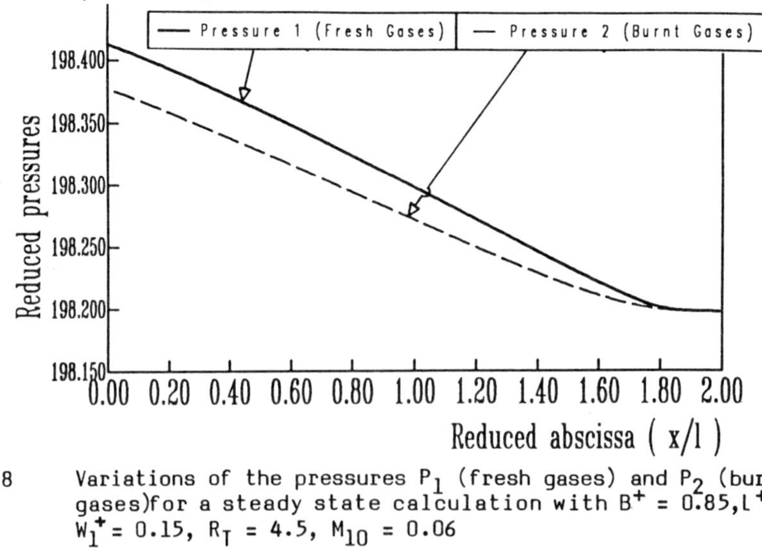

Fig. 8 Variations of the pressures P_1 (fresh gases) and P_2 (burnt gases)for a steady state calculation with $B^+ = 0.85$, $L^+ = 2$, $W_1^+ = 0.15$, $R_T = 4.5$, $M_{10} = 0.06$

Another interesting feature is illustrated by the pressure variations associated with the steady flame spreading (Fig. 8). As indicated before, pressure $\overline{p}_1$ (in the fresh gases) and $\overline{p}_2$ (in the burned gases) are calculated exactly at each duct abscissa by entropy conservation laws (30) and interfacial relations (27) and (28). As long as the flame speed is not zero, $\overline{p}_1$ and $\overline{p}_2$ are different (for $0 < x^+ < 1.7$). When the flame reaches the wall, $\overline{p}_1$ and $\overline{p}_2$ become equal and constant because there are no viscous or friction losses and therefore no pressure drop. The pressure variations between $x = 0$ and $x/\ell = 1.7$ are due to heat addition and to the associated gas acceleration.

5. ANALYSIS OF COMBUSTION OSCILLATIONS

The simulation of combustion oscillations and related instabilities consists of exciting a given steady flow by superposing a chosen perturbation. The numerical solution of system (4-1) is obtained by a finite differences method based on the TRAC code developped for two phase flow problems. This method is fully implicit and involves a pressure solver and two levels of Newton iterations. A complete description of this method is given in (Poinsot and Candel 1987) as well as numerical tests performed on simple situations to determine the stability, dispersionand dissipation characteristics of the scheme. A spatial excitation method is used to introduce perturbations on the steady flow. This method consists of changing some flow variables (usually the pressure) at time t = 0 and then letting the system evolve by itself.

The response of the ducted flame depends on many parameters

1) The steady flow regime characterized by the blockage B^+, the reduced length L^+, the initial normal flame speed W_1^+, the temperature ratio R_T and the upstream Mach number M_{10}.

2) The boundary conditions at the duct ends. Usually the duct outlet is an open end and may be described as a pressure release boundary with an impedance $Z_s = 0$. Two conditions are of special interest for the duct inlet. In one case the inlet may be described as a pressure release boundary ($Z_i = 0$). In the other case the inlet may be treated as a rigid surface (for the perturbations only) : $Z_i = \infty$
In that second case the velocity perturbations vanish at the inlet section. The second case is a simplified representation of some practical situations in which the combustor is a part of a longer system in which the resonant modes exhibit a pressure antinode (a velocity node) at the combustor inlet.

Pressure release boundaries have a vanishing impedance $Z = 0$ while rigid surfaces have an infinite impedance $Z = \infty$ (the classical concept of acoustic impedance which only applies to small perturbations is used here in a wider sense).

3) The perturbation waveform. The perturbation may be of the transient type or it may take the form of a spatial excitation imposed at t = 0. The present analysis is limited to this last case. Sinusoidal oscillations corresponding to a specific duct mode are used to start the calculations. Some transient cases may be found in Poinsot (1987).

4) The amplitude of the perturbation. The amplitude is an important parameter in nonlineat calculations.

5) The interaction index N. This parameter describes the local coupling between the pressure and the combustion process. Because this index is

not known the calculations must be carried for different values of N.

For a given initial perturbation the flame response may take different forms :

(1) the amplitude of the perturbation may decrease and the flow progressively returns to the steady state. This is a stable case.

(2) the amplitude of the perturbation may grow and a self amplifying oscillation may be obtained. This is an unstable case.

These two situations are illustrated below. An essentially stable case corresponding to a short duct excited in the fundamental open-open mode is first described. An unstable case corresponding to a long duct excited in the quarter wave mode is then examined and the mechanism of instability is then discussed in some detail.

5.1 Numerical simulations of the flame response

Among the many possible simulations that may be performed it is of interest to follow guidelines based on experimental studies of instabilities and results derived from linear models.

It is known from ramjet instability studies that three characteristic lengths control the flame response : the duct length L, the flame length L_f and the characteristic wave length λ_c of the convective mode propagating on the flame front. Experiments (see for example Putnam 1971, Mugridge 1980, Heitor et al. 1984, Poinsot et al. 1987) indicate that in low frequency acoustic instabilities the oscillation is coupled to an eigenmode of the system. Quarter-wave ($L = \lambda/4$) and three quarter-wave ($L = 3\lambda/4$) modes are often observed, half-wave ($L = \lambda/2$) modes are also found in ducts with open ends. Half-wave and quarter-wave perturbations will be successively used as initial excitations.

The response of the flame to a half wave perturbation of the pressure is displayed in Fig. 9. The flame front position and pressure oscillate at a reduced frequency $f_1^+ = 8.64$. The value of this oscillation frequency is here directly determined from the numerical simulation. A close estimate of this frequency may be obtained if one uses a local acoustic velocity

$$c_\varphi(x) = \left[\left(\frac{1-\eta}{\bar{\rho}_1} + \frac{\eta}{\bar{\rho}_2}\right) \Big/ \left(\frac{1-\eta}{\bar{\rho}_1 c_1^2} + \frac{\eta}{\bar{\rho}_2 c_2^2}\right) \right]^{1/2} \tag{42}$$

to evaluate the wave travel time from the duct inlet to the duct outlet and back. One then finds that

$$f_1 \simeq 1 \Big/ \int_0^L dx / c_\varphi(x)$$

The local velocity c_φ is derived from two-phase flow studies (for example Stewart and Wendroff 1984) and it is exactly valid when there is no interfacial mass transfer (i.e. when $W_1^+ = 0$).

However the frequency estimates obtained from this expression are within 1 % of the exact values deduced from the numerical calculations. The local wave speed (42) is also useful in the definition of a reference CFL number

$$CFL = (\Delta t/\Delta x)\ Max\,[c_\varphi(x)] \tag{43}$$

As indicated (Poinsot and Candel 1987), this number must be kept sufficiently small to reduce the numerical dissipation.

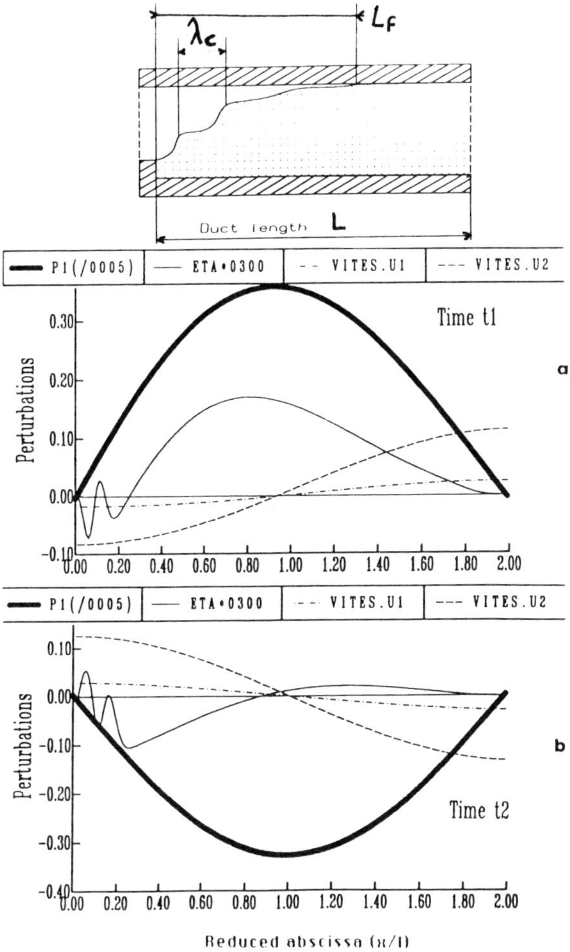

Fig. 9 Spatial excitation of a flame in a short duct.
Steady state characteristics : $B^+ = 0.85$, $L^+ = 2$, $W_1^+ = 0.1$, $R_T = 4.5$, $M_{10} = 0.06$
Initial excitation characteristics : $Z_i = 0.$, $Z_s = 0.$, Sinus of order 1 ($L = \lambda/2$), Amplitude 0.01. Interaction index N = 1
a) variable perturbations at time $t_1^+ = 0.18$
b) variable perturbations at time $t_2^+ = 0.24$

Consider now the perturbations in the flow variables as plotted in Fig. 9 at two time instants $t_1^+ = 0.18$ and $t_2^+ = 0.24$ (the oscillation period is 0.116). The pressure and velocity perturbations p_1 and u_1 exhibit distributions of a standard open-open duct mode. The pressure is nearly sinusoidal and vanishes at both ends of the duct. The velocity perturbation in the fresh gases u_1 reaches its maximum absolute values at the duct ends. The flame front movement is more complex and may be separated in two components. First, the flame moves up and down with the pressure. At time t_1 (Fig. 9a) both perturbations are positive while at time t_2 they are negative (Fig. 9b). A second effect of pressure oscillations on the flame front movement is observed at the duct inlet.

At each acoustic cycle a flame front distortion is generated and then propagates downstream with a wave speed close to the fresh gas velocity u_1. At time t_1, after one acoustic cycle, a single spatial period of this distortion is observed (Fig. 9a). At time t_2 after two acoustic cycles the front perturbation exhibits two full periods (Fig. 9b). The wavelength of this convective mode λ_c is the second characteristic length of the problem.

$$\lambda_c \simeq u_{10}/f_1 \tag{44}$$

Convective modes are observed in many recent studies of unstable combustion (for example Keller et al. 1981, Campbell et al. 1983). The convective wavelength may be assimilated to the typical length scale of the flame front motion. Its value is fixed by the frequency of the resonant acoustic mode which excites the flame.

In the linear stability models of Marble and Candel (1978), and Subbaiah (1983), it is suggested that the flame length L_f constitutes another important scale of the problem. These authors indicate that a strong coupling between the longitudinal acoustic modes and the flame convective modes occurs when the flame length is a multiple of the convective wavelength i.e. when

$$L_f/\lambda_c = n \text{ where n is an integer or equivalently when } fL_f/u_{10} = n \tag{45}$$

This condition may be interpreted as a resonance condition between acoustic and convective perturbations.

These considerations provide useful guidelines for the present analysis. It appears particularly instructive to compare the flame response in two generic situations. In the first case the resonance condition (45) is not satisfied while in the second it is nearly fulfilled. For this comparison the flame length is kept nearly constant while the duct length is changed. The flow characteristics for these two cases are given in Table 1, the characteristic lengths scales appear in Table 2 and the steady flow configurations are displayed in Fig. 10. In case 1 the flame length is much longer than the convective wavelength ($L_f/\lambda_c \simeq 13.5$) and a resonant coupling between the acoustic and convective modes is improbable. In case 2 the linear criterion (45) is nearly satisfied ($L_f/\lambda_c \simeq 2.05$ for the half-wave mode and $L_f/\lambda_c \simeq 1.02$ for the quarter-wave oscillation). This case is expected to be more unstable than the first configuration.

	Case 1	Case 2
Blockage B^+	0.85	0.85
Reference flame speed w_I^+	0.1	0.09
Mach number M_{10}	0.06	0.15
Duct length L^+	2	6
Frequency of mode half-wave	8.64	1.18
Frequency of quarter-wave mode	4.32	0.59

Table 1 Characteristics of the two cases examined in the numerical simulation

		Case 1	Case 2
Half wave mode	Convective wavelength	0.116	0.849
	L_f/λ_c	13.53	2.05
Quarter wave mode	Convective wavelength	0.232	1.70
	L_f/λ_c	6.76	1.02

Table 2 Characteristic lengths scales for the two test cases examined in the simulation

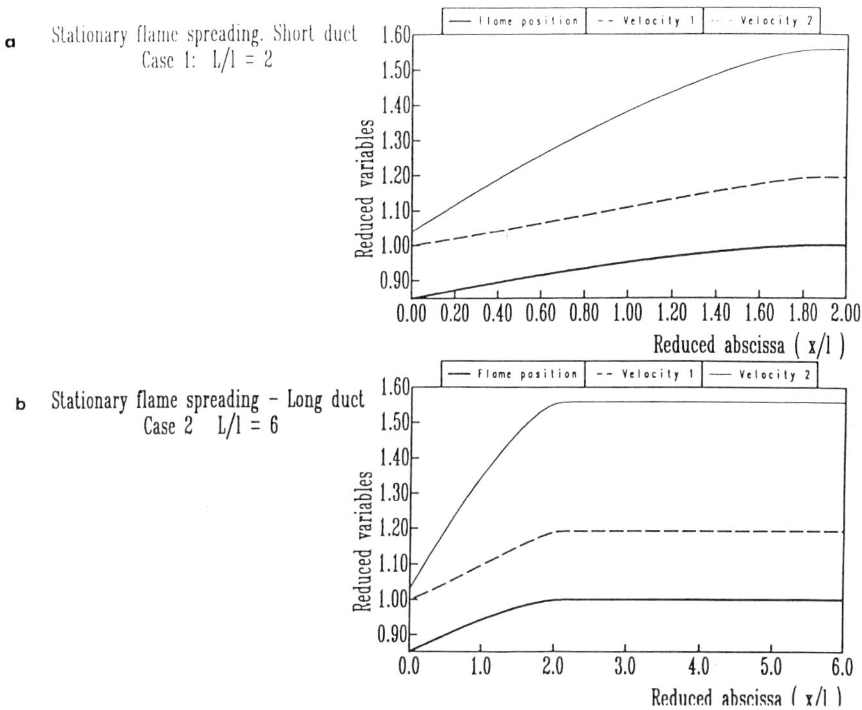

Fig. 10 Flow configuration for two typical flames :
a) case 1 : $B^+ = 0.85$, $L^+ = 2$, $W_1^+ = 0.1$, $R_T = 4.5$, $M_{10} = 0.06$ (short duct criterion (45) not fulfilled)
b) case 2 : $B^+ = 0.85$, $L^+ = 6$, $W_1^+ = 0.09$, $R_T = 4.5$ $M_{10} = 0.15$ (long duct criterion (57) fulfilled)

To rate the stability characteristics of the system one may consider the growth or decay of various flow perturbations. If the perturbations are small compared to the mean it is natural to examine the evolution of the acoustic energy stored in the system. For uniform flows this energy density takes the form

$$E = \frac{1}{2}\frac{p'^2}{\rho c^2} + \frac{1}{2}\rho u'^2 + \frac{\bar{u}\,u'p'}{c^2} \qquad (46)$$

where p', u' are the pressure and velocity perturbations p' = p-p, u' = u'-u. For the present compound flow one may take

$$\bar{E}(t) = \frac{1}{L\ell}\int_0^L dx\,[(\ell-\eta)E_1 + \eta E_2] \qquad (47)$$

This average density is a function of time. If E(t) decreases after the initial excitation the flame is stable, if it increases it is unstable. The amplification factor per cycle may be taken to be

$$a = E(t_0 + T) / E(to) \qquad (48)$$

where T is the acoustic period and to is an arbitrary time instant.

Amplification factors corresponding to half-wave and quarter-wave excitations are respectively given in Tables 3 and 4. The calculations correspond to an initial excitation amplitude $Max(p'/p_{10}) = 0.04$.

Interaction Index :	N = 0	N = 1	N = 2	N = 3
Case 1	0.84	0.95	1	1.05
Case 2	0.80	0.84	0.89	0.94

Table 3 Acoustic amplification for a flame excited in the fundamental open-open mode ($Z_i = 0$, $Z_s = 0$)

Interaction Index :	N = 0	N = 1	N = 2	N = 3
Case 1	0.88	0.99	1.11	1.23
Case 2	0.85	1.24	1.90	2.75

Table 4 Acoustic amplification for a flame excited in the quarter wave-mode ($Z_i = \infty$, $Z_s = 0$)

The results obtained deserve some comments.

(1) It is first noted that the energy amplification factors given in Tables 3 and 4 do not change if the initial excitation amplitude is lowered. This indicates that the coupling between the flame and the duct resonant mode is still in the linear regime.

(2) The amplification factor remains less then one when the interaction index N is zero. When this index is increased the amplication factor also increases and in certain cases it takes values larger than unity.

(3) As expected a strong coupling is observed when the resonant condition (45) is satisfied. In case 1 this condition is not fulfilled and an unstable response is only obtained for large values of N (in excess of 2). As the amplification factor remains close to one even when N > 2 it

is likely that the corresponding mode of oscillation would be damped if dissipation effects were added to the model.

Consider now case 2. The linear criterion (45) is now satisfied. The half-wave mode remains stable even for high values of N but the quarter-wave mode becomes unstable when N only exceeds 0.4.
The flow configuration examined in case 2 resembles that of many recent experiments (Smith and Zukoski 1985, Heitor et al. 1984, Poinsot et al. 1987) which all reveal highly unstable quarter-wave modes. The sensitivity of ramjet combustors to quarter-wave oscillations is also found in the linear analysis of Yang and Culick (1986).

From this discussion it appears that the mean flow configuration and the interaction index strongly influence the flame response. When N = 0 the present model indicates that the flame is stable. Instability is observed when N differs from zero and it is strongest when a resonant interaction between acoustic and convective modes develops. The quarter-wave mode is more unstable than the half-wave oscillation.
Numerical simulations indicate that the coupling is particularly strong for odd values of n in the criterion (45).
Some differences are also found between the earlier studies of Marble and Candel (1978) and Subbaiah (1983) and the present work. Instability was predicted in these references even for zero values of the interaction index while the present calculations indicate that the flame is stable for N = 0. The difference may be due to some changes in the basic model such as the inclusion ofmean compressibility effects and the representation of flame extinction at the wall which are contained in the present model and were absent in the previous descriptions.

5.2 Instability mechanism for the quarter-wave mode

The instability mechanism corresponding to the flow regime of case 2 under a quarter-wave initial excitation is now considered in more detail. The initial amplitude Max (p'/p_{10}) = 0.03 and the interaction index N = 1. Figure 11 shows that the mean energy $\bar{E}(T)$ increases. The perturbed flame structure is now described at five instants t_1 ... t_5 of

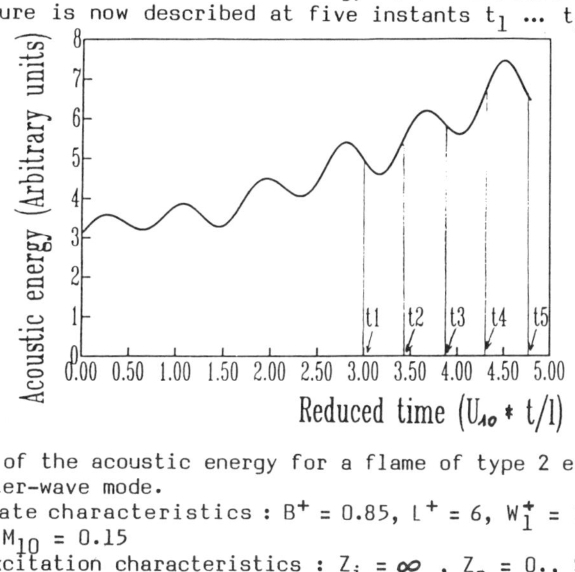

Fig. 11 Variation of the acoustic energy for a flame of type 2 excited by a quarter-wave mode.
Steady state characteristics : $B^+ = 0.85$, $L^+ = 6$, $W_1^+ = 0.09$ $R_T = 4.5$, $M_{10} = 0.15$
Initial excitation characteristics : $Z_i = \infty$, $Z_s = 0.$, Sinus order 1/2 ($L = \lambda/4$), Amplitude 0.03, Interaction index N = 1

the third oscillation cycle. As shown in Fig. 12, time instants t_1, t_3, t_5 correspond to zero values of the pressure perturbation in the combustor at the point $x/\ell = 1$. While instants t_2 and t_4 are relative to minimum and maximum pressure values. The global nonsteady rate of reaction in the duct is also shown in this Figure. This quantity is obtained by integrating the local mass flow rates across the flame

$$\dot{\omega} = \int_0^L \rho_1 w_1 \, dx/\cos\theta$$

Note that the nonsteady heat release in the duct is directly proportional to this expression

$$\dot{Q}(t) = (-\Delta h_f^o)\,\dot{\omega}$$

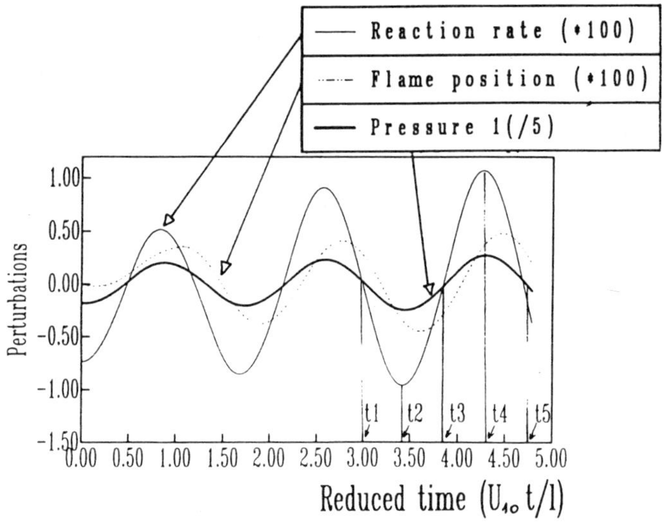

Fig. 12 Time variation of pressure, flame position at the combustor point ($x/\ell = 1$) and of the global heat release $\dot{Q}(t)$ (Same conditions as Fig. 11)

The perturbations in pressure, flame position and velocities corresponding to the five selected instants of time are displayed in Fig. 13 to 18. At time t_1 (Fig. 13), the pressure is almost zero at all points in the duct. The flame front movement results from a combination of acoustic and convective modes. From $x/\ell = 0$ to $x/\ell = 0.5$ the perturbation is negative indicating that this part of the flame has been lowered in the duct. From $x/\ell = 0.5$ to $x/\ell \simeq 1.7$ the perturbation is positive indicating that the flame is located on the upstream side of its steady location. The wall is reached at $x/\ell \simeq 1.8$.

At instant t_1 the pressure is low and the interaction between the reaction rate and the pressure wave is small according to the Rayleigh criterion (the applicability of this criterion to the present situation is questionable. It is used here to obtain an intuitive understanding of the mechanism).

At instant t_2 the pressure is negative everywhere in the duct (Fig. 14). The velocity perturbations are nearly zero. The flame is in its lowest position. The global reaction rate is minimum and the acoustic energy is increasing (as shown in Fig. 12). This growth is due to a pumping mechanism whereby pressure and heat release perturbations are both negative and satisfy the Rayleigh criterion.

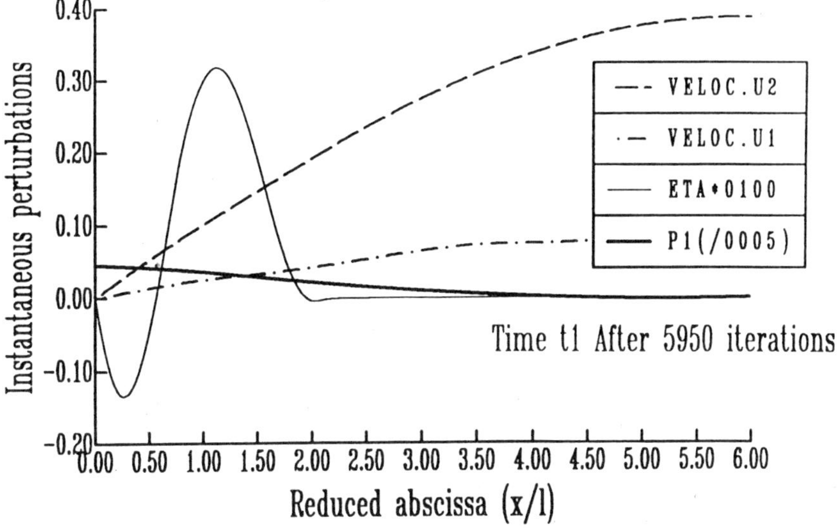

Fig. 13 Pressure, flame front position and velocities at time t_1 (Same conditions as Fig. 11)

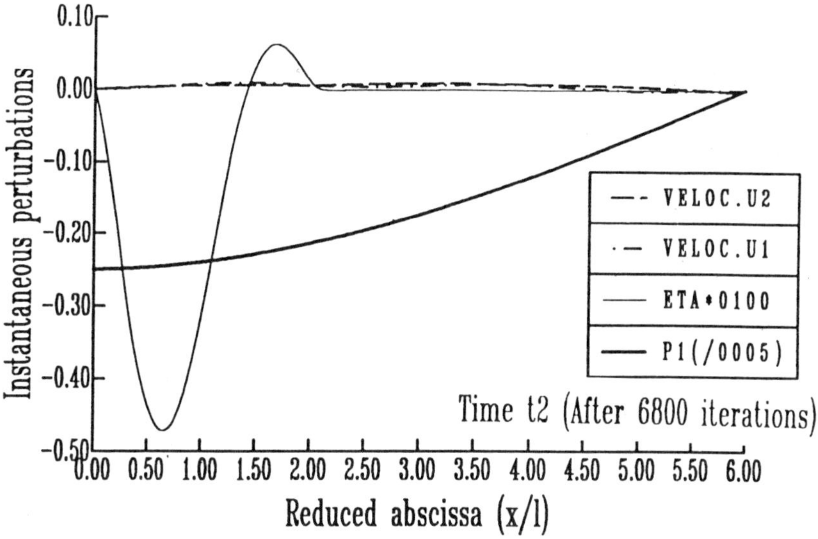

Fig. 14 Pressure, flame front position and velocities at time t_2 (Same conditions as Fig. 11)

At instant t_3 (Fig. 15), the pressure is again close to zero and the flame has begun to move above its steady position in the duct. This motion has a convective character and starts at the duct inlet. Between x/ℓ = 0.6 and x/ℓ = 1.8 the flame front is below its steady position and moving upwards. Energy addition to the acoustic oscillation is small in this situation.

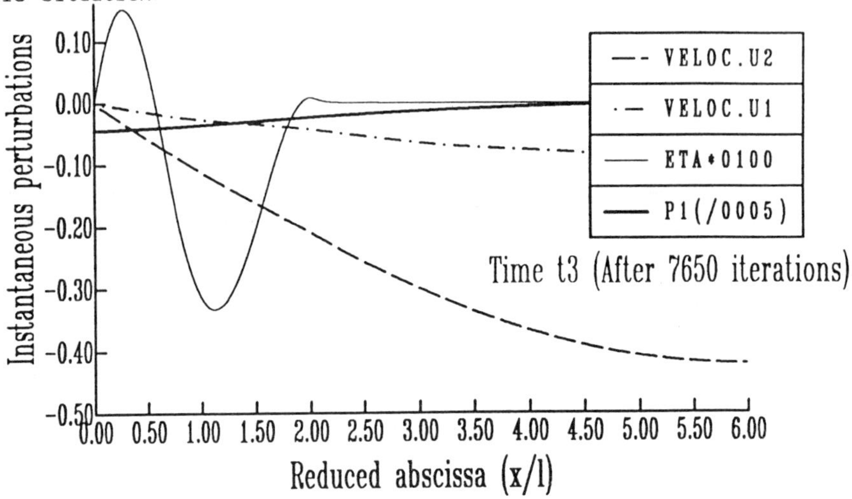

Fig. 15 Pressure, flame front position and velocities at time t_3 (Same conditions as Fig. 11)

At instant t_4 (Fig. 16) the pressure is maximum in the chamber. The flame front has moved upstream and reached its limit location. As the pressure is high the coupling between the acoustic field and the heat release is significant and the acoustic energy is increased.

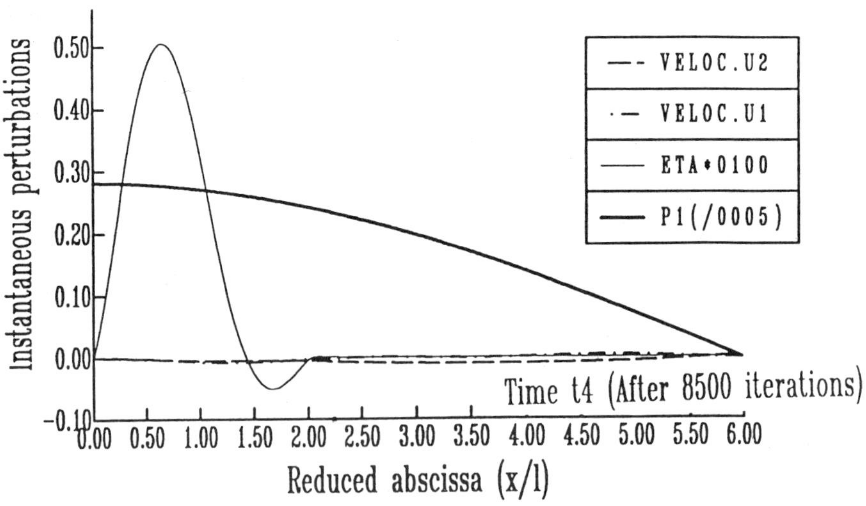

Fig. 16 Pressure, flame front position and velocities at time t_4 (Same conditions as Fig. 11)

Finally the flame configuration at instant t_5 is similar to that found at t_1 but the perturbation amplitude is higher (Fig. 17).

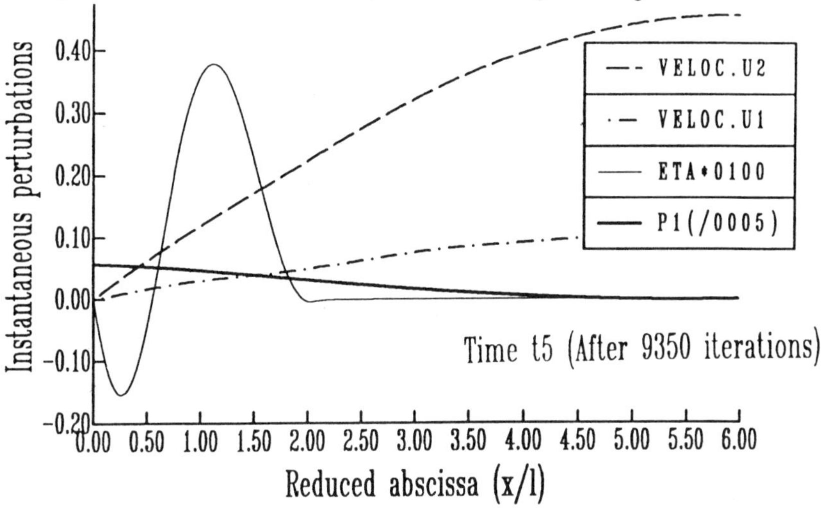

Fig. 17 Pressure, flame front position and velocities at time t_5 (Same conditions as Fig. 11)

These results are summarized in Fig. 18 which displays the instantaneous position of the front along with the pressure and global rate of reaction.

5.3 Discussion

It is worth noting that the instability mechanism described above is not associated with large mass flow perturbations at the duct inlet. The inlet is a velocity node and the mass flow perturbations are only due to relatively small density oscillations. The two cases described in this paper (zero and infinite impedance conditions at the duct inlet) are clearly simplified situations where the pressure and velocity effects are decoupled. However these limit cases give a basis for studies of more general configurations. The present calculations indicate how a quarter-wave pressure perturbation may trigger an unstable flame response. The sensitivity of combustion to quarter-wave acoustic modes has been known for a long time (see for example Putnam 1971, Heitor et al. 1984, Mugridge 1980). It is found here that a strong combustion-acoustic coupling is obtained if the convective wavelength coincides with the flame length itself

$$\lambda_c = u_{10}/f = L_f \tag{49}$$

This result confirms and refines an earlier finding of Marble and Candel (1978). Their linear analysis indicated that strong flame responses could be obtained for

$$L_f = n\lambda_c \tag{50}$$

Our calculations show that even values of n lead to stable flame responses while odd values of n (and in particular n = 1) lead to

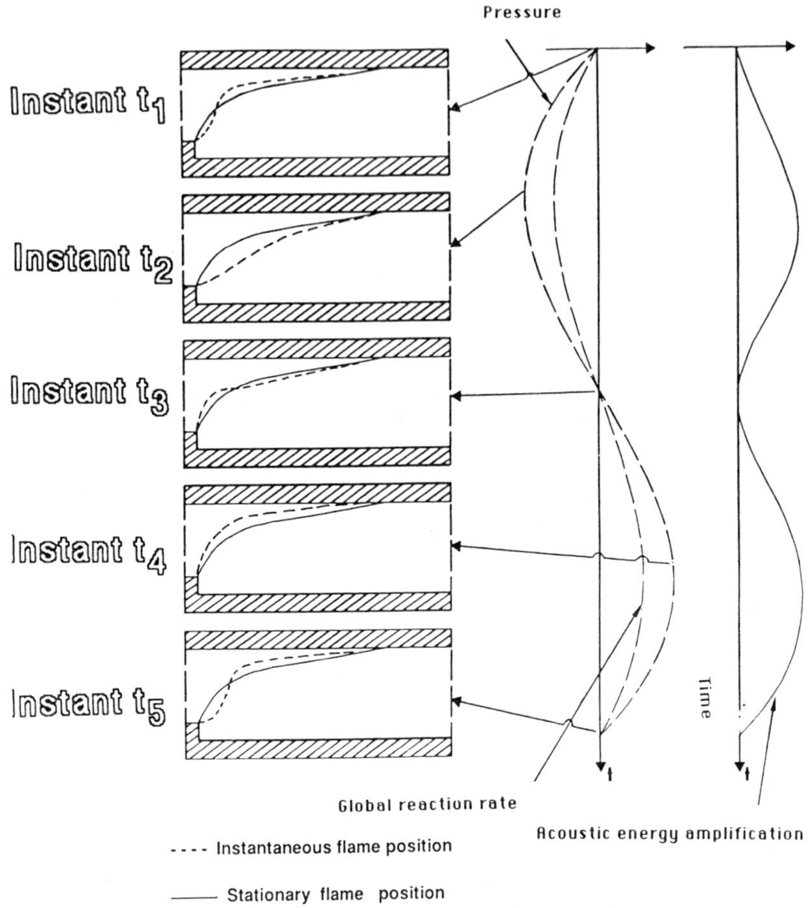

Fig. 18 Quarter wave combustion instability mechanism

unstable behavior. Expression (49) may be used in a simple way to predict certain ramjet instabilities. The frequency of the quarter-wave mode may be obtained from the approximate wave speed (42). One finds that

$$f = (1/4)/\left(\int_0^L dx/c_\varphi\right)$$

The fresh gas velocity u_{10} is usually known and the flame length may be obtained from measurements or numerical calculations. It is then a simple matter to see if condition (49) is satisfied.

A rough test of this condition may be carried on the basis of some recent experiments. The convective wavelength and the flame length corresponding to three combustor configurations are given in Table 5. The flame lengths in the studies of Smith (1985) and Keller et al. (1982) were determined from Schlieren views. The flame length in the work of Poinsot et al. (1987) was obtained from computerized gas analysis (see Darabiha et al. 1986). The "**humming**" mode described by Keller et al. (1982) is chosen because it is clearly acoustically coupled. The

	Mode	Frequency Hz	Velocity u_{10}(m/s)	λ_c (m)	L_f (m)
Smith 1985	Quarter-wave	457	70	0.15	0.15
Keller et al.1982	"Humming"	175	13.3	0.075	0.10
Poinsot et al.	Quarter-wave	440	41	0.093	0.10

Table 5 Convective length λ_c and flame length L_f for three different experimental configurations

estimated convective wavelengths and flame lengths coincide in the three cases indicating that certain ramjet instabilities may be predicted on the basis of expression (49). Of course this is only a necessary condition and it cannot represent all the possible oscillations found in practice.

5.4 Some nonlinear features

In principle, direct simulation methods allow the prediction of linear growth and limit cycles of unstable oscillations. The calculation of limit cycles is in fact a difficult problem which is not yet fully solved with the present model. Nevertheless certain nonlinear features of large amplitude oscillations may be exhibited with the TFS model. Consider once more case 2 (see Table 1) under a quarter-wave excitation with a high initial amplitude (Max $p'/p_{01} = 0.1$). In this situation the evolution of the average energy (47) differs markedly from that found in the linear regime. The energy increases only once in a cycle when the flame reaches its limit position upstream. This behavior may be understood if one examines the evolution of the global rate of reaction and of the pressure and flame position at the duct midpoint. The global rate of reaction is now strongly nonlinear while the pressure perturbation remains sinusoidal (Fig. 19). This phenomenon is found in many experiments (see Lang 1986, Poinsot 1987). It is due in the present model to a "**saturation**" of combustion when the flame travels upstream. The flame front reaches the wall at $t^+ = 1$ and remains at this location until $t^+ = 1.6$. This explains why the global rate of reaction diminishes rapidly during that time. This calculation indicates that the model is able to deal with nonlinear mechanisms. However the closure rule used to describe the influence of pressure on the normal flame speed is certainly invalid in the case of very large amplitudes because the flame speed is bounded by physical mechanisms. There are also many circumstances where the flame front rolls up into large scale patterns. This situation cannot be represented with the TFS model.

6. CONCLUSION

A direct simulation of low frequency combustion instabilities is described in this paper. The simulation is based on the thin flame sheet model derived by Marble and Candel in their investigation of ramjet and afterburner instabilities. While earlier studies based on this description were performed in the linear range (for small perturbations) the present model may be used to study large amplitude perturbations. The numerical calculations are performed with a fully implicit finite differences method.

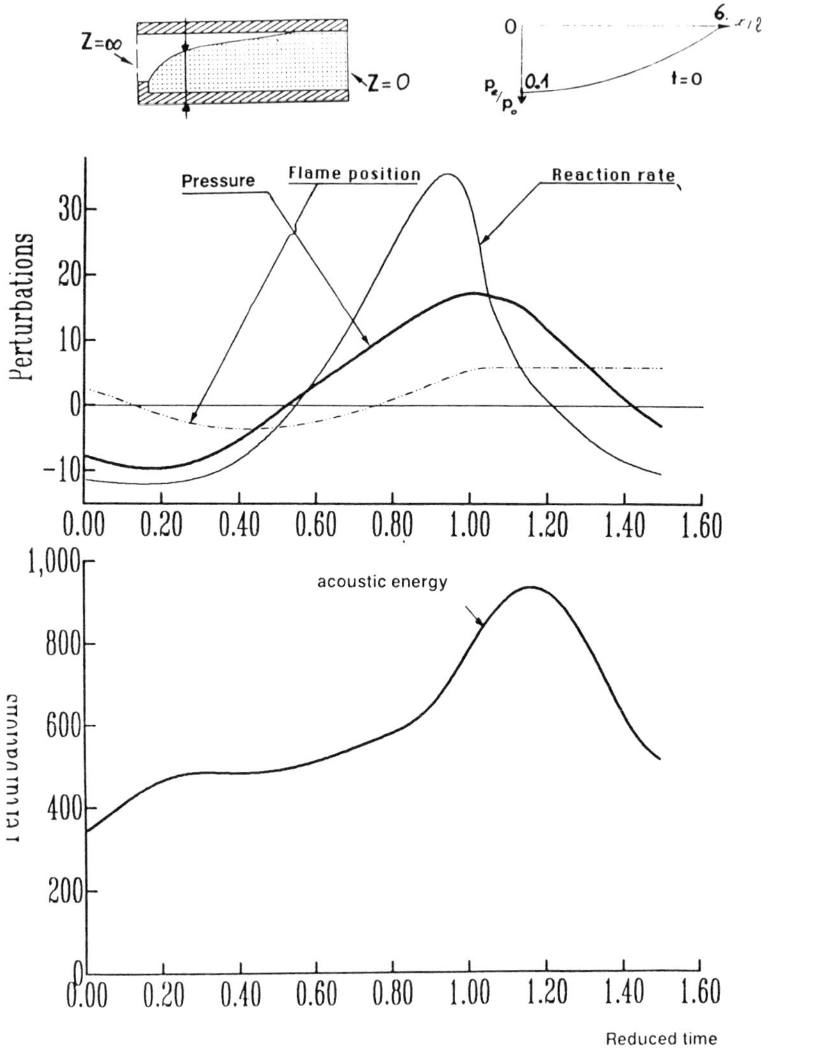

Fig. 19 Time variations of pressure, flame front position at the combustor point located at x/ℓ = 1 and of the global heat release $\dot{Q}(t)$. Time variation of the acoustic energy.

Results of calculations show that stability is mainly determined by the steady flow configuration and by the interaction parameter introduced to represent the effects of pressure on the local normal flame speed. Quarter-wave excitations induce strong instabilities when convective perturbations travelling on the flame front couple with the acoustic oscillation. In this circumstance the convective wavelength coincides with the flame length.

REFERENCES

Barrère M. and Williams F.A. (1968) . Comparison of combustion instabilities found in various types of combustion chambers. 11th Int. Symp. on Comb. 169-181.

Blackshear P.L. (1956). Growth of disturbances in a flame generated shear region. NACA Report 1360

Blackshear P.L., Rayle W.D. and Tower L.K. (1955) . Study of screeching combustion in a 6-in simulated afterburner. NACA TN 3567

Bray K.N.C., Campbell I.G., Lee O.K.L and Moss J.B. (1983). An investigation of reheat buzz instabilities. AASU Memo 83/2.

Brown G.L. and Roshko A. (1974). On density effects and large structure in turbulent mixing layers. J. of Fluid Mech. 64, 775-816.

Burgess J.C. (1981). Active adaptive sound control in a duct: A computer simulation. J. Acous. Soc. Am. 70, 3, 715-726.

Campbell I.G., Bray K.N.C. and Moss J.B. (1983). Combustion oscillations in a ducted premixed flame. IMechE C69/83, 85-93.

Crocco L. (1965). Theoretical studies on liquid propellant rocket instability. 10th Symp. Int. on Comb. 1101-1128.

Crocco L. and Cheng S.I. (1956). " Theory of combustion instability in liquid propellant rocket motors. " Agardograph No8. Butterworths Sci. Pub.

Crocco L. (1951). Aspects of combustion instability in liquid propellant rocket motors. Part I. J. of Am. Rocket Soc. 21, 163-178.

Crocco L. (1952). Aspects of combustion instability in liquid propellant rocket motors. Part II. J. of Am. Rocket Soc. 22, 7-16.

Crocco L. (1969). Research on combustion instability in liquid propellant rockets. 12th Symp. (Int) on Combustion

Cumpsty N.A. and Marble F.E. (1977). The interaction of entropy fluctuations with turbine blade rows; a mechanism of turbojet engine noise. Proc. Royal Soc. London 357, 323-344.

Delhaye J.M., Giot M. and Riethmuller M.L. (1984). "Thermohydraulics of two phase systems applied to industrial design and nuclear engineering." Hemisphere Mc Graw-Hill.

Harrje D.T. and Reardon F.H. (1972). Liquid propellant rocket combustion instability. NASA SP-194.

Heitor M.V., Taylor A.M. and Whitelaw J.H. (1984). Influence of confinement on combustion instabilities of premixed flames stabilized on axisymmetric baffles. Comb. and Flame 57, 109-121.

Kailasanath K., Gardner J., Boris J. and Oran E. (1985). Acoustic vortex interactions in a idealized ramjet combustor. 22nd JANNAF Comb. Meeting October.

Keller J.O. and Westbrook C.K. (1986). Response of a pulse combustor to changes in fuel composition. 21st Symp. (Int.) on Combustion Munich 3-8 August.

Kourta A., Braza M., Chassaing P. and Haminh H. (1987). Numerical analysis of natural and excited 2D mixing layers. AIAA J. 25, 2, 279-286.

Lang W. (1986). Dynamik und Stabilitat selbsterregter Verbrennungsschwingungen beim Auftreten mehrerer Frequenzen. Thèse Dr Ing. Technische Universitat Munich. 30 Juin.

Marble F.E. and Candel S. (1978). An analytical study of the non steady behavior of large combustors. 17th Symp. Int. on Comb., Pittsburgh 761-769.

Marble F.E. and Candel S.M. (1977). Acoustic disturbances from gas nonuniformities convected through a nozzle. J. Sound Vib. 55, 225-243.

Marble F.E. and Cox D.W. (1953). Servo stabilization of low frequency oscillations in a liquid bipropellant rocket motor. J. of Am. Rocket Soc. 23, 2, 63-81.

Marble F.E., Subbaiah M.V. and Candel S.M. (1979). Analysis of low frequency disturbances in afterburners. Proc. Spec. Meeting on Comb. Modelling 17.1-17.11.

Mugridge B.D. (1980). Combustion driven oscillations. J. of Sound and Vibration 70, 3, 437-452.

Pitz R.W. and Daily J.W. (1983). Combustion in aturbulent mixing layer formed at a rearward facing step. AIAA J. 21, 11, 1565-1570.

Poinsot T. and Candel S. (1986). The influence of differencing and CFL number on implicit time dependant non linear calculations. J. of Comp. Physics 62, 2, 282-296.

Poinsot T., Trouve A., Veynante D., Candel S. and Esposito E. (1987). Vortex driven acoustically coupled combustion instabilities. J. of Fluid Mechanics. 177, 4, 265-292.

Putnam A.A. (1971). "Combustion driven oscillations in industry " Fuel and Energy Science Series. J.M. Beer Ed. American Elsevier N-Y.

Rogers D.E. and Marble F.E. (1956). A mechanism for high-frequency oscillations in ramjet combustors and afterburners. Jet Prop. 26, 456-462.

Schmitt D. and Lourme D. (1982). A model of instability of liquid propellant engine with radial injection. Acta Astronautica 9, 445-454.

Scurlock A.C. (1948). Flame stabilization and propagation in high velocity gas streams. MIT Cambridge Mass., Meteor Report 19.

Smith D.A. and Zukoski E.E. (1985). Combustion instability sustained by unsteady vortex combustion. AIAA/SAE/ASME/ASEE 21st Joint Prop. Conf. July, Monterey.

Stewart H.B. and Wendroff B.B. (1984). Two-phase flow: models and methods. LA-UR 83 2124.

Subbaiah M.V. (1983). Non-steady flame spreading in two dimensional ducts. AIAA J. 21, 11, 1557-1564.

Tsien H.S. (1952). Servo Stabilization of combustion in rocket motors. J. Amer. Rocket Soc. 22.

Tsien H.S. (1951). Influence of flame front on the flow field. J. of Applied Mechanics 6, 188-194.

Winant C.D. and Browand F.K. (1974). Vortex pairing, the mechanism of turbulent mixing layer growth at moderate Reynolds numbers. J. of Fluid Mech. 63, 237-255.

Yang V. and Culick F.E.C. (1986). Analysis of low-frequency combustion instabilities in a laboratory ramjet combustor. Comb. Sci. Tech. 45, 1-25.

Zukoski E.E. (1978). Afterburners ('The aerothermodynamics of aircraft turbines'). edited by G.G. Oates, AFAPL-TR-78-52.

NON LINEAR EFFECTS OF BLOW UP AND LOCALIZATION PROCESSES IN BURNING PROBLEMS

A.A. SAMARSKII

KELDYSH INSTITUTE OF APPLIED MATHEMATICS, USSR ACADEMY OF SCIENCES, MOSCOW

When complex physical processes of burning are studied some major problems arise due to a nonlinear nature of mathematical models. This nonlinearity disables the superposition principle, which is inherent to linear problems. It results in a variety of possible evolution paths for a dissipative process and determines discrete spatial-temporal scales in continuum and so the properties of a nonlinear medium, which do not depend on an external perturbation. Nonlinear dissipative media with bulk heat sources and sinks may possess a certain intrinsic ordering which is characterized by spontaneous development of complex dissipative structures in a medium. In the course of evolution the self-organization process arises.

Even the simplest nonlinear burning models have these properties. The evolution principles and the spatial-temporal "architecture" of dissipative structures that arise during burning can be studied in detail on simple models extracted from comprehensive mathematical models by revealing the basic mechanisms that are responsible for the phenomena under consideration.

In this report a description is given of the nonstationary burning process properties in continuum by employing the mathematical models formulated in the parabolic approximation. An emphasis is made on essentially nonstationary burning regimes. The nonlinear evolution problems that allow unbounded solutions (the blow up regimes) were long considered as some exotic examples with a "pathological" behavior, however such singularity in time solutions is connected with the thermal explosion in the theory of burning, the laser beam self-focusing in nonlinear media, the T-layer effect in magnetic hydrodynamics, etc. The investigation of strong nonstationary effects, even in the framework of simple nonlinear parabolic models, require considerable efforts in both the theory of nonlinear equations and the construction of proper numerical algorithms because the burning blow up regime has a singularity which disables traditional techniques. Such simple nonlinear parabolic models allow the study of features of a blow up regime, for example the localization of burning in a dissipative medium, when during the whole time of burning heat conduction cannot spread temperature outside a certain fixed region.

In this report a number of new methods used in the theory of parabolic equations are discussed, including the group analysis of nonlinear equations, which provides detection of unusual temperature waves in nonlinear media, and construction of stable numerical algorithms for extremely nonlinear equations.

1. THE MODEL : NONLINEAR THERMAL CONDUCTIVITY IN A MEDIUM WITH BULK ENERGY RELEASE

The below results are summarized in the review [1] where a comprehensive list of references is presented.

C.-M. Brauner and C. Schmidt-Lainé (eds.), Mathematical Modeling in Combustion and Related Topics, 217–231.

1. The model under consideration is based on the quasilinear parabolic equation which in the nondimensional form is

$$u(t) = \nabla (u^{\sigma}\nabla u) + u^{\beta}, \; u = u(t,x), \tag{1}$$

with a perturbed background as initial data

$$u(0,x) = u_o(x) \geq 0, \; x \in R^N \; ; \; u_o^{\sigma+1} \in C^1(R^N), \tag{2}$$

where u is temperature and $\sigma > 0$, $\beta > 0$ ($k(u) = u^{\sigma}$ is typical of the electron thermal conductivity in plasma). The strong nonlinearity of (2) results in a possible blow up effect with unlimited growth of temperature perturbation

$$\limsup_{t \to T_o} \sup_x u(t,x) = +\infty \tag{3}$$

in the finite blow up time $T_o > 0$.

A behavior of the thermal wave front essentially depends on a relation between σ and β.

If $\beta < \sigma+1$, the wave propagates with an increasing velocity, the temperature growing all over the medium

$$|X_{\phi}(t)| \sim (T_o - t)^m, \quad m = \frac{\beta - (\sigma+1)}{2\,(\beta - 1)}, \tag{4}$$

so that $|X_{\phi}(t)| \to +\infty$ when $t \to T_o$. It is the so called HS burning regime (Fig. 1).

The case $\beta \geq \sigma+1$, when the burning localization takes place, is most interesting.

Theorem : Let $\beta = \sigma+1$ (S-regime) and $N = 1$. If $X_-(t)$ and $X_+(t)$ are the left and the right front coordinates of the spatially finite thermal waves then the estimates

$$X_+(t) \leq X_+(0) + L_S, \quad X_-(t) \geq X_-(0) - L_S \tag{5}$$

are valid. Here $L_S = 2\pi\sqrt{\sigma+1}/\sigma$ is the basic length of the S-regime.

Thus, according to (5) the thermal wave front can propagate over a distance not exceeding L_S. That is the burning localization takes place (Fig. 2). The localization arises even more clearly when $\beta > \sigma+1$ (the LS-regime of burning). In this case an unlimited growth of temperature usually occurs at a single point only. If this point is $x = 0$ an asymptotic temperature distribution $u(T_o, x)$ is formed near it (Fig. 3). By using special comparison techniques we succeed in obtaining accurate estimates for $u(T_o, x)$ near $x = 0$, which characterize the energy output of exothermal reaction. Under an assumption of symmetry $u = u(t,|x|)$ and provide that $\sigma+1 < \beta < (\sigma+1)(N+2)/(N-2)_+$ (there we denote $(\cdot)_+ = \max\{0,(\cdot)\}$) we obtained near $x = 0$

$$u(T_o, x) \geq C_o|x|^{-2/[\beta-(\sigma+1)]} \quad \text{where} \quad x \neq 0 \tag{6}$$

Similar upper estimate is obtained for a certain class of initial temperature perturbations :

$$u(T_o, x) \le c_o |x|^{-2/[\beta-(\sigma+1)]} \quad \text{where} \quad x \neq 0 \tag{7}$$

In (6) and (7) C_o and c_o are constants that depend only on σ , β and N .

Consider the spatial-temporal evolution of the burning wave near the blow up time. It is found that in some cases the stable asymptotics of the burning wave is described by self-similar solutions of the specific form

$$u_a(t,x) = (T_o-t)^{-\frac{1}{\beta-1}} \theta(\xi) \quad , \quad \xi = \frac{x}{(T_o-t)^m} \quad , \tag{8}$$

where the function $\theta(\xi) \ge 0$ satisfies the nonlinear elliptic equation

$$\begin{aligned} &\nabla_\xi \cdot (\theta^\sigma \nabla_\xi \theta) - m \nabla_\xi \theta \cdot \xi - \frac{1}{\beta-1} \theta + \theta^\beta = 0 , \\ &\theta(\xi) \to 0 \quad , \quad |\xi| \to +\infty \quad . \end{aligned} \tag{9}$$

To establish the solvability of (9) and to find the number of various solutions is a rather hard problem. It is shown that when $\beta \geq \sigma + 1$ the problem (9) has a set of different solutions which describe possible types of the burning self-organization in continuum [2] .

Note that determining the asymptotic of evolution processes that result in the development of a number of types of metastable stationary dissipative structures with the spatial-temporal distribution described by invariant-group solutions (or self-similar solutions in a specific case) can be considered as an analog of the second law of thermodynamics for open nonlinear systems.

2. Properties of nonlinear (implicit) difference schemes. Computer simulation is an important part of investigation of blow up regimes. Since burning is an essentially nonstationary process during which the spatial-temporal singularities arise the quality of difference schemes is of particular concern. Here we present the results [3] of investigation of implicit schemes applied to the equation

$$u'(t) = (u^{\sigma+1})_{xx} + u^\beta \quad , \quad t > 0 \quad , \quad 0 < x < \ell \quad , \tag{10}$$

for which we consider the boundary value problem

$$\begin{aligned} &u(0, x) = u_o(x) \ge 0 \quad , \quad 0 < x < \ell \quad ; \quad u_o^{\sigma+1} \in C^1 \quad , \\ &u(t, 0) = u(t,\ell) = 0 \quad , \quad t \ge 0 \quad . \end{aligned} \tag{11}$$

We introduce in a usual way the uniform grid

$$\omega_h = \{x_k = kh \ , \ k = 1, 2,\dots, M-1 \ ; \ h = \ell/M\} \ ,$$

and variable time step $\tau_j \to 0$ when $j \to +\infty$. Consider the implicit conservative scheme

$$\frac{\hat{u}-u}{\tau_j} = (\hat{u}^{\sigma+1})_{\bar{x}x} + \hat{u}^\beta \quad , \quad (t,x) \in \omega_\tau \times \omega_h \quad , \tag{12}$$

$$u^{o} = u_{oh} \geq 0 \quad , \quad x \in \omega_h \quad ; \quad u = 0 \quad , \quad x \in \partial\omega_h \quad , \tag{13}$$

where the designations are usual [4] and u_{oh} is the projection of u_o on the grid ω_h. There are local and global problems concerning a behavior of the difference solution. The local prolems involve questions of solvability of difference equations on a time layer and uniqueness of the local solution while the global problems are connected with the development of grid blow up regimes.

From this viewpoint the case $\beta < \sigma+1$ is very simple, the scheme being solvable on any time layer, the solution being unique for sufficiently small τ_j and being convergent (when $\tau_j = 0(h^2) \to 0$) to a global unique generalized solution of the original differential problem (10), (11).

The situation is quite different when $\beta \geq \sigma+1$ and the operator of (12) is not coercitive. As a result, the following unusual effects, which must be taken into account in computations, are possible :

- the scheme (12) may prove to be locally unsolvable ;
- a nonunique solution may exist ;
- the global solvability is absent.

As an example let us consider some specific estimates when the above effects take place.

Theorem : Let $\beta > \sigma+1$. Then if

$$|u|_c \geq a_o \tau_j (h^2)^{-\beta/[\beta-(\sigma+1)]} + b_o \tau_j^{-1/(\beta-1)} \tag{14}$$

where the constants a_o , $b_o > 0$ depend only on σ and β the scheme is unsolvable on a given time layer. For any sufficiently small τ_j there are at least two solutions (the "greater" solution, which has no physical sense, is unstable). If

$$1 - \frac{4}{h^2} \sin^2 \frac{\pi h}{2\ell} (u_{oh}, \psi_h)_{L^2(\omega_h)} > 0$$

where $\psi_h > 0$ is the first eigenfunction of the problem $(\psi_h)_{\bar{x}x} + \lambda\psi_h = 0$, $\psi_h = 0$ on $\partial\omega_h$, $\|\psi_h\|_{L^1(\omega_h)} = 1$ then there is the grid in time ω_τ , $T_{oh} = \Sigma\, \tau_j \neq T_o$ such that

$$\|u^j\|_{C(\omega_h)} \to +\infty \quad , \quad j \to +\infty \quad ,$$

i.e. the global solution is absent.

Let us note that (14) provides a rather unusual condition for the time step, which is necessary to ensure stable computations of the blow up. It has the form

$$\tau_{solv} = \text{const}\,(h^2)^{\frac{\beta-1}{\beta-(\sigma+1)}} \qquad (\beta > \sigma+1)$$

and essentially differs from the traditional restrictions $\tau = 0(h^2)$ for parabolic equations.

Now consider an example of the blow up difference solution for the case $\sigma = 2$, $\beta = 3$ (the S-regime), which is obtained in separable variables :

$$u_k^j = S^j \theta_k , \quad \tau_j = A \rho^{-2j} ,$$

where $S^j = \rho^j$, $j = 0, 1,...$; $A = 2 \rho^{-3}(\rho-1)$, $\rho = \text{const} > 1$,

and the spatial function θ_k has the form

$$\theta_k = \{2 [3(1 - \frac{4}{h^2} \text{Sin}^2 \frac{a_h h}{2})]^{-1}\}^{1/2} \text{Sin}(a_h kh) ,$$

$$a_h = \frac{\pi}{\ell_h} , \quad \ell_h = \frac{3\pi h}{2} (\arcsin \frac{h}{2})^{-1} , \quad h = 2 \text{ Sin} \frac{3\pi}{2(M+1)} .$$

This solution determines the "difference" basic length which approximates the differential length $\ell_o = 3\pi$.

2. THE GROUP ANALYSIS IS A NECESSARY TOOL OF MATHEMATICAL MODELING

The group analysis enables us to establish a symmetry of the model under consideration, i.e. to find the classical Lie groups and the Lie-Bäcklund groups admissible by a system of differential equations. Knowing the symmetry allows constructing new solutions from the available ones and determining a set of particular solutions -invariant or partially invariant. For variational problems the equation symmetry enables one to establish a full set of conservation laws and to find their basis ; if there exists no variational principle the symmetry makes it possible to "generate" the conservation laws from the known ones. The symmetry allows the classification of equations and searching for equations with an infinite group or infinite series of conservation laws, as well as the possibility of coupling the solutions of various equations (through the Bäcklund transformations) with the linearizing replacements of variables.

The promising fields of group analysis are :

1) extension of the Lie-Bäcklund group theory ;

2) group analysis of discrete (finite-difference) manifolds ;

3) nonlocal symmetries (the transformation groups containing integral - type variables) ;

4) approximate group analysis ;

5) computerization of analytic calculations of transformation groups and investigation of group properties of various mathematical models.

By this time a lot of models have been studied (three are hundreds of works), including the following.

(1) The 1-3D heat conduction equation for isotropic and anisotropic media [9] :

$$u_t = \sum_{i=1}^{N} (k_i(u)u_{x_i})_{x_i} + Q(u), \quad N = 1,2,3 \tag{1}$$

(2) The heat conduction equations of hyperbolic type :

$$T_t = - W_x ,$$
$$W_t + \lambda(T)W + k(T)T_x = 0.$$

(3) The double diffusion system :

$$u_t = (k_1(u)u_x)_x + Q_1(u,v),$$
$$u_t = (k_2(v)v_x)_x + Q_2(u,v).$$

(4) The MGD equations with the buik source term and with the finite conductivity and thermal conductivity.

As an example let us consider an invariant solution of the 2D equation of nonlinear anisotropic thermal conductivity with a source term

$$u_t = (u^{\sigma_1}u_x)_x + (u^{\sigma_2}u_y)_y + u^\beta, \qquad \sigma_1 > 0,\ \sigma_2 > 0,\ \beta > 0. \tag{2}$$

The symmetry of this equation is determined by the four operators (the infinitesimal transformations) :

$$X_1 = \frac{\partial}{\partial t},\ X_2 = \frac{\partial}{\partial x},\ X_3 = \frac{\partial}{\partial y},$$
$$X_4 = 2(1-\beta)t\frac{\partial}{\partial t} + (\sigma_1+1-\beta)x\frac{\partial}{\partial x} + (\sigma_2+1-\beta)y\frac{\partial}{\partial y} + 2u\frac{\partial}{\partial u}$$

These operators determine the classical four-parametric Lie group of point transformations. (Eq.(2) does not allow the nontrivial Lie-Bäcklund group that is different from the Lie group).

We consider the invariant solution for

$$(X_4 - t_o X_1) = 2(1-\beta)(t_o-t)\frac{\partial}{\partial(t_o-t)} + (\sigma_1+1-\beta)\,x\frac{\partial}{\partial x} + (\sigma_2+1-\beta)y\frac{\partial}{\partial y} + 2u\frac{\partial}{\partial u},$$

where the parameter $t_o > 0$ determines the time interval $(0,t_o)$ on which the solution exists.

Then the invariant solutions has the form :

$$u(t,x,y) = \frac{\tilde{u}(\xi,\eta)}{(t_o-t)^{1/\beta-1}}, \qquad \xi = x(t_o-t)^{(\sigma_1+1-\beta)/2(\beta-1)},\ \eta = y(t-t_o)^{(\sigma_2+1-\beta)/2(\beta-1)} \tag{3}$$

where the function $\tilde{u}(\xi,\eta)$ is given by the solution of the equation

$$(\tilde{u}^{\sigma_1}\tilde{u}_\xi)_\xi + (\tilde{u}^{\sigma_2}\tilde{u}_\eta)_\eta + \frac{\sigma_1+1-\beta}{2(\beta-1)}\xi\tilde{u}_\xi + \frac{\sigma_2+1-\beta}{2(\beta-1)}\eta\tilde{u}_\eta + \tilde{u}^\beta - \frac{\tilde{u}}{\beta-1} = 0 \tag{4}$$

The parameters σ_1, σ_2 and β determine a direction of heat propagation and the time history of $u(t,x,y)$.

Equation (1) was solved numerically by means of two different techniques: the finite element method and the finite difference method.

As it is seen from the invariant solution (3) the heat propagation behavior depends on the relation between β, σ_1 and σ_2. Three types of the solution can be distinguished qualitatively[1]: the S-S regime, the HS-S regime, and the LS-S regime. The computer simulation of these regimes is given in Fig. 4-6 .

The group analysis of (1) enables us to separate the equations admitting an infinite group. In addition to all the linear equations of (1), the two nonlinear equations are distinguished :

$$u_t = (u^{-2}u_x)_x + \alpha u + \beta \quad , \quad \alpha\beta = 0 \ ,$$

$$u_t = \Delta(\ell n u) + \gamma u \equiv \left(\frac{u_x}{u}\right)_x + \left(\frac{u_y}{u}\right)_y + \gamma u \quad .$$

These equations allow an infinite succession of essentially different invariant solutions to be constructed, the first one among them is linearized by the Bäcklund transformation.

Applying the group analysis we could obtain the invariant solutions that describe a spiral propagation of inhomogeneities in a homogeneous nonlinear medium.

Equation (2) written in the polar coordinates has the form

$$u_t = \frac{1}{\sigma+1}\left\{\frac{1}{r}\frac{\partial}{\partial r}\left(r\frac{\partial u^{\sigma+1}}{\partial r}\right) + \frac{1}{r^2}\left(\frac{\partial^2 u^{\sigma+1}}{\partial\phi^2}\right)\right\} + u^\beta \quad . \tag{5}$$

It has the invariant solution

$$\begin{aligned} u(t,r,\phi) &= \tilde{u}(R,\Phi)\,(t_o-t)^{-1/\beta-1} \quad , \\ R &= r(t_o-t)^{-\frac{\beta-(\sigma+1)}{2(\beta-1)}} \\ \Phi &= \phi - c_o\,\ell n(t_o-t) \quad , \quad c_o = \text{const} \quad , \end{aligned} \tag{6}$$

where the function $\tilde{u}(R,\Phi)$ is determined by the equation

$$\begin{aligned} &\frac{1}{\sigma+1}\left\{\frac{1}{R}\frac{\partial}{\partial R}\left(R\frac{\partial\tilde{u}^{\sigma+1}}{\partial R}\right) + \frac{1}{R^2}\left(\frac{\partial^2\tilde{u}^{\sigma+1}}{\partial\Phi^2}\right)\right\} = \\ &= \frac{\beta-(\sigma+1)}{2}\,R\,\frac{\partial\tilde{u}}{\partial R} - c_o\frac{\partial\tilde{u}}{\partial\Phi} + \tilde{u} - \tilde{u}^\beta \quad . \end{aligned} \tag{7}$$

It is clear that the invariant solution (6) describes the propagation of the temperature field inhomogeneities (the local maxima, weak discontinuities, etc.) along the logarithmic spiral.

The trajectories of such inhomogeneities are sketched in Fig. 7 for : $\beta < \sigma+1$ - untwisting spirals ; $\beta > \sigma+1$ - twisting spirals ; $\beta = \sigma+1$ - circles. The arrow indicates the direction of motion along the spiral with time (when $c_o > 0$) . The growth of $u(t,r,\phi)$ in time and the spiral motion of inhomogeneities occur in the blow up regime.

In the three-dimensional case we can construct similar solutions that describe propagation of the inhomogeneities u along the spirals applied on cones.

3. THE MODEL : EXOTHERMIC REACTION OF THE METALLIC SURFACE OXIDATION UNDER THE LASER RADIATION

The investigation of the metal flaming and burning under the laser action is of great scientific and practical interest. In the simplest parabolic formulation, provided that a metallic oxid is thin film, the problem can be reduced to the nonlinear boundary value problem for a parabolic thermal conductivity equation. In the dimensionless formulation we have [5]:

$$\frac{\partial T}{\partial t} = \Delta T \equiv \frac{1}{r}\frac{\partial}{\partial r}\left(r\frac{\partial T}{\partial r}\right) + \frac{\partial^2 T}{\partial z^2}, \quad (r,z) \in \Omega, \ \Omega = \{r \geq 0, z > 0\}, \tag{1}$$

$$-\left.\frac{\partial T}{\partial z}\right|_{z=0} = I(r) + g(T) \ ; \ T(0,x), \ x \in \Omega , \tag{2}$$

where $I(r) = I_o \exp(-r^2/r_o^2)$, I_o and r_o are the intensity and effective radius of the radiation beam. In the nonlinear boundary condition (2) the source $g(T) \geq 0$ describes the energy production in the oxidation reaction.

A typical example of the energy release approximation is the function $g(T) = \exp(-1/T)$. In this case it has been established that in the region $\{I_o, r_o\}$ there is the monotonic curve $I_o = h(r_o) > 0$ such that an infinitely propagating thermal wave arise when $I_o > h(r_o)$ (the high intensity of the beam).
But if $I_o < h(r_o)$ the problem has the stationary state, and a stationary burning wave arise (Fig. 8).

Near the zero stability curve $I_o \simeq h(r_o)$ various instabilities may appear, a perculiar "thermal noise" may arise. The situation becomes more complicated if the system has a nonunique stationary state, near which the temperature field evolution becomes even more unstable so the amplitude oscillations may develop. To investigate these effects a model was proposed that was obtained by the energy averaging of the initial evolution problem :

$$\begin{aligned} \frac{d}{dt}(\psi\,\phi^3) &= M_1 + \frac{\mu_1}{2}\phi^2 \exp\left(-\frac{1}{\psi}\right) - \delta\,\psi\,\phi , \\ \frac{d}{dt}(\psi\,\phi^4) &= M_2 + \frac{\mu_2}{2}\phi^3 \exp\left(-\frac{1}{\psi}\right) - \kappa\,\psi\,\phi^2 , \end{aligned} \tag{3}$$

where μ_1, μ_2, δ, $\kappa > 0$ are constants, $M_1 \sim I_o\, r_o^2$ and $M_2 \sim I_o\, r_o^3$.
Here $\phi(t) > 0$ and $\psi(t) > 0$ are, respectively, the half-width and amplitude of the thermal structure.

Figure 9 shows a sketch of phase trajectories of the system (3) for its four different stationary states. Only the first ("lower") state is stable, the orthers influence the evolution. This case is shown in Fig. 10 where the picture of the solution branching in the stationary problem is presented. Four stationary states correspond to the interval of the intensity parameter $I_o \in (I_o^-, I_o^+)$.

Another description usually used for the oxidation reaction is provided by the function $g(T) = T^\alpha$, where $\alpha = \text{const} > 0$. In this case we have a critical value of the reaction parameter $\alpha = 2$. When $1 < \alpha \leq 2$ there are no stationary states, the burning wave develops in the blow up regime and forms a perculiar thermal high-intensity layer near a limited central part of the metallic surface (the skin-effect, the burning localization). If $\alpha > 2$ the evelution brings the system (1), (2) into a stationary state and the temperature field becomes stabilized.

4. SIMULATION OF THE SURFACE EVAPORATION

The intensive surface evaporation of solids under the action of concentrated energy fluxes is similar to the medium burning process. When analyzing the surface evaporation we should generally consider kinetics of the phase transition together with the gasdynamics of vapors and the processes in solids.

A detailed mathematical description of the problem can be found in [7,8] . By ignoring hydrodynamic processes in the solids and taking into account the possibility of the plasma formation in a vapour we can describe the mathematical model by a system, involving the quasilinear heat conduction equation for a solid

$$c_p\,\rho\left(\frac{\partial T}{\partial t} - v_s\,\frac{\partial T}{\partial z}\right) = \frac{\partial}{\partial z}\,\lambda\,\frac{\partial T}{\partial z}\quad,\qquad -\ell < z < 0\quad, \tag{1}$$

$$z = -\ell\quad,\qquad \lambda\,\frac{\partial T}{\partial z} = 0\quad .$$

and the gasdynamics equations for the vapor with the equation for radiation transport in a plasma :

$$\frac{\partial \rho}{\partial t} + \frac{\partial(\rho u)}{\partial z} = 0\quad,$$

$$\frac{\partial(u-v_s)}{\partial t} + u\,\frac{\partial u}{\partial z} = -\frac{1}{\rho}\,\frac{\partial p}{\partial z}\quad,$$

$$\frac{\partial}{\partial t}\left[\varepsilon + \frac{(u-v_s)^2}{2}\right] + u\,\frac{\partial}{\partial z}\left(\varepsilon + \frac{u^2}{2}\right) = \tag{2}$$

$$= -\frac{1}{\rho}\left[\frac{\partial}{\partial z}(pu) + \frac{\partial W}{\partial z} + \frac{\partial G}{\partial z}\right]\quad,$$

$$\mu\,\frac{dI_\nu}{dz} + \kappa_\nu\,I_\nu = \kappa_\nu\,I_{eq}\quad,\qquad W = \int_0^{+\infty} d\nu \int_{-1}^{1} \mu I_\nu\, d\mu\quad,$$

where p, u, ρ, T are the pressure, the velocity, the density and the temperature, respectively ; W is the radiation flux, G is the source intensity ; the indices S and O reffered to solids and gas, respectively ; and $M = u_o/u_c$ is the Mach number.

The evaporation process at the phase interface must be considered as a strong gasdynamic discontinuity, on which three conservation laws are formulated :

$$\rho_s\,v_s = \rho_o\,v_o,\; p_s + \rho_s\,v_s^2 = p_o + \rho_o\,v_o^2\;,\; \lambda\,\frac{\partial T}{\partial z} = L_v\,\rho_s\,v_s - G_o \tag{3}$$

In addition two relations are used to describe the kinetics of the phase transition :

$$p\,(0,t) = p_o\,(T_s\,,\,M)\;,\quad u(0,t) = u_o(T_s\,,\,M) \tag{4}$$

The specific feature of this problem is that the nonequilibrium degree of the evaporation process (i.e. the mass flow rate) is not given a priori ; it is determined by the parameter M through the boundary conditions (4).

Computer simulation of the nonlinear evaporation problem showed that in the subsonic region $M < 1$ the quasistationary variations of surface temperature T_s result in changes of the nonequilibrium degree of evaporation process. The problem solution qualitatively changes in the transonic region of the vapour flow. In particular, it shows in a sub-definite character of the linearized formulation of the problem (1) - (4), which requires an additional boundary condition at the phase interface for the system closure when $M \geqslant 1$. The T_s variation with M is shown in Fig. 11.

In the case of evaporation into vacuum a change in M may be produced also by vapor heating. In this case the M values may be negative near the boundary (Fig. 12).

REFERENCES

1. Galaktionov V.A., Doroduitsin V.A., Elenin G.G., Kurdyumov S.P., Samarskii A.A. A quasilinear heat conduction equation with a source term : aggrevation, localization, symmetry, point solutions, asymptotics, structures. In : Modern Problems of Mathematics. Latest Advances. V.28 (Science and Technology Reviews, VINITI AN SSSR), M., 1986, p. 95-206.
2. Kurdyumov S.P., Kurkina E.S., Potapov A.B., Samarskii A.A. Architecture of multidimensional thermal sturctures. Dokl. AN SSSR, 1984, v. 274, N 5, p. 1071-1075.
3. Galaktionov V.A., Samarskii A.A. On difference solutions of one class of quasilinear parabolic equations. I : JVM i MF, 1983, v. 23, N 3, p. 646-659 ; II : ibid, N 4, p. 831-838.
4. Samarskii A.A. Theory of Difference Schemes. M., Nauka, 1977, 656 p.
5. Bunkin F.V., Kirichenko N.A., Luk'janchuk B.S. UFN, 1982, v. 138, N 1, p. 45-94.
6. Bunkin F.V., Galaktionov V.A., Kirichenko N.A., Kurdyumov S.P., Samarskii A.A. On a nonlinear problem of laser thermochemistry. I : Diff. equat., 1985, v. 21, N 11, p. 1947-1958. II : ibid, N 12, p. 2097-2105.
7. Mazhukin V.I., Samokhin A.A. On some peculiarities of a mathematical model of the intensive surface evaporation of a matter. Dokl. AN SSSR, 1985, v. 281, N 4, p. 830-833.
8. Mazhukin V.I., Pestryakova G.A. Mathematical modeling of the surface evaporation by the laser radiation. Dokl. AN SSSR, 1984, v. 278, N 4, p. 843-847.
9. Dorodnitsyn V.A., Knyazeva I.V., Svirshchevskii Group properties of nonlinear heat conduction equation with a source. Preprint Inst. Appl. Math. N 134, 1982.

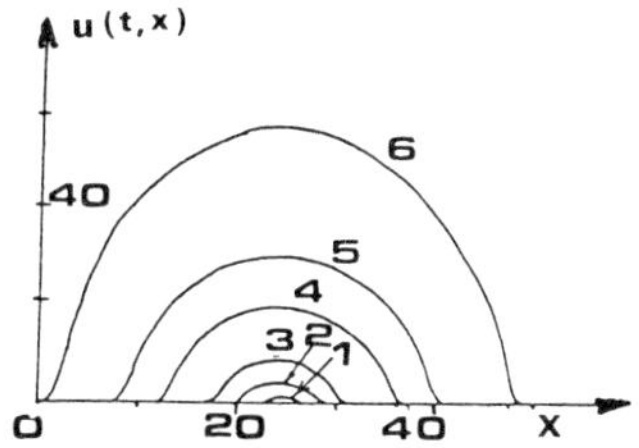

Figure 1

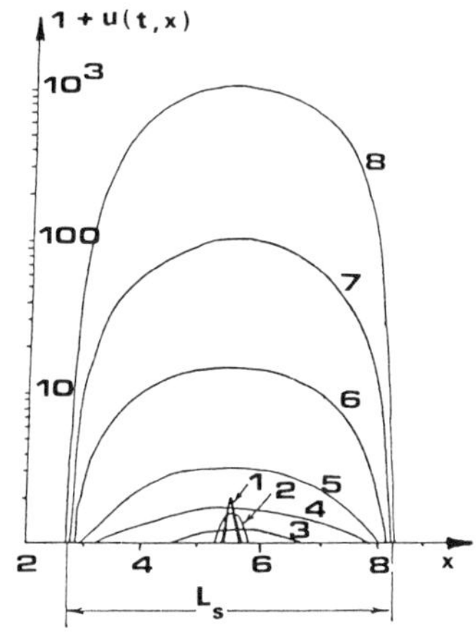

Figure 2

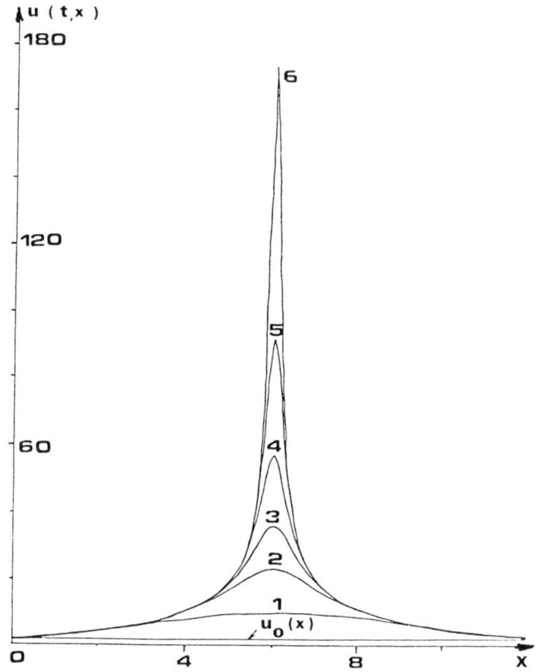

Figure 3

Figure 4

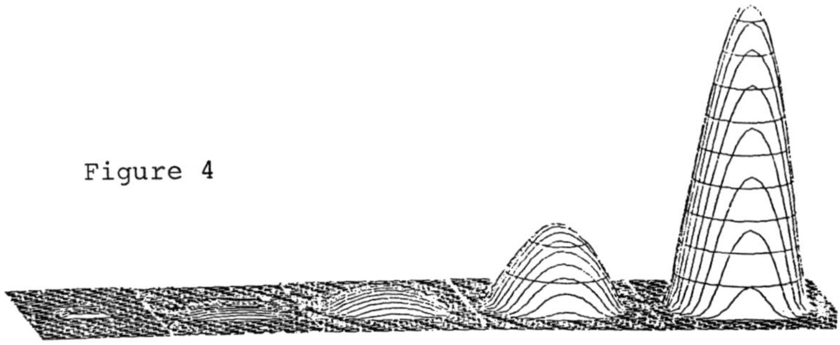

Figure 5

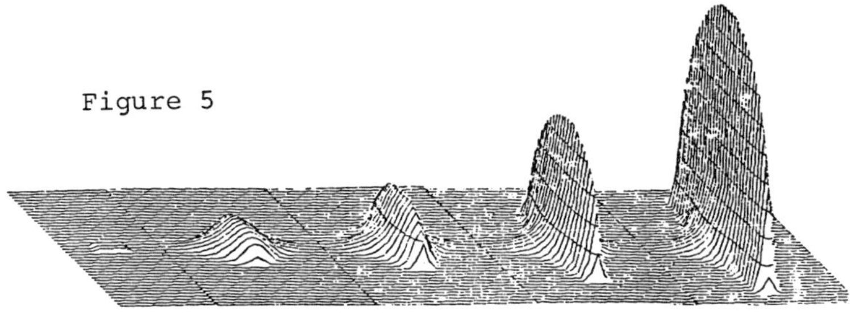

Figure 6

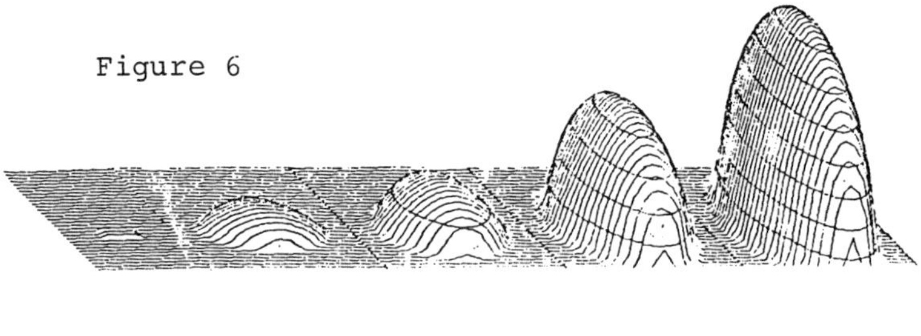

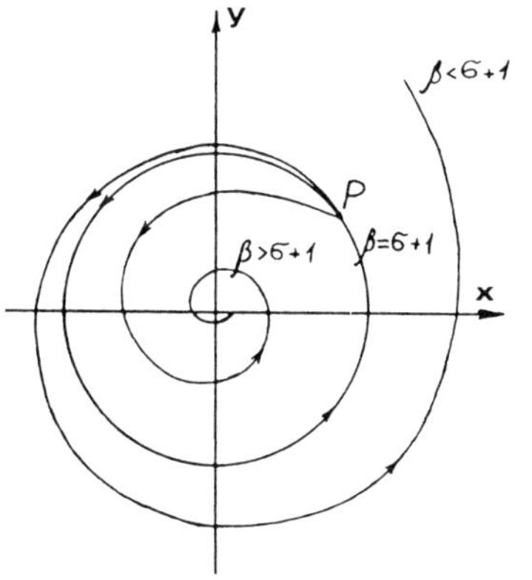

Figure 7

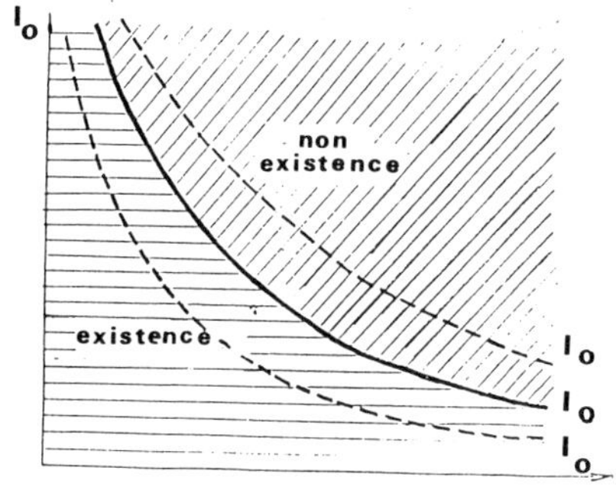

Figure 8

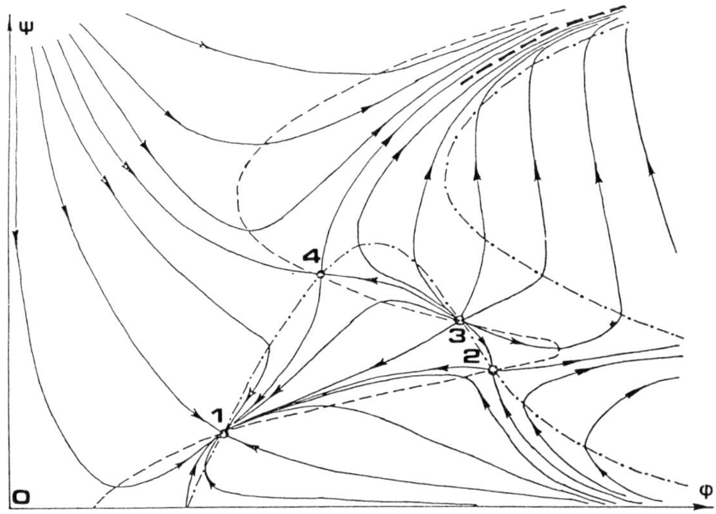

Figure 9

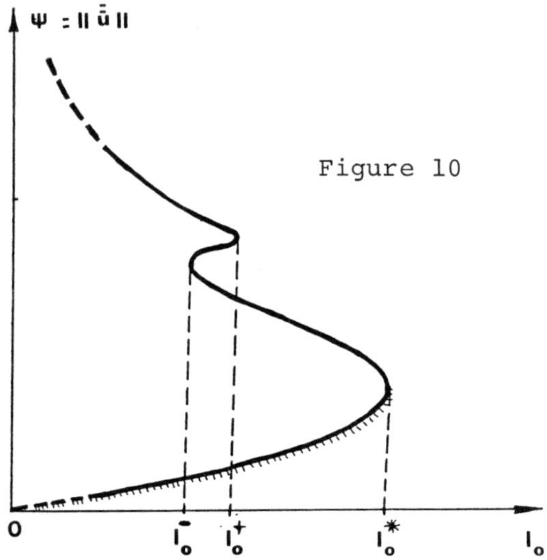

Figure 10

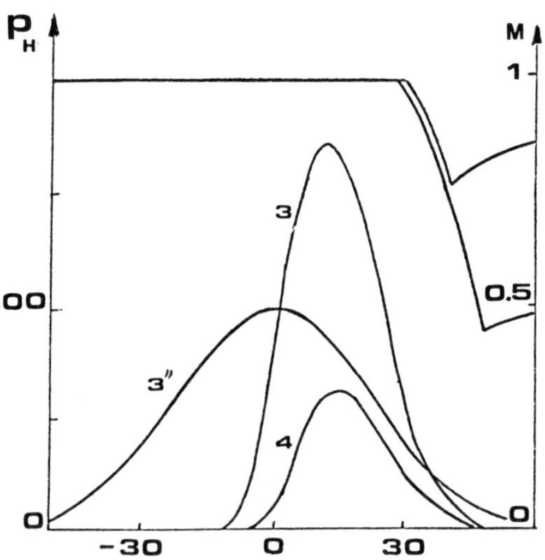

Figure 11

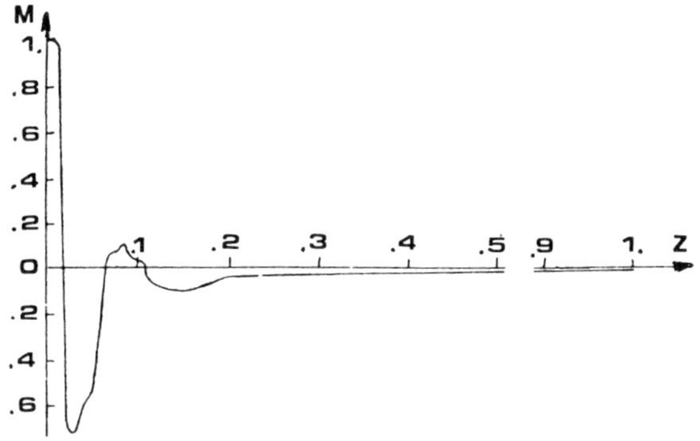

Figure 12

REMARKS ON THE STABILITY ANALYSIS OF REACTIVE FLOWS

B. SCHEURER

CEA, Centre d'Etudes de Limeil - Valenton - BP 27, 94190 Villeneuve Saint Georges, FRANCE.

ABSTRACT

We study a simple model of compressible reacting flow. First, we derive a dispersion relation for the linearized problem, making a distinction between frozen and equilibrium sound speed. Second, we study the stability of the Von Neumann-Richtmyer scheme applied to this model. One finds a natural generalization of the C.F.L. condition.

INTRODUCTION

A simple model--one irreversible reaction--of compressible reactive flow is presented in this report. Equations of gas dynamics are augmented by one equation for the progress variable of the chemical reaction and the appropriate equations of state. The resulting set of equations is a nonlinear hyperbolic system with source terms. However, acoustic waves, i.e., small perturbation of a given constant state, will not be described any more by the wave equation but by a more general dispersive equation. Only in the so-called frozen equilibrium limit will the wave equation be recovered with the appropriate sound speed. Such a classical analysis (see |8|, |1| for instance) is extended to the discrete set of equations obtained by using the Von Neumann-Richtmyer scheme (see |9|). Such a scheme is known for its simplicity and its robustness to handle strong shocks. The known stability results (see |9|, |7|) for such a scheme are generalized in our context. Both frozen and equilibrium limits appear again, and we make a distinction between implicit and explicit schemes used to discretize the reaction equation.

1. THE DISPERSION RELATION FOR A REACTIVE FLOW

We consider 1-D reactive flows involving a single irreversible chemical reaction $A \to B$. The flows will be compressible, viscous, but heat conduction and chemical diffusion effects are neglected.

Let us introduce the pressure p, the specific volume v, the internal energy e, the velocity u, and the progress variable λ of the reaction $A \to B$, changing from 0 for no reaction to 1 for complete reaction. Using Lagrangian coordinates (a,t), equations of motion are given by :

This research was supported in part by NAS1-18017 while the author was in residence at the Institute for Computer Applications in Science and Engineering (ICASE), NASA Langley Research Center, Hampton, VA 23665.

C.-M. Brauner and C. Schmidt-Lainé (eds.), Mathematical Modeling in Combustion and Related Topics, 233–242.

$$(1.1)\quad \begin{cases} \dfrac{\partial v}{\partial t} - \dfrac{\partial u}{\partial a} = 0 \\[2mm] \dfrac{\partial u}{\partial t} + \dfrac{\partial \bar{p}}{\partial a} = 0 \\[2mm] \dfrac{\partial}{\partial t}\left(e + \dfrac{u^2}{2}\right) + \dfrac{\partial}{\partial a}(\bar{p}u) = 0 \\[2mm] \dfrac{\partial \lambda}{\partial t} = r. \end{cases}$$

We recall that (a,t) are related to the Eulerian coordinates (x,t), where x = x(a,t) gives the position, at time t, of a fluid element that was initially at position a. We complete the system (1.1) by prescribing $\bar{p}$, the internal energy (per unit mass) e, and the reaction rate r ; precisely we have

$$(1.2)\qquad \bar{p} = p + \omega = p - \mu \frac{u}{a} \qquad\qquad \mu > 0$$

where $\bar{p}$ is the artificial viscosity

$$(1.3)\qquad e = e(p,v,\lambda)$$

$$(1.4)\qquad r = r(p,v,\lambda).$$

In (1.1), we rewrite the third equation in a nonconservative form as

$$(1.5)\qquad \frac{\partial e}{\partial t} + \bar{p}\,\frac{\partial v}{\partial t} = 0.$$

Using (1.1), (1.3) and (1.5), we also obtain

$$(1.6)\qquad \frac{\partial e}{\partial p}\frac{\partial p}{\partial t} + \frac{\partial e}{\partial v}\frac{\partial v}{\partial t} + \frac{\partial e}{\partial \lambda}\frac{\partial \lambda}{\partial t} + \bar{p}\,\frac{\partial v}{\partial t} = 0,$$

hence a relation between $\frac{\partial p}{\partial t}$ and $\frac{\partial v}{\partial t}$

$$(1.7)\qquad \frac{\partial e}{\partial p}\left(\frac{\partial p}{\partial t} + \frac{\frac{\partial e}{\partial v} + \bar{p}}{\frac{\partial e}{\partial p}}\,\frac{\partial v}{\partial t} + \frac{\frac{\partial e}{\partial \lambda}}{\frac{\partial e}{\partial p}}\,r\right) = 0.$$

We first notice (see |3| for instance) that

$$(1.8)\qquad \frac{\frac{\partial e}{\partial \lambda}}{\frac{\partial e}{\partial p}} = -\left(\frac{\partial p}{\partial \lambda}\right)_{e,v}.$$

Next, we set

$$(1.9)\qquad \bar{a}^2 = \frac{\frac{\partial e}{\partial v} + \bar{p}}{\frac{\partial e}{\partial p}}$$

$$(1.10)\qquad \bar{b} = \left(\frac{\partial p}{\partial \lambda}\right)_{e,v}.$$

The first coefficient will be interpreted (see the Appendix below) as the mass speed of sound ; the second one is related to the themicity (see |3|). Therefore, (1.7) reads

$$(1.11)\qquad \frac{\partial p}{\partial t} + \bar{a}^{-2}\,\frac{\partial v}{\partial t} = \bar{b}r.$$

In order to investigate the stability of the solution of (1.1)-(1.4), we shall "linearize" the equation for λ around some local equilibrium state $\lambda^* = \lambda^*(p,v)$ (see the Appendix). We have

$$(1.12)\qquad \frac{\partial \lambda}{\partial t} = \frac{\lambda^*(p,v) - \lambda}{\tau},$$

where τ is a "relaxation time", assumed given. Next, we linearize the equilibrium state by taking the undisturbed fluid denoted by the subscript 0--as reference state

$$(1.13)\qquad \lambda^*(p,v) = \lambda^*_{p_0}\,p + \lambda^*_{v_0}\,v.$$

Finally, we obtain for the equation governing λ

$$(1.14)\qquad \frac{\partial \lambda}{\partial t} = \frac{1}{\tau}\,(\lambda^*_{p_0}\,p + \lambda^*_{v_0}\,v - \lambda).$$

The linearized system associated to (1.1) is obtained classically by writing $v = v_0 + v'$, $p = p_0 + p'$, etc. .., in (1.1), (1.11), and (1.14), where $v_0, p_0, \ldots$ is a constant reference state. Neglecting second order terms and performing the approximation of frozen coefficients, we finally find (dropping the ')

$$(1.15)\qquad \left[\begin{aligned} &\frac{\partial v}{\partial t} - \frac{\partial u}{\partial a} = 0\\ &\frac{\partial u}{\partial t} + \frac{\partial p}{\partial a} = \frac{\partial}{\partial a}\left(\bar{\mu}\,\frac{\partial u}{\partial a}\right)\\ &\frac{\partial p}{\partial t} + \bar{a}^{-2}\,\frac{\partial v}{\partial t} = \bar{b}\,\frac{\partial \lambda}{\partial t}\\ &\frac{\partial \lambda}{\partial t} = \frac{1}{\tau}\,(\lambda^*_{p_0}\,p + \lambda^*_{v_0}\,v - \lambda), \end{aligned}\right.$$

where $\bar{\mu}$, $\bar{a}^{-2}$, $\bar{b}$, τ, $\lambda^*_{p_0}$, $\lambda^*_{v_0}$ are now given constants. An equivalent form of (1.15), which is more convenient for our analysis, is

$$(1.16)\qquad \left[\begin{aligned} &\frac{\partial v}{\partial t} - \frac{\partial u}{\partial a} = 0\\ &\frac{\partial u}{\partial t} + \frac{\partial p}{\partial a} = \frac{\partial}{\partial a}\left(\bar{\mu}\,\frac{\partial u}{\partial a}\right)\\ &\frac{\partial p}{\partial t} + \bar{a}^{-2}\,\frac{\partial u}{\partial a} = \bar{b}\,\frac{\partial \lambda}{\partial t}\\ &\frac{\partial \lambda}{\partial t} = \frac{1}{\tau}\,(\lambda^*_{p_0}\,p + \lambda^*_{v_0}\,v - \lambda). \end{aligned}\right.$$

The dispersion relation associated with (1.16) is classically obtained (see |10| for instance) by representing v, u, p, λ in (1.16) as a harmonic

$$(1.17)\qquad v = V_0\,\exp(\omega t + ika)\ \ldots,\ \text{etc.}$$

Substituting (1.17) into (1.16), we obtain a linear system for V_0, U_0, P_0, l_0

$$(1.18)\qquad \begin{cases} \omega V_0 - ik\, U_0 = 0 \\ \omega U_0 + ik\, P_0 = -\bar{\mu}\, k^2\, U_0 \\ \omega P_0 + \bar{a}^{-2}\, ik\, U_0 = \bar{b}\, \omega\, l_0 \\ \omega\, l_0 = \frac{1}{\tau}\, (\lambda^*_{p_0}\, P_0 + \lambda^*_{v_0}\, V_0 - l_0). \end{cases}$$

To have a nontrivial solution, the determinant associated with (1.18) should be zero, and this gives the dispersion relation, a relation between ω and k. A simple computation gives

$$(1.19)\qquad \omega\Big[\tau\omega(\omega^2 + \bar{\mu}\, k^2\omega + \bar{a}^{-2}\, k^2) + (1 - \bar{b}\, \lambda^*_{p_0})(\omega^2 + \bar{\mu}\, k^2\, \omega) + (\bar{a}^{-2} - \bar{b}\, \lambda^*_{v_0})k^2\Big] = 0.$$

We rearrange this equation as

$$(1.20)\qquad \omega\left[\frac{\tau\, \bar{a}^{-2}}{\bar{a}^{-2} - \bar{b}\lambda^*_{v_0}}\, \omega\Big(\frac{1}{\bar{a}^{-2}}\, \omega^2 + k^2 + \frac{\bar{\mu}}{\bar{a}^{-2}}\, k^2\omega\Big) + \frac{1 - \bar{b}\, \lambda^*_{p_0}}{\bar{a}^{-2} - \bar{b}\, \lambda^*_{v_0}}\, \omega^2 + k^2 + \frac{1 - \bar{b}\, \lambda^*_{p_0}}{\bar{a}^{-2} - \bar{b}\, \lambda^*_{v_0}}\, \bar{\mu}\, k^2\omega\right] = 0.$$

Of course the roots $\omega = \omega(k)$ of (1.20) characterize the stability of (1.16) according to the sign of their real parts. In view of the results given in the Appendix, the coefficients in (1.20) do have a simple interpretation in terms of the frozen and equilibrium mass sound speed $\bar{a}_f$ and $\bar{a}_e$. We refer to the Appendix for the definition ; here we just note that

$$(1.21)\qquad \bar{a}_f^{-2} \geq \bar{a}_e^{-2}.$$

We also introduce a new relaxation time :

$$(1.22)\qquad \bar{\tau} = \frac{\bar{a}_f^{-2}}{\bar{a}_f^{-2} - \bar{b}\, \lambda^*_{v_0}}\, \tau.$$

Finally, we rewrite the characteristic equation (1.20) in terms of τ, $\bar{a}_f^{-2}$, and $\bar{a}_e^{-2}$

$$(1.23)\qquad \omega\left[\bar{\tau}\, \omega\Big(\frac{1}{\bar{a}_f^{-2}}\, \omega^2 + k^2 + \frac{\bar{\mu}}{\bar{a}_f^{-2}}\, k^2\omega\Big) + \frac{1}{\bar{a}_e^{-2}}\, \omega^2 + k^2 + \frac{\bar{\mu}}{\bar{a}_e^{-2}}\, k^2\omega\right] = 0.$$

Of course, as $\bar{\tau} \to \infty$ (resp. $\bar{\tau} \to 0$), one recovers the limit of the frozen (resp. equilibrium) flow. Both limits are dispersion relations associated to a wave operator perturbed by a damping term $\Delta \frac{\partial u}{\partial t}$, with the appropriate coefficient, corresponding to the viscous term in (1.2).

Due to (1.21), one can easily check, using Routh-Hurwicz criterium (see |5| for instance and the next section)

(1.24) $\qquad \mathrm{Re}\ \omega(k) \leq 0,$

where $\omega = \omega(k)$ is any root of (1.23). Therefore, the solutions of (1.16) decay with time, i.e., the solutions of (1.1) are linearly stable.

2. VON NEUMANN-RICHTMYER SCHEME FOR REACTIVE FLOWS

Now we investigate the (linear) stability of the Von Neumann-Richtmyer scheme used to discretize (1.1) or (1.16) and completed by a simple Euler scheme for the λ-equation. Let $\Delta t, \Delta a$ denote the mesh scale in time and space. For (1.1), we have, using standard convention

$$(2.1)\quad \left\{\begin{array}{l}
\dfrac{v^{n+1}_{j+\frac{1}{2}} - v^{n}_{j+\frac{1}{2}}}{\Delta t} - \dfrac{u^{n+\frac{1}{2}}_{j+1} - u^{n+\frac{1}{2}}_{j}}{\Delta a} = 0 \\[2ex]
\dfrac{u^{n+\frac{1}{2}}_{j} - u^{n-\frac{1}{2}}_{j}}{\Delta t} + \dfrac{\bar{p}^{n}_{j+\frac{1}{2}} - \bar{p}^{n}_{j-\frac{1}{2}}}{\Delta a} \\[2ex]
\dfrac{e^{n+1}_{j+\frac{1}{2}} - e^{n}_{j+\frac{1}{2}}}{\Delta t} + \dfrac{1}{2}\left(\bar{p}^{n+1}_{j+\frac{1}{2}} + \bar{p}^{n}_{j+\frac{1}{2}}\right) \dfrac{v^{n+1}_{j+\frac{1}{2}} - v^{n}_{j+\frac{1}{2}}}{\Delta t} = 0 \\[2ex]
\dfrac{\lambda^{n+1}_{j+\frac{1}{2}} - \lambda^{n}_{j+\frac{1}{2}}}{\Delta t} = r^{n+\theta}_{j+\frac{1}{2}},
\end{array}\right.$$

where $r^{n+\theta}_{j+\frac{1}{2}} = r(\theta p^{n+1}_{j+\frac{1}{2}} + (1-\theta) p^{n}_{j+\frac{1}{2}}, \ldots)$, $0 \leq \theta \leq 1$. Note that we used the energy equation under the form (1.5). Second order accuracy corresponds to the choice $\theta = \frac{1}{2}$.

Such a sheme is now applied to the linearized version of (1.1), that is (1.16), again using the assumption of frozen coefficients. To avoid fractional indexes, we make the shift $j+\frac{1}{2} \to j$, $n+\frac{1}{2} \to n+1$, etc.

We find

$$(2.2)\quad \left[\begin{array}{l} \dfrac{v_j^{n+1} - v_j^n}{\Delta t} - \dfrac{u_{j+1}^{n+1} - u_j^{n+1}}{\Delta a} = 0 \\[2ex] \dfrac{u_j^{n+1} - u_j^n}{\Delta t} + \dfrac{p_j^n - p_{j-1}^n}{\Delta a} = \bar{\mu}\, \dfrac{u_{j+1}^n - 2u_j^n + u_{j-1}^n}{(\Delta a)^2} \\[2ex] \dfrac{p_j^{n+1} - p_j^n}{\Delta t} + \bar{a}^{2}\, \dfrac{u_{j+1}^{n+1} - u_j^{n+1}}{\Delta a} = \bar{b}\, \dfrac{\lambda_j^{n+1} - \lambda_j^n}{\Delta t} \\[2ex] \dfrac{\lambda_j^{n+1} - \lambda_j^n}{\Delta t} = \dfrac{1}{\tau} (\lambda^*_{p_0} p_j^{n+\theta} + \lambda^*_{v_0} v_j^{n+\theta} - \lambda_j^{n+\theta}), \end{array}\right.$$

where $p_j^{n+\theta} = \theta p_j^{n+1} + (1-\theta)p_j^n$, etc. .., $0 \leq \theta \leq 1$. To carry out a stability analysis for the scheme (2.2), we set

$$(2.3)\qquad v_j^n = V_0\, r^n\, \phi^j = V_0 (e^{\omega\Delta t})^n\, (e^{ik\Delta a})^j$$

and similar relation for u_j^n, p_j^n, λ_j^n. Substituting (2.3) into (2.2), we find a set of linear equations for V_0, U_0, P_0, l_0 :

$$(2.4)\quad \left[\begin{array}{l} \dfrac{r-1}{\Delta t} V_0 - \dfrac{e^{ik\Delta a} - 1}{\Delta a}\, r\, U_0 = 0 \\[2ex] \left(\dfrac{r-1}{\Delta t} + \dfrac{4\bar{\mu}}{(\Delta t)^2} \sin^2 k \dfrac{\Delta a}{2}\right) U_0 + \dfrac{1 - e^{-ik\Delta a}}{\Delta a} P_0 = 0 \\[2ex] \dfrac{r-1}{\Delta t} P_0 + \bar{a}^{2}\, \dfrac{e^{ik\Delta a} - 1}{\Delta a}\, r\, U_0 = \bar{b}\, \dfrac{r-1}{\Delta t} l_0 \\[2ex] \dfrac{r-1}{\Delta t} l_0 = \dfrac{1}{\tau} (\theta r + (1-\theta))(\lambda^*_{p_0} P_0 + \lambda^*_{v_0} V_0 - l_0). \end{array}\right.$$

The characteristic equation associated to this system is obtained by computing its determinant. We find, for $\beta \equiv \sin^2 \dfrac{k\Delta a}{2}$

$$(2.5)\quad \begin{array}{l} \dfrac{r-1}{\Delta t} \left\{\bar{\tau}\, \dfrac{r-1}{\Delta t} \left[\dfrac{1}{\bar{a}_f^{2}} \left(\dfrac{r-1}{\Delta t}\right)^2 + \dfrac{4\bar{\mu}}{\bar{a}_f^{2}} \dfrac{1}{(\Delta a)^2}\, \beta\, \dfrac{r-1}{\Delta t} + \dfrac{4}{(\Delta a)^2}\, \beta r\right]\right. \\[2ex] \left. + (\theta r + (1-\theta)) \left[\dfrac{1}{\bar{a}_e^{2}} \left(\dfrac{r-1}{\Delta t}\right)^2 + \dfrac{4\bar{\mu}}{\bar{a}_e^{2}} \dfrac{1}{(\Delta a)^2}\, \beta\, \dfrac{r-1}{\Delta t} + \dfrac{4}{(\Delta a)^2}\, \beta r\right]\right\} = 0, \end{array}$$

where we used the definition of the frozen and equilibrium mass sound speeds (see the Appendix) and the definition (1.22) of $\bar{\tau}$.

Equation (2.5) is to compare to its continuous analog (1.23). Of course, $r = 1$ is one trivial root. To insure stability, we have to check that the remaining roots of (2.5) do have modulus smaller or equal than 1. This will be achieved assuming conditions on Δt and Δa.

To simplify notations, let us introduce

$$(2.6)\qquad A = 4\,\beta\,\bar{\mu}\,\frac{\Delta t}{(\Delta a)^2}\,,\qquad B = 4\,(\bar{a}_f\,\frac{\Delta t}{\Delta a})^2\beta$$

$$B' = 4\,\bar{a}_e\,(\frac{\Delta t}{\Delta a})^2\qquad C = (\frac{\bar{a}_f}{\bar{a}_e})^2\,\frac{\Delta t}{\bar{\tau}}$$

Therefore (2.5), up to a factor $\frac{r-1}{\Delta t}$, can be rewritten as

$$(2.7)\qquad \begin{aligned}&(r-1)\left[(r-1)^2 + A(r-1) + Br\right]\\ &+ C(\theta(r-1) + 1)\left[(r-1)^2 + A(r-1) + B'r\right] = 0.\end{aligned}$$

We are now looking for conditions insuring that the roots r of (2.7) satisfy $|r| \leqslant 1$. Classically, we perform the transformation $r = \frac{1+z}{1-z}$ ($|r| \leqslant 1 \iff \mathrm{Re}\, z \leqslant 0$) and apply the Routh-Hurwicz criterium. From (2.7), we deduce

$$(2.8)\qquad \sum_{j=0}^{3} \alpha_j z^j = 0.$$

The coefficients α_j in (2.8) are given by

$$(2.9)\qquad \alpha_3 = 2(4 - 2A - B) + (2\theta - 1)\,C\,(4 - 2A - B')$$

$$(2.10)\qquad \alpha_2 = 4A + 2\,(2\theta - 1)\,AC + C(4 - 2A - B')$$

$$(2.11)\qquad \alpha_1 = 2B + (2\theta - 1)\,B'C + 2AC$$

$$(2.12)\qquad \alpha_0 = B'C.$$

The Routh-Hurwicz criterium insures $\mathrm{Re}\, z \leqslant 0$, for any root z of (2.8), as soon as (see |5|)

$$(2.13)\qquad \alpha_2,\ \alpha_3 \geqslant 0$$

$$(2.14)\qquad \alpha_1\alpha_2 - \alpha_0\alpha_3 \geqslant 0.$$

By the definitions (2.6) (we recall $\beta \equiv \sin^2 \frac{k\Delta a}{2})$ and the property (1.21), we have

$$(2.15)\qquad A,\ B,\ B',\ C \geqslant 0 \qquad \text{and} \qquad B \geqslant B'.$$

To study (2.13) and (2.14), we discuss according to the sign of $2\theta - 1$ ($0 \leqslant \theta \leqslant 1$).

i) <u>The case</u> $2\theta - 1 \geqslant 0$.

Here the scheme for the λ-equation is unconditionally stable. In view of (2.9), (2.10), and (2.15), the condition (2.13) is satisfied as soon as

$$(2.16)\qquad 4 - 2A - B \geqslant 0 \qquad \text{and} \qquad 4 - 2A - B' \geqslant 0.$$

Due to (2.6) and (1.21), we finally obtain the sufficient condition

$$(2.17)\qquad (\bar{a}_f \frac{\Delta t}{\Delta a})^2 + 2\bar{\mu} \frac{\Delta t}{(\Delta a)^2} \leqslant 1.$$

Thus, we recover the usual condition of stability for the Von Neumann-Richtmyer scheme (see |9|) but with the appropriate sound speed. Next, we look at the condition (2.14), which reads

$$(2.18)\qquad \begin{aligned} &(4A + 2(2\theta - 1)AC)\ (2B + (2\theta - 1)B'C + 2AC) \\ &+ 2AC^2(4 - 2A - B') + 2(4 - 2A)(B - B')C \geqslant 0. \end{aligned}$$

Since $2\theta - 1 \geqslant 0$, (2.18) is satisfied as soon as (2.17) ; hence, (2.16) holds.

<u>In summary, (2.17) insures (linear) stability for (2.2) if $1/2 \leqslant \theta \leqslant 1$.</u>

ii) <u>The case</u> $2\theta - 1 \leqslant 0$.

Here the scheme for the λ-equation becomes explicit. Again, using (2.6), we rewrite (2.13) as

$$(2.19)\qquad 2(4 - 2A - B - 4(1 - 2\theta)C) + (1 - 2\theta)\ C\ (2A + B') \geqslant 0$$

$$(2.20)\qquad 2A(2 - (1 - 2\theta)C) + C(4 - 2A - B') \geqslant 0.$$

Thus, due to (2.15), (2.19) is satisfied as soon as

$$(2.21)\qquad 4 - 2A - B - 4(1 - 2\theta)C \geqslant 0.$$

With (2.6), we thus obtain the sufficient condition

$$(2.22)\qquad (\bar{a}_f \frac{\Delta t}{\Delta a})^2 + 2\bar{\mu} \frac{\Delta t}{(\Delta a)^2} + (1 - 2\theta)\ (\frac{\bar{a}_f}{\bar{a}_e})^2 \frac{\Delta t}{\bar{\tau}} \leqslant 1\ ;$$

that is the condition (2.17) enforced by an extra term coming from the explicit scheme used for the λ-equation. We remark that this extra term vanishes in the frozen limit $\bar{\tau} \to +\infty$.

Of course, (2.21) implies (2.16), since $B \geqslant B'$; therefore, (2.20) is satisfied as soon as

$$(2.23)\qquad 2 - (1 - 2\theta)C \geqslant 0,$$

and we find the natural condition

$$(2.24)\qquad (\frac{\bar{a}_f}{\bar{a}_e})^2 \frac{1 - 2\theta}{2\bar{\tau}} \Delta t \leqslant 1.$$

It remains to check (2.14), i.e., (2.18). Due to (2.23), it is sufficient to satisfy

$$(2.25)\qquad 2B - (1 - 2\theta)B'C > 0$$

or

$$(2.26)\qquad \frac{1 - 2\theta}{\bar{\tau}} \Delta t \leqslant 1.$$

But, since $\bar{a}_f^2 \geqslant \bar{a}_e^2$, (2.26) is a consequence of (2.24).

In summary, (2.22) and (2.24) insure (linear) stability for (2.2), if $0 \leq \theta \leq 1/2$.

The above results could be easily generalized to similar 2-D reactive flows discretized by the natural extension of the Von Neumann-Richtmyer scheme (see |6|, for instance). Similar results are anticipated for the Godunov scheme, which reduces, after linearization, to the Courant-Isaacson-Rees scheme (|7|). They will be published elsewhere.

REFERENCES

1. Clarke J.F. and McChesney M. : The Dynamics of Real Gases, Butterworths, 1964.
2. Dautray R. and Lions J.L. : Analyse mathématique et calcul numérique pour les sciences et les techniques. Vol. 3, Masson, 1985.
3. Fickett W. and Davis W.C. : Detonation, University of California Press, 1979.
4. Godunov S. and al. : Résolution numérique des problèmes multidimensionnels de la dynamique des gaz. Edition Mir, 1979.
5. Henrici P. : Applied and Computational Complex Analysis. John Wiley, 1974.
6. Lascaux P. : Numerical methods for time dependent equations. Applications to fluid flows problems. Tata Institute, 1976.
7. Richtmyer R.D. and Morton K.W. : Difference Methods for Initial Value Problems. John Wiley, 1967.
8. Vincenti W.G. and Kruger C.H. : Introduction to Physical Gas Dynamics. Krieger Publishing Co., 1975.
9. Von Neumann J. and Richtmyer R.D. : A Method for Numerical calculation of Hydrodynamics Shocks. J. Appl. Phys., Vol. 21, 1950, pp. 232-237.
10. Whitham G.B. : Linear and Nonlinear Waves, John Wiley, 1974.

APPENDIX

Here, for the convenience of the reader, we recall some classical results concerning the definition of the sound speed in reactive flows (see for instance |3|, |8|, |1|). Here we take $\bar{p} = p$, that is (see (1.2)) ; we suppose $\mu = \bar{\mu} = 0$.

According to the laws of thermodynamics,

(A.1) $$TdS = de + pdv - \nu d\lambda,$$

where T, S, ν are the temperature, the entropy, and the chemical potential. From (A.1) and (1.3), we deduce

(A.2) $$TdS = \frac{\partial e}{\partial p}\, dp + \left(\frac{\partial e}{\partial v} + p\right) dv + \left(\frac{\partial e}{\partial \lambda} - \nu\right) d\lambda.$$

Now, we consider two cases.

i) The flow is frozen : $dS = d\lambda = 0$.

Then, (A.2) becomes

(A.3) $$\left(\frac{\partial p}{\partial v}\right)_{S,\lambda} = \frac{\frac{\partial e}{\partial v} + p}{\frac{\partial e}{\partial p}}$$

or

$$(A.4) \qquad \left(\frac{\partial p}{\partial \rho}\right)_{S,\lambda} = v^2 \frac{\dfrac{\partial e}{\partial v} + p}{\dfrac{\partial e}{\partial p}} \equiv c_f^2.$$

Therefore, the definition (1.9) of $\bar{a}^{-2}$ (where $\bar{p} = p$) is that of the frozen mass sound speed $\rho^2 c_f^2$. We shall use the standard notation $a_f^2 \equiv \rho^2 c_f^2$.

ii) <u>The flow is in chemical equilibrium : $dS = 0$ and $\lambda = \lambda^*(p,v)$.</u>

In particular $\nu = 0$ and $d\lambda = \frac{\partial \lambda^*}{\partial p} dp + \frac{\partial \lambda^*}{\partial v} dv$ in (A.2) ; collecting terms we find

$$(A.5) \qquad \left(\frac{\partial p}{\partial v}\right)_{S,\lambda=\lambda^*(p,v)} = - \frac{\dfrac{\partial e}{\partial v} + p \dfrac{\partial \lambda^*}{\partial v} \dfrac{\partial e}{\partial \lambda}}{\dfrac{\partial e}{\partial p} + \dfrac{\partial \lambda^*}{\partial p} \dfrac{\partial e}{\partial \lambda}}.$$

But $\left(\frac{\partial p}{\partial v}\right)_{S,\lambda=\lambda^*} = - \frac{1}{v^2} \left(\frac{\partial p}{\partial \rho}\right)_{S,\lambda=\lambda^*}$, and thus we find the equilibrium sound speed :

$$(A.6) \qquad c_e^2 \equiv \left(\frac{\partial p}{\partial \rho}\right)_{S,\lambda=\lambda^*} = v^2 \frac{\dfrac{\partial e}{\partial v} = p \dfrac{\partial \lambda^*}{\partial v} \dfrac{\partial e}{\partial \lambda}}{\dfrac{\partial e}{\partial p} + \dfrac{\partial \lambda^*}{\partial p} \dfrac{\partial e}{\partial \lambda}}.$$

The equilibrium mass sound speed is then $a_e^2 \equiv \rho^2 c_e^2$, and we easily check (see (1.20), (1.18), (1.10)) that

$$(A.7) \qquad \frac{\bar{a}^{-2} - \bar{b}\, \lambda^*_{v_0}}{1 - \bar{b}\, \lambda^*_{p_0}} = a_e^2 .$$

Similarly, we find (see (1.20)) that

$$(A.8) \qquad \frac{\bar{a}^{-2}}{\bar{a}^{-2} - b\lambda^*_0} = \frac{\dfrac{\partial e}{\partial v} + p}{\dfrac{\partial e}{\partial v} + p + \dfrac{\partial \lambda^*}{\partial v} \dfrac{\partial e}{\partial \lambda}}$$

Finally, in view of (A.4) and (A.6), we have

$$(A.9) \qquad a_f^2 \geqslant a_e^2.$$

EXPERIMENTS WITH PREMIXED FLAMES

S.SIVASEGARAM and J.H.WHITELAW
Imperial College of Science and Technology
London SW7 2BX, England

1. INTRODUCTION

A series of experiments with bluff-body stabilised premixed flames has quantified the flammability and stability limits of a wide range of geometric arrangements [1 to 6] and has provided confirmation of the existence of non-gradient diffusion [7, 8]. This paper presents some additional measurements and examines the implications of the findings of the overall investigation for engineering practice and for the modelling of combusting flows.

The discussion is restricted to axisymmetric flows with baffle-stabilised flames. Where the baffle is confined within a pipe of sufficient length to ensure energy release within it, the general nature of the flammability and stability limits is as shown on figure 1 which shows that unstable combustion can exist over a wide range of equivalence ratios between lean and rich extinction. The nature of the instability can be fluid-dynamic or acoustic and the following section examines the relative importance of the two as a function of geometry and equivalence ratio.

The implications of the results obtained with confined flames are considered in the third section in terms of calculation methods and the discussion extended to the unconfined flame of references 7 and 8 which confirm the need for a Reynolds-stress transport model. These arguments can be viewed in the broader context of modelling of steady premixed flames and the transient flames associated with internal-combustion engines. The former are addressed in the papers of [9, 10] and the latter in [11, 12].

2. INFLUENCE OF ACOUSTIC AND SHEDDING FREQUENCIES

Examination of the nature of the fluctuations in the combusting flow within a duct, where the flame is stabilised by a bluff-body, reveals the existence of discrete frequencies associated with aerodynamics and acoustics. In the

C.-M. Brauner and C. Schmidt-Lainé (eds.), Mathematical Modeling in Combustion and Related Topics, 243–259.

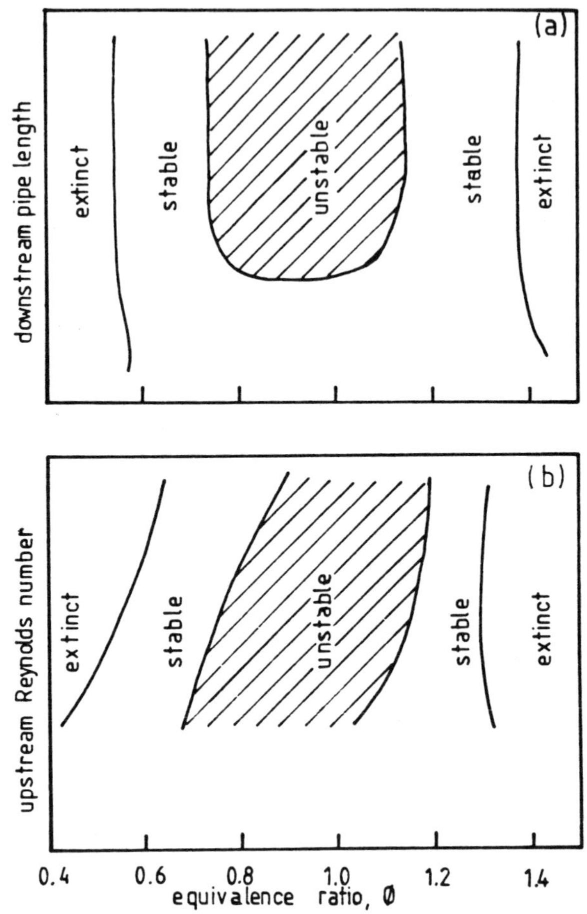

Figure 1. Flammability and stability characteristics for a confined disk stabilised flame.
(a) effect of confinement (b) effect of flow rate.

arrangement of figure 2, for example, the shear layers downstream of the disk give rise to shedding frequencies and the existence of a pipe with one end acoustically closed and the other open ensures the presence of a quarter-wave acoustic frequency and its harmonics. The measurements of references 1 to 6 show that the acoustic frequency dominates in the range of equivalence ratios defined and enclosed by the stability limits except for cases where the length of duct downstream of the disk stabiliser is insufficient to ensure energy release within the duct. The amplitude of the acoustic oscillations is dependent upon the energy release and is much larger than that associated with stable combustion.

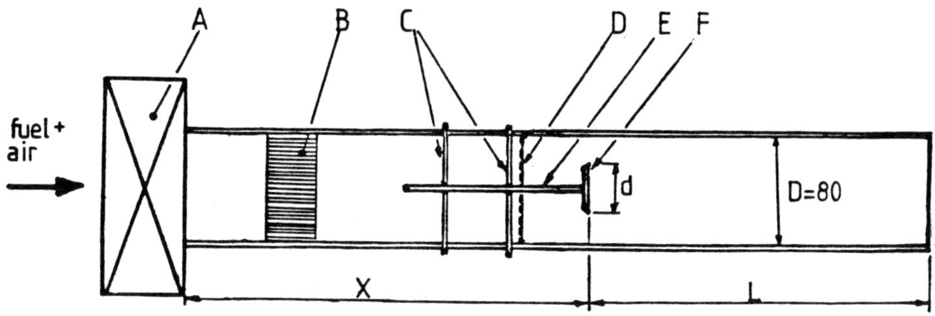

Figure 2. Burner geometry.
A - swirl register, B - flow straightener, C - pylons,
D - flame arrestor, E - sting, F - baffle.

It is not clear why there are boundaries of equivalence ratio across which the acoustic effects do not extend and experiments have been performed to investigate the nature of the flow close to the stability limits. Since it is simpler to consider the possible relationship between aerodynamic and acoustic effects, the nature of the flow downsteam of confined axisymmetric baffles was examined first in isothermal flow. As a consequence, 40 and 56 mm disks were arranged in an 80 mm diameter pipe and measurements of local velocity characteristics obtained in the isothermal flow and of free field sound intensity and wall pressure characteristics in the combusting flow downstream of the disks. A hot-wire anemometer was used for the velocity measurements and spectral information obtained from an FFT analyser; a condenser microphone and the FFT analyser were used to process the pressure signals.

2.1 Isothermal flow measurements

Power spectra of hot-wire signals obtained at stations downsteam of the disks showed distinct peaks similar to those of Fig. 3. The relative amplitude of the peak, corresponding to the energy associated with velocity fluctuations at the preferred frequency, was stronger for the smaller disk. In the case of the larger disk at 0^{0}, a preferred frequency was observed only in the immediate vicinity of the shear layer surrounding the region of recirculation, and with a peak of small amplitude. At angles of 30 and 40^{0}, this disk gave rise to easily observed preferred frequencies up to four pipe diameters downstream of the disk. With the smaller disk, the peaks were clearly noticable six pipe diameters downstream although they could barely be distinguished at ten pipe diameters.

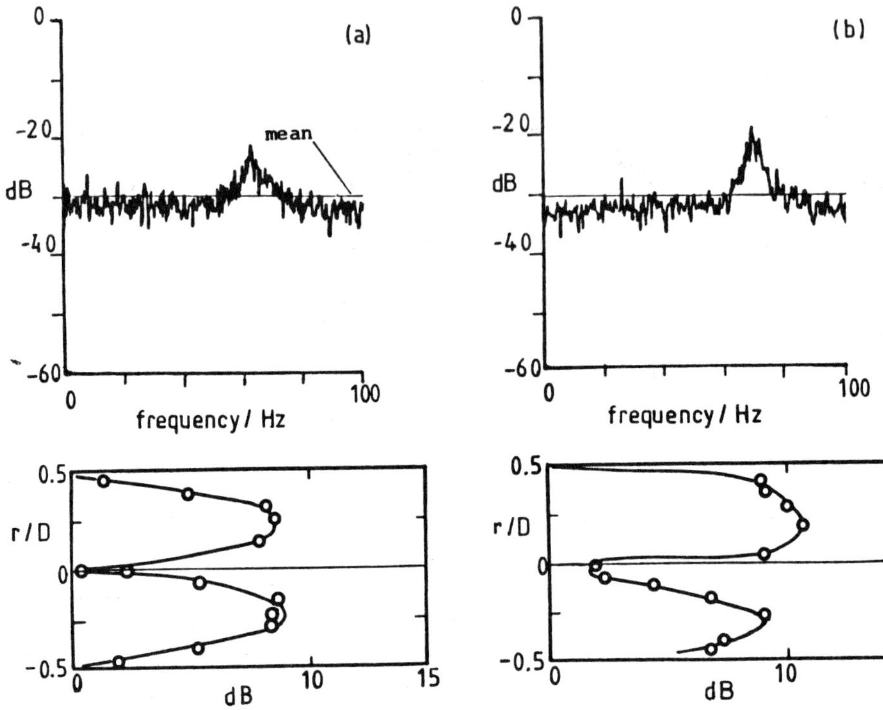

Figure 3. Hot-wire signal power spectra at r/D = 0.25, x = 1.25D and radial variation of amplitude of preferred frequency **relative** to mean
(a) disk d/D = 0.5 normal to flow (b) disk d/D = 0.5 inclined at 30^0

The radial profiles of the ratio of power spectral density at the preferred frequency to the mean density, measured 1.25 pipe diameters downstream of the small disk at 0^0 and 30^0 angles are also shown in Fig. 3. As can be seen, the distribution is symmetric for the disk at zero angle and asymmetric for the 30^0 angle with maxima in the shear layer generated from the edge of the disk, as previously observed in references [13] and [14] with disks held in a uniform flowfield in a large wind tunnel. It appears that confinement of the disks decreases the sensitivity of the preferred frequency to angle, since the difference in Strouhal number for the flow over a disk at zero and 40^0 is almost a factor of two in the experiments of [14] in contrast to the 15% difference observed here. The larger disk gave rise to larger frequencies as can be seen from Fig 4a, but to lower amplitudes.

Vortex shedding frequencies for disks at angles varying from zero to 40^0

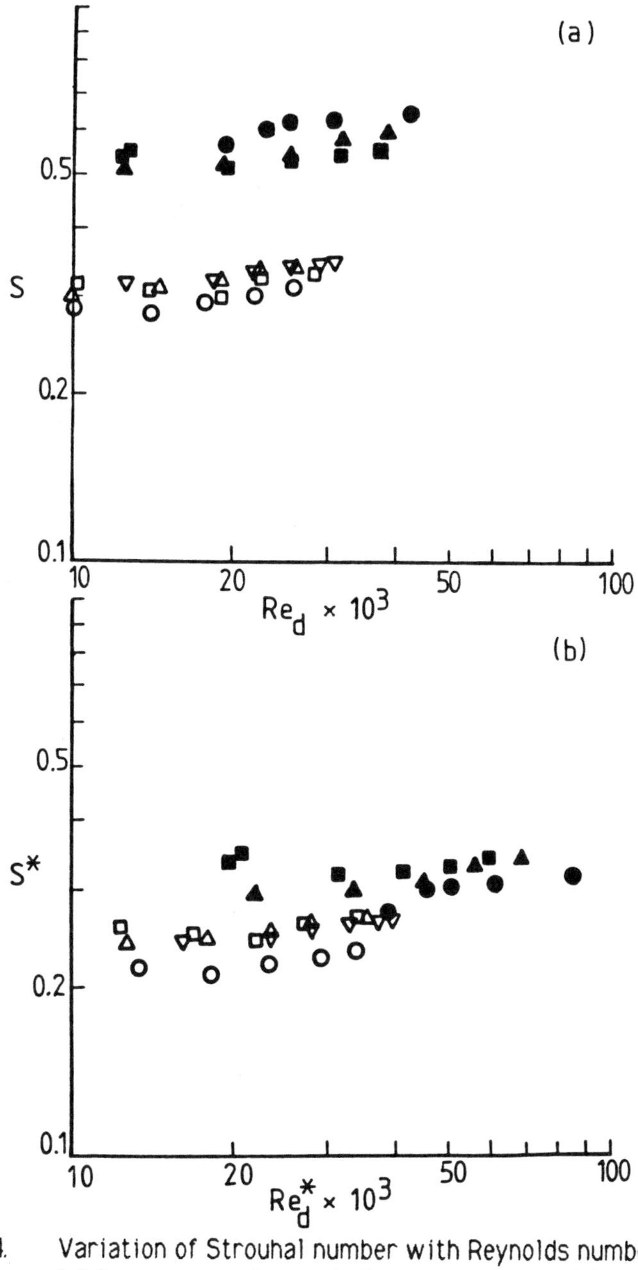

Figure 4. Variation of Strouhal number with Reynolds number.
(a) based on upstream bulk mean velocity (b) based on mean axial velocity in plane of disk.
o,● disk normal to flow; ▽ inclined at 20°;
△,▲ inclined at 30°; □,■ inclined at 40°.
Open symbols - d/D = 0.5; solid symbols - d/D = 0.7.

may be correlated by a modified Strouhal number defined in terms of the preferred frequency and a characteristic velocity and diameter which were defined in terms of the mean velocity profile across a plane normal to the axis at an axial station where the mean static pressure was minimum. Fig. 4b shows the present data in terms of Strouhal and Reynolds numbers based on the mean axial velocity across the plane of the disk and that its variation with disk diameter and angle is small but considerably larger than with Reynolds number.

Combustion alters the properties of the flowfield downstream of the disk and both the amplitude and the value of vortex shedding frequency are expected to be strongly dependent on the equivalence ratio of the mixture. Direct comparison of frequencies in isothermal flow and combusting flow will therefore not be possible but the influence of disk geometry and upstream flow conditions on shedding frequency in combusting flow should be consistent with those in isothermal flow. The proneness of a given disk geometry to vortex shedding in isothermal flow is likely to influence the strength of vortex shedding in combusting flow.

2.2 Combusting flow measurements

Figure 5 shows power spectra for wall static pressures 0.1 pipe diameter upstream of the disk and for radiated sound intensity at a location 20 pipe diameters from the axis and in the exit plane of the duct for flames stabilised behind the 40 mm disk at 0^0, located 12 pipe diameters from the exit and at an upstream Reynolds number of 44,000. The spectra in Fig. 5a to d are for apparently stable combustion and those in Fig. 4e and f for unstable combustion. Comparison of the power spectra for wall static pressure and sound level reveals that moderately strong pressure fluctuations occur over frequency bands around 200 Hz and 300 Hz, in addition to the sharp peaks observed at the four preferred frequencies marked f_1, f_2, f_3 and f_4. These frequency bands are not discrete and they were not significantly influenced by changes in equivalence ratio. Discrete frequencies close to 30 Hz and 90 Hz were observed at equivalence ratios 0.05 to 0.1 away from the extinction limits, but their amplitudes were small.

The frequency f_2 was first observed in the power spectra for wall pressure at about an equivalence ratio of 0.1 greater than the lean extinction limit, and a further increase of 0.05 in equivalence ratio was needed before frequency f_1 could be identified as a discrete frequency peak. Both frequencies increase with equivalence ratio as may be seen from Fig. 6 and,

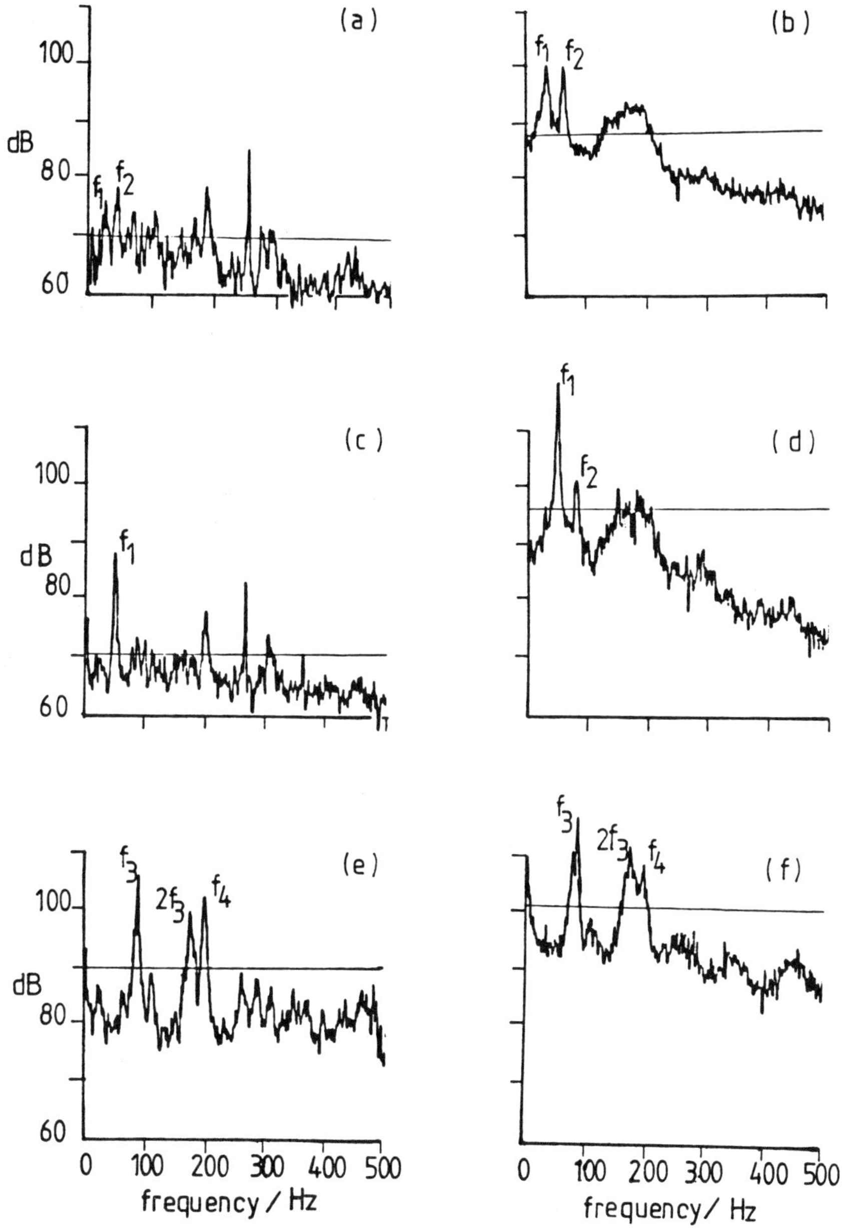

Figure 5. Power spectra for radiated sound and wall static pressure. Disk d/D = 0.5 normal to flow, L = 12 D, Re_D = 44,000.

(a) (b) ϕ = 0.67; (c) (d) ϕ = 0.70; (e) (f) ϕ = 0.80.

(a), (c) and (e) are sound spectra and the rest, pressure spectra.

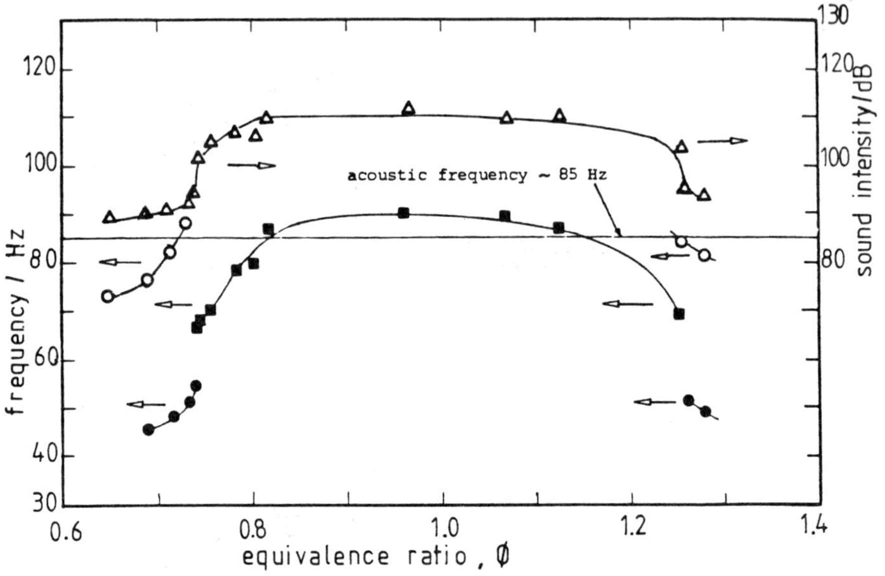

Figure 6. Variation of preferred frequencies and sound intensity with equivalence ratio.
Geometry as in figure 5, Re = 44,000.
● - f_1; ○ - f_2; ■ - f_3; Δ-sound intensity.

as the stability limit is approached, the amplitude and value of frequency f_1 rise sharply, while the amplitude of frequency f_2 tends to decline relative to the mean power spectral density. The onset of instability is marked by a sharp increase in sound level and the disappearance of f_1 which seems to be related to shedding. These events occur in reverse sequence as the equivalence ratio is increased through the rich stability limit to extinction. Strong oscillations in the range of equivalence ratios which give rise to unstable combustion were investigated in considerable detail in references 1 to 6 and were identified as acoustic frequencies, for example f_3, which varies from around 15 Hz less than the frequency of a standing acoustic quarter wave between the swirl register and the disk to nearly 5 Hz more than this frequency. The lower values of the frequency were observed close to the stability limit and the maximum in the middle of the unstable zone. The higher values of frequency were always associated with higher sound intensities, generally 5 to 10 dB more than the sound levels measured at the onset of instability. The variation in frequency is as a result of the variation in the position about which the flame front is oscillating and this position tends to move from downstream of the disk to a little upstream of the disk as the equivalence ratio approaches unity. This was confirmed by the lengthwise variation of the amplitude of wall static pressure fluctuations filtered at the preferred frequency. The position of the pressure node was just upstream of the disk at the highest frequency and at lower frequencies the position of the node could not be determined with precision although it was certainly downstream of the disk. Other acoustic frequencies, for example f_4, which is distinct from the harmonics of f_3, can result from acoustic three-quarter wave resonance with a node at the exit and an antinode at the swirl register, as from quarter waves based on the downstream duct length as in the case of the dump diffusors and disks with a large blockage ratio.

The distribution of amplitudes of fluctuating pressure along the wall for frequency f_1, at different equivalence ratios, indicated that the oscillations were strongest near the disk and did not suggest an acoustic standing wave. The dependence of the frequency on disk geometry and Reynolds number distinguishes f_1 from the three frequencies discussed earlier and was examined further. Both the frequency and amplitude of f_1 vary with equivalence ratio with a sharp increase as the stability limit is approached. As the equivalence ratio is increased from the extinction limit the flame front moves closer to the disk and this results in an increase in mean flow velocity around the recirculation bubble and a corresponding increase in the frequency of oscillation of the recirculation bubble. The sharp increase in the strength of the oscillations may be explained in terms of the heat release in the vicinity of the recirculation bubble as the flame front moves closer to the disk. It may be seen from Fig. 6 that the sound intensity reaches 100 dB, compared to around 90 dB in most of the stable combustion zone, before the onset of instability leads to a further rise to around 110 dB. Just prior to the onset of instability, the amplitude at frequency f_1 in the sound spectra rises to 20 dB over the mean. These observations demonstrate that strong oscillations are possible in apparently stable combustion and that the mechanism of the oscillations is distinct from that in unstable flames in that the dominant mode f_1 is not related to acoustic phenomena. They are consistent with the findings in dump combustors with a constricted exit where oscillations associated with shedding have been identified in smooth combustion for flames stabilised behind the rearward facing step of an axisymmetric combustor [15]. The shedding frequencies, however, tended to lock onto the nearest acoustic frequency of the combustor duct.

Figure 7 shows the variation of frequency f_1 with equivalence ratio for three

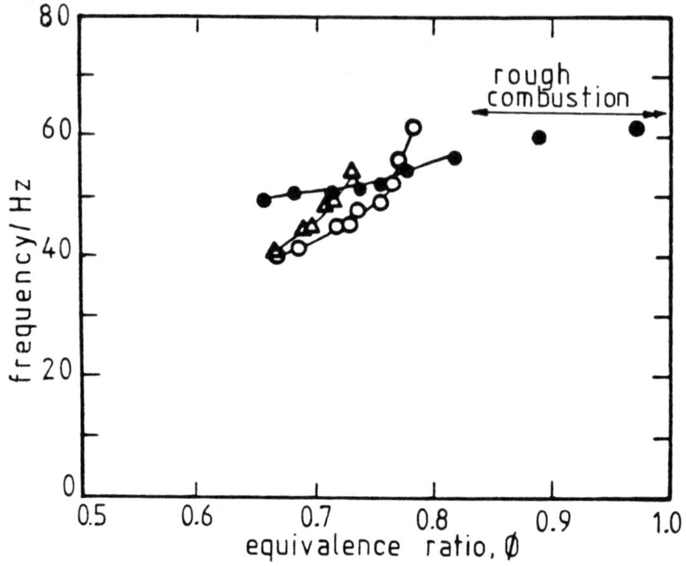

Figure 7. Variation of frequency f_1 with disk location. Disk d/D = 0.5 held normal to flow, Re = 44000. ● - L = 5.3 D; ○ - L = 9.5 D; Δ - L = 13.5 D.

different positions of the 40 mm disk held at 0°, at an upstream Reynolds number of 44,000. With the disk located far downstream (L/D ~ 5.3), f_1 was not distinguishable from f_2 and interaction between them cannot be ruled out. For the two upstream positions the curves show greater similarity. It should however be noted that when the disk was placed further upstream (L/D>11) the transition to instability occurred before f_2 exceeded 60 Hz. For Reynolds numbers between 31,000 and 49,000, f_1 had the widest range of values distributed over a moderately wide range of equivalence ratios when the disk was located between 9 and 10 pipe diameters from the exit. The frequencies are compared on the basis of the range over which they show gradual increase before the sharp rise close to the stability limit. This criterion has its limitations in that the fluid properties inside the recirculation bubble and flow characteristics around it are not exactly comparable, and detailed measurements of the flow in the region downstream of the disk are desirable. It nevertheless permits us to check whether the expected trends are observed, and examine the extent to which they are in agreement with measurements in isothermal flow.

Figure 8 shows the variation of f_1 with equivalence ratio for the 40 mm disk

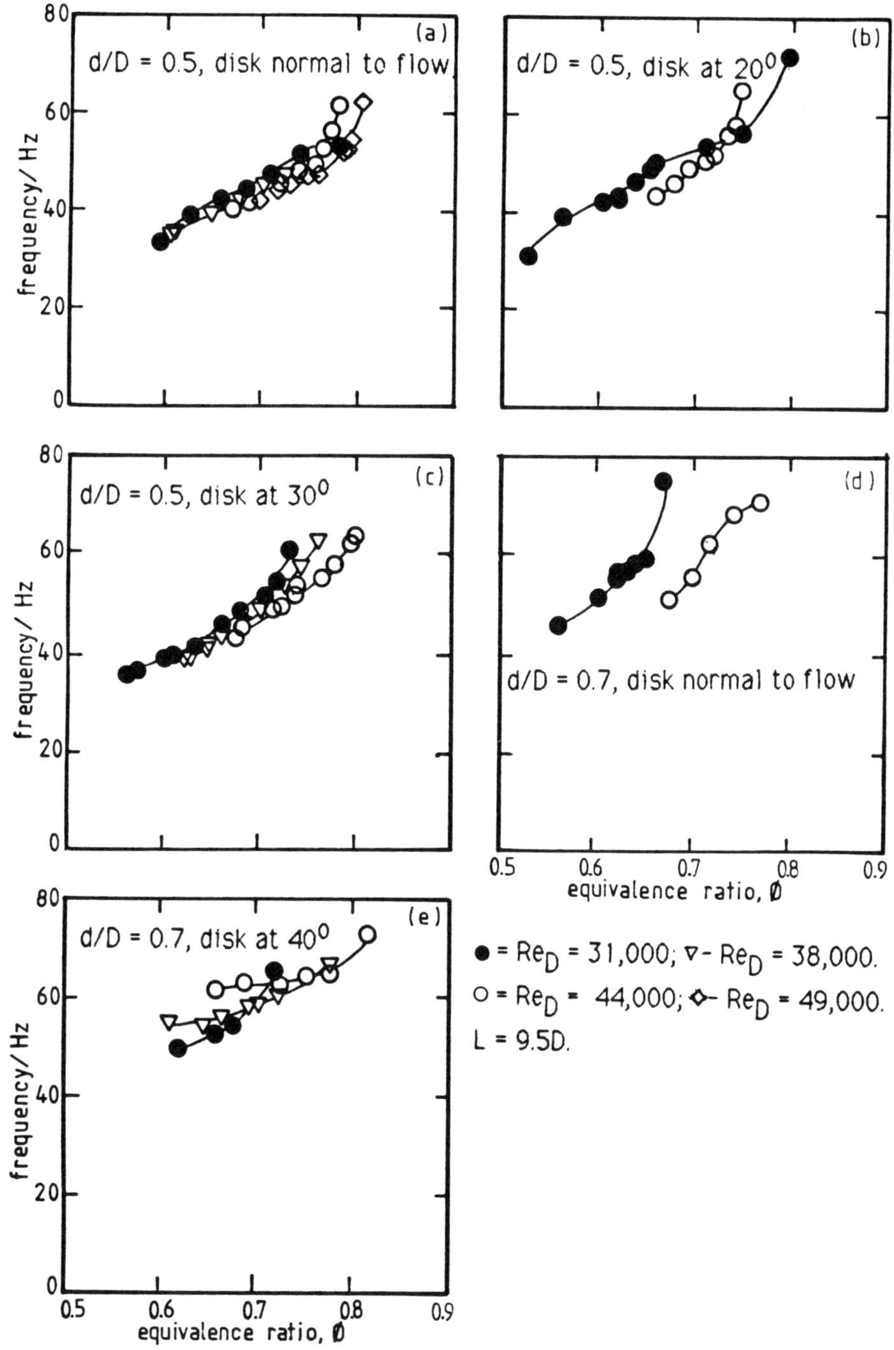

Figure 8. Influence of disk geometry and Reynolds number on frequency f_1.

held at zero, 20 and 30^0 and the 56 mm disk held at zero and 40^0, located 9.5 pipe diameters from the exit, at Reynolds numbers ranging from 31,000 to 49,000. With the 40 mm disk, the frequencies range from 30 to 50 Hz at an upstream Reynolds number of 31,000 and from 40 to 60 Hz at a Reynolds number of 44,000. The influence of the angle of the disk is not strong enough to be identified. These frequency ranges correspond to Strouhal numbers from 0.2 to 0.33 based on mean upstream velocity at the smaller Reynolds number, and 0.19 to 0.29 at the larger Reynolds number. The corresponding Strouhal numbers in isothermal flow are around 0.285 and 0.30, respectively. It appears that f_1 tends to increase while the Strouhal number tends to decrease with Reynolds number and this difference between combusting and isothermal flows is consistent with the increase in the lean blow-off and stability limits with Reynolds number. The ranges of equivalence ratio over which frequency variation is considered differ for the two Reynolds numbers and the equivalence ratio at either end of the range is around 10% higher for the larger Reynolds number than that for the smaller Reynolds number. A higher temperature in the recirculation bubble results in a lowering of the relevant Reynolds number and therefore the frequency. The influence of Reynolds number was similar for the larger disk. The Strouhal number in combusting flow for the disk held normal to the flow varies from 0.35 to 0.55 at an upstream Reynolds number of 31,000 and from 0.33 to 0.47 at a Reynolds number of 44,000. The corresponding Strouhal numbers in isothermal flow are 0.57 and 0.62, respectively. The Strouhal numbers for the larger disk held at an angle of 40^0 appear to be up to 10% higher than those for the disk held normal to the flow, in contrast to isothermal flow where the Strouhal number for the disk held normal to the flow is 10% higher. This difference is small compared to the variation in frequency with equivalence ratio, and may be partly attributed to the fact that the blockage effect of the larger disk at 0^0 (blockage ratio 0.5) is much larger than that for the inclined disk (blockage ratio 0.38). The stability characteristics are significantly different for the two disk angles with the stable operating range much narrower for the disk held normal to the flow than for the inclined disk: this result is consistent with the expected influence of blockage ratio.

Vortex shedding frequencies for the larger disk held normal to the isothermal flow are nearly 1.5 times those for the smaller disk held normal to the flow at the same upstream Reynolds number. In combusting flow the corresponding ratio for the value of frequency f_1 is 1.4 and shows that the effect of disk size on frequency is in qualitative agreement with that for vortex shedding in isothermal flow.

3. Modelling

The representation of the acoustically-dominated flames by calculation methods based on the solution of conservation equations in differential form, requires the solution of unsteady equations and appropriate modelling of acoustic, turbulence, heat transfer and combusting features. The acoustic frequencies are always comparatively low so that, as for reciprocating engines, useful numerical solutions of appropriate unsteady equations can be obtained. The representation of acoustic effects can be achieved by making use of experimentally determined information to provide the acoustic frequency and ensuring that all energy above that associated with a normal turbulent flow is assigned to it. It is still, of course, necessary to model the essential physical and chemical features of the combustion.

In principle, the acoustic nature of the flow can be determined from a knowledge of the geometry and of the impedences offered by the upstream swirl register and downstream baffle and pipe exit, together with a compressible form of the equations of conservation of mass, momentum and energy. This more general approach is dependent on information about local mean and fluctuating flow properties including heat release rate, especially in the region immediately downstream of the baffle, and, in view of the complexity of the problem, theoretical studies have preferred an integral approach. Studies based on one-dimensional flow approximations, assumption of complete combustion within a flame front of negligible thickness, and nearly invariant flame speed have successfully demonstrated the existence of stability limits and identified acoustic frequencies which would be associated with combustion oscillations [16-18]. Successful prediction of stability limits and the strength of the oscillations would depend on the development of theoretical models capable of calculating combusting flows without preferred frequency oscillations.

The modelling of premixed turbulent combustion, in the absence of preferred-frequency oscillations is difficult, particularly since, unlike diffusion flames, it requires consideration of turbulence-chemistry interactions [19, 20]. Simple approaches where a choice is made between reactions limited by chemical and diffusion control [21, 22] do not necessarily give rise to unique solutions and do not take adequate account of the interactions. More advanced combustion models assume fast chemistry and chemical equilibrium so that it is not necessary to introduce more than one equation for each of a conserved scalar and its fluctuations. The experiments of [23], concerned with the combustion of premixed methane and air in a duct and stabilised by a disk, suggest that these assumptions, are adequate for lean mixtures but inadequate in high temperature regions with higher equivalence ratios which cause concentrations of carbon monoxide higher than those associated with equilibrium. It should be emphasised that these considerations are based on the measurements of mean quantities, though they are similar to those chosen for non-premixed flames in a range of flows. To overcome the likely limitations of these assumptions will require the development of models based on joint probability functions involving more than one scalar and possibly their calculation by Monte Carlo methods as, for example, in [24].

In addition to those complexitites introduced by oscillations and by the essential turbulence-chemistry interaction, it is evident from the experiments of [7] and [8] that counter-gradient diffusion can exist over a substantial region of a premixed flame. This concept has been discussed in [25] and demonstrates that the gradient assumptions of eddy-viscosity approaches to turbulence are inappropriate. The magnitudes of the corresponding effects are large and it is necessary, therefore, to model turbulence transport by equations based on turbulent fluxes. Thus a Reynolds-stress transport model is required. The theoretical analyses of [9, 10] show that the interaction of mean pressure gradient with turbulent quantities, in particular the normal stresses and concentration fluctuations, can dominate the production of turbulence through non-gradient transport of turbulent heat flux. The effect can be expected to increase in importance with pressure gradient and, therefore, with flow confinement.

Nomenclature

d	diameter of baffle
D	diameter of pipe
f	frequency
L	pipe length downstream of baffle
r	radial distance from pipe axis
Re_d	Ud/ν
Re_D	UD/ν
S	fd/U
U	bulk mean velocity upstream of baffle
x	axial distance downstream of baffle
X	pipe length upstream of baffle
ν	kinematic viscosity

ϕ equivalence ratio = (fuel: air ratio of mixture)/(fuel: air ratio at stoichiometry)

superscript

* based on mean axial velocity in the plane of the baffle.

Acknowledgement

The authors are glad to acknowledge financial support from the US Office of Naval Research through Grant N00014-84-6-0185.

REFERENCES

1. Heitor M V, Taylor A M K P and Whitelaw J H: Influence of confinement on combustion instabilities of premixed flames stabilised on axisymmetric baffles. Comb and Flame 57, 109, 1984.

2. Katsuki M and Whitelaw J H: The influence of duct geometry on unsteady premixed flames. Comb and Flame 63, 87, 1986.

3. Sivasegaram S and Whitelaw J H: Oscillations in confined disk-stabilised flames. Comb and Flame, to appear in 1987.

4. Sivasegaram S and Whitelaw J H: Suppression of oscillations in disk-stabilised flames. J Propulsion and Power, to appear in 1987.

5 Sivasegaram S and Whitelaw J H: Oscillations in axisymmetric dump combustors. Comb Sci and Tech, to appear in 1987.

6. Sivasegaram S and Whitelaw J H: Combustion oscillations of relevance to augmentors. To be presented at Propulsion Conference of AIAA, San Diego, 1987.

7. Heitor M V, Taylor A M K P and Whitelaw J H: The interaction of turbulence and pressure greadients in a baffle-stabilised premixed flame. J Fluid Mech, to appear in 1987.

8. Heitor M V, Taylor A M K P and Whitelaw J H: Simultaneous velocity and temperature measurements in a premixed flame. Experiments in Fluids 3, 323, 1985.

9. Bray K N C, Libby P A and Moss J B: Unified modelling approach for premixed turbulent combustion. Part 1: General formulations. Comb and Flame 61, 87, 1983.

10. Libby P A: Theory of normal premixed turbulent flames revisited. Prog Energy Comb Sci, 11, 83, 1985.

11. Argueyrolles B, Southaite P, Gauffie S and Borghi R: Application of a "presumed pdf" model of turbulent combustion to reciprocating engines - part 2. Proc Int Conference on Supercomputer Application in Automotive Industry, Zurich, Computer Mechanics Publication, 303, 1986.

12. Ahmed-Befrui B, Gosman A D, Lockwood F C and Watkins A P: Multidimensional calculation of combustion in an idealised homogenous charge engine. SAE Paper 810151, 1981.

13. Calvert J R: Experiments in the low speed flow past cones. J Fluid Mech, 27, 273, 1967.

14. Calvert J R: Experiments in the flow past an inclined disk. J Fluid Mech, 29, 691, 1967.

15. Schadow K C, Crump J E, Mahan V A, Nabity J A, Wilson K J and Gutmark E: Large-scale coherent structures as drivers of ramjet combustion instabilities. Proc JANNAF Propulsion Meeting, CPIA Publication 425, San Diego, California, 1985.

16. Bloxidge G J and Dowling A P: Reheat buzz: an acoustically driven combustion instability. AIAA paper 84-2321, 1984.

17. Bloxidge G J, Dowling A P, Hooper N and Langhorne N: Active control of reheat buzz. AIAA paper 87-0433, 1987.

18. Yang V and Culick F E C: Analysis of low frequency combustion instabilites in a laboratory ramjet combustor. Comb Sci Tech, 45, 1, 1986.

19. Jones W P and Whitelaw J H: Modelling and measurements in turbulent combustion. Twentieth Symposium (Int) on Combustion, 233, 1984.

20. Libby P A and Williams F A: Some implications of recent theoretical studies in turbulent combustion. AIAA J, 19, 261, 1981.

21. Spalding D B: Mixing and chemical reaction in steady confined turbulent flames. Thirteenth Symposium (Int) on Combustion, 649, 1971.

22. Magnussen B F and Hjertager H: On mathematical modelling of turbulent combustion with special emphasis on soot formation and combustion. Sixteenth Symposium (Int) on Combustion, 719, 1976.

23. Heitor M V, Taylor A M K P and Whitelaw J H: Velocity and scalar characteristics of turbulent premixed flames stabilised on confined axisymmetric baffles. Comb Sci and Tech. Under consideration.

24. Pope S B: Monte Carlo calculations of premixed turbulent flames. Eighteenth Symposium (Int) on Combustion, 1001, 1981.

25. Libby P A and Bray K N C: Counter gradient diffusion in premixed turbulent flames. AIAA J, 19, 205, 1981.

SOLUTION OF TWO-DIMENSIONAL AXISYMMETRIC LAMINAR DIFFUSION FLAMES BY ADAPTIVE BOUNDARY VALUE METHODS

M. D. Smooke
Department of Mechanical Engineering
Yale University
New Haven, CT

A. A. Turnbull
Teradyne Inc.
Agoura Hills, CA

R. E. Mitchell
Sandia National Laboratories
Livermore, CA

D. E. Keyes
Department of Mechanical Engineering
Yale University
New Haven, CT

1. INTRODUCTION

The flame type of most practical combustion devices is the diffusion flame. These flames are important in the interaction of heat and mass transfer with chemical reactions in ram jets, jet turbines and commercial burners. The ability to predict the coupled effects of complex transport phenomena with detailed chemical kinetics in these systems is critical in the modeling of turbulent reacting flows, in improving engine efficiency and in investigating the effects of facility related contaminants on the flammability limits of combustion under supersonic conditions. Specifically, the ability to understand those factors affecting flame extinction is critical in improving engine efficiency. Combustor geometry and flow field patterns may produce flames with sharp edges. Extinction along these edges results from incomplete combustion and, hence, produces a decrease in fuel efficiency. The opening of such flames corresponds to the local extinction of a specific highly curved portion of the flame front. An understanding of the mechanisms that cause this behavior is important in flame stability and in determining flammability limits for flames exhibiting strong curvature [1]. In addition,

C.-M. Brauner and C. Schmidt-Lainé (eds.), Mathematical Modeling in Combustion and Related Topics, 261–300.

most commercial power generating units employ diffusion flames as their primary flame type. Emphasis on improving the efficiency of these units combined with environmental issues dealing with the production of nitrogen based pollutants helps to motivate the study of this type of diffusion flame. Most of the oxides of nitrogen are formed during combustion when part of the oxygen combines with atmospheric nitrogen rather than with the fuel. The burning of hydrocarbon fuels in these units can produce large quantities of nitrogen dioxide and nitric oxide. Both compounds are considered toxic and nitric oxide is related to the formation of photochemical smog [2-4]. To be able to determine the emissions of these flames as well as to be able to reduce the emission levels, a basic flame study is required.

Most of the detailed chemistry, computational combustion studies that have appeared in the literature have focused on essentially one-dimensional configurations, i.e., freely propagating or burner-stabilized premixed flames and counterflow premixed or diffusion flames [5-31]. The interaction of heat and mass transfer and chemical reaction in practical combustion systems, however, requires a multidimensional study. Three-dimensional models combining both fluid dynamical effects with finite rate chemistry are as yet computationally infeasible. As a result, the modeling of chemically reacting flows generally proceeds along two independent paths. In one case chemistry is given priority over fluid mechanical effects and these models are used to assess the important elementary reaction paths in, for example, hydrocarbon fuels. In the other case, multidimensional fluid dynamical effects are emphasized with chemistry receiving little or no priority.

In this paper we apply a detailed chemistry, complex transport combustion model to a two-dimensional, axisymmetric laminar diffusion flame in which a cylindrical fuel stream is surrounded by a coflowing oxidizer jet (see also [32-35]). In this configuration we can study the interaction of fluid flow and chemical reaction in diffusion flames while obtaining a computationally feasible problem. Unlike some models in which diffusion in the axial direction is neglected (see, e.g., [36]), we will treat the fully elliptic problem.

In the next section we formulate the problem using both a primitive variable and a stream function-vorticity development and we describe the transport and chemistry models used in the calculations. A flame sheet starting estimate is discussed in Section 3 and the computational method is developed in Section 4. Numerical results for a methane-air flame are presented in Section 5.

2. PROBLEM FORMULATION

2.1 Governing Equations

Conclusions derived from studies of laminar flames are important in the characterization of the combustion processes occurring in turbulent flames, in the formation of engine pollutants and in the determination of flammability limits for combustion systems. By studying laminar flames, we can identify the important reactions controlling extinction and we can identify the important species involved in pollutant formation while providing information on the fluid mechanics of the flames. One of the simplest two-dimensional flame configurations with practical importance is the axisymmetric diffusion flame. Although axisymmetric flames are important in combustion applications, they have received relatively little attention in theoretical flame studies. Part of this neglect is due to the two-dimensional nature of the problem coupled with the complexities associated with the combined effects of transport phenomena and chemical processes. In the axisymmetric diffusion flame we consider, a fuel jet discharges into a laminar air stream. The tubes through which the fuel and the oxidizer flow are concentric and have radii R_I and R_O, respectively. The two gases make contact at the outlet of the inner tube and a flame that resembles a candle results (see Figure 1).

Our model of axisymmetric laminar diffusion flames considers the full set of two-dimensional governing equations. In primitive variables (r and z denote the radial and axial coordinates, respectively) the governing equations can be written in the form

Continuity:

$$\frac{1}{r}\frac{\partial}{\partial r}(r\rho v_r) + \frac{\partial(\rho v_z)}{\partial z} = 0, \tag{2.1}$$

Radial Momentum:

$$\begin{aligned}\left[r\rho v_r\frac{\partial v_r}{\partial r} + r\rho v_z\frac{\partial v_r}{\partial z}\right] - 2\frac{\partial}{\partial r}\left(r\mu\frac{\partial v_r}{\partial r}\right) - \frac{\partial}{\partial z}\left(r\mu\frac{\partial v_r}{\partial z}\right)\\ +\frac{2}{3}\frac{\partial}{\partial r}\left(\mu\frac{\partial(rv_r)}{\partial r}\right) + \frac{2}{3}\frac{\partial}{\partial r}\left(r\mu\frac{\partial v_z}{\partial z}\right) - \frac{\partial}{\partial z}\left(r\mu\frac{\partial v_z}{\partial r}\right)\\ +2\mu\frac{v_r}{r} - \frac{2}{3}\frac{\mu}{r}\frac{\partial}{\partial r}(rv_r) - \frac{2}{3}\mu\frac{\partial v_z}{\partial z} + r\frac{\partial p}{\partial r} = 0,\end{aligned} \tag{2.2}$$

Axial Momentum:

$$\begin{aligned}\left[r\rho v_r\frac{\partial v_z}{\partial r} + r\rho v_z\frac{\partial v_z}{\partial z}\right] - \frac{\partial}{\partial r}\left(r\mu\frac{\partial v_z}{\partial r}\right) - 2\frac{\partial}{\partial z}\left(r\mu\frac{\partial v_z}{\partial z}\right)\\ +\frac{2}{3}\frac{\partial}{\partial z}\left(\mu\frac{\partial(rv_r)}{\partial r}\right) + \frac{2}{3}\frac{\partial}{\partial z}\left(r\mu\frac{\partial v_z}{\partial z}\right) - \frac{\partial}{\partial r}\left(r\mu\frac{\partial v_r}{\partial z}\right) + r\frac{\partial p}{\partial z} - r\rho g = 0,\end{aligned} \tag{2.3}$$

Species:

$$\left[r\rho v_r\frac{\partial Y_k}{\partial r} + r\rho v_z\frac{\partial Y_k}{\partial z}\right] + \frac{\partial}{\partial r}(r\rho Y_k V_{k_r}) + \frac{\partial}{\partial z}(r\rho Y_k V_{k_z}) - rW_k\dot{w}_k = 0,\ k = 1,2,\ldots,K, \tag{2.4}$$

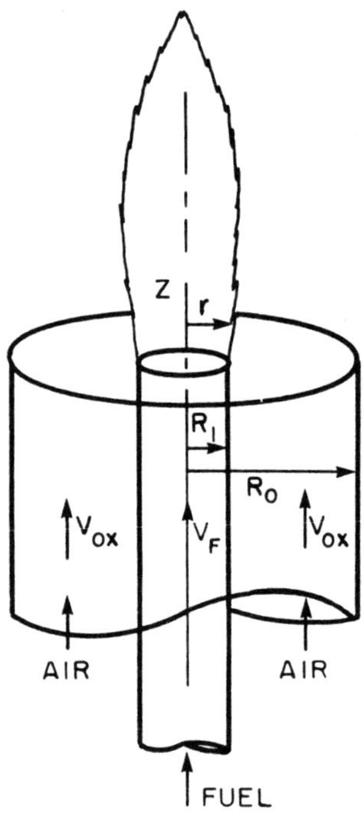

Figure 1. Schematic of an unconfined axisymmetric laminar diffusion flame.

Energy:

$$c_p \left[r\rho v_r \frac{\partial T}{\partial r} + r\rho v_z \frac{\partial T}{\partial z} \right] - \frac{\partial}{\partial r}\left(r\lambda \frac{\partial T}{\partial r} \right) - \frac{\partial}{\partial z}\left(r\lambda \frac{\partial T}{\partial z} \right)$$
$$+ r \sum_{k=1}^{K} \left\{ \rho c_{pk} Y_k \left(V_{k_r} \frac{\partial T}{\partial r} + V_{k_z} \frac{\partial T}{\partial z} \right) \right\} + r \sum_{k=1}^{K} h_k W_k \dot{w}_k = 0. \tag{2.5}$$

The system is closed with the ideal gas law

$$\rho = \frac{p\overline{W}}{RT}. \tag{2.6}$$

Although both confined and unconfined flames are of importance, we will only focus our study on unconfined flames. The boundary conditions are given by

Axis of Symmetry ($r = 0$):

$$\frac{\partial \rho}{\partial r} = v_r = \frac{\partial v_z}{\partial r} = \frac{\partial Y_k}{\partial r} = \frac{\partial T}{\partial r} = 0, \quad k = 1, 2, \ldots, K, \tag{2.7}$$

Exit ($z \to \infty$):

$$\frac{\partial v_r}{\partial z} = \frac{\partial v_z}{\partial z} = \frac{\partial Y_k}{\partial z} = \frac{\partial T}{\partial z} = 0, \quad k = 1, 2, \ldots, K, \tag{2.8}$$

Inlet ($z = 0$):
$r < R_I$

$$\begin{gathered} \rho = \rho_I, \\ v_r = 0, \\ v_z = v_I, \\ Y_k = Y_{k_I}, \quad k = 1, 2, \ldots, K, \\ T = T_I, \end{gathered} \tag{2.9}$$

$R_I < r < R_O$

$$\begin{gathered} \rho = \rho_O, \\ v_r = 0, \\ v_z = v_O, \\ Y_k = Y_{k_O}, \quad k = 1, 2, \ldots, K, \\ T = T_O, \end{gathered} \tag{2.10}$$

Outer Zone ($r = R_O$):

$$\frac{\partial v_r}{\partial r} = \frac{\partial v_z}{\partial r} = \frac{\partial Y_k}{\partial r} = \frac{\partial T}{\partial r} = 0, \; k = 1, 2, \ldots, K, \tag{2.11}$$

where the subscripts I and O refer to the inner jet and the outer jet, respectively, and where $\rho_I, \rho_O, v_I, v_O, Y_{k_I}, Y_{k_O}, T_I$ and T_O are specified quantities. Similar boundary conditions can be derived for a confined flame.

We can reduce the size of the system to be solved by introducing the vorticity and the stream function [33]. The vorticity is a measure of the counterclockwise rotation in the flow. In particular, formulation of the vorticity transport equation serves to eliminate the pressure as one of the dependent variables. We define the vorticity such that

$$\omega = \frac{\partial v_r}{\partial z} - \frac{\partial v_z}{\partial r}. \tag{2.12}$$

The stream function ψ is used to replace the radial and axial components of the velocity vector by a single function. It is defined in such a way that the continuity equation is satisfied identically. We have

$$\rho r v_r = -\frac{\partial \psi}{\partial z}, \tag{2.13a}$$

$$\rho r v_z = \frac{\partial \psi}{\partial r}. \tag{2.13b}$$

With the definitions in (2.12) and (2.13), the governing equations become

Stream Function:

$$\frac{\partial}{\partial z}\left(\frac{1}{r\rho}\frac{\partial \psi}{\partial z}\right) + \frac{\partial}{\partial r}\left(\frac{1}{r\rho}\frac{\partial \psi}{\partial r}\right) + \omega = 0, \tag{2.14}$$

Vorticity:

$$r^2\left[\frac{\partial}{\partial z}\left(\frac{\omega}{r}\frac{\partial \psi}{\partial r}\right) - \frac{\partial}{\partial r}\left(\frac{\omega}{r}\frac{\partial \psi}{\partial z}\right)\right]$$
$$-\frac{\partial}{\partial z}\left(r^3\frac{\partial}{\partial z}\left(\frac{\mu}{r}\omega\right)\right) - \frac{\partial}{\partial r}\left(r^3\frac{\partial}{\partial r}\left(\frac{\mu}{r}\omega\right)\right) + r^2 g\frac{\partial \rho}{\partial r} + r^2\nabla\left(\frac{v_r^2 + v_z^2}{2}\right) \cdot \text{iso } \rho = 0, \tag{2.15}$$

Species Conservation:

$$\left[\frac{\partial}{\partial z}\left(Y_k\frac{\partial \psi}{\partial r}\right) - \frac{\partial}{\partial r}\left(Y_k\frac{\partial \psi}{\partial z}\right)\right] + \frac{\partial}{\partial r}(r\rho Y_k V_{k_r})$$
$$+\frac{\partial}{\partial z}(r\rho Y_k V_{k_z}) - rW_k\dot{w}_k = 0, \quad k = 1, 2, \ldots, K, \tag{2.16}$$

Energy:

$$c_p\left[\frac{\partial}{\partial z}\left(T\frac{\partial\psi}{\partial r}\right)-\frac{\partial}{\partial r}\left(T\frac{\partial\psi}{\partial z}\right)\right]-\frac{\partial}{\partial r}\left(r\lambda\frac{\partial T}{\partial r}\right)-\frac{\partial}{\partial z}\left(r\lambda\frac{\partial T}{\partial z}\right)$$
$$+r\sum_{k=1}^{K}\left\{\rho c_{pk}Y_k\left(V_{k_r}\frac{\partial T}{\partial r}+V_{k_z}\frac{\partial T}{\partial z}\right)\right\}+r\sum_{k=1}^{K}h_kW_k\dot{w}_k=0. \tag{2.17}$$

The boundary conditions for the unconfined flame in the stream function-vorticity formulation are given by

Axis of Symmetry $(r=0)$:

$$\psi=\omega=\frac{\partial Y_k}{\partial r}=\frac{\partial T}{\partial r}=0,\quad k=1,2,\ldots,K, \tag{2.18}$$

Exit $(z\to\infty)$:

$$\frac{\partial\psi}{\partial z}=\frac{\partial\omega}{\partial z}=\frac{\partial Y_k}{\partial z}=\frac{\partial T}{\partial z}=0,\quad k=1,2,\ldots,K, \tag{2.19}$$

Inlet $(z=0)$:
$r<R_I$

$$\begin{gathered}\psi=\rho_I v_I r^2/2,\\ \omega=0,\\ Y_k=Y_{k_I},\quad k=1,2,\ldots,K,\\ T=T_I,\end{gathered} \tag{2.20}$$

$R_I<r<R_O$

$$\begin{gathered}\psi=\frac{1}{2}\rho_I v_I R_I^2+\frac{1}{2}\rho_O v_O(r^2-R_I^2),\\ \omega=0,\\ Y_k=Y_{k_O},\quad k=1,2,\ldots,K,\\ T=T_O,\end{gathered} \tag{2.21}$$

Outer Zone $(r=R_O)$:

$$\begin{gathered}\psi=\frac{1}{2}\rho_I v_I R_I^2+\frac{1}{2}\rho_O v_O(R_O^2-R_I^2),\\ \frac{\partial\omega}{\partial r}=0,\\ \frac{\partial Y_k}{\partial r}=0,\ k=1,2,\ldots,K,\end{gathered} \tag{2.22}$$

$$\frac{\partial T}{\partial r} = 0.$$

In addition to the variables already defined, T, denotes the temperature; Y_k, the mass fraction of the kth species; p, the pressure; v_r and v_z, the velocities of the fluid mixture in the radial and axial directions, respectively; ρ, the mass density; W_k, the molecular weight of the kth species; $\overline{W}$, the mean molecular weight of the mixture; R, the universal gas constant; λ, the thermal conductivity of the mixture; c_p, the constant pressure heat capacity of the mixture; c_{pk}, the constant pressure heat capacity of the kth species; $\dot{w}_k$, the molar rate of production of the kth species per unit volume; h_k, the specific enthalpy of the kth species; g, the gravitational constant; μ the viscosity of the mixture and V_{k_r} and V_{k_z}, the diffusion velocities of the kth species in the radial and axial directions, respectively.

2.2 Transport Model

The transport of momentum, species mass fractions and energy in a chemically reacting flow requires the evaluation of the viscosity, the thermal conductivity and the diffusion coefficients of the species with respect to the mixture. Accurate and efficient evaluation of these quantities is an important aspect of combustion calculations. The diffusion velocities V_{k_r} and V_{k_z} are each divided into three parts. If we consider the radial velocity and omit the subscript r (a similar analysis holds for V_{k_z}), then

$$V_k = v_k + w_k + v_c, \quad k = 1, 2, \ldots, K, \tag{2.23}$$

where v_k is the diffusion velocity due to mole fraction gradients, w_k is the thermal diffusion velocity and v_c is a constant diffusion velocity (independent of species).

We approximate v_k by the Curtiss-Hirschfelder [37] approximation. Specifically, we write

$$v_k = -(1/X_k) D_k \frac{\partial X_k}{\partial r}, \quad k = 1, 2, \ldots, K, \tag{2.24}$$

where X_k is the mole fraction of the kth species and D_k is related to the binary diffusion coefficients through the relation

$$D_k = \frac{(1 - Y_k)}{\sum_{j \neq k}^{K} X_j / \mathcal{D}_{jk}}. \tag{2.25}$$

The thermal diffusion velocity can be written in terms of a thermal diffusion ratio Θ_k. We have

$$w_k = \frac{D_k \Theta_k}{X_k} \frac{1}{T} \frac{\partial T}{\partial r}, \tag{2.26}$$

where the quantities D_k have already been defined. We will only consider thermal diffusion for the light species having molecular weight less than five and only in the trace limit.

The constant diffusion velocity v_c is introduced to satisfy the condition

$$\sum_{k=1}^{K} V_k Y_k = 0, \tag{2.27}$$

which must be satisfied if mass is guaranteed to be conserved. We can write

$$v_c = -\sum_{k=1}^{K} Y_k(w_k + v_k). \tag{2.28}$$

The diffusion coefficients, the viscosity and the thermal conductivity of the mixture are evaluated using the procedures described in [38].

2.3. Chemistry Model

The burning of hydrocarbon fuels proceeds through a complex sequence of elementary reversible reactions. The intermediate species in such mechanisms are of extreme importance. They are involved in the chain initiation, chain propagation, chain inhibition and chain breaking reactions that describe the oxidation of a given fuel. We model the kinetic aspects of axisymmetric diffusion flames by allowing the chemical production and destruction of a species to occur through a sequence of m reversible elementary chemical reactions. We can represent these reactions in the form

$$\sum_{k=1}^{K} \nu_{jk}^{R} Z_k \rightleftharpoons \sum_{k=1}^{K} \nu_{jk}^{P} Z_k, \quad j = 1, 2, \ldots, m, \tag{2.29}$$

where $\nu_{jk}^{R}(\nu_{jk}^{P})$ is the stoichiometric coefficient of species k appearing as a reactant (product) in reversible reaction $j = 1, 2, \ldots, m$ and where Z_k represents the chemical symbol for the kth species.

The production rate $\dot{w}_k$ for the kth species can be written in the form

$$\dot{w}_k = \sum_{j=1}^{m} (\nu_{jk}^{P} - \nu_{jk}^{R}) \left[k_j^f(T) \prod_{n=1}^{K} \left(\frac{\rho Y_n}{W_n} \right)^{\nu_{jn}^{R}} - k_j^r(T) \prod_{n=1}^{K} \left(\frac{\rho Y_n}{W_n} \right)^{\nu_{jn}^{P}} \right]. \tag{2.30}$$

The function $k_j^f(k_j^r)$ is the rate constant for the forward (reverse) path of reaction j. We assume k_j^f has the following modified Arrhenius temperature dependence

$$k_j^f = A_j^f T^{\beta_j^f} \exp(-E_j^f/RT), \tag{2.31}$$

with a similar expression for k_j^r. The reverse rate constant can be written in terms of the forward rate constant and the equilibrium constants K_j^c by

$$k_j^r = k_j^f / K_j^c. \tag{2.32}$$

The pre-exponential factor A_j^f, the temperature exponent β_j^f and the activation energy E_j^f can be compiled from published experimental work. The chemical production rates as well as thermodynamic quantities such as c_p, c_{p_k} and h_k are evaluated using CHEMKIN [39].

3. FLAME SHEET STARTING ESTIMATES

The governing equations in (2.1-2.11) or (2.14-2.22) are nonlinear and require a starting estimate for the finite difference method described in Section 4. The determination of a sufficiently "good" initial solution estimate in two-dimensional problems can be difficult. The difficulty is due to the exponential dependence of the chemistry terms on the temperature and to the nonlinear coupling of the fluid and the thermochemistry solution fields. In previous work, we focused our efforts on the solution of adiabatic and nonadiabatic premixed laminar flames by adaptive finite difference methods [20-24]. In these problems the governing differential equations were discretized and the resulting nonlinear difference equations were solved by Newton's method. Cubic polynomials and Gaussian shaped profiles were used as starting estimates for the major and minor species on an initial coarse grid. In adiabatic and nonadiabatic premixed laminar flame problems the conservation of mass and momentum reduced to the specification of a constant mass flow rate and a constant thermodynamic pressure. Hence, thermochemical considerations played a more important role in these problems than did fluid dynamical aspects. This is not the case in axisymmetric laminar diffusion flames. There is a strong coupling between the fluid dynamic and the thermochemistry solution fields in these flames. We have found that the solution procedure used in premixed laminar flame problems does not provide a sufficiently robust or efficient starting estimate from which Newton's method will converge. In addition, the relaxation of these starting estimates by time-dependent methods to steady-state (or at least until the solution is within the convergence domain of Newton's method) is extremely slow. The importance of these flames in combustion modeling, however, necessitates the development of an efficient starting procedure.

The burning rate in a diffusion flame is controlled by the rate at which the fuel and the oxidizer are brought together in the proper proportions. This is in distinction to premixed flames where the burning rate is controlled by chemical reactions. In the limit of infinitely fast kinetics, the fuel and the oxidizer are separated by a thin exothermic reaction zone. In this zone the fuel and the oxidizer are in stoichiometric proportion and the temperature and products of combustion are maximized. In such an ideal situation, no oxidizer is present on the fuel side and no fuel is present on the oxidizer side. The fuel and oxidizer diffuse towards the reaction zone as a result of concentration gradients in the flow. In diffusion flames of practical interest, the oxidation of the fuel to form intermediates and products proceeds through a detailed kinetics mechanism. In these problems combustion takes place at a finite rate and some fuel and oxidizer co-exist on either side of the reaction zone. Nevertheless, the use of a thin, infinitely fast, global reaction model is a natural starting point for the determination of a "good" initial solution estimate for our finite rate axisymmetric model (see also [32-35]).

We assume that the fuel and the oxidizer obey a single overall irreversible reaction of the type

$$\text{Fuel } (F) + \text{Oxidizer } (X) \rightarrow \text{Products } (P), \tag{3.1}$$

in the presence of an inert gas (N). We have

$$\nu_F F + \nu_X X \rightarrow \nu_P P, \tag{3.2}$$

where ν_F, ν_X and ν_P are the stoichiometric coefficients of the fuel, the oxidizer and the product, respectively. In addition, we neglect thermal diffusion and assume that c_p and c_{p_k} are constant and that the ordinary mass diffusion velocities can be written in terms of Fick's law. With these approximations we can write

Stream Function:

$$\frac{\partial}{\partial z}\left(\frac{1}{r\rho}\frac{\partial\psi}{\partial z}\right) + \frac{\partial}{\partial r}\left(\frac{1}{r\rho}\frac{\partial\psi}{\partial r}\right) + \omega = 0, \tag{3.3}$$

Vorticity:

$$r^2\left[\frac{\partial}{\partial z}\left(\frac{\omega}{r}\frac{\partial\psi}{\partial r}\right) - \frac{\partial}{\partial r}\left(\frac{\omega}{r}\frac{\partial\psi}{\partial z}\right)\right]$$
$$-\frac{\partial}{\partial z}\left(r^3\frac{\partial}{\partial z}\left(\frac{\mu}{r}\omega\right)\right) - \frac{\partial}{\partial r}\left(r^3\frac{\partial}{\partial r}\left(\frac{\mu}{r}\omega\right)\right) + r^2 g\frac{\partial\rho}{\partial r} + r^2\nabla\left(\frac{v_r^2+v_z^2}{2}\right)\cdot \text{iso } \rho = 0, \tag{3.4}$$

Species:

$$-\left[\frac{\partial}{\partial r}\left(Y_F\frac{\partial\psi}{\partial z}\right) + \frac{\partial}{\partial z}\left(Y_F\frac{\partial\psi}{\partial r}\right)\right] - \frac{\partial}{\partial r}\left(r\rho D_F\frac{\partial Y_F}{\partial r}\right)$$
$$-\frac{\partial}{\partial z}\left(r\rho D_F\frac{\partial Y_F}{\partial z}\right) - rW_F\nu_F\dot{w} = 0, \tag{3.5}$$

$$-\left[\frac{\partial}{\partial r}\left(Y_X\frac{\partial\psi}{\partial z}\right) + \frac{\partial}{\partial z}\left(Y_X\frac{\partial\psi}{\partial r}\right)\right] - \frac{\partial}{\partial r}\left(r\rho D_X\frac{\partial Y_X}{\partial r}\right)$$
$$-\frac{\partial}{\partial z}\left(r\rho D_X\frac{\partial Y_X}{\partial z}\right) - rW_X\nu_X\dot{w} = 0, \tag{3.6}$$

$$-\left[\frac{\partial}{\partial r}\left(Y_P\frac{\partial\psi}{\partial z}\right) + \frac{\partial}{\partial z}\left(\frac{Y_P\partial\psi}{\partial r}\right)\right] - \frac{\partial}{\partial r}\left(r\rho D_P\frac{\partial Y_P}{\partial r}\right)$$
$$-\frac{\partial}{\partial z}\left(r\rho D_P\frac{\partial Y_P}{\partial z}\right) + rW_P\nu_P\dot{w} = 0, \tag{3.7}$$

$$-\left[\frac{\partial}{\partial r}\left(Y_N\frac{\partial\psi}{\partial z}\right)+\frac{\partial}{\partial z}\left(Y_N\frac{\partial\psi}{\partial r}\right)\right]-\frac{\partial}{\partial r}\left(r\rho D_N\frac{\partial Y_N}{\partial r}\right)$$
$$-\frac{\partial}{\partial z}\left(r\rho D_N\frac{\partial Y_N}{\partial z}\right)=0, \tag{3.8}$$

and

Energy:

$$c_p\left[\frac{\partial}{\partial z}\left(T\frac{\partial\psi}{\partial r}\right)-\frac{\partial}{\partial r}\left(T\frac{\partial\psi}{\partial z}\right)\right]-\frac{\partial}{\partial r}\left(r\lambda\frac{\partial T}{\partial r}\right)-\frac{\partial}{\partial z}\left(r\lambda\frac{\partial T}{\partial z}\right)$$
$$-r\frac{W_F\nu_F h_F+W_X\nu_X h_X-W_P\nu_P h_P}{c_p}\dot{w}=0, \tag{3.9}$$

where

$$\dot{w}=-\frac{\dot{w}_F}{\nu_F}=-\frac{\dot{w}_X}{\nu_X}=\frac{\dot{w}_P}{\nu_P}, \tag{3.10}$$

is the rate of progress of the reaction.

If we introduce the heat release per unit mass of the fuel Q where

$$Q=h_F+\frac{W_X\nu_X}{W_F\nu_F}h_X-\frac{W_P\nu_P}{W_F\nu_F}h_P, \tag{3.11}$$

and if we assume that the Lewis numbers

$$\mathrm{Le}_F=\frac{\lambda}{\rho D_F c_p}\ ,\mathrm{Le}_X=\frac{\lambda}{\rho D_X c_p}, \tag{3.12a}$$

$$\mathrm{Le}_P=\frac{\lambda}{\rho D_P c_p}\ ,\mathrm{Le}_N=\frac{\lambda}{\rho D_N c_p}, \tag{3.12b}$$

are all equal to one, then each of the Shvab-Zeldovich variables

$$Z_F=Y_F-Y_{F_O}+\frac{c_p}{Q}(T-T_O), \tag{3.13}$$

$$Z_X=Y_X-Y_{X_O}+\frac{c_p}{Q}\frac{W_X\nu_X}{W_F\nu_F}(T-T_O), \tag{3.14}$$

$$Z_P=Y_P-Y_{P_O}-\frac{c_p}{Q}\frac{W_P\nu_P}{W_F\nu_F}(T-T_O), \tag{3.15}$$

$$Z_N=Y_N-Y_{N_O}, \tag{3.16}$$

satisfies the differential equation

$$-\left[\frac{\partial}{\partial r}\left(Z_k\frac{\partial\psi}{\partial z}\right)+\frac{\partial}{\partial z}\left(Z_k\frac{\partial\psi}{\partial r}\right)\right]-\frac{\partial}{\partial r}\left(r\rho D_k\frac{\partial Z_k}{\partial r}\right)$$
$$-\frac{\partial}{\partial z}\left(r\rho D_k\frac{\partial Z_k}{\partial z}\right)=0,\quad k=F,X,P,N. \tag{3.17}$$

One can show that all of the Z_k are proportional to each other and to a conserved scalar S which satisfies an equation similar in form to (3.17).

To complete the specification of the starting estimate, we must be able to recover the temperature and the major species profiles from the conserved scalar. Of critical importance to this procedure is an estimate of the location of the flame front. In the Shvab-Zeldovich formulation fuel and oxidizer cannot co-exist. Hence, on the fuel side of the flame $Y_X = 0$ and on the oxidizer side $Y_F = 0$. If we denote variables at the flame front with the subscript f, then it can be shown (see also [40]) that, for a fixed value of the axial coordinate z, the location of the flame front is defined such that

$$S(r_f)|_{fixed\ z} = S_f = \frac{Y_{X_O}}{Y_{X_O} + \frac{W_X \nu_X}{W_F \nu_F} Y_{F_I}}. \tag{3.18}$$

The location of the flame front can be obtained by solving (3.18) at each axial coordinate level.

If we utilize the proportionality of the Z_k's to S along with the expressions in (3.13-3.16), we can derive expressions for the temperature and species on the fuel and oxidizer sides of the flame. On the fuel side, we have

$$T = T_I S + \left[T_O + Y_{X_O} \frac{Q}{c_p} \frac{W_F \nu_F}{W_X \nu_X} \right] (1 - S), \tag{3.19}$$

$$Y_F = Y_{F_I} S + Y_{X_O} \frac{W_F \nu_F}{W_X \nu_X} (S - 1), \tag{3.20}$$

$$Y_X = 0, \tag{3.21}$$

$$Y_P = Y_{X_O} \frac{W_P \nu_P}{W_X \nu_X} (1 - S), \tag{3.22}$$

and

$$Y_N = Y_{N_O}(1 - S) + Y_{N_I} S. \tag{3.23}$$

On the oxidizer side, we have

$$T = T_O(1 - S) + \left[\frac{Q}{c_p} Y_{F_I} + T_I \right] S, \tag{3.24}$$

$$Y_F = 0, \tag{3.25}$$

$$Y_X = Y_{X_O}(1 - S) - Y_{F_I} \frac{W_X \nu_X}{W_F \nu_F} S, \tag{3.26}$$

$$Y_P = \frac{W_P \nu_P}{W_F \nu_F} Y_{F_I} S, \tag{3.27}$$

and

$$Y_N = Y_{N_O}(1 - S) + Y_{N_I} S. \tag{3.28}$$

The equations in (3.3), (3.4) and (3.17) (corresponding results can be derived for the primitive variable analogues) are solved for the stream function, the vorticity, the temperature and the major species in the flame. For a given profile of the conserved scalar, we solve (3.18) for the location of the flame front on each axial level. We then utilize the relations in (3.19-3.28) to obtain expressions for T, Y_F, Y_O, Y_P and Y_N. The recovered temperature profile is used in the ideal gas law to evaluate the density. The temperature is also needed in forming the viscosity and the diffusion coefficient (D_k is replaced by D). If we introduce the Prandtl number

$$\mathrm{Pr} = \frac{\mu c_p}{\lambda}, \tag{3.29}$$

and recall that all of the Lewis numbers are equal to one, we can write

$$\rho D = \frac{\lambda}{c_p} = \frac{\mu}{(\mathrm{Pr})_{ref}}, \tag{3.30}$$

where $(\mathrm{Pr})_{ref}$ is a reference Prandtl number. Specifically, we use an approximate value for air, $(\mathrm{Pr})_{ref} = 0.75$. Hence, determination of ρD is reduced to the specification of a transport relation for the viscosity. We use the simple power law

$$\mu = \mu_0 \left(\frac{T}{T_0}\right)^r, \tag{3.31}$$

where $r = 0.7$, $T_0 = 298$ K and $\mu_0 = 1.85\times 10^{-4}$ gm/cm-sec is again a reference value for air [41]. The temperature exponent was determined by fitting the equation in (3.31) to the mixture viscosity and temperature data of a representative one-dimensional finite rate chemistry calculation. The scaled heat release parameter $Q/c_p = \Delta T$ can be determined from an estimate of the peak temperature (e.g., from an experiment) or from the heat of combustion of the system under consideration and a representative heat capacity.

4. METHOD OF SOLUTION

Our goal is to obtain a discrete solution of the governing equations in two dimensions on the mesh $\mathcal{M}_2$ the initial nodes of which are formed by the intersection of the lines of the mesh $\mathcal{M}_r$

$$\mathcal{M}_r = \{0 = r_0 < r_1 < \ldots < r_i \ldots < r_{M_r} = R_O\}, \tag{4.1}$$

and the mesh $\mathcal{M}_z$

$$\mathcal{M}_z = \{0 = z_0 < z_1 < \ldots < z_j \ldots < z_{M_z} = L\}. \tag{4.2}$$

Computationally, we combine a steady-state and a time-dependent solution method. A time-dependent approach will be used to help obtain a converged numerical solution on an initial coarse grid using the flame sheet starting estimate. Grid points will then be inserted adaptively and the steady-state solution procedure will be used to complete the problem. Later in the section we will relax the restriction that the grid lines connect opposite sides of the computational domain and discuss the implementation of a locally defined grid.

4.1 Newton's Method

We approximate the spatial operators in the governing partial differential equations by finite difference expressions. Diffusion terms are approximated by centered differences and convective terms by upwind approximations. The problem of finding an analytic solution of the equations is then converted into one of finding an approximation to this solution at each point (r_i, z_j) of the mesh in two dimensions. With the difference equations written in residual form, we seek the solution U^* of the system of nonlinear equations

$$F(U) = 0. \tag{4.3}$$

For an initial solution estimate U^0 which is sufficiently close to U^*, the system of nonlinear equations in (4.3) can be solved by Newton's method. This leads to the iteration

$$J(U^n)(U^{n+1} - U^n) = -\lambda^n F(U^n), \quad n = 0, 1, 2, \ldots. \tag{4.4}$$

$J(U^n) = \partial F(U^n)/\partial U$ is the Jacobian matrix and λ^n $(0 < \lambda \leq 1)$ is the nth damping parameter [42].

We point out that with the spatial discretizations used in forming (4.3), the Jacobian matrix in (4.4) can be written in block nine diagonal form. For problems involving detailed transport and complex chemistry, it is often more efficient to evaluate the Jacobian matrix numerically as opposed to analytically. The numerical procedure we implement extends the ideas outlined by Curtis, Powell and Reid [43]. We form several columns of the Jacobian simultaneously using vector function evaluations and the Jacobian's given sparsity structure. If to each column of the Jacobian we associate the i and j values of the node corresponding to the column's diagonal block, then all columns of the Jacobian having the same value of the parameter

$$\alpha = (i + 3j) \text{mod } 9, \tag{4.5}$$

can be evaluated simultaneously. Ideas along these lines have also been explored in [44-45]. Once the Jacobian is formed we solve the Newton equations with a block-line SOR method. The Newton iteration continues until the size of $||U^{n+1} - U^n||_2$ is reduced appropriately.

In the axisymmetric problem the formation of the Jacobian and its partial factorization in the block-line SOR method accounts for a substantial part of the cost of the calculation. As a result, the use of a modified Newton method in which the Jacobian is re-evaluated only periodically is indicated. The immediate implication of applying the modified Newton method is that the partial factorization of the Jacobian can be stored and each modified Newton iteration can be obtained by performing relatively inexpensive block-line SOR back substitutions. If in the course of a calculation, we determine that the size of the $(n+1)^{\text{st}}$ modified Newton step is larger than the value predicted by the theorem, we form a new Jacobian and restart the iteration count. We have found that the error estimate can reduce the cost of premixed flame calculations by as much as 50 percent over a full Newton method [46].

4.2 Adaptive Gridding

The solution of the governing equations in the axisymmetric problem contains regions in each coordinate direction in which the dependent variables exhibit high spatial activity (steep fronts and sharp peaks). Efficient solution of these problems requires that the high activity regions be resolved adaptively. Techniques that attempt to equidistribute positive weight functions have been used with a great deal of success in premixed and counterflow flame problems. Flames with 30 to 40 chemical species and over 100 chemical reactions can be solved efficiently by adaptively placing grid points in the high activity regions. We will follow a similar approach in the axisymmetric problem.

Adaptive mesh refinement in two dimensions can proceed along several different paths. The simplest procedure involves determining the grid points of the mesh $\mathcal{M}_2$ by equidistributing positive weight functions over mesh intervals in both the r and z directions (see, e.g., [47-48]). Specifically, we attempt to equidistribute the mesh $\mathcal{M}_r$ with respect to the non-negative function W_r and constant C_r for each of the M_z+1 horizontal grid lines. We write

$$\int_{r_i}^{r_{i+1}} W_r \;\, dr < C_r, \quad i=0,1,\ldots,M_r-1, \tag{4.6}$$

for $j=0,1,\ldots,M_z$. Similarly, we attempt to equidistribute the mesh $\mathcal{M}_z$ with respect to the non-negative function W_z and constant C_z for each of the M_r+1 vertical grid lines. We have

$$\int_{z_j}^{z_{j+1}} W_z \;\, dz < C_z, \quad j=0,1,\ldots,M_z-1, \tag{4.7}$$

for $i=0,1,\ldots,M_r$.

In implementing the two-dimensional adaptive grid strategy, we first solve the boundary value problem on an initial coarse grid. We then test the inequality in (4.6) one r subinterval at a time for all the j grid lines and all the dependent solution components. If the inequality is not satisfied, a grid point is inserted at the midpoint of the r subinterval in question for $j=0,1,\ldots,M_z$. Once this procedure has been carried out in the r direction, we reverse the process and begin again in the z direction. This procedure produces an orthogonal tensor product grid in which the coordinate lines connect opposite boundaries of the computational domain. A typical grid determined by this procedure is illustrated in Figure 2. The weight functions in the equidistribution procedure are chosen such that the grid points are placed in regions of high spatial activity with the goal of reducing the local discretization error. We use a combination of first and second derivatives of the solution profiles (see also [20], [24], [30-31] and [49]). The particular combinations of function and slope and the values of C_r and C_z can be changed to produce a solution to a desired level of accuracy. Typically, grids with 50 to 70 adaptively chosen points are required to obtain accurate one-dimensional flame profiles.

Although an adaptive tensor product gridding strategy is straightforward to implement, it can produce a mesh with a large number of "unwanted" grid points. In particular, the high activity regions in a two-dimensional axisymmetric diffusion flame occur typically at the bottom left-hand corner of the computational domain–near the fuel inlet (see Section 5). It is in this region that the temperature and species gradients are largest. The activity in the remainder of the domain is generally small. However, the global gridding requirements in the small "flame region" produce large numbers of essentially "wasted" points in the rest of the computational domain. This results in a significant loss of efficiency due to the additional CPU time and the extra storage needed to solve the problem. An alternative procedure to the global tensor product gridding in (4.6) and (4.7) is to employ a local mesh refinement technique [50-52]. In such a method only the regions of high activity are refined. In this way grid points can be saved and the efficiency of the method increased.

In our local mesh refinement method each grid point is assigned a set of four pointers. These pointers are used in determining the neighbors of the given grid point–the points above, below, to the left and to the right. This information is stored in a set of four arrays. The new mesh is determined by employing the equidistribution procedure using a line integral around the boundary of each rectangular cell. Specifically, if we let $\mathbf{W}$ denote a vector of appropriate positive weight functions, then we require

$$\oint_{\Omega} \mathbf{W} \cdot d\mathbf{s} < C, \tag{4.8}$$

where Ω denotes the boundary of each grid cell. When the integral is not satisfied, the cell is divided into four smaller cells by bisecting opposite boundaries of the cell. A typical grid obtained in a flame sheet calculation is illustrated in Figure 3 with an expanded view of the bottom left-hand corner in Figure 4. The grid in Figure 3 contained 330 points while a corresponding tensor product mesh using the smallest grid spacings from the local mesh would contain 1075 points–a factor of 3.5 more. Based upon the results presented in Section 5, it is clear that local mesh refinement techniques must ultimately be employed in two-dimensional, complex chemistry, axisymmetric diffusion flame calculations.

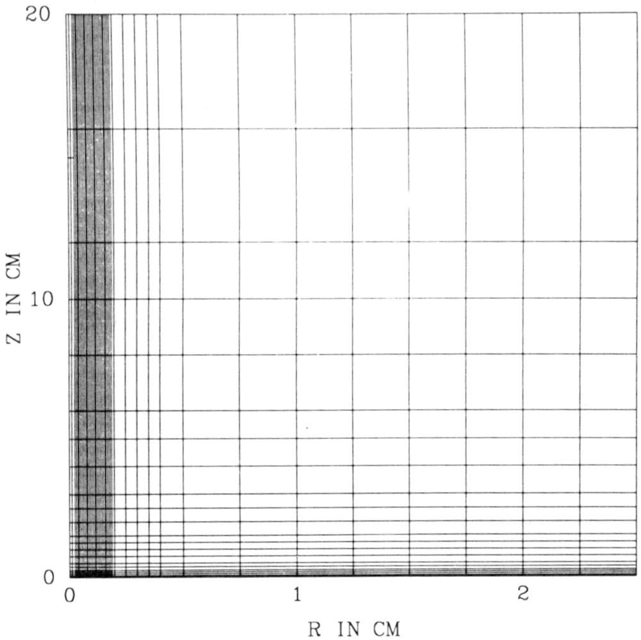

Figure 2. A typical tensor product grid.

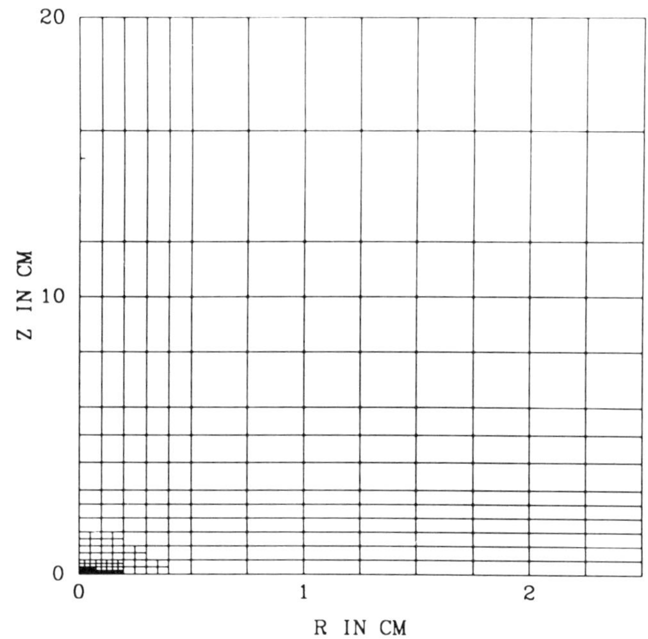

Figure 3. A local grid used in the flame sheet calculations.

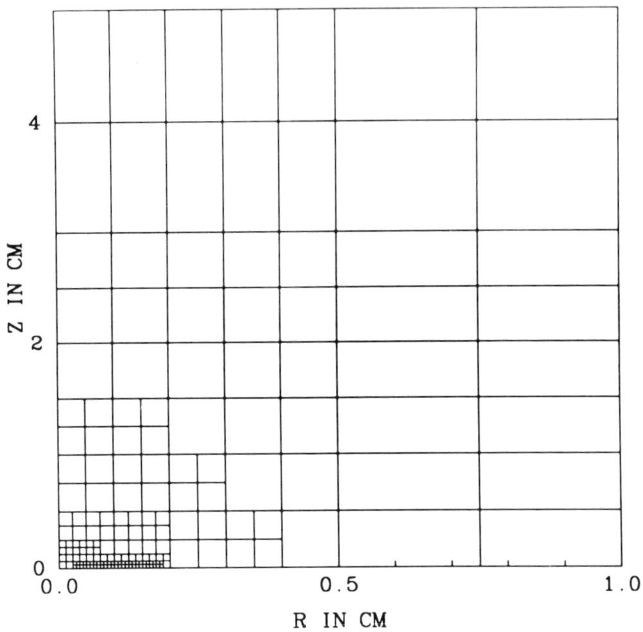

Figure 4. An expanded view of the bottom left-hand corner of the local grid in Figure 3.

4.3 Coarse to Fine Grid Methodology

In both our adaptive grid strategies, the equidistribution condition is checked one mesh interval (cell) at a time and grid points are added appropriately. The coarse grid solution is then interpolated linearly onto the new finer grid. The interpolated result serves as an initial solution estimate for the iteration procedure on the finer grid. The process is continued on successively finer and finer grids until several termination criteria are satisfied. We anticipate that as the size of the mesh spacing gets smaller, the interpolated solution should become a better starting estimate for Newton's method on the next finer grid. For a class of nonlinear boundary value problems, Smooke and Mattheij [53] have shown that there exists a critical mesh spacing such that the interpolated solution lies in the domain of convergence of Newton's method on the next grid. As a result, the hypotheses of the Kantorovich theorem [54] are satisfied and the sequence of successive modified Newton iterates can be shown to satisfy a recurrence relation scaled by the first Newton step [46].

4.4 Time-Dependent Starting Estimates

The coarse to fine grid strategy and the flame sheet starting estimate helps eliminate many of the convergence difficulties associated with solving the governing equations directly. Nevertheless, to obtain a starting estimate on the initial grid that lies in the convergence domain of Newton's method we apply a time-dependent iteration to the flame sheet solution. We remark that fundamentally there are two mathematical approaches for solving flame problems – one uses a transient method and the other solves the steady-state boundary value problems directly. Generally speaking, the transient methods are robust but computationally inefficient compared to the boundary value methods, which are efficient but have less desirable convergence properties. Most of the numerical techniques that have been used to solve one-dimensional flame problems have employed a time-dependent method. Variations of this approach have been considered by a variety of researchers (see, e.g., [5-8], [10-19]). In these methods, the original nonlinear two-point boundary value problem is converted into a nonlinear parabolic mixed initial-boundary value problem. This is accomplished by appending the term $\partial(\bullet)/\partial t$ to the left-hand side of the conservation equations. This same procedure can be employed in our two-dimensional calculations. We obtain

$$\frac{\partial U}{\partial t} = F(U), \tag{4.9}$$

with appropriate initial conditions. If the time derivative is replaced, for example, by a backward Euler approximation, the governing equations can be written in the form

$$\mathcal{F}(U^{n+1}) = F(U^{n+1}) - \frac{(U^{n+1} - U^n)}{\tau^{n+1}} = 0, \tag{4.10}$$

where for a function $g(t)$ we define $g^n = g(t^n)$ and where the time step $\tau^{n+1} = t^{n+1} - t^n$. At each time level we must solve a system of nonlinear equations that look very similar to the nonlinear equations in (4.3). Newton's method can again be used to solve this system. The important difference between the system in (4.3) and (4.10) is that the diagonal of the steady-state Jacobian is weighted by the quantity $1/\tau^{n+1}$. This produces a better conditioned system and the solution from the nth time step ordinarily provides an excellent starting guess to the solution at the $(n+1)^{\text{st}}$ time level. The work per time step is similar to that for the modified Newton iteration, but the timelike continuation of the numerical solution produces an iteration strategy that will, in general, be less sensitive to the initial starting estimate than if Newton's method were applied to (4.3) directly. As a result, when we ultimately implement Newton's method on the steady-state equations directly, we obtain a converged numerical solution with only a few additional iterations. This time-dependent starting procedure can also be used on grids other than the initial one.

5. NUMERICAL RESULTS

In this section we illustrate the results of applying the flame sheet starting estimate and the computational method discussed in Sections 3 and 4 to a methane-air, axisymmetric laminar diffusion flame. Detailed transport coefficients and a 42 reaction, fifteen species reaction mechanism (see Table I, [28]) were used in the calculations. The computational configuration was such that the radius of the inner fuel jet $R_I = 0.2$ cm, the radius of the outer oxidizer jet $R_O = 2.54$ cm and the length of the axial domain $L = 20.0$ cm. The boundary conditions at the inlet were given by

Inlet $(z = 0)$:

$r < R_I$

$$T = 298 \text{ K},$$
$$Y_{CH_4} = 1.0,\ Y_k = 0,\ k \neq CH_4, \tag{2.20}$$
$$v_r = 0.0 \text{ cm/sec},$$
$$v_z = 5.0 \text{ cm/sec},$$

$R_I < r < R_O$

$$T = 298 \text{ K},$$
$$Y_{O_2} = .21,\ Y_{N_2} = .79,\ Y_k = 0,\ k \neq O_2, N_2, \tag{2.21}$$
$$v_r = 0.0 \text{ cm/sec},$$
$$v_z = 25.0 \text{ cm/sec}.$$

The appropriate boundary conditions for the stream function-vorticity system can be formed using the definition in (2.12) along with (2.20)-(2.22).

TABLE I

Methane-Air Reaction Mechanism
Rate Coefficients in the Form $k_f = AT^{\beta}\exp(-E_0/RT)$.
Units are moles, cubic centimeters, seconds, Kelvins and calories/mole.

	REACTION	A	β	E
1.	$CH_4 + M \rightleftharpoons CH_3 + H + M$	1.00E+17	0.000	86000.
2.	$CH_4 + O_2 \rightleftharpoons CH_3 + HO_2$	7.90E+13	0.000	56000.
3.	$CH_4 + H \rightleftharpoons CH_3 + H_2$	2.20E+04	3.000	8750.
4.	$CH_4 + O \rightleftharpoons CH_3 + OH$	1.60E+06	2.360	7400.
5.	$CH_4 + OH \rightleftharpoons CH_3 + H_2O$	1.60E+06	2.100	2460.
6.	$CH_2O + OH \rightleftharpoons HCO + H_2O$	7.53E+12	0.000	167.
7.	$CH_2O + H \rightleftharpoons HCO + H_2$	3.31E+14	0.000	10500.
8.	$CH_2O + M \rightleftharpoons HCO + H + M$	3.31E+16	0.000	81000.
9.	$CH_2O + O \rightleftharpoons HCO + OH$	1.81E+13	0.000	3082.
10.	$HCO + OH \rightleftharpoons CO + H_2O$	5.00E+12	0.000	0.
11.	$HCO + M \rightleftharpoons H + CO + M$	1.60E+14	0.000	14700.
12.	$HCO + H \rightleftharpoons CO + H_2$	4.00E+13	0.000	0.
13.	$HCO + O \rightleftharpoons OH + CO$	1.00E+13	0.000	0.
14.	$HCO + O_2 \rightleftharpoons HO_2 + CO$	3.00E+12	0.000	0.
15.	$CO + O + M \rightleftharpoons CO_2 + M$	3.20E+13	0.000	-4200.
16.	$CO + OH \rightleftharpoons CO_2 + H$	1.51E+07	1.300	-758.
17.	$CO + O_2 \rightleftharpoons CO_2 + O$	1.60E+13	0.000	41000.
18.	$CH_3 + O_2 \rightleftharpoons CH_3O + O$	7.00E+12	0.000	25652.
19.	$CH_3O + M \rightleftharpoons CH_2O + H + M$	2.40E+13	0.000	28812.
20.	$CH_3O + H \rightleftharpoons CH_2O + H_2$	2.00E+13	0.000	0.
21.	$CH_3O + OH \rightleftharpoons CH_2O + H_2O$	1.00E+13	0.000	0.
22.	$CH_3O + O \rightleftharpoons CH_2O + OH$	1.00E+13	0.000	0.
23.	$CH_3O + O_2 \rightleftharpoons CH_2O + HO_2$	6.30E+10	0.000	2600.
24.	$CH_3 + O_2 \rightleftharpoons CH_2O + OH$	5.20E+13	0.000	34574.
25.	$CH_3 + O \rightleftharpoons CH_2O + H$	6.80E+13	0.000	0.
26.	$CH_3 + OH \rightleftharpoons CH_2O + H_2$	7.50E+12	0.000	0.
27.	$HO_2 + CO \rightleftharpoons CO_2 + OH$	5.80E+13	0.000	22934.
28.	$H_2 + O_2 \rightleftharpoons 2OH$	1.70E+13	0.000	47780.
29.	$OH + H_2 \rightleftharpoons H_2O + H$	1.17E+09	1.300	3626.
30.	$H + O_2 \rightleftharpoons OH + O$	5.13E+16	-0.816	16507.
31.	$O + H_2 \rightleftharpoons OH + H$	1.80E+10	1.000	8826.
32.	$H + O_2 + M \rightleftharpoons HO_2 + M^a$	2.10E+18	-1.000	0.
33.	$H + O_2 + O_2 \rightleftharpoons HO_2 + O_2$	6.70E+19	-1.420	0.
34.	$H + O_2 + N_2 \rightleftharpoons HO_2 + N_2$	6.70E+19	-1.420	0.
35.	$OH + HO_2 \rightleftharpoons H_2O + O_2$	5.00E+13	0.000	1000.
36.	$H + HO_2 \rightleftharpoons 2OH$	2.50E+14	0.000	1900.
37.	$O + HO_2 \rightleftharpoons O_2 + OH$	4.80E+13	0.000	1000.
38.	$2OH \rightleftharpoons O + H_2O$	6.00E+08	1.300	0.
39.	$H_2 + M \rightleftharpoons H + H + M^b$	2.23E+12	0.500	92600.
40.	$O_2 + M \rightleftharpoons O + O + M$	1.85E+11	0.500	95560.
41.	$H + OH + M \rightleftharpoons H_2O + M^c$	7.50E+23	-2.600	0.
42.	$H + HO_2 \rightleftharpoons H_2 + O_2$	2.50E+13	0.000	700.

[a] Third body efficiencies: $k_{32}(H_2O) = 21k_{32}(Ar)$, $k_{32}(H_2) = 3.3k_{32}(Ar)$, $k_{32}(N_2) = k_{32}(O_2) = 0$.

[b] Third body efficiencies: $k_{39}(H_2O) = 6k_{39}(Ar)$, $k_{39}(H) = 2k_{39}(Ar)$, $k_{39}(H_2) = 3k_{39}(Ar)$.

[c] Third body efficiency: $k_{41}(H_2O) = 20k_{41}(Ar)$.

The flame sheet model provided initial solution profiles for the stream function, the vorticity, the temperature and the major species, i.e., CH_4, O_2, N_2, CO_2 and H_2O. The starting estimates for the minor species in the full chemistry solution were approximated by Gaussian profiles that were centered at the location of the flame sheet on each axial level. They had peak heights of at most a few percent. To conserve mass in the starting estimate, the N_2 mass fraction was reduced accordingly. The flame sheet starting estimate required approximately 150 adaptive time steps and five Newton iterations to converge. Once the flame sheet estimate was calculated, we solved the full set of governing equations in a two-step procedure. We first determined a solution to the stream function, vorticity and species equations (2.8-2.10) based on the flame sheet temperature profile. This fixed flame sheet temperature solution (T_{OUT}) was then used as input to the full fluid dynamic-thermochemistry model (2.8-2.11) in which the energy equation was included (T_{IN}). This procedure helped to reduce both convergence difficulties and the total CPU time and is similar to the two-pass solution method used in the solution of adiabatic premixed laminar flames [23] and counterflow diffusion flames [40].

The flame sheet and the first T_{OUT} calculation were performed on a 31×19 tensor product grid. One hundred fifty five adaptive time steps were required to reduce the norm of the steady-state residuals below 1.0×10^{-3}. This was sufficient to bring the numerical solution within the convergence domain of Newton's method. After the time steps, Newton's method converged with only seven iterations. Once this solution was obtained, the mesh was refined and a solution was calculated on a finer grid. This procedure continued until a refined grid containing 1395 adaptively placed grid points was obtained. The refined fixed temperature solution was then used as the starting estimate for the complete fluid dynamic-thermochemistry solution. As the computational mesh was refined, Newton's method typically converged with a smaller number of time steps than on the coarser grids. The mesh spacing was such that 32000 equispaced points would have been needed to obtain comparable accuracy. The total CPU time for the entire procedure was approximately 60 hours on an FPS-264 ($\approx$ 10 hours on a Cray-XMP).

In Figure 5 we illustrate the temperature isotherms and in Figures 6 and 7 we plot the methane and oxygen isopleths as functions of the radial and axial coordinates. We note immediately the high temperature region extending from the boundary of the fuel and oxidizer jets to the axis of symmetry. The figure also illustrates the extremely high temperature gradients directly above the inlet. It is in this region that the fuel and the oxidizer first meet in stoichiometric proportion. In particular, it is clear from these figures that combustion occurs only in a thin region above the inlet. The methane and oxygen co-exist in only a very small region. Most of the methane disappears within 1.0 mm of the fuel jet. The resulting heat release produces an extremely rapid rise in the temperature. As one moves in the direction of increasing r, we observe that the temperature rapidly decreases and approaches its inlet value approximately 1.0 cm

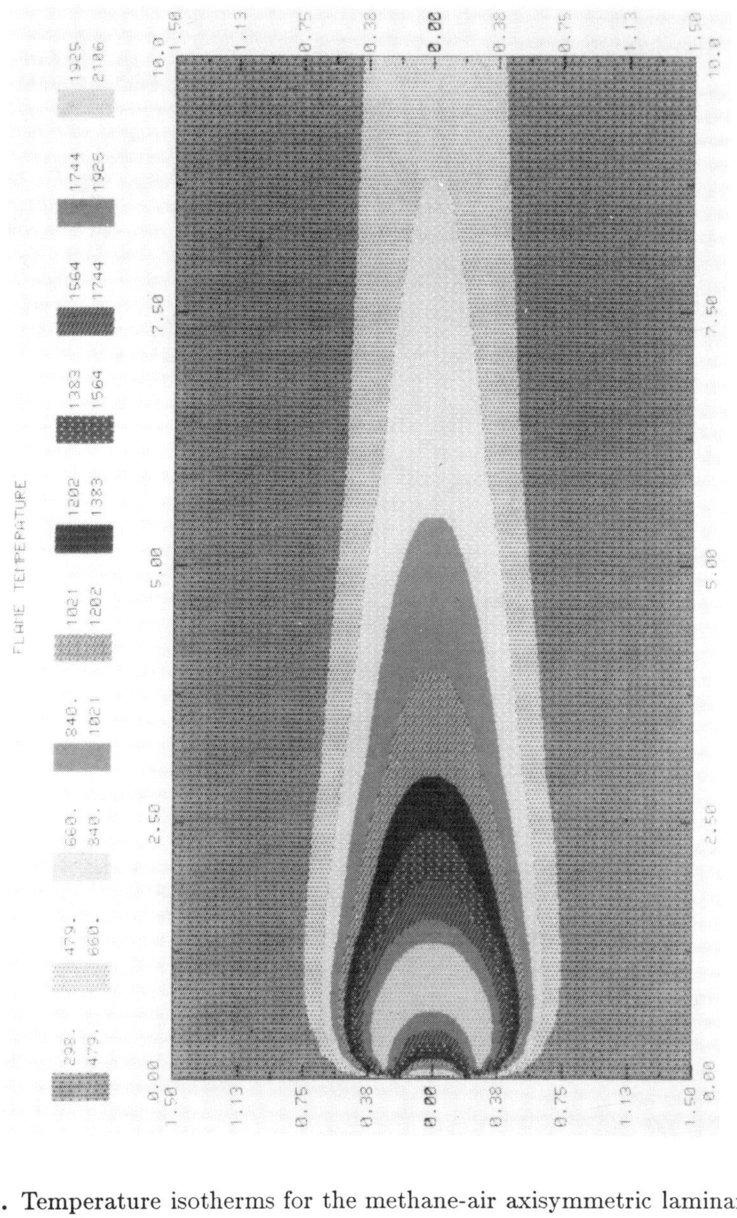

Figure 5. Temperature isotherms for the methane-air axisymmetric laminar diffusion flame.

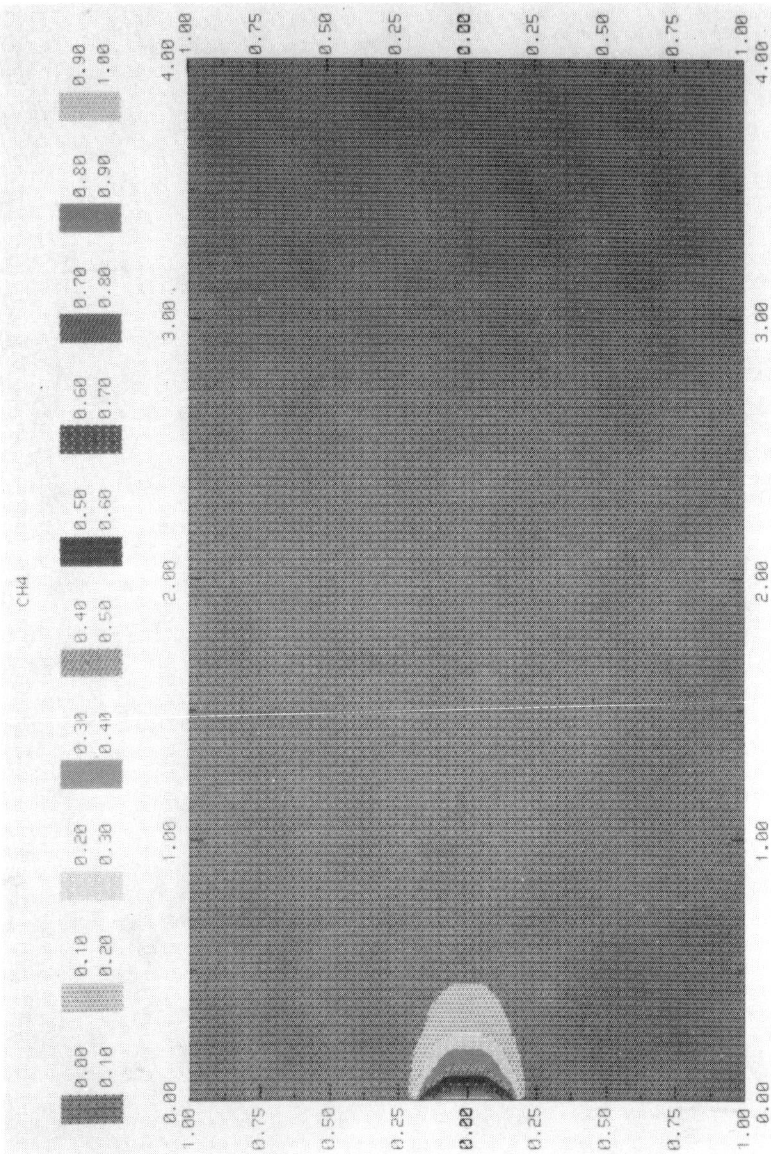

Figure 6. Methane (CH_4) isopleths for the methane-air axisymmetric laminar diffusion flame.

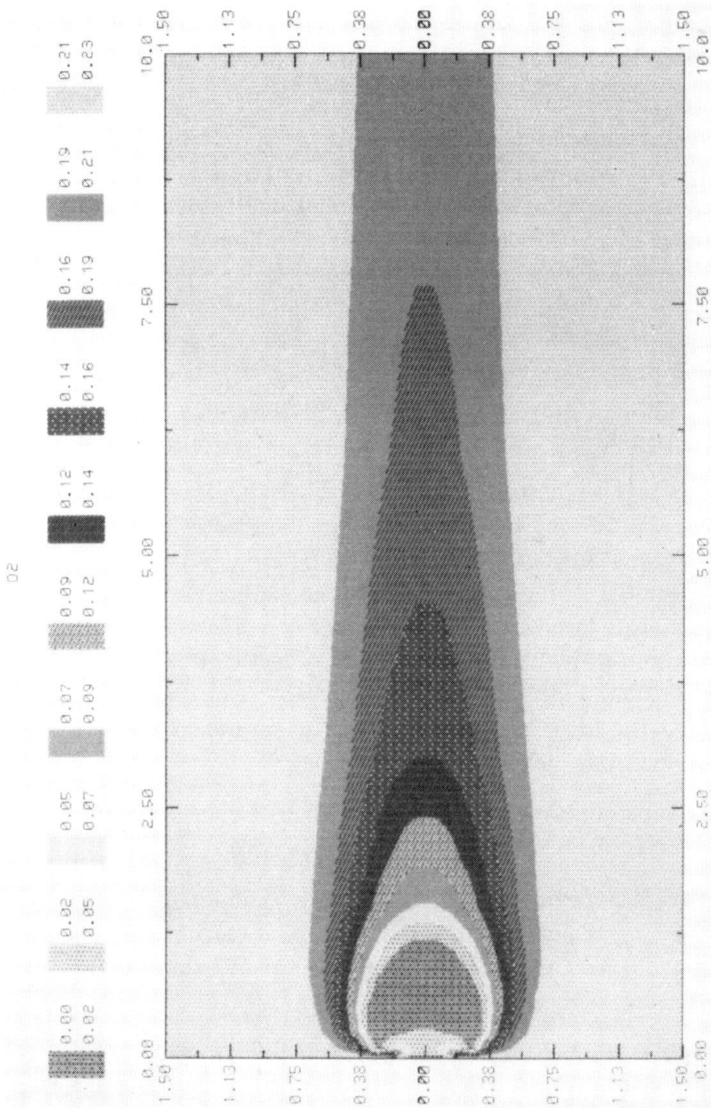

Figure 7. Oxygen (O_2) isopleths for the methane-air axisymmetric laminar diffusion flame.

from the center of the fuel jet. In addition, if we define the flame height as the first axial location where the maximum temperature occurs on the axis of symmetry, we obtain a flame height of 1.25 cm.

From a structural viewpoint, methane is the simplest hydrocarbon–it is the only hydrocarbon without a carbon-carbon bond. As a result, its oxidation differs significantly from other hydrocarbon fuels. Westbrook and Dryer [55] and Warnatz [56] have postulated that the oxidation of methane occurs through roughly two parallel paths. In one path carbon-hydrogen bonds are broken to form methyl radicals. The methyl radicals are then oxidized to methoxy radicals and/or formaldehyde. This is followed by the formation of the formal group which is then oxidized to form carbon monoxide. In the second path, methyl radical recombination is followed by the oxidation of the resulting C_2 species. In our calculations, the former path was chosen and the latter path was neglected.

In Figures 8-11 we illustrate isopleths for H_2O, CO, H_2 and CO_2, respectively. Similar plots for CH_3, OH and CH_2O are illustrated in Figures 12-14. It is clear from these figures that large quantities of H_2O, CO and H_2 are produced soon after the methane has been consumed. It is in this region that the methane is attacked by O, H and OH radicals and CH_3 is formed. In this region only small amounts of OH, H and O exist (see, e.g., Figure 13) due to the high affinity of methane for these radicals. The peak values of both CH_3 and CH_2O also occur after the methane has disappeared. The oxidation of CH_2O to HCO and the subsequent formation of CO occurs in regions of high methyl and formaldehyde concentrations.

The oxidation of CO to CO_2 proceeds primarily via the reaction

$$CO + OH \rightarrow CO_2 + H.$$

Hence, the rate of CO oxidation depends on the availability of OH radicals. However, as Westbrook and Dryer point out [55] the presence of most hydrocarbon species inhibits the oxidation of CO. This can be attributed to the fact that the rate of the reaction

$$H + O_2 \rightarrow OH + O,$$

is considerably smaller than the reaction rates of H atoms with hydrocarbon species and the rate of the CO oxidation reaction is also smaller than the reaction rates of hydrocarbon species with OH. As a result, small quantities of hydrocarbons can effectively restrict the oxidation of CO to CO_2. Although carbon monoxide and hydrogen are found during the oxidation of the hydrocarbon species, it is not until after the hydrocarbons and the hydrocarbon fragments have been consumed that the OH level rises and CO_2 is formed. In Figure 13 we observe that in the axial direction the OH radical pool increases after the disappearance of the methane and the formation of CO. The CO is then oxidized to form CO_2. Hence, in Figure 11 we observe that carbon dioxide forms downstream of the regions of high CH_3, CH_2O and CO concentration.

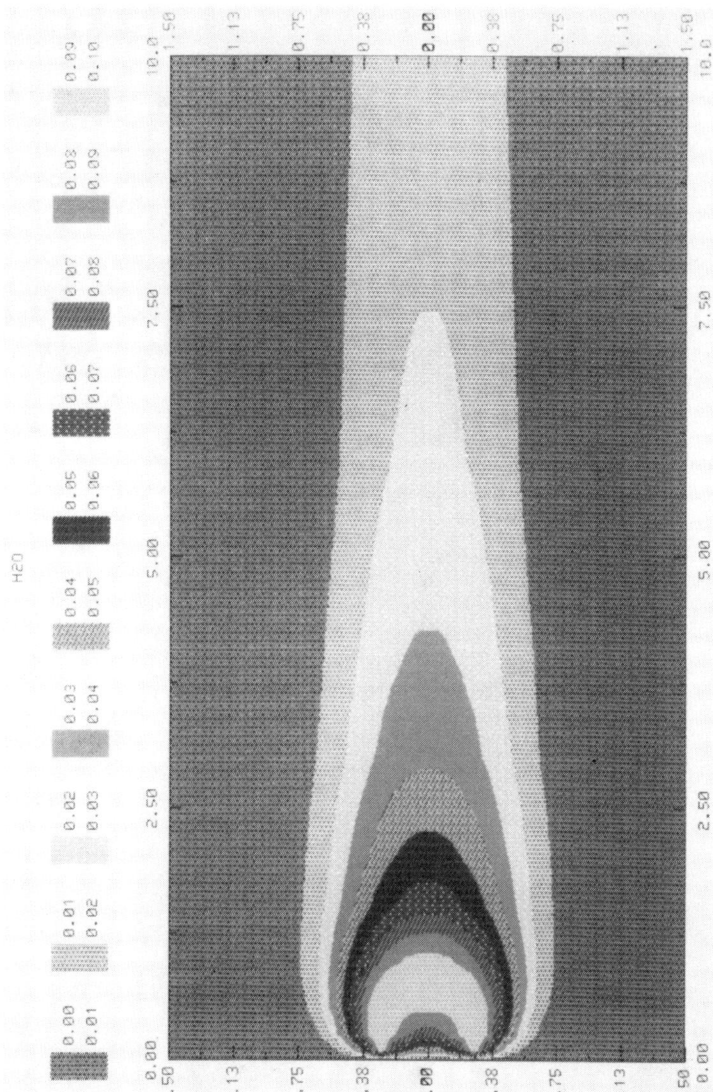

Figure 8. Water (H_2O) isopleths for the methane-air axisymmetric laminar diffusion flame.

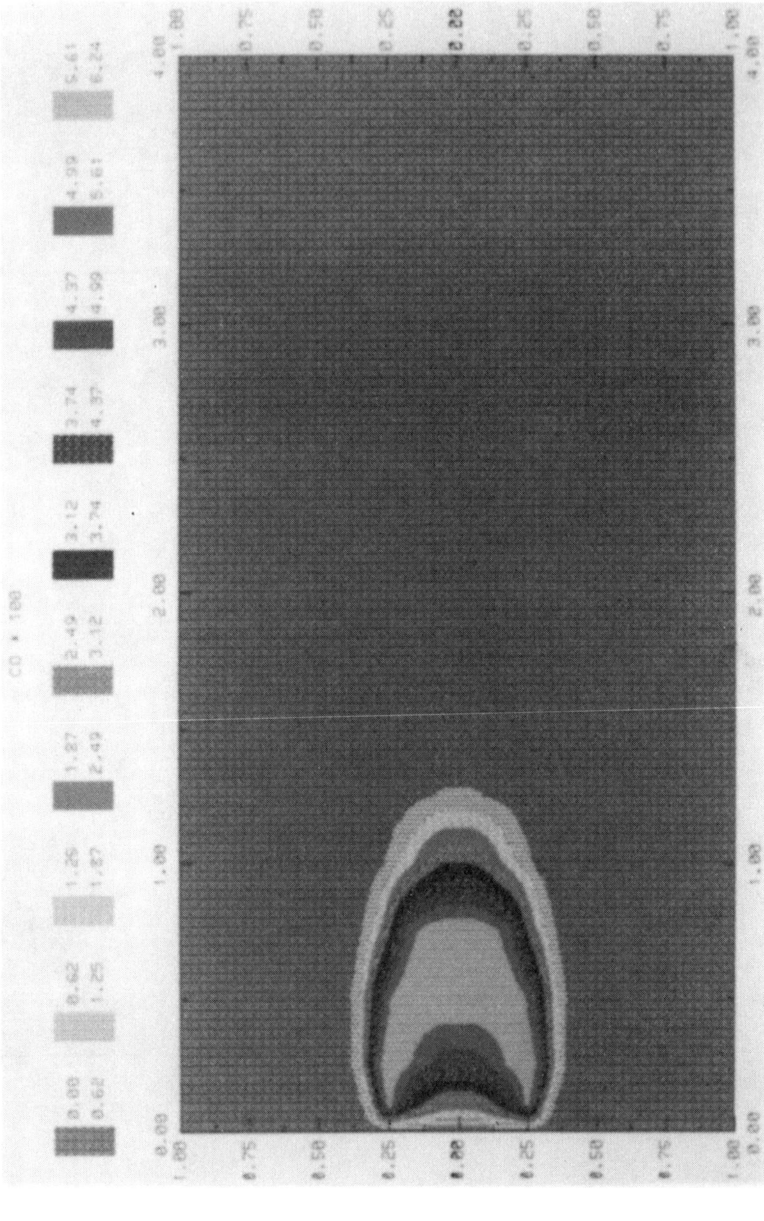

Figure 9. Carbon monoxide (CO) isopleths for the methane-air axisymmetric laminar diffusion flame.

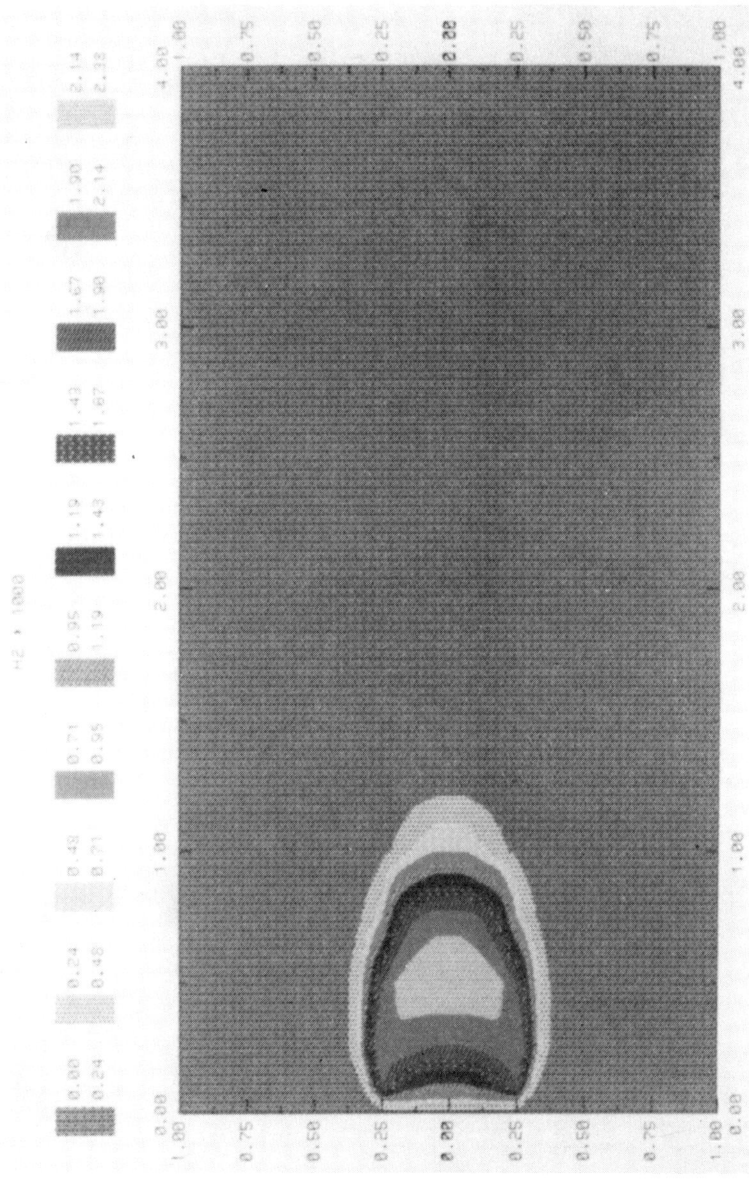

Figure 10. Molecular hydrogen (H_2) isopleths for the methane-air axisymmetric laminar diffusion flame.

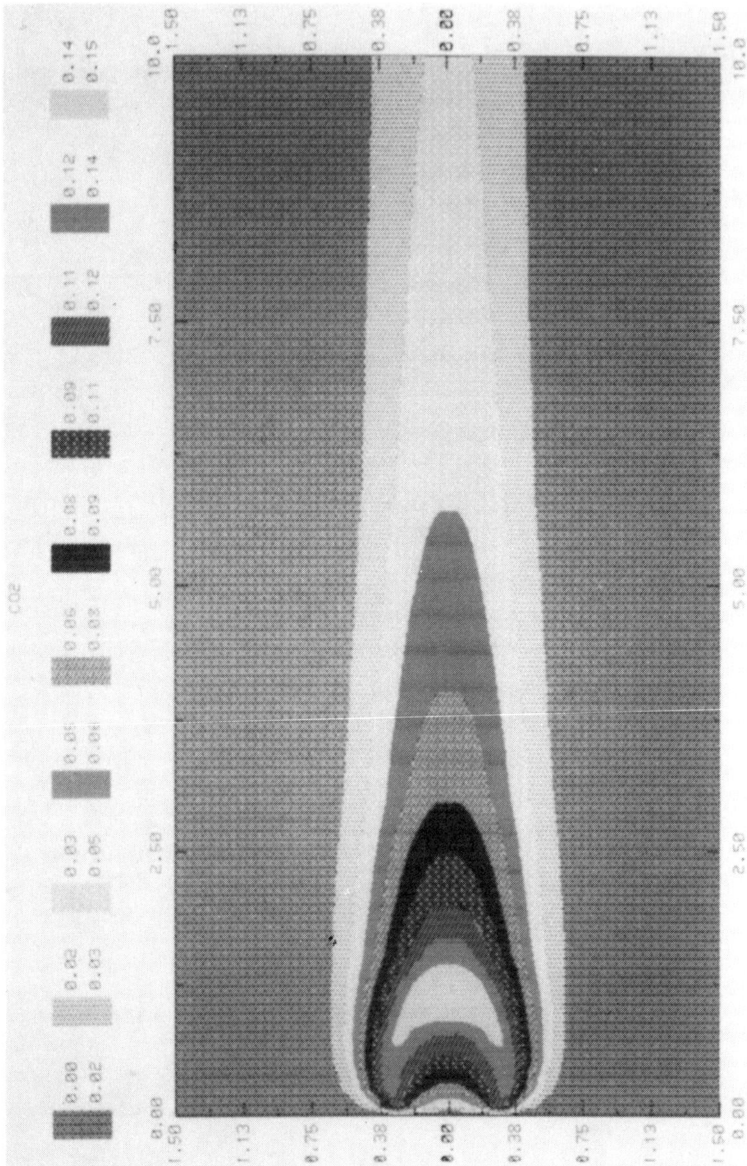

Figure 11. Carbon dioxide (CO_2) isopleths for the methane-air axisymmetric laminar diffusion flame.

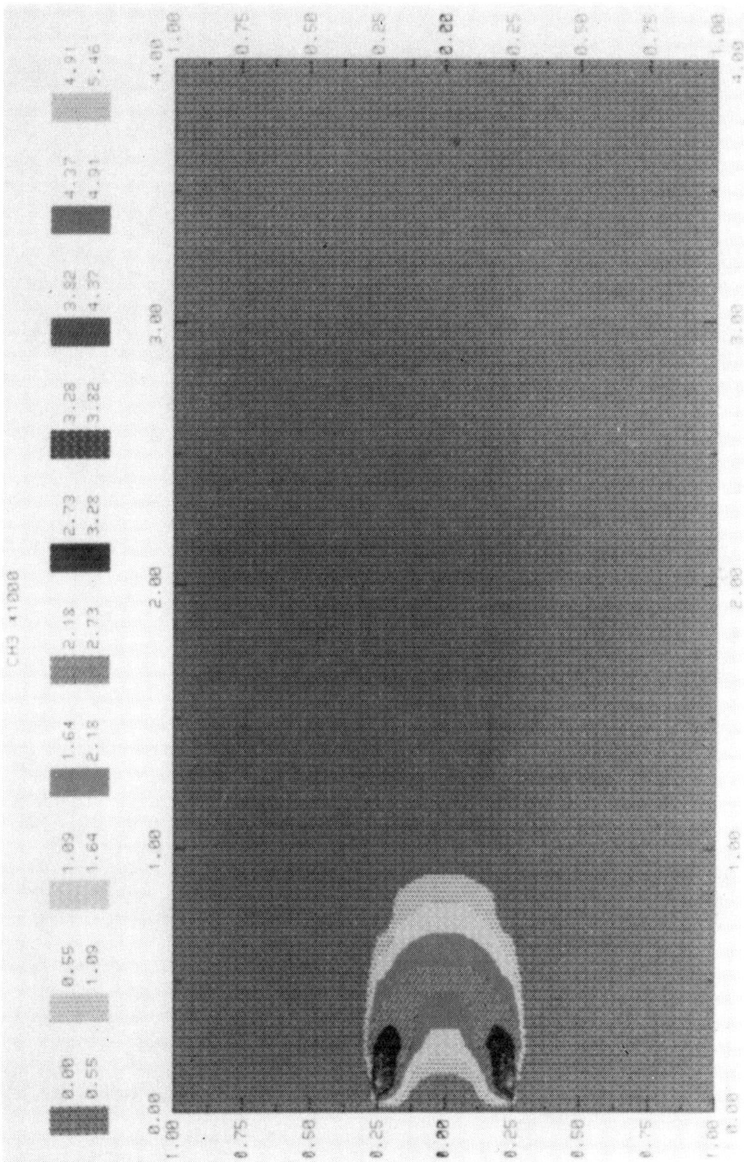

Figure 12. Methyl radical (CH_3) isopleths for the methane-air axisymmetric laminar diffusion flame.

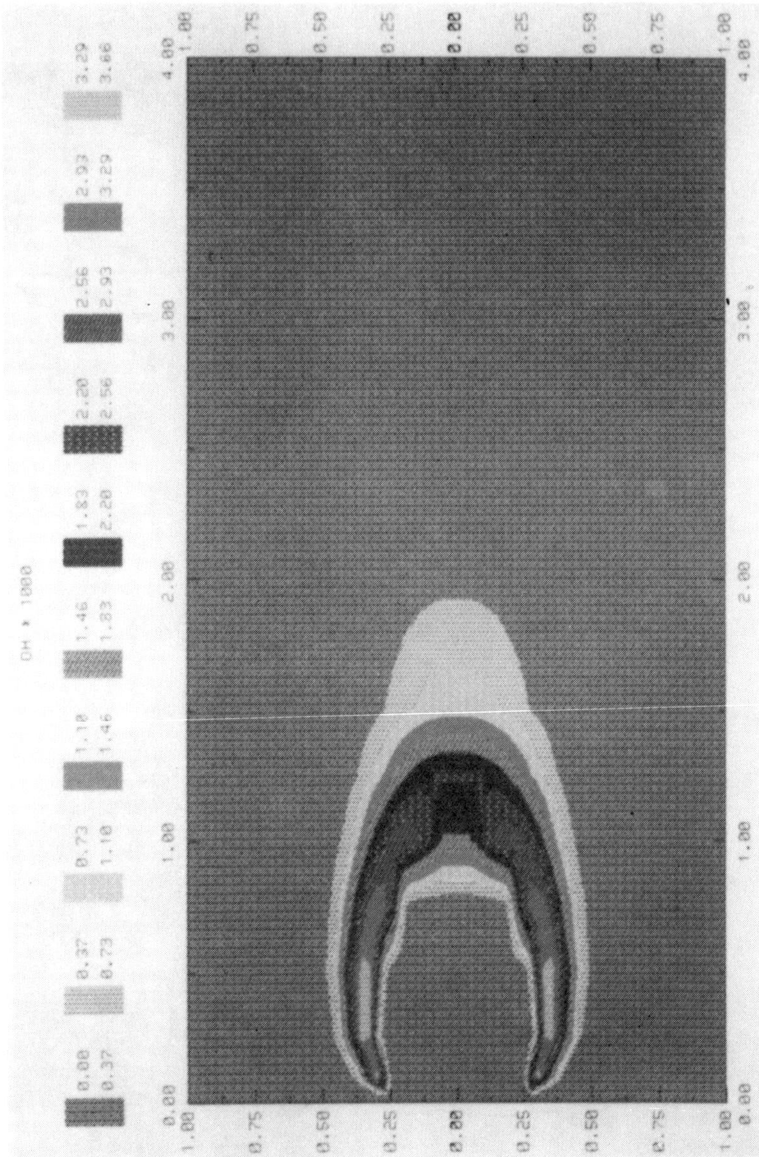

Figure 13. Hydroxyl radical (OH) isopleths for the methane-air axisymmetric laminar diffusion flame.

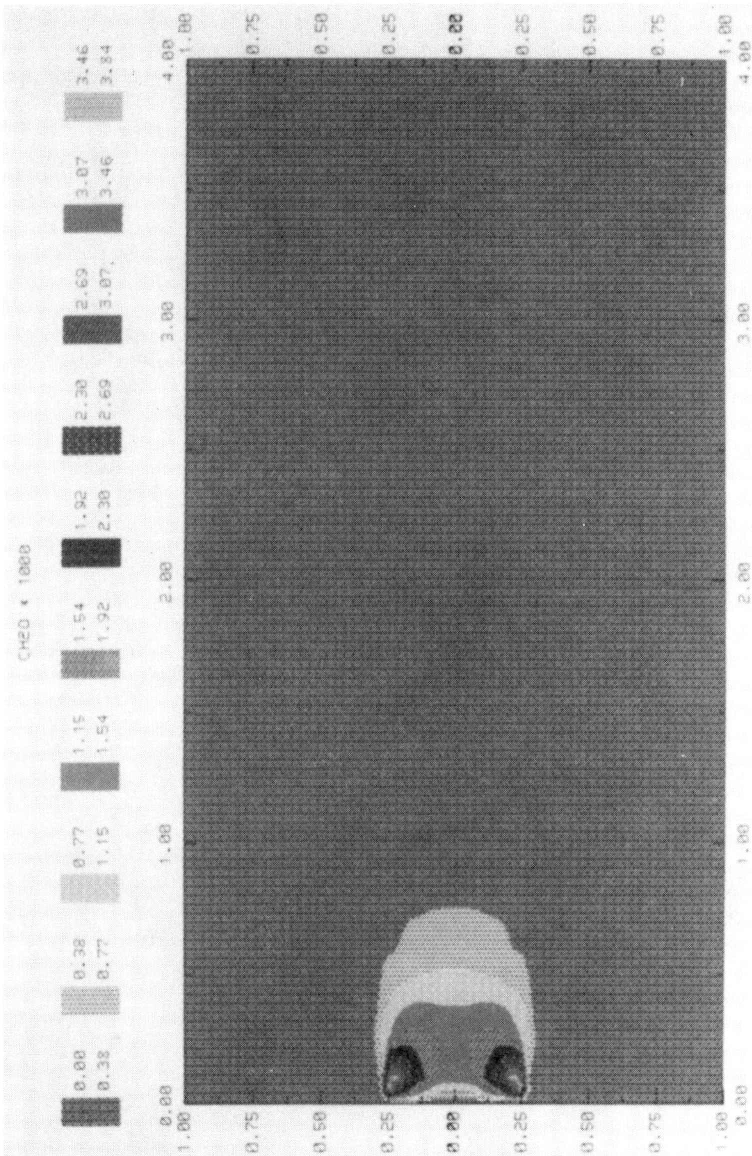

Figure 14. Formaldehyde (CH_2O) isopleths for the methane-air axisymmetric laminar diffusion flame.

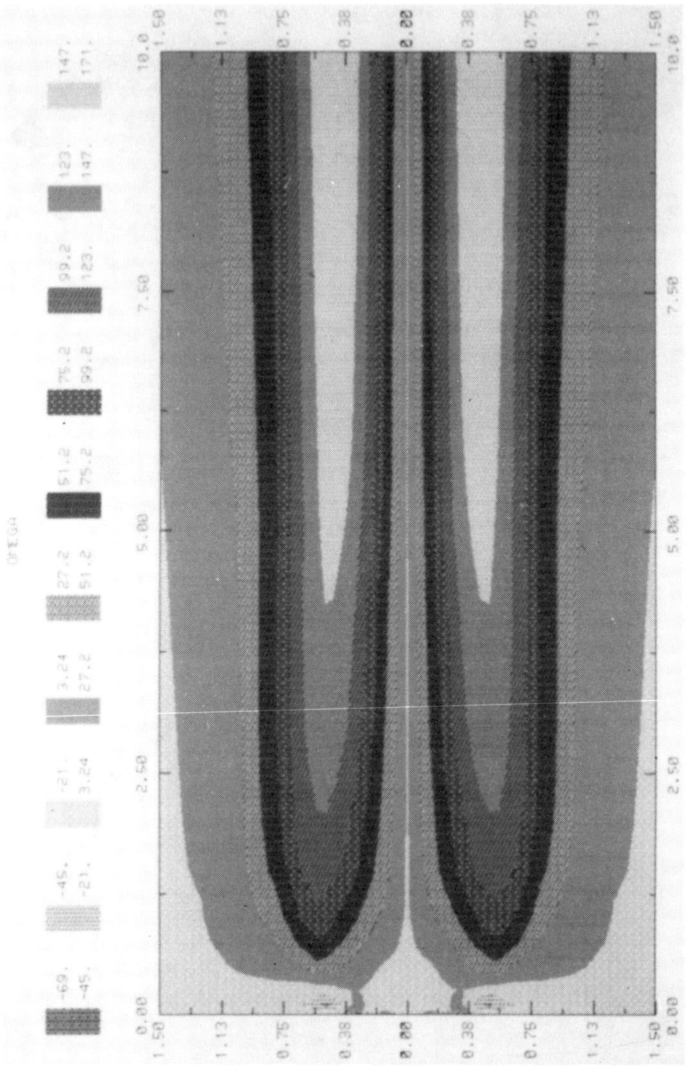

Figure 15. Vorticity isopleths for the methane-air axisymmetric laminar diffusion flame.

Figure 13 also illustrates that the peak values of the OH radical (similar results hold for the H and O radicals) are observed on the outer lean boundary of the flame. Based upon our studies of counterflow diffusion flames, the highest levels of OH radicals occur where the CH radical peaks. CH emission gives the methane flame its characteristic bluish color and the brightest blue regions of the flame occur along the outer boundary. This is precisely where the OH levels are highest.

Finally, in Figure 15 we plot the vorticity contours in the flame. We observe two large vorticity cells that begin approximately 4.0 cm above the inlet and are centered approximately 0.5 cm from the axis of symmetry. These are the regions of highest counterclockwise rotation in the flame. Here the flow has a strong velocity component towards the axis of symmetry. This is in contrast to the region directly above the inlet where the vorticity is negative and the flow direction is such that there is a substantial radial component of the velocity causing the flame to "bow out". The negative vorticity region is limited to the area directly above the jets. At a height of approximately 1.0 mm above the inlet, the vorticity changes sign and the inward radial component of the velocity begins to increase.

ACKNOWLEDGEMENT

This work was supported in part by the Office of Naval Research and the National Science Foundation. Computations supporting this research were performed on the Cornell National Supercomputer Facility, which is supported in part by the National Science Foundation, New York State and the IBM Corporation.

REFERENCES

1. C. K. Law, S. Ishizuka and P. Cho, *Comb. Sci. and Tech.*, **28** (1982), p. 89.
2. N. R. Sax, *Dangerous Properties of Industrial Materials*, Reinhold Publishing Co., New York, (1968).
3. P. K. Mueller and M. Hitchcock, *J. of Air Poll. Cont. Ass.*, **19** (1969), p. 670.
4. L. C. Harding, *J. of Air Poll. Cont. Ass.*, **20** (1970), p. 377.
5. D. B. Spalding, *Phil. Trans. Roy. Soc. London*, **249A** (1956), p. 1.
6. G. K. Adams and G. B. Cook, *Comb. and Flame*, **4**, (1960), p. 9.
7. G. Dixon-Lewis, *Proc. Roy. Soc. London*, **298A** (1967), p. 495.
8. D. B. Spalding, D. L. Stephenson and R. G. Taylor, *Comb. and Flame.*, **17** (1971), p. 55.
9. K. A. Wilde, *Comb. and Flame*, **18** (1972), p. 43.
10. L. Bledjian, *Comb. and Flame*, **20** (1973), p. 5.

11. S. B. Margolis, *J. Comp. Phys.*, **27** (1978), p. 410.

12. J. Warnatz, *Ber. Bunsenges. Phys. Chem.*, **82** (1978), p. 193.

13. J. Warnatz, *Ber. Bunsenges. Phys. Chem.*, **82** (1978), p. 643.

14. J. Warnatz, *Ber. Bunsenges. Phys. Chem.*, **82** (1978), p. 834.

15. G. Dixon-Lewis, *Phil. Trans. of the Royal Soc. London*, **292** (1979), p. 45.

16. C. K. Westbrook and F. L. Dryer, *Comb. Sci. and Tech.*, **20** (1979), p. 125.

17. C. K. Westbrook and F. L. Dryer, *Comb. and Flame*, **37** (1980), p. 171.

18. T. P. Coffee and J. M. Heimerl, *Comb. and Flame*, **39** (1980), p. 301.

19. T. P. Coffee and J. M. Heimerl, *Comb. and Flame*, **43** (1981), p. 273.

20. J. A. Miller, R. E. Mitchell, M. D. Smooke and R. J. Kee, Nineteenth Symposium (International) on Combustion, Reinhold, New York, (1981), p. 181.

21. M. D. Smooke, *J. Comp. Phys.*, **48** (1982), p. 72.

22. M. D. Smooke, J. A. Miller and R. J. Kee, Numerical Methods in Laminar Flame Propagation, N. Peters and J. Warnatz, eds., Friedr. Vieweg and Sohn, Wiesbaden, (1982).

23. M. D. Smooke, J. A. Miller and R. J. Kee, *Comb. Sci. and Tech.*, **34** (1983), p. 79.

24. J. A. Miller, M. D. Smooke, R. M. Green and R. J. Kee, *Comb. Sci. and Tech.*, **34** (1983), p. 149.

25. J. Sato and H. Tsuji, Sixteenth Symposium on Combustion, Japan, (1978), p. 13.

26. W. A. Hahn, and J. O. L. Wendt, Eighteenth Symposium (International) on Combustion, Reinhold, New York, (1981), p. 121.

27. J. Sato and H. Tsuji, Comb. Sci. and Tech., **33** (1983), p. 193.

28. J. A. Miller, R. J. Kee, M. D. Smooke, and J. F. Grcar, Paper # WSS/CI 84-20 presented at the 1984 Spring Meeting of the Western States Section of the Combustion Institute, University of Colorado, Boulder, CO, (1984).

29. G. Dixon-Lewis, T. David, P. H. Gaskell, S. Fukutani, H. Jinno, J. A. Miller, R. J. Kee, M. D. Smooke, N. Peters, E. Effelsberg, J. Warnatz, and F. Behrendt, Twentieth Symposium (International) on Combustion, Rheinhold, New York, (1985), p. 1893.

30. M. D. Smooke, I. K. Puri, and K. Seshadri, to be published in the Twenty-First Symposium (International) on Combustion, (1986).

31. V. Giovangigli and M. D. Smooke, *Comb. Sci. and Tech.* **53** (1987), p. 23.

32. S. P. Burke and T. E. W. Schumann, *Industrial Engineering Chemistry*, **29** (1928), p. 998.

33. R. E. Mitchell, Sc.D. Thesis, MIT, (1975).

34. R. E. Mitchell, A. F. Sarofim and L. A. Clomburg, *Comb. and Flame*, **37** (1980), p. 227.

35. M. D. Smooke, R. E. Mitchell and J. F. Grcar, *Elliptic Problem Solvers II*, G. Birkhoff and A. Schoenstadt, eds., Academic Press, New York, (1984).

36. J. A. Miller and R. J. Kee, *J. Phys. Chem.*, **81** (1977), p. 2534.

37. C. F. Curtiss and J. O. Hirschfelder, *J. Chem. Phys.*, **17** (1949), p. 550.

38. R. J. Kee, J. Warnatz and J. A. Miller, Sandia National Laboratories Report, SAND83-8209, (1983).

39. R. J. Kee, J. A. Miller and T. H. Jefferson, Sandia National Laboratories Report, SAND80-8003, (1980).

40. D. E. Keyes and M. D. Smooke, to be published in *J. Comp. Phys.*, (1987).

41. A. M. Kanury, *Combustion Phenomena*, Gordon and Breach, New York, (1982).

42. P. Deuflhard, *Numer. Math.*, **22** (1974), p. 289.

43. A. R. Curtis, M. J. Powell, and J. K. Reid, *J. Inst. Math. Appl.*, **13** (1974), p. 117.

44. G. N. Newsam and J. D. Ramsdell, Harvard University Report TR-17-81, (1981).

45. T. F. Coleman and J. J. More, Argonne National Laboratory Report ANL-81-39, (1981).

46. M. D. Smooke, *J. Opt. Theory and Appl.*, **39** (1983), p. 489.

47. J. Kautsky and N. K. Nichols, *SIAM J. Sci. Stat. Comput.*, **1** (1980), pp. 499.

48. R. D. Russell, *Working Codes for Boundary Value Problems in ODE's*, B. Childs et al., eds., Springer-Verlag, New York, (1979).

49. M. D. Smooke, *AIChE J.*, **32** (1986), p. 1233.

50. M. Berger, W. D. Gropp, and J. Oliger, *Numerical Grid Generation Techniques*, NASA Conference Publication 2166, (1981).

51. M. Berger, Ph. D. Thesis, Stanford University, (1982).

52. W. D. Gropp, *SIAM J. Sci. and Stat. Comp.* **1** (1980), p. 191.

53. M. D. Smooke and R. M. M. Mattheij, *Appl. Num. Math.*, **1** (1985), p. 463.

54. L. V. Kantorovich and G. P. Akilov, *Functional Analysis in Normed Spaces*, Pergamon Press, (1964).

55. C. K. Westbrook, and F. L. Dryer, *Prog. Energy Combust. Sci.* **10** (1984), p. 1.

56. J. Warnatz, *Comb. Sci. and Tech.* **34** (1983), p. 177.

SHOCK INDUCED THERMAL EXPLOSION

D. S. STEWART

Cornell Mathematical Science Institute
Cornell University, Ithaca, New York and
Department of Theoretical and Applied Mechanics
University of Illinois, Urbana, Illinois 61801

1. INTRODUCTION

Shock initiation experiments on condensed phase solid and liquid explosives as well as gases have suggested the strong role of hot spots (also called reaction centers) as sites where ignition of the explosive first takes place. Shock initiation studies in liquid Nitromethane have provided representative evidence. In these experiments, an inert piston is driven into the explosive and after the elapse of an "induction period", an ignition event occurs in a reaction center, (thought to be located at the piston face for Nitromethane). The continued growth and development of the center ultimately gives rise to a detonation.

The concept of a single, isolated hot spot as the primary ignition site is clearly an idealization. Even with initiation by a piston driven shock (nominally plane), instabilities of an inherent fluid dynamical nature soon develop and result in the formation of multiple, distributed hot spots or reaction centers. Estimates of the sizes of single and multiple reaction centers and their distribution in plane and multidimensional geometry are necessary to assess the initiation environment of shocked explosive materials.

This paper will give a basic model for piston driven shock initiation in Section 2. In section 3, we will consider some recent results developed by Jackson, Kapila and Stewart [1] that describe the formation of a single hot spot. Using these results we estimate the size and the length of time required to form a shock initiated hot spot for Nitromethane. Section 4 considers plane geometry with multiple hot spots where the multiple 1-d hot spots are due to variation in the initial (unshocked) material properties. In particular, when the time for the leading shock to cross a characteristic imperfection is small compared to the homogeneous induction time for the pure material, we show (numerically) that nonlinear feedback results in an "effective" induction time which is different by a sizable factor from the homogeneous induction time. Based on reasonable modeling concepts for the reaction rate for condensed materials, we pose a problem for the induction phase that will determine this effective time in a limit where the ratio of the imperfection acoustic passage time to the homogeneous induction time goes to zero. Finally in Section 5 we consider multiple hot spot formation by deriving a 3-d model for the induction phase for a single phase explosive material. We pose stability problems that should shed light on the early ignition environment and the distribution and spacing of ignition hot spots.

C.-M. Brauner and C. Schmidt-Lainé (eds.), Mathematical Modeling in Combustion and Related Topics, 301–314.

2. BASIC MODEL

The basic model equations for an explosive material and a discussion of their derivation can be found in [2], [3] and more recently in [1]. (The formulation in all these papers are consistent with only a minor difference the definition of the characteristic time scale.) We mainly follow Stewart's presentation in [3].

A prominent feature of the explosives we consider is that the energy release rate of the explosive is sensitive in that a slight change in the local thermodynamic state for an explosive particle generates a large change in the energy release rate. Thus the explosive model used in the study assumes Arrhenius kinetics with a large (dimensionless) activation energy and it has this feature. As a consequence, in the initial phase of the reaction center development before the elapse of the induction time, only a small amount of energy is released into the shocked state of the explosive.

The initial environment of a shocked, but nominally unreacted explosive. The complete explosive model is given by the equations of reactive gasdynamics (reactive Euler equations) with a simple (polytropic) equation of state. The energy release is modeled after a single, first-order, irreversible chemical reaction of the Arrehnius form. Initial and boundary conditions can correspond to a piston or for simplicity, to confinement between two parallel planes. The initial (shocked) state possesses a slight spatial nonuniformity of the same order as that produced by the combustion of the explosive.

The mathematical problem, (in dimensionless form) is given below:

$$\frac{D\tilde{u}}{Dt} + \rho^{-1}\,\tilde{\nabla}(P) = 0 \tag{1}$$

$$\frac{DP}{Dt} + \gamma\, c^2\, \tilde{\nabla}\cdot\tilde{u} = \gamma\,\beta\, w \tag{2}$$

$$\frac{DP}{Dt} - \gamma\, c^2 \frac{D\rho}{Dt} = \gamma\,\beta\, w \tag{3}$$

$$\rho\,\frac{D\lambda}{Dt} = w. \tag{4}$$

For an ideal, polytropic fluid

$$p = \rho\, T, \quad c^2 = P/\rho.$$

w is given by

$$w = [1/(\beta\,\theta)]\;\rho(1-\lambda)\,\exp\,(\theta - \theta/T). \tag{5}$$

In the above $\rho, \underset{\sim}{u} = (u,v,w)$, P, T, λ are respectively the dimensionless density, particle velocity, pressure, temperature and reaction variable. The parameters that appear are γ, β and θ which are respectively the polytropic exponent, a heat release parameter and the dimensionless activation energy. The dependent variables (except u) are one at the unreacted shocked state. The independent variables $\tilde{x} = (x,y,z)$ and t are distance and time and the time scale is chosen to be the spatially homogeneous induction time for the shocked state, i.e.

$$\tilde{t}_i = \tilde{k}^{-1} e^{\theta}/\beta\,\theta. \tag{6}$$

The length scale is the sound speed in the unreacted shocked state times the reference induction time.

In a one dimensional system the equations (1) - (4) can be written conveniently in characteristic form as

$$P_t + (u + c)\, P_x + \gamma\rho c\, [u_t + (u + c)\, u_x] = \gamma\beta\, w \tag{7}$$

$$P_t + (u - c)\, P_x - \gamma\rho c\, [u_t + (u - c)\, u_x] = \gamma\beta\, w \tag{8}$$

$$P_t + u\, P_x - \gamma\, [\rho_t + u\, \rho_x] = \gamma\beta\, w \tag{9}$$

$$\lambda_t + u\, \lambda_x = w/\rho.$$

3. THE TIME FOR FORMATION AND THE SIZE OF REACTION CENTERS IN NITROMETHANE

Recently in [1], we considered the problem of the formation of a single hot spot in a one dimensional geometry. It was assumed that the material was confined between two parallel planes and that the initial shocked state possessed a slight spatial nonuniformity of the same order as that produced by the combustion of the explosive. Hence the boundary condition are given by

$$u(0,t) = u(\ell,t) = 0.$$

The initial boundary conditions are given by

$$T = 1 + \theta^{-1} T_1(x,0),\; p = 1 + \theta^{-1} p_1(x,0),\; \lambda = 1 + \theta^{-1} \lambda_1(x,0)$$

$$\rho = 1 + \theta^{-1} \rho_1(x,0) \quad \text{and} \quad u = \theta^{-1} u_1(x,0).$$

Detailed examples were worked out for the specific initial data

$$p_1(x,0) = \lambda_1(x,0) = u_1(x,0) = 0 \text{ and}$$

$$T_1(x,0) = a[1 - (x/\ell)], \quad \text{or} \quad a[1 - (x/\ell)^2],$$

where a and ℓ are prescribed constants. Only the main features of the solution relevant to the current discussion are given here. The interested reader is referred to [1] for additional details.

The subsequent analysis in the paper shows that the slight perturbations to the otherwise uniform (shocked) state causes the reaction center to form at a particular location where the initial energy release rate (although small) has a maximum. For the examples in the paper these are chosen to be at $x = 0$. The asymptotic solution is carried as far as the end of the birth of the reaction center where the temperatures and pressures within the center become larger than their original (shocked) values and are first limited by reactant depletion.

The evolution of the reaction center occurs in two stages. In the induction period, the state of the material is a small perturbation of the initial state and the underlying physical processes are those of linearized acoustics coupled to a weak but nonlinear reaction. The reduced equations require numerical solution which exhibits thermal runaway, characterized by a local unboundedness of the perturbation quantities in a finite time.

Induction is followed by the explosion stage, whose temporal extent is exceedingly short compared to the induction period, and which consists of several distinct spatial zones. There is an outer "frozen zone", in which the state of the explosive remains essentially the same as that at the end of the induction stage. This is distinct from the innermost core of the reaction center. This region is a rapidly shrinking inner zone, or "hot spot", in which intense chemical activity leads to an explosive growth of temperature and pressure, ultimately limited only by the consumption of reactant. To leading order, the "hot spot" is a constant-density, spatially homogeneous thermal explosion.

As the "hot spot" layer shrinks, it recedes away from the "frozen zone", thereby creating an expanding zone of "transition". This new zone is also effectively frozen in time on the very short time scale of the "hot spot" formation; but in contrast with the outer "frozen zone", the temperature and pressure variation across it are substantial.

The solution found for the growth of the reaction center is essentially self similar and is reasonably insensitive to the initial data with the following exceptions. First the actual induction time, t_e, for this spatially inhomogeneous ignition process is different but on the same order as the spatially homogeneous induction time. The induction time does depend to first order on the perturbations to the initial (shocked) state. One must always solve the induction zone problem to find the leading order estimate for t_e and the point in the region where the localized ignition occurs.

Secondly, the form of the reaction center is a self similar structure of one of two types; either an "internal type" or a "boundary type" reaction center. The "internal type" occurs away from boundary with a zero temperature derivative at the temperature peak, while the "boundary type" reaction center occurs at boundaries with a nonzero temperature derivative at the temperature peak. The thickness of the "internal type" reaction center is slightly greater than that of the "boundary type".

The main structural features of the evolution of the reaction center can be summarized in a simple (and practical) way by giving only the leading-order results in the explosion stage. Close to the reaction center ($x \to 0$) and close to the induction time t_e, ($t \to t_e$) the evolution of the center is described approximately by the formulas

$$T = p = \frac{1}{[1 + \theta^{-1} \ln(t_e - t)]}, \quad \rho = 1, \ u = 0, \tag{10}$$

$$\text{in the interval} \quad 0 < x < (t_e - t)^{(1 - \mu)},$$

and

$$T = p = \frac{1}{[1 + \theta^{-1} (\ln x)/(1 - \mu)]}, \quad \rho = 1, \ u = 0, \tag{11}$$

$$\text{in the interval} \quad x > (t_e - t)^{(1 - \mu)},$$

for the approximate time interval from the initial instant to times exponentially close to the induction time

$$0 < t < t_e - e^{[-\frac{\theta \ \beta\gamma}{1+\beta\gamma}]}. \tag{12}$$

The value of μ is $(\gamma-1)/(2\gamma-1)$ for a boundary type reaction center and 1/2 for an internal type reaction center. A sketch of the evolution of the temperature and pressure according to these formulas is shown in Figure 1. The formulas are graphed for five equal time intervals. The last curves shown have the maximum temperature and pressure within the flat "hot spot" region of $1 + \beta\gamma$.

It is worth noting that formulas (10) and (11) can be shown to hold for the case of piston and leading shock boundary conditions as well.

Having generated these results for the early generation of hot spot structure, it is a reasonable question to ask, what is the time scale relevant to the generation of the hot spot and what are its dimensions? The answer is relevant to the question of direct numerical simulation as it suggests the smallest spatial and temporal scales that need to be represented. The dimension of the reaction center is important information for the experimentalist.

Nitromethane has been modeled successfully in part by Bdzil and his coworkers [4], [5] as a polytropic fluid. Also Nitromethane's shock initiation behavior has been studied and to some extent its thermomechanical properties are thought to have been determined. A recent representative experimental study is due to Hardesty [6]. Table 1 shows the properties of Nitromethane used to determine the hot spot size as a function of the initial shock input pressure.

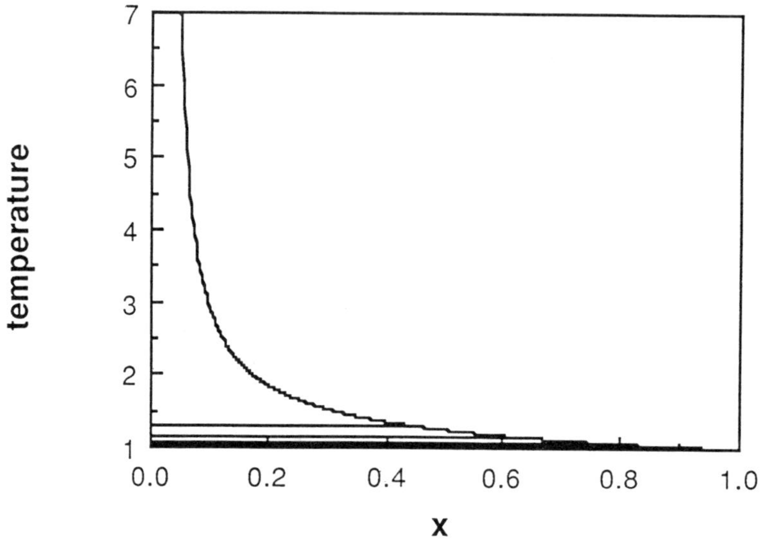

Figure 1. An "internal type" reaction center, $t_e = 1$, $\gamma = 3$, $\beta = 2$, $\theta = 7$ and $\mu = 1/2$.

Table 1

Thermomechanical properties of Nitromethane

ideal gas constant: $\tilde{R} = 1.987$ cal/(mole deg K)

activation energy: $\tilde{E} = 23$ kilocal/mole[6]

frequency factor: $\tilde{k} = 2.6 \times 10^9$/sec

heat of combustion: $q = 1.1$ kilocal/mole[6]

unshocked density: $\tilde{\rho}_0 = 1.1278$ gm/cm^3

inert shock density at 9.47 GPa: $\tilde{\rho}_s = 1.861$ gm/cm^3

unshocked shock density vrs. P (GPa): $\tilde{\rho}_s = 0.0299\ \tilde{P}_s + 1.5779$

inert shock temperature (deg K): $\tilde{T}_s = 332.5 + 75.3\ \tilde{P}_s + .9\ \tilde{P}_s^2$ [5]

polytropic exponent: $\gamma = -\ .1189\ \tilde{P}_s + 5.2157$, $\tilde{P}_s$ (GPa)

estimate of sound speed: $\tilde{c}_s = [\gamma\ \tilde{P}_s/\tilde{\rho}_s]^{1/2}$

Table 2

Parameter definitions

$$\theta = \tilde{E}/(\tilde{R}\ \tilde{T}_s)$$

$$\beta = \tilde{q}(\gamma - 1)\ /\ \tilde{c}_s^2$$

$$\mu = 1/2$$

$$\tilde{t}_i \approx \tilde{k}^{-1}\ e^{\theta}/\beta\ \theta.$$

Three quantities with dimension were calculated that characterize the hot spot: $\tilde{t}_e$ as defined in Table 2 as an order of magnitude estimate of the induction time or time of formation, $\tilde{I}$ is width of the constant pressure and temperature region of the reaction center, i.e. the edge of the hot spot at the end of the formation period when the temperature and pressure are maximum with $T = 1 + \beta\ \gamma$, $\tilde{L}$ is a width characterizing the hot spot such that 1 percent of the maximum temperature rise from the shock state is attained (i.e. the location where $T + 1 = .01\ \beta\ \gamma$). Table 3 shows $\tilde{t}_e$, $\tilde{I}$ and $\tilde{L}$ versus $\tilde{P}_s$. A few comments: The results are to be regarded as an order of magnitude estimates of times and widths. The value of t_e has only been determined up to an O(1) factor which in a piston problem may be as small as 1/10 - 1/100. $\tilde{I}$ and $\tilde{L}$ scale accordingly. The value of γ was between 4 and 5. The value of $\tilde{c}_s$ (comparable to the shock velocity) is perhaps low compared to that may be inferred from Hardesty's data but it was found that changing γ to increase $\tilde{c}_s$ did not significantly change the estimates. Also the estimates for the widths were not sensitive to the precise value of μ and it was taken to be 1/2. Thus the values given in Table 3 are deemed reasonable estimates for the stated purpose.

Table 3

$\tilde{P}_s$ (GPa)	$\tilde{T}_s$ (deg K)	θ	$\tilde{c}_s$ (mm/μsec)	β	$\tilde{t}_e$ (10^{-4}sec)	$\tilde{I}$ (10^{-4}m)	$\tilde{L}$(m) (10^{-4}m)
5.0	731.5	15.82	3.80	1.27	1.42	5.81	3375
5.5	773.9	14.96	3.97	1.17	0.691	4.63	1817
6.0	816.7	14.17	4.13	1.08	0.360	3.77	1035
6.5	860.0	13.46	4.28	1.00	0.199	3.12	619.3
7.0	903.7	12.81	4.43	.94	0.167	2.63	387.4
7.5	947.9	12.21	4.56	.88	0.072	2.24	252.1
8.0	992.5	11.66	4.69	.84	0.045	1.94	170.1
8.5	1037.6	11.16	4.82	.79	0.030	1.70	118.4
9.0	1083.1	10.69	4.96	.76	0.021	1.51	84.7
9.5	1129.1	10.25	5.05	.72	0.015	1.34	62.2
10.0	1175.5	9.85	5.16	.69	0.011	1.21	46.78

Hardesty's experiments ranged over shock input pressures from approximately 7.5 GPa to 9.5 GPa. The striking aspects of these results is that the inner width of the hot spot ($\tilde{\Gamma}$) is nealy constant, but for the same pressure range the outer edge of the hot spot $\tilde{L}$ decreases by two orders of magnitude as $\tilde{P}_s$ is increased from 5 to 10 GPa. This indicates that the physical structure of the hot spot is very broad for low pressures and hence its evolution in isolation of the initiating shock as stimulus may be in question for low pressures.

4. MULTIPLE 1-d HOT SPOTS AND "EFFECTIVE" INDUCTION BEHAVIOR

Practical condensed phase explosives materials are made of granular composites. Generally speaking, the time for an inert shock to pass over a grain or void in the composite is very short compared to the time for substantial burning to occur. In this limiting case many interactions must be accounted for and in some sense average to describe the "effective" induction behavior of the explosive material. Clearly the macroscopic effective properties are the important ones for engineering purposes.

A simple model for describing the interaction of multiple 1-d hot spots in the induction phase due to heterogeneities in the unshocked material was previously introduced by Stewart in [3] and the detailed derivation is found there. (Note that in the following discussion the time scale is now chosen to be $\tilde{k}^{-1} e^{\theta}/\theta$, following [3].

Variation of the local thermodynamic state of the particle is allowed through the Lagrangian coordinate ζ. The relation between the lab and lagrangian coordinate is given by

$$x = \zeta + \int_0^t u[x(\zeta,\hat{t}),\ \hat{t}]\ d\hat{t}.$$

The governing equations are still those of (6) - (9). The dependence on the Lagrangian coordinates comes in explicitly through the shock relations. For example for a polytropic fluid

$$E(\tilde{P},\ \tilde{\rho},\ \lambda;\ \zeta) = \tilde{P}/\ (\tilde{\gamma} - 1) - \tilde{q}\lambda$$

where the explicit particle dependence is given by material functions such as $\tilde{\gamma}(\zeta)$ and $\tilde{q}(\zeta)$.

In the following, the 1 subscript (which is later dropped) refers to the perturbations to the constant inert shock state. Additionally, D_1 is the perturbed shock velocity, and the strong approximation is employed for convenience. Then the following induction-zone model behind the induction shock is derived

$$P_{1t} + P_{1\xi} + \gamma[u_{1t} + u_{1\xi}] = \gamma\beta\ h \tag{13}$$

$$P_{1t} - P_{1\xi} - \gamma[u_{1t} - u_{1\xi}] = \gamma\beta\ h \tag{14}$$

$$P_{1t} - \gamma[\rho_{1t}] = \gamma\beta\, h \tag{15}$$

$$\lambda_{1t} = h \tag{16}$$

where ξ is now a piston centered coordinate. Note that to leading order the Lagrangian coordinate ζ and the piston coordinate ξ are identical. The piston boundary condition is simply

$$u_1 = 0 \quad \text{at} \quad \xi = 0. \tag{17}$$

The function $h(P_1,\rho_1,\lambda_1)$ corresponds to the perturbation of the reaction rate. When the ideal equation of state and Arehnius kinetics are chosen it follows that

$$h = \exp\,[T_1], \quad \text{where } T_1 = P_1 - \rho_1. \tag{18}$$

The perturbed normal shock relations represents the shock boundary conditions which become

$$\rho_1 = -R\,\Omega, \quad u_1 = [2/(\gamma + 1)]\, D_1 - U\,\Omega,$$

$$P_1 = \{2[2\gamma(\gamma - 1)]^{1/2}/(\gamma - 1)\}\, D_1 - Q\,\Omega, \; \lambda_1 = 0.$$

In the above R U and Q are constants that depend on the assumptions of character of the inhomogeneity. Here Ω is a spatial distribution $\Omega(\xi)$ which when evaluated at the leading shock is simply an explicit function of time. Note then that the shock is simply an explicit function of time. Note then that the shock conditions are those of a simple homogeneous materials with additional contributions that reflect the acoustic "noise" generated by the passage of the leading shock over the heterogeneities. The presence of this noise changes the induction time, sometimes dramatically.

In what follows some results of numerical experiments are presented that reflect the following choices of R, U, Q and $\Omega(\xi)$. The former were chosen corresponding to allowing variations in $\tilde{\gamma}$ following the modeling in [2]. Thus

$$R = [2\gamma(\gamma-1)/(\gamma+1)], \; Q = -\,[\gamma(\gamma-1)^2/(\gamma+1)],$$

$$U = (\gamma-1)[2\gamma(\gamma-1)]^{1/2}/(\gamma+1)$$

Then Ω was chosen such that

$$\Omega(\xi) = \alpha \sin(\pi\, \omega\, \xi)$$

and h as in equation (18). Experiments were carried out (for $\gamma = 2.86$, $\beta = 1$) holding γ and β fixed while varying α and ω. The results are shown in Figures 2 and 3.

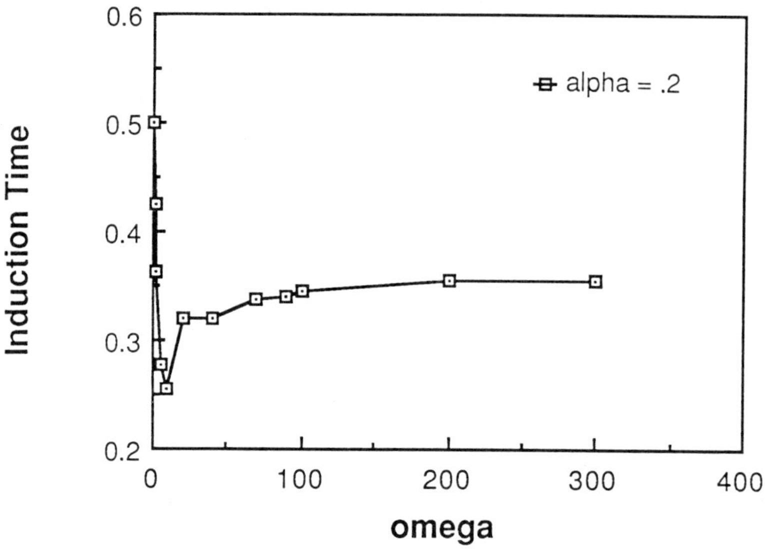

Figure 2. Variation of the dimensionless induction time t_e for a fixed value of α while varying ω.

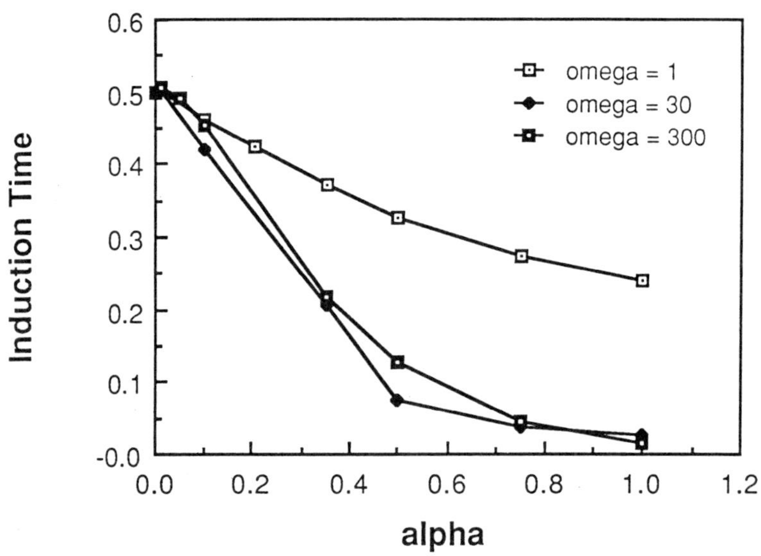

Figure 3. Variation of the dimensionless induction time t_e for a fixed value of ω while varying α.

The results do show some interesting effects. In particular when $\omega \to \infty$ (corresponding to the crossing of the shock over many inhomogeneities prior to thermal runaway) the induction time is observed numerically to tend to a definite limit. This result suggests the induction environment is behaving is an averaged way. Indeed we expect that the induction behavior in this limit is controlled by two physically distinct scales. The first is the induction time of the homogeneous (without heterogeneities) explosive. This time is presumed to be a long scale of the evolution. The final effective induction time, although modulated is still expected to be measured on this scale, see Figure 4.

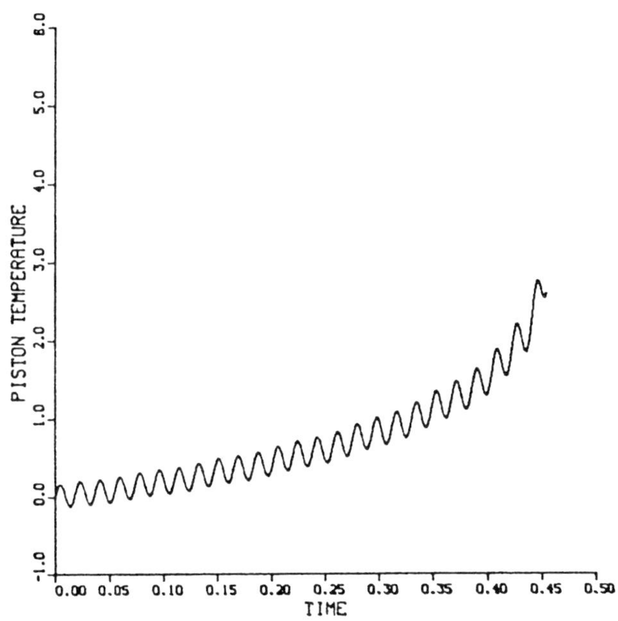

Figure 4. Shows a typical plot of the temperature at the piston when ω is large, here $\alpha = .1$, $\omega = 300$.

The other time is the acoustic passage time across the heterogeneities which represents a fast time. That fast time, high frequency interactions can modulate long time effective averaged behavior is well known and for example is treated in Kevorkian and Cole's book [6] in chapter 5.

The observation of two time scales as $\omega \to \infty$ and the motivation of trying to describe effective behavior, motivates some further modeling assumptions in order to get to a simpler problem. For example, if one assumes highly pressure sensitive kinetics for the reaction rate (not unreasonable for condensed phase materials) then it can be shown, as in [3] that h can be replaced by

$$h = \exp[P_1]$$

Then from the governing equations (13) - (16) and the boundary condition (17) - (18) it is a simple matter to derive the following nonlinear wave equation and boundary conditions for P_1 (now set equal to P) namely

$$P_{tt} - P_{\xi\xi} = \gamma \; \beta(h)_t \tag{19}$$

with the piston boundary condition

$$P_\xi = 0 \quad \text{at} \quad \xi = 0, \tag{20}$$

and the shock boundary condition

$$P_t + A\,P_\xi = B\,\beta\,h + C\,\Omega(\xi) \quad \text{at} \quad \xi = V\,t, \tag{21}$$

where A, B, C and V are constants that depend on γ as derived from [3]

$$A = 3\gamma(\gamma-1)^{1/2}/[(2\gamma)^{1/2}(2\gamma-1)],\; B = [2\gamma(\gamma-1)]^{1/2}(\gamma-1)^2/(\gamma+1),$$

$$C = \gamma(\gamma-1)/(2\gamma-1),\; V = [(\gamma-1)/2\gamma].$$

If the distribution function Ω reflects the fact that there is a high density of imperfections, then it can be written as

$$\Omega = \Omega\,(\varepsilon^{-1}\,\xi), \quad \varepsilon << 1.$$

In order to cast this problem in the form discussed at length in [6] we introduce new scales that are appropriate for acoustics on the scales of the imperfections. Thus define new scales χ and τ such that

$$\xi = \varepsilon\,\chi \quad \text{and} \quad t = \varepsilon\,\tau.$$

The above problem becomes simply

$$P_{\tau\tau} - P_{\chi\chi} = \varepsilon\,\beta\,(e^P)_\tau$$

subject to

$$P_\chi = 0 \quad \text{at} \quad \chi = 0 \quad \text{and}$$

$$P_\tau + A\,P_\chi = \varepsilon\,B\,\beta\,e^P + C\,\Omega(\chi) \quad \text{at} \quad \chi = V\,\tau.$$

For $\varepsilon << 1$, the leading order solution is simply linear acoustics. In the standard way, the higher order corrections become secular and the resolution is to allow the solution to be modulated on the "slow" time scale (which for this problem is homogeneous induction time scale). For finite ε the thermal runaway time t_e is a function of γ, β and ε, represented as $t_e(\gamma, \beta; \varepsilon)$. The effective induction time (which should be regarded as a property of the composite explosive) is then found in the limit as $\varepsilon \to 0$ i.e. $t_e(\gamma, \beta; 0)$.

This problem is still under study and promises to shed light on the effective behavior of explosive materials.

5. THREE DIMENSIONAL DISTRIBUTION OF HOT SPOTS AND THE STABILITY OF THE HOMOGENEOUS STATE

As mentioned in the introduction, the concept of a single isolated reaction center as an ignition site is clearly an idealization. In section 4, we showed some of the consequences of multiple-phase hot spots for a heterogeneous material and effective properties. We also suggested the analysis of a simple model which deserves further study.

One can also consider formation of multiple hot spots that form due to transverse wave disturbances to the one dimensional or the spatially homogeneous solution. It should be possible, again by analyzing a simpler model problem in the induction phase, to determine characteristic spacing of the hot spots that may arise in the transverse direction. We close the paper by introducing such a problem.

The simplest problem that we can pose uses the rate idealization of section 4, namely letting $h = \exp[P]$. Then it is a simple matter to show that the following wave equation for P holds

$$P_{tt} - \tilde{\nabla} \cdot \tilde{\nabla} (P) = \gamma \beta (\exp[P])_t .$$

$$P = - \ln (1 - \gamma \beta t). \tag{22}$$

Introducing a normal mode expansion of the form

$$P = - \ln (1 - \gamma \beta t) + \hat{P}(t) e^{i}(k x + l y + m z)$$

leads to the following evolution equation for $\hat{P}$, namely

$$\hat{P}_{tt} + K^2 \hat{P} = \frac{\gamma\beta}{(1-\beta\gamma t)} \hat{P}_t , \quad K^2 = (k^2 + l^2 + m^2),$$

which can be recognized as a Bessel's equation.

At very early times the stability problem is the stability of the (early time) inert shocked state P = 0 with $\hat{P}$ replaced by a constant times $e^{\lambda t}$. This leads to the dispersion relation

$$\lambda^2 - \gamma \beta \lambda + k^2 = 0;$$

and we find

$$\lambda = (\gamma\ \beta/2)\ [1 \pm \sqrt{1 - 4\ K^2/(\gamma\ \beta)}]$$

The maximum growth rate is then confined to disturbances with the lowest wave numbers. Also for very low wave numbers λ is strictly real and there is no propagation of the growing disturbance. This result is consistent with observations made in numerical studies that the solutions tend to localize at the maxima of broad disturbances. At any event these results are only preliminary and are still under investigation by the author.

REFERENCES

1. Jackson, T. L., Kapila, A. and Stewart, D. S., Evolution of a reaction center in an explosive material (1987), T.&A.M. Report No. 484, Department of Theortical and Applied Mechanics, University of Illinois.

2. Jackson, T. L. and Kapila, A., Shocked induced thermal runaway (1985), SIAM Journal of Applied Math. 45, 130-137.

3. Stewart, D. S., Plane shock initiation of homogeneous and heterogeneous condensed phase explosives with a sensitive rate (1986), Combust. Science and Tech. Vol. 48, pp 309-330.

4. Bdzil, J. B., Steady state two dimensional detonation (1981), Journal Fluid Mech. Vol. 108, pp 195-226.

5. Bdzil, J. B., Engelke, R. and Christenson, D., Kinetics study of a condensed detonating explosive, J. Chem. Phys. 75 (10) 5694-5699.

6. Hardesty, D., An investigation of the shock initiation of liquid nitromethane, Combust. and Flame, 27, 229-251.

7. Kevorkian, J. and Cole, J., Perturbation Methods in Applied Mathematics, Springer Verlag.

ACKNOWLEDGMENTS

This work was completed while the author was a Visiting Scientist at the Cornell Mathematical Sciences Institute. The work is supported by Los Alamos National Laboratory under Contract DOE-LANL-9XR6-5128C1.

INFLUENCES OF DETAILED CHEMISTRY ON ASYMPTOTIC APPROXIMATIONS FOR FLAME STRUCTURE

F. A. Williams
Department of Mechanical and Aerospace Engineering
Princeton University, Princeton, NJ 08544, USA

This paper is dedicated to G.S.S. Ludford, in memory of his numerous contributions to flame theory.

1. INTRODUCTION

Traditionally there have been two distinct approaches to the calculation of flame structure. The earliest approach employed analytical methods in which approximations to the solutions of the ordinary differential equations describing steady, planar flames were obtained - initially through physical reasoning, without even identifying the differential equations. In more recent years, systematic development of the analytical methods has been provided by asymptotic anlaysis, which also enabled error estimates to be made. The second approach, which awaited the advent of computers, involved numerical integration of the differential equations. Today it is generally accepted that, if truly accurate descriptions of flame structures are sought for real flames, then numerical methods must be employed.

Unfortunately, numerical integrations with full chemistry are expensive and time-consuming. Moreover, unless considerable effort is expended after the integrations are obtained, the resulting curves do not contribute greatly to our understanding of flame structure. Results of analytical methods facilitate understanding - indeed, it must be there before the simplifications that enable analytical methods to be pursued can be initiated. Yet, these methods have been deemed manageable only if highly simplified descriptions of the chemistry in flames are introduced.

Recent advances in the development of asymptotic methods have carried their applicability well beyond the constraints traditionally visualized. For example, asymptotic descriptions of methane-air flames with full chemistry have been obtained, and there is no obvious impedement to extending these approaches to other hydrocarbon-air systems, alcohols, etc. The purpose of the present paper is to describe these new advances in asymptotics and to appraise their potential future contributions by identifying outstanding problems.

2. THE CONSERVATION EQUATIONS

Attention will be restricted to steady, planar flames in ideal gas mixtures, without distributed losses, radiant energy transfer, Soret or Dufour effects. One of the principal interests in the simplifications provided by asymptotic descriptions is the relative ease with which they

C.-M. Brauner and C. Schmidt-Lainé (eds.), Mathematical Modeling in Combustion and Related Topics, 315–341.

lend themselves to addressing nonsteady and nonplanar phenomena. Totally numerical approaches with full chemistry become intractable for three-dimensional, time-dependent flows even with the largest computers. Simplifications through asymptotics can enable these flows to be considered. Relevant ideas involve reasoning about effects of departures from steady, planar configurations on diffusion of intermediate species. These aspects of contributions of asymptotic methods will not be discussed here.

Attention is further restricted to Mach numbers low enough for the pressure p to be constant in a close approximation. The conservation of energy and of chemical species can then be written as (1)

$$\lambda dT/dx = m\left(\sum_{i=1}^{N} h_i \epsilon_i - h_o\right) \tag{1}$$

and

$$m d\epsilon_i/dx = W_i \sum_{k=1}^{M} \left[\nu_{ik} B_k T^{\alpha_k} e^{-E_k/R^o T} \prod_{j=1}^{N} (X_j p/R^o T)^{n_{jk}}\right], i=1,\ldots,N, \tag{2}$$

where x is distance, T temperature and ϵ_i the mass-flux fraction of chemical species i.

Here the mass flux is $m = \rho v$, where $\rho = (p/T) \sum_{i=1}^{N} W_i X_i$ is density and v velocity, and m is a constant by mass conservation. In Eq. (1) λ is the thermal conductivity, h_i the (thermal plus chemical) enthalpy per unit mass for species i, and h_o a constant enthalpy per unit mass determined by boundary conditions. In Eq. (2) X_j is the mole fraction of species j, R^o the universal gas constant, W_i the molecular weight of species i, and ν_{ik} the increase in the number of moles of species i in reaction k. Reaction-rate parameters are n_{jk}, which is the order of reaction k with respect to species j, and the specific reaction-rate constant for reaction k, namely $k_k = B_k T^{\alpha_k} \exp(-E_k/R^o T)$, the temperature dependence of which is assigned a three-parameter representation through the constant prefactor B_k, the constant temperature exponent α_k and the constant activation energy E_k. There are N species in the system, which experiences M reactions.

Under the stated assumptions the Stefan-Maxwell form of the diffusion equations is appropriate and can be written as

$$dX_i/dx = (m/\rho) \sum_{j=1}^{N} [(X_iX_j/D_{ij})(\epsilon_j/Y_j-\epsilon_i/Y_i)], i=1,\dots,N, \quad (3)$$

where D_{ij} represents the binary diffusion coefficient for the pair. The mass fraction of species i is $Y_i=W_iX_i/\sum_{j=1}^{N} W_jX_j$.

The quantities h_i, λ and ρD_{ij} are functions of T alone, which are considered known from thermodynamic and transport properties. Since $\sum_{i=1}^{N} \epsilon_i = \sum_{i=1}^{N} X_i=1$, Eqs. (1), (2) and (3) constitute 2N-1 independent equations for 2N-1 independent unknowns, T, ϵ_i and X_i, i=1,...,N-1.

This formulation excludes nonplanar flow; contributions from dm/dx would have to be taken into account in counterflow or stagnation-flow configurations, for example. These contributions modify flame structure by introducing "flame stretch". Qualitative influences of flame stretch nevertheless can be extracted from the formulation by introducing suitable two-point boundary conditions at two finite values of x. For example, a stagnant diffusion layer can be described by letting $m \longrightarrow 0$ and imposing appropriate boundary conditions at $x = x_1$ and at $x = x_2 > x_1$; all of the essential characteristics of stretched diffusion flames are then recovered, with an increase in stretch corresponding to a decrease in $x_2 - x_1$.

3. ACTIVATION-ENERGY ASYMPTOTICS FOR PREMIXED FLAMES WITH ONE-STEP CHEMISTRY

The physical ideas underlying the most well-known asymptotic analysis for premixed flames extend back to Zel'dovich and Frank-Kamenetskii (2). As an approximation, in Eq. (2) take M = 1, and seek an asymptotic expansion for large values of the Zel'dovich number

$$\beta_1 = E_1(T_\infty-T_o)/R^o T_\infty^{\;2} \quad , \qquad (4)$$

where T_o is the initial temperature and T_∞ the equilibrium temperature of the burnt gas. From Eq. (2) it is then evident that asymptotically $\epsilon_i = \epsilon_{io}$ = constant until T is near T_∞. This leads to the existence of the convective-diffusive or "preheat" zone followed by the thin reaction zone, as illustrated in Fig. 1.

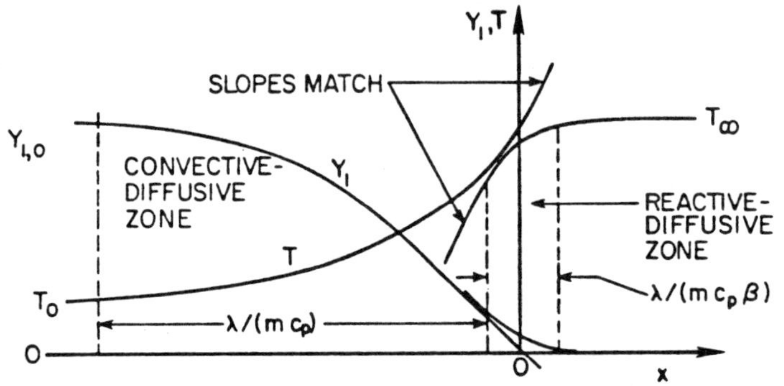

FIGURE 1. Illustration of premixed flame structure for one-step chemistry according to activation-energy asymptotics.

A representative specific heat at constant pressure c_p may be introduced and a nondimensional distance variable defined as

$$\xi = \int_0^x (mc_p/\lambda)dx \ , \tag{5}$$

along with a nondimensional temperature

$$\tau = (T-T_o)/(T_\infty-T_o) \ . \tag{6}$$

Solutions for τ and X_i tend to be exponential in ξ in the preheat zone. From the translational invariance the reaction zone may be placed at $\xi = 0$ in the outer variable, and then appropriate inner-variable scalings identify $\beta_1\xi$ and $\beta_1(1-\tau)$ as variables of order unity in the reaction zone, whereas ξ and τ are of order unity in the preheat zone. Differential equations in the inner variables are derived and solved subject to matching conditions to determine profiles in the reaction zone as well as the burning velocity v_o, which is an eigenvalue of these equations.

Basic aspects of this asymptotic approach are well known and may be found in books (1,3,4). Since $\beta_1 = 10$ is a representative value for the large parameter with the present approximation, two-term expansions are needed to reduce errors in v_o to a few percent. Errors in numerical integrations often tend to be a few percent; therefore two-term expansions are needed if close comparisons between numerics and asymptotics are to be made. Experimental errors typically are of this same order. Hence in all respects, the 10% to 20% errors anticipated from activation-energy asymptotics at leading order are somewhat larger than desired. This was recognized even in the first formal application of the method (5), which obtained two-term expansions.

There have been continuing reasons to pursue activation-energy asymptotics for flames with one-step chemistry. The reasons stem from interest in testing the predictions against those for more complex chemistry, to see how good one-step approximations can be. For this purpose it has been necessary to take into account phenomena that were neglected in the earliest analyses. Initially, thermodynamic and transport coefficients had been set constant, and only one-reactant, one-product flames had been addressed. For one-reactant flames, general two-term results are now available, taking into account variable thermodynamic and transport properties and arbitrary reaction orders n_{11} (6,7). Correspondingly general results have now been obtained for two-reactant flames with arbitrary orders n_{11} and n_{21} (8). The earliest analyses of two-reactant flames, summarized by Mitani (9), for example, employed Fick's approximation for diffusion, which is accurate only if reactants are dilute or if binary diffusion coefficients are nearly equal for all species. The more recent work (8,10) incorporates the improved description of diffusion provided by the Stefan-Maxwell equations.

Very little research remains to be pursued in activation-energy asymptotics for premixed flames with one-step chemistry. Only two avenues appear to remain open. One is to investigate the influences of Soret and Dufour effects. Since coefficients of thermal diffusion typically are small, numerically of order β^{-1} compared with binary diffusion coefficients, the Soret and Dufour effects can conveniently be introduced as second-order terms to obtain improvements to the most general results currently available (8). Resulting corrections to burning velocities could be large enough to be tested experimentally.

The second avenue involves calculations for rate functions different from the Arrhenius functions and not having the power dependences on reactant concentrations shown in Eq. (2). One-step descriptions are obtained formally from full chemistry by introducing suitable steady-state and partial-equilibrium approximations into the rate equations. When this is done, the one-step rates usually are more complicated than those that have been considered; for example, the resulting concentration dependence is not simply proportional to $X_j^{n_{j1}}$. The analyses are no more difficult in principle, since only different integrals need to be performed, and the results will be qualitatively the same in the most important respects. However, the number of different functions that may arise is so great that it would seem to be of interest to pursue such analyses only for specific kinetic schemes that arise in problems with real chemistry.

4. THE OZONE DECOMPOSITION FLAME

The ozone flame involves only one element, O, and has N = 3 and M = 6. Since it is the simplest example of a real flame, it has often served as a testing ground for numerical and analytical methods. Asymptotic methods have been applied to the ozone decomposition flame in a series of recent papers (11,12,13,14), which have resulted in a complete classification of its propagation regimes.

The relevant reaction scheme for the ozone flame is $O_3 + M \longrightarrow O_2 + O + M$, $O_3 + O \longrightarrow 2O_2$, $2O + M \longrightarrow O_2 + M$, $O_2 + O + M \longrightarrow O_3 + M$, where M represents any of the three species, O, O_2 or O_3. The total of six reactions is obtained by adding the reverses of the second and third of these steps to the mechanism. These two reverse reactions involve dissociation of the very stable product O_2, and although they are important for approaching the final equilibrium composition at sufficiently high final temperatures T_∞, there is no regime in which they have a significant influence on the burning velocity. Equilibrium dissociation begins to become important at a T_∞ that increases with p and that has a value of about 2000 K at atmospheric pressure, but dissociation rates are so slow that the approach to equilibrium always occurs in a downstream convective-reactive zone that is too wide to influence the burning velocity. Thus, the two reverse reactions that are neglected here must be included if accurate profiles near the hot boundary are to be obtained at sufficiently large values of T_∞, while they play no role in the classification of regimes relevant to the burning velocity. The problem reduces to one with N = 3 and M = 4; i = 1,2,3 will refer to O, O_2 and O_3, respectively, and M = 1,2,3,4 will identify the reactions in the order listed above.

The first reaction is endothermic, requiring an energy $\mu = 0.7$ times that released in the overall combustion process, $O_3 \longrightarrow (3/2)\ O_2$. Its measured activation energy is $E_1 \approx 22$ kcal/mol. The other three steps are exothermic with activation energies near zero; that for the second has been quoted to be between 4 and 6 kcal/mol, so that its ratio to E_1 is $b \approx 0.2$, while both recombination steps have $E_k = 0$. Equation (4) therefore may define a large expansion parameter for activation-energy asymptotics, to be used with the full kinetics.

With β_1 large, since the first reaction must occur to initiate combustion, a preheat zone may be expected to exist, like that for one-step chemistry as illustrated in Fig. 1. At sufficiently high pressures, in the reaction zone all four of the reactions may have comparable rates, so that a structure like Fig. 1 is obtained with the four steps occurring in the reaction zone. Three independent differential equations then remain to be solved in the reaction zone – those for T, Y_1 and Y_3, say. The steady-state approximation for O atoms,

$$X_1[k_2X_3+(p/R^oT)(2k_3X_1+k_4X_2)] = k_1X_3 \quad , \qquad (7)$$

can be justified if T_∞ is small enough, thereby reducing the number of independent differential equations to two, giving a problem with the same attributes as that for one-step kinetics, but with a more complicated reaction-rate expression. However, for roughly $p \lesssim 100$ atm at most values of X_{3o} (for $p \lesssim 10$ atm if $X_{3o} \lesssim 0.2$), the two recombination steps are slow in the reaction zone, and a more interesting structure is obtained. The remaining discussion is restricted to these lower pressures, which clearly encompass the major conditions of practical importance.

When recombination is negligible in the reaction zone, it occurs in a larger downstream zone that typically maintains a convective-reactive balance, as illustrated in Fig. 2. The reaction-zone solutions must then be matched to those for the recombination zone which result in zero gradients as reaction-zone matching conditions. The temperature at the downstream end of the reaction zone is reduced below T_∞ because of the energetic consequences of the nonzero O concentration there and is given by

$$T_f = T_\infty - 3(1+\mu)(T_\infty-T_o)(Y_{1f}/Y_{3o}) \quad , \qquad (8)$$

where the subscript f identifies conditions there. In the reaction zone, either the steady-state approximation for O may apply (now simply $X_1 = k_1/k_2$), or it may be necessary to retain separate differential equations for T, Y_1 and Y_3. The former situation is termed the steady-state regime and the latter the merged regime, both of which have the zone arrangement shown in Fig. 2.

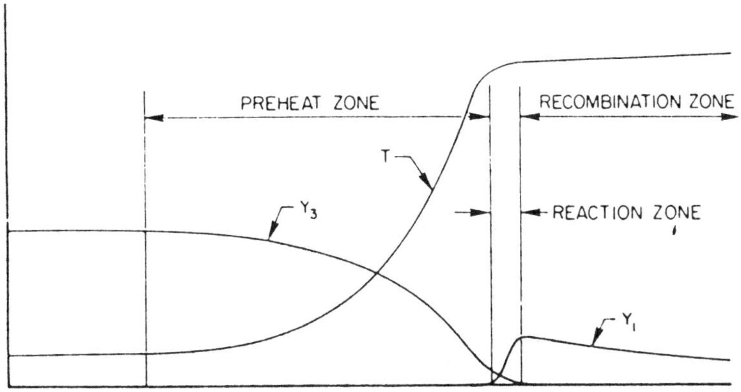

FIGURE 2. **Illustration of ozone flame structure for the steady-state and merged regimes.**

In the reaction zone, $\beta X_3/X_{3o}$ is the reactant-concentration variable of order unity, where β is defined by Eq. (4) with T_∞ replaced by T_f, and the scalings show through Eqs. (2) and (3) for ϵ_1 and X_1 that the steady state applies if X_1 is small compared with X_3. Since $X_1 = k_1/k_2$ in the steady state, this occurs if T_f is not too large. At larger values of T_f, X_1 becomes of the same order of magnitude as X_3 in the reaction zone, and the merged regime occurs.

Burning velocities predicted for these two regimes are shown in Fig. 3. In the steady-state regime, it is found that, through fortuitous cancellations, variable-property effects have negligible influences on the burning velocity, and one-term and two-term expansions in β^{-1} give essentially the same predictions. These predictions are seen to be in excellent agreement with those of full numerical integrations of the original set of conservation equations, shown by the squares in the figure. In the merged regime β becomes smaller ($\beta \approx 5$ at $X_{3o} \approx 0.8$), and appreciable differences develop between burning velocities predicted by one-term and two-term expansions, as seen in the figure. An heuristic one-third rule has been proposed (7) for estimating true burning velocities when results of one-term and two-term expansions in β^{-1} differ significantly, and this approximation is seen in Fig. 3 to provide good agreement with the predictions of numerical integrations.

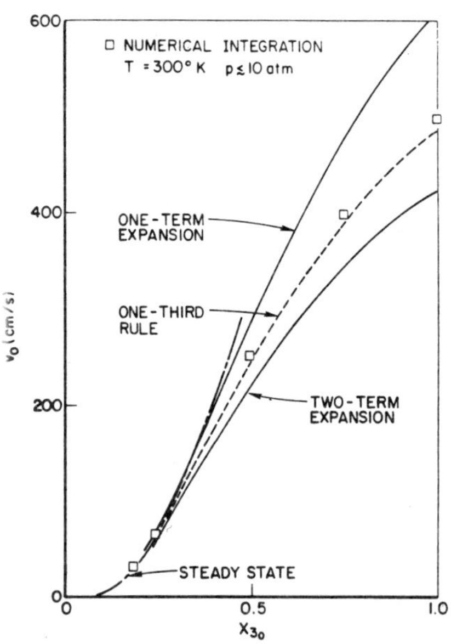

FIGURE 3. Comparison between asymptotic and numerical results for the burning velocity as a function of the initial ozone mole fraction.

With X_3/X_{3o} of order β^{-1} in the reaction zone, consistency prevents X_1/X_{3o} from becoming larger than order β^{-1}. However, it is possible for X_3/X_{3o} and X_1/X_{3o} both to be of order unity in the reaction zone (13) if b is small. This condition never occurs for the ozone flame but could occur for flames having the same kinetic scheme but different values of rate parameters. A radically different flame structure then is obtained from activation-energy asymptotics, as illustrated in Fig. 4. In the reaction zone ξ becomes of order unity, so that convection becomes important at leading order, and a convective-reactive-diffusive balance is maintained. A steady-state approximation becomes applicable in the differential equation for thermal enthalpy in this zone. The O concentration in this reaction zone is now so large that the intermediate diffuses into the preheat zone, where it liberates heat by the second reaction. The preheat zone therefore becomes an exothermic zone, and what was formerly the only reaction zone may now be called a generation zone (see Fig. 4), the zone in which the active intermediate is generated. Because of the two (convective-reactive-diffusive) reaction zones, this propagation regime has been termed the two-zone regime. There is a transition layer between the two reaction zones, with thickness of order $\beta^{-1/2}$ compared with the reaction zones.

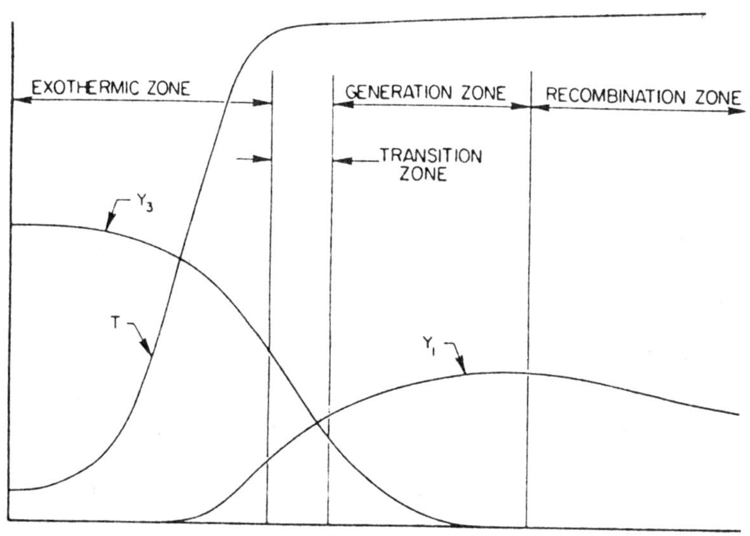

FIGURE 4. Illustration of flame structure in the two-zone regime for flames modeled after the ozone flame.

The two-zone structure provides a mechanism whereby diffusion of an active intermediate can play an important role in the flame structure and dynamics. Diffusion coefficients for intermediates affect burning velocities already in merged regimes by influencing radical distributions within the thin reaction zone, but direct influences of moderate strain or curvature on structure through modification of intermediate diffusion are not encountered until the two-zone regime. If the activation energy of the second step is too high, so that b is of order unity, then the preheat zone must remain inert, but for b = 0 the two-zone structure exhibits exothermicity all the way to the cold boundary (without compromising activation-energy asymptotics). Although there have often been discussions of radical diffusion upstream potentially influencing flame structure, the two-zone regime represents the first time that this effect has been obtained from activation-energy asymptotics. The energetic influences of the intermediate increase as the heat of dissociation, μ, increases. For μ of order β or larger with b small, there could be significant heat release in the preheat zone from intermediate diffusion even though Y_1/Y_{3o} and Y_3/Y_{3o} are of order β^{-1} in a thin reaction zone that (for $\beta Y_1/Y_{3o}$ small) may maintain a steady state for the intermediate; the relevant asymptotic analyses have not yet been initiated.

5. PREMIXED HYDROGEN-HALOGEN FLAMES

Hydrogen-halogen flames involve two elements, hydrogen (H) and the halogen (F,Cℓ,Br or I, denoted here by Ha). The species are Ha,H,Ha_2,H_2 and HHa, so that N = 5; the subcripts i = 1,...,5 will identify the species in the order shown. The overall combustion reaction is $H_2 + Ha_2 \longrightarrow 2HHa$, but, except possibly for Ha = I at low temperatures, this four-center step is too slow to be part of the kinetic mechanism. The mechanism generally is assumed to have M = 5, the steps being $Ha_2 + M \longrightarrow 2Ha + M$, $Ha + H_2 \longrightarrow HHa + H$, $H + Ha_2 \longrightarrow HHa + Ha$, $H + HHa \longrightarrow H_2 + Ha$, and $2Ha + M \longrightarrow Ha_2 + M$, identified by k = 1,...,5 in the order given. Here the fifth is the reverse of the first and the fourth the reverse of the second; the reverse of the third is deemed too slow to be important. In addition, H_2 is considered to be too stable for its dissociation, $H_2 + M \longrightarrow 2H + M$, or the reverse thereof, to be important. If these additional steps were included, then the value M = 8 would apply; if the four-center step and its reverse also were included, then M = 10. The mechanism is more complicated than that of the ozone flame, but much less complicated than that of the hydrogen-oxygen flame or of hydrocarbon flames.

Attention will be restricted to the mechanism with M = 5. Once the flame structure is obtained by asymptotic methods for this mechanism, it is straightforward to test whether the steps that have been neglected are important in any of the regimes. These tests remain to be performed. In fact, thorough identifications of regimes have not yet been obtained. However, a number of ideas on structures have been developed from asymptotic concepts, many through analysis of model problems (3,15,16,17,18).

Within the context of the mechanism with M = 5, the steady-state approximation for H generally is introduced, giving

$$X_2 = k_2X_1X_4/(k_3X_3 + k_4X_5) \quad , \tag{9}$$

which is used to eliminate the H concentration. This reduces the order of the system of differential equations, and since atoms are conserved in chemical reactions, only two independent chemical rate equations remain; in Eq. (2) these may be taken to apply to ϵ_3 and to $\epsilon_1 + (W_1/W_2)\epsilon_2$. An overall representation of the two steps is $Ha_2 \rightleftarrows 2Ha$, $H_2 + Ha_2 \longrightarrow 2HHa$, the first, corresponding to k = 1 and k = 5, in general having finite rates in both directions, and the second, the result of k = 2,3,4, being irreversible with a rate proportional to the intermediate concentration X_1. Model studies have addressed the detailed straight-chain mechanism $A_2 + M \longrightarrow 2A + M$, $A_2 + A \longrightarrow B + A$, $2A + M \longrightarrow C$, which is readily seen to be related to this reduced mechanism if the identifications A = Ha, B = 2HHa − H_2, C = Ha_2 + M are made. There are many similarities between this model mechanism and the ozone-flame mechanism; the main difference is the regeneration of the intermediate A in the straight-chain step. Thus, similarities with the regimes for the ozone flame are to be anticipated. The range of validity of the steady state for H seems wide but has not been explored thoroughly and deserves to be checked for each halogen.

Analyses through activation-energy asymptotics may be based on large values of β_1 of Eq. (4), with the step k = 1 being Ha_2 dissociation. The reverse (recombination) step has zero activation energy, while the effective activation energy for the propagation step (overall $H_2 + Ha_2 \longrightarrow 2HHa$) is that for k = 2 of the original mechanism, which is denoted by bE_1, so that b is its ratio to that for dissociation. With the ratio of the heat of dissociation of Ha_2 to the heat released in $H_2 + Ha_2 \longrightarrow 2HHa$ denoted by μ, a table of values of b and μ can be prepared for each of the four halogens, as shown in Table I. The table indicates that b ranges from small values

TABLE I. Activation-Energy Ratios and Recombination-Energy Ratios for Halogens

Halogen	F_2	$C\ell_2$	Br_2	I_2
b	0.04	0.12	0.45	1.11
μ	0.29	1.31	1.85	15.62

(nearly zero) to values of order unity, while μ ranges from small to large values. These values may be compared with those quoted previously for the ozone flame.

The sequence of possible regimes parallels those for ozone but exhibits certain differences caused by the chain mechanism. At low enough temperatures and high enough pressures, the dissociation will maintain partial equilibrium, so that

$$X_1 = X_3 k_1/[k_5(p/R^o T)] \quad . \tag{10}$$

In the literature this condition seems to be universally termed a steady state for Ha, but that terminology is improper because a correct application of the steady-state approximation for Ha entails retaining contributions from the steps k = 2,3,4, which may be significant for Ha even when the steady state for H, Eq. (9), is accurate. Conditions for equivalence of the partial equilibrium and the steady state need to be checked. Since it is Eq. (10) that has been employed and tested in practice, where the phrase "steady state for Ha" appears in the literature, it should be translated as "partial equilibrium for Ha_2 dissociation". In this context, Ha = Br almost exclusively. When Eqs. (9) and (10) apply, the problem reduces to that of one-step chemistry, and the flame structure is as described previously for steady-state regimes.

When the rates of dissociation and recombination are of the same order of magnitude as the propagation rate in the reaction zone, the situation is analogons to the merged regime of the ozone flame and has been termed (18) the merged regime of the hydrogen-halogen flame. The corresponding structure is illustrated in Fig. 5 and differs from that of Fig. 2 in that the recombination zone is absent.

At still lower pressures and higher temperatures, the recombination becomes negligible in the reaction zone but occurs in broader zones on each side of the reaction zone, as illustrated in Fig. 6. This situation may be termed the slow-recombination regime, and in its distinguished limit the thickness of each recombination zone is the same as that of the preheat zone, and each maintain a convective-reactive-diffusive balance. With μ of order unity or larger (and b of order unity), these recombinations influence temperatures to order β_1^{-1} in the distinguished limit, and it becomes important to allow for $T_f < T_\infty$, as with the ozone flame [compare Eq. (8)].

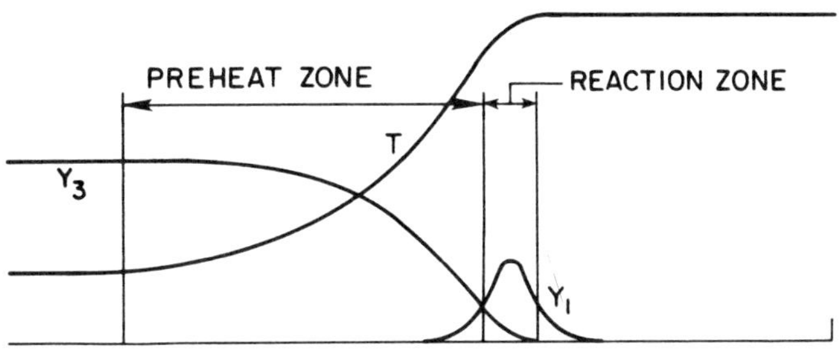

FIGURE 5. Illustration of premixed hydrogen-halogen flame structure for the steady-state and merged regimes.

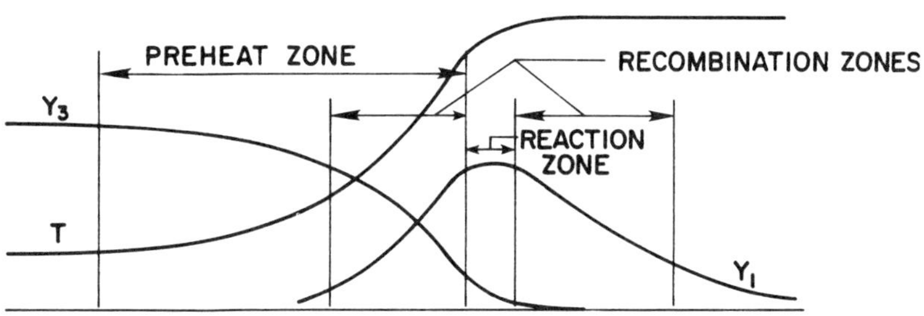

FIGURE 6. Illustration of premixed hydrogen-halogen flame structure for slow-recombination regimes.

At still slower rates of recombination, the recombination becomes important only in a broad downstream zone that maintains a convective-reactive balance, and the structure resembles that in Fig. 2, except that the Y_1 profile extends to the beginning of the preheat zone. Reasonable structures have been obtained with the recombination neglected entirely (3,16). For halogen-rich mixtures, inclusion of the recombination affords the opportunity to investigate influences of equilibrium dissociation of Ha_2 at the final state.

For small b there are possibilities of two-zone structures, resembling Fig. 4, with orderings like those that have been cited for the ozone flame. The two-zone regime can arise only if the recombination rate is not too large. From Table I it is seen that the two-zone regime might be encountered for F_2 or for $C\ell_2$ but not for Br_2 or I_2.

Continuing research is needed to obtain complete classifications of regimes for premixed hydrogen-halogen flames.

6. PREMIXED METHANE-AIR FLAMES

The remaining discussion will be concerned with methane-air flames, first the premixed flame (19,20,21), then the diffusion flames (22,23). Asymptotic methods different from activation-energy asymptotics play roles in the description of these flames. Methane-air flames are representative of flames in general hydrocarbon-oxygen-nitrogen mixtures, and there is every reason to believe that the asymptotic methods being employed can readily be extended to them.

In comparison with the flames considered thus far, hydrocarbon flames have very complicated chemistry. Since the chain branching of the hydrogen-oxygen system is but one subelement of the mechanism, it may be thought easier to resolve the hydrogen-air flame structure first. However, attempts to apply asymptotic methods to hydrogen flames with real kinetics have continually encountered complexities (1). Asymptotics have helped to clarify structures of flames with the model branched-chain kinetics, $A + B \longrightarrow 2B$, $2B + M \longrightarrow 2C + M$ (3,16,17,24), including influences of stretch (25,26) and unsteadiness (27); structures and stability also recently have been investigated for the mechanism $A + B \longrightarrow 2B$, $A + B \longrightarrow C$ (28). These model studies are of help in understanding some aspects of hydrogen-oxygen flame propagation but have not played roles for hydrocarbon flames. Simplifications contributing to understanding of hydrocarbon flames can be achieved without awaiting full resolution of the hydrogen-oxygen problem.

A minimal set of elementary steps for the methane flame, with a representative set of rate constants, is given in Table II. The reactions will be numbered as indicated at the left in the table. Addition of steps (2), (3), (4), (6), (9b) and (10b) gives $CH_4 + 2H + H_2O \longrightarrow CO + 4H_2$, an overall fuel-consumption step, which through steady-state approximations for the intermediates may be assumed to proceed at the rate of step (2). Steps (8) and (9) together produce the water-gas shift, $CO + H_2O \rightleftarrows CO_2 + H_2$, which is reversible, of quantitative importance, and, with a steady state for OH, has the rate of step (8a) for its forward rate. The qualitative essence of the structure can be obtained by assuming water-gas equilibrium in the reaction zones, which has some approximate experimental support, at least for diffusion flames. A combination of steps (9), (10) and (11), with steady states for OH and O, gives $O_2 + 3H_2 \rightleftarrows 2H_2O + 2H$, the overall oxygen-consumption, radical-production process, reversible but with the forward rate of step (11a). Finally, if a steady state is assumed for HO_2, steps (9b), (12) and (13) give the overall recombination process $2H + M \longrightarrow H_2 + M$, at the rate of step (13). Step (11a) is taken to be the slowest in the hydrogen-oxygen chain, and partial equilibria for steps (9) and (10) are closely equivalent to steady states for OH and O and are employed for simplification.

TABLE II ELEMENTARY REACTION STEPS AND RATE CONSTANTS

	Reaction			B*	α	E*
1	CH_4+OH	⟶	CH_3+H_2O	1.6×10^{6}	2.1	2460
2	CH_4+H	⟶	CH_3+H_2	2.2×10^{4}	3	8750
3	CH_3+O	⟶	CH_2O+H	6.8×10^{13}	0	0
4	CH_2O+H	⟶	$HCO+H_2$	3.3×10^{14}	0	10500
5	CH_2O+OH	⟶	$HCO+H_2O$	7.5×10^{12}	0	167
6	$HCO+M$	⟶	$CO+H+M$	1.6×10^{14}	0	14700
7	$HCO+H$	⟶	$CO+H_2$	4×10^{13}	0	0
8a	$CO+OH$	⟶	CO_2+H	1.5×10^{7}	1.3	-758
8b	CO_2+H	⟶	$CO+OH$	1.47×10^{12}	0.51	24211
9a	H_2+OH	⟶	H_2O+H	1.17×10^{9}	1.3	3626
9b	H_2O+H	⟶	H_2+OH	4.4×10^{9}	1.32	18756
10a	H_2+O	⟶	$OH+H$	1.8×10^{10}	1	8826
10b	$OH+H$	⟶	H_2+O	7.93×10^{9}	1	6963
11a	$H+O_2$	⟶	$OH+O$	5.13×10^{16}	-0.816	16507
11b	$OH+O$	⟶	$H+O_2$	1.71×10^{14}	-0.444	-503
12	$OH+HO_2$	⟶	H_2O+O_2	5×10^{13}	0	1000
13	$H+O_2+M$	⟶	HO_2+M	2.1×10^{18}	-1	0

* Units: mol, cm^3, s, K, cal/mol

By the reasoning just outlined, a test mechanism with four overall steps is obtained, termed the four-step mechanism. If, furthermore, a steady-state is assumed for H, then a three-step mechanism is obtained, $CH_4 + O_2 \longrightarrow CO + H_2 + H_2O$ (fuel consumption), $CO + H_2O \rightleftarrows CO_2 + H_2$ (water-gas), and $O_2 + 2H_2 \longrightarrow 2H_2O$ (oxidation), with rates readily derivable from those indicated above by introducing the new steady state. If X_{1e} is the H mole fraction corresponding to full partial equilibrium of steps (9), (10) and (11), then the steady-state H mole fraction is

$$X_1 = X_{1e}\sqrt{F}, \quad F \equiv 1 - (k_2/k_{11a})(X_3/X_2)-(k_{13}/k_{11a})(p/R^{o}T), \quad (11)$$

where X_2 and X_3 are the O_2 and CH_4 mole fractions, respectively. Introduction of equilibrium for the water-gas shift removes this reaction from the system and results in what is called the two-step mechanism,

$$CH_4 + O_2 \longrightarrow \frac{2}{1+r}(H_2+rCO) + \left(\frac{2r}{1+r}\right) H_2O + \left(\frac{1-r}{1+r}\right) CO_2 ,$$

$$\left(\frac{2}{1+r}\right)(H_2+rCO) + O_2 \longrightarrow \left(\frac{2}{1+r}\right) H_2O + \left(\frac{2r}{1+r}\right) CO_2 ,$$

where r is the water-gas-equilibrium ratio of CO to H_2 concentrations. In the two-step mechanism, the fuel consumption produces a specified combination of H_2 and CO intermediates, which in turn are oxidized in the second step. Of course, if the second step of the two-step mechanism is fast, then the one-step, overall combustion mechanism, $CH_4 + 2O_2 \longrightarrow CO_2 + 2H_2O$, is obtained.

The asymptotic analyses of methane flames relate to the four-step, three-step or two-step mechanism. For the premixed flame, the structure according to the three-step mechanism is illustrated in Fig. 7. Following the preheat zone there is a thin layer in which the fuel-consumption step occurs, and this is followed by a broader (but still reactive-diffusive) layer of oxidation of intermediates, in the upstream part of which the water-gas reaction departs from equilibrium. The ratios of the thicknesses of these three reaction layers to that of the preheat zone are denoted by δ, ϵ and ν, respectively. For discussion purposes, the value $\nu = 0$ may be employed, giving the two-step mechanism.

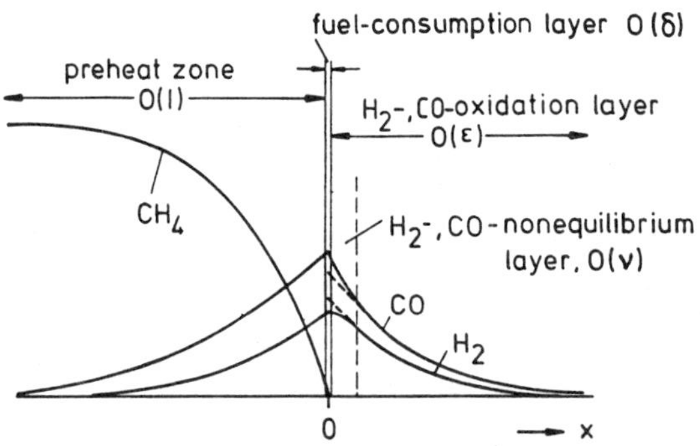

FIGURE 7. Illustration of premixed methane-air flame structure according to asymptotic analysis.

The temperature at the fuel-consumption zone (where conditions will be identified by the subscript f) turns out to be determined largely by the kinetics in the oxidation zone and (for stoichiometric flames under many conditions of practical interest) results in

$$\delta \equiv (k_{11a}/k_2)_f(X_{2f}/X_{3o})(\bar{W}_o/\bar{W}_f)/L_3 \tag{12}$$

being a small parameter, where $\bar{W}$ is the average molecular weight, and L denotes the Lewis number. The differential equation for the suitably scaled CH_4 concentration y in the fuel-consumption zone is found to be

$$d^2y/d\varsigma^2 = \Lambda y\sqrt{F} \tag{13}$$

(subject to $y \longrightarrow 0$ as $\varsigma \longrightarrow \infty$ and $y \longrightarrow -\varsigma$ as $\varsigma \longrightarrow -1$), a quadrature of which determines the burning-rate eigenvalue Λ, proportional to $(k_{11a}^2/k_2)_f/v_o^2$ by its definition. The last term in Eq. (11) for F typically is negligible (away from flammability limits), and Eq. (11) then gives $F = 1-y$. The factor $\sqrt{F}$ in Eq. (13) serves the function that the activation energy serves in activation-energy asymptotics; the reaction ceases in the preheat zone because of radical depletion by CH_4, not because of any decrease in rate with decreasing temperature.

The small parameter ϵ for the oxidation zone is proportional to $v_o^{1/2}/k_{13f}^{1/4}$ and through energetics determines T_f for stoichiometric flames as

$$T_f = T_\infty - (T_\infty - T_o)\epsilon\, z_f \quad , \tag{14}$$

where $z_f \approx 2^{1/4}$ is obtained from the solution to a suitably scaled differential equation describing the structure of this zone. Activation-energy asymptotics are irrelevant to this structure which, in the first approximation, is simply that for a third-order, temperature-independent reaction. The product $\Lambda\epsilon^4$ is independent of v_o, and from results of analyses of each zone [Eq. (14) with the stated z_f, and $\Lambda \approx$ 15/8] the expression

$$(15/16)k_{2f}k_{13f}(p/R^oT)_f = k_{11af}^2(L_2/L_3) \tag{15}$$

is obtained for calculating T_f. Equation (15) describes the competition between the branching effect of step (11a) and the chain-breaking effects of steps (2) and (13). It gives a T_f that increases with increasing p, from about 1600 K at 1 atm to about 2300 K at 30 atm. After T_f is evaluated, the burning velocity v_o is readily calculated from Eq. (14) and the definition of ϵ.

In view of the apparent overall activation energy of about 20 kcal/mol, associated with k_{11a}^2/k_2 in the definition of Λ, it may be thought that the temperature dependence of v_o arises directly through these Arrhenius factors. In fact, the variation of these factors with T is less than that of the concentration factors that also appear in the definition of Λ. Since the rate of CO and H_2 oxidation decreases as the reactant concentrations decrease, from Eq. (14) and the definition of ϵ it is found that for stoichiometric flames v_o is approximately proportional to $(T_\infty - T_f)^4$. This same dependence of v_o is obtained from the concentration factors in Λ so that, because of the specific chemistry involved, the concentration factors surprisingly cancel and do not appear in Eq. (15). The temperature dependence of the burning velocity is dominated by the factor $(T_\infty - T_f)^4$, so that the effective overall activation energy E is roughly $4R^0T_\infty^2/(T_\infty - T_f)$, which typically is in the range of 40 to 60 kcal/mol. It is true enough that the activation energies in k_{11a} and k_2 influence T_f in Eq. (15), but this is fixed when v_o is varied by varying T_o or T_∞, so that in fact the rate ratios at T_f and the energetics control the E deduced from T dependences of v_o. It is purely by chance that the resulting E is in the range for a typical elementary step.

The quantitative agreement between predicted and measured burning velocities for stoichiometric flames, obtained by the analysis describes with the three-step mechanism, is not very good. The difficulty lies not in the zone selection but rather in the specified chemical mechanism for the fuel-consumption zone. For example, reaction (1) of Table II has not been included in the kinetic scheme. By beginning with the mechanism defined, it is straightforward to systematically test for the importance of new steps in the hydrocarbon chemistry (21) when their rates are given. As a result, it is found that the steps listed in Table II are likely to be important. Since the new chemistry is confined to the fuel-consumption layer, it becomes straightforward to revise the burning-velocity computations (21). Essentially, only the definition of F in Eq. (11) needs to be changed, and the first integral of Eq. (13) then involves a more complicated integrand, so that the quadrature is performed numerically. Resulting revised burning velocities (21) agree well with results of full numerical integrations employing the same rate coefficients. The sensitivities of the predictions to poorly known rates have identified elementary steps for which better rate data are needed.

7. METHANE-AIR DIFFUSION FLAMES

The three-step mechanism defined above has recently been remarkably successful in describing the structure and extinction of diffusion flames (22). This is of special interest because there has always been an open question as to whether the kinetic mechanisms are the same in premixed and diffusion flames with complex kinetics. The extent of success (22) is an indication of the extent to which the mechanisms are the same.

For diffusion flames, it is convenient to transform from x to the mixture fraction Z as the independent variable. By definition, Z = 0 in the oxidizer stream, and Z = 1 in the fuel stream. In premixed flames Z remains essentially constant and cannot serve as an independent variable. In the one-step approximation for infinite rates, there is an infinitesimally thin reaction zone in the diffusion flame located at the stoichiometric value $Z = Z_s$, as illustrated by the dashed lines in Fig. 8. For methane-air flames, Z_s = 0.054, which is small enough to serve as a small parameter. The dashed lines show fuel and oxygen diffusing into the reaction zone from opposite sides and the temperature peaking there at the stoichiometric adiabatic flame temperature T_s (here and in Fig. 9 y_i is a reduced mole fraction of species i). With one-step chemistry, activation-energy asymptotics have given the structure of the reaction zone and formulas specifying conditions for extinction. These results have been extended to take into account reversibility of the one-step reaction and dissociation (29), (30).

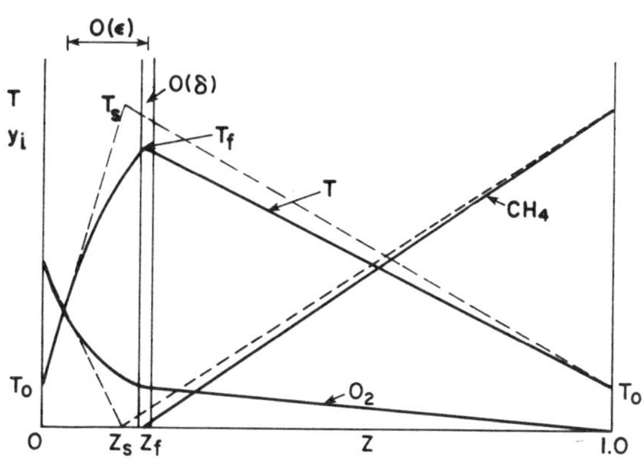

FIGURE 8. Illustration of methane-air diffusion-flame structure according to a two-step mechanism derived by asymptotic methods.

Unlike unstrained premixed flames, diffusion flames possess a one-parameter family of structures. The parameter can be taken to be a Damköhler number, the ratio of a diffusion time to a chemical time. The diffusion time is

$$t = (D_{12}|\nabla Z|^2)^{-1} = [D_{12}(dZ/dx)^2]^{-1} = (x_2-x_1)^2/D_{12}, \qquad (16)$$

where D_{12} is a representative diffusion coefficient. Here the first equality is general, the second applies for a flame perpendicular to the x axis, and the third for the stagnant diffusion layer. The strain rate in a counterflow and the rate of scalar dissipation in a turbulent flow are inversely proportional to t. Extinction occurs if t becomes small enough to decrease the Damköhler number below its critical value. For convenience, reduction of t here will be termed increase of strain.

Measurements and full numerical integrations of the conservation equations for the structure with detailed chemistry show that increasing strain produces increasing leakage of O_2 through the reaction zone but no CH_4 leakage. This behavior is contrary to predictions of activation-energy asymptotics with one-step chemistry. However, the two-step mechanism qualitatively yields the correct behavior. The structure with the two-step mechanism is illustrated by the solid curves in Fig. 8. There is a thin fuel-consumption zone, with thickness in Z of order δ, on the fuel-rich side of a thicker (but still thin) zone of H_2 and CO oxidation, with thickness of order ϵ. The intermediates H_2 and CO are produced along with H_2O and CO_2 in the fuel-consumption zone, whose structure is essentially identical to that for the premixed flame, so that δ is given by Eq. (12). For the oxidation zone, ϵ is like that of the premixed flame but is proportional to $t^{-1/4}$; the structure of this zone differs from that of the premixed flame only in that in Fig. 8 the temperature decreases and the O_2 concentration increases with increasing distance from the fuel-consumption zone. An increase in ϵ (a decrease in t) broadens the oxidation zone, moves the fuel-consumption zone ($Z=Z_f$) to larger values of Z, and increases the O_2 leakage, according to the asymptotic analysis. These predictions agree qualitatively with reality.

Quantitative predictions of structure and extinction from the two-step mechanism are not very good. Substantial improvements are achieved by introducing corrections for water-gas nonequilibrium. This gives the three-step mechanism and results in a thin water-gas nonequilibrium layer on the rich side of the oxidation zone, adjacent to the fuel-consumption zone. The zone arrangement then is entirely analogous to that of the premixed flame, illustrated in Fig. 7; the kinetic schemes are the same, and activation energies are equally irrelevant.

An alternative viewpoint suggests that there may be significant differences between premixed-flame and diffusion-flame kinetics (23). Estimates indicate that the full steady-state approximations for H and OH may not be very accurate and that partial equilibrium for step (11) of Table II in fact may be poor over a certain range of Z. It therefore may be necessary to revert to a four-step mechanism to obtain an improved approximation to the flame structure. However, simplifications motivated by asymptotic analysis may be introduced into the four-step kinetics to obtain a readily manageable formulation from which extinction conditions are easily computed. The structure is illustrated in Fig. 9, which focuses on the region of Z where the chemistry occurs.

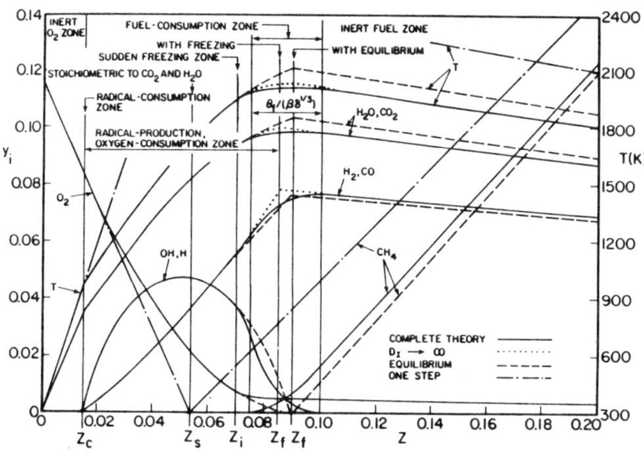

FIGURE 9. Illustration of methane-air diffusion-flame structure according to a reduced four-step mechanism derived by asymptotic methods.

If w_k denotes the rate of step k, then with the numbering of Table II, the steady states and partial equilibria introduced lead to $w_1 + w_2 = w_3 = w_4 + w_5 = w_6 + w_7$, $w_{12} = w_{13}$, $w_{9a} = w_{9b}$ and $w_{10a} = w_{10b}$. The four-step mechanism then is slightly revised compared with that introduced for the premixed flame and can be written as

$$CH_4 + (\frac{2}{1+s})OH + (\frac{2s}{1+s})H \longrightarrow CO + (\frac{2+4s}{1+s})H_2 + (\frac{1-s}{1+s})H_2O \ ,$$

$$CO + H_2O \rightleftarrows CO_2 + H_2 \ ,$$

$$O_2 + (\frac{1+3s}{1+s})H_2 \rightleftarrows (\frac{2}{1+s})OH + (\frac{2s}{1+s})H + (\frac{2s}{1+s})H_2O \ ,$$

$$(\frac{2}{1+s})OH + (\frac{2s}{1+s})H + (\frac{1-s}{1+s})H_2 \longrightarrow (\frac{2}{1+s})H_2O \ ,$$

where s is the ratio of H to OH, defined by the equilibrium

constant for step (9). For simplicity the analysis introduces the further partial equilibrium $w_{8a} = w_{8b}$, so that water-gas equilibrium prevails, and the second step indicated above may be deleted. It is likely that, in future work, this condition should be relaxed near $Z=Z_f$ for improved accuracy, but as found previously, the approximation has no substantial qualitative influence on the flame structure. A reduced four-step mechanism then is obtained which, with the r defined previously, can be written as

$$CH_4 + (\frac{2}{1+s})OH + (\frac{2s}{1+s})H \rightarrow \frac{r(3+5s)}{(1+r)(1+s)} CO$$

$$+ \frac{(3+5s)}{(1+r)(1+s)} H_2 + \frac{3r(1+s)-2s}{(1+r)(1+s)} H_2O + \frac{(1+s)-2r(1+2s)}{(1+r)(1+s)} CO_2, \quad (I)$$

$$O_2 + \frac{(1+3s)}{(1+r)(1+s)} H_2 + \frac{r(1+3s)}{(1+r)(1+s)} CO \rightleftarrows (\frac{2}{1+s}) OH + (\frac{2s}{1+s}) H$$

$$+ \frac{2s-r(1+s)}{(1+r)(1+s)} H_2O + \frac{r(1+3s)}{(1+r)(1+s)} CO_2 \quad , \quad (II)$$

$$(\frac{2}{1+s})OH + (\frac{2s}{1+s})H + \frac{(1-s)}{(1+r)(1+s)} H_2 + \frac{r(1-s)}{(1+r)(1+s)} CO \rightarrow$$

$$\frac{2+r(1+s)}{(1+r)(1+s)} H_2O + \frac{r(1-s)}{(1+r)(1+s)} CO_2 \quad . \quad (III)$$

Step (I) represents fuel consumption by radicals, step (II) oxidation of H_2 and CO to consume oxygen and produce radicals, and step (III) radical consumption by recombination. All steps are bimolecular except for (III), which is termolecular. The rate of (I) involves k_1 and k_2, that of (II) k_{11}, and that of (III) k_{13}.

Step (I) occurs in the narrow fuel-consumption zone at $Z=Z_f$, shown in Fig. 9, and involves a diffusion flame between CH_4 and radicals. Step (III) occurs in a narrow recombination zone at a temperature T_c defined by $k_{11a} = k_{13}(p/R^oT)$ and shown at $Z=Z_c$ in Fig. 9; for $p < 1$ atm, it is found that $T_c < 1000$ K, and so three-body processes are negligible throughout the main high-temperature reaction zones, although they still influence flame structure through the boundary conditions that they impose at $Z=Z_c$. It is the appreciable activation energy of step (11a) that makes the recombination zone thin and places it at $Z=Z_c$; in this respect, the structure is quite different from that of Fig. 8, for which the rate of the recombination step remains important throughout the zone of thickness ϵ. Between $Z=Z_c$

and $Z = Z_f$ in Fig. 9 is the radical-production, oxygen-consumption zone in which step (II) occurs. This is the only reversible step of the three, and it is found to maintain partial equilibrium at low T while having a negligible reverse rate of high T. To approximate this behavior, a point of sudden freezing for (II) is introduced, $Z=Z_i$, such that there is equilibrium for $Z < Z_i$, while the reverse reaction is negligible for $Z > Z_i$. The relatively high activation energy of step (11a), compared with (11b), is an aspect of the motivation for the sudden-freezing approximation (as opposed to a near-equilibrium approximation, for example).

Figure 9 differs from Fig. 8 first in that radicals not in partial equilibrium play a significant role and second in that activation energies influence the structure. In addition to defining Z_c and Z_i, a nonnegligible activation energy is involved in the structure of the fuel-consumption zone. Activation-energy asymptotics are in fact applied to this zone in obtaining predictions of extinction conditions (23), which occur in Liñán's premixed-flame regime because Z_f is small. Overall activation energies deduced from the dependence of t at extinction on T_s agree well with experiment.

The roles of the radicals in Fig. 9 differ appreciably from their roles in the premixed Flame. For the premixed flame with the four-step mechanism, the partial equilibrium for H is poorest in the upstream part of the fuel-consumption zone, where a distinct radical-consumption layer exists (21). This would correspond to the fuel-rich side of Z_f in Fig. 9. Contrary to this behavior, in the mechanism of Fig. 9 departures from this full partial equilibrium are encountered in general in two regions, a thin layer at $Z=Z_c$ where the recombination occurs, and a thicker region in the oxidation layer between $Z=Z_i$ and $Z=Z_f$. In Fig. 9, this latter nonequilibrium additionally extends throughout the fuel-consumption zone. Thus, if Fig. 9 applies then there are substantial differences in kinetics for premixed flames and diffusion flames.

The successes of models of the types represented by Figs. 8 and 9 may be illustrated by the results shown in Fig. 10, taken from the work of Seshadri and Peters (22). The curve represents the theoretical predictions of the maximum temperature according to the three-step mechanism of the augmented Fig. 8, while the triangles are results of a full numerical integration with the four-step mechanism of the premixed flame, and the squares measured values for counterflow diffusion flames. The agreement here is excellent, and the agreement between the observed and predicted value of t for extinction (about 0.05 s) isn't too bad.

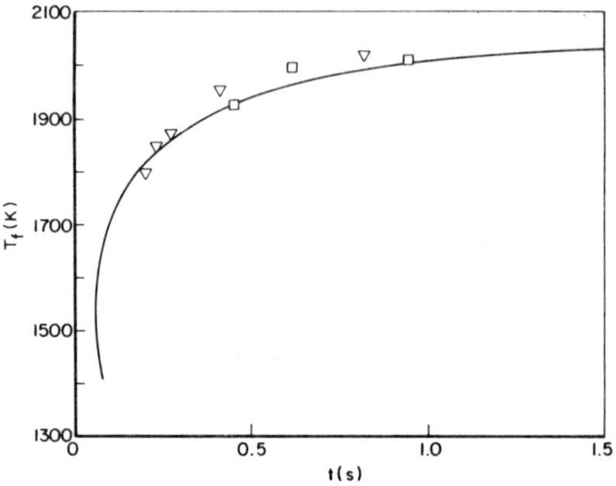

FIGURE 10. Dependence of the maximum temperature on the effective inverse strain rate for methane-air diffusion flames, according to asymptotic analysis with a three-step mechanism, numerical integration with a four-step mechanism, and experiment.

These diffusion-flame studies by asymptotic methods are in their infancies. The substantial differences between Figs. 8 and 9 illustrate the large uncertainties that remain concerning the structures and the extent to which methods of activation-energy asymptotics may play a role. Further research is needed to better identify accuracies and ranges of validities of the two models. Despite their great differences, in some respects the numerical predictions of the models are similar - for example concerning extinction conditions at atmospheric pressure. The extreme differences in the roles of the three-body recombination (being localized at $Z=Z_c$ in Fig. 9 but extending through the oxidation layer in Fig. 8) suggest appreciable differences in predictions of pressure dependences of extinction conditions. Experimental (or full numerical) studies of pressure dependences of extinction therefore could help to distinguish between the two views.

The temperature agreement in Fig. 10 is a result of the oxidation-zone kinetics with water-gas nonequilibrium and the chemistry of the fuel-consumption zone reducing the peak temperature by an appropriate amount below the adiabatic flame temperature. By contrast, with the model of Fig. 9, activation-energy asymptotics in the fuel-consumption layer are needed to obtain a suitable temperature reduction. Neither model has addressed the further chemistry that occurs on the fuel-rich side of Z_f and that is responsible for the well-known "pyrolysis dip" of the temperature profiles there, ultimately leading to soot production at sufficiently large values of t. Perturbation methods can be introduced in both models in the region $Z > Z_f$ to approach these questions. It is thus seen that much remains to be done in asymptotic analyses of diffusion-flame structures.

8. CONCLUDING REMARKS

When the full chemistry of real flames is taken into account, problems of flame-structure calculations at first seem hopelessly complicated for any but completely numerical approaches. However, with the recent progress of asymptotic methods, it is beginning to appear that no flame is too tough for an asymptotic approach. Activation-energy asymptotics do not always afford the appropriate vehicle. Parameters involving reaction-rate ratios, stoichiometry, energetics, etc., arise as large or small parameters of expansion in the asymptotic methods. By taking advantage of the magnitudes of these parameters for real flames, results of asymptotic analyses for flames with complex chemistry afford great simplifications over methods of full numerical integration and are contributing appreciably to our understanding of flame structure.

ACKNOWLEDGEMENTS

This work was supported in part by the National Science Foundation through Grant No. CBT-8603989. H.K. Chelliah helped in the research.

REFERENCES

1. Williams, F.A.: Combustion Theory. 2nd ed., Benjamin/Cummings, Menlo Park, CA, 1985.

2. Zel'dovich, Ya.B. and Frank-Kamenetskii, D.A.: Zhur. Fiz. Khim. 12, 100 (1938).

3. Zel'dovich, Ya.B., Barenblatt, G.I., Librovich, V.B. and Makhviladze, G.M.: The Mathematical Theory of Combustion and Explosion. Nauka, Moscow, 1980.

4. Buckmaster, J.D. and Ludford, G.S.S.: Theory of Laminar Flames. Cambridge, Cambridge, 1982.

5. Bush, W.B. and Fendell, F.E.: Comb. Sci. and Tech. 1, 421 (1970).

6. Rogg, B. and Williams, F.A.: Comb. Sci. and Tech. 42, 301 (1985).

7. Rogg, B.: Comb. Sci. and Tech. 45, 317 (1985).

8. Chelliah, H.K. and Williams, F.A.: Comb. Sci. and Tech. 51, 129 (1987).

9. Mitani, T.: Comb. Sci. and Tech. 21, 175 (1980).

10. Sen, A.K. and Ludford, G.S.S.: Eighteenth Symposium (International) on Combustion. The Combustion Institute, Pittsburgh, 1981, p. 417.

11. Rogg, B. and Wichman, I.S.: Combust. Flame 62, 271 (1985).

12. Liñán, A. and Rodriguez, M.: Combustion and Non Linear Phenomena. Clavin, P., Larrouturou, B. and Pelcé, P., eds., Les Editions de Physique, Les Ulis, 1986, p. 51.

13. Rogg, B., Liñán, A. and Williams, F.A.: Combust. Flame 65, 79 (1986).

14. Rogg, B.: Combust. Flame 65, 113 (1986).

15. Zel'dovich, Ya.B.: Kinetika i Kataliz 2. 305 (1961).

16. Istratov, A.G. and Librovich, V.B.: Zhur. Prikl. Mekh. Tekhn. Fiz. No. 1, 68 (1962).

17. Liñán, A.: Tech. Rept. No. 1, AFOSR Contract No. EOOAR 68-0031, INTA, Madrid, 1971.

18. Liñán, A. and Clavin, P.: Premixed Flames with Non-Branching Chain-Reactions. Laboratoire de Recherche en Combustion, Université de Provence, Centre St. Jérôme, Marseille, 1986.

19. Peters, N.: Numerical Simulation of Combustion Phenomena. Glowinski, R., Larrouturou, B. and Teman, R., eds., Springer-Verlag, New York, 1986, p. 90.

20. Paczko, G., Lefdal, P.M. and Peters, N.: Twenty-First Symposium (International) on Combustion. The Combustion Institute, Pittsburgh, 1987.

21. Peters, N. and Williams, F.A.: Combust. Flame, to appear, 1987.

22. Seshadri, K. and Peters, N.: Asymptotic Structure and Extinction of Methane-Air Diffusion Flames. Submitted to Combust. Flame, 1987.

23. Treviño, C. and Williams, F.A.: An Asymptotic Analysis of the Structure and Extinction of Methane-Air Diffusion Flames. In preparation, 1987.

24. Zel'dovich, Ya.B.: Zhur. Fiz. Khim. 22, 27 (1948).

25. Seshadri, K. and Peters, N.: Combust. Sci. and Tech. 33, 35 (1983).

26. Tam, R. and Ludford, G.S.S.: Combust. Sci. and Tech. 40, 303 (1984); 43, 227 (1985).

27. Ludford, G.S.S. and Peters, N.: Dynamics of Flames and Reactive Systems. Bowen, J.R., Manson, N., Oppenheim, A.K. and Soloukhin, R.I., eds., Vol. 95, Progress in Astronautics and Aeronautics, American Institute of Aeronautics and Astronautics, New York, 1984, p. 75.

28. Clavin, P., Fife, P. and Nicolaenko, B.: SIAM J. Appl. Math. 47, 296 (1987).

29. Peters, N. and Williams, F.A.: Dynamics of Flames and Reactive Systems. Bowen, J.R., Manson, N., Oppenheim, A.K. and Soloukhin, R.I., eds., Vol. 95, Progress in Astronautics and Aeronautics, American Institute of Aeronautics and Astronautics, New York, 1984, p. 37.

30. Williams, F.A. and Peters, N.: Dynamics of Reactive Systems Part I: Flames and Configurations. Bowen, J.R., Leyer, J.-C. and Soloukhin, R.I., eds., Vol. 105, Progress in Astronautics and Aeronautics, American Institute of Aeronautics and Astronautics, New York, 1986, p. 152.

PART II
SHORTER PAPERS

NUMERICAL STUDY OF PARTICLE-LADEN JETS: A LAGRANGIAN APPROACH*

J. Anagnostopoulos, G. Bergeles.
Laboratory of Aerodynamics,
National Technical University of Athens

NOMENCLATURE

C_d	= drag coefficient	Γ_t	= turbulent diffussivity
D	= jet diameter	ε	= rate of k dissipation
d_p	= particle diameter	ϱ	= density
g	= accel.of gravity	σ_t	= Prandtl/Schmidt number
G	= particle mass flux		
k	= turb. kinetic energy	Subscripts	
Le	= dissip. length scale	c	= centerline quantity
m	= particle mass	p	= particle property
r	= radial distance	o	= initial condition
Re	= Reynolds number		
S	= source term	Superscripts	
t	= time	$(\ )'$	= fluctuating quantity
t_e	= eddy lifetime	$(\vec{\ })$	= vector quantity
u.v	= velocity component	$(\bar{\ })$	= mean value
x	= axial distance		

1. INTRODUCTION

In the last few years,there has been a tendency towards the use of coal instead of petrol for energy production. This fact has led to increased efforts in order to improve the performances of coal burners.

This study is a first step of an effort aimed at the development of a procedure, which would accurately predict the flow field properties within a lignite burner. For this purpose, a stochastic flow model was developed, based on the Lagrangian approach, for the treatment of the non-evaporating, solid-phase.

The continuous phase equations were solved with the use of finite difference techniques [10]. Effects of interphase slip and turbulence on particle motion were taken into account with the help of the Monte-Carlo technique. Results obtained are favorably compared with available experimental data.

Several models have been proposed, based on the Lagrangian

* The research was financially supported by EEC/DGXII, contract number EN3F-022-GR(T-T).

C.-M. Brauner and C. Schmidt-Lainé (eds.), Mathematical Modeling in Combustion and Related Topics, 345–354.

approach, such as those by Migdal and Agosta [1], Gosman et al. [2], Durst et al. [3], Shuen et al. [4], Abbas et al. [5]. A second way for handling two-phase flows is by treating the descrete phase as a continuum, by means of an Eulerian approach, as in the models by Elghobashi et al. [6], Melvill and Bray [7], Shuen et al. [4], Durst et al. [3].

In the following there is a theoretical presentation of the Lagrangian model, along with the mathematical formulation of the basic equations. The algorithm for the numerical solution of the equations is then described. The performance of the model compared to existing experimantal data of Shuen [4] and the results of a sensitivity study, concerning the effects of initial conditions are also reported.

2. THEORETICAL MODEL

2.1. Gas-phase

The conservation equations for the mean flow quantities expressed in cylindrical coordinates, take the form:

$$\partial/\partial x(\rho\bar{u}\Phi)+(1/r)\partial/\partial r(r\rho\bar{v}\Phi) = (1/r)\partial/\partial r(r\Gamma_t\partial\Phi/\partial r)+S_\varphi+S_p \tag{1}$$

Turbulence is accounted for by a two-equation model for the turbulent kinetic energy, (k) and its rate of dissipation, (ε), (Launder and Jones [9]). The turbulent diffusivity Γ_t and turbulent viscosity μ_t are obtained from:

$$\Gamma_t = \mu_t/\sigma_t \tag{2}$$

$$\mu_t = C_\mu\rho k^2/\varepsilon \tag{3}$$

where σ_t is the turbulent Prandtl/Schmidt number, taken as a constant.

The parameter S_p represents the additional sources due to the presence of particles. Expressions for Φ, S_φ and S_p and also the empirical constants of the model are given in Table 1.

At the jet exit, variations of kinetic energy, viscous dissipation and gas density were neglected since the injector exit Mach number was less than 0.3 for the present case. Also, since the examined spray was too dilute, the volume fraction of the solid phase was neglected.

Since fully developed pipe flow was assumed at the injector exit, initial values of k and ε were obtained from Hinze [11], for the corresponding mean exit velocity, which was obtained from power law expressions (Schlichting [12]).

The boundary conditions at the jet axis were:

$$r=0,\ \partial\Phi/\partial r = 0 \tag{4}$$

At all the free boundaries, entrainment boundary conditions were used to account for the entering fluxes.

2.2. Particle-phase

Particles were treated as sources of momentum for the continuous phase, by a Lagrangian approach. A finite number of

TABLE 1. Source terms and constants of k-ε model

Φ	S_φ	S_p
1	0	0
u	0	$\bar{S}_{pu}$
k	$\mu_t (\vartheta\bar{u}/\vartheta r)^2 - \varrho\varepsilon$	$\overline{uS_{pu}} - \bar{u}\bar{S}_{pu}$
ε	$C_1\mu_t \varepsilon/k(\vartheta\bar{u}/\vartheta r)^2 - C_2\varrho\varepsilon^2/k$	$-2C_3\mu_t \varepsilon/k(\vartheta\bar{S}_{pu}/\vartheta r)$
C_μ = .09, C_1 = 1.44, C_2 = 1.87, C_3 = .1-5., σ_k = 1.. σ_ε = 1.3		

particles were introduced from a finite number of starting locations into the flow field and their trajectories were calculated by solving Lagrangian equations for the instantaneous motion, of the form [13]:

$$du_{pi}/dt = (3\varrho C_d/4d_p\varrho_p)(u_i - u_{pi})|\bar{u}-\bar{u}_p| + g(1-\varrho/\varrho_p) \quad (5)$$

$$dx_{pi}/dt = u_{pi}, \qquad i = x,y,z \quad (6)$$

which were obtained after neglecting effects of Bassett and Magnus forces, (since $\varrho/\varrho_p > 200$), particle collisions and particle rotation.

The drag coefficient C_d for a solid sphere was calculated according to Wallis [14] from the equations:

$$C_d = 24/Re_p(1+.15Re_p^{.687}) \qquad \text{for } Re_p > 1000 \quad (7)$$

$$C_d = 0.44 \qquad \text{for } Re_p \leqq 1000 \quad (8)$$

where Re_p the particle Reynolds number defined as:

$$Re_p = \varrho\,|\bar{u} - \bar{u}_p|\,d_p/\mu \quad (9)$$

Since the actual particles used in experiments are not exactly spherical, for each particle diameter, the drag coefficient was multiplied by a factor ≧ 1, (DR, Table 2), as it is experimentally derived by Shuen et al. [4].

Effects of turbulence on the particle motion were simulated by a stochastic technique as follows: The turbulence is assumed to be isotropic and the fluctuating velocities of the gas phase to follow a Gaussian distribution, having standard deviation given by:

$$\sigma_{ij} = (2k/3)^{.5} \quad (10)$$

The instantaneous flow field, assosiated with a turbulent eddy, was then estimated by random selection of the fluctuating velocities at any arbitrary point of the flow domain.

The characteristic size of an eddy was estimated as equal to the dissipation length scale of the flow [2]:

$$Le = C_\mu^{3/4}k^{3/2}/\varepsilon \quad (11)$$

and the eddy lifetime as:

$$t_e = Le/(2k/3)^{.5} \quad (12)$$

A particle assumed to interact with an eddy as long as the eddy lifetime, or until it transversed the eddy. Hence the **turbulent** particle dispersion can be represented by **tracking,** in this way, a statistically satisfactory number of particles and averaging their trajectories.

Effects of solid-phase on the gas-phase properties were accounted for, by appropriate source terms, introduced into the gas-phase governing equations. The change in momentum as a particle passes through a control volume can be formed as, [5]:

$$S_p = m_p(u_{pin} - u_{pout}) + m_p g(1 - \varrho/\varrho_p)(t_{out} - t_{in}) \qquad (13)$$

where the subscripts in and out represent particle properties when it enters and leaves the control volume. At each computational cell, the total source term is the summation of the momentum change of all the particles transversing the cell.

Similar source terms were derived by Shuen et al.[4] for the k and ε equations, in order to account for the effects of solid-phase on turbulence modulation (Table 1). These terms are not used here, since the flow was too dilute.

3. SOLUTION PROCEDURE

3.1. Gas-phase

The set of the differential equations with the associated boundary conditions was numerically solved with the SIMPLE algorithm [10]. For this purpose, the generally available TEACH-T computer code, developed in Imperial College, was appropriately modified for the case. During the solution procedure, the particle source terms are kept constant.

3.2. Particle-phase

For the solution of the equations of particle motion, a 4th order Runge-Kutta integration method was used, since it was found to be faster than a finite-differences technique, with the same accuracy [8].

Initial conditions for the particle-phase are stochastically prescribed at the jet exit, accorting to the data of Shuen [4].

The trajectory of each particle is followed until the particle leaves the predetermined computational area.

The gas properties appearing in the equations, are kept constant as long as the particle remains into an eddy.

Particle-phase was subdivided into groups having the same mass flow rate. Each group was defined by its initial position, velocity and the number and diameter of individual particles it contained.

The trajectories of about 5000 particle groups were computed and averaged, in order to get a statistically representative solution. The evaluation of a smaller number of particle trajectories, produced no significant change to the gas-phase results, while the predicted values of particle properties showed increased scattering.

The optimum time step for the integration of the Lagrangian

equations cannot be kept constant, when they are expressed in cylindrical coordinates since it depends on the radial distance. To overcome this difficulty, the equations were solved in their Cartesian form and the results and inputs were transformed wherever necessary. The additional computation time required for these transformations was estimated to be less than 10% of total time requirements.

The properties of each particle when it enters and leaves a computational cell were found using a prediction-correction technique.

3.3. Overall procedure

The overall solution algorithm consisted of the following steps: First a solution is obtained for the gas-phase; the resulting flow field is used for the calculation of particle trajectories and source terms. These terms are then introduced into the gas-phase equations which are solved again, producing a new flow field. The procedure is repeated until the last introduction, through the source terms, of the solid particles in gas phase does not further change the computed flow field.

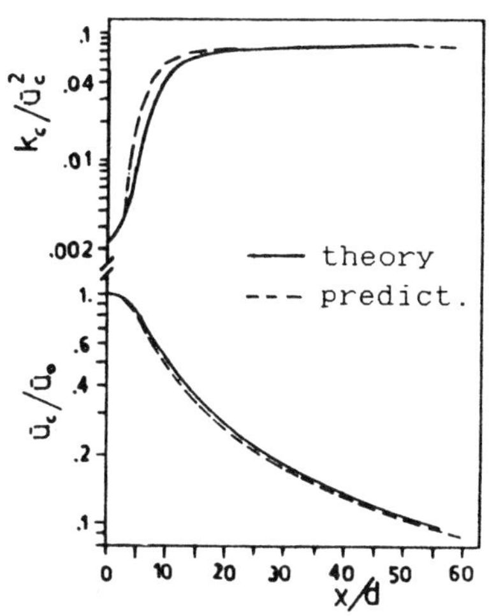

FIGURE 1. Axial variation of gas-phase quantities, single-phase jet.

TABLE 2.. Test Conditions

Particle Sauter Mean Diameter (μm)	119
Diam. standart deviation (μm)	19
DR (Drag Ratio)	1.25
Loading Ratio	0.66
Centerline Exit Velocity (m/s)	
Air	25.22
Particle	21.89
Centerline Exit Particle Mass Flux (kg/m^{2}s)	18.89
Amb. Temper. (K)	296
Amb. press. (kPa)	97
Injector Inside Diameter (mm)	10.9
Particle Density (kg/m^3)	2620

4. RESULTS AND DISCUSSION

The developed model, as it was introduced into the TEACH-T computer code, was first validated against theoretical results of Shuen et al. [4], for a single phase round jet having Re ≅ 22000. The agreement is very good (Fig. 1,2).

After that the developed model was adjusted to predict the

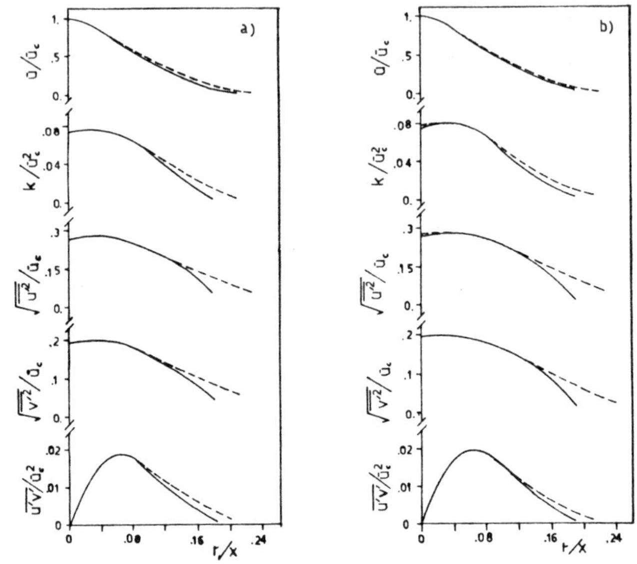

FIGURE 2. Radial variation of gas quantities, single-phase jet. a) x=20*D. b) x=40*D. ——— theory; - - - - predictions.

experiments of Shuen et al. [4], for a vertical, particle - laden, free, round jet, issuing into stagnant air. The test conditions are summarized in Table 2.

The experimental error of the measurements was estimated by Shuen to be within 10% for mean and fluctuating gas velocities, whithin 15% for mean and fluctuating particle velocities and whithin 13% for the particle mass flux.

The predictions of the gas fluctuating velocity components were acquired using the relation $\overline{u'^2}:\overline{v'^2}:\overline{w'^2} = k:0.5k:0.5k$, which is typical for single-phase jets [4].

In order to investigate the effects of initial parameters on the predictions, the change of the results due to a 100% increase of the values of k_o, ε_o, C_d, and d_p and a 50% decrease of k_{po} is plotted in Fig. 3 to 7, along with the described basic model predictions. From the diagramms it is apparent that a change in the gas-phase initial parameters doesn't have significant effect on the predictions, which on the other hand seem to be very sensitive to particle-phase parameters variations.

Measured and predicted profiles of centerline mean gas velocity are shown in Fig. 3. The rate of the predicted velocity decay along the axis is smaller than that of the single - phase jet but larger compared to the experimentally derived rate.

The predicted axial particle velocity is in very good

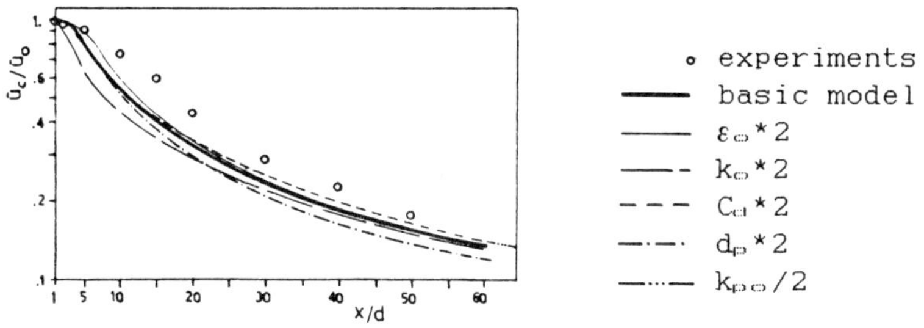

FIGURE 3. Centerline gas-phase velocity.

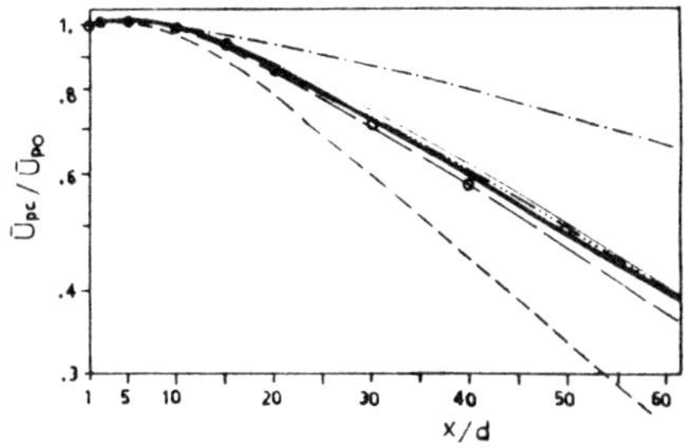

FIGURE 4. Centerline solid-phase velocity.

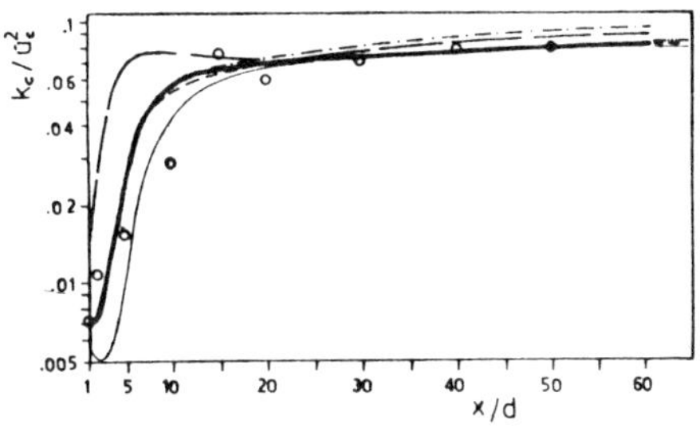

FIGURE 5. Centerl. gas-phase turb. kinetic energy.

agreement whith the experiments (Fig. 4), as well as the axial profile of gas kinetic energy (Fig. 5).

Predicted centerline particle mass flux is underestimated at distances greater than 10D downstream, but is improved when the initial particle kinetic energy is decreased by 50% (Fig. 6).

Axial particle turbulence level predictions are shown in Fig. 7 to be slightly decreasing with increasing distance, while the experimental values show an opposite behaviour. This is probably due to the fact that particle diameter distribution was ignored and the Sauter mean diameter used instead, to represent each particle group.

Radial variations of gas-phase properties are drawn in Fig.8; the model appears to overestimate the values for all these properties, which are approaching anyway the air-jet results, due to the low loading ratio, as well as to the fact that the constants used in the turbulence model were fixed for single-phase jet.

Radial particle velocity decay is underestimated at x=20D, but is well predicted at larger distances (Fig. 9).

Radial particle turbulence level is underestimated, as the axial one, and especially near the axis (Fig. 9).

The model produced good results for the particle mass flux at x=20D and predicted somewhat greater spreading rate at x=40D (Fig. 9).

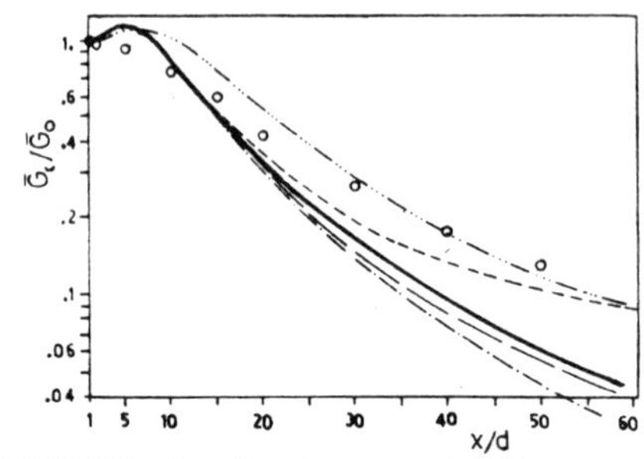

FIGURE 6. Cent. particle mass flux.

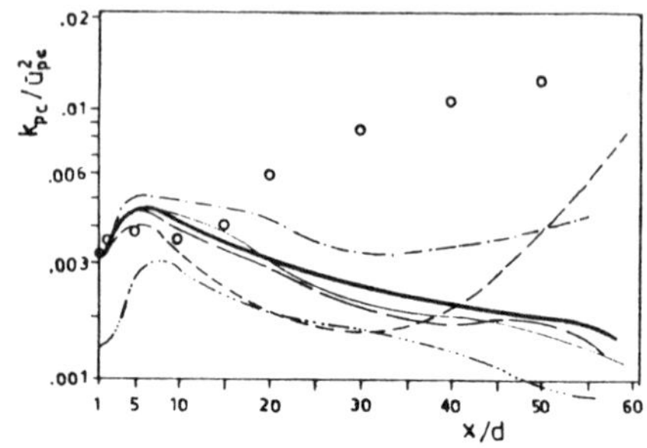

FIGURE 7. Cent. solid-phase turb. kinetic energy.

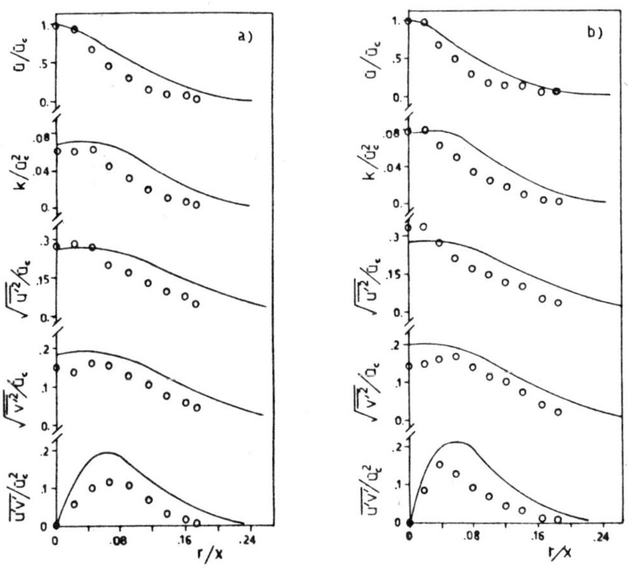

FIGURE 8. Radial variation of gas-phase quantities. a) x=20*D. b) x=40*D. o experiments; ——— predictions.

In order to determine the particle dispersion as a function of particle diameter, we plotted the axial variation of concentration of particles of different diameter, but same initial turbulence level. Fig. 10 shows that the concentration of particles having diameter smaller than the Sauter mean diameter, increases with the axial distance. This can explain the underestimation of axial mass flux, since the model ignores the diameter variation.

5. CONCLUSIONS

A stochastic flow model for the prediction of two - phase, particle -laden flows, was developed and tested using existing experimental data. The obtained results showed reasonably good agreement with the experiments.

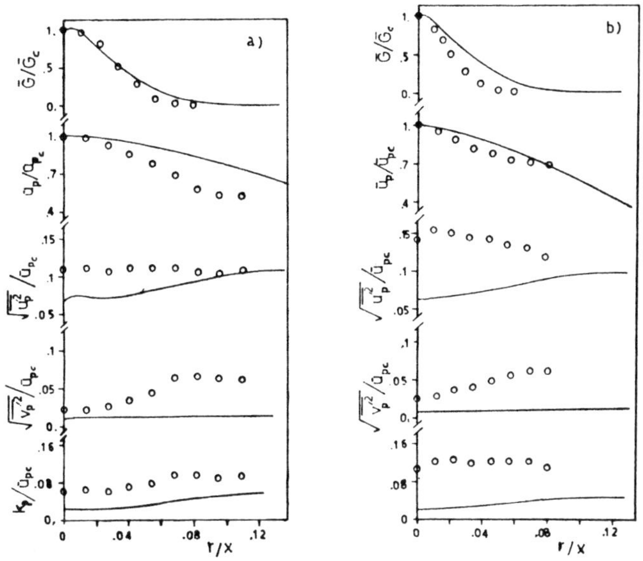

FIGURE 9. Radial variation of solid-phase quantities. a) x=20*D. b) x=40*D. o experiments; ——— predictions.

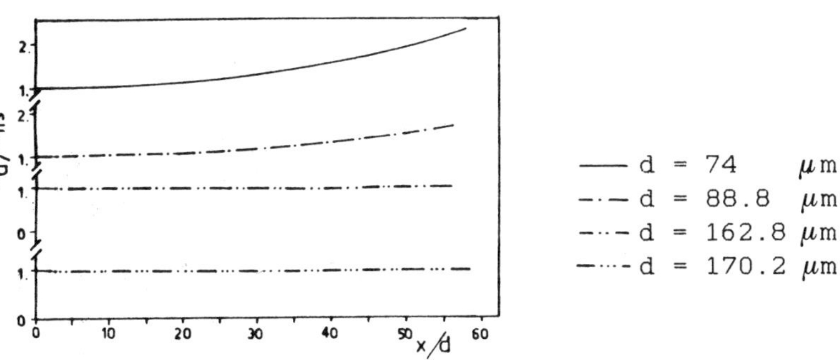

FIGURE 10. Axial variation of particle consentration. N_d = number of particles of diameter d (μm).

The observed deviations in the particle mass flux and particle turbulence level are mainly due to the fact that the particle diameter was taken constant while the assumption of isotropic gas turbulence seems to produce no significant error for our case .

Furthermore, the usage of constant turbulence coefficients, may cause the observed differences between experimental and predicted gas-phase properties.

A sensitivity study was conducted in order to determine the effect of initial parameters on the predictions. The results of this study indicated that the definition of the initial properties of the solid-phase is of great importance for the resulting predictions.

Hence, the evaluation of such a model requires experimental data with accurate measurements for the values of the initial parameters of the solid-phase.

REFERENCES

1. Migdal D, Agosta DV: A Source Flow Model for Continuum Gas-Particle Flow. Trans. ASME, J.Appl. Mech., Dec.1967, p.860.
2. Gosman AD, Ioannides E: Aspects of Computer Simulation of Liquid-Fueled Combustors. J. Energy, v7,6, 1983, p.482.
3. Durst F, Milojevic D, Schoenung B: Eulerian and Lagrangian Predictions of Particulate Two-phase Flows: a Numerical Study. Appl. Math. Modelling, v8, April 1984, p. 101.
4. Shuen JS, Solomon ASP, Zhang QF, Faeth GM: A Theoretical and Experimental Study of Turbulent Particle-Laden Jets. NASA CR-168293, November 1983.
5. Abbas AS, Koussa SS, Lockwood FC: The Prediction of the Particle-Laden Gas Flows. Eighteenth Symposium on Combustion, The Combustion Institute, 1981
6. Elghobashi S, Abou-Arab T, Rizk M, Mostafa A: Prediction of the Particle-Laden Jet with a Two-equation Turbulence Model. Int. J. Multiphase Flow, v10,6, 1984, p.697.
7. Melville WK, Bray KNC: A Model of the Two-phase Turbulent Jet. J. Heat Mass Transfer, v22, 1979, p.647.
8. Diakoumakos E, Anagnostopoulos J, Bergeles G: A Theoretical Study of Solid-Air, Two-Phase Flow. To be presented at the Math. Modeling in Combustion Conf., E.C.L., 1987.
9. Jones WP., Launder BE: The Prediction of Laminarization of with a Two-equation Model of Turbulence.
10. Patankar SV, Spalding DB: A Calculation Procedure for Heat Mass and Momentum Transfer in Three-Dimensional Parabolic Flows. Int. J. Heat Mass Transfer, v15, 1972, p.1787.
11. Hinze JO: Turbulence. 2nd Edition, McGraw-Hill Book Co., New York, 1975, p.427, p.724.
12. Schlichting H: Boundary Layer Theory. McGraw-Hill Book Co., New York, 1979, p.599.
13. Tchen CM: Mean Value and Correlation Problems Connected with the Motion of Small Particles in a Turbulent Fluid. Ph.D. thesis, University of Delft, 1947.
14. Wallis CB: One Dimensional and Two Phase Flow. Mcgraw-Hill Book Co., New York, 1969.

MATHEMATICAL MODELISATION of ENCLOSED COMBUSTION at CONSTANT or VARIABLE PRESSURE by VIBE LAW

Philippe ARQUES
Ecole Centrale de LYON .BP 163 . 69131 ECULLY. FRANCE

For the modelisation of enclosed combustion, different empiric or semi-empiric relations have been proposed (tab : 1). Among them, the VIBE law is applied to the combustion in a reciprocating engine /1/.

Tab :1

	Relations	Conditions
NEUMANN	$x = (2-t/t_f).(t/t_f)$	t_f : final combustion time
or	$x = (3-2.(t/t_f).(t/t_f)^2$	
GONCAR	$x = 1 - (1+t/t_m).\exp(-t/t_m)$	t_m: time where combustion velocity is maximal
AKULOV	$dx/dt = (1/\tau).(x+(dx/dt).t).(1-x)$	τ : ignition delay
KOSKIN	$t/\tau = -(\mathrm{Log}(1-x))/(x+ax^2+bx^3)$	a and b : constants from engine
SEMENOV	$dx/dt = K.(1-x).x.\exp(-E/RT)$	E: activation energy R: constant of gas T: temperature
VIBE	$x = 1 - \exp(-6,908.(t/t_f)^{m+1}$	m : exponent

1.VIBE LAW

The burnt fraction $x = M(t)/M(t_f)$ of combustible relative to time t is defined by this relation :

$$x = 1 - \exp(\ -6.908.(t/t_f)^{(m+1)}\ )$$

where x = 0,999 for $t = t_f$. In this relation t_f is the final combustion time.

The combustion velocity is expressed thus :

C.-M. Brauner and C. Schmidt-Lainé (eds.), Mathematical Modeling in Combustion and Related Topics, 355–363.

$$w = dx/d(t/t_f) = 6{,}908.(m+1).(t/t_f)^m.\exp(-6.908.(t/t_f)^{(m+1)})$$

and the combustion mean velocity w_m is proportional to $1/t_f$.

The burnt fraction $x(t/t_f)$ and the combustion velocity $w(t/t_f) = dx/d(t/t_f)$ define a group of curves (Fig : 1) for each value of parameter m .

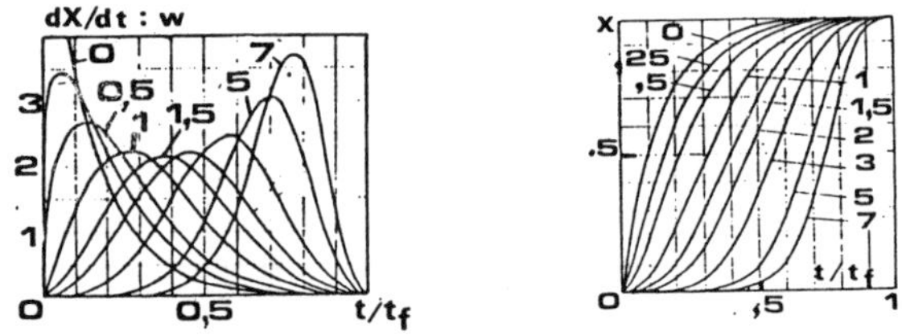

FIGURE 1. Burnt fraction and combustion velocity for some values of parameters m.

Between final combustion time t_f, time t_M of maximal combustion velocity w_{Max} and parameter m, there is this relation (Fig : 2) :

$$t_M/t_f = (m/(6.908.(m+1)))^{1/(m+1)}$$

and the maximal combustion velocity has for equation:

$$w_{Max} = 6.908.(m+1).(m/(6.908.(m+1).e)^{m/(m+1)}$$

2. DIESEL ENGINE

Values of exponent m and final combustion time t_f for some combustion parameters are given by Vibe :

* Intake pressure (bar)	:	2	2,5
Exponent m	:	0,33	0,73
Final combustion time t_f (s)	:	0,0135	0,0107
Final combustion angle θ_f (V)	:	162	129

* Rotation per minute N (t/mn)	:	1600	1200	1000
Exponent m	:	-0,16	-0,13	-0,08
Final combustion time t_f (s)	:	0,0378	0,0383	0,033
Final combustion angle θ_f (V)	:	364	276	200

FIGURE: 2
Maximal velocity W_{Max} of combustion

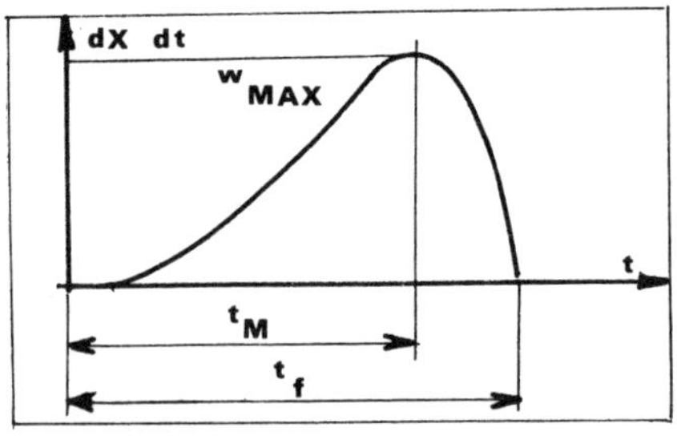

* Engine load (%)	:	100	25
Exponent m	:	-0,16	-0,16
Final combustion time t_f (s)	:	0,0379	0,02615
Final combustion angle θ_f (V)	:	364	251

* Injection timing (V)	:	18	13
exponent m	:	0,08	-0,06
Final combustion time t_f (s)	:	0,0221	0,0301
Final combustion angle θ_f (V)	:	212	289

* Carburant	Kérozène	fuel	"heavy"
Exponent m	:-0,16	-0,16	0,16
Final combustion time t_f (s)	: 0,0315	0,0378	0,0301
Final combustion angle θ_f (V)	: 302	364	400

For DIESEL engine, the final combustion time is connected to the engine rotation per minute and to the mixture ratio by this relation /2/ :

$$t_f/t_{f0} = (Ri/Ri_0)^{0,6}.(N/N_0)^{0,5}$$

(0 is a reference state of the engine)

and the combustion exponent m :

$$m/m_0 = (\tau_0/\tau)^{0,5}.(P/P_0).(T_0/T).(N_0/N)^{0,3}$$

P et T are the engine intake pressure and temperature and τ the ignition delay.

3. SPARK PLUG ENGINE /1,3/

* Intake pressure (bar)	:	1,277	1,044
Exponent m	:	3,6	3,6
t_f (s)	:	0,045	0,052

* Air excess Θ	: 0,865	0,93	1,07	1,22	1,27
Exponent m	: 2,6	3,6	4,3	4,2	3,4
Final comb. time t_f (s)	: 0,00907	0,01205	0,0081	0,0154	0,0127
Final comb.angle θ_f (V)	:49	94	43,7	120	68,6
	(1)	(2)	(1)	(2)	(1)

Moteur :	bore mm	stroke mm	N t/mn	ϵ	Ignition timing (V)
(1)	73	120,65	900	4,6	25
(2)	152	102	1300	5	55

ϵ : engine volumetric ratio

For an aircraft engine (stroke : 190 mm, bore : 160 mm, rotation per minute : 1600t/mn, volumetric ratio ϵ = 6,35, light fuel), the combustion exponent m is constant and equal to 0.38, while the combustion final time varies with the air excess and is minimal for an air/combustible ratio of approximately 0,85 .At this ratio, the turbulent flame velocity is maximal :

Air excess ⊘ :	0,572	0,627	0,760	0,849	0,980	1,076	1,13
Final combustion angle θ_f (V) :	54,7	52	48,4	46,1	50,1	59,6	70,8

For spark plug ignition engine, with a weak mixture combustion ($Ri_g < 1$), exponent m of VIBE law reads :

$$m = 0,24.Ri_g^{-0,7}.(N/1000)^{0,37}$$

and final combustion time t_f :

$$t_f = 9,77.Ri_g^{-1,1}.(N/1000)^{-1,27}$$

or

$$t_f = 55,6.Ri_g^{-1,4}$$

The final combustion time increases with the air excess and decreases when the engine rotation per minute increases and, with it, the " turbulence " of the mixture in combustion.

Thus m and t_f define unambiguously the combustion period in a reciprocating engine. The optimal values m and t_f, or for this last parameter θ_f, may be computed as follows :
- the indicated mean pressure (imp) and the indicated efficiency are as high as possible.

- the pressure and the pressure gradient relative to the time are the lowest possible.

For that very reason, it is desirable that the final combustion time θ_f is equal to 50 ± 10 V and that the combustion exponent m is little different from 1.5, for that reason, it is necessary that the time t_M/t_f ratio is equal to 0.376 .

4. KOUCHOUL ENGINE

The engine and the cycle imagined by KOUCHOUL present the particular feature of enabling technological separation of the two functions of aërodynamic mass transfer and heat release during combustion. We have used this property to calculate the relationship between a parameter characterising the aerodynamic turbulence and the physical and chemical characteristics of the unburned mixture.

We have determined the swirl velocity per crankshaft angle and the flow characteristics of what is conventionally designated as "turbulence" in an engine motored (velocity anemometry method) and in a compression mahine (schlieren techniques). Taking into account the development of combustion, it is possible to investigate the relationship between a parameter characterising the mixture ratio and the flame velocity /4,5,6/.

The KOUCHOUL engine contains two cylinders in which move two pistons out of phase by 25 crank angle (Fig : 3).

The kammers are connected by a channel in the cylinder head. The combustion kammer (b), limited by the advanced piston, is filled with a rich mixture and has a volumetric ratio near to 8. The other cylinder (P) has a very high volumetric ratio and is filled with a weak mixture or air. The turbulent combination of these two masses is obtained by the production of opposite variation of the cylinder volumes due to the phase difference between the piston motions. Ignition takes place in the rich mixture in combustion chamber, turbulence is provoked by the air entering via the channel from the high compression cylinder. The flame velocity in the weak mixture, resulting from the association of rich mixture with air coming from the other cylinder, is increased by this flow and the associated turbulence. The gas characteristics during the combustion period in the cylinder are defined by an equation system deduced from energy, continuity, impulsion and state equations.

We have constructed a KOUCHOUL engine with a crankshaft with two cranks out of phase by 25 . The high compression cylinder has a volumetric ratio of 50 and is filled with air. The global volumetric ratio of the engine is 8. We have calculated and measured with an anemometry apparatus the flow velocity in the combustion chamber and transfer canal, the engine being motored. With a schieleren apparatus, in a special compression machine, we have observed the swirl created in combustion chamber by the air coming /3/ (Fig : 4).

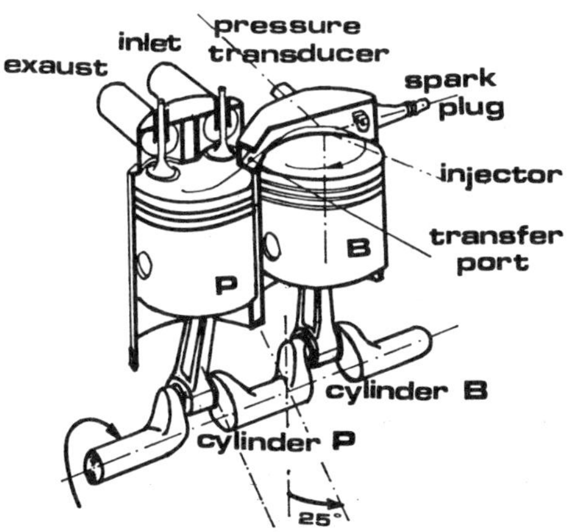

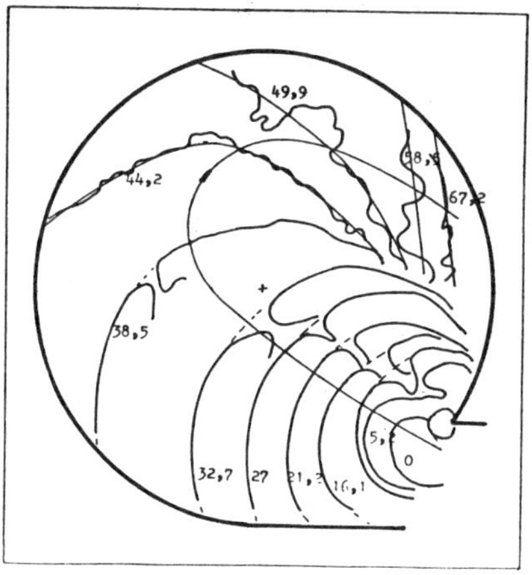

FIGURE: 3 FIGURE: 4 Swirl in combustion chamber
KOUCHOUL Engine
B : combustion cylinder
P : compression cylinder, C : Crankshaft.

In this engine, when air is coming from the transfer port in the combustion chamber, the flame velocity is accelerated and the combustion chamber pressure increase, then the air velocity and the flame velocity decrease. Consequently, the flame velocity is a function of the differential volumes variation and has alow dependence from mixture ratio.

* The relation between final combustion time t_f, global mixture ratio Rig and air flow velocity U_i in the transfer port is :

$$t_f = 24.3.Rig^{0.2}.(U_i(N)/U_i(1000))^{0.2}$$

with Rig < 1. $U_i(N)$ is the air velocity in the transfer port at the ignition angle for a N rotation per minute of the engine representative of air turbulence in the combustion chamber.

In this relation, the exponents of the global mixture ratio and the air velocity ratio are low. This show the insenbility of KOUCHOUL engine to the combustion of the weak mixture.

Exponent m of VIBE law is connected at the same parameters by this relation :

$$m = 13.3.Rig^{0.5}.(U(N)/U(1000))^{-0.5}$$

- When the global mixture ratio is decreasing to 1.2 or when the gas velocity in the transfer port is increasing, the flame velocity increases, as does the combustion velocity dx/dt with the consequence of decreasing the law VIBE exponent m .
- variation of air intake pressure and temperature is too low to appear in this relation.

5. ENCLOSED FIRES AT CONSTANT PRESSURE

To understand better the working of fire hearths inside homes it is necessary to study large volume combustion under well defined conditions.

M. LEBEY /7/ chose to try to improve the methods of analysis applied to fires by studying the evolution of the combustion of single fires in a room of 40 m^3 (4x4x2.5). He has shown through experimentation that the semi-empirical VIBE enables the description of the evolution of single enclosed fires under constant pressure conditions .
This relation which gives the evolution of burnt mass of fuel Mb(t) as a function of time t is written as :

$$M(t) = M(t_f).(1 - \exp(t/t_f)^{(m+1)})$$

M(t) is the combustible furnace mass at the time t measured with a weighing machine. t_f is the final combustion time which is extrapolled from the datas.

He has studied the combustion stages of several loads of wood comprising oak battens which were placed at regular intervals and in a fixed parallelepipedic volume inside a room with a controlled outlet. The length of sticks depending on the wood crib sizes. Ignition is created by a combustion of 40 cm^3 of alcohol located under the furnace.

The experimentations show :

- VIBE's law is defined from the first moments of combustion. The final period of combustion is not really defined because it is a cinder of wood combustion .

- exponent m is a function of the levels number N_e

$$m = 0,045.(N_e - 3.91)^2 +0.75$$

- final combustion (t_f) is a function of the initial ratio of wood and air in the room with the coefficient k :

$$t_f = (6.908.(m+1)/k)^{1/(m+1)}$$

Coefficient k , experimentally determined, is connected at the air/wood ratio R_i with the équation :

$$k = 1.9.(R_i - 0.2)^{-0.6}$$

R_i is the air-wood ratio calculated as this :

$$R_i = \frac{(\text{wood mass/room air mass})}{(\text{wood mass/room air mass})_{\text{stoechiometric}}}$$

- the temperature inside the room follow a law similar to that of VIBE.
- the lowest figure for exposant m is determined from the combustion of a sample of wood subjected to a constant radiation flux /8/.

Tab : 2

Combustion	m	t_f mn	W_M /mn	t_M mn	W_{Max} /mn
enclosed	1,254	73,98	0,0135	24,19	2,19
free	1,841	46,95	0,02129	20,41	2,21

The parameters m : VIBE law exponent and the combustion final time t_f are caracteristics of the combustion evolution of single wood crib fires located in given situations. The combustion of two furnaces free and enclosed has for consequences in the free combustion relative to the enclosed fire (Tab : 2)):

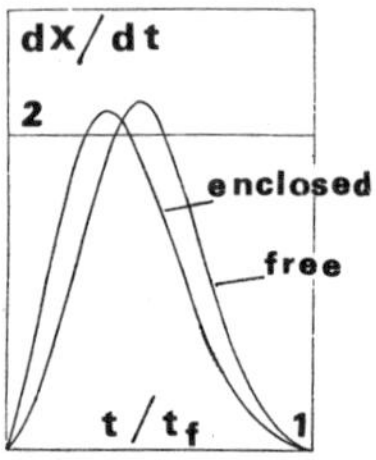

FIGURE 5 Combustion velocity for free or enclosed fire

- exponent m is higher and final combustion time t_f is lower,
- mean velocity of combustion W_M is higher,
- maximal combustion velocity W_{Max} is the same (Fig : 5).

Thus, it is possible, knowing at each moment, the burnt fraction or the radiative temperature evolution of the furnace, to predict :
- if the fire is in an extension period or in a extinguishing combustion period,
- the probable final combustion time on the measurement date.

6. CONCLUSIONS

The combustion of mixtures with air excess, enclosed in volume, may be defined by a VIBE law for which the exponent m and the final combustion time t_f is expressed by functions relative to the combustion physical parameters :

* Optimal values of m and t_f (1.5 and 50V) are rarely reached in DIESEL engine ($0 < m < 1.5$ and $\theta_f > 80V$) or in spark plug engine ($2.5 < m < 4.5$).

* Exponent m is connected to the mass of combustible parameter at the ignition time and after:

- for DIESEL engine : one parameter characteristic of the combustion chamber and one other of the "turbulence" in the engine , for example the rotation per minute,

- For spark plug engine : mixture homogeneous ratio and, as DIESEL engine, a "turbulence" parameter,
- for enclosed fire in room : the furnace height.

* the final combustion time is connected to the mixture ratio and to the engine rotation per minute.

In the case of KOUCHOUL engine, the influence of mixture ratio and "turbulence" is low.

REFERENCES

1. VIBE, Von I. Brennverlauf und Kreisprozess von Verbrennungs-motoren. VEB VERLAG TECHNIK. BERLIN
2.WOSCHNI G,ANISITIS F, Eine Methode zur Vorausberechnung des Anderung des Brennverlaufs mittelschnelllaufender Dieselmotoren bei geänderten Betriebsbedingungen.M.T.Z.,1973,34,n4,page 106,115.
3.ARQUES Ph. Les moteurs alternatifs à combustion interne. Editeur Masson. PARIS 1987.
4.ARQUES Ph. Flow and Flame propagation in the KOUCHOUL engine. Conf: Stratified charge automotive engine. The Inst. Mech.Engin.London. Grande Bretagne. november,25.1980.
5.ARQUES Ph.Swirl development in cylindrical combustion chamber of reciprocating engine.Soc.of Engin. Sciences 21st an. meeting.17pm9 1984.
6.ARQUES Ph. Cyclic pressure variation and flow in a KOUCHOUL engine. Preprint XXIIIst meeting Society of Engineering Sciences.Buffalo (New York) .USA. August 25,27 1986.
7.LEBEY M. and ARQUES Ph. A study of the combustion of single fires aiming at a better knowledge of accidental fires.XXIst Meeting Society of Engineering Science Inc. Blacksburg. USA. October,15.1984.
8.VOYELLE C,AKRICH C,DELBOURGO R. Comportement du bois dans un incendie. Entropie. N 87, 1979.

EXPERIMENTAL AND NUMERICAL STUDY OF A HEATED TURBULENT ROUND JET

E.M. BAHRAOUI*, G. BRUN***, D. JEANDEL**, L. FULACHIER*
* Institut de Mécanique Statistique de la Turbulence
12, avenue du Général Leclerc - 13003 MARSEILLE - FRANCE
** Lab. de Mécanique des Fluides & d'Acoustique - Ecole Centrale de Lyon
36 Avenue Guy de Collongue - 69131 ECULLY CEDEX - FRANCE
*** Société METRAFLU -64 Chemin des Mouilles - 69134 ECULLY CEDEX - FRANCE

1. INTRODUCTION

A comparison between experimental and numerical results is presented for a heated turbulent round jet.

The main purpose of the experiments is to study the effects of strong temperature gradients on the structure of a turbulent jet. Experiments are also carried out on the slightly heated turbulent jet for comparison. The mean velocity, the mean temperature, and the temperature fluctuations are measured. The Reynolds stress and the turbulent lateral heat flux are inferred from these profiles by integration of the momentum and enthalpy equations. The eddy diffusivities for momentum and heat are also deduced.

Numerical predictions are obtained by a finite element simulation of the flow. The k-ε model is based on Favre decomposition of velocity and temperature. A constant turbulent Prandtl number is used. A semi-implicit time discretization is used and the continuity equation is satisfied by an Uzawa algorithm coupled with a conjugate gradient method. These numerical procedures are described in the second part of this paper.

Comparison between experimental and numerical results are presented in the last part.

2. EXPERIMENTAL ARRANGEMENTS

2.1. Flow facility

Experiments are carried out in an axisymmetric, vertical turbulent air jet issuing from an annular nozzle (fig.1 ; outside diameter, D = 25,3 mm ; inside diameter 18,2 mm ; air flow rate, Q = 15 m^3/hr ; Reynolds number Re = 8 200). A weak flow rate of an air (q_a) - propane (q_g = 0,04 m^3/hr) mixture can be injected into the central pipe (exit diameter d' = 7 mm). For q_a = 0,73 m^3/hr, stoichiometric proportions are almost realized : the premixed flame appears to be laminar (L.F). It is attached to the inside edge of the central tube R_{ed} = 1 080 or $R_{ed'}$ = 2 800 ; $\theta_m \simeq$ 1 850 K at x/D = 1,4). In the case of the laminar flame, the annular turbulent jet does not appear to contribute to the combustion. Indeed, if the annular flow is stopped, the laminar flame remains almost unchanged (fig.2). Thus, the external turbulent flow is only heated by the flame (F.C).

In a second experiment (P.C), considered as a reference, the annular flow is slightly heated by means of an electrical resistance, without combustion (Q $\simeq$ 15 m^3/h, q_a = q_g = 0). The nominal value of the mean temperature at the jet exit is 22 K above the ambiant. Heat can be considered as a passive contaminant in this case (P.C).

C.-M. Brauner and C. Schmidt-Lainé (eds.), Mathematical Modeling in Combustion and Related Topics, 365–373.

More details can be found in Bahraoui (1981).

2.2. Instrumentation

The mean longitudinal velocity* is measured with a special Pitot tube (mean temperature < 900 K ; inside diameter of total pressure tube = 0,6 mm). A reference static pressure tube is located outside the jet flow to take into account the radial variation of the static pressure. The density variation is obtained from the mean temperature profiles. We obtain with a good approximation (Bahraoui 1981) : $\hat{u} = (2\tilde{T}\Delta P/\rho_e T_e)^{\frac{1}{2}}$; ΔP, pressure difference ; ρ , air density at T ; subscript e, refers to the exterior of the jet.

The mean temperature $\tilde{T}$ is measured with thermocouples (Ch.Chr.Const. d = 0.15) and cold wires (d = 0.63 μm, Pt). It is quite difficult to know if these thermometers measure $\tilde{T}$ or $\bar{T}$, but the difference between the two temperatures has been found less than 8% at 900 K in our experiments.

Two types of thermocouples are used : a 1 mm diameter Platinum/Platinum 10% Rhodium thermocouple, in order to obtain a rough measurement of the flame temperature, and a more sensitive 0.15 mm diameter Chromel-Chromel-Constantan thermocouple for temperature less than 1 300 K. The catalysis and radiation effects have been accounted for (Trinite 1980 and Bahraoui 1981). The mean temperature $\tilde{T}$ is also measured with Platinum cold wires operated with a constant intensity. A 0.6 μm diameter, 0.4 mm long wire operated with a 0.12 mA current intensity could be used until the station x/d = 2.9. The temperature profiles obtained with wires and thermocouples are practically identical.

The temperature fluctuations are measured with the cold wire (0,63 diameter). Both the time constant (15.10^{-6} s) and the contamination by the velocity fluctuations are negligible.

At low frequencies (n $\leqslant$ 10 Hz) the contribution to the variance is diminished because of the end effects : the error on $(\overline{\theta'^2})^{\frac{1}{2}}$ is about 10 % for 0,63 μm diameter cold wire.

2.3. Reynolds stress and turbulent heat transfer

The Reynolds stress and the lateral component of the turbulent heat transfer are deduced from momentum and enthalpy equations, taking into account the continuity equation. These equations are written for sections above the flame where the fluid can be considered as heated air with no chemical reaction. For the present experiments, the momentum and enthalpy equations (Favre 1975) may be written, to a good approximation (Bahraoui 1985), as follows.

$$\bar{\rho}\,\tilde{u}\,\frac{\partial \tilde{u}}{\partial x} + \bar{\rho}\,\tilde{v}\,\frac{\partial \tilde{u}}{\partial r} = -\frac{1}{r}\frac{\partial}{\partial r}\left(r\,\overline{\rho u''v''}\right) - (\bar{\rho}-\rho_e)g$$

$$\bar{\rho}\,\tilde{u}\,\frac{\partial \tilde{\theta}}{\partial x} + \bar{\rho}\,\tilde{v}\,\frac{\partial \tilde{\theta}}{\partial r} = -\frac{1}{r}\frac{\partial}{\partial r}\left(r\,\overline{\rho\theta''v''}\right)$$

V is the radial velocity and g the acceleration due to gravity.

The integrations are performed at the distances x/D = 4,5 ; x/D = 8 and x/D = 11). Reynolds stress and heat flux profiles are obtained in both experiments (F.C and P.C). The variations close to the axis are deduced from expansions around $r = 0$.

---

* We use Favre averaging (i.e. mass weighted averaged quantites) defined by $\xi = \tilde{\xi} + \xi''$ with $\overline{\rho\xi''} = 0$.

From the previous results, the eddy viscosity ν_t and the eddy thermal diffusivity α_t which are defined by :

$$-\overline{\rho u''v''} = \bar{\rho}\nu_t \frac{\partial \tilde{u}}{\partial r} \qquad -\overline{\rho\theta''v''} = \bar{\rho}\alpha_t \frac{\partial \tilde{\theta}}{\partial r}$$

may be deduced

These definitions generalize the classical ones used when heat can be considered as a passive contaminant.

3. NUMERICAL METHOD

3.1. Governing equations

The numerical simulation of the jet has been obtained by a finite element technique.

The classical conservation equations are written in a dimensionless form by introduction of the flow characteristic scales. A state equation is deduced from the perfect gas equation, supposing that pressure variations have no influence on density. In a dimensionless form, this equation reduces to a direct relation between temperature and density.

$$\rho = \frac{1}{T}$$

The classical Reynolds decomposition is applied to pressure and density, whereas Favre decomposition is applied to velocity and temperature. For these last variables, we then write (cf[3])

$$\xi(\vec{x},t) = \tilde{\xi}(\vec{x},t) + \xi''(\vec{x},t)$$

$$\text{with} \quad \tilde{\xi}(x,t) = \frac{\overline{\rho\,\xi}}{\bar{\rho}}$$

The use of these decompositions leads to mean equations where the density weighted Reynolds stress tensor $\widetilde{u''\otimes u''}$ and the turbulent temperature fluxes $\widetilde{u''\theta''}$ are appearing. These quantities are expressed, following Jones [5], as :

$$\bar{\rho}\,\widetilde{\vec{u}''*\vec{u}''} = \frac{2}{3}\left(\bar{\rho}k + \frac{1}{Rt}\operatorname{div}\vec{\tilde{u}}\right)I - \frac{1}{Rt}\left(\overrightarrow{\text{grad}}\,\vec{\tilde{u}} + \overrightarrow{\text{grad}}^{t}\,\vec{\tilde{u}}\right)$$

$$\bar{\rho}\,\widetilde{\vec{u}''*\theta''} = -\frac{1}{Rt\,\sigma_T}\,\overrightarrow{\text{grad}}\,\tilde{T}$$

The turbulent Reynolds number R_t is given by :

$$\frac{1}{Rt} = C_\mu\bar{\rho}k^2/\varepsilon$$

and the turbulent Prandtl number Pr_t is supposed constant. k is the turbulent kinetic energy defined as : $k = \frac{1}{2}\widetilde{u_i''u_i''}$ and ε is its dissipation rate. Modelled equations for k and ε are obtained from equations used in incompressible case.

A closed system of equations may then be written for the mean momentum $\vec{Q} = \bar{\rho}\tilde{\vec{u}}$, a generalised pressure $\bar{p}^*$, the mean temperature $\tilde{T}$, the mean density $\bar{\rho}$, and the scalar turbulent scales k and ε.

This closed system is :

$$\bar{\rho} = \frac{1}{\tilde{T}}$$

$$\frac{\partial \bar{\rho}}{\partial t} + \mathrm{div}\ \vec{Q} = 0$$

$$\frac{\partial \vec{Q}}{\partial t} + \mathrm{div}\left(\frac{\vec{Q}}{\bar{\rho}} * \vec{Q}\right) = -\vec{\mathrm{grad}}(\bar{p}^{*}) + \mathrm{div}\left(\left(\frac{1}{Re} + \frac{1}{Rt}\right)\left(\vec{\mathrm{grad}}\ \frac{\vec{Q}}{\bar{\rho}} + \vec{\mathrm{grad}}^{t}\ \frac{\vec{Q}}{\bar{\rho}}\right)\right)$$

$$\text{with} \quad \bar{p}^{*} = \bar{p} - \frac{2}{3}\left(\left(\frac{1}{Re} + \frac{1}{Rt}\right)\mathrm{div}\ \frac{\vec{Q}}{\bar{\rho}} + \bar{\rho}k\right)$$

$$\frac{\partial \tilde{T}}{\partial t} + \frac{\vec{Q}}{\bar{\rho}} \cdot \vec{\mathrm{grad}}(\tilde{T}) = \frac{1}{\bar{\rho}}\ \mathrm{div}\left(\left(\frac{1}{RePr} + \frac{1}{Rt\sigma_T}\right) \vec{\mathrm{grad}}\ \tilde{T}\right)$$

$$\frac{\partial k}{\partial t} + \frac{\vec{Q}}{\bar{\rho}} \cdot \vec{\mathrm{grad}}(k) = \frac{1}{\bar{\rho}}\ \mathrm{div}\left(\left(\frac{1}{Re} + \frac{1}{Rt\sigma_k}\right) \vec{\mathrm{grad}}\ k\right) + P - \varepsilon$$

$$\frac{\partial \varepsilon}{\partial t} + \frac{\vec{Q}}{\bar{\rho}} \cdot \vec{\mathrm{grad}}(\varepsilon) = \frac{1}{\bar{\rho}}\ \mathrm{div}\left(\left(\frac{1}{Re} + \frac{1}{Rt\sigma_\varepsilon}\right) \vec{\mathrm{grad}}\ \varepsilon\right) + C_1 \frac{\varepsilon}{k} P - C_2 \frac{\varepsilon^2}{k}$$

with

$$P = \frac{1}{\bar{\rho}}\left(\frac{1}{Rt}\left(\vec{\mathrm{grad}}\ \frac{\vec{Q}}{\bar{\rho}} + \vec{\mathrm{grad}}^{t}\ \frac{\vec{Q}}{\bar{\rho}}\right) - \frac{2}{3}\left(\frac{1}{Rt}\ \mathrm{div}\ \frac{\vec{Q}}{\bar{\rho}} + \bar{\rho}k\right)\vec{I}\right) * \vec{\mathrm{grad}}\ \frac{\vec{Q}}{\bar{\rho}}$$

and where the values of the constants are :

$C_\mu = 0.09 \quad C_1 = 1.44 \quad C_2 = 1.92 \quad \sigma_k = 1.0 \quad \sigma_\varepsilon = 1.3$

Temperature fluctuations, namely $\widetilde{\theta''^2}$, are calculated by equation :

$$\frac{\partial \widetilde{\theta''^2}}{\partial t} + \frac{\vec{Q}}{\bar{\rho}} \cdot \vec{\mathrm{grad}}(\widetilde{\theta''^2}) = \frac{1}{\bar{\rho}}\ \mathrm{div}\left(\left(\frac{1}{RePr} + \frac{1}{Rt\sigma_{\theta_2}}\right) \vec{\mathrm{grad}}\ \widetilde{\theta''^2}\right) + C_{\theta_1} P_\theta - C_{\theta_2} \frac{\varepsilon}{k} \widetilde{\theta''^2}$$

where $P_\theta = \frac{1}{\bar{\rho}}\left(\frac{1}{R_t\sigma_t}\left(\vec{\mathrm{grad}}\ \tilde{T} \cdot \vec{\mathrm{grad}}^{t}\ \tilde{T}\right)\right)$

As for incompressible case, the values of the constants are :

$C_{\theta_1} = 1. \quad C_{\theta_2} = 2. \quad \sigma_{\theta_2} = 1.$

3.2. Numerical method

The equation described above are discretised using a semi implicit time scheme where non linear terms are linearised. A weak formulation of the problem is obtained by multiplication by test functions, integration over the calculation domain and application of Green's Theorem to the highest order terms.

Two triangulations of the domain are then used : a coarse one where pressure is approximated by piecewise linear fonctions and a fine one where other variables are approximated linearly.

These two triangulations are related by the following scheme :

The Galerkin formulation of the problem leads to matrix systems. For scalar quantities, these systems are solved using L.D.U decomposition of the matrix.
The momentum equation is coupled with the continuity equation. This is satisfied by minimising the integral quantity :

$$\int (r_h)^2 \, d\Omega \qquad \text{with } r_h = \sum_{j=1}^{nn2} r_j \, N2_j$$

where r_h is defined by :

$$\int r_h \, N2_i \, d\Omega = \int (\mathrm{div}\vec{Q}_h - S) \, N2_i \, d\Omega \qquad \forall i = 1, nn2$$

(N2i : basic function and nn2 total number of nodes of coarse mesh).
This minimisation is performed by a conjugate gradient method proposed by Buffat [6].

3.3. Calculation of the heated turbulent jet

To simulate the experimental jet described in part I, a calculation domain, beginning at x/D = 4.5 and extending to x/D = 12.5 has been choosen. In the initial section, experimental profiles are imposed as boundary conditions. The maximum velocity is about 15 m/s. For weakly heated jet, the maximum temperature is about 15°C. Thus, the passive scalar hypothesis is applied and density is taken as a constant. The strongly heated jet corresponds to a maximum temperature variation of 270°C. Results presented here have been obtained with a turbulent Prandtl number value of 0.5. This relatively weak value seems to give better numerical results than a higher one.

4. EXPERIMENTAL AND NUMERICAL RESULTS

Comparisons between experimental and numerical results are presented on figures (3) to (16).
Figures (3) to (8), show comparisons of mean quantities. The velocity and temperature axial decays are correctly predicted. Nevertheless, a deviation is noted on temperature decay for both the wealky heated and the strongly heated cases. For this last one, the deviation has a influence on velocity axial decay. Velocity and temperature radial profiles show also an almost good agreement between experimental and numerical data. The temperature behaviour close to the axis seems to be better predicted for the weakly heated jet.

Results for turbulent quantities are presented on figures (9) to (16). They concern radial profiles of turbulent viscosity, Reynolds stress, turbulent temperature flux and fluctuations of temperature.
Turbulent viscosity profiles are underpredicted by numerical simulation. The deviation is much greater in the strongly heated case despite the correct prediction of the profile global shape. Simultaneously there is some discrepancy between predicted and measured Reynolds stress.
For turbulent temperature flux, the profile shape general behaviour is quite good. The location of the maximum is accurately predicted but its value is not in agreement with the measured one.
The temperature fluctuations $\sqrt{\overline{\theta^2}}$ are shown on figures (15) and (16). For station x/D = 8, many features of the measured data are well predicted (behaviour near the axis), but, far downstream, there are large descripancies on the levels and on the profile global shapes. It seems that the equilibrium between production and dissipation terms is badly modelled.

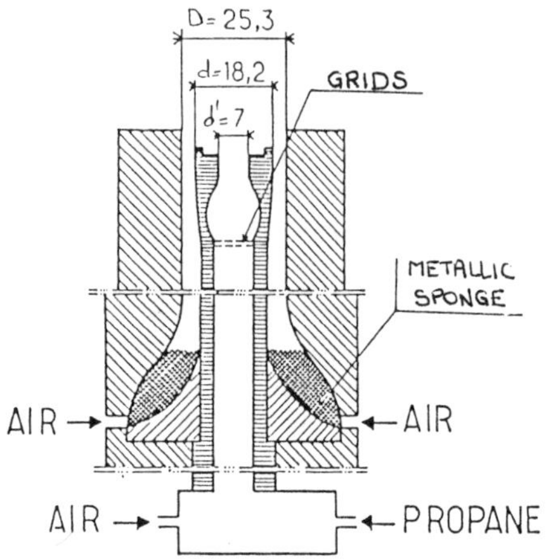

Fig. 1 : Burner

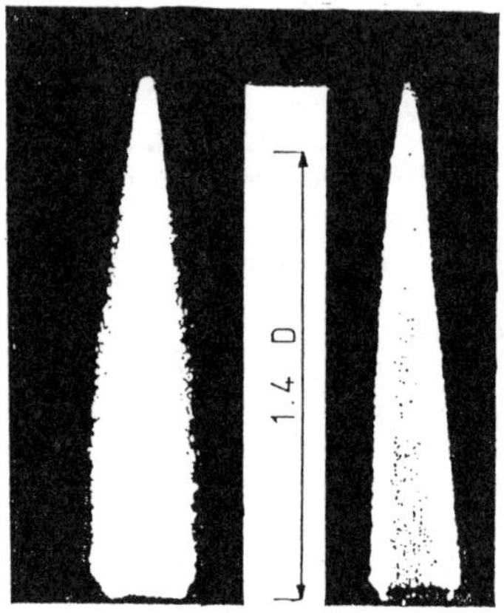

Fig. 2 : LAMINAR FLAME LAMINAR FLAME WITHOUT ANNULAR FLOW

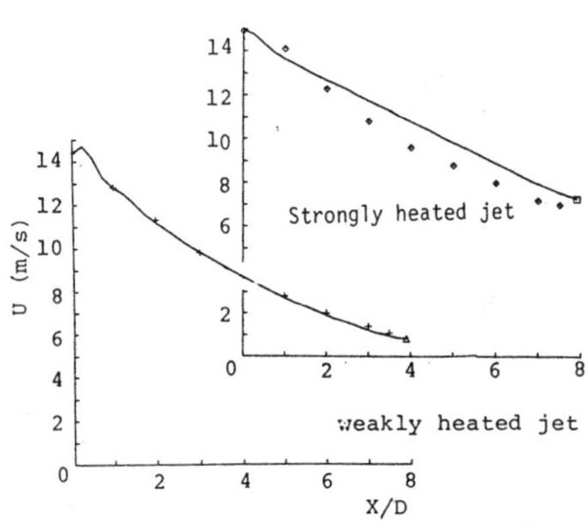

Fig 3: AXIAL DECAY OF VELOCITY

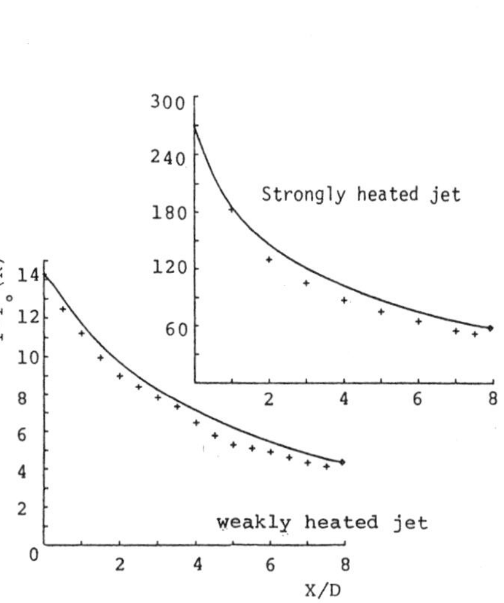

Fig 4 : AXIAL DECAY OF TEMPERATURE

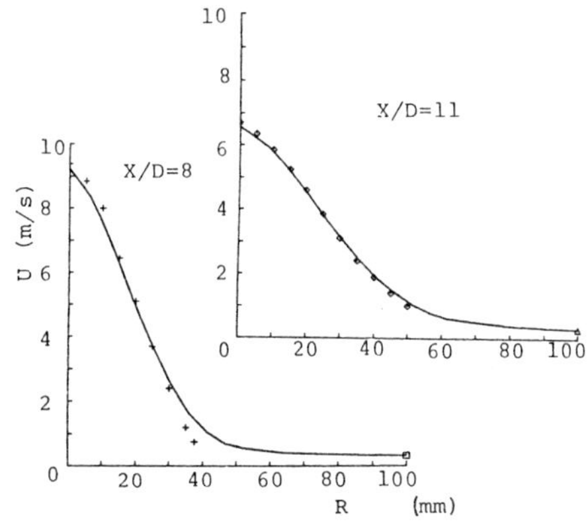

Fig 5: Weakly heated jet

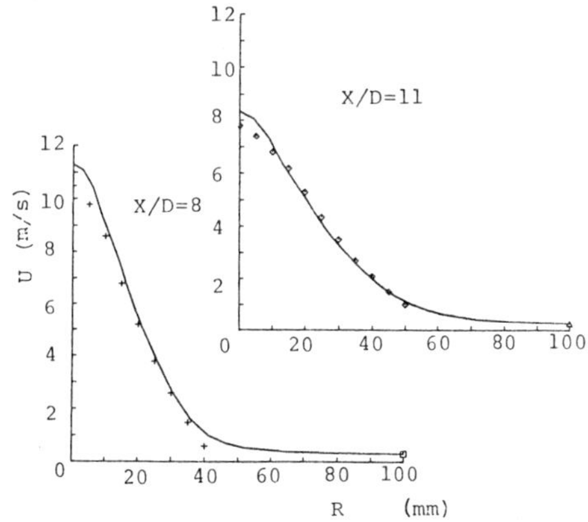

Fig 6: Strongly heated jet

MEAN VELOCITY PROFILES

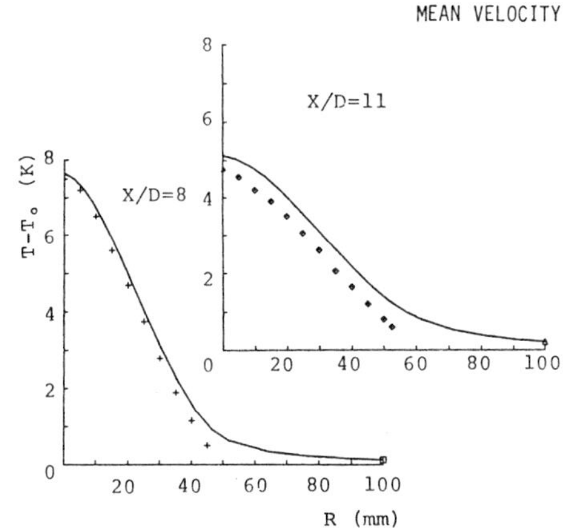

Fig 7: Weakly heated jet

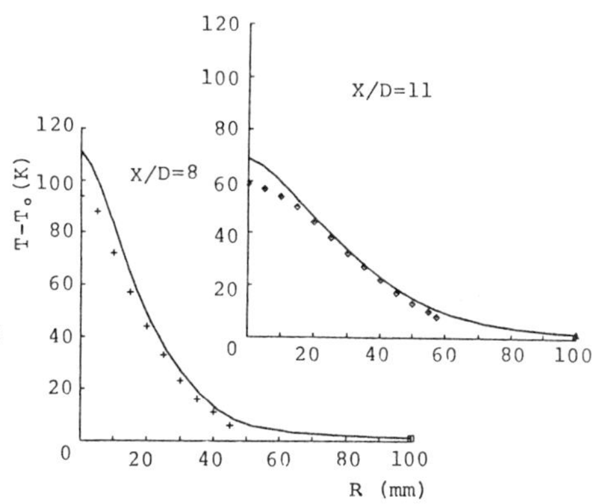

Fig 8: Strongly heated jet

MEAN TEMPERATURE PROFILES

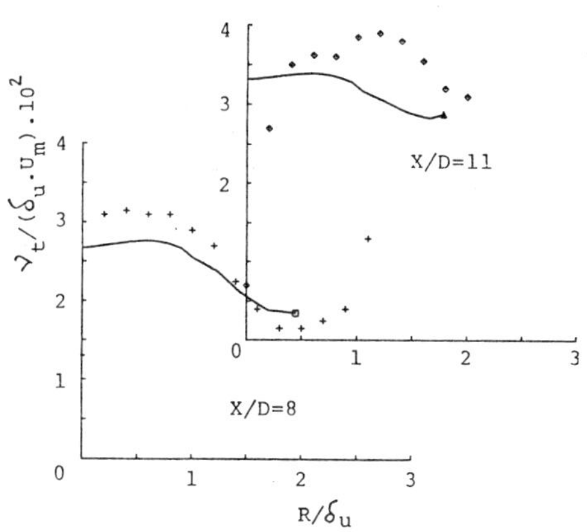

Fig 9: Weakly heated jet

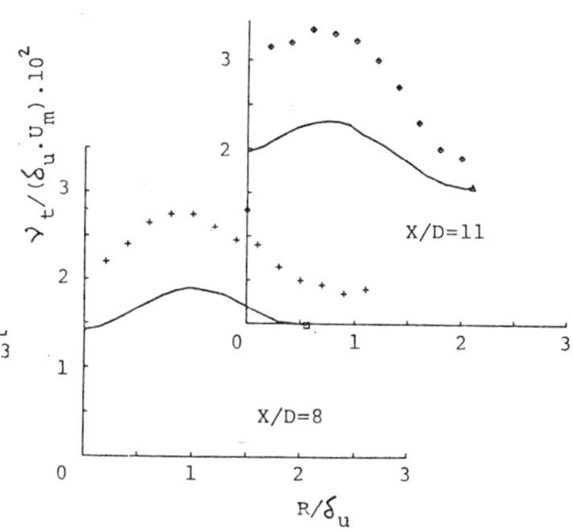

Fig 10: Strongly heated jet

TURBULENT VISCOSITY PROFILES

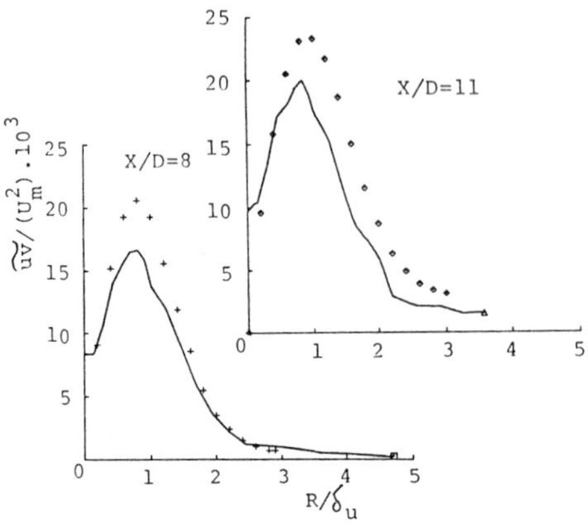

Fig 11: Weakly heated jet

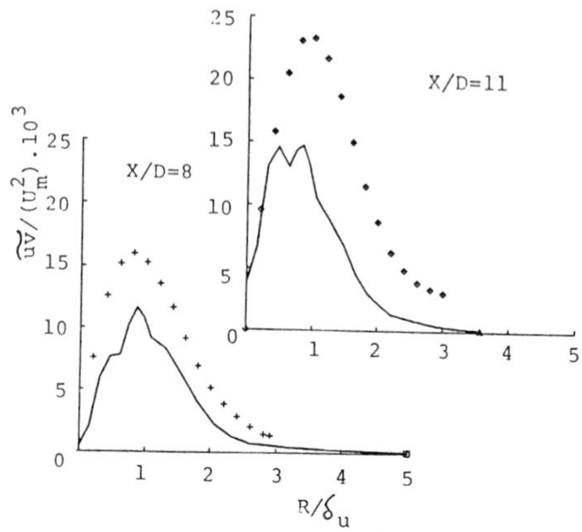

Fig 12: Strongly heated jet

REYNOLDS STRESS PROFILES

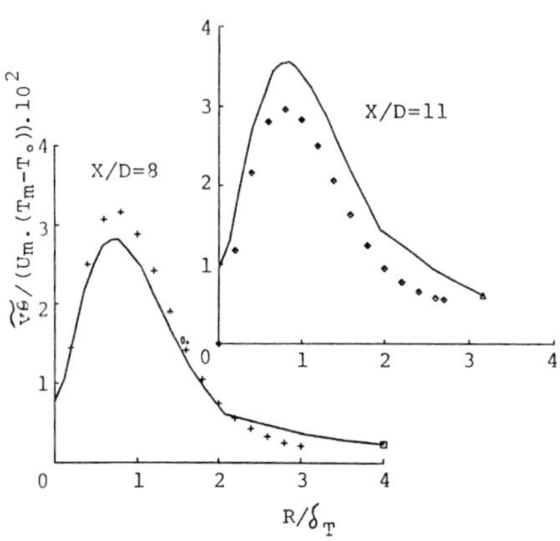

Fig 13: Weakly heated jet

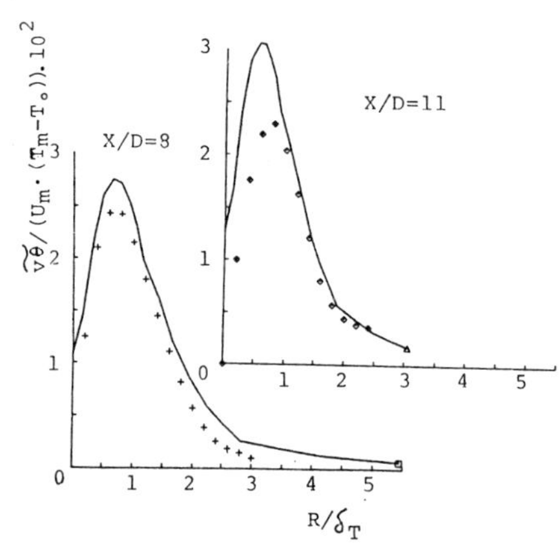

Fig 14: Strongly heated jet

TURBULENT TEMPERATURE FLUX PROFILES

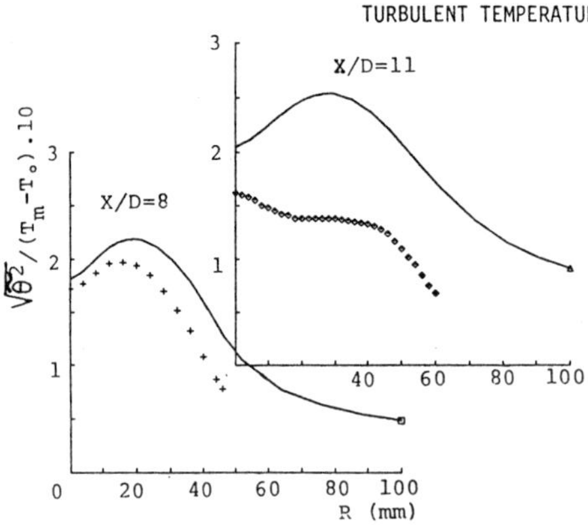

Fig 15: Weakly heated jet

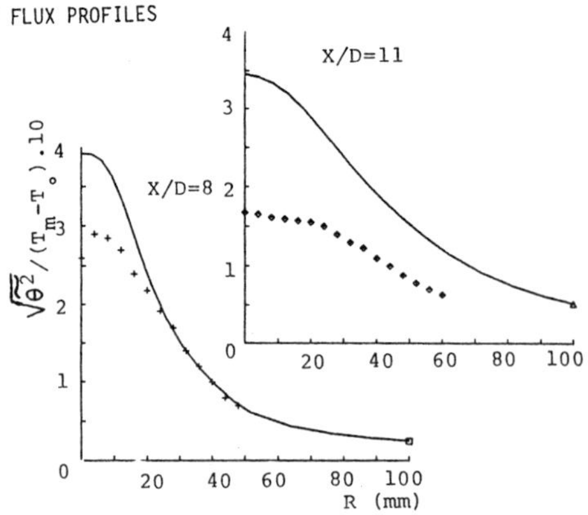

Fig 16: Strongly heated jet

FLUCTUATIONS OF TEMPERATURE

5. CONCLUSION

A finite element method has been applied to test classical turbulent assumptions when they are extended to variable density flows. The first results obtained in the heated jet are encouraging. The model involves a large number of assumptions. Two kinds of errors may arise : those encountered in isothermal flows and those introduced by variable density.

The results seem to be in good agreement with data for mean variables but discrepancies arise for turbulent quantities.

The use of a better modelisation of turbulent temperature flux seems to be necessary to obtain better results. In particular experimental results (cf [4]) show that the constant turbulent Prandtl number hypothesis is not a good hypothesis in this case. The use of the calculation procedure in other experimental configurations such as heated boundary layer (cf Cheng [7]) will give us other indications on the validity of the assumptions.

REFERENCES

1. BAHRAOUI EM :" Contribution à l'étude d'un écoulement turbulent en présence de combustion".Thèse de 3ème cycle, I.M.S.T, université d'Aix-Marseille II FRANCE, 108 pp, 1987.
2. TRINITE M :" A method for simultaneous measurements of concentration and temperature fluctuations in air-hydrogen flame". Rep. Labo. Méc. des Fluides de l'Ecole Centrale de Lyon, Ecully, France, Euromech 132, VI 5-VI 5C, 1980.
3. FAVRE A :" Equations statistiques des fluides turbulents compressibles". Proceeding of the fifth canadians congress of applied mechanics, Fredericton, G3 - 34, 1975.
4. BAHRAOUI EM, FULACHIER L : "Quelques résultats sur un jet turbulent axisymétrique fortement chauffé". Journées d'étude "Ecoulements turbulents à masse volumique variable". I.M.S.T, Marseille, FRANCE, p15-37, 1985.
5. JONES WP, Mc GUIRK JJ :" Mathematical modelling of gas-turbine combustion chambers". Agard CP 275, 1979.
6. BUFFAT M :" Finite element analysis on microcomputer of laminar flow". 2nd International conference on microcomputer in engng. Development and application of software.Swansea, 1986.
7. CHENG RK, NG TT : "Some aspects of strongly heated turbulent boundary layer flow". Physics of fluid 25 (8), August 1982.

RELEVANCE OF COMPUTATION METHODS IN COMBUSTION TO THE DEVELOPMENT OF TURBOREACTOR FURNACES.

G. BAYLE-LABOURE

Département combustion SNECMA - Centre de Villaroche

Summary:

Developments in turboreactor cycles, the need to conform to pollution standards, the improvement in manoevrability, life, and the reduction in the size and mass of combustion chambers have lead to increasingly expensive partial tests.

This is why engine designers have worked out computational methods for sizing and optimising the various elements with a limited number of tests, and for extrapolating the results obtained into the whole sphere of flights (Plate 1).

As regards combustion the physical phenomena at work are particularly complex: the flow is compressible, turbulent and recirculating, and in most cases purely three-dimensional; the local richness of the air-fuel mixture is very variable in the furnace, and the chemical reactions which bring many molecular species into play have very different speeds. Diphase phenomena (the atomization and vaporization of the fuel) cannot be overlooked. Moreover, the considerable interaction between turbulent and chemical kinetics is poorly understood from the fundamental point of view (plate 2).

In spite of these difficulties, manufacturers and SNECMA in particular have perfected computational methods which are now operational dimensioning tools. The relevance of these tools is illustrated by the three examples of methods presented below.

C.-M. Brauner and C. Schmidt-Lainé (eds.), Mathematical Modeling in Combustion and Related Topics, 375–382.

INTERET DES METHODES DE CALCULS EN COMBUSTION DANS LE DEVELOPPEMENT DES FOYERS DE TURBOREACTEURS

G. BAYLE-LABOURE (*)

Département Combustion SNECMA - Centre de Villaroche

L'évolution des cycles des turboréacteurs, la nécessité de satisfaire des normes de pollution, l'amélioration de la pilotabilité et de la durée de vie ainsi que la réduction de l'encombrement et de la masse des chambres de combustion conduisent à des essais partiels de plus en plus coûteux.

C'est pourquoi les motoristes ont développé des méthodes de calculs afin de dimensionner et d'optimiser les différents éléments avec un nombre limité d'essais et d'extrapoler les résultats obtenus dans l'ensemble du domaine de vol (planche 1).

Dans le cas de la combustion, les phénomènes physiques mis en jeu sont particulièrement complexes : l'écoulement est compressible, turbulent recirculant et dans la plupart des cas purement tridimensionnel, la richesse locale du mélange air-carburant est très variable dans le foyer, et les réactions chimiques qui mettent en jeu de nombreuses espèces moléculaires ont des vitesses très dispersées. Les phénomènes diphasiques (pulvérisation, vaporisation du carburant) ne peuvent pas être négligés. Par ailleurs, la forte interaction entre la turbulence et la cinétique chimique est encore mal connue sur le plan fondamental (planche 2).

Malgré ces difficultés, les constructeurs et la SNECMA en particulier ont mis au point des méthodes de calculs qui constituent aujourd'hui des outils de dimensionnement opérationnels. L'intérêt de ces outils est illustré par 3 exemples de méthodes présentées ci-dessous.

1er EXEMPLE

CALCUL DE L'INTERACTION DIFFUSEUR / TETE DE CHAMBRE

Pour garantir une bonne alimentation des orifices du tube à flamme, la qualité de l'écoulement dans le diffuseur et au contournement du tube à flamme doit être assurée.

Une méthode de calcul bidimensionnelle aux éléments finis mise au point par l'Ecole Centrale Lyonnaise est utilisée à la SNECMA pour efféctuer cette vérification. La planche 3 présente l'application de cette méthode à l'optimisation de la forme des casquettes d'une chambre de combustion.

La validation de cette méthode a été effectuée par comparaison à des résultats obtenus en analogie hydraulique et par des mesures aérodynamiques.

(*) Contributor to the Panel on the modeling of industrial problems.

La version 1 à casquettes courtes présente un décollement alors que la version 2 à casquettes longues optimisées donne une aérodynamique saine qui permet en particulier de garantir une bonne alimentation des films de refroidissement.

L'utilisation d'une telle méthode au stade de la mise au point permet d'éviter des essais coûteux et générateurs de délais.

2ème EXEMPLE

CALCUL DES PERFORMANCES DE COMBUSTION

Les méthodes purement tridimensionnelles en combustion ne peuvent pas encore aujourd'hui être considérées comme réellement prédictives. Au stade du dimensionnement d'une chambre de combustion, une méthode monodimensionnelle d'assemblage de foyers homogènes est maintenant couramment utilisée. Elle offre l'avantage de rendre possible l'utilisation d'une modélisation détaillée de la cinétique chimique (60 réactions - 25 espèces) indispensable pour prédire correctement les performances de rendement et de stabilité dans la très large plage de richesses rencontrée à l'intérieur des foyers. La comparaison de cette méthode à des résultats expérimentaux est présentée sur les planches 5 et 6.

Cette méthode permet une optimisation du volume global de la chambre ainsi que de la répartition des débits par les différents orifices. Elle permet également après recoupement d'un certain nombre de points d'essais d'extrapoler les résultats dans l'ensemble du domaine de vol et des régimes "moteur".

3ème EXEMPLE

OPTIMISATION DE LA CARBURATION D'UNE RECHAUFFE

Dans le cas d'une réchauffe, l'approximation bidimensionnelle axisymétrique est acceptable en première approche.

Une méthode aux volumes finis mise au point par l'ONERA est utilisée (planche 4). La turbulence est décrite par un modèle $k - \varepsilon$ et la combustion turbulente par un modèle Eddy Break up. Cette méthode est en particulier, utilisée pour optimiser la carburation de la veine. Le carburant est injecté ponctuellement dans le maillage à l'emplacement des rampes et le calcul permet de décrire sa dispersion dans le canal. La répartition du débit de carburant entre les rampes d'injection peut être ainsi ajustée grâce à cette méthode évitant ainsi de nombreux essais d'optimisation.

CONCLUSION

Malgré la complexité des phénomènes mis en jeu, des méthodes de calculs en combustion pouvant être considérées comme prédictives sont aujourd'hui opérationnelles à la SNECMA.

L'amélioration de ce type de méthodes passe par une description de plus en plus fine de phénomènes physiques élémentaires tels que la turbulence, la cinétique chimique ou les écoulements diphasiques. Au-delà des problèmes physiques et numériques posés par ce type de modélisation, les méthodes de calculs en combustion nécessitent pour leur validation, la mise au point de méthodes de mesures optiques non intrusives utilisables à l'intérieur des foyers (planche 6).

L'utilisation industrielle de méthodes de calculs conduit à l'intégration de chaque code dans un environnement informatique plus vaste permettant d'assurer des conditions d'utilisation souples et rapides. Enfin, une validation approfondie de chaque code doit être réalisée par confrontation à des résultats expérimentaux afin de définir leur domaine d'utilisation et leur niveau de fiabilité (planche 7).

INTÉRÊTS DES MÉTHODES DE CALCULS EN COMBUSTION dans le développement des foyers de turboréacteurs 1

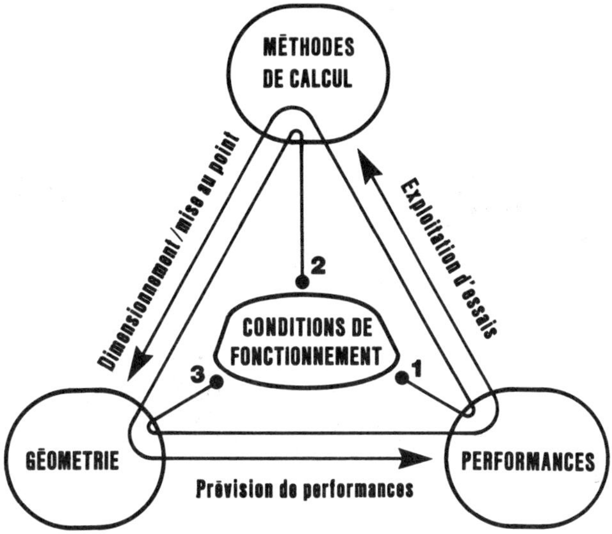

MODÉLISATION DE LA COMBUSTION DANS LES FOYERS DE TURBORÉACTEURS difficultés du problème 2

- CONDITIONS DE FONCTIONNEMENT TRÈS VARIABLES
  - **RÉGIME MOTEUR**
  - **CONDITIONS DE VOL**
- GÉOMETRIE
  - **FORTEMENT TRIDIMENSIONNELLE**
  - **COMPLEXE**
- DIFFICULTÉ DES MESURES INTERNES AUX FOYERS
- COMPLEXITÉ DES PHÉNOMÈNES PHYSIQUES MIS EN JEU
  - **ÉCOULEMENT TURBULENT ET RECIRCULANT**
  - **CINÉTIQUE CHIMIQUE**
  - **PHÉNOMÈNES DIPHASIQUES**
  - **PHÉNOMÈNES THERMIQUES**
  - **PHÉNOMÈNES TRANSITOIRES ET INSTATIONNAIRES**

OPTIMISATION DE LA TÊTE D'UNE CHAMBRE DE COMBUSTION

3

LIGNES DE COURANT

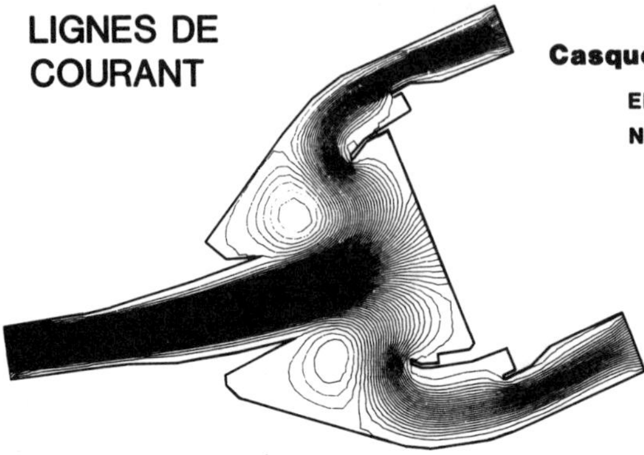

Casquettes courtes

ELEMENTS : 2524

NOEUDS : 1404

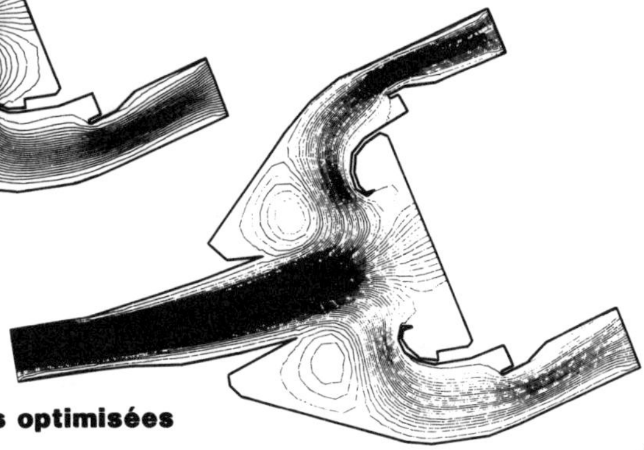

ELEMENTS : 2308

NOEUDS : 1296

Casquettes longues optimisées

CALCUL D'UNE RÊCHAUFFE

4

CHAMP DE VITESSES

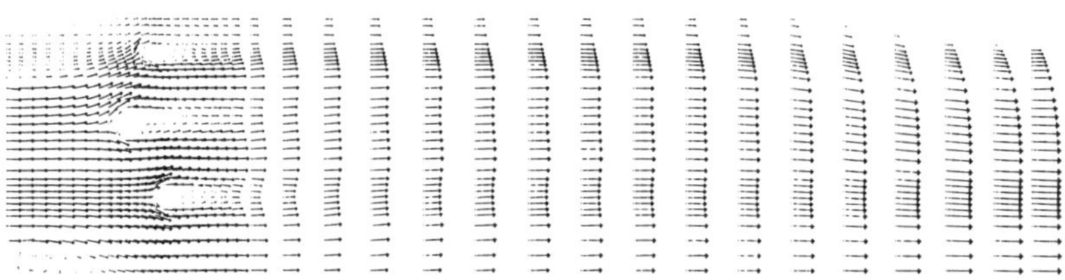

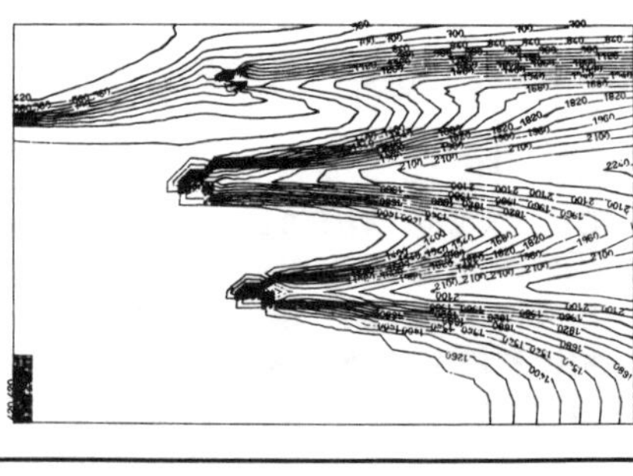

ISOTHERMES

CALCUL D'UNE CHAMBRE DE COMBUSTION PAR ASSEMBLAGE DE FOYERS HOMOGĒNES 5

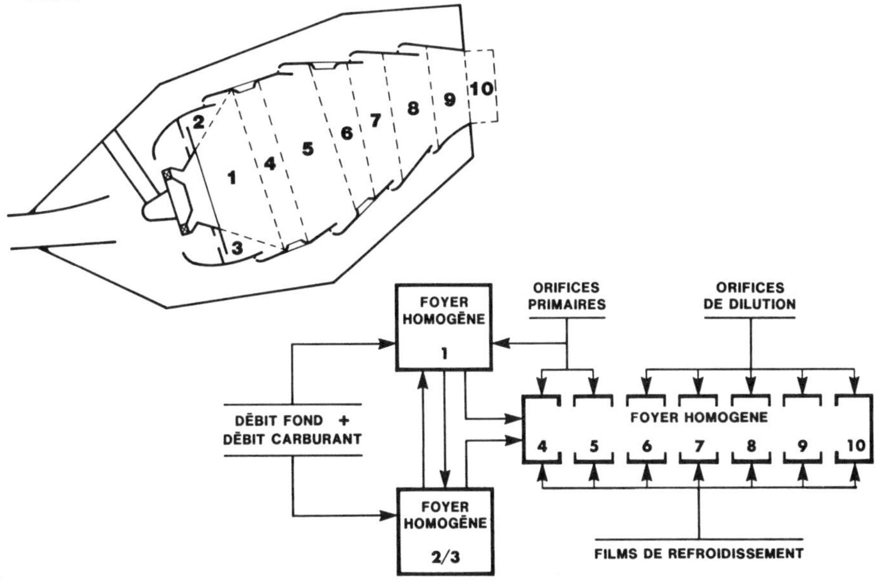

CALCUL D'UNE CHAMBRE DE COMBUSTION PAR ASSEMBLAGE DE FOYERS HOMOGĒNES 6

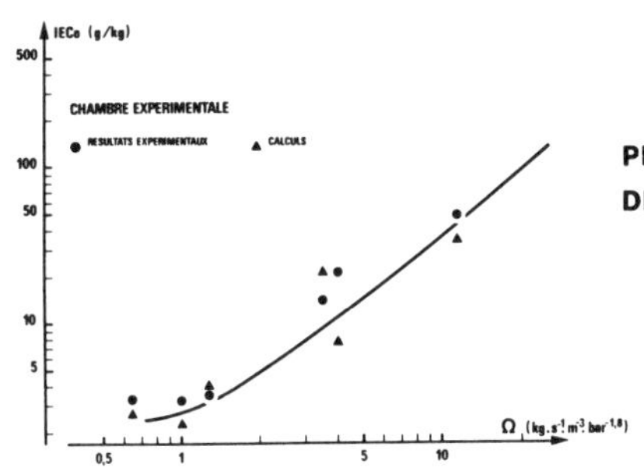

PRĒVISION DES PERFORMANCES
DE RENDEMENT/POLLUTION

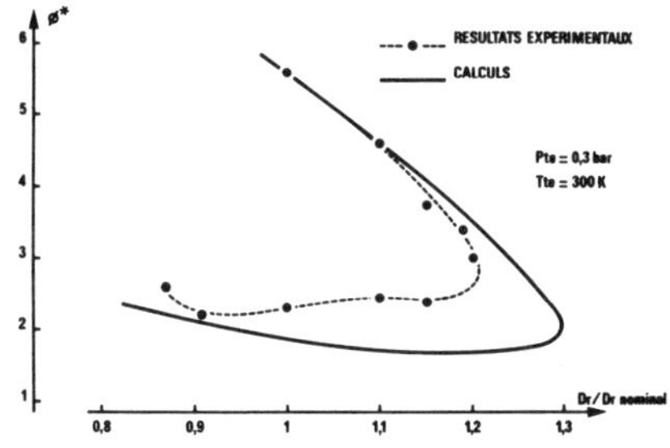

PRĒVISION DES PERFORMANCES
DE STABILITĒ

BUTÉES ACTUELLES DES MÉTHODES 7

● MODÉLISATION DES PHÉNOMENES PHYSIQUES ÉLÉMENTAIRES

- **COMBUSTION TURBULENTE**
- **CINÉTIQUE CHIMIQUE**
- **FRAGMENTATION DE NAPPES LIQUIDES**

● MÉTHODES NUMÉRIQUES

- **DIFFUSION NUMÉRIQUE**
- **REPRÉSENTATIVITÉ DES MAILLAGES**

● MÉTHODES DE MESURES A DÉVELOPPER

- **PONCTUELLES**
- **IN SITU . NON INTRUSIVES**

● VALIDATION DES MÉTHODES

- **EXPÉRIENCES FONDAMENTALES**
  VALIDATION DES MODÈLES ÉLÉMENTAIRES
- **MESURES SUR FOYERS RÉELS**
  VALIDATION DES CODES INDUSTRIELS

CRITÈRES D'UNE MÉTHODE INDUSTRIELLE 8

● ENVIRONNEMENT INFORMATIQUE

- **NOMBREUSES MÉTHODES COUPLÉES**
- **NOMBREUX FICHIERS / BASES DE DONNÉES**

● CONDITIONS D'UTILISATION

- **SOUPLESSE ET RAPIDITÉ DE MISE EN OEUVRE**
- **ROBUSTESSE**

● VALIDATION

- **PLAGE DE VALIDATION IDENTIFIÉE**
  / DOMAINE D'UTILISATION
- **PRÉCISION / FIABILITÉ DES RÉSULTATS**
- **AMÉLIORATIONS SUCCESSIVES A PRENDRE EN COMPTE**

CHARACTERIZING SELF-SIMILAR BLOW-UP

J. BEBERNES
Department of Mathematics
University of Colorado
Boulder, Colorado 80309

and

D. KASSOY
Department of Mechanical Engineering
University of Colorado
Boulder, Colorado 80309

1. INTRODUCTION

The purpose of this paper is to give a precise description of the asymptotic behavior of the radially symmetric solutions $u(x,t)$ of

$$u_t - \Delta u = f(u) \tag{1}$$

which blow up in finite positive time T and to determine where blow-up occurs for the perturbed semilinear equation

$$u_t - \Delta u = f(u) + g(t) \ . \tag{2}$$

When a fuel in a bounded container is ignited, a large amount of heat can be produced in the ensuing chemical reaction. The further evolution of the system is determined by the competition between the heat addition due to the chemical reaction and the heat loss through the boundary of the container due to thermal conduction. If the heat loss is suitably large, the system rapidly approaches an equilibrium state which is retained until all the fuel is consumed. In contrast, if the chemical reaction is sufficiently vigorous, the temperature in the interior of the container increases very rapidly and an explosive event will occur.

The single semilinear parabolic equation of reaction and diffusion (1) with $f(u) = \delta e^u$ is a mathematical model for the combustion of a solid fuel with u being a measure of the temperature. This relatively simple model contains a single parameter. The model predicts a critical value δ_{FK}, a mild equilibrium state is reached. If instead the parameter δ is greater than δ_{FK}, a supercritical event occurs and an explosion is observed [3]. From a mathematical point of view, this means that the solution of the reaction-diffusion equation becomes infinite within a finite time.

In other reaction-diffusion systems of biological interest, u can be interpreted as a concentration with $f(u) = u^p$, $p > 1$, describing the reactant consumption, a similar behavior for u occurs depending on the size of the container.

With $f(u) = \delta e^u$ and $g(t) = \frac{\gamma - 1}{\gamma \cdot \mathrm{vol}\Omega} \int_\Omega u_t(x,t)dx$, the integro-partial differential equation (2) serves as a model for a perfect reactive gas with gas constant γ in a bounded container. Again u is a measure of the temperature and a critical value δ_{FK} of δ exists for which blow-up occurs if $\delta > \delta_{FK}$. This ignition model is developed in [18] and the theoretical results are proven in [6].

Besides the computation of the critical parameter δ_{FK}, it is important to know how an explosion develops. Recent results for (1) by Weissler [22], [23], [24] and Friedman and McLeod [10] indicate that, for radially symmetric solutions, when the blow-up time is

C.-M. Brauner and C. Schmidt-Lainé (eds.), Mathematical Modeling in Combustion and Related Topics, 383–392.

approached, the solution u becomes infinite only at the center of the container, remaining bounded at all other points.

A more detailed description of the temperature profile around the center spot as the explosion time is approached can be obtained by the introduction of similarity variables – in other words the change of both dependent and independent variables. One such set of coordinates, the hot-spot variable grouping, was considered by Kassoy and Poland [17] and Kapila [16]. A vigorous mathematical study of the behavior of the reaction-diffusion equation (1) in these new coordinates was completed by A. Bressan, D. Eberly, and the first author of this paper in [2] for $f(u) = \delta e^u$ and with spatial dimension $n = 1$ or 2. Similar results, for the nonlinearity $f(u) = u^p$ with $n = 1,2$ or $p < \frac{n+2}{n-2}$, where obtained by Giga and Kohn [12]. We showed that, as the blow-up time T is approached, the temperature profile tends to be a limit, in these new coordinates, which is spatially constant. Essentially, this means that the hot spot variable grouping provides a good description of the temporal evolution of the temperature as the blow-up time is approached but that the space variable is stretched too much so that no information concerning the spatial behavior is retained in the limit.

Dold [25] has suggested an alternative rescaling of variables in which the space coordinates are less stretched. This appears to be a promising approach to this problem, but the computations in [25] are purely formal. They do suggest that a spatially nontrivial, nonsingular limit should be observed. The expansion can be carried forth to higher order terms as we indicate at the end of this paper.

2. KNOWN FACTS AND STATEMENT OF PROBLEMS

Throughout the paper we will consider the two problems

(1) $$u_t - \Delta u = f(u)$$

(3) $$u(x,0) = \phi(x) \geq 0, \quad x \in \Omega$$

$$u(x,t) = 0, \qquad x \in \partial\Omega,\ t \geq 0$$

and

(2) $$u_t - \Delta u = f(u)$$

(3) $$u(x,0) = \phi(x), \quad x \in \Omega$$

$$u(x,t) = 0, \qquad x \in \partial\Omega,\ t \geq 0$$

where $\Omega = B_R = \{x \in \mathbb{R}^n\colon\ |x| < R\}$, ϕ is nonnegative, radially symmetric, and radially decreasing, with $\Delta\phi + f(\phi) \geq 0$, f is C^1 and nonnegative for $u \geq 0$, and g is continuous and nonnegative on its domain of existence. The typical prototypes which we will consider are $f(u) = e^u$ or u^p with $p > 1$ and in some cases $g(t) = \frac{K}{\operatorname{vol}\Omega} \int_\Omega u_t(x,t)dx$ where $0 \leq K < 1$.

The following results which we now summarize are known – certainly they are standard for (1)-(3) and their proofs for (2)-(3) can be found in [6].

2.1. <u>Facts</u>

2.1.1. A unique solution $u(x,t)$ exists on $\Omega \times [0,\sigma)$, $\sigma > 0$ sufficiently small (Local Existence).

2.1.2. $u(x,t) \geq 0$.

2.1.3. If $g'(t) \geq 0$ or if $g(t) = \frac{K}{\text{vol}\,\Omega} \int_\Omega u_t(x,t)dx$, $K < 1$, then $u_t \geq 0$ and $U(t) = \sup_\Omega u(x,t)$ is increasing.

2.1.4. If $u(x,t)$ exists on $0 \leq t < \sigma < \infty$ and $U(\sigma^-) < \infty$, then the solution u can be uniquely extended to $0 < t < \sigma + \epsilon$, $\epsilon > 0$.

Let $T = \sup\{\sigma:\ u(x,t) \text{ exists on } 0 \leq t < \sigma\}$.

2.1.5. $u(x,t)$ is radially symmetric and $u(\cdot,t)$ is radially decreasing for $t \in [0,t)$.

2.1.6. If $T < \infty$, then $U(T^-) = \sup_\Omega u(x, T^-) = +\infty$.

ASSUME $T < \infty$, then $u(x,t)$ blows-up in finite time.

2.2. Definition

A point $x \in \Omega$ is a blow-up point if there exists $\{(x_n, t_n)\}$ such that $t_n \to T^-$, $x_n \to x$, and $u(x_n, t_n) \to \infty$ as $n \to \infty$.

2.3. Problems

2.3.1. Where. Describe the set of blow-up points.

2.3.2. How. Describe the behavior of $u(x,t)$ near blow-up points as $t \to T^-$.

3. WHERE?

3.1. $f(u) = e^u$

THEOREM 1. [10]

a) $x = 0$ is the only blow-up point in B_R (SPBU).

b) Given $\alpha \in (0,1)$, there exists $K_\alpha > 0$ such that

$$u(x,t) \leq -\frac{2}{\alpha}\, ln|x| + K_\alpha \quad \text{on } \overline{B}_R \times [0,T] .$$

IDEA OF PROOF. For $n = 1$, find $F(u)$ such that

$$J = u_x + \epsilon x F(u), \quad \epsilon > 0 .$$

Then show by use of the maximum principle that

$$J_t - J_{xx} + bJ \leq 0 \text{ implies } J \leq 0 .$$

This is true if $F(u) = e^{\alpha u}$, $\alpha \in (0,1)$, so

$$u_x \leq -\epsilon x F(u) .$$

Set $H(u) = e^{\alpha u}$, $\alpha \in (0,1)$, then

$$H_x = -e^{-\alpha u} u_x \geq \epsilon x$$

$$e^{-\alpha u(x,t)} \geq \frac{\alpha \epsilon x^2}{2}$$

$$u(x,t) \leq -\frac{1}{\alpha}\, ln\, \frac{\alpha\epsilon x^2}{2} = \frac{2}{\alpha}\, ln\, \frac{1}{x} - \frac{1}{\alpha}\, ln\, \frac{\alpha\epsilon}{2}$$

Thus, blow-up occurs at a single point and

$$u(x,t) \leq \frac{2}{\alpha}\, ln\, \frac{1}{x} + K_\alpha$$

The following extension can be made:

THEOREM 2.

$$u(x,t) \leq 2ln\,|x|^{-1} + ln(ln\,|x|^{-1}) + C$$

for $x \in \overline{B}_R$, $t \in [0,T]$.

Proof. Similar to previous theorem, letting

$$F(u) = \frac{e^u}{u+1}$$

3.2. $f(u) + g(t)$, $g(t) \geq 0$.

THEOREM 3. If $\int_0^T g(t)dt = +\infty$, then the solution $u(x,t)$ of IBVP (3)-(2) satisfies

$$\lim_{t \to T^-} u(x,t) = +\infty$$

for all $x \in B_R$ and hence blow-up occurs everywhere.

Proof. Fix $\overline{x} \in B_R$ and let $\rho = R - |\overline{x}|$. On the ball $B_\rho(\overline{x}) \leq B_R$, the solution $u(x,t)$ is an upper solution for

$$(4) \qquad \begin{cases} v_t = \Delta v + g(t) \\ v(x,0) = 0, \quad |x-\overline{x}| < \rho \\ v(x,t) = 0, \quad x \in \partial B_\rho(\overline{x}),\ t \in [0,T). \end{cases}$$

The solutions of (4), at $(\overline{x}, t)$, can be expressed as

$$v(\overline{x}, t) = \int_0^t \int_{B_\rho(\overline{x})} G(\overline{x}, y, t-s)g(s)dyds$$

$$\geq \int_0^t g(s) \int_{B_\rho(\overline{x})} G(\overline{x}, y, T)dy\ .$$

where $K(\rho) = \int_{B_\rho(\overline{x})} G(\overline{x}, y, T)dy$.

As $t \to T^-$, $v(\overline{x}, t) \to +\infty$. Since $v(x,t) \leq u(x,t)$ on $B_\rho(\overline{x})$, $u(\overline{x}, t) \to \infty$ as $t \to T^-$. But $\overline{x} \in B_R$ was arbitrary so blow-up occurs everywhere.

THEOREM 4. If $g(t) = \frac{K}{\text{vol}\,\Omega} \int_\Omega u_t(x,t)dx$, $0 \leq K < 1$, and if blow-up occurs at some $\overline{x} \neq 0$, then it blows up everywhere in B_R.

Proof. Observe

$$\int_0^t g(s)ds = \frac{K}{\text{vol}\,\Omega} \int_0^t \int_\Omega u_t(x,s)dxds = \frac{K}{\text{vol}\,\Omega} \int_\Omega [u(x,t),-\phi(x)]dx$$

If $\lim_{t \to T^-} u(\overline{x},t) = +\infty$, then by radial monotonicity of u,

$$\int_\Omega u(x,t)dx \geq \int_{|x|\leq|\bar{x}|} u(x,t)dx \geq (\mathrm{vol}\,B_{|\bar{x}|})u(\bar{x},t) \to \infty$$

as $t \to T^-$. Hence, $\int_0^T g(s)ds = +\infty$ and, by THEOREM 3, blow-up occurs everywhere.

3.3 $\underline{f(u) = e^u,\ g(t)}$.

THEOREM 5. If $f(u) = e^u$, $g(t) = \frac{K}{\mathrm{vol}\,\Omega} \int_\Omega u_t dx$, then the solution $u(x,t)$ of (3)-(2) blows up at the single point $x = 0$.

Corollary. If $f(u) = e^u$, $\int_0^T g(s)ds < \infty$, then blow-up occurs at $x = 0$ only.

Proof. For $n = 1$, take, as in Theorem 1,

$$J(x,t) = u_x + \epsilon x F(u,t)$$

where $F(u,t) = e^{\alpha(u-G(t))}$, $G(t) = \int_0^t g(s)ds$, $\alpha \in (0,1)$.

Then, by the maximum principle, $u_x \leq -\epsilon x F(u,t)$. Set $H(u,t) = \frac{1}{\alpha} e^{-\alpha[u-G(t)]}$, then

$$H_x = -e^{-\alpha[u-G(t)]}u_x \geq \epsilon x$$

$$H(u(x,t),t) - H(u(0,t),t) \geq \frac{\epsilon x^2}{2}$$

$$e^{-\alpha[u-G(t)]} \geq \frac{\epsilon \alpha x^2}{2}$$

$$u(x,t) \leq \frac{2}{\alpha}\, \ln x^{-1} - \frac{1}{\alpha}\, \ln \frac{\alpha\epsilon}{2} + G(t)$$

Thus,

$$\int_\Omega u(x,t)dx = \int_\Omega \left[\frac{2}{\alpha}\, \ln x^{-1} - \frac{1}{\alpha}\, \ln \frac{\alpha\epsilon}{2}\right]dx + \mathrm{vol}\,\Omega\cdot G(t)$$

But

$$G(t) = \int_0^t g(s)ds = \frac{K}{\mathrm{vol}\,\Omega} \int_0^t \int_\Omega u_t(x,s)dxds = \frac{K}{\mathrm{vol}\,\Omega} \int_\Omega [u(x,t)-u(x,0)]dx$$

$$\leq \frac{K}{\mathrm{vol}\,\Omega} \int_\Omega u(x,t)dx\ ,$$

so $\int_\Omega u(x,t)dx \leq \int_\Omega \left[\frac{2}{\alpha}\, \ln x^{-1} - \frac{1}{\alpha}\, \ln \frac{\alpha\epsilon}{2}\right]dx + K \int_\Omega u(x,t)dx$ which implies $(1-K) \int_\Omega u(x,t)dx < +\infty \Rightarrow$ blow-up at single point by THEOREM 4.

3.4 $\underline{f(u) = u^p,\ u^p + g(t),\ p > 1}$.

THEOREM 6. [10]. Let $f(u) = (u+\lambda)^p$, $p > 1$, $\lambda \geq 0$. Consider IBVP (1)-(3).

a) $x = 0$ is the only blow-up point in B_R.

b) Given $\alpha \in (1,p)$, there is $C > 0$ such that

$$u(x,t) \leq \frac{C}{|x|^{2/\alpha - 1}} .$$

THEOREM 7. Consider IBVP (2)-(3) with $f(u) = (u+\lambda)^p$, $\lambda \geq 0$, $p > 1$.

a) If $\int_0^T g(s)ds = +\infty$, blow-up occurs everywhere.

b) If $\int_0^T g(s)ds < +\infty$, then blow-up occurs at a single point.

THEOREM 8. Consider (2)-(3) with $g(t) = \frac{K}{\operatorname{vol}\Omega} \int_\Omega u_t(x,t)dx$, $0 \leq K < 1$, then:

a) if $1 < p < 1 + 2/n$, blow-up occurs everywhere.

b) if $p > 1 + 2/n$ and K is sufficiently small, then SPBU.

Conjecture. If $p > 1 + 2/n$ and $K < 1$, then SPBU.

4. HOW?

4.1. Here we consider only solid fuel ignition models with $f(u) = u^p$ or e^u.

THEOREM 9. [12]. Consider (1)-(3) with $f(u) = u^p$, $p > 1$. If $n = 1,2$, or $n \geq 3$ and $p \leq \frac{n+2}{n-2}$, then $(T-t)^\beta u(x,t) \to \beta^\beta$ as $t \to T^-$ uniformly on $0 \leq |x| \leq c(T-t)^{1/2}$, $c > 0$, where $\beta = \frac{1}{p-1}$.

Let $f(u) = e^u$. Since $u(x,t)$, the solution of IBVP (1)-(3) is radially symmetric, IBVP (1)-(2) can be reduced to a problem in one spatial dimension. Let

$$D = \{(r,t)\colon\ 0 \leq r \leq R,\ 0 \leq t < T\}$$

Then, if $r = |x|$, $v(r,t) = u(x,t)$ satisfies

$$v_t = v_{rr} + \frac{n-1}{r} v_r + e^v \tag{5}$$

$$\begin{cases} v(r,0) = \phi(r) \\ v_r(0,t) = 0, \quad v(R,t) = 0 \end{cases} \tag{6}$$

To study the asymptotic behavior of v as $t \to T^-$, make the change of variables

$$\left.\begin{aligned} \tau &= -ln(T-t) \\ \eta &= \frac{r}{(T-t)^{1/2}} \\ \theta &= v + ln(T-t) = v - \tau \end{aligned}\right\} \qquad \begin{cases} t = T - e^{-\tau} \\ r = \eta e^{-\tau/2} \\ v = \theta - ln(T-t) \end{cases}$$

Then D transforms to $D' = \{(\eta,\tau)\colon\ 0 \leq \eta \leq Re^{\tau/2},\ \tau \geq -ln T\}$ and $\theta(\eta,\tau) = v - \tau$ solves

(7) $$\theta_\tau = \theta_{\eta\eta} + \left(\frac{n-1}{\eta} - \frac{\eta}{2}\right)\theta_\eta + e^\theta - 1$$

(8) $$\begin{cases} \theta(\eta, -ln\mathrm{T}) = \phi(\eta \mathrm{T}^{1/2}) + ln\mathrm{T} \\ \theta_\eta(0, \tau) = 0 \qquad \theta(\mathrm{R}e^{\tau/2}, \tau) = -\tau \end{cases}$$

The proof of the following theorem is based on the ideas of [12].

THEOREM 10. If there is a unique function $y(\eta)$ which is a nonincreasing globally Lipschitz continuous solution of

(9) $$\begin{cases} y'' + \left(\frac{n-1}{\eta} - \frac{\eta}{2}\right)y' + e^y - 1 = 0 \\ y'(0) = 0, \;\; y(0) = \alpha \geq 0, \end{cases}$$

then as $\tau \to \infty$ the solution $\theta(\eta,\tau)$ tend uniformly to $y(\eta)$ on compact subsets of $\mathbb{R}^+$.

THEOREM 11. For $n = 1,2$ the only nonincreasing globally Lipschitz continuous solution of (9) is

$$y(\eta) \equiv 0 .$$

The proof of this theorem can be found in [2]. As a consequence of the previous two theorems, we have:

THEOREM 12. For $n = 1,2$, the solution $u(x,t)$ of IBVP (1)-(3) satisfies

$$u(x,t) - ln(T-t)^{-1} \to 0$$

as $t \to T^-$ uniformly on

$$0 \leq |x| \leq C(T-t)^{1/2}, \;\; C > 0 .$$

4.2. Question

What happens, as $t \to T^-$, to the solution $u(x,t)$ of IBVP (1)-(3) when i) $f(u) = e^u$, $n \geq 3$? ii) $f(u) = u^p$, $n \geq 3$, with $p > \frac{n+2}{n-2}$?

THEOREM 13. For each $n \in (2,10)$, there is an increasing unbounded sequence of positive initial values $\{\alpha_m(n)\}_{m=1}^{\infty}$ such that the solutions $y(x,\alpha_m)$ of

(9) $$\begin{cases} y'' + \left(\frac{n-1}{x} - \frac{x}{2}\right)y' + e^y - 1 = 0 \\ y(0) = \alpha_m, \;\; y'(0) = 0 \end{cases}$$

satisfy

$$\lim_{x\to\infty} \left(1 + \frac{x}{2}\, y'(x)\right) = 0 .$$

IDEA OF PROOF. Equation (9) has a singular solution:

$$S_e(x) = ln\, \frac{2(n-2)}{x^2}$$

Set $h(x) = y(x,\alpha) = S_e(x)$, then h satisfies

$$h'' + \left(\frac{n-1}{x} - \frac{x}{2}\right)' + \frac{2(n-2)}{x^2}(e^h - 1) = 0$$

Since the linearized version, for x near 0, is

$$L'' + \frac{n-1}{x} + \frac{2(n-2)}{x^2} L = 0$$

and its solutions have zeros which accumulate at 0, we use this idea to show that there is an unbounded increasing sequence $\{\alpha_k\}$ such that the number of zeros of $h(x)$ on $(0, 2(n-2)]^{1/2})$ increases as $\alpha_k \to \infty$

THEOREM 14. Let $n > 2$, $p > \frac{n+2}{n-2}$, then there exists an unbounded increasing sequence $\{a_k\}$ such that:

$$(10) \qquad \begin{cases} w'' + (\frac{n-1}{x} - \frac{x}{2})w' + |w|^{p-1}w - \frac{1}{p-1} w = 0 \\ w(0) = \alpha_k, \ w'(0) = 0 \end{cases}$$

has a bounded solution $w(x)$ on $[0,\infty)$ with

$$\lim_{x \to \infty} w(x) = 0$$

Idea is similar. Use singular solution:

$$S_p(x) = \left\{ \frac{-4\beta[\beta + \frac{2-n}{2}]}{x^2} \right\}^{\beta}, \quad \beta = (p-1)$$

This multiple existence of solutions complicates the stability analysis required to precisely describe the evolution of time-dependent solutions $u(x,t)$ of (1)-(3) near the blow-up singularity.

THEOREM 15. Let $n \geq 3$.

a) For $f(u) = e^u$, the solution $u(x,t)$ of (1)-(3) satisfies $u(x,t) + ln(T-t) \to 0$ uniformly on $|x| \leq C(T-t)^{1/2}$ as $t \to T^-$.

b) For $f(u) = u^p$, the solution $u(x,t)$ of (1)-(3) satisfies $(T-t)^\beta u(x,t) \to \beta^\beta$ uniformly as $t \to T^-$.

The proof of this theorem can be found in [26].

5. SPECULATIONS

As we have observed, we now have a precise description of the temporal evolution as the blow-up time is approached, but the spatial variable is stretched too much and we have no information concerning the spatial variable. Different scalings have been suggested ([25], [14]).

To illustrate a matched asymptotic analysis which gives additional information, let us consider the one-dimensional case where

$$u_t = u_{xx} + e^u$$

$$u(x,0) = \phi(x) \geq 0$$
$$u(\pm R,t) = 0\ .$$

Set $u + -ln\, w$. Let $\eta = x\tau^{-1/2}$, $\tau = T-t$, then in the inner asymptotic region:

$$w \sim \tau\left[1 + \frac{1/2}{ln\,\tau}\,(1 - \frac{\eta^2}{2}) + \frac{1/2 + D_1(1-\frac{\eta^2}{2})}{ln^2\,\tau^{-1}} + 0(\frac{1}{ln^3\,\frac{1}{\tau}})\right].$$

The nonuniformity in the inner layer solution for $\eta = 0((ln\,\frac{1}{\tau})^{1/2})$ implies that a new variable must be defined. Let $\xi = \frac{x}{(\tau\,ln\,\tau^{-1})^{1/2}}$, then by matching we find a second layer solution

$$w \sim \tau\left\{(\frac{\xi^2}{4} + 1) + \frac{1}{ln\,\tau^{-1}}\left[\frac{\xi^2}{4}\,ln(\frac{\xi^2}{4} + 1) - \frac{D_1}{2}\,\xi^2 - \frac{1}{2}\right] + 0(\frac{1}{ln^2\,\frac{1}{\tau})}\right\}.$$

Again a nonuniformity will appear when $\frac{ln\,\xi^2}{ln\,\tau^{-1}} = 0(1)$. Defining a new variable $s = \frac{ln\,\xi^2}{ln\,\tau^{-1}}$, we can find a third layer solution which expressed in terms of the original solution gives

$$u(x,t) \sim 2\,ln\,\frac{1}{x} + ln(ln\,\frac{1}{x}) + ln\,8 - \frac{3}{4}\,\frac{ln\,ln\,\frac{1}{x}}{ln\,\frac{1}{x}} + 0\left(\frac{ln^2\,ln\,x^{-1}}{ln^2 x^{-1}}\right).$$

Note the similarity with the upper bound obtained in Theorem 2.

REFERENCES

1. Baras, P. and L. Cohen: Sur l'explosion totale après T_{max} de la solution d'une equation de la chaleur semi-linéaire, C.R. Acad. Sci. Paris, t.300 (1985), 295-298.
2. Bebernes, J., Bressan, A., and Eberly, D.: A Description of Blow-up for the Solid Fuel Ignition Model, Indiana J. Math., to appear.
3. Bebernes, J. and Kassoy, D.: A Mathematical Analysis of Blow-up for Thermal Reactions, SIAM J. Appl. Math. 40 (1981), 476-484.
4. Bebernes, J. and Troy, W.: On the Existence of Solutions to the Kassoy Problem in Dimension 1, SIAM J. Math. Anal., to appear.
5. Bebernes, J., Bressan, A., and Lacey, A.: Total Blow-up Versus Single Point Blow-up, J. Differential Equations, to appear.
6. Bebernes, J. and Bressan, A.: Thermal Behavior for a Confined Reactive Gas, J. Differential Equations, 44 (1982), 118-133.
7. Bellout, H.: A Criterion for Blow-up of Solutions to Semilinear Heat Equations, preprint.
8. Eberly, D.: On the Existence of Solutions to the Kassoy Problem on Dimensions 1 and 2, J. Math. Anal. Appl., to appear.
9. Eberly, D. and Troy, W.: On the Existence of Logarithmic-type Solutions to the Kassoy Problem in Dimension 3, J. Differential Equations, to appear.

10. Friedman, A. and McLeod, B.: Blow-up of Positive Solutions of Semilinear Heat Equations, Indiana Univ. Math. J. 34, No. 2 (1985), 425-447.

11. Giga, Y.: On Elliptic Equations Related to Self-Similar Solutions for Nonlinear Heat Equations, Hiroshima Math. J. 16 (1986), 541-554.

12. ga, Y. and Kohn, R.V.: Asymptotically Self-Similar Blow-up of Semilinear Heat Equations, Comm. Pure and Appl. Math. 38 (1985), 297-319.

13. Giga, Y. and Kohn, Robert V.: Characterizing Blowup Using Similarity Variables, Indiana Univ. Math. J., to appear.

14. Galaktionov, V.A. and Posashkov, S.A.: The equation $u_t = u_{xx} + u^\beta$. Localization and Asymptotic Behavior of Unbounded Solutions, Diff. Uravnen 22 (1986), 1165-1173.

15. Hocking, L.M., Stewartson, K., and Stuart, J.T.: A Nonlinear Instability Burst in Plane Parallel Flow, J. Fluid Mech. 51 (1972), 705-735.

16. Kapila, A.: Reactive-Diffusive Systems with Arrhenius Kinetics: Dynamics of Ignition, SIAM J. Appl. Math. 39 (1980), 21-36.

17. Kassoy, D. and Poland, J.: The Thermal Explosion Confined by a Constant Temperature Boundary: I. The Induction Period Solution, SIAM J. Appl. Math. 39 (1980), 412-430.

18. Kassoy, D. and Poland, J.: The Induction Period of a Thermal Explosion in a Gas Between Infinite Parallel Plates, Combustion and Flame, 50 (1983), 259-274.

19. Lacey, A.: Mathematical Analysis of Thermal Runaway for Spatially Inhomogeneous Reactions, SIAM J. Appl. Math. 43 (1983), 1350-1366.

20. Lacey, A. and Tzanetis, D.: Global Existence and Convergence to a Singular Steady State for a Semilinear Heat Equation, preprint.

21. Troy, W.: The Existence of Bounded Solutions for a Semilinear Heat Equation, SIAM J. Math. Anal., to appear.

22. Weissler, F.B.: Single Point Blowup of Semilinear Initial Value Problems, J. Diff. Eq. 55 (1984), 204-224.

23. Weissler, F.B.: An L^∞ Blowup Estimate for a Nonlinear Heat Equation, Comm. Pure Appl. Math. 38 (1985), 291-296.

24. Weissler, F. and Mueller, C.E.: Single Point Blow-up for a General Semilinear Heat Equation, Indiana Univ. Math. J. 34 (1985), 881-913.

25. Dold, J.: Analysis of the Early Stage of Thermal Runaway, Quat. J. Mech. Appl. Math. 38 (1985), 361-387.

26. Bebernes, J. and Eberly, D.: A Description of Self-Similar Blow-up for Dimensions $n \geq 3$, Analyse Nonlinear, to appear.

SOME FINITE-ELEMENT INVESTIGATIONS OF STIFF COMBUSTION PROBLEMS: MESH ADAPTION AND IMPLICIT TIME-STEPPING

F. BENKHALDOUN, A. DERVIEUX, G. FERNANDEZ,
H. GUILLARD, B. LARROUTUROU

INRIA, Sophia-Antipolis, 06560 Valbonne, FRANCE

1. INTRODUCTION

The purpose of this paper is to present some preliminary numerical experiments illustrating a global approach to the solution of multidimensional compressible reactive flows. The emphasis is put on solutions to a class of stiff combustion problems, namely thin flame propagation phenomena in medium or low Mach number flows. In combustion problems, stiffness is due both to the presence of disparate characteristic length scales and disparate characteristic time scales.
The spatial stiffness essentially results from the small length scales in the flame region: the reaction zone thickness is much smaller than the flame thickness, which is itself small compared to the size of the vessel in which the flame propagates (see [3]). This may result in inaccurate or even unstable calculations (see [12]). The main numerical tools used to face this difficulty are the adaptive mesh techniques, whose application to computational fluid dynamics and to combustion has received increasing attention in the past ten years. We propose in this paper two different adaptive gridding procedures for steady laminar flames, namely local element divisions and nodes movement.
The temporal stiffness appears in low Mach number reacting flows, such as those occurring in internal combustion engines: the time scale associated to the acoustic wave propagation is then much smaller than the flame propagation time scale. The numerical solution of temporally stiff problems often requires implicit time integration schemes. In this study, we use a linearized implicit scheme which can be employed in a wide range of Mach numbers, since stable and consistent results are obtained with small or large Courant numbers.

The approach described in this paper therefore combines adaptive gridding techniques and non factored implicit non oscillatory schemes. These ingredients are associated in the context of a finite-element formulation on unstructured meshes (triangles in two dimensions). This method is applied to the calculation of thin flames propagating in a compressible laminar gaseous flow.

Section 1 is devoted to the spatial approximation; some numerical examples are presented using uniform meshes. In Section 2, we describe the mesh adaption methods and their application to representative steady problems. The implicit time stepping and other numerical results are presented in Section 3.

C.-M. Brauner and C. Schmidt-Lainé (eds.), Mathematical Modeling in Combustion and Related Topics, 393–409.

2. SPATIAL APPROXIMATION

We first recall the main features of a finite-element upwind scheme developed for the Euler equations by F. FEZOUI [5], and then describe its extension to the solution of reactive flow problems.

2.1. Euler scheme

2.1.1. The system of Euler equations for perfect gas flows writes:

$$\begin{cases} W_t + F(W)_x + G(W)_y = 0, \\ W = \begin{pmatrix} \rho \\ \rho u \\ \rho v \\ E \end{pmatrix}, \quad F(W) = \begin{pmatrix} \rho u \\ \rho u^2 + p \\ \rho uv \\ u(E+p) \end{pmatrix}, \quad G(W) = \begin{pmatrix} \rho v \\ \rho uv \\ \rho v^2 + p \\ v(E+p) \end{pmatrix}, \end{cases} \tag{2.1}$$

with

$$p = (\gamma - 1)\left[E - \frac{1}{2}\rho(u^2 + v^2)\right], \qquad (\gamma = 1.4). \tag{2.2}$$

2.1.2. The spatial discretization uses a finite-element triangulation of the flow domain $\mathcal{D}$. A finite-volume scheme, which is a simplified version of the usual Galerkin approximation, is then introduced using a dual partition in control volumes or cells: a cell C_i is constructed around each vertex i by means of the medians of neighboring triangles (see Fig. 1).

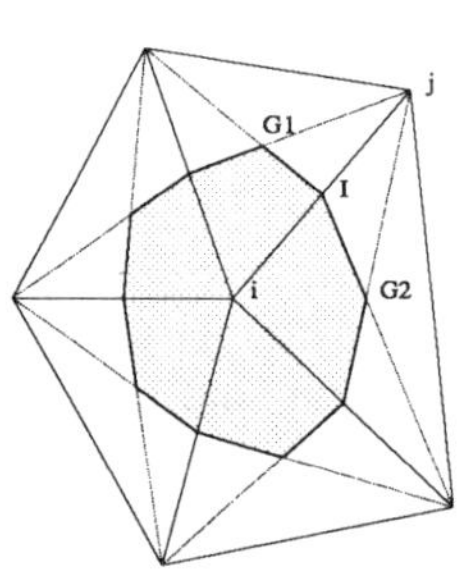

Fig. 1 : Cell C_i

Let us consider two neighbor vertices i and j; let I be the middle of the segment ij, and let G_1 and G_2 be the centroids of the two triangles having ij as a common side. Lastly, let $\vec{n}_{ij} = (n^x_{ij}, n^y_{ij})$ be the outward unit normal to the cell C_i on the bisegment G_1IG_2. We define the vector $\vec{\eta}_{ij}$ by:

$$\vec{\eta_{ij}} = (\eta^x_{ij}, \eta^y_{ij}) \; ; \quad \eta^x_{ij} = \int_{G_1IG_2} n^x_{ij} d\sigma \; ; \quad \eta^y_{ij} = \int_{G_1IG_2} n^y_{ij} d\sigma \, . \tag{2.3}$$

Furthermore, using two values W_R and W_L of W and a vector $\vec{\eta}$, we define the numerical flux splitting function Φ as (see for example [9]):

$$\Phi(W_R, W_L, \vec{\eta}) = \frac{1}{2}[P(W_R) + P(W_L)] + \frac{1}{2}|Q(\overline{W})|(W_R - W_L) \tag{2.4}$$

with $\overline{W} = \frac{1}{2}(W_R + W_L)$, $P(W) = \eta^x F(W) + \eta^y G(W)$, $Q(\overline{W}) = \frac{\partial P}{\partial W}(\overline{W})$, and $|Q|$ being defined by:

$$\begin{cases} Q = T\Lambda T^{-1}, \ \Lambda \ \text{diagonal}: \Lambda = (\lambda_k)_{k=1,4} \ , \\ |Q| = T|\Lambda|T^{-1}, \ |\Lambda| = (|\lambda_k|)_{k=1,4} \ . \end{cases} \tag{2.5}$$

2.1.3 A first-order upwind non oscillatory scheme can now be defined using the following approximation of the fluxes at vertex i:

$$(F(W)_x + G(W)_y)_i \simeq \mathcal{F}(W) = \frac{1}{area(C_i)} \sum_{j \in \kappa(i)} \Phi(W_i, W_j, \vec{\eta_{ij}}) \tag{2.6}$$

where $\kappa(i)$ is the set of neighboring nodes of i.

2.1.4. A second-order extension of the previous scheme can be derived using a MUSCL-type-interpolation; we first define an approximate gradient of W at the vertex i by the formulas:

$$\begin{cases} W_x(i) = \int\int_{supp\ (i)} \frac{\partial W}{\partial x} \, dxdy \ / \ area[supp(i)] \ , \\ W_y(i) = \int\int_{supp\ (i)} \frac{\partial W}{\partial y} \, dxdy \ / \ area[supp(i)] \ , \end{cases} \tag{2.7}$$

where $supp\ (i)$ denotes the union of those triangles which have i as a vertex, and where the integrand $\frac{\partial W}{\partial x}$ is the derivative of the usual P-1 Galerkin interpolant of the variable W. We can then define W_{ij} and W_{ji} by:

$$\begin{cases} W_{ij} = w_i + \frac{1}{2} \begin{pmatrix} W_x(i) \\ W_y(i) \end{pmatrix} . \vec{ij} \\ W_{ji} = w_j + \frac{1}{2} \begin{pmatrix} W_x(j) \\ W_y(j) \end{pmatrix} . \vec{ji} \end{cases} \tag{2.8}$$

and obtain a second-order accurate approximation of the fluxes at vertex i by the expression:

$$(F(W)_x + G(W)_y)_i \simeq \frac{1}{area(C_i)} \sum_{j \in \kappa(i)} \Phi(W_{ij}, W_{ji}, \vec{\eta_{ij}}) \tag{2.9}$$

This fully second-order accurate approximation may produce oscillatory results, and thus will not be employed in the sequel; we use instead a TVD-like scheme derived by using "limited" extrapolated values W^l_{ij} and W^l_{ji} in place of W_{ij} and W_{ji} in (2.9); we refer to [5] for the details.

2.2. A numerical example

The ability of this scheme to capture traveling shock waves with accuracy and without oscillations is illustrated by the following example. The flow domain is a rectangular vessel [0,0.1] x [0,0.05], with adiabatic slipping walls; different initial conditions are used in a subdomain $\mathcal{D}_1$ of dimensions 0.03 x 0.03 and in the remaining part $\mathcal{D}_0$ of $\mathcal{D}$ (see Fig. 2). The initial conditions are the following:

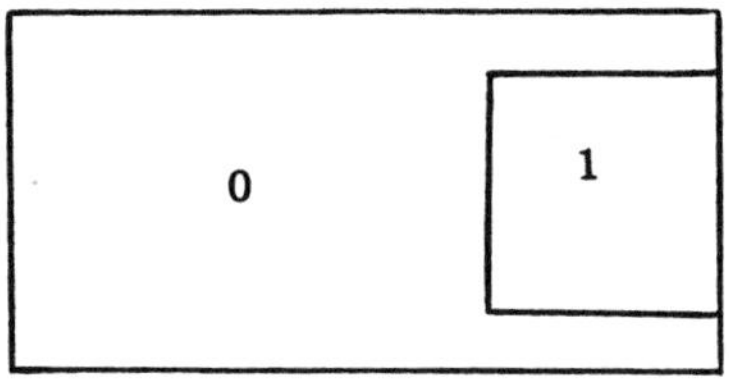

Figure 2.

$$\begin{cases} P_0 = 30\ bars\ , & \rho_0 = 15\ kg.m^{-3}\ , \quad u_0 = v_0 = 0 \text{ in } \mathcal{D}_0\ , \\ P_1 = 80\ bars\ , & \rho_1 = 15\ kg.m^{-3}\ , \quad u_1 = v_1 = 0 \text{ in } \mathcal{D}_1\ . \end{cases} \tag{2.10}$$

This calculation gives a simplified representation of the strong acoustic waves that may occur during an engine knock. A two-step explicit time integration is combined to the slope-limited scheme defined above. The results of two calculations using a coarse 21 x 21 mesh and a finer 101 x 51 triangulation are shown on Figs. 3 and 4. They demonstrate the capability of this second-order scheme to capture strong pressure waves with no oscillation.

2.3. Reactive-Euler scheme

2.3.1 We now consider the following nondimensionalized "reactive-Euler" model, that describes the combustion of an inviscid gaseous mixture in which a single one-step chemical reaction takes place:

$$
\left\{
\begin{array}{l}
\tilde{W}_t + \tilde{F}(\tilde{W},\tilde{W}_x)_x + \tilde{G}(\tilde{W},\tilde{W}_y)_y = \tilde{S}(\tilde{W}), \quad \tilde{W} = \begin{pmatrix} \rho \\ \rho u \\ \rho v \\ E \\ \rho Y \end{pmatrix} = \begin{pmatrix} W \\ \rho Y \end{pmatrix}, \\[2ex]
\tilde{F}(\tilde{W},\tilde{W}_x) = \begin{pmatrix} F(W) \\ 0 \end{pmatrix} + \hat{F}(\tilde{W},\tilde{W}_x), \quad \hat{F}(\tilde{W},\tilde{W}_x) = \begin{pmatrix} 0 \\ 0 \\ 0 \\ -D_T T_x \\ \rho u Y - D_Y Y_x \end{pmatrix}, \\[2ex]
\tilde{G}(\tilde{W},\tilde{W}_y) = \begin{pmatrix} G(W) \\ 0 \end{pmatrix} + \hat{G}(\tilde{W},\tilde{W}_y), \quad \hat{G}(\tilde{W},\tilde{W}_y) = \begin{pmatrix} 0 \\ 0 \\ 0 \\ -D_T T_y \\ \rho v Y - D_Y Y_y \end{pmatrix}, \\[2ex]
\tilde{S}(\tilde{W}) = \begin{pmatrix} 0 \\ 0 \\ 0 \\ \mathcal{Q}\Omega \\ -\Omega \end{pmatrix}.
\end{array}
\right. \tag{2.11}
$$

In these equations, T is the mixture temperature, Y is the reactant mass fraction. The additional flux terms $\hat{F}$ and $\hat{G}$ represent the convection of reactant and the thermal and molecular diffusions (D_T and D_Y are the non dimensional thermal and molecular diffusion coefficients). The source term $\tilde{S}$ contains the reactive terms: Ω is the reaction rate, given by:

$$
\Omega = A\rho Y \, exp\left(-\frac{N}{T}\right) \tag{2.12}
$$

where A is the non dimensional Arrhenius prefactor and N is the normalized activation energy.

2.3.2. For the discrete spatial approximation of system (2.11), we discretize the Euler terms as in Section 2.1.. The additional terms are approximated using a standard Galerkin formulation, except for the time-derivative and source terms integrals, which are "mass-lumped" for the sake of simplicity and positiveness:

$$
\int_D \frac{\partial \tilde{W}}{\partial t}\phi_i \simeq area(C_i)\left[\frac{\partial \tilde{W}_i}{\partial t}\right], \quad \int_D \Omega\phi_i \simeq area(C_i)\Omega_i\,, \tag{2.13}
$$

where ϕ_i is the basis finite-element function associated to vertex i (see [8] for more details).

2.4. Numerical illustration

The evolution of a flame in a square vessel is shown on Fig. 5. A sudden heating at the top of the vessel first induces strong acoustic waves; then the flame front is initiated and propagates through the vessel. This computation proves that the discretization of the Euler terms and of the diffusive and reactive terms do combine well, producing no noticeable oscillations.

3. MESH ADAPTION

The above scheme should combine well with mesh adaption techniques: from a general point of view, monotone schemes tolerate well non regular meshes since they tend to avoid the oscillations which irregular meshes can generate. Moreover, many mesh refinement techniques can be applied with no modification of the discrete scheme since it is defined on arbitrary finite-element non structured triangulations. Lastly mesh deformations are also well accepted by finite-element-like formulations because no preferred direction appears in the discrete domain.

3.1. Mesh enrichment

The ability of the scheme to handle non structured meshes is here illustrated by the application of an adaptive mesh-enrichment technique based on element divisions (the procedure is due to B. PALMERIO, see [10] for a detailed description). To show the efficiency of this adaptive method, we consider the calculation of a flame stabilized in a divergent nozzle. The boundary conditions are chosen to obtain a steady flame: the inflow gas velocity is greater than the steady planar flame speed, while the outflow velocity is less than the flame speed. In this example, a first calculation is performed with a regular mesh of 287 nodes (Fig. 6.a). Then the enrichment procedure is applied, using the first mesh and solution: any triangle in which the reaction rate Ω is larger than a given value is divided into four new triangles. A new solution is then computed and the enrichment process is applied a second time. The resulting mesh with 536 nodes is displayed on Fig 6.b. The comparison (Fig 7.) of the reaction rate values on the coarser and finer mesh shows a dramatic improvement. In particular, the errors in the reaction rate values on the coarse mesh are shown to result in an unaccurate location of the flame.

3.2. Node movement

The above enrichment process has the main drawback of adding points in an isotropic non directional manner, which can be costly. We now wish to obtain a <u>directional</u> stretching, i.e. much smaller mesh sizes in one direction than in the orthogonal direction. A possible procedure consists in moving the nodes along some trajectories. In the calculation of an anchored flame in an open rectangular tube reported below, the nodes are displaced along straight lines parallel to the tube walls. At each adaption, the new node locations are determined using a one-dimensional analysis of the local truncation error, and the variable values are interpolated from the old mesh to the new one in a conservative manner (see [1], [2] for more details).

As an example, we study the reacting flow of a gaseous mixture in a rectangular open channel with a flame thermally anchored at a hot point H located on the upper wall of the channel. A first calculation is done with a regular 61 x 11 mesh (see Fig. 8).

A second calculation uses a 31 x 11 adaptive mesh (see Fig. 9). The results obtained are clearly much better.

4. IMPLICIT TIME-STEPPING

In many reactive flows of industrial interest, the pressure waves propagate much faster than the entropy wave. For the numerical simulation of these flows, explicit schemes would result in very severe stability restrictions on the time step and therefore in very high computational costs. It is then interesting to employ implicit schemes operating with larger time steps. The implicit method presented in this section is essentially derived from the implicit Euler solver described in [7]. A systematic study of its adaptation to reactive flows is the object of [4]. This scheme is fully conservative, and is designed to avoid numerical oscillations; it is based on a non factored (fully multidimensional) implicit formulation, and allows the use of large Courant numbers. Therefore, the presented scheme should be able to apply for a large range of reactive flow calculations, from supersonic to low Mach flows.

4.1. Description of the scheme

The (non reactive) implicit Euler scheme is described in [7]; we now briefly describe a semi-implicit reactive Euler scheme, that is implicit for Euler terms and explicit for the other terms. Using the notations of Section 2., we write the global scheme as:

$$\frac{\tilde{W}^{n+1} - \tilde{W}^n}{\Delta t} + \begin{pmatrix} \mathcal{F}(W^n) \\ 0 \end{pmatrix} + \begin{pmatrix} \mathcal{A}(W^n)(W^{n+1} - W^n) \\ 0 \end{pmatrix} + \hat{\mathcal{F}}(\tilde{W}^n, \tilde{W}^n_x) = \mathcal{S}(\tilde{W}^n) \quad (4.1)$$

where $\mathcal{F}$ is the first-order discrete nonlinear operator for $F(W)_x + G(W)_y$ introduced in (2.6), $\hat{\mathcal{F}}$ is the discrete nonlinear operator for the additional flux $\hat{F}(\tilde{W}, \tilde{W}_x)_x + \hat{G}(\tilde{W}, \tilde{W}_y)_y$, and $\mathcal{S}$ is the discrete nonlinear operator for the source term $\tilde{S}$.

The linear operator $\mathcal{A}(W)$, which approximates $\frac{\partial \mathcal{F}}{\partial W}$, is based on a first-order flux-splitting similar to (2.6) (more precisely, it uses ROE's averaging [11]; we refer to [6] for more details). Because of this upwind evaluation of $\mathcal{A}$, the linear system to be solved at each time-step is diagonally dominant and can be solved with a few iterations of a node-wise Gauss-Seidel relaxation algorithm.

4.2. Numerical experiments

4.2.1. In order to compare the explicit and implicit versions, we consider the simple case of a planar flame propagating in a closed rectangular vessel; the domain is discretized with 61 x 3 mesh points. The ratio of the sound velocity to the flame speed is about 10^3. Over 110,000 time steps are necessary to obtain the solution at time $t = 0.14$ with the explicit version, while the same computation requires only 1,200 implicit time steps with a Courant number of about 100. We wish to emphasize that with such a low Mach number (10^{-3}), the spatial pressure variations are very small ; this illustrates that the classical isobaric model is relevant for these very low Mach number flows. Furthermore, the evolution of the acoustic waves does not interact with the slower evolutions, as shown by the flame speed histories and variable profiles presented in Fig 10. It can also be noticed that no fluctuation resulting from transient

acoustic waves appears during the calculation. Lastly, we have to mention the low frequency fluctuations appearing at the end of the calculation : several computations using different grids have shown that these oscillations occur because the reaction zone become thinner (and thus is not captured well enough by the coarse mesh) at the end of the computation due to the temperature rise into the constant volume vessel.
This experiment shows the ability of the code to operate with large Courant numbers. The resulting CPU gain is greater than 5. and could be even larger with a better adjustment of the number of linear iterations. Higher Courant numbers cannot be used in this example because of the stability restriction due to the explicit treatment of the diffusive terms.

4.2.2. The calculation of a flame propagating in a square vessel is now used to show the performance of the code in a two-dimensional context. The Mach number is again of the order of 10^{-3}. The examination of contours and velocity field (Figs 11 and 12) proves the stability of the scheme; the non homogeneity of the regular (Union Jack like) mesh due to alternating clusters only slightly appears in the numerical results.

5. CONCLUSION

This paper deals with stiff problems in combustion. A combination of mesh adaption techniques and upwind finite-element schemes, including a highly stable implicit formulation, is proposed. These features interact to allow a robust and satisfactory behavior of the resulting code. We wish to emphasize that this is mainly due to the following facts :

(i) - Upwind Finite Elements are well adapted to mesh adaption.

(ii) - Upwinding is useful for constructing implicit schemes involving simple linear algebra, because of diagonal dominance.

(iii) - Implicit schemes can be very useful when very small computational cells exist, as in an adaptive mesh.

We think that the results presented in this paper eloquently prove (i) and (ii). The point (iii) is currently investigated. Several other necessary improvements such as the extension to second-order accuracy or the use of different methods to solve the implicit system are also in progress.

REFERENCES

[1] F. BENKHALDOUN, Thesis, to appear.

[2] F. BENKHALDOUN, B. LARROUTUROU, “Explicit adaptive calculations of wrinkled flame propagation”, to appear in Int. J. Num. Meth. Fluids.

[3] P. CLAVIN, “Dynamic behavior of premixed flame fronts in laminar and turbulent flows”, Prog. Energy Combust. Sci. **11**, p. 1, (1985).

[4] G. FERNANDEZ, Thesis, to appear.

[5] F. FEZOUI, ”Résolution des équations d’Euler par un schéma de Van Leer en éléments finis”, INRIA Report 358, (1985).

[6] F. FEZOUI, G. FERNANDEZ, “Etude comparative sur la stabilité des schémas implicites décentrés : le cas du tube à choc”, INRIA report, to appear.

[7] F. FEZOUI, B. STOUFFLET, ”A class of implicit upwind schemes for Euler simulation with unstructured meshes”, submitted to J. Comp. Phys..

[8] A. HABBAL, A. DERVIEUX, H. GUILLARD, B. LARROUTUROU, “Explicit calculations of reactive flows with an upwind finite-element hydrodynamical code”, INRIA Report, to appear.

[9] A. HARTEN, P. D. LAX, B. VAN LEER, “On upstream differencing and Godunov type schemes for hyperbolic conservation laws”, SIAM Rewiew, **25**, p.35, (1983).

[10] B. PALMERIO, ”Self adaptive FEM algorithms for the Euler equations”, INRIA Report 338, (1984).

[11] P. L. ROE, “Approximate Riemann solvers, Parameter vectors and difference schemes”, J. Comp. Phys, **43**, p. 357, (1981).

[12] M. D. SMOOKE, “On the use of adaptive grids for premixed combustion”, in “Combustion and non-linear phenomena”, Clavin, Larrouturou, Pelcé Eds., Editions de Physique, Paris, (1985).

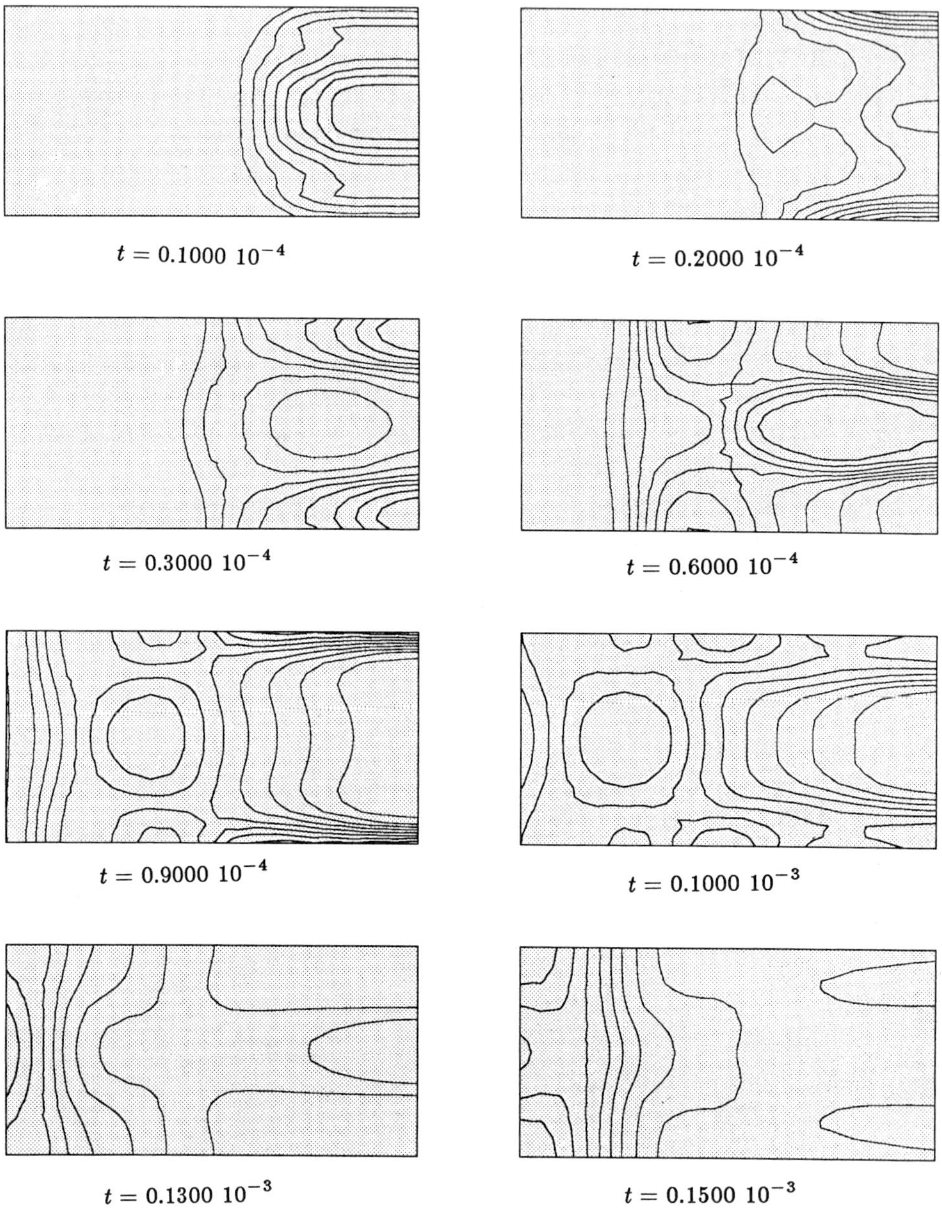

Figure 3 :: Strong acoustics calculation ; pressure contours at different times.
(coarse 21x21 mesh)

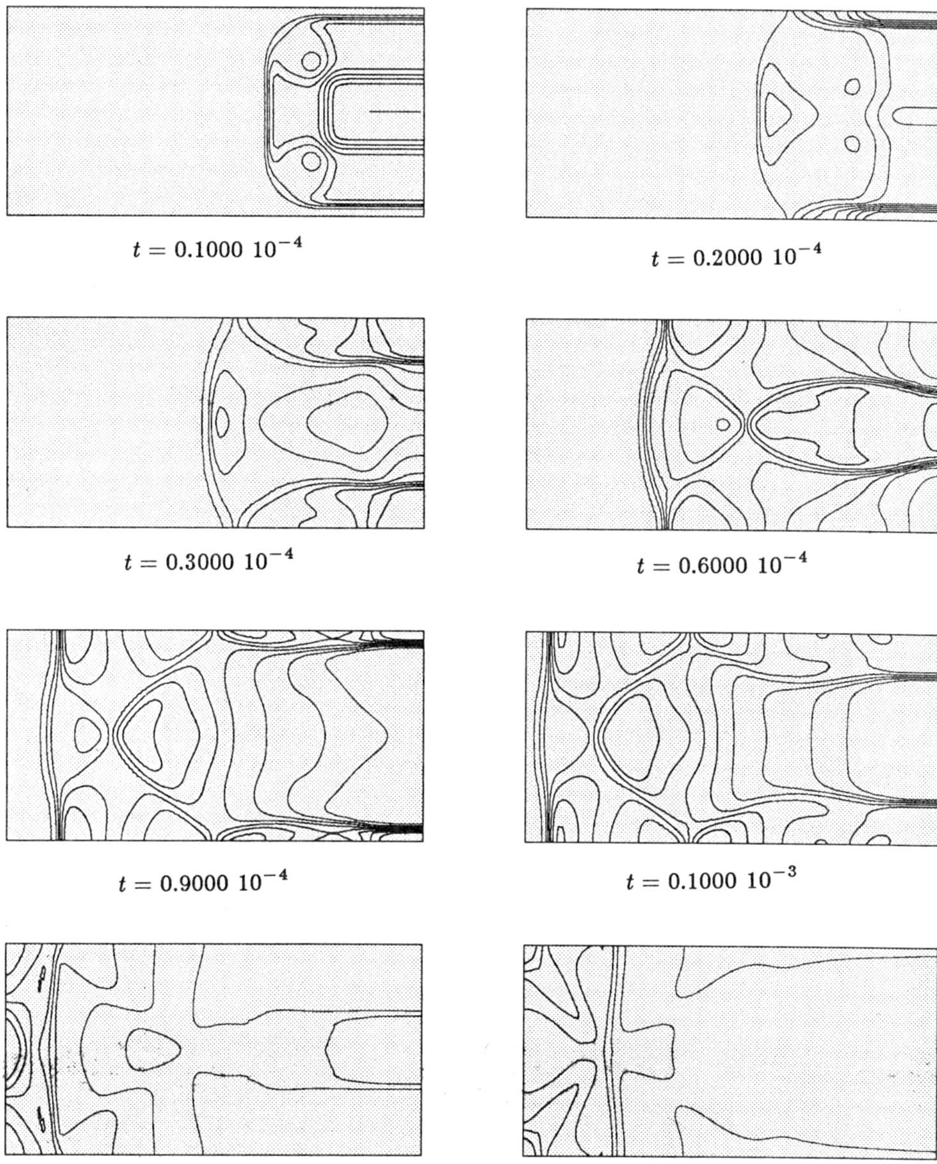

$t = 0.1000\ 10^{-4}$ $t = 0.2000\ 10^{-4}$

$t = 0.3000\ 10^{-4}$ $t = 0.6000\ 10^{-4}$

$t = 0.9000\ 10^{-4}$ $t = 0.1000\ 10^{-3}$

$t = 0.1300\ 10^{-3}$ $t = 0.1500\ 10^{-3}$

Figure 4 : Strong acoustics calculation ; pressure contours at different times.
(finer 101x51 mesh).

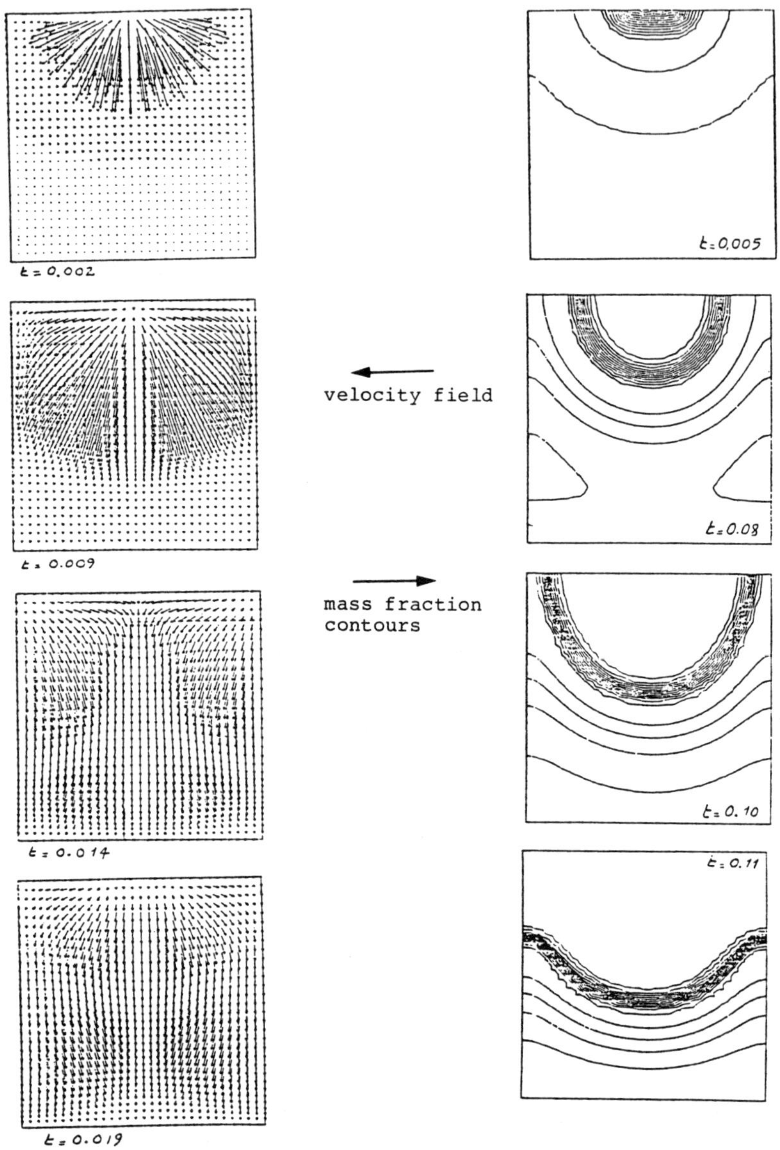

Figure 5 : Reactive flow in a closed vessel.

Figure 6a :
(287 vertices)

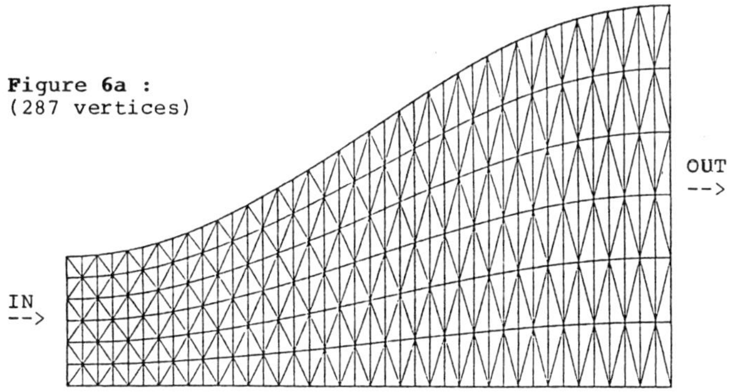

Figure 6b :
(536 vertices)

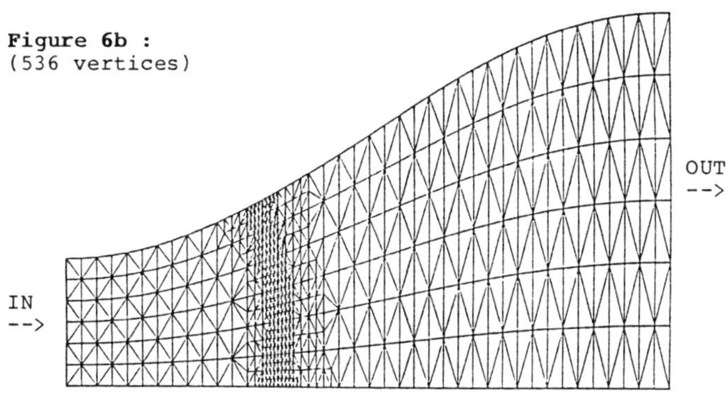

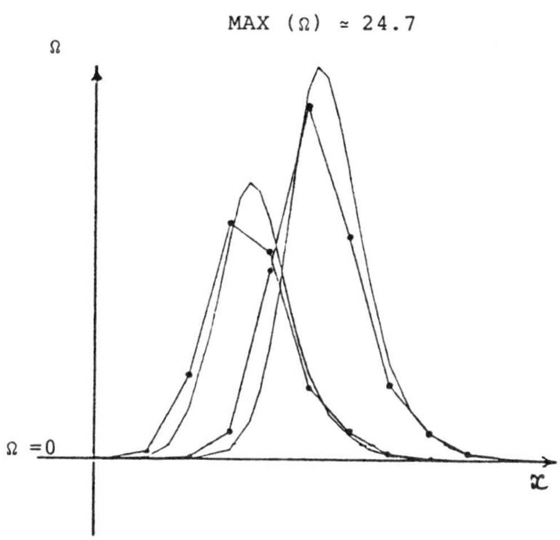

Figure 7 :
Reaction rate distribution along top and bottom slipping boundaries

—•—•—•— using the coarse mesh

______ using the refined mesh

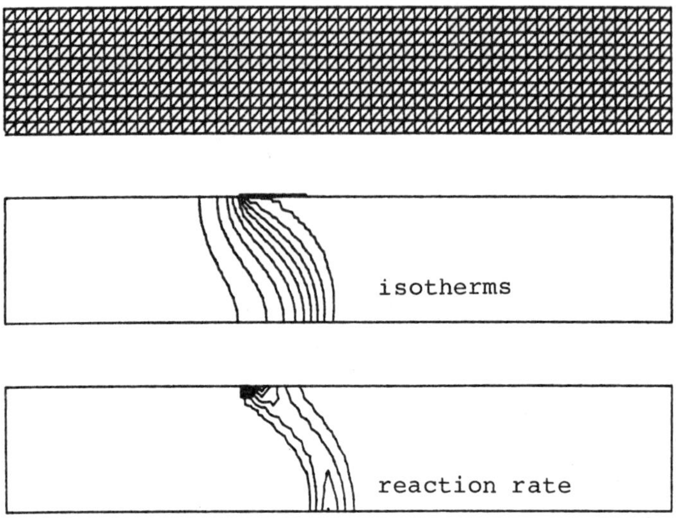

Figure 8 : Anchored flame. 61x11 regular mesh.

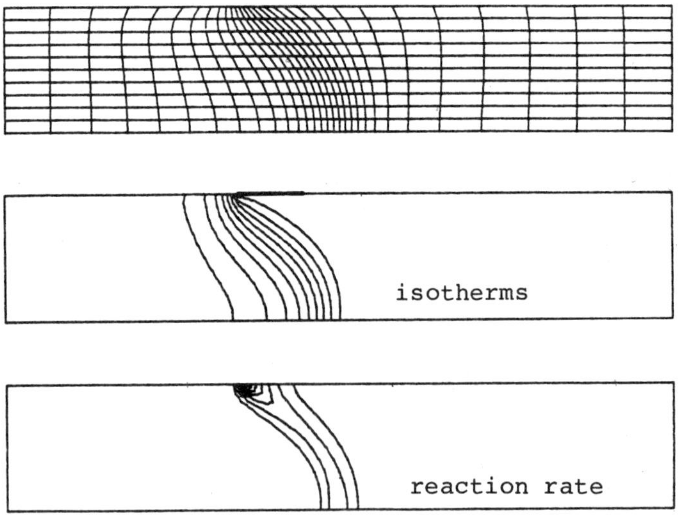

Figure 9 : Anchored flame. 31x11 adapted mesh.

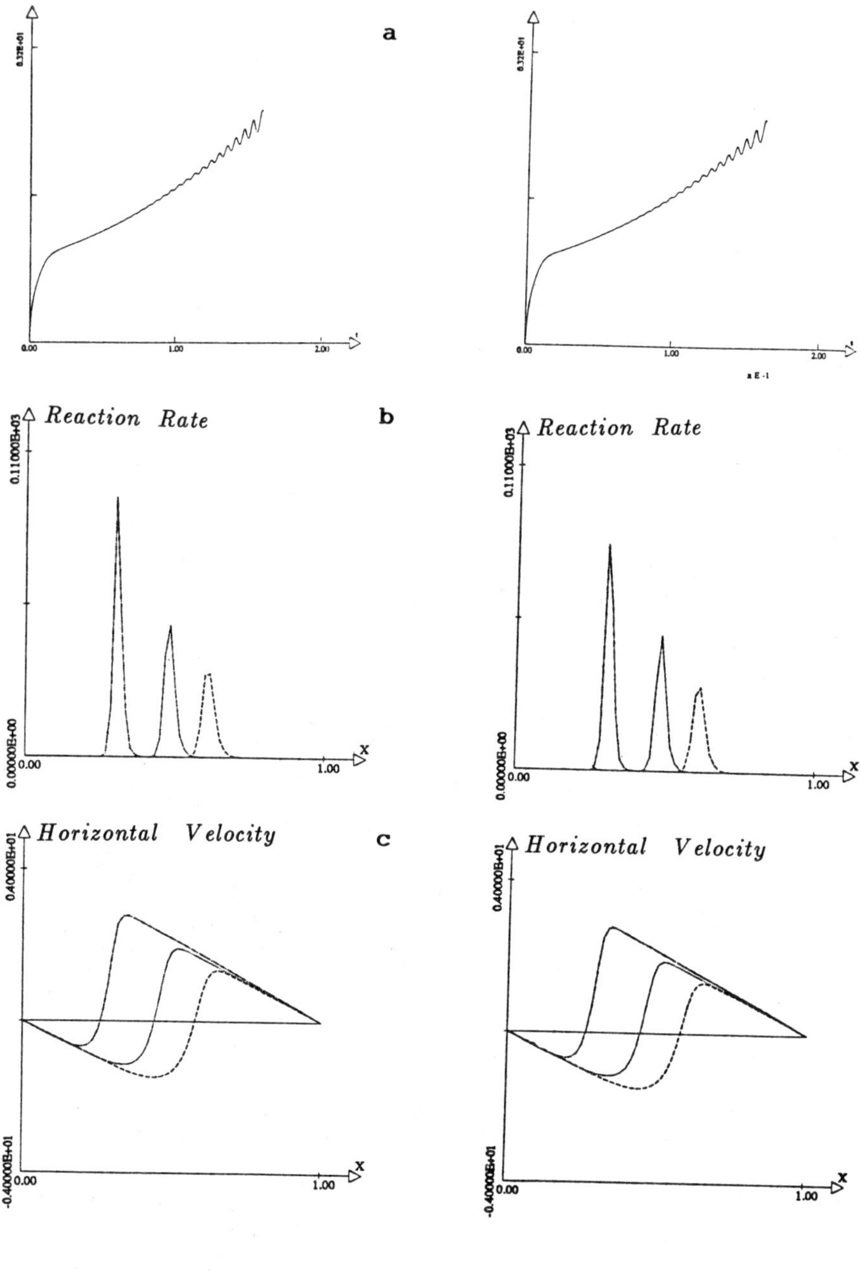

Figure 10 : Explicit (left) versus Implicit (right) computations of 1-d flame. flame speed history (a) reaction rate (b) and velocity (c) at different times.

VELOCITY FIELD

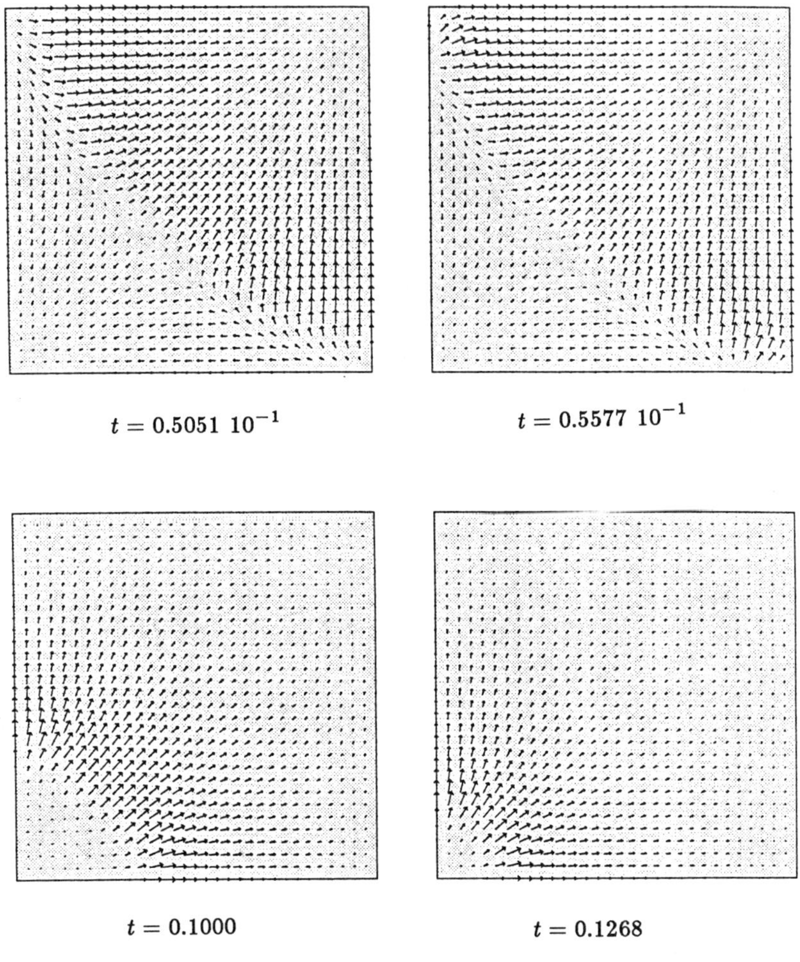

Figures 11

MASS FRACTION CONTOURS

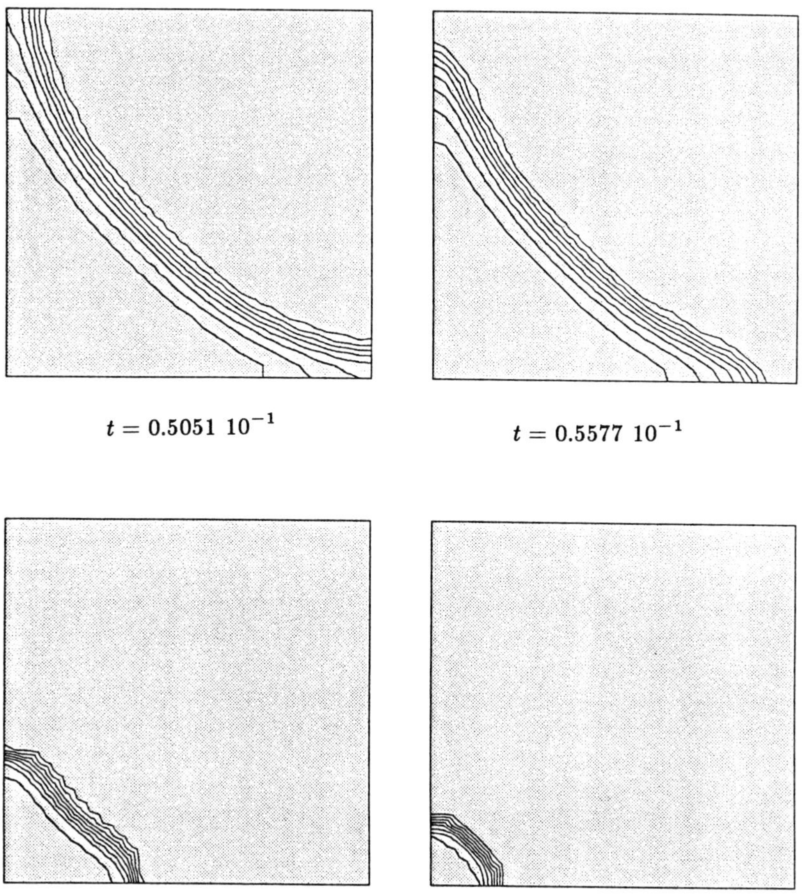

$t = 0.5051\ 10^{-1}$ $t = 0.5577\ 10^{-1}$

$t = 0.1000$ $t = 0.1268$

Figures 12

MODELING OF TURBULENT DIFFUSION FLAMES WITH DETAILED CHEMISTRY

H. BOCKHORN

Physikalisch Chemisches Institut, Universität Heidelberg
Im Neuenheimer Feld 253, D-6900 Heidelberg

1. INTRODUCTION

Because of their technical relevance chemically reacting turbulent flows have attracted increased interest in recent years. For instance, the need of high combustion efficiency and decreased pollutant emission from turbulent combustion devices has made great demands upon methods of prediction and calculation of these flows. As a consequence, models are required that are able not only to predict the turbulent flow field and global flame properties like the flame length or the overall heat release, but that allow equally well for an accurate and confident prediction of the temperature and the species concentrations throughout the flame. In particular, the detailed numerical prediction of the latter quantities requires complex chemistry to be taken into account, an essential shortcoming of many of the models of turbulent combustion currently available.

Besides pdf-methods (pdf = probability density function) [1], several practical possibilities exist to link chemical reactions to models of turbulent flow field predictions and to compute the turbulent mean values of the thermochemical variables of interest. One method is to solve modeled transport equations for the turbulent flow field and a conserved scalar quantity. The mean values of the other quantities of interest are then obtained by applying a laminar-flamelet model [2, 3]. In such a model the structure of a turbulent diffusion flame is thought to consist of an ensemble of laminar diffusion flames. The mean thermochemical variables in the turbulent flow are determined by the local mean values of the conserved scalar and the scalar dissipation rate. The relations between those quantities are derived, for example, from laminar-flamelet computations. This model is applicable if the length scales of turbulent eddies are larger than the reaction zone thickness of the flamelets and has successfully been applied to systems with fast chemistry, see e.g. [4, 5] . Problems that still remain are concerned with the closure of the averaged transport equations, e.g., the modeling of higher moments[1] or the finding of realistic relations between the conserved scalar and the thermochemical variables.

Another approach is to solve the averaged equations for the turbulent flow achieving closure of the system of differential equations by modeling the higher moments and to look at certain fluid particles from a Lagrangian point of view [6, 7] . The chemical reaction rate of an individual fluid particle is derived from the instantaneous values of the respective thermochemical variables, these values being determined by the "history" of the particle, e.g. by its turbulent exchange of mass and heat with other particles.

Finally, a third approach implies the solution of the averaged transport equations for the turbulent flow field including conservation equations for the enthalpy and mass fractions of the chemical species. Again, the higher moments have to be modeled to achieve closure of the problem. Furthermore, the chemical source terms in the species mass fractions transport equations must be modeled to describe the chemistry-turbulence interactions occuring in the flow. Here often a direct closure approach with a pdf-closure of the chemical production terms is being used [8,9] . This approach is not subjected to any restriction with regard to the length scales of the turbulent eddies and the reaction zone.

In this paper the latter approach is being pursued. In particular, flames in simple flow

[1]If not stated otherwise, throughout this paper by "higher moments" we mean moments higher than the mean, i.e., the first moment, of the respective variables.

C.-M. Brauner and C. Schmidt-Lainé (eds.), Mathematical Modeling in Combustion and Related Topics, 411–420.

geometries are numerically simulated, viz., turbulent confined H_2/air and CO/air jet diffusion flames, the reaction mechanisms of these systems comprising 44 elementary reactions of 11 species and 72 elementary reactions of 15 species, respectively. The numerical simulations are compared with experimental data available in the literature [10, 11].

2. MATHEMATICAL MODELING

2.1. The Flow – Field Model

In the present paper the turbulent mixing of a fuel jet with air is described by the Favre averaged conservation equations for a swirl-free, axisymmetric flow. A conventional eddy viscosity gradient model of turbulent transport and a $k-\epsilon$ turbulence model are adopted, because the focus of interest is the chemical source term closure. The flow field model thus comprises the transport equations for total mass, for momentum in longitudinal and radial direction, and the transport equations for turbulence kinetic energy k and dissipation rate ϵ (for details see e.g. [8, 12, 13, 14, 23]). For the turbulent viscosity the relationship

$$\mu_t = C_\mu \bar{\rho} k^2/\epsilon \tag{1}$$

is adopted. In Eq. (1), and below, $\bar{\rho}$ denotes the time averaged density and C_μ is an empirical constant, $C_\mu = 0.09$.

2.2. The Turbulent Reaction Model

The classical approach to link complex chemistry to models of turbulent flow field predictions involves modeling of the conservation equations of the chemical species. By virtue of an elliptic turbulent-transport operator L,

$$L(\tilde{\phi}) \equiv \frac{\partial(\bar{\rho}\tilde{u}\tilde{\phi})}{\partial x} + \frac{1}{r}\frac{\partial(r\bar{\rho}\tilde{v}\tilde{\phi})}{\partial r} - \frac{\partial}{\partial x}\left(\frac{\mu_t}{\sigma_\phi}\frac{\partial\tilde{\phi}}{\partial x}\right) - \frac{1}{r}\frac{\partial}{\partial r}\left(r\frac{\mu_t}{\sigma_\phi}\frac{\partial\tilde{\phi}}{\partial r}\right) \tag{2}$$

for any Favre averaged variable $\tilde{\phi}$, these equations may be written as

$$L(\tilde{Y}_i) = \overline{w_i}, \quad i = 1, \ldots N, \tag{3}$$

where $\overline{w_i}$ is the mean rate of production of species i.[1] Taking detailed chemistry into account, the local instantaneous value of the rate of production of species i is given by

$$w_i = M_i \sum_{k=1}^{M} (\nu''_{i,k} - \nu'_{i,k})(AT^\alpha)_k exp(-\frac{T_{a,k}}{T}) \prod_{j=1}^{N} \left(\frac{\rho Y_j}{M_j}\right)^{\nu'_{j,k}}. \tag{4}$$

Here M_i is the molar mass of species i, and $\nu'_{i,k}$ and $\nu''_{i,k}$ are its stoichiometric coefficients in reaction k, representing there reactant and product, respectively; T is the temperature, $(AT^\alpha)_k$ and $T_{a,k}$ are the pre-exponential factor in the specific reaction-rate constant and the activation temperature of reaction k, respectively. Note that Eq. (4) implies that forward and backward chemical reactions are treated separately.

According to the principles of pdf closure, the local mean rate of production of species i then is

$$\overline{w_i} = \int_0^1 \cdots \int_0^1 \int_0^1 \int_0^\infty \int_0^\infty w_i P(\rho, T, Y_1, Y_2, \ldots, Y_N; x, r) d\rho dT dY_1 dY_2 \ldots dY_N. \tag{5}$$

Here $P(\ldots)$ is a joint probability density function which in this work was empirically specified to be a normal distribution. With this specification the pdf-closure of the chemical production

[1] For details of the governing equations, including the numerical values adopted for various model constants such as the σ_ϕ's appearing in Eq. (2), reference is given to [23].

terms may be achieved by using a procedure similar to that developed by *Gutheil* [15] modeling the CO/air system by an overall reaction involving a three-variables rate expression.

In consistency with the applied turbulence model, where pressure fluctuations are neglected, the joint pdf appearing in Eq. (5) is represented by a $N+1$ dimensional clipped normal distribution, viz.,

$$P(\rho, T, Y_1, \ldots, Y_N) = \delta(\rho - \bar{\rho}) \cdot P(T, Y_1, \ldots, Y_N) \tag{6}$$

with

$$P(T, Y_1, \ldots, Y_N) = C\ exp\left[-\frac{1}{2}\,Q\right]. \tag{7}$$

In Eq. (7) the exponent Q as well as the constant C of the clipped normal distribution are functions, yet to be determined, of the mean values, variances and covariances of $T, Y_1, \ldots, Y_N$. Clipping of the normal distribution is necessary, because mass fractions and temperature vary between 0 and 1, and T_0 and T_{ad}, respectively, where T_{ad} denotes the adiabatic flame temperature.

The effect of weighting the local, instantaneous chemical production terms according to Eq. (5) is illustrated in Fig. 1 for the simple case of a two-variable function $w(\phi, \psi) = A\phi^2 exp(-B/\psi)$ which is characteristic for a bimolecular chemical-reaction-rate expression. The left hand side of this figure shows the logarithm of the ratio $\overline{w_\phi}/w(\tilde{\phi}, \tilde{\psi})$ and $\overline{w_\psi}/w(\tilde{\phi}, \tilde{\psi})$, respectively, in dependence on the turbulence intensities of ϕ and ψ. The $\overline{w_\phi}$ and $\overline{w_\psi}$ are obtained by weighting $w(\phi, \psi)$ with the marginal distributions of a two dimensional normal distribution, i.e. statistical independence of ϕ and ψ has been assumed. The graph clearly shows the effect of the exponential term of the rate expression: This term weighted by the marginal distribution brings about an acceleration of the reaction rate by some orders of magnitude depending on the value of the exponent whereas the quadratic term weighted in the same manner produces only a slight reaction rate enhancement. The right hand side of Fig. 1 elucidates the effect of the correlation of the two variables. Here the ratio of the reaction rate weighted by a two dimensional normal distribution to the marginal distribution weighted rate in dependence on the correlation

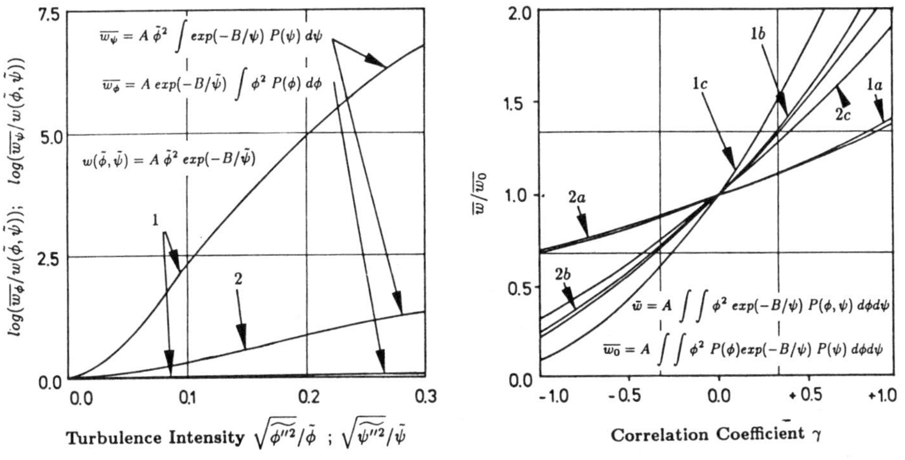

FIGURE 1. Effect of pdf-closure on the mean reaction rate for a two-variable rate expression. Case 1: $\tilde{\phi} = 0.1$, $\tilde{\psi} = 1200$, $B = 51600$; Case 2: $\tilde{\phi} = 0.5$, $\tilde{\psi} = 2500$, $B = 35000$; a: $\sqrt{\widetilde{\phi''^2}}/\tilde{\phi} = 0.1$, $\sqrt{\widetilde{\psi''^2}}/\tilde{\psi} = 0.1$, b: $\sqrt{\widetilde{\phi''^2}}/\tilde{\phi} = 0.3$, $\sqrt{\widetilde{\psi''^2}}/\tilde{\psi} = 0.1$, c: $\sqrt{\widetilde{\phi''^2}}/\tilde{\phi} = 0.3$, $\sqrt{\widetilde{\psi''^2}}/\tilde{\psi} = 0.3$. Further explanation see text.

coefficient $\gamma = \widetilde{\phi''\psi''}/\sqrt{\widetilde{\phi''^2}}\sqrt{\widetilde{\psi''^2}}$ is displayed. Note that for $\gamma = 0$ the variables ϕ and ψ are statistically independent, for $|\gamma| = 1$ they depend linearly on each other. The graphs illustrate that, compared with statistically independent variables, a positive correlation accelerates the rate depending on the turbulence intensities and the absolute values of the variables. The effect of the exponential term of the rate expression again is more pronounced.

For the general case considered in the present paper the pdf-closure must be applied to each of the individual terms of the sum appearing in Eq. (4). For instance, for a bimolecular reaction k in which species i is being produced the respective terms, say $\overline{w_{i,k}}$, may be written as

$$\overline{w_{i,k}} = \int_0^1 \cdots \int_0^1 \int_0^1 \int_0^\infty M_i (AT^\alpha)_k exp(-\frac{T_{a,k}}{T}) \left(\frac{\rho Y_l}{M_l}\right) \left(\frac{\rho Y_m}{M_m}\right)$$

$$P(T, Y_1, Y_2, \ldots, Y_l, Y_m \ldots, Y_N) dT dY_1 dY_2 \ldots dY_N. \quad (8)$$

The findings for $w(\phi, \psi)$ discussed above have been generalized by evaluating rate expressions like that given by Eq. (8) for a number of selected mean values, turbulence intensities and correlation coefficients of T, $Y_1, \ldots, Y_N$ and for all possible combinations of $\nu_{i,k}$. The mean rates of chemical production have then been subjected to a multiple regression analysis resulting in a "library" of relations for the mean rates as functions of these quantities. During the computation of a turbulent flame these relations are then drawn from the "correlation library" and they are used, according to Eq. (8), for the pdf-closure of the chemical production terms appearing in Eq. (3).

2.3. Further Modeling Requirements

Pdf-closure of the mean chemical production rates according to Eq. (8) necessitates the solution of transport equations for variances and covariances, respectively, of various scalar quantities. Assuming gradient transport and "dissipation hypothesis" [8, 16] analogously to the turbulent flow field model the transport equations for the covariances of two scalars ϕ and ψ (and hence for the variances of ϕ or ψ) can be written as

$$L(\widetilde{\phi''\psi''}) = C_1 \mu_t \left[(\frac{\partial \tilde{\phi}}{\partial x} \frac{\partial \tilde{\psi}}{\partial x}) + (\frac{\partial \tilde{\phi}}{\partial r} \frac{\partial \tilde{\psi}}{\partial r}) \right] - C_2 \bar{\rho} \epsilon \frac{\widetilde{\phi''\psi''}}{k} + \overline{\phi'' w_\psi} + \overline{\psi'' w_\phi} \,, \quad (9)$$

with empirical constants $C_1 = 2.8$ and $C_2 = 2.0$. Again, the last two terms on the right hand side of Eq. (9) are obtained by pdf-closure. The main difference of the Eq. (9) to the averaged conservation equations for momentum, enthalpy, k and ϵ are the additional chemical production terms and the terms containing the gradients of $\tilde{\phi}$ and $\tilde{\psi}$. Thus, a correlation coefficient does not differ very much from unity for scalar quantities with similar profiles, e.g. mixture fraction and enthalpy, the source terms of which in the respective transport equations for are identically zero. For chemical species which are present in either one of the two feeds and are both reactants, products or intermediates of the chemical reaction, the correlation coefficient is nearly 1 in the region of weak chemical reaction and decreases in the main reaction zone of the flow field depending on the strength of the chemical source or sink. Similarly, the correlation coefficients for scalars the axial or radial gradients of which attain different sign, e.g. for reactants present in both of the feeds or for reactants and products etc., tend to small values depending on the course of the gradients and the respective source terms (for further discussion see, e.g., [15]). This behaviour is used in this work to estimate the correlation coefficients entering the pdf-closure corresponding to Eq.(8) rather than to solve $(N^2 - N)/2$ additional equations for the covariances.

The mean temperature $\tilde{T}$ has been calculated from the mean specific enthalpy $\tilde{h}$ the latter quantity being obtained by solving the transport equation

$$L(\tilde{h}) = 0, \quad (10)$$

where $\tilde{h} = \sum_{1}^{N} \widetilde{Y_i h_i}$. In particular, expressing the specific enthalpies h_i of the chemical components as polynomials [17], $h_i = \sum_{k=1}^{p} a_{i,k} T^k$, the caloric equation of state can be written as

$$\tilde{h} = \sum_{i=1}^{N} \sum_{k=0}^{p} a_{i,k} \left[\tilde{Y}_i \tilde{T}^k + \frac{k(k-1)}{2} \tilde{Y}_i \tilde{T}^{k-2} \widetilde{T''^2} + k \tilde{T}^{k-1} \widetilde{Y_i'' T''} \right], \tag{11}$$

which, for given $\tilde{h}$, represents an implicit equation for $\tilde{T}$. Note that in Eq. (11) moments of orders higher than two have been neglected.

Similarly, the temperature variance $\widetilde{T''^2}$ can be obtained from the enthalpy variance $\widetilde{h''^2}$, the transport equation of which is [23]

$$L(\widetilde{h''^2}) = C_1 \mu_t \left[\left(\frac{\partial \tilde{h}}{\partial x} \right)^2 + \left(\frac{\partial \tilde{h}}{\partial r} \right)^2 \right] - C_2 \bar{\rho} \widetilde{h''^2} \frac{\tilde{\epsilon}}{\tilde{k}}. \tag{12}$$

Closure of the set of governing equations is completed through the ideal-gas equation of state,

$$\bar{p} = \bar{\rho} R \sum_{i=1}^{N} (\tilde{T} \tilde{Y}_i + \widetilde{Y_i'' T''}) / M_i, \tag{13}$$

where $\bar{p}$ means the mean pressure and R is the universal gas constant.

The reaction mechanisms of the CO/air flame and the H_2/air flame are given in detail in [18]. In particular, for the CO/air flame 72 elementary reactions of 15 species, viz. $N_2, CO, CO_2, O_2, H_2, H_2O, H, O, OH, HO_2, H_2O_2, CHO, CH_2O, N$ and NO have been taken into account whereas for the H_2/air flame 44 elementary reactions between the 11 species $N_2, O_2, H_2, H_2O, H, O, OH, H_2O_2, HO_2, N$ and NO constitute a comprehensive mechanism.

2.4. Numerical Method and Boundary Conditions

The computations were performed using the method developed by *Spalding* and co-workers [19, 20]. A non-uniform grid with 20 times 24 grid points was found to be sufficient for producing numerically accurate results. The appropiate boundary conditions for density, mass fluxes and enthalpy were computed according to the experimental conditions given in [10] and [11]. A suitable power law was assumed as initial velocity profile in the fuel nozzle. At the inlet of the combustion device radial profiles were specified for all dependent variables, at the outlet zero gradients in axial direction. The boundary conditions for k, ϵ and μ_t at the walls were calculated using the wall functions described in refs. [19, 20], the walls themselfs assumed as being impermeable for matter. Since no information on wall temperatures and heat transfer properties are given in [10] and [11], the combustion devices were treated as adiabatic systems.

3. RESULTS AND DISCUSSION

3.1. Calculation of Variances

As has been discussed in the previous section, the applied pdf-closure requires the calculation of the variances of temperature and mass fractions. Figure 2 exemplarily gives profiles of the turbulence intensities of temperature, $\sqrt{\widetilde{T''^2}}/\tilde{T}$, and carbon dioxide mass fraction, $\sqrt{\widetilde{Y''^2_{CO_2}}}/\widetilde{Y_{CO_2}}$, for the CO/air flame. The figure shows that the turbulence intensity of the temperature initially raises to about 45%, the maximum being attained in the main reaction zone which is located at $20 \leq x/D \geq 50$,compare Fig. 3, and then decreases. Similarly, in radial direction the turbulence intensity of the temperature decreases and attains low values outside the main reaction zone. The turbulence intensity of the carbon dioxide mass fraction varies between 10% and 50% in the reaction zone. At locations in the flame with very low carbon dioxide mass fractions and negligible conversion these turbulence intensities increase to even higher values.

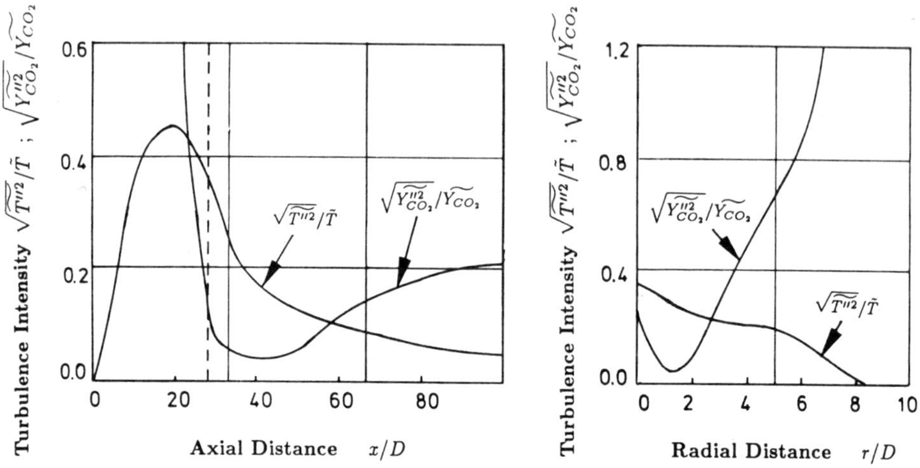

FIGURE 2. Calculated turbulence intensities of temperature and carbon dioxide mass fraction for the carbon monoxide air diffusion flame. Radial profiles are given at the axial distance indicated by the dashed line.

In general the turbulence intensities in the region of the flame with non-negligible reaction rates do not exceed about 80%.

3.2. Mean Turbulent Reaction Rates

Profiles of the mean reaction rate for the reaction $CO + OH \longrightarrow CO_2 + H$, which is one of the essential reactions of the CO/air system, are shown in Fig. 3. The three cases indicated correspond to different averaging procedures. Case 1 involves the mean values of temperature and mass fractions, in case 2 the instantaneous rates are weighted by the respective marginal distributions of the more dimensional normal distribution assuming statistically independent variables. In case 3 the mean reaction rate is calculated according to Eq. (8). This figure clearly shows the effect of the pdf-closure on the rate of this reaction which has an activation temperature of 4030 K [18] for temperatures between 1800 K and 2700 K. Weighting the instantaneous reaction rate by the marginal distributions doubles the reaction rate in the main reaction zone of the flame. This reflects the effect of the temperature fluctuations on the exponential term of the rate expression of this reaction. Furthermore, the weighting procedure according to Eq. (8) leads to an enhancement of the mean turbulent reaction rate. Simultaneously, the thickness of the zone where this reaction occurs is slightly shrinked.

3.3. Temperature and Species Concentrations

In Figs 4 and 5 profiles of the mean temperature and the mean concentrations of the major species are compared with various experimental data, viz., data on the CO/air flame by *Razdan* and *Stevens* [10] and on the H_2/air flame by *Drake et al.*[11].

For both flames it is seen that the agreement between measured and computed profiles for both systems generally is good. The observed disagreement between computed and measured temperatures for the CO/air flame are due to the boundary conditions that are not unambiguously defined in ref. [10] for the respective experimental system. Furthermore, in ref. [10] temperatures have been measured by unshielded thermocouples and, therefore, some of the discrepancies may be due to inconsistencies in ref. [10] in calculating radiation corrected values for the measured temperature data.

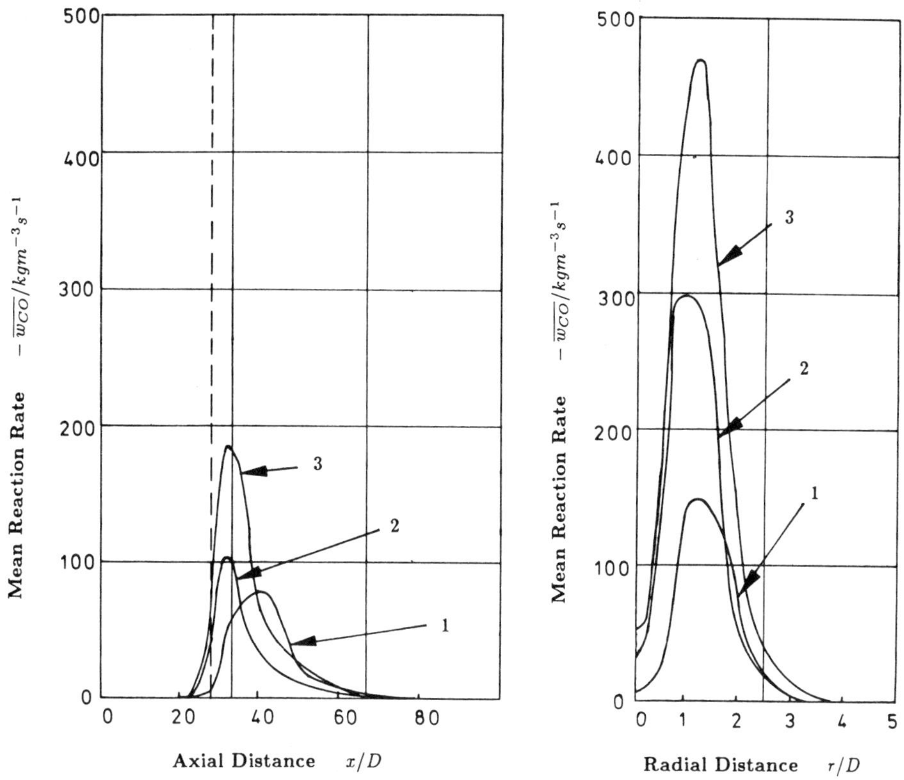

FIGURE 3. Calculated mean reaction rate for the reaction $CO + OH \longrightarrow CO_2 + H$. Radial profiles are given at the axial distance indicated by the dashed line. Further explanation see text.

Compared with the measured profiles, in radial direction a less extended reaction zone can be observed from the calculated radial profiles of both flames. This may be due to shortages of the assumed pdfs which may fail to describe the correlation of the thermochemical variables in the region of the jet edges because there the variables determining the reaction rate of the oxidation of carbon monoxide or hydrogen are no longer normally distributed. As a consequence, the effect of turbulent fluctuations on the chemical reaction rate is not properly predicted leading to overestimated mean chemical reaction rates. Furthermore, the chemical reactions may be influenced by "coherent structures" that may exist in this region and that, as well-known, cannot be predicted properly by the k-ϵ model of turbulence. In addition, for the H_2/air flame it might be necessary to improve the turbulent transport model by taking into account effects of preferential diffusion of hydrogen molecules or atoms (cf.[21]).

Some other results of this work will be discussed elsewhere [22].

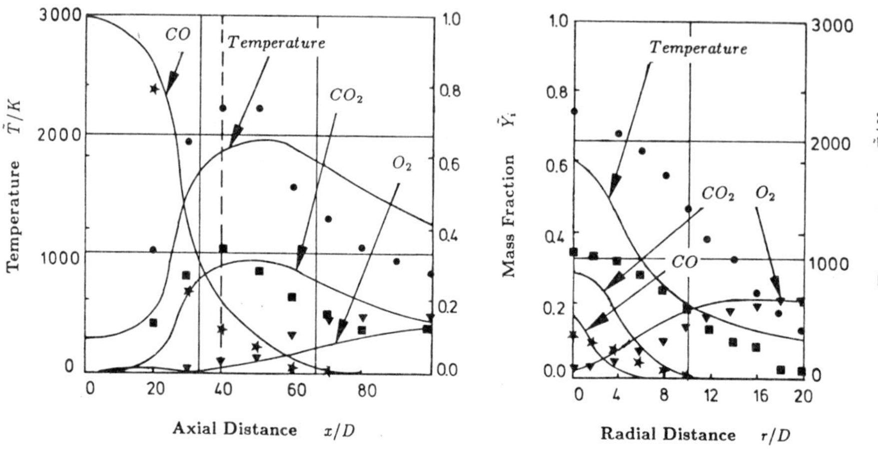

FIGURE 4. Profiles of mean temperature and mean mass fractions of major species for the carbon monoxide air diffusion flame. Solid lines represent calculations; symbols represent measurements reported by *Razdan* and *Stevens*[10]. (●) : temperature, (★) : carbon monoxide, (▼) : oxygen, (■) : carbon dioxide. Radial profiles are given at the axial distance indicated by the dashed line.

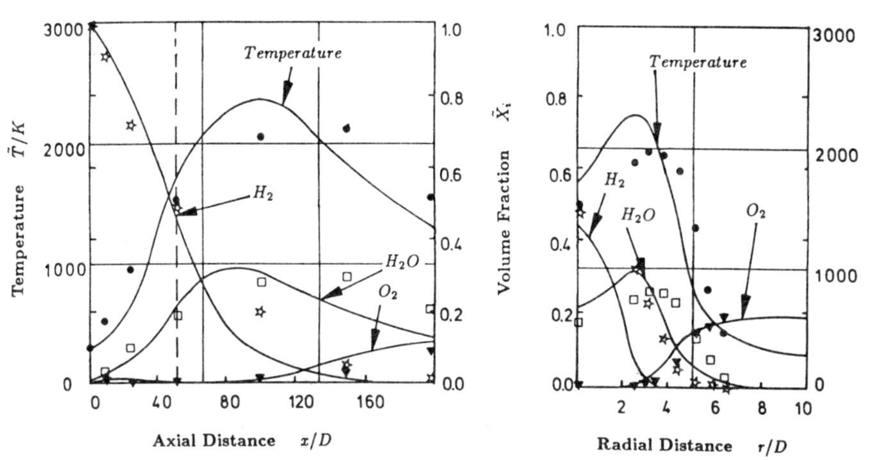

FIGURE 5. Profiles of mean temperature and mean mole fractions of major species for the hydrogen air diffusion flame. Solid lines represent calculations; symbols represent measurements reported by *Drake et al.*[11]. (●) : temperature, (☆) : hydrogen, (▼) : oxygen, (□) : water. Radial profiles are given at the axial distance indicated by the dashed line.

4. CONCLUSIONS

In this work a pdf-closure for the chemical production terms has been adopted to link complex chemistry to turbulent flow field predictions. In particular, a convenient method to handle more dimensional pdfs involved in the applied pdf-closure of the chemical source terms is suggested.

As first examples turbulent H_2/air and CO/air jet diffusion flames were numerically simulated. The agreement between the calculations and measurements for these flames available in the literature show reasonably good agreement though some aspects of the modeling, e.g., the modeling of variances and covariances, need experimental verification.

The author wishes to thank his colleagues E. Gutheil and B. Rogg for many helpful discussions.

REFERENCES

1. *Pope.S.B.*, PDF Methods for Turbulent Reactive Flows, Prog. Energy Combust. Sci. 11, 119 (1985)
2. *Williams,F.A.*, Recent Advances in Theoretical Descriptions of Turbulent Diffusion Flames, in: Turbulent Mixing in Nonreactive and Reactive Flows, ed S. N. B. Murthy, p. 189, Plenum Press, New York (1975).
3. *Peters,N.*, Laminar Diffusion Flamelet Models in Non-Premixed Turbulent Combustion, Prog. Energy Combust. Sci. 10, 319 (1984)
4. *Liew,S.K., Bray,K.N.C and Moss,J.B.*, A Stretched Laminar Flamelet Model of Turbulent Non-Premixed Combustion, Combust. Flame 56, 199 (1984)
5. *Rogg,B., Behrendt,F. and Warnatz,J.*, Turbulent Non-Premixed Combustion in Partially Premixed Diffusion Flamelets with Detailed Chemistry, Twenty First Symposium (International) on Combustion, The Combustion Institute, Pittsburgh 1987
6. *Ma,A.S.C., Spalding,D.B and Sun,R.L.T*, Application of "Escimo" to the Turbulent Hydrogen-Air Diffusion Flame, Nineteenth Symposium (International) on Combustion, p. 393, The Combustion Institute, Pittsburgh, 1982
7. *Borghi,R. and Pourbaix,E.*, Lagrangian Models for Turbulent Combustion, in: Turbulent Shear Flows 4, Selected Papers from the Fourth International Symposium on Turbulent Shear Flows,ed. L.J.S. Bradbury, F.Durst, B.E.Launder, F.W.Schmidt and J.H.Whitelaw, p.369, Springer Verlag, Heidelberg 1985
8. *Bilger,R.W.*, Turbulent Flows with Nonpremixed Reactants, in: Turbulent Reacting Flows, ed. P. A. Libby, F. A. Williams , p.65, Springer Verlag, Heidelberg 1980
9. *Bray,K.N.C.*, The Interaction between Turbulence and Combustion, Seventeenth Symposium (International) on Combustion, p. 223, The Combustion Institute, Pittsburgh, 1979
10. *Razdan, M.A. and Stevens,J.G.*, CO/Air Turbulent Diffusion Flame: Measurements and Modeling, Combust. Flame 59, 289 (1985)
11. *Drake, M.C., Pitz, R.W., Lapp. M., Fenimore, C.P., Lucht, R.P., Sweeney, D.W. and Laurendeau, N.M.*, Measurements of Superequilibrium Hydroxyl Concentrations in Turbulent Nonpremixed Flames Using Saturated Fluorescence, Twentieth Symposium (International) on Combustion, p. 327, The Combustion Institute, Pittsburgh, 1984
12. *Libby,P.A. and Williams,F.A.*,Fundamental Aspects, in: Turbulent Reacting Flows, ed. P. A. Libby, F. A. Williams, p. 1, Springer Verlag, Heidelberg 1980
13. *Launder,B.E. and Spalding,D.B.*, Mathematical Models of Turbulence, Academic Press, New York, 1972

14. *Kent, J.H. and Bilger, R.W.*,Fluctuations in a Round Sixteenth Symposium (International) on Ccombustion, p. 1643, The Combustion Institute, Pittsburg, 1977
15. *Gutheil,E. and Bockhorn,H.*, The Effect of More Dimensional PDFs on the Turbulent Reaction Rate in Turbulent Reacting Flows at Moderate Damköhler Numbers, Physico Chemical Hydrodynamics, 9 (1987), to appear
16. *Spalding,D.B*, Concentration Fluctuations in a Round Turbulent Free Jet, Chem. Eng. Sci. 26, 95 (1971)
17. *Stull,D.R. and Prophet, H.*, JANAF Thermochemical Tables, Office of Standard Reference Data, NBS, Washington, D.C., 1971
18. *Warnatz, J.*, Rate Coefficients in the C/H/O System, in: Combustion Chemistry, ed. W. C. Gardiner, Jr. , p. 197, Springer Verlag, New York (1984).
19. *Gosman, A. D., Pun, W. M., Runchal, A. K., Spalding, D. B., Wolfshtein, M.*, Heat and Mass Transfer in Recirculating Flows, Academic Press, London (1969).
20. *Srivatsa, S. K.*: Cham Computer Code 201, CORA2 - A Computer Code for Axi–Symmetrical Combustion Chambers, CHAM Technical Report 201/1 (1977).
21 *Drake, M.C.*, Stretched Laminar Flamelet Analysis of Turbulent H_2 and $CO/H_2/N_2$ Diffusion Flames, Twenty First Symposium (International) on Combustion, The Combustion Institute, Pittsburh 1987
22. *Bockhorn, H., Rogg, B. and Warnatz, J.*, A Direct closure Approach for Modeling of Turbulent Diffusion Flames with Detailed Chemistry, Paper presented at the Sixth Symposium on Turbulent Shear Flows, Toulouse 1987
23. *Behrendt, F., Bockhorn, H., Rogg, B. and Warnatz. J.*, Modeling of Turbulent CO/Air Diffusion Flames with detailed Chemistry, in: Proc. 2nd Int. Workshop on Modeling Chemical Reaction Systems, ed. J.Warnatz, Springer Verlag, Heidelberg, in Press

KINETIC MODELLING OF LIGHT HYDROCARBONS COMBUSTION

J.C. Boettner, M. Cathonnet, P. Dagaut and F. Gaillard

Centre de Recherches sur la Chimie de la Combustion et des Hautes Températures (CNRS) 45071 ORLEANS Cedex 2 - FRANCE

The use of detailed mechanisms to model the oxidation of light hydrocarbons is now possible because of the availability of kinetic data for the elementary reaction steps (Warnatz 1984). Such comprehensive mechanisms are built by comparison with experimental data obtained for a wide range of physical conditions in flames, plug-flow reactors, perfectly-stirred reactors and shock tubes. In the three last experimental techniques, the transport effects are negligible and the model equations are substantially simplified.

The oxidation of ethylene, propene, and propane in a jet stirred reactor has been studied experimentally (Dagaut 1986). A wide range of conditions were investigated : temperature varying from 900K to 1200K, pressure extending from 1 atm to 10 atm, and fuel-oxygen equivalence ratios including very lean to very rich mixtures (0.15 to 4.0) . The concentrations of molecular species were determined by gas chromatography . By varying the nominal residence time, it was possible to describe the evolution of the reaction from the beginning up to complete conversion .

KINETIC MODELLING

Chemical kinetics problems and atmospheric modelling often lead to initial value ordinary differential equations (ODE) :

$$dy/dt = f(t,y) \qquad a \le t \le b \quad , \ y(a) \text{ given} .$$

A local solution u(t) to this ODE is found by using a numerical method and the question which arises is how accurate is the approximation.

Problems of great practical importance deal with large Lipschitz constants and even when the mathematical problem is well conditioned and stable in the sense that small perturbations in initial conditions lead to small changes in the true solution , an attempt to integrate this problem with standard methods (Runge Kutta, Adams) leads to a severe step-size reduction due to numerical stability : this is what is called "stiffness".

C.-M. Brauner and C. Schmidt-Lainé (eds.), Mathematical Modeling in Combustion and Related Topics, 421–429.

The description of combustion phenomena involves transport together with chemical reaction, and stiffness occurs because :

- The time scale of the reactions is much smaller than transport times.
- The time scale between species is quite different.

In diffusion-convection problems, discretization of partial differential equations by the method of lines can lead also to stiff ODE.

By far the most popular numerical technique to reduce stiffness is the use of implicit techniques, and the development of ODE solver in the last fifteen years had lead to a variety of packages and automatic black-boxes.

Four classes of solvers are in wide use :

- backward differentiation formulae (BDF)
Gear (1971), Hindmarsh (1974,1977), Cash (1980)
- implicit Runge Kutta methods
Butcher (1979a,1979b), Alexander (1977)
- semi implicit Runge Kutta methods
Rosenbrock (1963) , Verwer (1981), Kaps (1979,1981)
- extrapolation methods
Bader (1981)

Packages exist for mass-action kinetics : some of them can handle large systems of equations through the use of sparse matrix techniques.

- HAVCHEM	Stabler (1978)
- MACKSIM	Carver (1979)
- CHEMKIN	Kee (1980)
- LARKIN	Deuflhard (1981)
- CREK1D	Pratt (1984)

Chemical kinetics are by far the largest stiff application area : stiffness arises mainly from large differences among the reaction rate constants. The law of mass-action states a linear or quadratic form in the concentration variable but if temperature is also varying, very severe exponential non-linearities are introduced through the Arrhenius term ($A.T^n.\exp(-E/RT)$) leading to a dramatic acceleration of the global reaction rate, especially in chain branching mechanisms. Kineticists soon turned to the quasi steady state approximation (QSSA) in order to overcome stiffness problems : it is effective when used by a skilled expert but leads to unreliable accuracy and efficiency in other cases. Fortunately most stiff ODE solvers are now able to integrate safely the great majority of problems without any assumption of QSSA in the model.

METHOD USED

It is described in CASH (1983) as a simple case of split linear multistep methods and uses :

- an implicit θ predictor

$$y^*(t+h)=y(t)+h.[(1-\theta).f(t)+\theta.f^*(t+h)]$$

- a modified trapezoidal rule corrector rewritten as

$$y^{**}(t+h)=y(t)+h.[f(t)+f^{**}(t+h)]/2+h.(\theta-\tfrac{1}{2}).[f^{**}(t+h)-f^*(t+h)]$$

The free parameter θ is choosen to make the method L-stable (the ratio y(t+h)/y(t) --> 0 when applied on the scalar test equation f = L.y with h.L --> infinity) and minimize local truncation error. The order of the method is 2 and an error estimate is available ($y^{**}-y^*$).

Both equations share the same coefficient matrix (I-h.θ.J where J denote the jacobian matrix $\delta f/\delta y$) and this property is essential for good numerical efficiency. The system is solved using a modified Newton method where the coefficient matrix is decomposed into triangular factors : repeated substitution processes to solve y(t+h) are very cheap. An other important cost saving device is to use an approximate jacobian for as many steps as convergence is adequate. Jacobian matrix can be estimated numerically (many ODE solvers have this opportunity) but it is relatively expensive and leads to instabilities in the computed solution especially under near equilibrium conditions. Fortunately the particular structure of our chemical systems enables an easy calculation of time derivative function f and jacobian matrix J.

This method has been used intensively in our Laboratory under various parametric conditions and experimental techniques (plug-flow reactor and reflected shock-waves).

Of great importance is the scaling of the local error estimate especially at low tolerance requirement. Scaling is performed by using different weights in the norm of the error estimate. We choosed the strategy used in LSODE (Hindmarsh 1980), a mixed relative and absolute tolerance :

$W_i=RTOL.Y_i+ATOL_i$ with :

- RTOL in the range 0.001-0.01
- $ATOL_i = 10^{-6}.Y_i(0)$ for initial reactives
  $=10^{-20}$ for products and intermediates
  (units in mole/cm^3)

We have been using LSODE and other Gear-like methods for many years and found them very reliable but rather difficult to interface on existing software especially under non isothermal conditions and mixed differential-algebraic equations.

Adiabatic reactor and reflected shock tube calculation are examples of non isothermal conditions where numerical stability enforces a relationship between step-size and temperature increase per step : we found empirically that an increase of 5 Kelvin was an upper limit condition for stability, and step-size depends heavily on it in the heat release zone.

PERFECTLY-STIRRED REACTOR

Description of the process occurring within a perfectly stirred reactor is obtained by relating the conservation of mass and energy to the net generation of chemical species in the volume (Jones 1968, Glarborg 1986).

The species conservation equations are given by

$$\phi(\sigma) \equiv \sigma - \sigma_e - f/dm = 0 \quad \text{with} \quad \phi'(\sigma) \equiv I - t_s.(\delta f/\delta y)$$

where

t_s : residence time
dm : mass flow rate
f : net chemical rate of each species
y : output molar concentration
σ, σ_e : output and input mass concentration
$I, \delta f/\delta y$: unity and jacobian matrix

Conservation of energy was not stated because experimental temperature (constant to within ± 3 Kelvin) is introduced in our isothermal model.

For a given set of residence times and initial estimated y, we used the same modified Newton method to solve these equations with positive constraints on σ. Surprisingly, such a computation is much more difficult than the "equivalent" differential problem and we had some troubles in tracking the correct trajectory which is not always attainable from low residence time : in fact we must start with near equilibrium conditions and go back along the trajectory.

The corresponding computation code shares a great number of routines with our ODE solver (chemical interpreter, linear system solver, thermodynamic routines, graphical output) and standardization of user interface is one of the key of transportability.

SENSITIVITY

Sensitivity analysis is becoming an effective mean of determining model details and quantifying the effects of parametric incertainty. Considerable work has been made in that field (Koda 1979, Dougherty 1980, Kramer 1982a,1982b).

Linear sensitivity coefficients are defined as $\delta y_i/\delta\alpha_j$ where y_i is usually a species concentration and α_j a rate constant. Second order sensitivities can be defined, providing information on non-linearities and parametric interactions. We have used a brute force method to compute first order sensitivities : it is very effective , easy to code and provides the necessary information to adjust safely unrealistic rate constants and optimize species profiles.

Complex chemical models frequently involve 50-100 species and 200-700 elementary steps. Modelling reactive dynamic flows including gas phase transport and gas-surface interaction necessitates the simultaneous solution of a very large number of points over a 3-D mesh. Solving the corresponding large stiff system of chemical kinetic equation is impracticable or at least very computer time consuming. Sensitivity analysis plays an important role in such problems by enabling simplification of the detailed mechanism to a realistic though much smaller mechanism.

COMPARISON WITH EXPERIMENTAL RESULTS IN P.S.R.

A detailed kinetic mechanism involving 277 reactions and 48 species has been developped (Dagaut et al. 1987) from existing litterature data (Warnatz 1984, Pitz and Westbrook 1986) to reproduce the experimental results. Sensitivity analysis was performed to refine the mechanism, and some rate constants were adjusted in their uncertainty range to fit our data.

Very good agreement was observed between computed results and experimental data for most of the molecular species determined experimentally (Dagaut et al. 1987). Examples of comparison between computation (lines) and experimental data (points) are given in figures 1 (propene oxidation), 2 (ethylene oxidation) and 3 (propane oxidation).

MODELLING OF IGNITION DELAYS

The same mechanism was used to simulate experiments of other authors on C_1 to C_3 hydrocarbon oxidation behind a reflected shock wave. In such conditions, we can make the constant density approximation, and the ignition delay was defined as the maximum rate of temperature rise.

The mechanism gives a good prediction of the ignition delays determined experimentally by several authors in the oxidation of methane, ethane, ethylene and propane. These experiments covered a wide range of operating conditions :
$900\ K <= T <= 2200\ K$ and $3\ atm <= P <= 12\ atm$.

Fig 4 shows an example of the comparison between the results of our computations (dotted lines) and the experimental data (full lines) of Burcat et al. (1971).

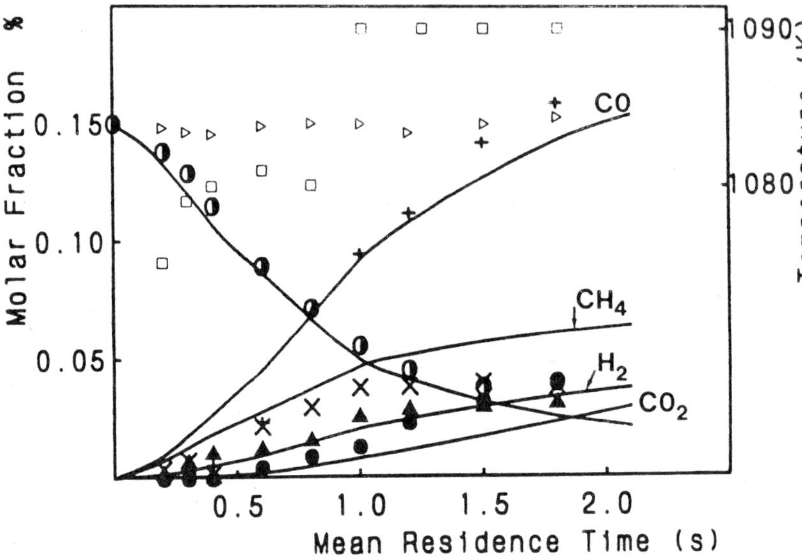

Figure 1 : Oxidation of propene in the jet stirred reactor.

P = 8 atm T = 1085 K = 1.5

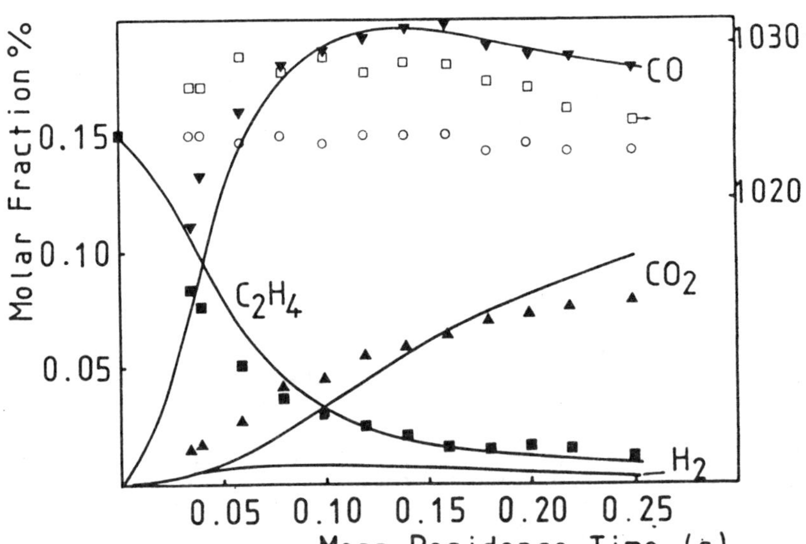

Figure 2 : Oxidation of ethylene in the jet stirred reactor.

P = 5 atm T = 1085 K = 2

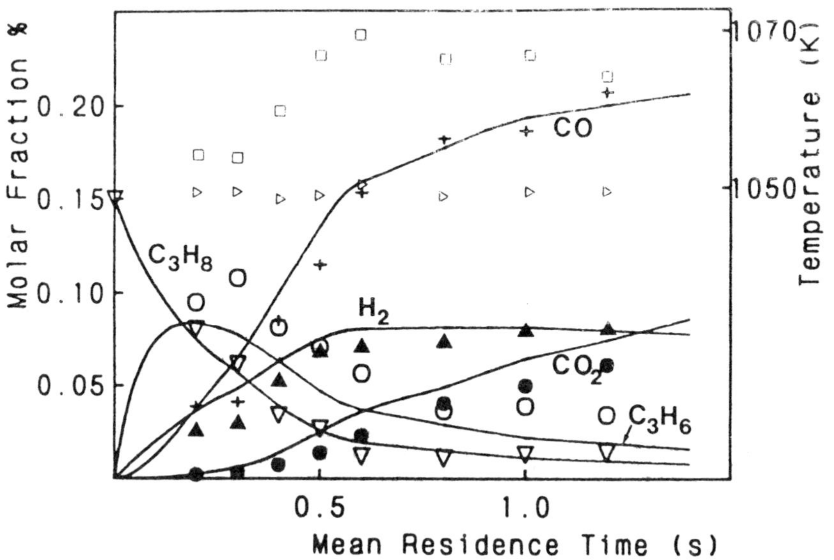

Figure 3 : Oxidation of propane in the jet stirred reactor.

P = 5 atm T = 1060 K = 1.5

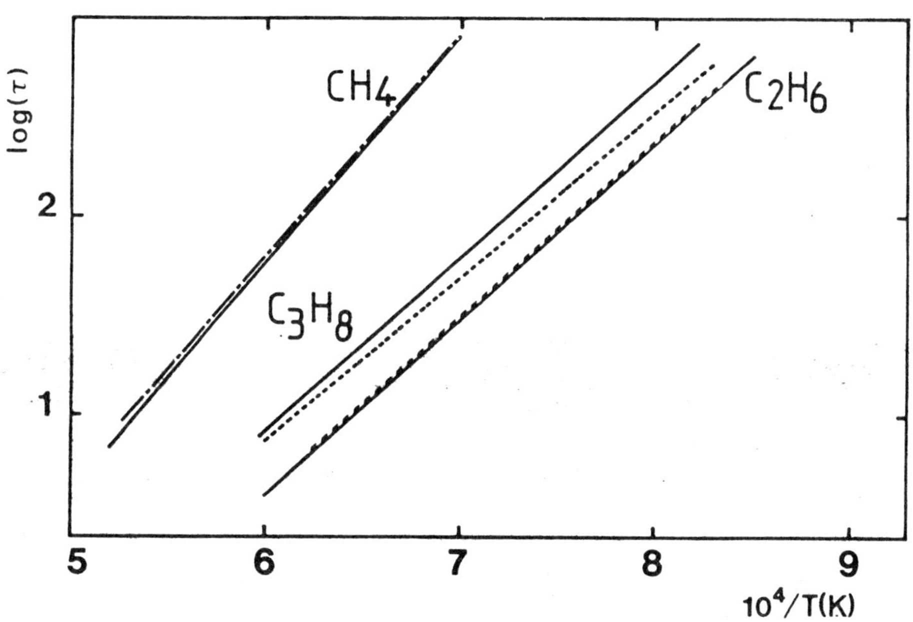

Figure 4 : Experiments of Burcat and Col. (Comb. and Flame, 1971, 16, 29) : Full lines . Computation : dotted lines.

REFERENCES

- Alexander, R. :
  Siam J. Numer. Anal.,14,1006 (1977)
- Bader, G.,Deuflhard :
  SFB 123, Rpt. 114, Heidelberg Univ. (1981)
- Burcat, A., Scheller, Lifshitz :
  Combust. and Flame ,16,29 (1971)
- Butcher, J.C. :
  J. ACM.,26,731 (1979a)
- Butcher, J.C., Burrage, Chipman :
  Rpt. 150, Dept of Computer Sci., Auckland Univ. (1979b)
- Carver, M.B., Boyd :
  Int. J. Chem. Kin.,11,1097 (1979)
- Cash, J.R. :
  Numer. Math.,34,219 (1980)
- Cash, J.R. :
  Numer. Math.,42,299 (1983)
- Dagaut, P. :
  Thèse de l'Université Pierre et Marie Curie, Paris (1986)
- Dagaut, P., Cathonnet, Boettner and Gaillard :
  submitted for publication (1987)
- Deuflhard, P., Bader, Nowak :
  "Modelling of chemical reaction systems" Series in Chemical Physics 18. Springer(1981)
- Dougherty, E.P., Rabitz :
  J. Chem. Phys.,72,6571 (1980)
- Gear, C.W. :
  "Numerical initial value problems in ODEs", Prentice-Hall, Englewood Cliffs, New Jersey (1971)
- Glarborg, P.,Kee,Grcar,Miller :
  Sandia Rpt.SAND86-8209 (1986)
- Hindmarsh, A.C. :
  L.L.L. Rpt. UCID-30001, Rev.3, (1974)
- Hindmarsh, A.C.,Byrne :
  L.L.L. Rpt. UCID-30112, Rev.1, (1977)
- Jones, A.,Prothero :
  Combust. and Flame ,12,457 (1968)
- Kaps, P.,Rentrop :
  Numer. Math.,33,55 (1979)
- Kaps, P.,Wanner :
  Numer. Math.,38,279 (1981)
- Kee, R.J., Miller ,Jefferson :
  Sandia Rpt. SAND80-8003 (1980)
- Koda, M.,McRae ,Seinfeld :
  Int. J. Chem. Kin.,11,427 (1979)
- Kramer, M.A.,Kee,Rabitz :
  Sandia Rpt. SAND82-8230 (1982a)
- Kramer, M.A.,Calo,Rabitz,Kee :
  Sandia Rpt. SAND82-8231 (1982b)
- Pitz, W.J., Westbrook :
  Combust. and Flame ,63,113 (1986)
- Pratt, D.T.,Radhakrishnan :
  NASA TM-83806 (1984)

- Rosenbrock, H.H. :
  Comp. J.,5,329 (1963)
- Stabler, R.N.,Chesick :
  Int. J. Chem. Kin.,10,461 (1978)
- Verwer, J.G. :
  Siam J. Numer. Anal.,19,155 (1981)
- Warnatz, J. :
  "Rate coefficients in the C/H/O system" in "Combustion Chemistry" (Gardiner, W.C., ed.) chap.5,pp.197 Springer Verlag, New York (1984)

APPLICATION OF VARIATIONAL METHODS TO PREMIXED, HOMOGENEOUS AND STEADY TURBULENT COMBUSTION.

J.D.CARTIGNY

Institut Français du Pétrole

Summary:

A study is made of steady, homogeneous and turbulent combustion. A single combustion reaction between the premixed reactants is considered. Quantity c represents the mass fraction of the product. A probability density is usually associated with this turbulent combustion, whose balance equation is shown. Compared to this probability, the average values of random variables relating to the physical scales of the flow are great enough to be worthy of interest. This study consists in applying variational methods to try to calculate precisely such average values, particularly the mean rate of chemical reaction, without having to solve precisely the equation of the previous balance. In order to do this, as the equation balance is linear, a steady function is introduced, whose value is equal to the average rate of chemical reaction. Because it is steady, it is possible, by first order approximations, particularly of the probability density, to obtain second degree accuracy for the value of this function.

Finally, the particular case of a single, immeasurably fast chemical reaction is analyzed. By making the appropriate choice, especially of the test function relating to the probability density, in estimating the preceding function, the mean rate of chemical reaction shows up as proportional to the quantity

$$-\tilde{c}\ (1 - \tilde{c})$$

where $\tilde{c}$ is the average value of variable c. This result is related to the classical formula of the eddy break up model, which is valid in the case of a single immeasurably rapid chemical reaction. The fact that the results obtained by the variational methods presented here are similar to those in the literature is encouraging.

C.-M. Brauner and C. Schmidt-Lainé (eds.), Mathematical Modeling in Combustion and Related Topics, 431–439.

APPLICATION DE METHODES VARIATIONNELLES A UNE COMBUSTION TURBULENTE PREMELANGEE, HOMOGENE ET STATIONNAIRE

J.D. CARTIGNY

INSTITUT FRANCAIS DU PETROLE

RESUME

Le cas d'une combustion turbulente, homogène et stationnaire est étudié. Une réaction unique de combustion entre des réactants prémélangés est prise en compte. La quantité c représente la fraction massique du produit. Une densité de probabilité est généralement associée à cette combustion turbulente, dont l'équation de bilan est présentée. Les valeurs moyennes, par rapport à cette probabilité, de variables aléatoires relatives aux grandeurs physiques de l'écoulement apparaissent comme des quantités dignes d'intérêt. Cette étude consiste en l'application de méthodes variationnelles pour essayer de calculer avec précision de telles valeurs moyennes, notamment le taux moyen de réaction chimique, sans avoir à résoudre exactement l'équation de bilan précédente. Pour ce faire, comme l'équation de bilan est linéaire, une fonctionnelle est introduite, stationnaire, et dont la valeur est égale au taux moyen de réaction chimique. Du fait de sa stationnarité, il est possible, à partir d'approximations du premier ordre, notamment de la densité de probabilité, d'obtenir une précision du second ordre sur la valeur de cette fonctionnelle.

Enfin, le cas particulier d'une seule réaction chimique, infiniment rapide, est analysé. Par un choix approprié, notamment de la fonction d'essai relative à la densité de probabilité, dans l'évaluation de la fonctionnelle précédente, le taux moyen de réaction chimique apparaît proportionnel à la quantité :

$$- \tilde{c}\ (1 - \tilde{c})$$

où $\tilde{c}$ désigne la valeur moyenne de la variable c. Ce résultat s'apparente à la formule classique du modèle de l'eddy break up, valable précisément dans le cas d'une seule réaction chimique infiniment rapide. Cette proximité de résultats entre le taux moyen de réaction chimique obtenu par les méthodes variationnelles présentées ici et la littérature demeure encourageante.

1. NOTATIONS

a,b	bornes d'une intégrale, ou quantités reliées à $\tilde{c}$
c	fraction massique des produits
D	quantité scalaire définie par l'équation (4.5)
f	fonction
F	fonctionnelle

g	fonction
H	opération linéaire
$\tilde{p}(c)$	densité de probabilité des particules
r,s	fonctions
t_K	temps caractéristique de la réaction chimique
t_R	temps de résidence
t_T	temps caractéristique de la turbulence
$\dot{w}(c)/t_K$	taux de réaction chimique
u,v	densité de probabilité
α	constante
ε	nombre positif très petit
λ	multiplicateur de Lagrange
δ	distribution de Dirac
σ	caractéristique des gaussiennes considérées

Indices

D	distribution
f	fonction
σ	dépendant du paramètre σ
$\tilde{c}$	valeur moyenne de c
H^+	opérateur adjoint de H
u^+	quantité duale de u

2. INTRODUCTION

Une manière d'aborder le problème de combustion est d'introduire une densité de probabilité. Les travaux de Pope [1.3] figurent parmi les meilleurs, réalisés au cours des dix dernières années. Il a, en outre, décrit une approche probabiliste pour la modélisation des écoulements réactifs turbulents [1].

Une réaction unique de combustion entre des réactants prémélangés est prise en compte. La quantité c représente la fraction massique du produit. Les réactants sont ajoutés au système à c=0. La densité volumique est supposée uniforme. La combustion est supposée stationnaire, avec une turbulence homogène.

Soit :

$$\tilde{p}(c)$$

la densité de probabilité associée à cet écoulement turbulent. C'est la probabilité d'avoir la fraction massique égale à c à l'intérieur du réacteur considéré. En considérant le volume global de ce réacteur, l'équation de bilan de la densité de probabilité s'écrit :

$$\frac{\delta(c=0)}{t_R} = \frac{\tilde{p}(c)}{t_R} + \frac{\partial}{\partial c} \left[\tilde{p}(c) \frac{\dot{w}(c)}{t_K} - \frac{\alpha}{t_T} (c-\tilde{c})\, \tilde{p}(c)\right] \qquad (2.1)$$

où t_R est défini comme le temps de résidence à l'intérieur du système, t_K comme le temps caractéristique de la réaction chimique et t_T comme celui de la turbulence. Cette équation est une application au cas stationnaire de celle démontrée par Bonniot [4].

Dans cet article est présentée l'application de méthodes variationnelles pour résoudre des problèmes liés à l'équation (2.1).

3. METHODES VARIATIONNELLES

3.1. Cadre mathématique général

Soient :

$$f_D = \frac{\delta(c=0)}{t_R} \qquad (3.1)$$

et l'opérateur H [5] donné par :

$$Hu = \frac{u}{t_R} + \frac{\partial}{\partial c} \left[\frac{\dot{w}(c)}{t_K}\, u(c) - \frac{\alpha}{t_T} (c-\tilde{c})\, u(c)\right] \qquad (3.2)$$

Si, en outre,

$$u(c) \equiv \tilde{p}(c) \qquad (3.3)$$

l'équation (2.1) s'écrit :

$$Hu = f_D \qquad (3.4)$$

L'espace considéré est un espace de fonctions pourvu d'un produit scalaire noté :

$$(r,t) = \int_a^b r(c)\, t(c)\, dc \qquad (3.5)$$

L'adjoint H^+ [6] a pour expression (Annexe) :

$$H^+v = \frac{v}{t_R} = \frac{\dot{w}(c)}{t_K} \frac{\partial v}{\partial c} + \frac{\alpha}{t_T} (c-\tilde{c}) \frac{\partial v}{\partial c} \qquad (3.6)$$

H^+ est différent de H.

Dans la partie 3, il est admis implicitement que

$$u(a) = u(b) = 0$$

Soit de plus l'équation suivante,

$$H^+u^+ = \frac{\dot{w}(c)}{t_K} \equiv g(c) \qquad (3.7)$$

g(c) étant une fonction.

Soit l'approximation gaussienne suivante,

$$h_\sigma(c) = \frac{e^{-\frac{c^2}{2\sigma^2}}}{\sigma\sqrt{2\pi}} \tag{3.8}$$

de la distribution $\delta(c=0)$. Quand σ tend vers zéro, $h_\sigma(c)$ tend vers la distribution de Dirac à l'origine au sens des distributions.

Considérons le problème "approché" suivant,

$$Hu = \frac{h_\sigma(c)}{t_R} = f_f \tag{3.9}$$

On [7] a résolu directement l'équation (3.9) et on a montré que lorsque σ tend vers zéro, la solution de (3.9) $u_{f,\sigma}$ tend vers une distribution u_D, solution de l'équation (3.4).

De plus,

$$\lim_{\sigma\to 0} \int \frac{\dot{w}(c)}{t_K} u_{f,\sigma}(c)\, dc = \langle u_D, \frac{\dot{w}(c)}{t_K} \rangle \tag{3.10}$$

où le nombre de droite de l'équation (3.10) représente l'effet de la distribution u_D, sur la fonction $\frac{\dot{w}(c)}{t_K}$ qui est de plus la définition de la valeur moyenne du taux de réaction chimique correspondant au problème originel,

$$\langle u_D, \frac{\dot{w}(c)}{t_K} \rangle \equiv \left(\widetilde{\frac{\dot{w}(c)}{t_K}}\right)_D \tag{3.11}$$

Parallèlement, la quantité

$$\int \frac{\dot{w}(c)}{t_K} u_{f,\sigma}(c)dc$$

est égale à la valeur moyenne du taux de réaction, relativement au problème approché dont $u_{f,\sigma}$ est une solution

$$\int \frac{\dot{w}(c)}{t_K} u_{f,\sigma}(c)\, dc \equiv \left(\widetilde{\frac{\dot{w}(c)}{t_K}}\right)_{f,\sigma} \tag{3.12}$$

L'équation (3.10) peut s'écrire :

$$\left(\widetilde{\frac{\dot{w}(c)}{t_K}}\right)_D = \lim_{\sigma\to 0} \left(\widetilde{\frac{\dot{w}(c)}{t_K}}\right)_{f,\sigma} \tag{3.13}$$

L'équation (3.13) nous permet d'approcher le taux moyen de réaction chimique, relatif à la distribution solution de l'équation de bilan de la densité de probabilité, par le taux moyen de réaction chimique relatif à une densité de probabilité, plus simple, puisqu'il s'agit d'une fonction, solution de l'équation entre fonctions (3.9).

L'introduction de techniques variationnelles permet donc de calculer directement des grandeurs moyennes du type

$$\left(\widetilde{\frac{\dot{w}(c)}{t_K}}\right)_{f,\sigma}$$

3.2 Fonctionnelle

La solution u de l'équation (3.9) est une densité de probabilité : elle doit donc vérifier l'équation

$$\int u(c)\ dc = 1 \tag{3.14}$$

Puisque u doit satisfaire la contrainte, exprimée par l'équation (3.14), on doit introduire un multiplicateur de Lagrange λ, dans la fonctionnelle à optimiser, définie par

$$F_\sigma(u,u^+,\lambda) = (g,u) + (u^+,f_f) - (u^+,Hu) + \lambda((u-1),1) \tag{3.15}$$

La bistationnarité de F_σ vis-à-vis des variations quelconques de u et de u^+ équivaut à u et u^+ et λ solutions du système suivant :

$$S^* \quad \begin{cases} H\ u_{f,\sigma} = f_f & (3.9) \\ H^+\ u^+ = g + \lambda & (3.16) \\ \int u(c)\ dc = 1 & (3.14) \end{cases}$$

Puis on procède de la manière suivante. Dans un premier temps, on tente d'obtenir des fonctions d'essais u_o, u_o^+ et λ_o, approximations du premier ordre des solutions du système S*. Dans un second temps, on calcule la fonctionnelle F_σ, en remplaçant dans l'équation (3.15) u et u^+, λ par leurs fonctions d'essais respectives. On obtient une approximation du second ordre de

$$\left(\frac{\tilde{\dot{w}}(c)}{t_K}\right)_{f,\sigma}$$

par la formule suivante :

$$\left(\frac{\tilde{\dot{w}}(c)}{t_K}\right)_{f,\sigma} = (g,u_{f\sigma}) \simeq F\ (u_o,u_o^+,\lambda_o) \tag{3.17}$$

On a donc obtenu une approximation du taux moyen de réaction chimique, relatif à la fonction, densité de probabilité, solution de l'équation (3.5) par le biais d'une fonctionnelle bistationnaire.

La méthode variationnelle, exposée ci-dessus, est appliquée au cas d'une réaction chimique infiniment rapide.

4. CAS D'UNE REACTION CHIMIQUE INFINIMENT RAPIDE

Jusqu'à ce point, il n'y a pas eu d'hypothèses restrictives, sur les valeurs des coefficients, t_R, t_K et t_T. Pour une réaction chimique infiniment rapide, t_K tend vers zéro, et le quotient $\frac{t_R}{t_T}$ tend lui vers une valeur finie.

Considérons comme fonction d'essais

$$u_o^+ = \lambda^+ + 1 + c\ (1-c) \tag{4.1}$$

$$u_o(c) = a \frac{e^{-\frac{c^2}{2\sigma^2}}}{\sigma\sqrt{2\Pi}} + b \frac{e^{-\frac{(c-1)^2}{2\sigma^2}}}{\sigma\sqrt{2\Pi}} \tag{4.2}$$

où a et b seront précisés ultérieurement,

et comme produit scalaire

$$(r,s) = \int_{-\varepsilon}^{1+\varepsilon} r(c)\ s(c)\ dc \tag{4.3}$$

où ε est une très petite constante positive.

La fonctionnelle précédente peut s'écrire :

$$F_\sigma = (g,u_o) + (\frac{h_\sigma\ (c)}{t_R},\ 1+c\ (1-c)) = (1+c)(1-c), Hu_o) + \lambda((u_o-1)1) \tag{4.4}$$

Soit D_σ égal à

$$D_\sigma \equiv (1+c(1-c), Hu_o) \tag{4.5}$$

D_σ est aussi égal à

$$D_\sigma = (H^+(1+c(1-c)), u_o) \tag{4.6}$$

La quantité H^+ (1+c(1-c)) peut s'écrire comme

$$H^+\ (1+c(1-c)) = \frac{1+c(1-c)}{t_R} - \frac{\dot{w}(c)}{t_K}\ (1-2c) + \frac{\alpha}{t_T}(c-\tilde{c})(1-2c) \tag{4.7}$$

Lorsque σ tend vers zéro, la limite de D_σ est, pour n'importe quel ε,

$$\lim_{\sigma \to 0} D_\sigma = \frac{1}{t_R} - \frac{a\dot{w}(0)}{t_K} + \frac{b\dot{w}(1)}{t_K} + \frac{\alpha}{t_K}\left[- a\tilde{c} - b(1-\tilde{c})\right] \tag{4.8}$$

mais comme, de plus

$$\lim_{\sigma \to 0} a = \tilde{c} \tag{4.9}$$

et

$$\lim_{\sigma \to 0} b = 1-\tilde{c} \tag{4.10}$$

La limite de F_σ vaut

$$\lim_{\sigma \to 0} F_\sigma = \frac{2a\dot{w}(0)}{t_K} + \frac{2a}{t_K}\ \tilde{c}(1-\tilde{c}) \tag{4.11}$$

or, dans les gaz imbrûlés, il est raisonnable de faire l'hypothèse que

$$\frac{\dot{w}(0)}{t_K} = 0 \tag{4.12}$$

pour tout t_K. Donc, la limite de la fonctionnelle s'écrit

$$\lim_{\sigma \to 0} F_\sigma = \frac{2\alpha}{t_T} \tilde{c}(1-\tilde{c}) \tag{4.13}$$

pour mémoire

$$\left[\frac{\widetilde{\dot{w}(c)}}{t_K}\right]_D = \lim_{\sigma \to 0} F_\sigma \tag{4.14}$$

donc :

$$\left[\frac{\widetilde{\dot{w}(c)}}{t_K}\right] = \frac{2\alpha}{t_T} \tilde{c}(1-\tilde{c}) \tag{4.15}$$

Pour une combustion stationnaire, avec une turbulence homogène, le taux moyen de réaction chimique, obtenu en rendant stationnaire la fonctionnelle introduite plus haut, aboutit, dans le cas d'une réaction chimique infiniment rapide, à une formule très proche du modèle classique de l'eddy break up de B. Spalding.

5. CONCLUSION

Une combustion stationnaire est étudiée, avec une turbulence homogène. Dans cet article, on présente :

- d'une part lc cadre mathématique, nécessaire pour calculer des valeurs moyennes de grandeurs physiques, lorsqu'on connaît l'équation de bilan de la densité de probabilité. Une fonctionnelle, ad hoc, est introduite, dont la bistationnarité permet, à partir d'approximations de premier ordre, notamment de cette densité de probabilité, d'obtenir une approximation de second ordre sur la valeur de cette fonctionnelle, à savoir le taux moyen de réaction chimique ;

- d'autre part, l'utilisation de la méthode variationnelle présentée ci-dessus, au cas d'une réaction chimique, infiniment rapide. Le taux moyen de réaction chimique obtenu en exprimant la stationnarité de la fonctionnelle précédente, est très semblable à celui du modèle classique de l'eddy break up imaginé par B. Spalding, justement pour une réaction chimique infiniment rapide. Cette proximité de résultats est réconfortante.

REFERENCES

1. S.B. Pope, "The Probability Approach to the Modelling of Turbulent Reacting Flows", Combustion in Flame, 27, 299 (1976).
2. S.B. Pope, "The Relationship between the Probability Approach and Particle Models for Reaction in Homogeneous Turbulence", Combustion and Flame, 35, 41 (1979).
3. S.B. Pope, "A Monte-Carlo Method for the Probability Density Functions. Equations of Turbulent Flow". MIT (1980).
4. C. Bonniot, "Influence de la Turbulence sur la Combustion d'Hydrocarbures dans un Foyer Homogène", ONERA, 1978-7 (1978).
5. Schechter, "The Variational Method in Engineering", Mac-Graw Hill.

6. J. Lewins, "Importance, The Adjoint Function", Pergamon Press (1965).
7. L. Tartar, Communication Privée.

ANNEXE

Calcul de l'Adjoint de l'Opérateur H

L'adjoint H^+ de l'opérateur est défini par :

$$(Hu,v) = (u,H^+v) \quad u \text{ et } v \tag{A.1}$$

Pour mémoire,

$$Hu = \frac{u}{t_R} + \frac{\partial}{\partial c}\left[\frac{\dot{w}(c)\ u}{t_K} - \frac{\alpha}{t_T}(c-\tilde{c})\ u\right] \tag{A.2}$$

d'où :

$$(Hu,v) = \int_a^b \frac{uv}{t_R} + \left\{ v\left[\frac{\dot{w}(c)}{t_K} - \frac{\alpha(c-\tilde{c})u}{t_T}\right]\right\}_a^b - \int_b^a \left[\frac{\dot{w}(c)}{t_K} - \frac{\alpha(c-\tilde{c})}{t_T}\right] u \frac{\partial v}{\partial c}\, dc \tag{A.3}$$

Si u est tel que :

$$u\ (a) = u\ (b) = 0 \tag{A.4}$$

aucune condition aux limites pour v n'est nécessaire et si H^+ est de plus défini par :

$$H^+_v = \frac{v}{t_R} - \frac{\dot{w}(c)}{t_K}\frac{\partial v}{dc} + \frac{\alpha(c-\tilde{c})}{t_T}\frac{\partial v}{dc} \tag{A.5}$$

d'une part, H^+ est bien l'adjoint de H, car vérifiant l'équation (A.1), d'autre part, H^+ est différent de H (non autoadjoint).

A MINIMAL MODEL FOR TURBULENT FLAME FRONTS

Hugues Chaté*
Department of Engineering, University of Cambridge,
Trumpington Street, Cambridge CB2 1PZ, England.

1 Motivation

The problem of premixed turbulent combustion is known to be very difficult to handle from the points of view of both theory and experiment [1,2]. Moreover, the proposed models often rely on assumptions which are beyond the reach of current experimental techniques. In this context, it appears very useful to have a simple model that contains the minimum number of physical hypotheses, can be treated both analytically and numerically, and allows visualization of the flame fronts.

The model presented here addresses turbulent premixed combustion in the corrugated laminar flamelet regime, a case intensively studied, in particular by the more conventional Bray-Moss-Libby (BML) model [3]. More precisely, the proposed system focuses on the problem of the evolution of an initially plane flame front in a given (artificial) turbulent flow field. It attempts to give visual information, to reveal the most important physical phenomena and to provide qualitative, quantitative and statistical results to compare with theoretical assumptions and experimental data.

It is worth noting that this "minimal model" may not be of interest simply for turbulent combustion. It can be cast into the more general case of the motion of an interface in a turbulent medium as only basic physical assumptions are retained from the full problem.

2 Description

In the regime considered, the characteristic reaction time is taken to be much smaller than the characteristic turbulent time (high Dämkohler number limit). The flame is then confined to a very small region of space and the streamwise variations of the temperature and the concentrations are essentially two-valued functions [3]. All

*Permanent address: DPh-G/PSRM, CEN Saclay, 91191 Gif-sur-Yvette Cedex, France.

C.-M. Brauner and C. Schmidt-Lainé (eds.), Mathematical Modeling in Combustion and Related Topics, 441–448.

chemical reactions take place within such flamelets whose distortion by the turbulent flow field is believed largely to determine the behaviour of the system. Together with the requirement of simple simulations, this motivated the choice of cellular-automata-like techniques, namely a discrete space and discrete time universe, where a site of the lattice is a logical variable whose state is either burnt or unburnt.

The combustion process itself is also represented in a very simple way, reproducing normal propagation of the front on the discrete space lattice. This is realized by the contaminative spreading of the burnt sites to their nearest neighbours each time a "combustion-step" is needed.

The most drastic simplifications are made in the modelling of turbulence. In fact, an"external" turbulence is applied to the front, as our purpose is not to include detailed hydrodynamics in the system, but rather to show how a laminar flamelet advances and is distorted in a given turbulent unburnt mixture. The eddy concept is used in a geometrical (real-space) rather than in the usual (Fourier-space) meaning of classical representations of developed turbulence. The turbulent unburnt mixture is seen as a collection of localized vorticity structures having a well defined statistical distribution of sizes.

The form of this distribution is suggested in a related paper by Kerstein [4]. His argument, though not entirely satisfactory, provides a reasonable basis on which to proceed. The model proposed here takes a power-law distribution of eddy sizes:

$$d(l) \sim l^{-5/3} \tag{1}$$

where $d(l)$ is the density of eddies of size l.

The scaling of the circumferential velocities arises from the ordinary Kolmogorov theory (K41) [5]:

$$v_l \sim l^{1/3} \tag{2}$$

3 Practical implementation

Although in principle such a model can be built in any number of dimensions, the one presented here takes place in a 2D-universe which can be understood as a typical slice of a flame surface. A plane front appears then as a straight line between the burnt and the unburnt mixture. For the sake of numerical simplicity again, a square lattice was chosen. The lattice step then stands for the smallest scale of the system, and is not *a priori* linked to any physical scale such as the Kolmogorov scale of K41 theory.

Dynamics and combustion are separated, so that a time-step consists of a certain number of "dynamical-steps" N_{dyn} and "combustion-steps" N_{comb}. This allows a simple control of the laminar burning velocity u_L, which can be expressed in non-dimensional units as the number of combustion-steps per time-step:

$$u_L = N_{comb} \tag{3}$$

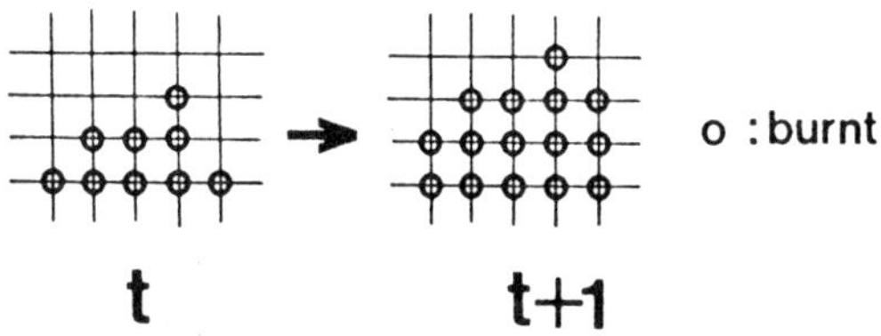

Figure 1: *Effect of a combustion-step using the Von Neumann neighborhood of each site on the front.*

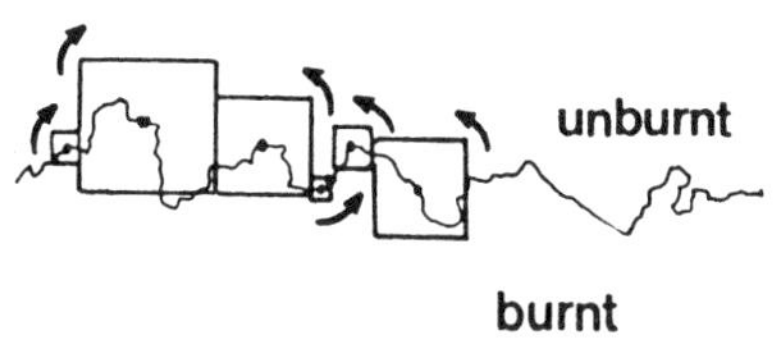

Figure 2: *Schematic representation of how "eddies" are applied to the front during a dynamical-step.*

As already mentioned, the burning process is realized at each combustion-step by changing the state of all unburnt sites in the Von Neumann neighborhood of a burnt site on the front (Fig. 1), an idea already used by Borghi [6]. The Von Neumann neighborhood was preferred to the Moore neighborhood (eight nearest neighbors) which allows propagation over distances topologically equivalent to two lattice steps.

A dynamical-step consists of the application of "eddies" along the front. They are square portions of the universe centered on a site at the limit between the burnt and the unburnt mixture (Fig. 2). Their sizes are chosen at random according to (1), between a minimum size l_{min} and a maximum size l_{max}. Lengths l_{min} and l_{max} are thus the smallest and largest dynamical scales in the system. That means that l_{min} can be seen as the Kolmogorov scale in the framework of K41 theory whereas l_{max} is an integral scale. It is worth noting that these physical scales are well defined and under direct control. The direction of rotation of each eddy is chosen at random and the rotation angles are scaled according to (2):

$$w_l = w_1 l^{-2/3} \text{ for } l = l_{min}, \dots, l_{max} \tag{4}$$

Finally, because of the progression of the front, its mean position is calculated when needed, so that it can be recentered in the middle of the universe in order to avoid going out of the limits.

It should be obvious now that the system is remarkably light from the point of view of numerical simulation. It is much easier to perform numerical simulations

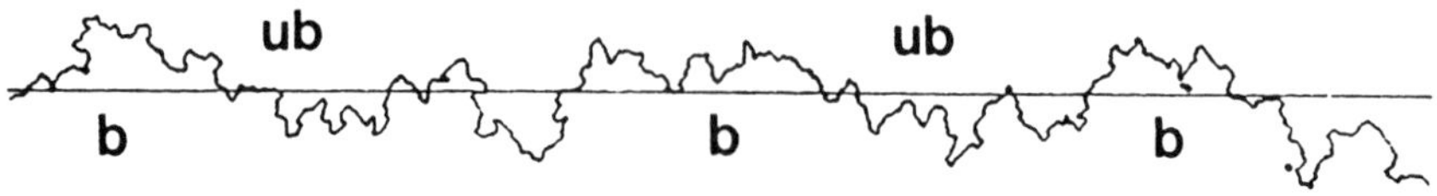

Figure 3: *A typical front after 160 time-steps from an initially plane front. The length of the universe is of 800 sites.* $l_{min} = 6$, $l_{max} = 64$, $w_1 = \pi/2$, $N_{dyn} = 1$, $N_{comb} = 1$. *The symbols* "**b**" *and* "**ub**" *identify regions of burnt and unburnt gas respectively. The front is moving upwards.*

with it than with the classical models or even with the realistic cellular automata of Clavin *et al* [7]. This is naturally achieved at the price of a few limitations. The main phenomena not taken into account are the flame stretching and the intrinsic instability mechanisms of a laminar flame front.

Nevertheless, the next section shows that this very simple model gives strikingly realistic results and achieves its intended aims, exhibiting qualitative, quantitative and statistical information of use to both experimenters and theoreticians.

4 Results

It follows from the above presentation that there are five main parameters in the system: the minimum and maximum sizes of eddies l_{min} and l_{max}, the numeric prefactor of the rotation angles w_1, and the numbers of dynamical- and combustion-steps per time-step, N_{dyn} and N_{comb}. Parameters l_{min}, l_{max}, w_1 and N_{dyn} control the turbulence precisely and meaningfully whereas N_{comb} defines the laminar burning velocity u_L.

Many quantities can be extracted easily so that the systematic exploration of their variation with the five parameters is out of reach of this article. Only a few examples will be presented here.

The first point to mention is that an initially plane front quickly reaches a statistically stable state after a short transient (Fig. 3). The system does not "explode" and does not show any instability. It possesses well defined characteristics such as a constant turbulent velocity u_T, a statistically constant mean width and burning rate (to be measured here as the mean length of the front). Clearly, the combustion is the limiting step in what would otherwise be a roughly diffusive process. The precise study of the transition between $u_L = 0$ (no combustion) and $u_L \neq 0$ will be reported elsewhere but this property of the system can already be quoted as *combustion-limited diffusion.*

The front shown in Figure 3 is very realistic, in particular in regard of the recent results obtained in IC engines [8,9] that show the same kind of highly corrugated interfaces between the burnt and the unburnt mixture. It must be stressed here

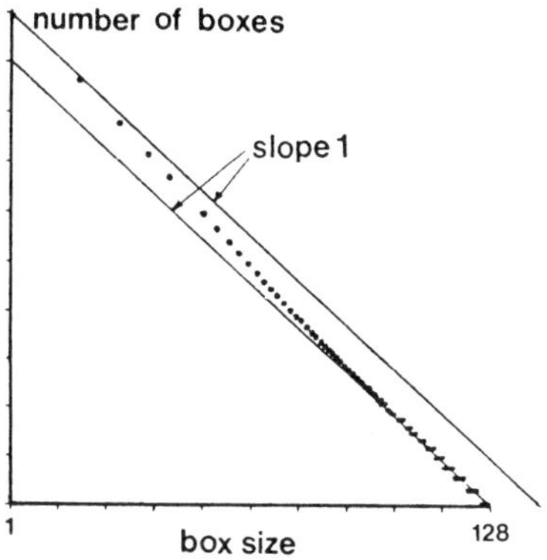

Figure 4: *Log-log plot of the the result of the box counting algorithm on the front of Fig. 3. The lower and upper cut-offs of the fractal range are approximately 4 and 50. A least-square fit between these values of the box-sizes gives a fractal dimension* $d_F = 1.16 \pm 0.01$.

that although the eddies are applied only once and then replaced by another random choice, they have an effect on the front over several time-steps. This "lifetime" is increasing with their size, in qualitative agreement with usual views on turbulence [10]. Moreover this effect helps to recover the hierarchical structure of the eddies.

The fronts have to be characterized by geometrical properties. This is achieved by a box-counting algorithm which exhibits a fractal range (Fig. 4). The fractal dimension weakly increases with the turbulence intensity and decreases with the laminar burning velocity u_L. The cut-offs limiting this fractal behaviour are of the order of l_{min} and l_{max}. This is not surprising since the dynamical process creates scales roughly within this range and tends to increase the wrinkling of the front as the combustion process tends to smooth it by burning out all circumvolutions smaller than u_L (in non-dimensional units).

Figure 5 shows the variation of both the effective fractal dimension and the turbulent velocity with the laminar burning velocity u_L. Both quantities are ill-defined for $u_L < 1$ (no combustion case), mainly because they are time dependent for $u_L = 0$ and thus have no meaning. For u_L very large (weak turbulence limit), $d_f - 1$ must go to zero because the front recovers its plane shape at each time-step. For the same reasons $u_T = u_L$ in this limit. Considering that the distorting process creates no wrinkles larger than l_{max}, this must occur for finite values of u_L.

Statistical results such as correlation functions and fluctuations can also be easily obtained. Here histograms of burnt and unburnt packet sizes at different locations from the mean position have been calculated (Fig. 6). The mean position is the line delimiting an equal area of burnt and unburnt mixture respectively above and

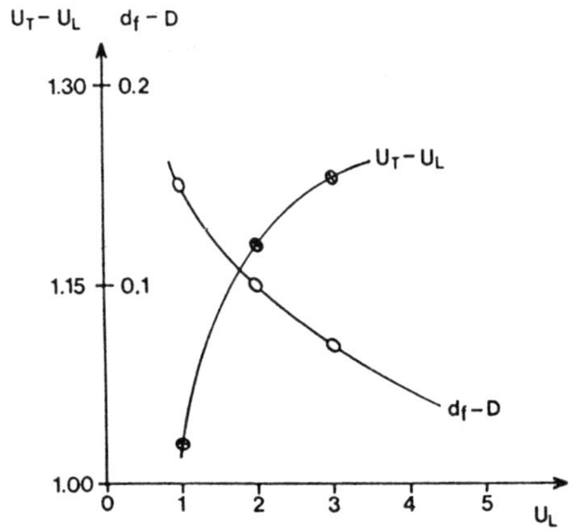

Figure 5: *Variation of the turbulent velocity u_T and the effective fractal dimension $d_f - D$ with the laminar burning velocity $u_L = N_{comb}$. D is the topological dimension of the front (here $D = 1$). $l_{min} = 2$, $l_{max} = 32$, $w_1 = \pi/2$, $N_{dyn} = 1$.*

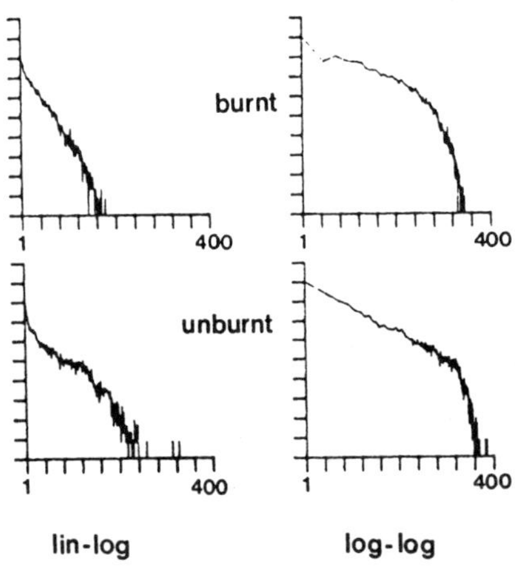

Figure 6: *Histograms of burnt and unburnt packets at 8 sites from the mean position of the front. Same parameter values as in Fig. 3 and 4. The results are cumulated over 2000 time-steps. This position corresponds to a mean progress variable $\tilde{c} = 0.399$.*

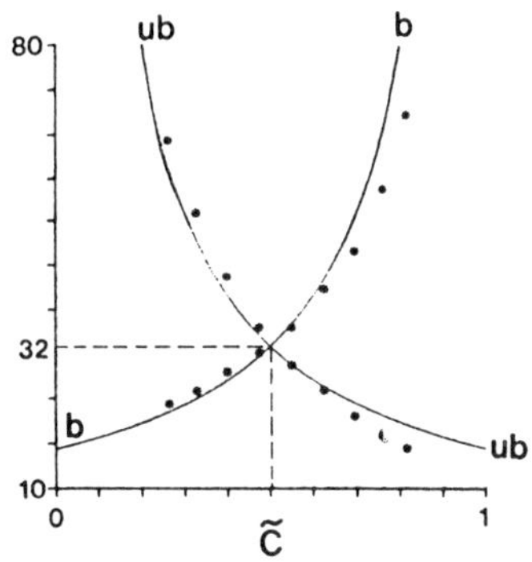

Figure 7: *Variation of the mean packet size with the mean progress variable $\tilde{c}$. Results extracted from the histograms of burnt and unburnt packets for the parameter values of Fig. 3. Solid lines show the theoretical prediction of the BML model.*

below itself (see Fig. 3). The histograms show a fractal range for small scales and an exponential tail for large packets. From each pair of histograms a mean burnt and unburnt packet size and a mean progress variable $\tilde{c}$ is then calculated. This allows to plot the variation of the mean packet size with $\tilde{c}$ (Fig. 7), a result to be compared with the prediction of the BML theory [3]. The agreement is good and this suggests that the essential physics is contained in very general features that do not depend of the details of the full problem, hence confirming the approach taken in the construction of the model.

5 Current work

Ongoing work is completing a systematic study of the qualitative and quantitative effects of the variation of the main parameters. The results are being compared with available experimental data. Analytical approaches try to derive the turbulent velocity u_T and the fractal dimension d_f. Furthermore, attempts are made to determine a satisfactory Reynolds number and a characteristic velocity of the "turbulence" modelled, though it is not believed that these quantities should describe the turbulent mixture better than the simple parameters of the model, l_{min}, l_{max}, w_1 and N_{dyn}.

A careful investigation of the relevant scales in the regime considered using the same numerical techniques [11] is in progress. This should enable a better representation of the involved scaling laws in a similar model. Finally, extensions of the system to include flame stretching and a non-trivial chemical kinetics are under consideration.

References

[1] P.Clavin, "Dynamical behavior of premixed flame fronts in laminar and turbulent flows", Prog. in En. and Comb. Sci., **11** (1985) 1-59.

[2] K.N.C.Bray, "Turbulent flows with premixed reactants" in *Turbulent Reacting Flows*, Topics in Applied Physics **44**, Springer 1980.

[3] K.N.C.Bray, P.A.Libby, J.B.Moss, "Scalar length scale variations in premixed turbulent flames" in: Proceedings of the 20th Symposium on Combustion, The Combustion Institute, pp 421-427 (1984).

[4] A.R.Kerstein, "Pair-exchange model of turbulent premixed flame propagation" in: Proceedings of the 21st Symposium on Combustion, The Combustion Institute, (1985).

[5] Landau and Lifschitz, *Fluid Dynamics*, Pergamon Press, London, 1959.

[6] R.Borghi, "Using a simulation by cellular automata as a help for turbulent combustion modelling" in: Proceedings of the 9th Australasian Fluid Mechanics Conference, pp 234-237 (1986).

[7] P.Clavin, P.Lallemand, Y.Pomeau, G.Searby, "Simulation of front dynamics in flows: a new proposal based on lattice-gas models", to be published in Journal of Fluid Mechanics, 1987.

[8] T.A.Baritaud, R.M.Green, "A 2D flame visualisation technique applied to the I.C. engine", SAE paper 860025 (1986).

[9] A.O.zur Loye, F.V.Bracco, "Two-Dimensional visualisation of premixed-charge flame structure in an I.C. engine", SAE Paper 870454 (1987).

[10] H.Tennekes and J.L.Lumley, *A First Course in Turbulence*, MIT Press, Cambridge Mass., 1972.

[11] H.Chaté, R.S.Cant, to be published.

A THEORETICAL STUDY OF AIR-SOLID TWO-PHASE FLOWS*

E. Diakoumakos, J. Anagnostopoulos, G. Bergeles
Laboratory of Aerodynamics,
National Technical University of Athens.

NOMENCLATURE

C_d	= drag coefficient		x	= axial distance
D_p	= particle diam.		Γ_t	= turbulent diffussivity
g	= accel.of gravity		ε	= rate of k dissipation
k	= turb. kinetic energy		ϱ	= density
L	= dissip. length scale			
Re	= Reynolds number		Subscripts	
t	= time		f	= fluid property
t_e	= eddy lifetime		p	= particle property
u	= velocity			

1. THE ENGINEERING PROBLEM

The two-phase flow and especially the dispersion of particles entrained in a turbulent flow, is a phenomenon of great importance in many fields of engineering.

Referring to the vital problem of energy production, it is understood that the use of coal instead of petrol has been rising since the first energy crisis. It is known that coal burning, produces amounts of fly ash in the atmosphere which are almost 500 times greater than those of the case of liquid fuel. This makes the improvement of coal burners essential, also for reasons of greater efficiency. That could be achieved by being able to understand and then predict the procedure of coal particle motion within the burner.

This work continues with a review of experimental and theoretical findings on the subject and the mathematical and numerical treatments of the two-phase model. Applications of the proposed model for several test cases are discussed.

2. PREVIOUS THEORETICAL AND EXPERIMENTAL WORK

In recent years a substantial amount of work concerning two-phase flows, has been conducted. Experiments have been done by many researchers as Snyder & Lumley (1971) and Wells & Stock (1983) for a grid generated turbulent flow. Measurements

* The work was financially supported by EEC/DCXII, Contract number EN3F-022-GR(T-T).

C.-M. Brauner and C. Schmidt-Lainé (eds.), Mathematical Modeling in Combustion and Related Topics, 449–459.

concerning flows in ducts, have been done by Tsuji et al. (1984), Lee & Durst (1982), Steimke & Dukler (1983).

The results have led to conclusions, which can be in general summarized as follows : 1. The probability density functions of particle velocities and radial particle dispersion are Gaussian. 2. The effects of crossing trajectories are important for particles with large relaxation times, resulting in loss of correlation and reduction of the dispersion coefficient. 3. The diffusion rate of the dispersed phase is less than that of the carrier fluid. 4. Small particles seem to be following the flow and thus disperse more than the bigger ones. 5. Large particles drag the flow as they move, causing a change of the fluid kinetic energy. Smaller particles, on the other hand, seem to be absorbing energy from the surrounding fluid, decreasing the intensity of the turbulence. 6. The mean slip velocity between the two phases increases with increasing particle diameter and decreasing mass loading. 7. The velocity fluctuations are reduced with increasing mass loading. 8. For small loadings the mean flow field is practically uninfluenced by the presence of the particles.

In the field of theoretical two-phase flow studies, much has been done lately, due mainly to the advances in computer systems and numerical methods. For the prediction of particulate two-phase flows, two approaches are possible. Both of these treat the fluid phase as a continuum. Their difference lies in the treatment of the particulate phase. The first approach, called the Lagrangian approach, treats the second phase as single particles; by assuming different starting positions of discrete particles and following their trajectories, a two-phase flow can be simulated.

The second approach, called the Eulerian approach, treats the particle phase as a kind of interacting continuum, and solves the appropriate equations for both phases.

The Eulerian approach has been employed by many researchers, including Elghobashi et al. (1983), Mostafa & Elghobashi (1986), Picart, Berlemont & Gouesbet (1986), Melville & Bray (1979). The Lagrangian approach, originaly proposed by Migdal & Agosta (1967), has been used among others by Sharma & Crowe (1978), Gosman & Ioannides (1983), Shuen et al. (1985).

There are arguments as to the advantages of each of the two approaches. Relative comparison of the two methods also done by Durst et al. (1984), Mostafa & Mongia (1986), shows the following :

The Lagrangian approach seems to be a little more economical in storage requirements, when particles of one size diameter are used. This relative advantage is much increased, when a particle size distribution is required for the flow definition, which is the case for most real systems.

This also reflects to increased CPU time requirements for the Eulerian approach, if detailed information for the particulate phase is needed, because the equations have to be solved for each particle group.

The Eulerian approach can easily incorporate diffusion effects by using an appropriate diffusion coefficient in the formulation. Also the Eulerian approach exhibits numerical diffusion, unlike the Lagrangian approach. Predetermining the diffusion coefficient for the particles, based on some assumption for their concentration at a specified point, would however change the particle motion equations from parabolic to elliptic.

An alternative way of simulating turbulent diffusion effects on the particles, while using the Lagrangian approach, is by the more expensive, but also accurate and realistic, Monte-Carlo method. In this stochastic method the instantaneous fluid flow field must be modeled. This was done by Gosman & Ioannides (1983), with the spliting of the isotropic turbulent field into two components, mean and fluctuating. The second component is randomly sampled during each droplet's trajectory. The overall particle cloud properties are obtained by averaging a statistically significant sample.

Owing to the small time increments used for the calculation, the Lagrangian approach gives better results when high particle velocity gradients are present.

The Lagrangian approach gives more detailed information on the dispersed phase, as it produces characteristics of the particle motion in certain areas. The Eulerian approach on the other hand, has the advantage where flows of high particle concentration are handled and the volume fraction becomes significant (volume fraction > 0.001, loading > 0.1)

3. ANALYSIS OF PRESENT WORK

3.1. Mathematical formulation

In this work the Lagrangian formulation was used in order to predict particle motion in a steady, homogeneous, isotropic turbulent flow field. The method used was based on the model proposed by Gosman & Ioannides (1983) and its modification by Shuen et al. (1985) on the L_e, t_e values, which are discussed later. The instantaneous flow field can be represented by introduction of the Monte-Carlo stochastic approach, with the division of the fluid velocity U_f into two components. The first is the average fluid velocity $\langle U_f \rangle$ and the second the velocity fluctuation u_f, which is caused by the turbulence possessing a Gaussian probability distribution. The standard deviation of the fluctuation is :

$$\sigma = (2k/3)^{1/2} \tag{1}$$

Random sampling gives the instantaneous velocity :

$$U_f = \langle U_f \rangle + u_f \tag{2}$$

The discrete phase motion can be determined through the use of the Basset-Boussinesq-Oseen (BBO) equation. The assumptions for the particles are the following : Dilute, unbounded flow was concidered. The particles are concidered spherical with a drag coefficient of :

$$C_D = (1 + 0.15\, Re_p^{0.687})/(Re_p/24), \qquad Re_p \leq 1000$$

$$C_D = 0.44, \qquad Re_p > 1000 \tag{3}$$

where $Re_p = \varrho_f\, |\vec{U}_p - \vec{U}_f|\, D_p/\mu$ the particle Reynolds number. (4)

Since the ϱ_p/ϱ_f ratio is quite large, effects of Magnus force, Bassset force, virtual mass, can be neglected. Lift forces induced by particle rotation or velocity gradients, are also neglected. As an external flow field, the gravity is taken into account.

The BBO equation in cartesian coordinates takes the form of :

$$dU_{pi}/dt = 3/4(\varrho_f/\varrho_p)(C_D/D_p)(U_{fi} - U_{pi})\,|\vec{U}_f - \vec{U}_p| + g_i\,(1-\varrho_f/\varrho_p)$$

$$i = x,y,z \tag{5}$$

The particle position can be found from :

$$dx_{pi}/dt = U_{pi}, \quad i = x,y,z \tag{6}$$

A problem of special interest is the particle-eddy interaction time, which is the time that a particle spends in a particular eddy. This was solved by assuming that the interaction takes place as long as both the following criteria are satisfied:

$$|\Delta t| \leq t_e \qquad \text{and} \qquad |\Delta x_p| \leq L_e \tag{7}$$

where Δt and Δx_p the interaction time and distance and

$$L_e = C_\mu^{3/4}\, k^{3/2}/\varepsilon\ , \quad t_e = L_e\ /(2k/3)^{1/2} \tag{8}$$

the eddy characteristic size and life time respectively.

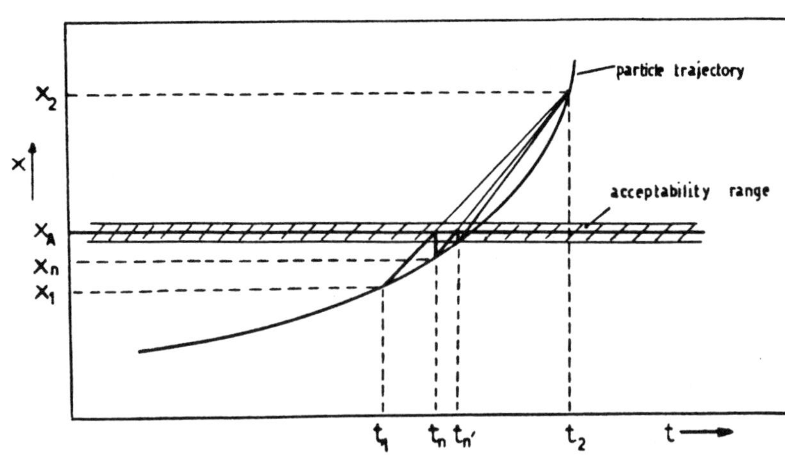

FIGURE 1. Prediction - correction scheme.

3.2. Numerical treatment

For the solution of the solid-phase motion equations a Runge-Kutta integration technique was used. A prediction-correction treatment for finding the exact position of the particle leaving the eddy and also for data sampling at predetermined positions was used. The procedure used is shown in figure 1.

This can be summarized in the following :

-Start at the position x_1 , the last position before the particle, with the predetermined time step, passes the x_A position, where we want to find the particle characteristics. Using the predetermined time step, the particle moves to position x_2. Estimate the time t_n for the particle to reach the given position, were the motion linear. This is done with the use of the expression :

$$(x_2 - x_1)/(t_2 - t_1) = (x_2 - x_A)/(t_2 - t_n) \quad (9)$$

-Using a time step $\Delta t = t_n - t_1$, and starting from the previous particle position, find a new x_n position and check whether or not it is whithin the given range of acceptability (convergenge).

-If it isn't, let $x_1 = x_n, t_1 = t_n$ (for the case where x_n is less than x_A - case of figure 1, accelerating motion) or $x_2 = x_n, t_2 = t_n$, (x_n larger than x_A, decelarating motion), find a new time step with the use of eq.(9), and recompute.

-Repeat until particle position is in the acceptability range.

The resulting computer programs, were written in FORTRAN and run on a CDC 171-8 computer.

3.3. Applications and results

3.3.1. Motion of discrete particle in a steady flow field

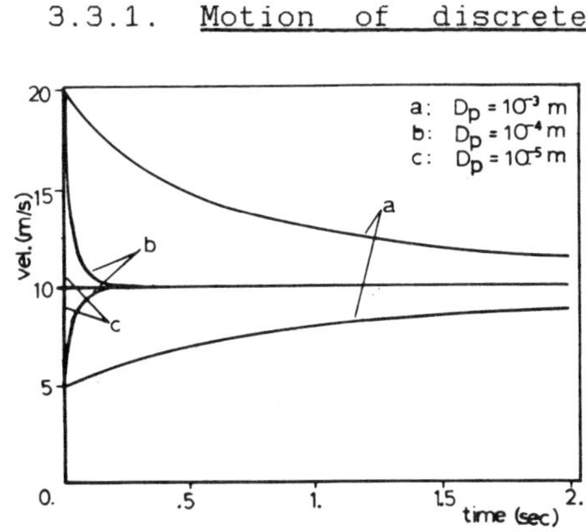

FIGURE 2. Effects of particle diameter on particle velocity changes; no gravity; ϱ_p=2500 kg/m³; ϱ_f=1.29kg/m³; U_p=10m/s

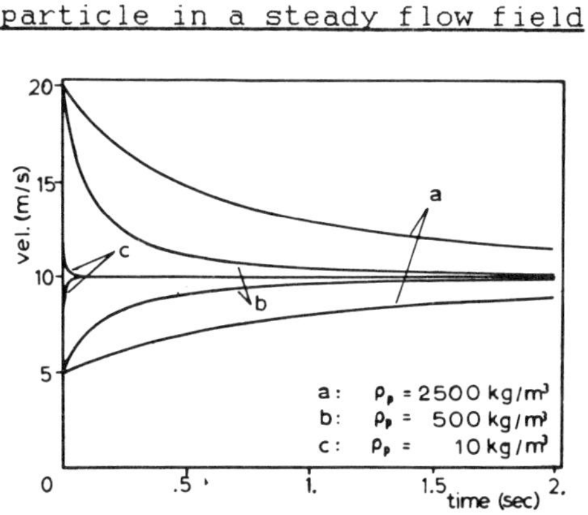

FIGURE 3. Effects of particle density on particle velocity changes; no gravity; D_p=.001m ϱ_f=1.29kg/m³; U_p=10m/sec.

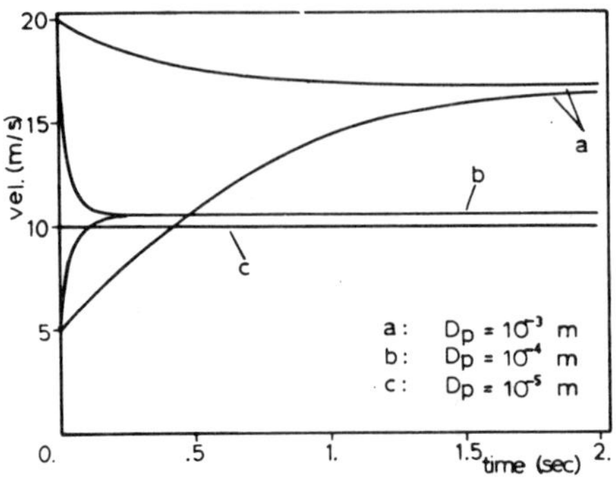

FIGURE 4. Effects of gravity on particle velocity changes; $\varrho_p=2500kg/m^3$; $\varrho_f=1.29kg/m^3$; $U_p=10m/sec$.

<u>without turbulent diffusion.</u> Cases were studied in order to observe the effect of several parameters on the motion of the particle, during the transition phase until the particle acquires its terminal velocity. These parameters were the particle diameter, the material density in comparison to that of the carrier fluid and the gravity field acting in the direction of the axis. Effects of swirl (W=wr) of the flow field on the particle radial motion, were also examined.

The results are shown in figures 2-5. These show that with increasing particle inertia, the time for the particle to react to the different field velocities is inrceased. Particle diameter and density play an equally significant role. Without the effect of the gravity field and the swirl the particle, after a sufficient transition time, follows the flow field.

If gravity is present, then the particle tends to acquire a

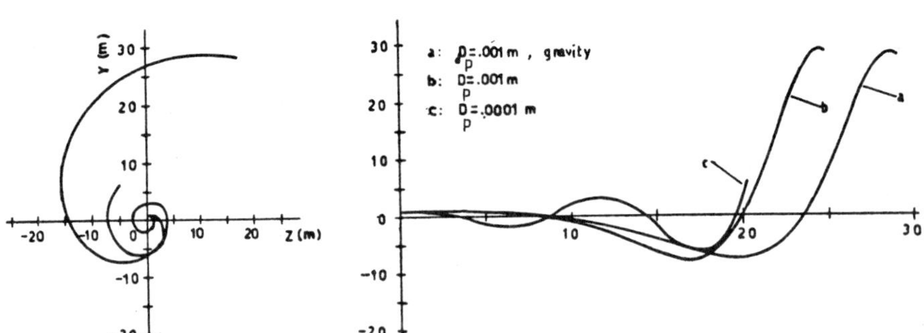

FIGURE 5. Particle motion in a flow field possessing swirl; total time 2s; $\varrho_p=2500kg/m^3$; $\varrho_f=1.29kg/m^3$; $U_p=10m/s$; $W_f=4\pi r$ m/s

velocity larger than that of the fluid. This is owed to the drag force required to compensate for the external field force. The velocity difference depends on the particle diameter and relative density. That is, the larger those parameters are, the larger the velocity difference. The effect of swirl can be seen in figure 5. The heavier particles do not react as quickly as the lighter ones to the velocity change, but after some time centrifugal effects seem to be of increasing importance. As a result, the lighter particles only have minor radial velocity.

3.3.2. Motion of fluid points in a homogeneous, isotropic, turbulent flow field. Comparison was made of the fluid point concentration acquired by using the different eddy characteristics (eddy length and eddy lifetime) definitions of Gosman-Ioannides and Shuen et al., and the analytic solution given by Hinze (1975). The distribution of the concentration of fluid particles introduced at a constant rate from a source into a homogeneous turbulent flow of the same fluid, is according to Hinze :

$$C(x,y,z)=S\ \exp[(-U_f(y^2+z^2)/(4\Gamma_t|x|)]/(4\pi\Gamma_t|x|) \quad (10)$$

where S the volumetric source strength and Γ_t the turbulent diffusivity.

Gosman & Ioannides evaluate the eddy lifetime as $t_e = L_e/|u_f|$ and the eddy size as $L_e = C_\mu^{1/2}\ k^{3/2}/\varepsilon$.

However Shuen et al. use the expressions (8) for the definition of the eddy characteristics. Calibration of the models was made by fixing the values of U_f, k and ε, in order to produce an S value relevant to the number of trajectories evaluated.

Using the Monte-Carlo method and following the trajectories of a number of "marked" fluid points, comparison has been made

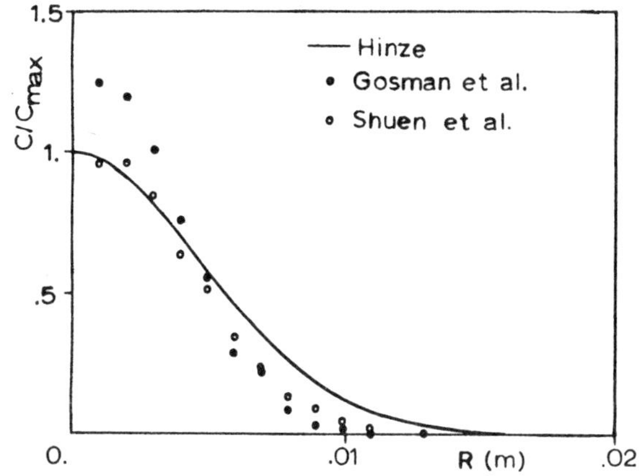

FIGURE 6. Comparison of concentration predictions of different models; Cmax is the maximum concentration value given by Hinze. Same axial location : 1 m after the source.

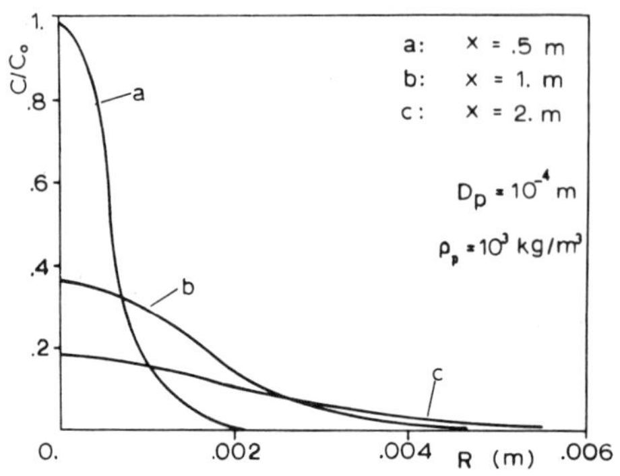

FIGURE 7. Particle dispersion at various axial locations; C_o is the particle concentr., at x=0.

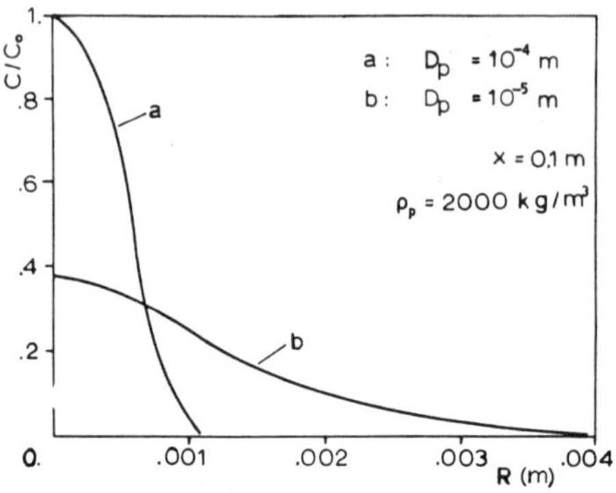

FIGURE 8. Effect of particle diameter on dispersion.

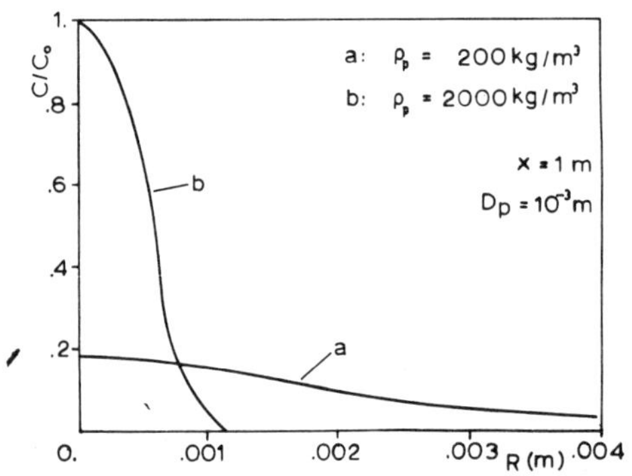

FIGURE 9. Effect of particle density on dispersion.

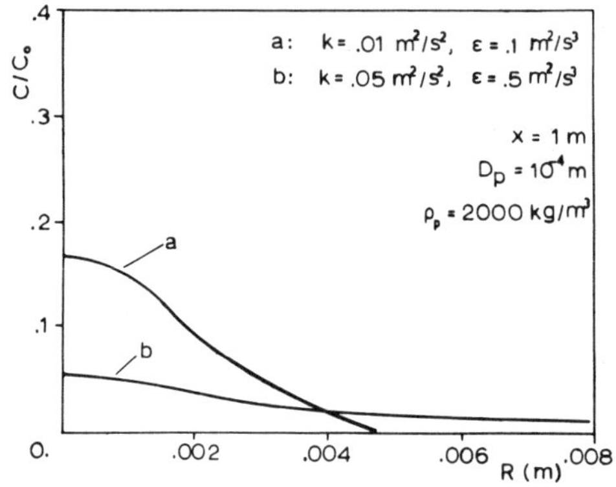

FIGURE 10. Particle dispersion for different k-ε.

of the models of Gosman & Ioannides and Shuen et al. The particles were assumed to be infinitely small and so to follow exactly the fluid motion. Figure 6 clearly shows that the model of Shuen et al. results in better agreement to Hinze's solution.

3.3.3. Motion of solid particles in a homogeneous, isotropic, turbulent flow field. Using the turbulence model of Shuen et al., and combining it with the particle motion equations and the Monte-Carlo method, resulted in the final proposition for the two-phase flow model. The effects of particle and flow field parameters were evaluated.

The statistic evaluation was done over 1000 trajectories for each case. The number was chosen as a compromise between computer time requirements and consistency in the results. Some typical results can be seen in figures 7 to 10; using the conclusions of the previous parts of the study, the predictions can be easily explained.

Heavier particles show inability to follow the flow fluctuations, being mainly affected by the mean flow characteristics. This leads to a crossing trajectories effect and these particles spend less time in a particular eddy than lighter particles. The result is reduced dispersion in comparison to small particles. It seems again that the particle inertia is the governing parameter for a certain turbulence level. It was also found that an increase of the dissipation rate ε and thus producing of "larger" eddies, resulted in an increase of the dispersion of particles (fig.10). Larger

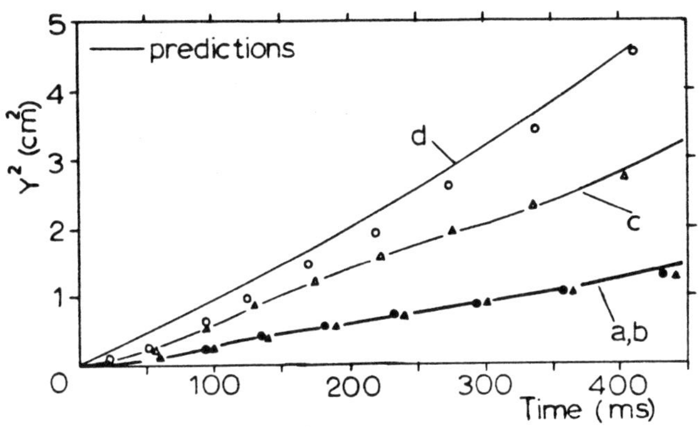

FIGURE 11. Particle dispersion predictions - experiments: a - • glass; b - ▲ copper; c - △ corn pollen; d - ○ hollow glass.

eddies produce grater interaction times with the particles, this meaning that a particle's trajectory is affected more by every single eddy. This can be understood by comparison to the results for the particle velocity change shown in figures 2,3 and 4.

Comparison has also been made with the experimental data given by Snyder & Lumley (1971). These data refer to grid generated turbulence and a variety of different particles. All of the particles are of diameter smaller than 100 μm and have densities which range from 260 kg/m^3 to 8900 kg/m^3. For each case 2000 trajectories were evaluated. Results of the comparison with particle radial dispersion data, can be seen in figure 11. Agreement with copper, glass and corn pollen data is quite satisfactory.

4. CONCLUSIONS

The present study was aimed at the proposal of a model for the prediction of particle motion characteristics in a two-phase flow. From the two widely used approaches to the problem, the Eulerian and the Lagrangian, the latter was selected as having advantages in giving detailed information about the dispersed phase. A simplified form of the BBO equation was used, having as acting forces only the drag and the force induced by the gravity field.

The effects of turbulent dispersion were evaluated with the help of the Monte-Carlo stochastic method. Turbulence models by Gosman-Ioannides and Shuen et al. were compared, the second showing better agreement with Hinze's analytic results.

Results of the conducted work show good agreement with experimental data for the simple flow field test cases selected. The larger particles seem to be affected by the mean fluid flow; on the other hand, small particles react quicker to velocity fluctuations, being therefore influenced by turbulence, and disperse more than than the heavier ones.

This preliminary evaluation is encouraging, but of course more testing is needed, with the use of proven experimental results, in order to show the capabilities of the model in quantitative predictions in more complex flow situations.

REFERENCES

1. Durst F, Milojevic D, Schoenung B: Eulerian and Lagrangian Predictions of Particulate Two-phase Flows: a Numerical Study. Appl. Math. Modelling, v8, April 1984, 101-115.
2. Elghobashi S, Abou-Arab T: A Two-equation Turbulence Model for Two-phase Flows. Phys. Fluids, v26, 4, 1983, 931-938.
3. Gosman AD, Ioannides E: Aspects of Computer Simulation of Liquid-Fueled Combustors. J. Energy, v7,6, 1983, 482-490.
4. Hinze JO: Turbulence. 2nd Edition, McGraw-Hill Book Co. New York, 1975.
5. Lee SL, Durst F: On the Motion of Particles in Turbulent Duct Flows. Int.J.Mult.Flow, v8, 2, 1982, 125-146.
6. Melville WK, Bray KNC: A Model of the Two-phase Turbulent Jet. J. Heat Mass Transfer, v22, 1979, 647-656.
7. Migdal D, Agosta DV: A Source Flow Model for Continuum Gas-Particle Flow. Trans. ASME, J.Appl. Mech., 1967, 860-865.
8. Mostafa AA, Elghobashi SE: Effect of Liquid Droplets on Turbulence in Round Gaseous Jet. NASA CR-175063, Feb.1986.
9. Mostafa AA, Mongia HC: Eulerian and Lagrangian predictions of Turbulent Evaporating Sprays. Paper 86-0452 at AIAA 24th Aerospace Sciences Meeting, 1986.
10. Picart A, Berlemont A, Gouesbet G: Modelling and Predicting Turbulence Fields and the Dispersion of Discrete Particles Transported by Turbulent Flows. Int.J.Mult.Flow, v12, 2, 1986, 237-261.

11. Sharma MP, Crowe CT: A Novel Physico-Computational Model for Quasi One-dimensional Gas-Particle Flows. J.Fl.Eng., v100, Sept. 1978, 343-349.
12. Shuen JS, Solomon ASP, Zhang QF, Faeth GM: Structure of Particle Laden Jets: Measurements and Predictions. AIAA J., v23, 3, 1985, 396-404.
13. Snyder WH, Lumley JL: Some Measurements of Particle Velocity Autocorrelation Functions in a Turbulent Flow. J.Fl.Mech., v48, 1, 1971, 41-71.
14. Steimke JL, Dukler AE: Laser Doppler Velocimeter Measurements of Aerosols in Turbulent Pipe Flow. Int.J.Mult.Flow, v9, 6, 1983, 751-754.
15. Tsuji Y, Morikawa Y, Shiomi H: LDV Measurements of Air-Solid Two-phase Flow in Vertical Pipe. J.Fl.Mech., v139, 1984, 417-434.
16. Wells MR, Stock DE: The Effects of Crossing Trajectories on the Dispersion of Particles in a Turbulent Flow. J.Fl.Mech, v136, 1983, 31-62.

Dynamic Transition of a Self-Igniting Region

J.W. Dold

School of Mathematics, University of Bristol,
University Walk, Bristol BS8 1TW, England.

1. *Introduction*

We consider a non-uniform thermal evolution in a self-igniting combustible gas. Typically [1,2], once a relatively hot kernel of reactants has been formed, the exponentially nonlinear dependence of the rate of heat production on temperature leads to a vigorous and progressive acceleration of the rate of peak temperature growth. This continues until the locally-available chemical reactants are almost consumed. Provided the ignition kernel is small enough or the remaining time to ignition is long enough, thermal expansion velocities are produced with little associated pressure rise. This paper examines the possibile transition from such a relatively gentle form of self-ignition to one in which the accelerating reaction produces substantial increases in pressure.

Since the hottest parts of a mixture tend to react fastest, the maximum growth-rate of the temperature also tends to focus itself on a smaller and smaller spatial region as time proceeds. Just as thermal conductivity normally acts to spread out any such concentration of temperature, its effect in the evolving ignition kernel is to work against this focussing of temperature growth. However, once thermal-runaway has begun, conduction fails to do more than reduce the rate at which the peak temperature region becomes more and more localised [2]. Along with this, conduction acts to progressively diminish its own importance as thermal runaway proceeds; it produces a larger, and so more vigorous, reacting region than would arise without conduction.

The rapid self-heating of the igniting region also tends to expand the gases locally. For slow enough processes, the accompanying pressure increase in the reacting region has time to push aside the surrounding gases, so producing an expansion velocity. It has been recognised (for example [3]) that, provided expansion velocities remain a small enough fraction of the speed of sound, the overall rise in pressure remains small compared with absolute pressure. Equivalently, this requires that a typical time to ignition (or induction time) for the hottest region should be long compared with the acoustic time for propagation of disturbances across the region. Should the reverse situation arise, there would then simply not be sufficient time for expansion effects to propagate outwards. Large pressure increases would result while density changes would remain small.

Indeed, because the shrinking of the region of fastest reaction, as the temperature increases, can be less rapid than the dramatic shortening of reaction times, a transition from one of these situations to the other

C.-M. Brauner and C. Schmidt-Lainé (eds.), Mathematical Modeling in Combustion and Related Topics, 461–470.

is possible. With continuous derivatives, cases of this sort will arise when the gradient of temperature at its maximum value is zero; non-zero gradients at the maximum are only possible when ignition occurs at a boundary of the medium, as examined by Jackson et al [4] and Stewart et al [5]. Correspondingly, the evolution of the temperature peak can change from taking place at almost constant pressure to almost constant density.

The occurrence of this transition makes a major qualitative difference between two types of thermal-runaway. Without it, the gas behaviour is fairly gentle and expansion Mach numbers are small. On the other hand, the transition marks the beginning of a large increase in pressure, with the consequent development of shock waves and possibly detonation. Kapila [6] has found that an elevated pressure quickly forms (as the reaction proceeds to completion) in the entire region where the acoustic time exceeds the reaction time. Such increases in pressure, and consequent propagation of strong shock waves, are observed as a part of "knock" in internal combustion engines (for example [7]). This happens through premature explosion of unburnt gases in combustion chambers, and can be very destructive. Building on the work of Clarke [8], Klein and Peters [9] have shown how multiple reflections within a closed cylinder can contribute to a thermal runaway which is in its transition stage throughout the cylinder.

This paper concentrates on a process of transition which is taken to happen on a small scale compared with the scale of any container. The region of interest may thus be taken to be unconfined. Because of the focussing effect of the thermal runaway, this is a reasonable assumption if the gases are taken to be initially cold enough for reaction times to be very long. Although some pressure increase will occur because of thermal heating in a closed vessel [9,10], as transition is approached a time will be reached when the remaining time to ignition is less than the time needed for acoustic waves to cross the vessel. This is still much longer than the time taken for acoustic waves to cross the region of greatest reaction-rate. Thus while the ignition kernel is no longer aware of any boundaries, the evolution is still well in advance of the transition.

A natural physical constraint on any real combustion process is that chemical reactions take place over a time-scale which is, on average, long compared with the mean collision time between molecules. To be consistent with this in the present context, conduction should not be very important in any situation where both the chemical induction time and the acoustic propagation time, across the region of significant reaction rate, are of the same magnitude [11]. To see this, it can be noted that conductive effects would only be comparable with the effects of acoustic transfer for changes occuring across a region the size of a few molecular collision distances; comparable chemical effects would then only be possible for reaction times of the order of the collision time, which is unrealistic.

It can therefore be argued that conduction and diffusion should be insignificant during and after any transition to almost constant density behaviour. However, it may be recalled that the relative importance of conduction diminishes as thermal runaway proceeds. Thus, given a situation in which conduction is initially important, it may still be realistic for a dynamical transition, to constant density behaviour, to happen in the later well-developed ignition kernel, when conductive effects have become much less significant. In modelling only the details immediately before and after transition, it remains consistent to neglect conductive and diffusive effects.

Considering a non-homogeneous thermal-runaway in an unconfined reactive gas, a point of maximum temperature may be taken to exist. The reaction-rate and temperature-growth will tend to be fastest, and so a hot ignition kernel will tend to evolve, primarily around this point. Given that, at some stage, a transition takes place in the nature of the accompanying gasdynamical effects, it is possible to place the transition in context by looking both forwards and backwards in time. As is shown below, the lowest order of spatial variation at the temperature peak tends to become the most significant as time proceeds. For smooth, continuous temperature changes, this lowest order will typically be quadratic in nature (although higher-order basic forms are possible) and a nondimensionalisation can be chosen which normalises the solution for the earlier bahaviour about this fundamentally quadratic spatial variation of the temperature peak. For "distant" initial conditions, it is only this order that is remembered by the temperature profile as transition is approached. In terms of this, the later solution is uniquely determined for a given ratio of principal specific heats γ.

2. *Model*

a) *Governing equations*

Neglecting conductive and diffusive effects, the following set of equations can be used to examine the effects of a model one-step reaction between υ_F fuel molecules and υ_X oxidant molecules, producing υ_P product molecules [12]. The superscript # is used to signify the dimensional values of the various physical quantities.

Energy: $$\rho^{\#}\mathscr{C}_P^{\#}\frac{DT^{\#}}{Dt^{\#}} = \upsilon_P W_P Q^{\#}\mathfrak{R}^{\#} + \frac{DP^{\#}}{Dt^{\#}} \tag{2.1}$$

Continuity: $$\frac{DP^{\#}}{Dt^{\#}} + \gamma P^{\#}\nabla^{\#}\cdot\mathbf{u}^{\#} = (\gamma-1)\upsilon_P W_P Q^{\#}\mathfrak{R}^{\#} \tag{2.2}$$

Momentum: $$\rho^{\#}\frac{D\mathbf{u}^{\#}}{Dt^{\#}} + \nabla^{\#}P^{\#} = \mathbf{0} \tag{2.3}$$

Gas-State: $$P^{\#} = \frac{\gamma-1}{\gamma}\,\mathscr{C}_P^{\#}\rho^{\#}T^{\#} \tag{2.4}$$

Species: $$\rho^{\#}\frac{Dc_\alpha}{Dt^{\#}} = -\upsilon_\alpha W_\alpha \mathfrak{R}^{\#} \qquad (\alpha = F \text{ or } X) \tag{2.5}$$

Reaction: $$\mathfrak{R}^{\#} = (\rho^{\#}c_X)^{\upsilon_X}(\rho^{\#}c_F)^{\upsilon_F}\,F^{\#}(T^{\#})\,e^{-E^{\#}/(R^{\#}T^{\#})} \tag{2.6}$$

In writing the gas state equation in the form (2.4) the relation $R^{\#}/W = \mathscr{C}_P^{\#}-\mathscr{C}_V^{\#} = (\gamma-1)\mathscr{C}_P^{\#}/\gamma$ has been used to substitute for $R^{\#}/W$ in the usual form of this equation; $R^{\#}$ is the universal gas constant, and W is the mean molecular weight of the mixture. Also, the usual form of the continuity equation, $D\rho^{\#}/Dt^{\#} + \rho^{\#}\nabla^{\#}\cdot\mathbf{u}^{\#} = 0$, has been modified to give the version (2.2) of this equation above, by using the gas-state equation to substitute for $D\rho^{\#}/Dt^{\#}$ and then using the energy equation to eliminate $DT^{\#}/Dt^{\#}$. The usefulness of this version of the equation, in combination with the momentum equation, lies mainly in showing how thermal expansion, associated with the chemical reaction rate $\mathfrak{R}^{\#}$, can produce gasdynamic disturbances of the pressure and velocity field.

It is assumed that the specific heat at constant pressure $\mathcal{C}_p^{\#}$ and the ratio of specific heats γ are constant, as are the heat of reaction $Q^{\#}$, activation energy $E^{\#}$ and the molecular weights W_α for the chemical species α. In the use of this model below, the activation energy will be taken to be large enough for the ratio $E^{\#}/(R^{\#}T^{\#})$ to remain large for all reasonable values of the absolute temperature $T^{\#}$. The pre-exponential function $F^{\#}(T^{\#})$ does not generally vary strongly with temperature, especially in comparison to the exponential term of the expression [12]. The operator $D/Dt^{\#}$ represents the Lagrangean time derivative $\partial/\partial t^{\#}+\mathbf{u}^{\#}\cdot\nabla^{\#}$, where $\mathbf{u}^{\#}$ is the velocity and $\nabla^{\#}$ is the spatial gradient operator. Other quantities used in equations (2.1) to (2.6) are density $\rho^{\#}$, pressure $P^{\#}$, and the mass fractions c_α of the species α.

b) *Non-Dimensionalisation*

Characteristic dimensional values (subscript c) for density, pressure, temperature, velocity, length and time scales can be related as follows, where $a_c^{\#}$ is a typical sound-speed given by $a_c^{\#} = \sqrt{(\gamma P_c^{\#}/\rho_c^{\#})} = \sqrt{[(\gamma-1)\mathcal{C}_p^{\#}T_c^{\#}]}$:

$$\rho_c^{\#} = \gamma P_c^{\#}/a_c^{\#2}, \qquad P_c^{\#} = \rho_c^{\#}a_c^{\#2}/\gamma,$$

$$T_c^{\#} = \frac{a_c^{\#2}/\mathcal{C}_p^{\#}}{\gamma-1}, \qquad u_c^{\#} = a_c^{\#}, \qquad x_c^{\#} = a_c^{\#}t_c^{\#}, \tag{2.7}$$

$$t_c^{\#} = \frac{R^{\#}\mathcal{C}_p^{\#}T_c^{\#2}}{\nu_P W_P Q^{\#}E^{\#}}\, e^{E^{\#}/(R^{\#}T_c^{\#})}\, \rho_c^{\#}/[(\rho_c^{\#}c_{FO})^{\nu_F}(\rho_c^{\#}c_{XO})^{\nu_X}F^{\#}(T_c^{\#})],$$

where the expression for $t_c^{\#}$ represents a typical time for thermal runaway to occur. Given that the ratio $E^{\#}/(R^{\#}T_c^{\#})$ is large, it is sensitively dependent on the choice of $T_c^{\#}$. The quantities c_{FO} and c_{XO} are typical reference values for the mass fractions of fuel and oxidant. In terms of these, dimensionless variables can be defined as follows

$$\rho = \rho^{\#}/\rho_c^{\#}, \quad P = P^{\#}/P_c^{\#} \quad T = T^{\#}/T_c^{\#},$$
$$\mathbf{u} = \mathbf{u}^{\#}/u_c^{\#}, \quad \boldsymbol{x} = \boldsymbol{x}^{\#}/x_c^{\#}, \quad t = t^{\#}/t_c^{\#}. \tag{2.8}$$

With α representing either F or X, we write $y_\alpha = c_\alpha/c_{\alpha O}$ so that y_α is a scaled mass fraction which will generally be of order one for representative choices of the reference mass fractions $c_{\alpha O}$. Measures of the nondimensional potential for heat release, available in these mass-fractions of reactants can be identified as Q_α where

$$Q_\alpha = \frac{\nu_P W_P}{\nu_\alpha W_\alpha}\frac{Q^{\#}}{\mathcal{C}_p^{\#}T_c^{\#}}\, c_{\alpha O}. \tag{2.9}$$

It is also convenient to define a nondimensional form F for the pre-exponential function $F^{\#}$ such that $F(T) = F^{\#}(T^{\#})/F^{\#}(T_c^{\#})$. This function will generally be an order-one function of the nondimensional absolute temperature T and, in particular, its value at $T = 1$ is unity. We also identify the small parameter $\epsilon = R^{\#}T_c^{\#}/E^{\#}$, representing the inverse of the large nondimensional activation temperature. The governing equations then become:

Energy: $$\rho\frac{DT}{Dt} = \epsilon\mathcal{R} + \frac{\gamma-1}{\gamma}\frac{DP}{Dt} \tag{2.10}$$

Continuity: $$\gamma^{-1}\frac{DP}{Dt} + P\nabla\cdot\mathbf{u} = \epsilon\mathcal{R} \tag{2.11}$$

Momentum: $$\rho\frac{D\mathbf{u}}{Dt} + \gamma^{-1}\nabla P = 0 \tag{2.12}$$

Gas-State: $$P = \rho T \tag{2.13}$$

Species: $$\rho\frac{Dy_\alpha}{Dt} = -\epsilon\mathcal{R}/Q_\alpha \tag{2.14}$$

Reaction: $$\mathcal{R} = (\rho y_F)^{\nu_F}(\rho y_X)^{\nu_X}\, F(T)\, e^{(1-1/T)/\epsilon}. \tag{2.15}$$

If we now assume symmetry about the plane $x = 0$, and define a one-dimensional Lagrangean coordinate s such that

$$s = \int_0^x \rho\, dx \tag{2.16}$$

then, in terms of s and t as independent variables, the relevant governing equations become

Energy $$\rho T_t = \epsilon\mathcal{R} + \frac{\gamma-1}{\gamma}P_t \tag{2.17}$$

Continuity $$\gamma^{-1}P_t + P\rho\, u_s = \epsilon\mathcal{R} \tag{2.18}$$

Momentum $$u_t + \gamma^{-1}P_s = 0 \tag{2.19}$$

Species $$\rho y_{\alpha t} + \epsilon\mathcal{R}/Q_\alpha = 0. \tag{2.20}$$

c) *Asymptotic Equations*

For disturbances of order ϵ, and assuming the nondimensionalisation is carried out appropriately, we can consider asymptotic solutions for these equations in the form

$$T \sim 1 + \epsilon\Gamma(s,t), \quad P \sim 1 + \epsilon\gamma p(s,t), \quad u \sim \epsilon v(s,t), \tag{2.21}$$

with ρ and y_α taking the forms $\rho = 1+O(\epsilon)$ and $y_\alpha = 1+O(\epsilon)$, as $\epsilon \to 0$ with s and t fixed. Equations for Γ, p and v then become

Energy: $$\Gamma_t = e^{\Gamma} + \mu p_t \qquad (2.22)$$

Continuity: $$p_t + v_s = e^{\Gamma} \qquad (2.23)$$

Momentum: $$v_t + p_s = 0, \qquad (2.24)$$

where μ represents the difference $\gamma-1$.

3. *Early-Time solution*

a) *Basic Behaviour*

Neglecting pressure changes in equation (2.22), the equation $\Gamma_t = e^{\Gamma}$ leads to the simple solution $\Gamma = -\ln[t_1-t + f_1(s)]$, where $f_1(s)$ is a function of integration. The highest temperature corresponds to a minimum value of f_1, which can be taken equal to zero by absorbing any non-zero part of the minimum value into the constant t_1 which then represents an ignition time, at which Γ becomes unbounded. If it is also assumed that f_1 is a smooth regular function of s and that the maximum temperature lies at $s = 0$ (which is not a rigid boundary of the domain of s), then the lowest-order variation of f_1 typically takes the quadratic form $f_1 = c_1 s^2 + O(s^4)$, for some non-zero constant c_1. In principle, it is possible for the lowest order variation of f_1 to be proportional to s^4, or some higher even order, but such cases would be exceptional and will not be studied here.

We can thus identify that, with negligible pressure variations, Γ typically takes the basic form

$$\Gamma = -\ln(t_1-t) - \ln[1 + c_1\frac{s^2}{t_1-t} + O(\frac{s^4}{t_1-t})]. \qquad (3.1)$$

This result illustrates the tendency for the most strongly-reacting region to focus on smaller values of s, as t approaches t_1. It can be seen that the reaction is most significant in the region where s is of order $\sqrt{(t_1-t)}$. Correspondingly, the terms of order s^4 or higher in the expansion of f_1 become decreasingly significant as the reaction proceeds so that, with t_1 and f_1 defined by "distant" initial conditions, only the lowest order needs to be considered as transition is approached.

Also, subtracting $\ln(c_1)$ from Γ and rescaling s and t by c_1^{-1}, it can be noted from equations (2.7) that a suitable choice of $T_c^{\#}$ can be made for which the constant $c_1(T_c^{\#})$ is set to unity. With this choice fixing the nondimensionalisation, disturbances in Γ can be studied in the fairly simple form

$$\Gamma = \phi - \ln(t_1-t + s^2), \qquad (3.2)$$

where $\phi(s,t)$ represents dynamic disturbances from the fundamental constant-pressure (or zero μ) solution $\Gamma \sim -\ln(t_1-t+s^2)$. For ϕ to arise purely through dynamic effects, contributions to ϕ of the form $-\ln[1+f(s)/(t_1-t+s^2)]$ are excluded; these would arise because of different representations for the function f_1.

b) *Dynamic Disturbances*

Using the definition (3.2), equations (2.22) to (2.24) become

$$\phi_t = \frac{e^{\phi}-1}{t_1-t+s^2} + \mu p_t \tag{3.3}$$

$$p_t + v_s = \frac{e^{\phi}}{t_1-t+s^2} \tag{3.4}$$

and

$$v_t + p_s = 0 \tag{3.5}$$

which need to be solved under the conditions of zero disturbance at infinity and symmetry:

$$\lim_{t_1-t \to \infty} (p,v,\phi) \equiv \mathbf{0} \equiv \lim_{|s| \to \infty} (p,v,\phi)$$

and

$$v(0,t) \equiv p_s(0,t) \equiv \phi_s(0,t) \equiv 0. \tag{3.6}$$

Examining these equations, it can be seen that disturbances in p, v and ϕ will only be small for large values of t_1-t. In this range, two distinct regions of solution can be distinguished: with s of order $\sqrt{(t_1-t)}$ the reaction is significant but acoustic effects are not properly modelled; with s of order t_1-t reaction effects are small and acoustic disturbances, arising from the thermal runaway in the reaction region, can be modelled. The full asymptotic solution for this stage has to be obtained by solving for both regions with appropriate matching between them. There is not enough space to present a full derivation here, and so we only present the composite solution, valid in both regions, to order $(t_1-t)^{-1}$. Writing τ for t_1-t and ζ for $s/\sqrt{\tau}$, the solution is

$$p \sim \frac{\pi}{2}(s+\tau)^{-\frac{1}{2}} + \tau^{-1}\frac{\zeta}{2}\left[\frac{\pi}{2} - \tan^{-1}\zeta\right], \tag{3.7}$$

$$v \sim \tau^{-\frac{1}{2}}\left\{\tan^{-1}\zeta - \frac{\pi}{2}\left[1 - \left(1 + \frac{s}{\tau}\right)^{-\frac{1}{2}}\right]\right\} - \tau^{-1}\mu\,\frac{\pi}{2}\,\frac{\zeta}{1+\zeta^2} \tag{3.8}$$

and

$$\frac{\phi}{\mu} \sim \frac{\pi}{2}(s+\tau)^{-\frac{1}{2}} - \tau^{-\frac{1}{2}}\frac{\pi}{1+\zeta^2} - \frac{\ln\tau}{\tau}\,\mu\,\frac{\pi^2/8}{1+\zeta^2} + \tag{3.9}$$

$$+ \tau^{-1}\left\{\frac{\zeta}{2}\left[\frac{\pi}{2} - \frac{3+\zeta^2}{1+\zeta^2}\tan^{-1}\zeta\right] + \frac{1}{2}\,\frac{\ln(1+\zeta^2)}{1+\zeta^2} - \mu\,\frac{\frac{1}{2}\pi^2\zeta^2}{(1+\zeta^2)^2}\right\},$$

as $\tau \to \infty$.

4. ***Transition***

The early-time solution clearly identifies two length-scales, the acoustic length scale s and the reaction length scale $\zeta = s/\sqrt{\tau}$. While acoustic disturbances propagate along characteristics parallel to $s = \tau$, for large values of τ the reaction occupies the relatively small region $s = O(\sqrt{\tau})$. This quadratic behaviour of the size of the reaction-rate region highlights the possibility of a transition in the relative scales of the acoustic and reaction lengths; namely that, for small values of τ, the acoustic length scale becomes much smaller than the reaction length scale as illustrated in figure 1.

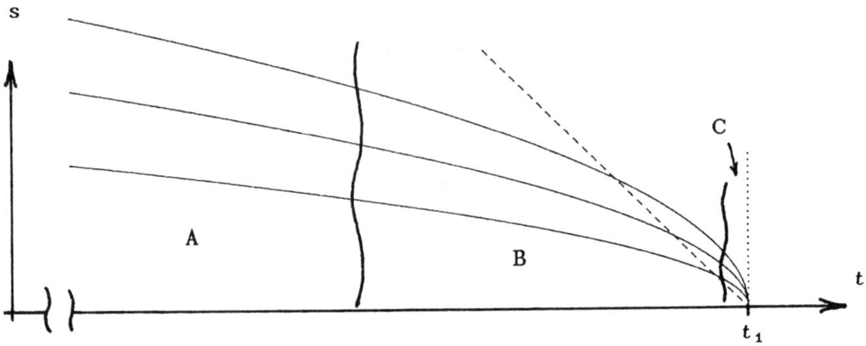

__Figure 1__. Lines of constant ζ and the acoustic characteristic $s = \tau$ (dashed). Region A, nearly-constant pressure; B, transition region; C, nearly-constant density.

For τ of order unity, the behaviour is influenced by significant interactions between acoustic and reaction effects. Using a method of lines, with mid-point interpolation to add points whenever the focussing of temperature bagan to cause a loss of spatial resolution, this stage of the evolution was solved numerically. Given the asymptotic solutions (3.7) to (3.9) for specified values of γ the full evolution could be solved well into the constant-density stage of the thermal runaway.

5. *Constant-Density Thermal Runaway*

After transition, the acoustic length-scale is much shorter than the length scale of the reaction. Thus acoustic effects are not able to propagate dynamic disturbances away from their sources, and reaction effects dominate. To model this stage it is convenient to identify a logarithmic time-scale ξ, such that $\tau = t_1 - t = e^{-\xi}$. It is also convenient to consider perturbations on the otherwise logarithmic growth of Γ in the form $\Gamma = \xi + \theta$. The governing equations then become

$$1 + \theta_\xi + \tfrac{1}{2}\zeta\theta_\zeta = e^{\theta} + \mu[p_\xi + \tfrac{1}{2}\zeta p_\zeta] \qquad (5.1)$$

$$p_\xi + \tfrac{1}{2}\zeta p_\zeta + e^{-\frac{1}{2}\xi} v_\zeta = e^{\theta} \qquad (5.2)$$

$$v_\xi + \tfrac{1}{2}\zeta v_\zeta + e^{-\frac{1}{2}\xi} p_\zeta = 0, \qquad (5.3)$$

in which the effect of letting ξ become large can be seen through the exponentially small terms.

Asymptotic solutions of these equations can be written as follows,

$$v \sim -\tau^{\frac{1}{2}} \ln\tau \, \frac{2a}{\gamma}\zeta - \tau^{\frac{1}{2}} \frac{2a}{\gamma}\zeta \ln(1+a\zeta^2) \qquad (5.4)$$

$$\Gamma \sim -\ln[\tau\gamma(1+a\zeta^2)] - \tau^{\frac{1}{2}}\ln\tau \, \frac{\gamma-1}{\gamma} \, a \, \frac{1+2a\zeta^2+3a^2\zeta^4}{1+a\zeta^2} \tag{5.5}$$

$$\gamma p \sim \Gamma + c - \tau^{\frac{1}{2}}\ln\tau \, \frac{2a}{\gamma}(1+3a\zeta^2), \tag{5.6}$$

where it can be noted that the expansion velocity v is small to first order as $\tau \to 0$. This relates to the fact that there is no longer time for expansion effects to propagate and indicates that density now remains nearly constant, for constant ζ, during the thermal runaway. The pressure begins to rise in tandem with the temperature, so that one can anticipate increases in absolute pressure of the same relative order as the remaining increase in absolute temperature through chemical reaction.

The constants a and c need to be determined by matching with the earlier solution for the transition-stage. Using the numerical results described in section 4 and the normalisation (3.2) of the early-time solution, these constants depend only on the specific-heats' ratio γ. Results of the matching, for γ between one and two, are plotted in figure 2. It can be noted that with γ equal to unity ($\mu = 0$), the equation for temperature (2.22) is decoupled from the momentum and continuity equations. The solution for Γ is thus independent of the transition so that a takes the value unity. The numerical matching confirms this trivial result, and confirms that c takes the value zero when $\gamma = 1$.

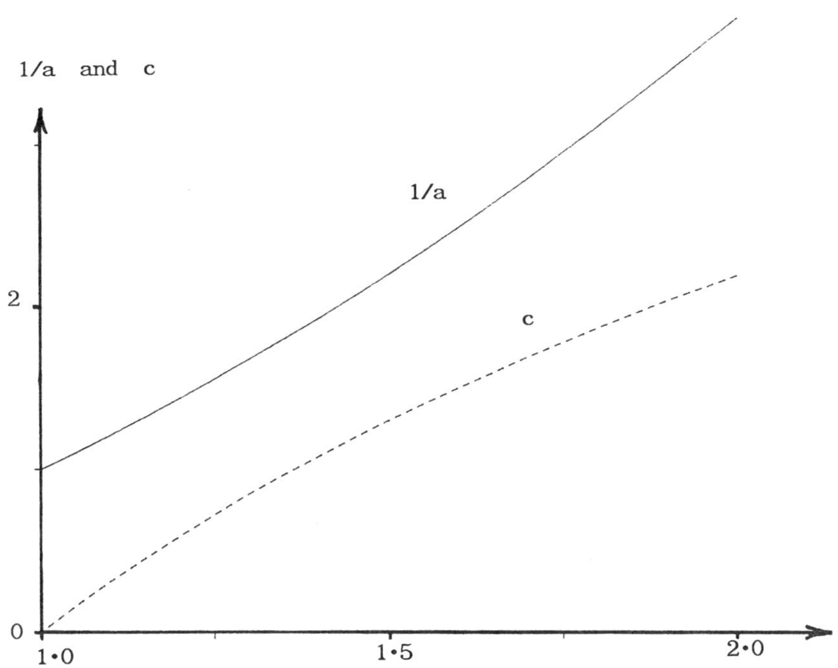

Figure 2. Results for $a(\gamma)$ [plotted as $1/a(\gamma)$] and $c(\gamma)$ [dashed].

With $a(\gamma)$ and $c(\gamma)$ determined in this way, the basic form of the evolution of the ignition kernel during its constant-density stage is known as a function of γ and the appropriate choice of nondimensionalisation which produces the normalisation (3.2).

6. *Conclusions*

A numerical solution has been used to follow the process of transition between almost-constant pressure and almost-constant density stages in the evolution of thermal runaway in an ignition kernel. With a suitable normalisation of the asymptotic solution at almost-constant pressure, the asymptotic behaviour in the near-constant density stage is determined as a function of the ratio of principal specific heats.

This transition forms a small but essential part in an overall process of thermal runaway and ignition which results in strong increases in pressure. It may be conjectured that igniting gases experience such a transition in an early stage of the production of engine knock.

References

1. Dold J.W., *Analysis of the Early Stage of Thermal Runaway*, Quart. J. Mech. Appl. Math. **38** (1985) 361-387
2. Dold J.W., *Analysis of Thermal Runaway in the Ignition Process*, submitted to SIAM J. Appl. Math., 1987
3. Buckmaster J.D., *An Introduction to Combustion Theory*, in *The Mathematics of Combustion*, J.D. Buckmaster (Ed.), SIAM Philadelphia 1985, 3-46
4. Jackson T.L., A.K. Kapila and D.S. Stewart, *Evolution of a Reaction Center in an Explosive Material*, manuscript 1987
5. Stewart D.S., T.L. Jackson and A.K. Kapila, *Evolution of a Localised Shock Induced Thermal Explosion*, Conference on Mathematical Modelling in Combustion, Lyon 27-30 April 1987
6. Kapila A.K., private communication, 1987
7. Griffiths J.F. and W. Nimmo, *Spontaneous Ignition and Engine Knock Under Rapid Compression*, Combustion and Flame **60** (1985) 215-218
8. Clarke J.F., *Finite Amplitude Waves in Combustible Gases*, in *The Mathematics of Combustion*, J.D. Buckmaster (Ed.), SIAM Philadelphia 1985, 183-245
9. Klein R. and N. Peters, *Cumulative Effects of Weak Pressure Waves in the Induction Period of a Thermal Explosion*, submitted to J. Fluid Mech., 1987
10. Poland J. and D.R. Kassoy, *The Induction Period of a Thermal Explosion in a Gas Between Infinite Parallel Plates*, Combustion and Flame **50** (1983) 259-275
11. Clarke J.F. and M. McChesney, *Dynamics of Relaxing Gases*, Butterworths London 1976, 79-90
12. Williams F.A., *Combustion Theory* (2nd Edition), Benjamin/Cummings California 1985

NUMERICAL MODEL FOR TURBULENT REACTIVE FLOWS WITH SWIRL

by F. DUPOIRIEUX and D. DUTOYA

Office National d'Etudes et de Recherches Aérospatiales
92320 Châtillon (FRANCE)

ABSTRACT

This paper presents a numerical method for the computation of turbulent reactive flows. In our case, the combustion is not stabilized by usual flame-holders behind which a recirculation takes place but by a swirler ring which gives a tangential velocity to the flow. Despite the existence of three velocity components, the flow is considered axisymmetric since it depends only on two space variables. In a first part, some points concerning the spatial discretization, the time integration and the ignition process are discussed. In a second part, some numerical results are displayed. The influence of some inlet parameters (stagnation pressure, tangential velocity) is investigated.

NOMENCLATURE

C_p : specific heat
h : stagnation enthalpy
K : kinetic energy
k : turbulent kinetic energy
p : pressure
q_r : radial momentum ($=\rho u$)
q_z : axial momentum ($=\rho w$)
r : radial coordinate
R_o : constant of the state equation
S_f : source term in the balance equation of the scalar f
t : time
T : temperature
u : radial velocity (along Or)
v : tangential velocity
w : axial velocity (along Oz)
z : axial coordinate
Δr : mesh size in the r direction
Δz : mesh size in the z direction
ε : dissipation rate of the turbulent kinetic energy
μ_t : turbulent viscosity
ρ : density
$\overline{(.)}$: usual turbulent average
$\widetilde{(.)}$: Favre average
$(.)'$: divergence to the usual average
$(.)''$: divergence to the Favre average

C.-M. Brauner and C. Schmidt-Lainé (eds.), Mathematical Modeling in Combustion and Related Topics, 471–480.

I. INTRODUCTION

When a new jet engine is designed, one of the main requirements is to reduce its dimensions. Concerning military engines, the after-burner part, which is fairly voluminous, must comply with this requirement. Thus a new idea has been recently examined: it consists in giving a tangential velocity to the flow, which has initially only one velocity component along the axis of the after-burner duct, in order to increase the length of the fluid particles path and thus to improve the heat transfer which pilots the combustion intensity. The tangential component is obtained through a swirler ring which is crossed by a part of the flow as indicated on figure 1. Although the velocity has at a given point three non zero components, the flow can be considered axisymmetric since the field of any quantity depends only on two space variables z and r (axial and radial coordinate respectively). In such an after-burner the combustion develops downstream of the swirler ring which operates like a flame holder. An experimental investigation of the flow is difficult since the axisymmetric geometry makes the installation of convenient windows for optical diagnostics impossible. Therefore some theoretical studies are needed and the purpose of this paper is to present the numerical computation of the flow downstream of the swirler without and with combustion and to examine the influence of the inlet conditions.

II. THE GOVERNING EQUATIONS

Since the flow is strongly turbulent with large density variations, the Favre averaging ($\tilde{f} = \overline{\rho f}/\bar{\rho}$) is used to prevent the correlations with dentity ($\overline{\rho' f'}$). The balance equations can be written as follows [1]:

$$\frac{\partial}{\partial t}(\bar{\rho} r) + \frac{\partial}{\partial r}(\bar{q}_r r) + \frac{\partial}{\partial z}(\bar{q}_z r) = 0 \tag{1}$$

$$\frac{\partial}{\partial t}(\bar{q}_r r) + \frac{\partial}{\partial r}\left| r\left(\frac{\bar{q}_r^2}{\bar{\rho}} + \bar{\rho}\,\widetilde{u''^2}\right)\right| + \frac{\partial}{\partial z}\left| r\left(\frac{\bar{q}_r \bar{q}_z}{\bar{\rho}} + \bar{\rho}\,\widetilde{u''w''}\right)\right| = -r\frac{\partial \bar{p}}{\partial r} + \bar{\rho}(\tilde{v}^2 + \widetilde{v''^2}) \tag{2}$$

$$\frac{\partial}{\partial t}(\bar{\rho}\tilde{v} r) + \frac{\partial}{\partial r}\left| r\left(\bar{q}_r \tilde{v} + \bar{\rho}\,\widetilde{u''w''}\right)\right| + \frac{\partial}{\partial z}\left| r\left(\bar{q}_z \tilde{v} + \bar{\rho}\,\widetilde{v''w''}\right)\right| = -\bar{\rho}(\tilde{u}\tilde{v} + \widetilde{u''v''}) \tag{3}$$

$$\frac{\partial}{\partial t}(\bar{q}_z r) + \frac{\partial}{\partial r}\left| r\left(\frac{\bar{q}_r \bar{q}_z}{\bar{\rho}} + \bar{\rho}\,\widetilde{u''w''}\right)\right| + \frac{\partial}{\partial z}\left| r\left(\frac{\bar{q}_z^2}{\bar{\rho}} + \bar{\rho}\,\widetilde{w''^2} + \bar{p}\right)\right| = 0 \tag{4}$$

$$\frac{\partial}{\partial t}(\bar{\rho}\tilde{f} r) + \frac{\partial}{\partial r}\left| r(\bar{q}_r \tilde{f} + \bar{\rho}\,\widetilde{u''f''})\right| + \frac{\partial}{\partial z}\left| r(\bar{q}_z \tilde{f} + \bar{\rho}\,\widetilde{w''f''})\right| = S_f r \tag{5}$$

where f represents all the scalars such as the enthalpy, the chemical species, k, ε and others (N.B.: if f is the total enthalpy, S_f includes the unsteady term $\partial\bar{p}/\partial t$).

These equations are completed by a state equation ($\bar{p} = \bar{\rho} R_0 \tilde{T}$) and a thermodynamic relationship ($\tilde{h} = \int_{T_0}^{T} C_p\, dT + \tilde{K}$).

All the statistic correlations are given by a usual k - ε model [2,3] whereas the source term S_f which includes the mean reaction effect is obtained through a turbulent combustion model which is described in [4]. We notice that the balance equations for the velocity components $\tilde{u}$ and $\tilde{v}$ cannot be written in a fully conservative way due to the terms $\bar{\rho}\widetilde{v^2}$ and $\bar{\rho}\widetilde{uv}$ stemming from the existence of the tangential component v. Although we are at first interested by the steady state, the unsteady terms ($\partial/\partial t$ terms) have been taken into account because a time iterative method is used. This method enables us to start from a physical and very well-known initial state, that is the gas at rest, and to converge rapidly by using an implicit scheme in time.

III. The Numerical Method

1) The spatial discretization

The above equations are discretized in space on a cartesian staggered grid (fig. 2) by means of a finite volume technique which ensures the conservativity of the scalars. The staggered grid prevents some wriggle in the pressure field. In addition to the conservativity, the spatial discretization must promote the stability and the accuracy of the method. Thus the discretization must be in compliance with the fact that any definite positive quadratic form made with the unknown quantities of the problem must keep a finite value, which is a sufficient and necessary condition of stability. To obtain this condition, let us consider the following quantity (for clearness the symbols of averaging are omitted):

$$E = \sum_{i,j} \frac{1}{2} \left| \frac{q_{ri,j}^2}{\rho_{i+\frac{1}{2},j}} + \rho_{i,j} v_{i,j}^2 + \frac{q_{zi,j}^2}{\rho_{i,j+\frac{1}{2}}} \right| r_i \Delta r_i \Delta z_j$$

This quantity can be considered as a positive definite quadratic form of the three velocity components since the density, which is imposed by the combustion but not by the Mach number (subsonic conditions), keeps a positive finite value. E is obviously very close to the total kinetic energy contained in the flow but is however slightly different because $q_{ri,j}$, $v_{i,j}$ and $q_{zi,j}$ are not expressed at the same point of the mesh (upper border, center and right border respectively). It can be easily established that the time dependent equation of E can be obtained from the discretized form of the equations (1), (2, (3) and (4):

$$\sum_{i,j} \Delta r_i \Delta z_j \left| u_{i+\frac{1}{2},j} \times (2) + v_{i,j} \times (3) + w_{i,j+\frac{1}{2}} \times (4) - \frac{1}{2}\left(u_{i+\frac{1}{2},j}^2 + v_{i,j}^2 + w_{i,j+\frac{1}{2}}^2 \right) \times (1) \right|$$

with:

$$u_{i+\frac{1}{2},j} = \frac{q_{ri,j}}{\rho_{i+\frac{1}{2},j}} \quad \text{and} \quad w_{i,j+\frac{1}{2}} = \frac{q_{zi,j}}{\rho_{i,j+\frac{1}{2}}}$$

This equation writes: $(\rho E)_t = W$, where the subscript t denotes the time derivative and W represents the source terms arising from the pressure, the viscous effects, the boundary conditions and on occasion from the spatial discretization. The three first mentioned source terms have a physical signification and balance each other when the steady state is reached (if a steady state does not exist, they do not make yet E infinite) whereas the term proceeding from the spatial discretization, if it exists, leads to inaccuracy and even to the non stability of the method since, if it is not balanced by the other physical terms, it may make E infinite. In order to illustrate how the spatial discretization is influenced by the accuracy/stability condition, let us consider the part of W (noted W_o) which is due to the source terms ρv^2 and $-\rho uv$ of equations (2) and (3) respectively.
It writes:

$$W_o = \sum_{i,j} \left| q_{ri,j} v_{i+\frac{1}{2},j}^2 - q_{ri-\frac{1}{2},j} v_{i,j}^2 \right|$$

For the sake of simplicity, the grid is assumed uniform and Δ_{ri} and Δ_{zj} have been dropped from that equation. Since it does not deal directly with the pressure and viscous effects, W_o must be made of terms which may be non zero only on the boundary. If $q_{ri-\frac{1}{2}j}$ is replaced by $\frac{1}{2}(q_{ri-1j} + q_{rij})$, the required estimate $v^2_{i+\frac{1}{2}j}$ is $\frac{1}{2}(v^2_{ij} + v^2_{i+1j})$, with which we have:

$$W_o = \sum_{i,j} \left| \frac{1}{2} q_{rij}(v^2_{ij} + v^2_{i+1j}) - \frac{1}{2}(q_{ri-1j} + q_{rij})v^2_{ij} \right|$$

$$= \sum_j \left| -\frac{1}{2} q_{roj}\, v^2_{1j} + \frac{1}{2} q_{rIj}\, v^2_{I+1j} \right|$$

where I is the number of cells in the r direction, q_{roj} is the transverse momentum component on the southern boundary and v_{I+1j} the value of v in the northern fictitious cells used to deal with the boundary conditions. It is clear that, with the above estimate of $v^2_{i+\frac{1}{2}j}$, W_o is made only of terms which are defined on the boundary and thus does not introduce non physical source terms in the equation of E. Actually W_o is zero because in our case the northern and southern boundaries are a wall and a symmetry axis, respectively, in such a way that q_{roj} and q_{rIj} vanish. An other estimate of $v^2_{i+\frac{1}{2}j}$ such $[\frac{1}{2}(v_{ij} + v_{i+1j})]^2$ would obviously not fulfil that condition. The same analysis for the other parts of the source term W leads to the following equations concerning the discretized convection terms of the equations (2) and (4):

$$q_{ri-\frac{1}{2}j} = \frac{r_{i-\frac{1}{2}}\, q_{ri-1j} + r_{i+\frac{1}{2}}\, q_{rij}}{r_i}$$

$$q_{rij-\frac{1}{2}} = \frac{\Delta z_{j-1}\, q_{rij-1} + \Delta z_j\, q_{rij}}{\Delta z_{j-1} + \Delta z_j}$$

$$q_{zij-\frac{1}{2}} = \frac{1}{2}(q_{zij-1} + q_{zij})$$

$$q_{zi-\frac{1}{2}j} = \frac{\Delta r_{i-1}\, q_{zi-1} + \Delta r_i\, q_{zij}}{\Delta r_{i-1} + \Delta r_i}$$

$$u_{ij} = \frac{q_{ri-1j}}{\rho_{i-1} + \rho_{ij}} + \frac{q_{rij}}{\rho_{ij} + \rho_{i+1j}}$$

$$w_{ij} = \frac{q_{zij-1}}{\rho_{j-1} + \rho_{ij}} + \frac{q_{zij}}{\rho_{ij} + \rho_{ij+1}}$$

N.B.: $q_{zi-\frac{1}{2}j}$ is estimated in a different way from previously for W_o.

2) The time integration

A time implicit integration using the Beam-Warming factored scheme [5-7] gives:

(6) $$(Id - A^r\,\Delta t)(Id - A^z\,\Delta t)\,\delta g = B(g^n), \quad g^{n+1} = g^n + \Delta t\,\delta g$$

where Id is the identity matrix, Δt is the time step, g is the vector of the unknown quantities consisting of $I \times J \times N$ components ($I \times J$: total number of cells, N: number of balance equations), B is a function proceeding from the spatial discretization. A^r and A^z are two matrices which have a block tridiagonal structure if the viscous terms are neglected. The superscript n denotes the time level.

This system is solved in two steps:

- r-step:

$$\delta g' = (Id - A^r \Delta t)^{-1} B(g^n) \quad with \quad \delta g' = (Id - A^z \Delta t)\, \delta g$$

- z-step:

$$\delta g = (Id - A^z \Delta t)^{-1}\, \delta g' \; and \; g^{n+1} = g^n + \Delta t\, \delta g$$

These two steps are relatively simple when A^r and A^z are block-tridiagonal but actually the viscous terms must be taken into account due to the strong turbulence and theReynolds stress $\bar{\rho}$ u" w" which is modeled in the form: $\mu_t[\partial/\partial r\,(q_z/\rho) + \partial/\partial z\,(q_r/\rho)]$ and thus introduces some crossed derivatives in the equations (2) and (4), makes A^r and A^z more complex. To turn around this difficulty, the equations of the system (6) are not solved altogether. During the r-step, a first set of equations relating to the quantities ρ, q_r, v and h is solved. This set proceeds from the fact that these four quantities are strongly coupled together due to the combustion and the source terms of the equations (2) and (3) but are independent of the other quantities. The corresponding system is block tridiagonal (4 × 4 blocks) since the crossed derivatives have an effect on A^r only through the equation of q_z. After solving, the results are introduced in the other equations, in particular that of q_z. The crossed derivative $\partial/\partial r\,[\mu_t\, \partial/\partial z(q_r/\rho)]$, of which the value is now well-known, can be included in the right handed term $B(g^n)$. The remaining equations can now be disposed to constitute N-4 simple tridiagonal systems which are solved easily. In the same way one block-tridiagonal system (3 × 3 blocks relating to ρ, q_z and h) and N-3 simple tridiagonal systems are successively solved during the z-step. Thus the time integration can be efficiently performed despite the crossed derivatives.

3) The ignition procedure

In a usual after-burner the ignition is operated in a cell located at the center of the recirculation appearing behind a flame holder. In this cell, the fuel mass fraction is suddenly canceled and the temperature becomes equal to the burnt gas temperature. The pressure is kept constant, which leads to a slight instantaneous loss of mass which has no influence on the results. Since the initial velocity in the center of the recirculation is close to zero, the fluid is still at mechanical equilibrium after the ignition. On the other hand, when the ignition is operated behind the swirler, the initial velocity is high, due to the high tangential component, and some terms in the steady state momentum equations become locally strongly unmatched. Two cases then occur: either the hot fluid is immediately carried away and the combustion cannot be stabilized or the numerical method diverges. To prevent that, the fuel mass fraction is progressively canceled within hundred time steps while the temperature is increased. Afterwards the conditions are kept constant in the cell in order to simulate the hot points generated by the swirler ring. Thus, a numerical process of "slow ignition" must be used for the flow including a swirl velocity component.

IV. Some Numerical Results

Some calculations have been performed with the following data:

- geometrical data:

- radius of the duct: 15 cm (noted R)
- inner radius of the swirler ring: 8 cm
- outer radius of the swirler ring: 12 cm

- initial conditions:
  - pressure : 1 atm.
  - temperature: > 1000 K

- physical data:
  . inlet turbulent kinetic energy: 150 m^2/s^2 just behind the swirler ring and 80 m^2/s^2 in the rest of the inlet section
  . inlet integral length scale of the turbulence : 1.5 cm
  . constants of the k-ε model: $C_1 = 1.44$, $C_2 = 1.92$, $C_\mu = 0.09$
  . mixture ratio: 0. or 0.76
  . C_p: complex function of the temperature.

The high initial temperature accounts for the fact that the gas has already burnt in the main combustion chamber of the engine. On the other hand, the pressure (1 atm.) is different from that existing in an aircraft engine but corresponds to the test facilities conditions. The inlet dissipation rate of the turbulent kinetic energy is given by:

$$\varepsilon_i = C_\mu^{3/4} \frac{k_i^{3/2}}{l_{ti}}$$

where l_t is the integral length scale of the turbulence and the subscript i indicates some inlet conditions. In the numerical procedure, the gas is initially at rest. The movement is obtained by a progressive increase of the inlet stagnation pressure. At first, two calculations without combustion have been performed, one with a final stagnation pressure which has a constant value p_A in the whole inlet section, the other with an inlet stagnation pressure p_B (r) accounting as follows for the effects of the centrifugal force:

$$p_B(r) = p_A + \int_0^r \rho \frac{v^2}{r'} dr'$$

For both cases, the value of the inlet tangential velocity is 90 m/s. The figures 3 and 4 give the inlet and outlet axial velocity profiles and the evolution of the static pressure along the duct. As expected the flow rate is larger with p_B (r) as inlet stagnation pressure and the static pressure is higher near the wall but above all it appears that some precise inlet data are required for the computation even if they are not easily accessible by experiment.

For the calculations with combustion, the inlet flow rate is imposed instead of the stagnation pressure in order to prevent reverse velocities in the inlet section during ignition. The initial state is the flow without combustion obtained by a previous calculation. The figure 5 gives the constant pressure lines for the initial flow (that is before ignition) and for the flow stabilized with combustion. We can see that the combustion makes the pressure gradients more parallel to the axis. The figure 7 gives the constant temperature and constant tangential velocity lines for the stabilized reactive flow. The combustion seems to propagate better to the wall than to the axis. This is merely due to the friction at the wall which increases both the residence time of the fluid particles and the transfer by turbulent diffusion. The tangential velocity is damped down by the viscous effects but still has a significant level in the outlet section, that is about 20 m/s.

Moreover, we have investigated the effect of the inlet tangential velocity on the combustion efficiency. Some calculations have been performed with three values of this tangential velocity: 45 m/s, 90 m/s and 135 m/s. The corresponding combustion efficiencies are given on figure 7 as function of the distance l from the swirler. This efficiency η(l) is defined as follows:

$$\eta(l) = \frac{\int_0^R \left(Y_o - Y(r,l) \right) q_z(r,l) r \, dr}{\int_0^R Y_o q_z(r,l) r \, dr}$$

where Y(r,l) is the fuel mass fraction at the point of coordinates (r,l) and Y_o is the initial fuel mass fraction. A gain which is probably due to the increase of the turbulent diffusion is obtained when the inlet tangential velocity v_t is raised from 45 m/s to 135 m/s but the gain is not significant at the end of the duct when v_t goes from 90 m/s to 135 m/s. Some more numerical tests are needed to determine the value of v_t which provides the best compromise between the combustion efficiency and the head loss Δp.

V. CONCLUSION

Some calculations of turbulent reactive flows with a tangential velocity component have been carried out in an axisymmetric geometry. Some precautions have been taken concerning the spatial discretization for obtaining the best conditions for stability and accuracy. A time implicit integration could be used despite the crossed derivatives proceeding from the Reynolds stress. The numerical results confirm that the tangential velocity improves the combustion efficiency.

REFERENCES

[1] Dutoya, D.: "Simulation of flow and heat transfer in annular cavities between disks in turbomachinery". AGARD-CP-390, p. 35-1.

[2] Launder, B.E.: "Turbulence", Ed. P. Bradshaw. Topics in Applied Physics, Springer-Verlag, 1976.

[3] Launder, B.E., Spalding, D.B.: "The numerical computation of turbulent flows". Computer Methods in Applied Mechanics and Engineering, vol. 3, nr. 2, p. 269, 1974.

[4] Dupoirieux, F.: "Numerical calculations of turbulent reactive flows and comparison with experimental results", Workshop on the gas flame structure, Novosibirsk, 1986.

[5] Beam, R.M., Warming, R.F.: "An implicit factored scheme for the compressible Navier-Stokes equations", AIAA Journal, vol. 16, April 1978, p. 398-402.

[6] Dutoya, D., Michard, P.J.: "An implicit type finite volume method for elliptical flow computation", La Recherche Aérospatiale, nr. 1980-2, p. 123-129.

[7] Dupoirieux, F.: "Méthodes numériques à convergence rapide utilisées pour le calcul des écoulements réactifs". Numerical Simulation of Combustion Phenomena, Sophia-Antipolis, 1985, Proceedings in Lecture Notes in Physics, nr. 241.

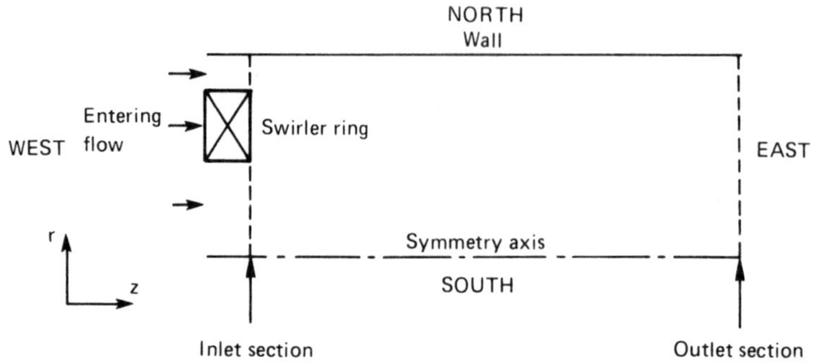

Fig. 1 – Geometry of the burner.

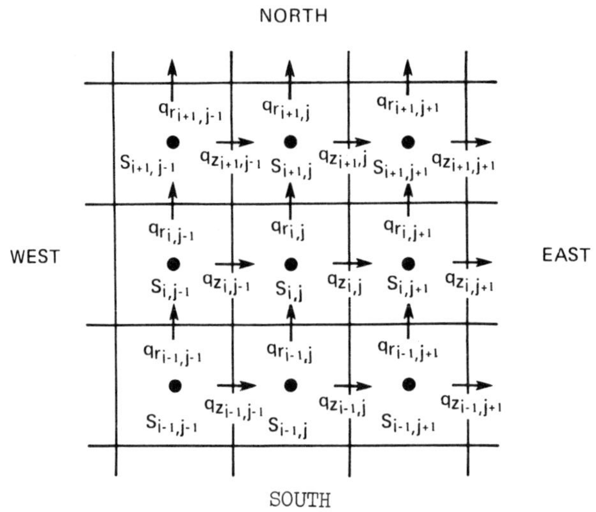

Fig. 2 – The staggered grid. $S_{l,m}$ represents at the mesh l, m all the scalars and the velocity components $u_{l,m}$, $v_{l,m}$, $w_{l,m}$.

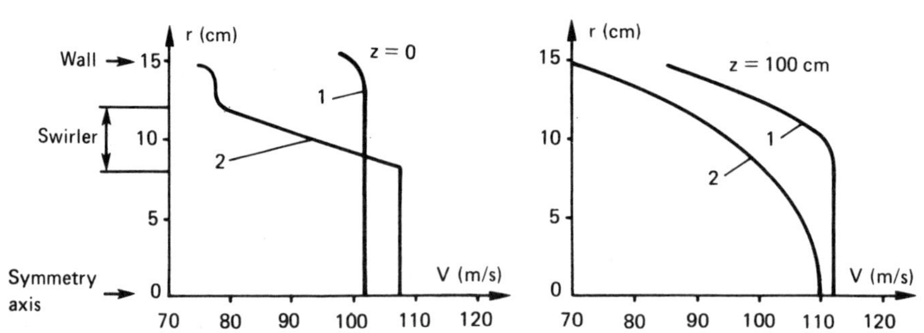

Fig. 3 – Axial velocity profiles in the inlet section (z = 0) and in the section z = 100 cm without combustion.
1 : The inlet stagnation pressure is corrected to take into account the effects of the centrifugal force.
2 : The inlet stagnation pressure is not corrected.

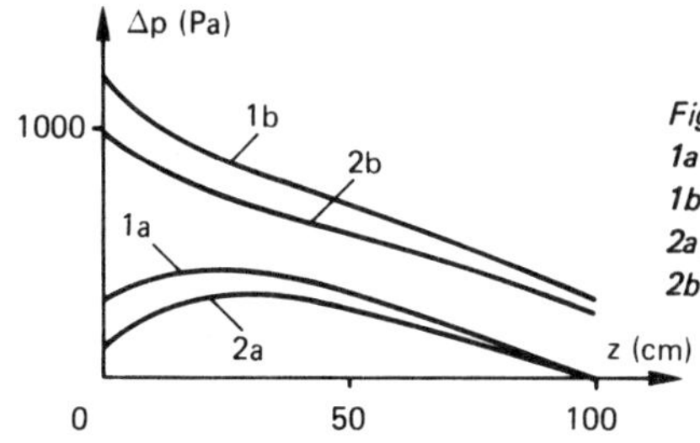

Fig. 4 — Evolution of the static pressure along the duct.
1a : axis pressure, 1b: wall pressure — when the inlet stagnation pressure is corrected
2a : axis pressure, 2b: wall pressure — when the inlet stagnation pressure is not corrected

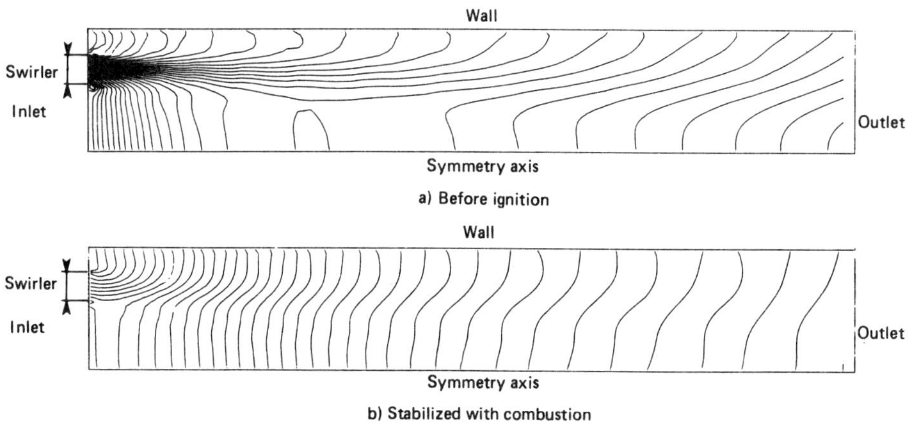

Fig. 5 — Constant pressure lines. Δp between two lines : a) 24 Pa, b) 29 Pa.

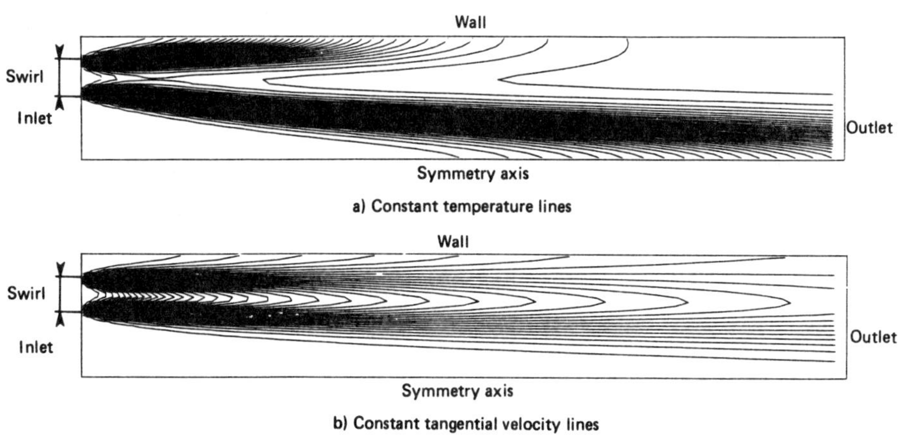

Fig. 6 — Isolines for a stabilized reactive flow. $\Delta T = 26$ K and $\Delta v = 2.25$ m/s between two lines.

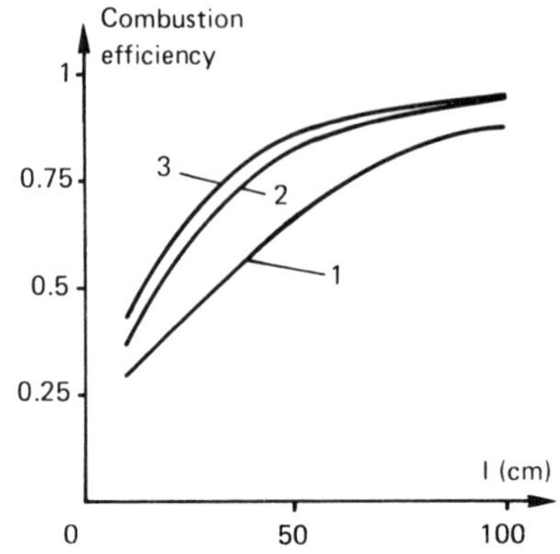

Fig. 7 — Combustion efficiency versus the distance from the swirler for three values of the inlet tangential velocity : 1) v = 45 m/s, 2) v = 90 m/s, 3) v = 135 m/s.

EXPERIMENTAL ANALYSIS ON THE STABILITY OF AN OBLIQUE FLAME FRONT

D. ESCUDIE

Laboratoire de Mécanique des Fluides - ECOLE CENTRALE DE LYON
36, Avenue Guy de Collongue - BP 163 69131 ECULLY Cedex FRANCE

1. INTRODUCTION

The field of Turbulent-Combustion includes many views of the interaction between the various scales present in a flow and a flame front. Only a few of them deal with the stability problem, particularly the stability analysis of a tilted front, and experimental studies in this area are still lacking. The various characteristics lengths must be known as in the case of pure turbulent combustion, and thus all the disturbance parameters such as amplitude and wavelength, can be defined.

Nevertheless, measurements in turbulent reacting flows involve significant difficulties and the frequency spectrum of interest in the analysis of turbulent flows is not easy to obtain. To avoid this problem a simpler configuration, in which the main flow is laminar and the velocity disturbances are generated by a well controlled phenomenon : the alternative vortices of a Karman vortex street, has been chosen.

So the purpose of this study is to describe the response of a stable premixed laminar V-shaped flame to perturbations created by a cylinder set upstream in the flow. To help our understanding of stability, the Reynolds numbers (based on the cylinder diameter) are such that the Karman vortex streets generated are laminar. In this way there are no turbulent velocity fluctuations accompanying the periodic formation of vortices and the fundamental scales to be considered are :

0. The flame front thickness "δ"
1. The vortex diameter "Λ"
2. The length between two successive vortices "a"
3. The width of the Karman vortex street "b"

Scales in the transversal direction of the flow (1,3) could be related to amplitude values, while the one in the axial direction (2) could be linked to a wavelength value. In addition the important kinematic parameters are :

- The mean axial velocity of the flow
- The laminar burning velocity of the obliquely propagating flame
- The rotation velocity of the vortices.

Such an interaction "flame front - Karman vortex street" was studied in 1980 by I. Namer |1|, by Hertzberg et al in 1984 |2|, and recently (1987) by D. Escudié and G. Charnay |3|. In all these works, investigations are essentially realized in terms of turbulent combustion, that is, to describe (by means of visualizations) the effect of coherent structures upon the shape and the structure of the flame front, and at the same time, to try to understand the influence of the flame on turbulence (Laser Doppler Anemometry, and Rayleigh measurements).

As regards stability, the well known investigations by Petersen and Emmons in 1961 |4| are still a reference in the V - flame case. They found for lean air fuel ratios, both natural and forced flame amplified oscillations ; near

C.-M. Brauner and C. Schmidt-Lainé (eds.), Mathematical Modeling in Combustion and Related Topics, 481–489.

stochiometric air fuel ratios no flame disturbances grow while for somewhat richer air fuel ratios only forced oscillations are possible. So according to the eddies size either the flame remains stable nor the waves, once excited, continue to grow and distord through both Huygens' propagation and through velocities induced in the unburned gases.

The experimental research presented in this paper is an attempt to supply more informations in connection whith the stability analysis of oblique flames generally hold by means of a bluff body. The results consist in two dimensional visualizations using the (tomography technique |5|) completed by some velocity measurements.

2. EXPERIMENTAL SET UP AND CONDITIONS.

The premixed flame studied is a lean hydrogen air flame (6,5 % H_2 volumic concentration) stabilized on a catalytic platinum wire (0,4 mm diameter). Experiments are performed in a quartz chamber in which the average upstream flow velocity is 5 m/s (cf. figure 1). Consequently the V-shaped flame angle is 4° and the resulting laminar burning velocity is about 18 cm/s. Moreover mean temperature measurements achieved with a 100 μm diameter coated chromel-alumel thermocouple enabled us to determine the flame front thickness $\delta \sim 0,7$ mm and $\gamma = (T_B - T_\infty) / T_B = 0,75$ in which T_B is the burned gases temperature and T_∞ the upstream gas flow temperature (density ratio n = 4)

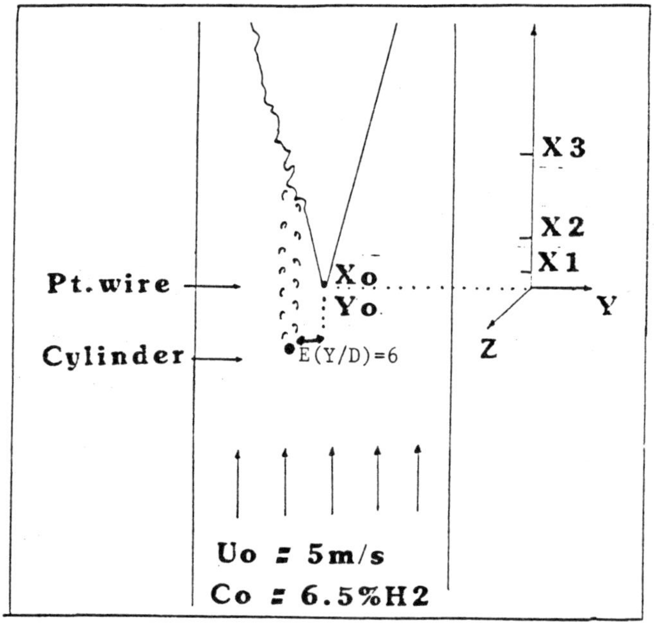

FIGURE 1. Experimental conditions

Vortices used to induce the disturbance, are generated by a cylinder set 1 cm upstream in the flow. Two cases have been investigated according to the diameter of the cylinder.

(A) - The fist one with a 0,4 mm cylinder diameter for which the laminar Karman vortex street generated is characterized by a Reynolds number Re = 133, and a strouhal frequency Fs = 2250 Hȝ. So in this part of the study, the vortices created will be, owing to Hamel Oseen theory, of the same order of magnitude as the flame front thickness (viscous vortex radius :0,425 mm).

(B) - In the second case the cylinder diameter is 1 mm and the strouhal frequency Fs = 1000 Hz. If the cylinder is fixed and for such a Reynolds number, the Karman vortex street is turbulent |6|. But following Berger observations |7|, when the cylinder is submitted to transverse vibrations at and near the natural vortex shedding frequency (Fs) the initiation of turbulence is delayed and stable vortex shedding range extended to Reynolds values about 300-400.

Therefore the rigid cylinder has been mounted elastically so that the periodic shedding of vortices due to the crossflow past the bluff obstacle might excite the body into resonant transverse vibrations, the vortices frequency and the body frequency being sufficiently close to one another (synchronization).

Griffin |8| emphasized the effects of this "lock-in" on the vortex street in comparison with the fixed cylinder case. For example in our study the oscillation frequency of the rod is slightly larger than Fs in the synchronization range and this implies a decrease of the formation length. There is also a substantial increase in vortex strength for all vibratory conditions investigated and an increased amplitude of cylinder vibrations.

So in this part the of the study, the characteristic length scale of the vortices (vortex radius : 1.2 mm) is greater than the flame front thickness, and as the vortex street still remains laminar, its effect on the flame front may be compared to the small cylinder previous case.

Experimental Techniques.

1 - The premixed air hydrogen flame is not luminous. For this reason a visualization technique, the tomography was first used. The mean flow is seeded with incense and as the particles pass through the flame front they evaporate and burn. When the combustion chamber is illuminated with a sheet of laser light (figure 2), particles present in the flow scattered the light and the flame appears as the interface between the bright and dark areas (burned regions without any particles remain dark).

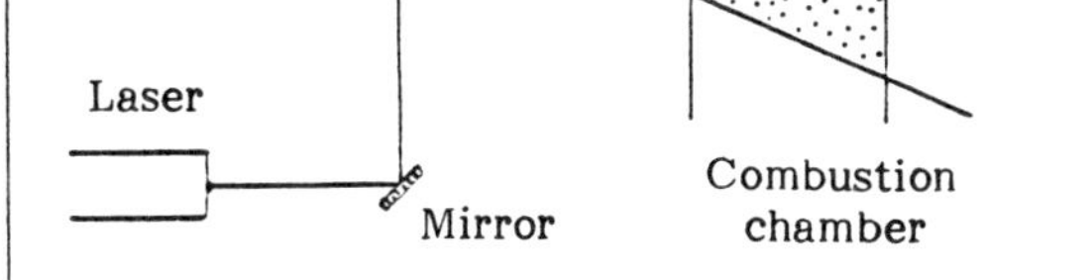

FIGURE 2. Experimental set up for tomographic visualization

2 - A description of the velocity field, essentially the axial velocity component, is realized by means of a forward scatter dual beam Laser Doppler Velocimetry technique. Then the whole flow has to be seeded with an alumina -silica mixture.

3. RESULTS AND DISCUSSION

Before studying the disturbed flames, the laminar premixed flame has been inspected by the tomographic method. The two V- branches are perfectly regular and there is no wrinkle or oscillation of the flame front. So even if it is in a lean premixed combustion case |9|, and probably because of the existence of a tangential velocity component in the gas flow, the flame is intrinsecally stable.

The mean axial velocity and temperature profiles are plotted in figure 3, versus the transverse coordinate Y, for three stations downstream the flame holder : at X1 = 13 mm, in the wire wake ; at X2 = 32 mm, where the wake is supposed to be balanced by combustion acceleration ; at X3 = 80 mm, where the velocity increases due to confinement associated with the important decrease of density.

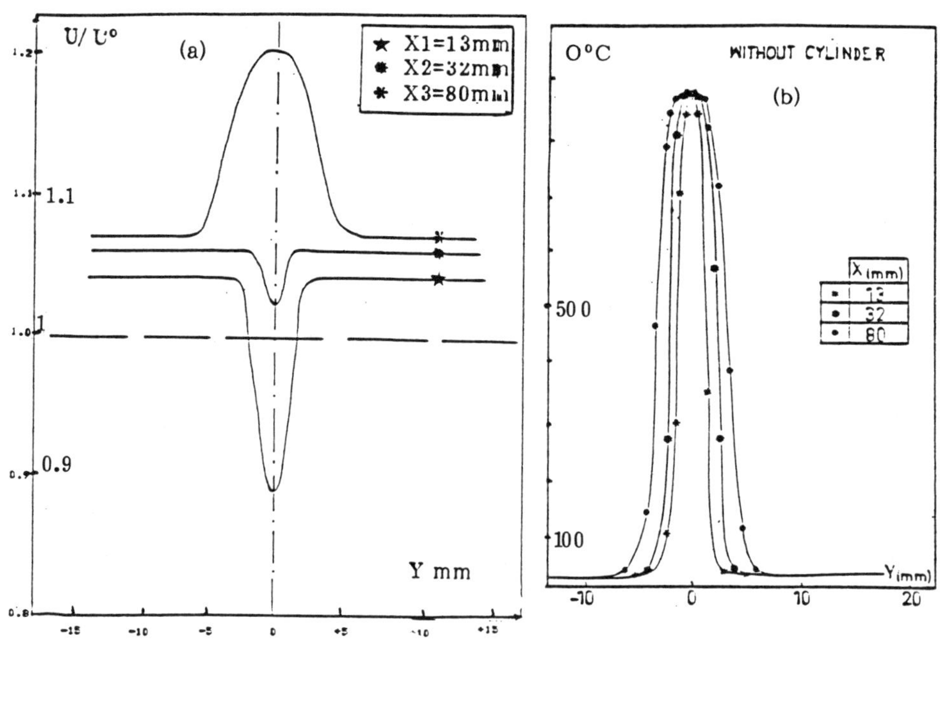

FIGURE 3. : a) Profile evolution of the mean axial velocity in the laminar combustion case.
b) Mean temperature profiles of the laminar flame.

3.1. Smallest eddies (as small as the flame thickness).

The flame is now disturbed by vortices, charateristic scales of which are of the same order of magnitude as the flame front thickness. In this situation, one generally expects an increase of the effective transport properties (for heat and mass) within the flame. For Petersen and Emmons |4| "the burned gas turbulence is probably about as energetic as the unburned gas exept that the passage through the flame will have stretched the normal (to the flame) component of vorticiy. Thus the short wavelength components are made non-isotropic by the flame".

In this study, mean velocity measurements with or without combustion are first investigated (cf. figure 4). They put into evidence a well known deflection of the mean streamlines |3||10| (the wake of the cylinder is deviated), and consequently the location at which the vortices will cross the flame will be further shifted.

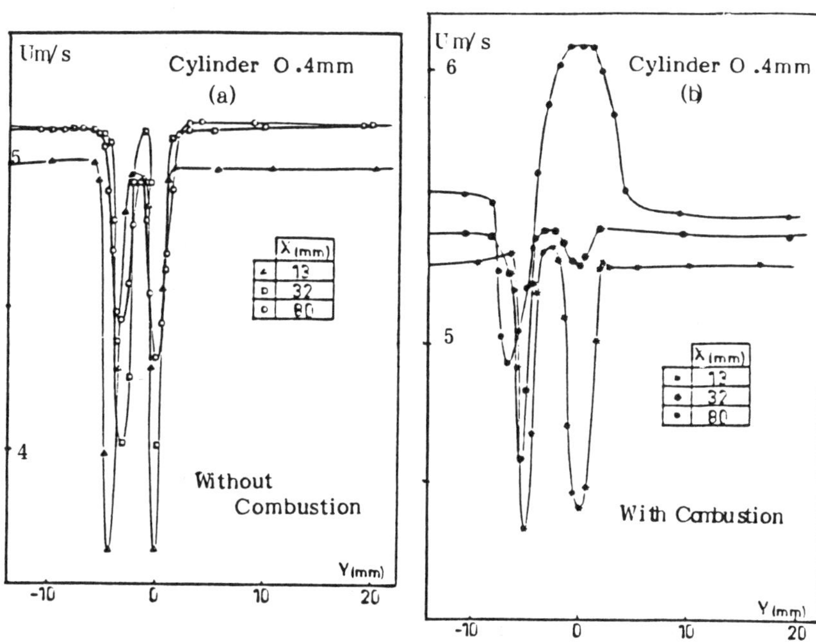

FIGURE 4. : Mean axial velocity profiles downstream the wire and rod system.
a) Without combustion - b) With combustion.

In addition, by assuming an approximate value of the vortex radius it is possible to calculate the mean tangential velocity (V_T) of the vortices |11|. In this case V_T = 0,8 m/s which is four times greater than the burning velocity. So although the vortex dimension is of the same order of magnitude as the flame front thickness, its velocity value is large compared to the velocity of the flame front.

Flow visualization helps ower understanding of the processes which take place when the laminar vortex street and the flame front interact. The first effect of interest is a wrinkle of the flame front in a well defined zone over the flame holder. At the same time some packs of burned gas are flowing away randomly along the flame front without any gradual increase of a node from the lower part of the flame.

This interaction is schematically summarized in figure 5 close to a photograph.

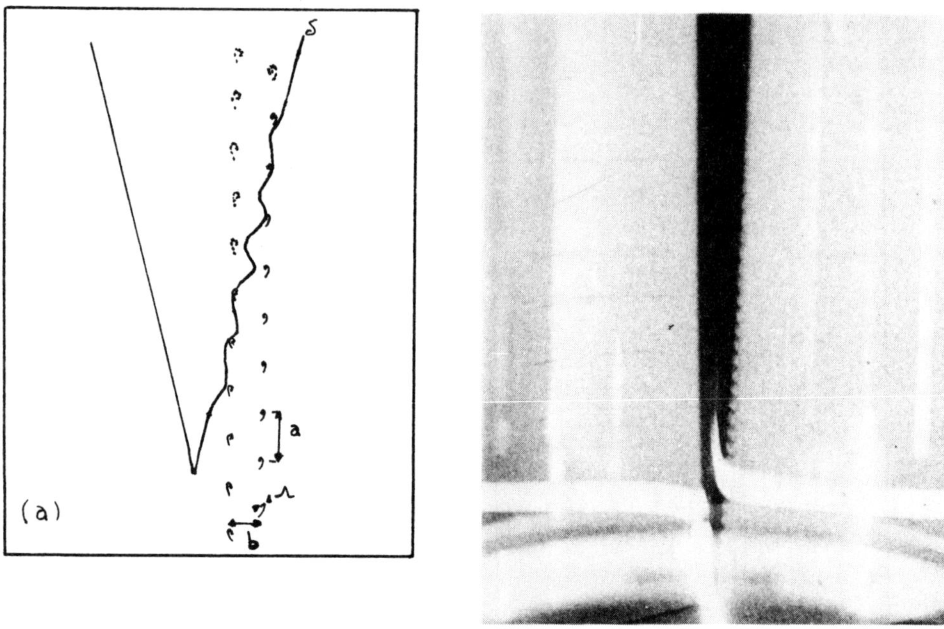

FIGURE 5. : Interaction between a **laminar** Karman vortex street and a laminar flame front. a) Schematic view - b) Tomographic visualization.

As pointed out by Hertzberg et al |2|, the distance between cups corresponds closely to the vortex spacing : the wavelength value is the "a" value of the Karman vortex street. An interesting point is the fact that the initial amplitude growth is followed by a gradual decay. In the lower part the perturbations are generated by the small vortices of the same side of the street ; their amplitude is equal to the vortex diameter. Then the vortices of the other side progressively become significant, the flame is swept away by the vortex street and the amplitude is the "b" value of the street. At the end of the interaction zone (upper part of the interaction zone), vortices of the first side of the street are "burned" at the flame front (due to an increase of viscosity effects) and the amplitude becomes again the one of the eddies.

It seems that there is no propagation of the disturbances. Each eddy is acting locally and immediatly damped down without affecting the dowmstream part of the flame front. Regarding the stability, this suggests the existence of a stable state, but this idea is inconsistent with the burned gaz pockets flowing away downstream. Up to now there is no straightforward explanation for this phenomena. Consequently the influence of the small eddies on the stability of the tilted flame front have to be improved by numerous experiments ; particularly with smallest diameter cylinder for which the Reynolds numbers would be far from a transition state (laminar --> turbulent Karman vortex street) and even if previous hot wire measurements indicate that the laminar vortex street is stable (perhaps this can be changed by the pressure drop due to combustion).

3.2. Large eddies

Keeping in mind the fact that the width of the laminar Karman vortex street "b" is important, and that the flame angle still remains small, it must be noted that the interaction length is much larger than in the previous case.

When the first kinematic effect in the small disturbances case was a deviation of the mean streamlines close to the flame front, here large vortices immediatly put into evidence a deflection of the flame. Figure 6 presents the photograph of this interaction between the laminar flame front and the stronger vortices.

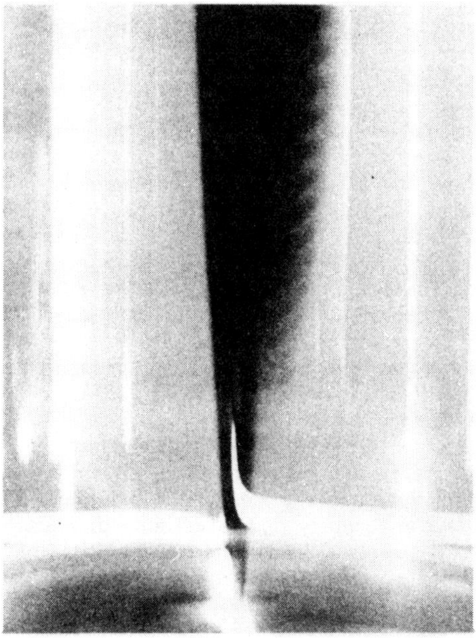

FIGURE 6. : Visualization of the interaction between the 1mm diameter vibrating rod and the laminar flame front.

A regular pattern appears with nodes and cups pointing towards the burned gas, greater than in the small disturbances case. Following a linear increasing zone these oscillations seems to be conserved and there is no longer a decrease of the wave amplitude. The latter remains approximately constant over the whole visualized region. The flame front is carried away inside the vortex street, but the perturbation is not amplified, and don't give rise to turbulent sheet.

Even if the shearing stresses developped are of importance and probably enable us to explain some local extinction in the cups, the reaction of the flame to these large disturbances has the same appearance that for the small ones. It can be imagined that a visualization of the flow far downstream will show a decrease of these oscillations at the end of the interaction zone. So the flame is supposed to be stable to this large eddies generated by a laminar Karman vortex street.

4. CONCLUSION

The interaction between a laminar Karman vortex street and a V-shaped flame can be a simplified situation for the stability study of a premixed laminar oblique flame.

When the characteristic scale of the vortices generated by the cylinder are of the same order of magnitude as the flame front thickness, they wrinkle the flame front all along the interaction zone. The wavelength of the propagating wave is the distance between two successive eddies and the amplitude grows from the vortex diameter value to the width "a" of the Karman vortex street, before decreasing again to the characteristic vortex scale. Along with this effect there is the added one of the random presence of burned gas pockets flowing away along the flame front.

Existence of this transition state could be due to a transition of the Karman vortex street from the laminar to the turbulent regime, but in our opinion and in the light of preliminary flow characterization without combustion, it is not the case. It seems more, in terms of stability analysis, to a particular response of the flame front to a small disturbance.

The part concerning the large eddies, exhibits the same kind of behaviour but without any instability problem, and that constitutes the essentiel difference.

To have a better understanding, of this difficult problem a more complete study has been realized |12|, with a comparison "laminar-turbulent" Karman vortex street. Nevertheless it would be worthwhile to briefly compare theoretical predictions with our experimental results. Unfortunately such a comparison is difficult. As a matter of fact only few researchers are concerned with the stability of the tilted flame |13| and results are in sharp disagreement with the experimental data. For this reason the attempt to analyze the stability of the V flame directly has been unsuccessful until the recent works of Garcia -Ybarra and Borghi |14|. They obtain a dispersion relation and a partial investigation of this relation shows the existence of structures such as polyhedral flames or travelling waves ; but at present, numerical investigation has not been completly performed.

In conclusion, despite many experimental and theoretical works on turbulent combustion, a great deal of effort has still now to be done, particularly concerning the stability analysis,in order to better understand all the phenomena involved in this interaction.

REFERENCES

|1| NAMER, I. (1980) : "An experimental investigation of the interaction between a Karman vortex street and a premixed laminar flame" ; Ph D. Thesis, Berkeley CA.

|2| HERTZBERG, J.R., NAMAZIAN, M., TALBOT, L. (1984) : "A laser tomographic study of a laminar flame in a Karman vortex street" ; Comb Science and Technol. 38,205.

|3| ESCUDIE, D. and CHARNAY, G. (1987) : "Experimental Study of the interaction between a premixed confined laminar flame and coherent structures" ; Turbulent Shear Flows 5, Springer-Verlag Berlin Heidelberg, pp 347-360

|4| PETERSEN, R.E., and EMMONS, H.W. (1961) : "Stability of laminar flames", The Physics of Fluids, vol. 4, 4, pp 456-464.

|5| BOYER, L. (1980) : "Laser tomographic method for flame front movements studies", Combust. and Flame, 39, 321.

|6| ROSHKO, A. (1953) : "On the development of turbulent wakes from vortex streets", NACA, Technical Note 2913.

|7| BERGER, E.W. (1967) - Phys-Fluids, 10, pp 8191-8193.

|8| GRIFFIN, O. M. and RAMBERG, S.E. (1974) : "The vortex street wakes of vibrating cylinders", J. Fluid. Moch., Vol. 66, 3, pp 553-576.

|9| QUINARD, J. (1984) : "Limite de stabilité et structures cellulaires dans les flammes de prémélange. Etude expérimentale". Thèse Docteur es-science-Université de Provence.

|10| ESCUDIE, D. et al (1983) : "Modification of turbulent flow field by an oblique premixed hydrogen-air flame", Prog. Astron. Aero, 88, p 147.

|11| LAMB, H. (1931) : "Hydrodynamics" 6th edition, Cambridge University Press.

|12| ESCUDIE, D. (1987) : "Stability of a premixed laminar V -Shaped flame", submitted to 11th ICDERS, Varsovie, Pologne.

|13| BLACKSHEAR, P.L., Jr. (1956) "Growth of disturbances in a flame generated shear region", NACA TN 3830.

|14| GARCIA-YBARRA, P and BORGHI, R. (1985) : "Stability study of an oblique flame front model", 10th ICDERS, Berkeley, CA.

VECTOR COMPUTERS AND COMPLEX CHEMISTRY COMBUSTION

V. GIOVANGIGLI †‡ and N. DARABIHA ‡

‡ Laboratoire EM2C du CNRS et de l'ECP, Ecole Centrale des Arts et Manufactures, 92295 Chatenay-Malabry cedex, France.
† Laboratoire d'Acoustique et Mécanique, UA868, Université Paris 6, T66 5Et., 4 place Jussieu, 75252 Paris cedex 05, France.

1. INTRODUCTION

Combustion chemistry usually involves many interacting species and many elementary reactions. Typically, tens of species and hundreds of reversible chemical reactions are used in hydrocarbon combustion kinetics. As a result, the governing conservation equations for total mass, species mass, momentum and energy, in a reactive flow, are large and stiff systems of partial differential equations. A consequence is that only zero and one dimensional models have formerly been used to look in detail at the coupling of a large number of species interactions [8] [13] [17] [19] [20] [22].

Among the numerical problems for computational solutions are those associated with multiple time and space scales and the costs due to complex chemistry. In addition, there are problems not directly related to chemistry such as complicated geometries or linear algebra which can be investigated separately. To overcome numerical difficulties arising from multiple time and space scales one has to use time-implicit algorithms and adaptive gridding techniques respectively [13] [17] [21]. In this paper we focus on computational costs inherent in complex chemistry due to multiple sums and products performed when evaluating thermodynamic properties, chemical production rates and multicomponent transport coefficients.

Use of a vector computer can significantly reduce these costs provided the corresponding algorithms vectorize. This problem is addressed in the present paper where high degrees of vectorization are obtained by structuring properly the corresponding multiple nested loops.

The formulas for evaluating chemical production rates, thermodynamic properties and diffusion coefficients are problem independent, and two packages, CHEMKIN and TRANSPORT, have been written at Sandia national laboratories by Kee et al. for evaluating these quantities [14] [15]. Since the vectorized algorithms are also problem independent, although the choice of an optimum algorithm may depend on various problem parameters like the number of species or the number of grid points, the resulting subroutines can easily be embeded in the CHEMKIN and TRANSPORT packages, which were originally written for scalar machines.

Finally, to test the resulting vectorized subroutines we have computed various flame structures on a Cray-1 computer, characterized by a vector/scalar speed ratio between 5.-10., and we have found vector/scalar performances around 6.-8. for the routines and 4. for the flame solver.

2. A DETAILED COMBUSTION MODEL

2.1 Governing equations

The governing equations of a gaseous laminar reacting flow are the equations for conservation of total mass, species mass, momemtum and energy, and the equation of state. The total mass conservation can be written :

$$\frac{\partial \rho}{\partial t} + \nabla.(\rho v) = 0, \tag{1}$$

where ρ is the density, v the mass averaged flow velocity, t the time, and the species mass conservation equation is :

$$\rho\frac{\partial Y_k}{\partial t} + \rho(v.\nabla)Y_k = -\nabla.(\rho Y_k V_k) + W_k\omega_k, \qquad k = 1,\ldots,K, \tag{2}$$

where Y_k is the mass fraction of the kth species, V_k the diffusion velocity of the kth species, W_k the molecular weight of the kth species, ω_k the molar production rate of the kth species and K the number

C.-M. Brauner and C. Schmidt-Lainé (eds.), Mathematical Modeling in Combustion and Related Topics, 491–503.

of species. The momentum conservation equation is :

$$\rho\frac{\partial v}{\partial t}+\rho(v.\nabla)v=-\nabla p+\nabla.\left\{\eta\Big(\nabla v+(\nabla v)^T-\tfrac{2}{3}(\nabla.v)\mathbf{I}\Big)\right\}+\sum_{k=1}^{K}\rho Y_k f_k, \tag{3}$$

where p is the pressure, η the viscosity and f_k the external force per unit mass of the kth species and the energy conservation equation is :

$$\rho c_p\frac{\partial T}{\partial t}+\rho c_p(v.\nabla)T=\nabla.(\lambda\nabla T)-\Big(\sum_{k=1}^{K}\rho Y_k V_k c_{pk}\Big).\nabla T-\sum_{k=1}^{K}h_k W_k\omega_k+\frac{\partial p}{\partial t}+(v.\nabla)p$$
$$+\eta\Big(\nabla v+(\nabla v)^T-\tfrac{2}{3}(\nabla.v)\mathbf{I}\Big):\nabla v+\sum_{k=1}^{K}\rho Y_k V_k.f_k, \tag{4}$$

where T is the temperature, c_p the constant pressure heat capacity of the mixture, c_{pk} the constant pressure heat capacity of the kth species, λ the thermal conductivity and h_k the specific enthalpy of the kth species. Finally the equation of state is

$$\rho=pW/RT, \tag{5}$$

where W is the molecular weight of the mixture and R the universal gas constant.

Equations (1)-(5), which may include other effects like radiative heat losses, etc., are usually further simplified according to the problem under study, e.g., steady flow, boundary layer flow, low mach number flow, zero dimensional model, etc.. These equations have to be completed by formulas expressing the transport coefficients λ and η, the diffusion velocities V_k, the thermodynamic properties c_p, c_{pk} and h_k and the chemical production rates ω_k in terms of the state variables T, p, Y_k, $k=1,\ldots,K$, and their gradients. These relations are written in the next sections.

2.2 Chemistry and thermodynamics

We consider I elementary reversible reactions involving K chemical species, which can be represented in the general form :

$$\sum_{k=1}^{K}\nu'_{ki}\mathcal{X}_k \rightleftharpoons \sum_{k=1}^{K}\nu''_{ki}\mathcal{X}_k, \qquad i=1,\ldots,I, \tag{6}$$

where the stoichiometric coefficients ν'_{ki} and ν''_{ki} are integers and $\mathcal{X}_k$ denotes the symbol of the kth species. The production rate of the kth species can be written as :

$$\omega_k=\sum_{i=1}^{I}\nu_{ki}q_i, \qquad \nu_{ki}=\nu''_{ki}-\nu'_{ki}, \tag{7}$$

where q_i is the rate of progress variable of the ith reaction. The rate of progress q_i is the difference between the forward and reverse rates :

$$q_i=\mathcal{K}_{fi}\prod_{k=1}^{K}[\mathcal{X}_k]^{\nu'_{ki}}-\mathcal{K}_{ri}\prod_{k=1}^{K}[\mathcal{X}_k]^{\nu''_{ki}}, \tag{8}$$

where $[\mathcal{X}_k]$ is the molar concentration of the kth species and $\mathcal{K}_{fi}$ and $\mathcal{K}_{ri}$ are the forward and reverse rate constants of the ith reaction. Denoting X_k the mole fraction of the kth species the quantities X_k, Y_k and $[\mathcal{X}_k]$ are related by :

$$X_k=Y_k\frac{W}{W_k}, \qquad [\mathcal{X}_k]=X_k\frac{p}{RT}, \qquad W=\Big(\sum_{k=1}^{K}\frac{Y_k}{W_k}\Big)^{-1} \tag{9}$$

The forward rate constant $\mathcal{K}_{fi}$ for the ith reaction is given by Arrhenius' law :

$$\mathcal{K}_{fi} = A_i T^{\beta_i} \exp\left(\frac{-E_i}{RT}\right), \tag{10}$$

where A_i is the pre-exponential factor, β_i the temperature exponent and E_i the activation energy. The reverse rate constant is related to the forward constant through the equilibrium constant $\mathcal{K}_{ei}$:

$$\mathcal{K}_{ri} = \frac{\mathcal{K}_{fi}}{\mathcal{K}_{ei}}, \qquad \mathcal{K}_{ei} = \left(\frac{p_{atm}}{RT}\right)^{\Delta\nu_i} \exp(\Delta Z_i), \tag{11}$$

$$\Delta\nu_i = \sum_{k=1}^{K} \nu_{ki}, \qquad \Delta Z_i = \sum_{k=1}^{K} \nu_{ki} Z_k, \qquad Z_k = \frac{S_k^0}{R} - \frac{H_k}{RT}, \tag{12}$$

where p_{atm} denotes the atmospheric pressure and H_k and S_k^0 the molar enthalpy and entropy of the kth species at the standard state of one atmosphere, respectively. In some reactions an arbitrary third body, usually denoted M, is required for the reaction to proceed. In this situation the rate of progress variable becomes

$$q_i = ctb_i\left(\mathcal{K}_{fi}\prod_{k=1}^{K}[\mathcal{X}_k]^{\nu'_{ki}} - \mathcal{K}_{ri}\prod_{k=1}^{K}[\mathcal{X}_k]^{\nu''_{ki}}\right), \qquad ctb_i = \left(\sum_{k=1}^{K}\alpha_{ki}\,[\mathcal{X}_k]\right) \tag{13}$$

where the α_{ki} are one if all the species contribute equally as third bodies, in which case the extra factor of Eq. (13), —the third body concentration ctb_i, is the total concentration $[M]$:

$$[M] = p/RT. \tag{14}$$

However, if the species act more or less efficiently as third bodies the α_{ki} may differ from one. Finally, some hydrocarbon reactions have pressure dependent rate coefficients. These rate coefficients can be estimated either from the Lindemann formula

$$\mathcal{K}_{fi} = A_i T^{\beta_i} \exp(-\frac{E_i}{RT}) \Big/ \left(1 + \frac{A'_i T^{\beta'_i} \exp(-\frac{E'_i}{RT})}{[M]}\right), \tag{15}$$

where A'_i, β'_i and E'_i are constants, or from more involved formulas like the quantum mechanical Kassel sum [24].

Thermodynamic properties are usually estimated from polynomial fits of the JANAF data. As a concrete exemple we will consider the Nasa fits of [9] also used in CHEMKIN, but any other type of fits would lead to similar vectorized algorithms. In these packages molar constant pressure heat capacities are taken in the form of fourth degree polynomial fits for two temperature ranges :

$$\frac{C_{pk}}{R} = \begin{cases} a_{1k} + a_{2k}T + a_{3k}T^2 + a_{4k}T^3 + a_{5k}T^4, & \text{if } T_{\text{low}} \le T \le T_{\text{mid}}, \\ a_{8k} + a_{9k}T + a_{10k}T^2 + a_{11k}T^3 + a_{12k}T^4, & \text{if } T_{\text{mid}} \le T \le T_{\text{high}}, \end{cases} \tag{16}$$

with similar expressions for the corresponding molar enthalpy H_k and molar standard state entropy at one atmosphere S_k^0 . Moreover the midpoint temperature is allowed to be species dependent in the CHEMKIN package [14] :

$$T_{\text{mid}} = T_{\text{mid}k}. \tag{17}$$

Finally, the thermodynamic properties in mass units and the mixture constant pressure heat capacity are easily obtained from :

$$c_{pk} = \frac{C_{pk}}{W_k}, \qquad h_k = \frac{H_k}{W_k}, \qquad c_p = \sum_{k=1}^{K} Y_k c_{pk}. \tag{18}$$

2.3 Transport properties

Comparisons between different mathematical approximations of the multicomponent transport properties have shown that the following expressions provide a good trade off between precision and computational costs [3] [15] [19] [25]. The diffusion velocity is divided into three parts :

$$V_k = \mathcal{V}_k + \mathcal{W}_k + V_{cor}, \tag{19}$$

where $\mathcal{V}_k$ is the diffusion velocity due to species gradients and is given in the Hirschfelder-Curtiss approximation [10] by

$$\mathcal{V}_k = -D_k \frac{1}{X_k} \nabla X_k, \qquad D_k = (1 - Y_k) \Big/ \sum_{l \neq k} X_l / \mathcal{D}_{kl}, \tag{20}$$

where the $\mathcal{D}_{kl}$, $k = 1, \ldots, K$, $l = 1, \ldots, K$, are the binary diffusion coefficients. The velocity $\mathcal{W}_k$ is due to temperature gradients and is included only for low molecular weight species like H and H_2. The trace light component limit is used for $\mathcal{W}_k$ and leads to

$$\mathcal{W}_k = \frac{D_k \Theta_k}{X_k} \frac{1}{T} \nabla T, \qquad \Theta_k = \sum_{l \neq k} \theta_{kl}, \tag{21}$$

where the θ_{kl} can be expressed in terms of collision integrals [2]. The velocity V_{cor} is a correction velocity included to insure that the mass is conserved [3] [15] [17] :

$$V_{cor} = -\sum_{k=1}^{K} Y_k (\mathcal{V}_k + \mathcal{W}_k). \tag{22}$$

The mixture thermal conductivity λ can be estimated either by the simple semi-empirical formula

$$\lambda = \frac{1}{2} \{ \sum_{k=1}^{K} X_k \lambda_k + (\sum_{k=1}^{K} X_k \lambda_k^{-1})^{-1} \}, \tag{23}$$

where λ_k is the thermal conductivity of the kth species, or by the more expensive Wilke formula :

$$\lambda = \sum_{k=1}^{K} \frac{X_k \lambda_k}{\sum_{l=1}^{K} X_l \phi_{kl}}, \qquad \phi_{kl} = \frac{1}{\sqrt{8}} \Big(1 + \frac{W_k}{W_l}\Big)^{-\frac{1}{2}} \Big(1 + (\frac{\eta_k}{\eta_l})^{\frac{1}{2}} (\frac{W_l}{W_k})^{\frac{1}{4}}\Big)^2, \tag{24}$$

where η_k is the viscosity of the kth species, and the mixture viscosity η can be estimated with similar expressions [3] [15] [17].

The transport coefficients η_k, λ_k, $\mathcal{D}_{kl}$ and θ_{kl} have to be expressed in terms of the state variables p, T, $Y_k, k = 1, \ldots, K$, and molecular parameters. The corresponding formulas, which can be found in [2] [11] [15], appear to be complex. However, while their dependence on the molecular parameters is indeed complex, their dependence on the temperature is relatively simple. As a consequence, it is possible to compute polynomial fits of the temperature dependent part of single component viscosities η_k and conductivities λ_k, binary diffusion coefficients $\mathcal{D}_{kl}$ and reduced thermal diffusion coefficients $\theta_{kl}/X_k X_l$ [3] [15] [17] [25]. This procedure leads to significant gains in computational efficiency with little or no compromise in accuracy.

As for the Thermodynamic properties, we will consider the polynomial fits of the TRANSPORT package [15] as a concrete exemple, but using any other type of fits would lead to similar vectorized algorithms. These fits are in the form

$$\log \eta_k = b_{1k} + b_{2k} \log T + b_{3k} (\log T)^2 + b_{4k} (\log T)^3, \qquad k = 1, \ldots, K, \tag{25}$$

$$\log \lambda_k = c_{1k} + c_{2k} \log T + c_{3k} (\log T)^2 + c_{4k} (\log T)^3, \qquad k = 1, \ldots, K, \tag{26}$$

$$\log \mathcal{D}_{kl} = d_{1kl} + d_{2kl} \log T + d_{3kl} (\log T)^2 + d_{4kl} (\log T)^3, \qquad \begin{cases} k = 1, \ldots, K, \\ l = 1, \ldots, K, \end{cases} \tag{27}$$

$$(\theta_{kl}/X_k X_l) = e_{1kl} + e_{2kl} T + e_{3kl} T^2 + e_{4kl} T^3, \qquad \begin{cases} k \in \mathcal{L}, \\ l = 1, \ldots, K, \end{cases} \tag{28}$$

where $\mathcal{L} = \{light(l);\, l = 1, \ldots, LIGHT\}$ denotes the indices of the light species, the fitting being done at the beginning of the calculation.

3. EVALUATION OF AERO-THERMO-CHEMISTRY EXPRESSIONS

3.1 Computational considerations

Examination of Eqs. (1)-(5) shows that, regardless of the particular problem under study, it is necessary to evaluate the thermodynamic properties such as c_p, c_{pk}, $k = 1, \ldots, K$, h_k, $k = 1, \ldots, K$, the chemical production rates ω_k, $k = 1, \ldots, K$, and the transport coefficients such as D_k, $k = 1, \ldots, K$, $\mathcal{D}_{kl}$, $k = 1, \ldots, K$, $l = 1, \ldots, K$, etc., in terms of the state variables T, p and Y_k, $k = 1, \ldots, K$. But evaluating these aero-thermo-chemistry quantities is expensive since they involve multiple sums and products as shown in the previous section. Typically, up to 70-95% of the CPU time may be spent for these evaluations in a one dimensional laminar flame solver [21].

Using a vector processor, e.g., Cray-1, CDC-205, Fujitsu VP200, NEC SX2, Hitachi S-820, FPS 264, etc., or a multiple vector processor, e.g., Cray X-MP, Cray-2, etc., can significantly reduce these computational costs, provided the corresponding algorithms vectorize. This problem is investigated in the following. We refer to [12] [16] [18] for more details on vectorization techniques. Altough different groups can have investigated this type of problem, there are few papers on the subject, e.g., Boris and Winsor [1]. The paper of Boris and Winsor, however, investigates the vectorization of reactive flow solvers based on timestep splitting, rather than the more specific problem of evaluating thermodynamic properties, chemical production rates and transport coefficients.

To vectorize subroutines for evaluating these aero-thermo-chemistry quantities, two fundamentally different situations have to be considered. Indeed, keeping in mind that in a flow field all these quantities only depend on the local state variables T, p and Y_k, $k = 1, \ldots, K$, and since most combustion problems are at least one dimensional, one has in general to evaluate these quantities for a large number of input data, each of these corresponding to a grid point, regardless of the grid structure. Nevertheless, the case of a single input data is still interesting for instance if an homogeneous problem is under study, say with a huge reaction mechanism, or if a grid point needs a special treatment. In particular, on a multiprocessor computer, both cases are of interest depending on how many grid points are treated per processor. These two different situations require different treatments and, to distinguish between them, the single imput data case is denoted SID and the multiple input data case MID. Moreover, if we note that the subscripts k or l are always used for the species and the subscript i for the chemical reactions, we introduce the subscript j, $j = 1, \ldots, J$, to index the different input conditions, J being the number of them. In the following, whenever the MID case is considered, we will add the subscript j to every quantity like c_{pj}, c_{pkj}, h_{kj}, ω_{kj}, D_{kj}, $\mathcal{D}_{klj}$, etc, which depend on the jth input data T_j, p_j and Y_{kj}, $k = 1, \ldots, K$. Further note that the CHEMKIN and TRANSPORT libraries only consider the SID case. Anticipating the results of the next sections, we point out that the best performances for the MID subroutines are obtained when inner loops are j loops, that is to say loops in j, and the resulting routines are significantly faster than the corresponding SID routines used J times, provided J is not too small.

3.2 Vectorization of chemistry and thermodynamics

3.2.1 Evaluation of thermodynamic properties In the SID case, the simplest way to compute single component thermodynamic properties, e.g., C_{pk}/R, $k = 1, \ldots, K$, is to do a loop in k and to use explicitly the polynomial fits, such that † :

$$\begin{array}{l} \textbf{do } k = 1,\ K \\ \quad \textbf{if } T \leq T_{\text{mid}k} \textbf{ then} \\ \quad\quad C_{pk}/R = a_{1k} + a_{2k}T + a_{3k}T^2 + a_{4k}T^3 + a_{5k}T^4 \\ \quad \textbf{else} \\ \quad\quad C_{pk}/R = a_{8k} + a_{9k}T + a_{10k}T^2 + a_{11k}T^3 + a_{12k}T^4 \\ \quad \textbf{endif} \\ \textbf{enddo} \end{array} \tag{29}$$

Although introducing an extra loop for the evaluation of the polynomial fits would allow the user-friendly possibility of easily changing the degree of the fits, the faster unrolled algorithm (29) seems more appropiate since changing of thermodynamic data base is infrequent. The unrolling of the fitting polynomial save loads and more operation can be overlaped. It also minimize the number of loop setups, even in scalar mode. We further note that it is the dependence of the intermediate range temperature $T_{\text{mid}k}$ on the species k (17) that introduces a conditional loop. With a species independent intermediate temperature T_{mid} [9], the if statements could be move outside the loop. We remark also that using only one fit over a wide temperature range would clearly be more convenient as pointed out by Boris and Winsor [1]. Conditional loops are vectorized on a Cray-1 with mask vectors, but (29) has to be rewritten using non ANSI FORTRAN statements.

† Loops are sketched in a symbolic way.

In the MID case, a conditional loop is still needed since T_j has to be compared to $T_{\text{mid}k}$. Since J is usually larger than K it is better to put the j loops as the inner loops :

$$
\begin{array}{l}
\textbf{do } k = 1,\ K \\
\quad \textbf{do } j = 1,\ J \\
\quad\quad \textbf{if } T_j \leq T_{\text{mid}k} \textbf{ then} \\
\quad\quad\quad C_{pkj}/R = a_{1k} + a_{2k}T_j + a_{3k}T_j^2 + a_{4k}T_j^3 + a_{5k}T_j^4 \\
\quad\quad \textbf{else} \\
\quad\quad\quad C_{pkj}/R = a_{8k} + a_{9k}T_j + a_{10k}T_j^2 + a_{11k}T_j^3 + a_{12k}T_j^4 \\
\quad\quad \textbf{endif} \\
\quad \textbf{enddo} \\
\textbf{enddo}
\end{array}
\tag{30}
$$

Of course the evaluation of c_{pk}, h_k, $Z_k = (S_k^0/R) - (H_k/RT)$, etc., is similar.

In the SID case the mixture thermodynamic properties, e.g., c_p, are then obtained with one inner product (18). Keeping in mind that the number of species K is generally lower than 100, this type of scalar reduction loop slows down the computer speed in vector mode, even if an optimized module is used. However, in the MID case, no scalar reduction loops are introduced so that no optimized modules are needed when the j loops are the inner loops and the k loops are the outer loops. Note also that in this situation the outer k loop could be partially unrolled into the inner j loop.

3.2.2 Evaluation of chemical production rates Before computing ω_k, $k = 1, \ldots, K$, one has to evaluate the rate of progress q_i, $i = 1, \ldots, I$, of each reaction. This can be done with an i loop. However, difficulties arise due to the products Π_{fi} and Π_{ri} in (8), to the sums ΔZ_i in (12), and to third body concentrations ctb_i in (13) :

$$
\Pi_{fi} = \prod_{k=1}^{K} [\mathcal{X}_k]^{\nu'_{ki}}, \quad \Pi_{ri} = \prod_{k=1}^{K} [\mathcal{X}_k]^{\nu''_{ki}}, \quad \Delta Z_i = \sum_{k=1}^{K} \nu_{ki} Z_k, \quad ctb_i = \sum_{k=1}^{K} \alpha_{ki} [\mathcal{X}_k]. \tag{31}
$$

Indeed the number of reactants or products in a chemical reaction never exeeds three and rarely two since most three body reactions are treated differently (13), so that the matrix ν_{ki}, $k = 1, \ldots, K$, $i = 1, \ldots, I$, is quite sparse. As a consequence, forming the products Π_{fi} and Π_{ri} or the sums ΔZ_i with (31) would be a great waste of time, regardless of the algorithm. The formulas (31) have to be replaced by

$$
\Pi_{fi} = \prod_{k \in \mathcal{R}(i)} [\mathcal{X}_k]^{\nu'_{ki}}, \qquad \Pi_{ri} = \prod_{k \in \mathcal{P}(i)} [\mathcal{X}_k]^{\nu''_{ki}}, \qquad \Delta Z_i = -\sum_{k \in R(i)} \nu'_{ki} Z_k + \sum_{k \in P(i)} \nu''_{ki} Z_k, \tag{32}
$$

where $\mathcal{R}(i) = \{kreac(i,l);\ l = 1, \ldots, KREAC(i)\}$ denote the indices of the reactants the ith reaction for which $\nu'_{ki} \neq 0$ and and $\mathcal{P}(i) = \{kprod(i,l);\ l = 1, \ldots, KPROD(i)\}$ the indices of the products the ith reaction for which $\nu''_{ki} \neq 0$ respectively. As already pointed out the number of reactants $KREAC(i)$ or products $KPROD(i)$ of the reactions rarely exeeds two and never three so that the k loops over $\mathcal{R}(i)$ and $\mathcal{P}(i)$ of (32) have to be unrolled in an i loop. However the reactant indices $kreac(i,l)$, $l = 1, \ldots, KREAC(i)$, and the product indices $kprod(i,l)$, $l = 1, \ldots, KPROD(i)$, of the ith reaction are arbitrary since kinetic mechanisms have no special structures, and their numbers $KREAC(i)$ and $KPROD(i)$ are varying with the reaction index i. Now on a CRAY-1 indirect addressing vector loops can only be obtained by using gathering-scattering macro-instructions, temporary arrays and loop splitting [18]. Furthermore variable-length loop unrolling can be achieved by introducing dummy species whose stoichiometric coefficients are zero in such a way that there are always three reactants and products in each reaction. Following these ideas, we introduce an extended set of stoichiometric coefficients n_1, n_2, n_3 and m_1, m_2, m_3 such that :

$$
n_{li} = \begin{cases} \nu'_{kreac(i,l)i}, & \text{if } 1 \leq l \leq KREAC(i), \\ 0, & \text{if } KREAC(i) < l \leq 3, \end{cases} \qquad m_{li} = \begin{cases} \nu''_{kprod(i,l)i}, & \text{if } 1 \leq l \leq KPROD(i), \\ 0, & \text{if } KPROD(i) < l \leq 3, \end{cases} \tag{33}
$$

and we set arbitrary values for $kreac(i,l)$ and $kprod(i,l)$ for $KREAC(i) < l \leq 3$ and $KPROD(i) < l \leq 3$ respectively. Then, if the molar concentrations of the reactants and products are gathered into the temporary arrays R_1, R_2, R_3, P_1, P_2 and P_3 :

$$
R_{li} = \left[\mathcal{X}_{kreac(i,l)}\right], \qquad P_{li} = \left[\mathcal{X}_{kprod(i,l)}\right], \qquad l = 1, 2, 3, \tag{34}
$$

and if the quantities $Z_k = (S_k^0/R) - (H_k/RT)$ of the reactants and products are gathered into the temporary arrays ZR_1, ZR_2, ZR_3, ZP_1, ZP_2 and ZP_3 :

$$ZR_{li} = Z_{kreac(i,l)}, \qquad ZP_{li} = Z_{kprod(i,l)}, \qquad l = 1,2,3, \tag{35}$$

then Π_{fi}, Π_{ri} and ΔZ_i can subsequently be estimated from

$$\Pi_{fi} = R_{1i}^{n_{1i}} R_{2i}^{n_{2i}} R_{3i}^{n_{3i}}, \qquad \Pi_{ri} = P_{1i}^{m_{1i}} P_{2i}^{m_{2i}} P_{3i}^{m_{3i}}, \tag{36}$$

$$\Delta Z_i = -n_{1i} ZR_{1i} - n_{2i} ZR_{2i} - n_{3i} ZR_{3i} + m_{1i} ZP_{1i} + m_{2i} ZP_{2i} + m_{3i} ZP_{3i}, \tag{37}$$

Note that (34)-(35) has to be rewritten with gathering non ANSI FORTRAN statements on a Cray-1. The extra cost due to the dummy variables in (36)-(37) is compensated by the vector speedup. Of course, the quantities $[\mathcal{X}_k]$, $k = 1, \ldots, K$, have to be evaluated with a simple k loop, and the quantities $Z_k = (S_k^0/R) - (H_k/RT)$, $k = 1, \ldots, K$, have to be evaluated like (29), before (34)-(35).

On the other hand, if $ithb$ is a flag concerning third bodies such that $ithb = 0$ if the ith reaction does not contain arbitrary third bodies, $ithb = 1$ if it does with unity efficiencies and $ithb = 2$ if some efficiency differs from unity and if we denote $kthb(i,l)$, $l = 1, \ldots, KTHB(i)$, the species whose efficiency α_{ki} in the ith reaction is not unity, then third body concentrations ctb_i can be obtained from

do $i = 1,\ I$
  $ctb_i = 1$
  if $ithb(i) \geq 1$ **then**
    $ctb_i = [M]$ (38)
  endif
  if $ithb(i) = 2$ **then**
    do $l = 1,\ KTHB(i)$
      $ctb_i = ctb_i + (\alpha_{kthb(i,l)i} - 1)\left[\mathcal{X}_{kthb(i,l)}\right]$
    enddo
  endif
enddo

Indeed, only a few third body efficiencies differ from one, so that vectorizing the innermost loop of (38) is not interesting. For the same reasons, only a small number of the chemical reactions are such that $ithb(i) = 2$, so that the true rate of the corresponding test is small and the use of a mask vector is not interesting. Only the two first statements are worth vectorizing by splitting the i loop.

Now the rate of progress q_i, $i = 1, \ldots, I$, can easily be obtained with an i loop :

do $i = 1,\ I$

$$\mathcal{K}_{fi} = A_i T^{\beta_i} \exp\left(-\frac{E_i}{RT}\right)$$

$$\mathcal{K}_{ri} = \mathcal{K}_{fi} \left(\frac{p_{atm}}{RT}\right)^{-\Delta\nu_i} \exp(+n_{1i}ZR_{1i} + n_{2i}ZR_{2i} + n_{3i}ZR_{3i} - m_{1i}ZP_{1i} - m_{2i}ZP_{2i} - m_{3i}ZP_{3i}) \tag{39}$$

$$q_i = ctb_i \left(\mathcal{K}_{fi} R_{1i}^{n_{1i}} R_{2i}^{n_{2i}} R_{3i}^{n_{3i}} - \mathcal{K}_{ri} P_{1i}^{m_{1i}} P_{2i}^{m_{2i}} P_{3i}^{m_{3i}}\right)$$

enddo

Note that the Lindeman form (15) or the irreversibility condition, i.e., $\mathcal{K}_{ri} = 0$, could easily be included. Finally, the chemical production rates ω_k, initialized to zero, are obtained with one scalar i loop :

do $i = 1,\ I$
  $\omega_{kreac(i,1)} = \omega_{kreac(i,1)} - n_{1i} q_i$
  $\omega_{kreac(i,2)} = \omega_{kreac(i,2)} - n_{2i} q_i$ (40)
  ....
  $\omega_{kprod(i,3)} = \omega_{kprod(i,3)} + m_{3i} q_i$
enddo

since none of the statements in (40) can vectorize [16] even by using scattering macro instructions, due to the fact that a species can be involved in different reactions. Note that only a few K and I temporary arrays are needed in the preceding algorithms.

In the MID case, everything is simpler. All the j loops have to be put as inner loops. In this situation, the number of operations is minimized, the vector lengths are maximized, and all the loops vectorize. The corresponding algorithm can be sketched as follows. First, the molar concentrations $[\mathcal{X}_{kj}]$, $k = 1, \ldots, K$, $j = 1, \ldots, J$, and the quantities Z_{kj}, $k = 1, \ldots, K$, $j = 1, \ldots, J$, are evaluated and stored in $K*J$ temporary arrays, and the production rates ω_{kj}, $k = 1, \ldots, K$, $j = 1, \ldots, J$, are initialized to zero. Then begins an outer loop over the chemical reactions, followed by the initialization to one, one and zero, of temporarry arrays of length J used to store the Π_{fij}, Π_{rij} and ΔZ_{ij}, respectively. The molar concentration products Π_{fij} of the reactants of the ith reaction are then evaluated together with the reactants' contribution to the sums ΔZ_{ij} for every $j = 1, \ldots, J$, :

do $l = 1,\ KREAC(i)$
  do $j = 1,\ J$
    $\Pi_{fij} = \Pi_{fij} \left[\mathcal{X}_{kreac(i,l)j}\right]^{\nu'_{kreac(i,l)i}}$ (41)
    $\Delta Z_{ij} = \Delta Z_{ij} - \nu'_{kreac(i,l)i} Z_{kreac(i,l)j}$
  enddo
enddo

The calculation of the Π_{rij} and of the contribution of the products of the ith reaction to the sums ΔZ_{ij} is then done as (41) for every $j = 1, \ldots, J$. Note that the number of operations for calculating Π_{fij}, Π_{rij} and ΔZ_{ij} is minimized, at variance with the SID case where dummy species are introduced. The third body concentrations for the ith reaction ctb_{ij} are evaluated for every $j = 1, \ldots, J$, using only vector loops, which contrasts with the SID case :

if $ithb(i) = 0$ **then**
  do $j = 1,\ J$
    $ctb_{ij} = 1$
  enddo
elseif $ithb(i) \geq 1$ **then**
  do $j = 1,\ J$
    $ctb_{ij} = [M_j]$ (42)
  enddo
endif
if $ithb(i) = 2$ **then**
  do $l = 1,\ KTHB(i)$
    do $j = 1,\ J$
      $ctb_{ij} = ctb_{ij} + (\alpha_{kthb(i,l)i} - 1)\left[\mathcal{X}_{kthb(i,l)j}\right]$
    enddo
  enddo
endif

The rate of progress of the ith reaction is then easily obtained for every $j = 1, \ldots, J$. Writing the corresponding algorithm is straigthforward. Finally the contribution of the ith reaction to the chemical production rates of its reactants is evaluated from :

do $l = 1,\ KREAC(i)$
  do $j = 1,\ J$
    $\omega_{kreac(i,1)j} = \omega_{kreac(i,1)j} - \nu'_{kreac(i,l)i} q_{ij}$ (43)
  enddo
enddo

followed by the analogous to (43) for the products of the ith reaction, and the loop over the chemical reactions ends. Note that the irreducible scalar kernel of (40) disappears in (43) and its analogous for the products, and that only two $K*J$ temporary arrays are needed, one for the $[\mathcal{X}_{kj}]$ and one for the Z_{kj}, together with some J temporary arrays.

3.3 Vectorization of transport properties

3.3.1 Evaluation of λ_k and η_k These evaluations are similar to those of c_{pk}, with an inner k loop in the SID case and an outer k loop with an inner j loop in the MID case, explicitely using the fits (25)-(26). Writing the corresponding algorithms is straightforward.

3.3.2 Evaluation of λ and η The vectorization of semi-empirical formulas like (23) is similar to that of (18), so that we will not discuss it again. The vectorization of the Wilke formula (24) leads to the following algorithm, where sum_k, $k = 1,\ldots,K$, denotes a temporary array initialized to zero :

$$
\begin{array}{l}
\textbf{do } l = 1,\ K \\
\quad \textbf{do } k = 1,\ K \\
\quad\quad sum_k = sum_k + X_l \frac{1}{\sqrt{8}}\Big(1 + \frac{W_k}{W_l}\Big)^{-\frac{1}{2}}\Big(1 + (\frac{\eta_k}{\eta_l})^{\frac{1}{2}}(\frac{W_l}{W_k})^{\frac{1}{4}}\Big)^2 \\
\quad \textbf{enddo} \\
\textbf{enddo} \\
\textbf{do } k = 1,\ K \\
\quad \lambda = \lambda + X_k\lambda_k/sum_k \\
\textbf{enddo}
\end{array}
\tag{44}
$$

An optimized modules can also be used for the last loop which is a scalar reduction loop. In the MID case, as usual, the j loops are put as inner loops and writing the corresponding algorithm is straightforward.

3.3.3 Evaluation of $\mathcal{D}_{kl}$ and D_k In the SID case the binary diffusion coefficients can be evaluated in a (k,l) loop :

$$
\begin{array}{l}
\textbf{do } (k,l) = (1,1),\ (K,K) \\
\quad \mathcal{D}_{kl} = \exp(d_{1kl} + d_{2kl}\log T + d_{3kl}(\log T)^2 + d_{4kl}(\log T)^3) \\
\textbf{enddo}
\end{array}
\tag{45}
$$

Although recent compilers, e.g., Fujitsu VP200, directly vectorize (45) as a (k,l) loop of length $K{*}K$, it is necessary to rewrite it as a one-dimensional loop on a Cray-1. On the other hand, exploiting the natural symmetry of the binary diffusion coefficients leads to the the following algorithm :

$$
\begin{array}{l}
\textbf{do } k = 1,\ K \\
\quad \textbf{do } l = k{+}1,\ K \\
\quad\quad \mathcal{D}_{kl} = \exp(d_{1kl} + d_{2kl}\log T + d_{3kl}(\log T)^2 + d_{4kl}(\log T)^3) \\
\quad\quad \mathcal{D}_{lk} = \mathcal{D}_{kl} \\
\quad \textbf{enddo} \\
\textbf{enddo}
\end{array}
\tag{46}
$$

which has to be used together with a compiler directive forcing the vectorization inhibited by a formal recursivity. Now, in scalar mode, exploiting the symmetry of the $\mathcal{D}_{kl}$ is always faster than (45) which approximately doubles the number of operations. This is true in vector mode if the number of interacting species K is large enough. On a Cray-1 this minimum number of species is between 20-30.

In the SID case, the mixture diffusion coefficients D_k, $k = 1,\ldots,K$, can then be obtained from the following algorithm, where sum_k, $k = 1,\ldots,K$, denotes a temporary array initialized to zero :

$$
\begin{array}{l}
\textbf{do } l = 1,\ K \\
\quad \textbf{do } k = 1,\ K \\
\quad\quad \textbf{if } k \neq l \textbf{ then} \\
\quad\quad\quad sum_k = sum_k + X_l/\mathcal{D}_{kl} \\
\quad\quad \textbf{endif} \\
\quad \textbf{enddo} \\
\textbf{enddo} \\
\textbf{do } k = 1,\ K \\
\quad D_k = (1 - Y_k)/sum_k \\
\textbf{enddo}
\end{array}
\tag{47}
$$

and where the conditional loop has to be rewritten with a non ANSI FORTRAN statement. Note also that D_k is not defined for a pure species mixture so that some care must be taken in this latter case. A procedure introduced by Kee, Warnatz and Miller [15] consists in computing

$$D_k = \frac{1}{W}\sum_{l\neq k}(X_l + \epsilon)W_l \Big/ \sum_{l\neq k}(X_l + \epsilon)/\mathcal{D}_{kl}, \tag{48}$$

where ϵ is a small positive constant. The corresponding extra cost is small.

In the MID case the evaluation of $\mathcal{D}_{klj}$ must be done with j loops as inner loops in (46), and that of D_{kj} with j loops as inner loops of (47). The extra cost introduced by (48) is small. Writing the corresponding algorithms is straightforward. Note however that if only the D_{kj}, $k = 1,\ldots,K$, $j = 1,\ldots,J$, are needed, it is not necessary to introduce a $K{*}K{*}J$ temporary array to store the $\mathcal{D}_{klj}$, which can be a problem for small memory computers and large chemical networks, since one can evaluate the $\mathcal{D}_{klj}$ only where they are needed. Note also that the choice of the fits (25)-(28) is not without consequences. For instance, the $(1/2)K*(K-1)*J$ exponentials needed for evaluating the $\mathcal{D}_{klj}$ are fairly expensive, especially in scalar mode. Finally, writing similar algorithms for the evaluation of the thermal diffusion ratio Θ_k is an easy task.

4. NUMERICAL EXPERIMENTS

In this section, performances of the SID and MID routines and of a laminar flame solver are presented. All tests are performed with Hydrogen-Air and Propane-Air mixtures. The kinetic scheme for Hydrogen-Air mixtures involves $K = 9$ species and $I = 19$ reactions and is taken from Miller and Kee [23]. For Propane-Air mixtures we have used a $K = 33$ species and $I = 126$ reactions reaction scheme due to Warnatz [24]. All tests have been done on a Cray-1 computer with the CFT 1.11 compiler.

Furthermore, the SID and MID routines have been written to use the work arrays of the CHEMKIN and TRANSPORT packages' subroutines. These work arrays store information about the species, the kinetic mechanism, the transport properties and they contain some work space. Larger work arrays were necessary however to store the extended set of stoichiometric coefficients and species names (33), and for the larger work space needed, especially for the MID routines. Up to these modifications, SID and MID routines may be embeded in the libraries of these packages. In particular, the excellent CHEMKIN FORTRAN code which reads symbolic descriptions of reaction mechanisms, the 'Interpreter', and the TRANSPORT FORTRAN code which computes the polynomial fits (25)-(28) can be used with the SID and MID routines. We refer to [14], [15] for more details.

4.1 Methodology

To measure the performance of a given program, we have used two types of tests. The first one is related to Amdhal's law which states that if P is the vector/scalar performance of a code,—the ratio of the execution times T_s and T_v with the vector mode inactive and active respectively, if S is the ratio of the average rates S_v and S_s at which the computer process the code in vector mode and scalar, i.e., sequential, mode respectively, —sometimes called the vector speedup :

$$P = T_s/T_v, \qquad S = S_v/S_s, \tag{49}$$

and if α is the the fraction of the code executed in vector mode, i.e., in the pipelined functional units, — also called the vectorization ratio, then we have :

$$P = 1\Big/\left(\frac{\alpha}{S} + 1 - \alpha\right). \tag{50}$$

This formula indicates that the higher the S and α values, the faster the program is processed. The vector/scalar speed ratio S depends of course on the computer but it also depends on the nature of the program. The program attributes affecting the value of S are for instance the vector lengths and the data reference method, i.e., the type of indexing. On a Cray-1 the workload dependent vector speedup S can range from 5. to 10.. Assuming an average value of $S = 7.5$, (50) gives that $\alpha = 0.00$ for $P = 1$, $\alpha = 0.58$ for $P = 2$, $\alpha = 0.77$ for $P = 3$, $\alpha = 0.87$ for $P = 4$, $\alpha = 0.92$ for $P = 5$, $\alpha = 0.96$ for $P = 6$, $\alpha = 0.99$ for $P = 7$, so that Amdhal's law provided an approximated method to measure the vectorization ratio of a code by measuring P.

On the other hand, it is also usefull to compare the execution times T_s and T_v of a program when the vector mode is inactive and active, respectively with the corresponding execution times $T_s^{\rm old}$ and $T_v^{\rm old}$ of different versions of the code :

$$\mathcal{P}_s = T_s^{\rm old}/T_s, \qquad \mathcal{P}_v = T_v^{\rm old}/T_v, \tag{51}$$

to measure the improvements of the new versions.

4.2 NUMERICAL RESULTS

4.2.1 SID subroutines The performances of the SID routines returning the species enthalpies h_k, the chemical production rates ω_k, the semi-empirical mixture conductivity λ, the Wilke mixture conductivity λ and the mixture diffusion coefficients D_k exploiting or not the symmetry of the $\mathcal{D}_{kl}$, are presented in Table 1 for a Hydrogen-Air mixture with $K = 9$ and $I = 19$, in Table 2 for a Propane-Air mixture with $K = 33$ and $I = 126$, and in the columns referenced by h_k, ω_k, λ_e, λ_w, D_k^{sym} and D_k, respectively. These subroutines represent quite well all the different types of algorithms considered in Section 3. In Tables 1 and 2, the first lines are the vector/scalar performances P of the SID routines, while the other two are the performance ratios $\mathcal{P}_s$ and $\mathcal{P}_v$ of the SID routines over the corresponding modules of the CHEMKIN and TRANSPORT libraries. The TRANSPORT subroutines [15] were modified in such a way that the mole fractions X_k are in the calling list instead of the mass fractions Y_k, which lead to faster routines. Mixture diffusion coefficients are estimated with the modified formula (48).

	h_k	ω_k	λ_e	λ_w	D_k^{sym}	D_k
P	2.10	1.95	1.62	4.06	2.04	3.89
$\mathcal{P}_s$	2.10	1.26	1.39	1.05	2.43	1.72
$\mathcal{P}_v$	6.05	2.51	1.67	1.43	3.64	4.90

TABLE 1
Performances of the SID routines for a Hydrogen-Air mixture.

	h_k	ω_k	λ_e	λ_w	D_k^{sym}	D_k
P	4.04	2.55	3.40	8.16	4.21	6.19
$\mathcal{P}_s$	2.32	1.39	1.53	1.03	2.87	1.81
$\mathcal{P}_v$	13.59	3.57	2.96	1.31	8.36	7.72

TABLE 2
Performances of the SID routines for a Propane-Air mixture.

The first important point is that the performances P of the SID routines depend on the reaction mechanism. Indeed, the larger are K and I, the longer are the vector lengths involved in the SID routines and therefore the larger is the speed ratio S, with a resulting larger P according to (50). The presence of scalar reduction loops in the algorithms for λ_e and the irreducible scalar kernel (40) in the evaluation of ω_k explain the corresponding performances. Note also the differences between D_k^{sym} and D_k due to the small vector lengths in (46).

In scalar mode, the performances over the CHEMKIN and TRANSPORT routines are sligthly larger than one. Indeed, a vectorized algorithm often run a little bit faster than a corresponding scalar version because fewer loop setups, vector loads, etc., are usually needed in the vectorized version. The higher values $\mathcal{P}_s = 2.43$–2.87 for D_k^{sym} are due to the smaller number of arithmetic operations.

Finally, in vector mode, the improvements over the CHEMKIN and TRANSPORT routines range from 1.31 to 13.5. The value $\mathcal{P}_v = 1.31$, for the evaluation of λ_w, one of the most expensive subroutines, shows that the corresponding TRANSPORT module is quite well vectorized. The unusual value $\mathcal{P}_v = 13.5$, for the evaluation of h_k $k = 1, \ldots, K$, is due in part to the bad vector/scalar performance $P = 0.69$ of the corresponding CHEMKIN module, due to scalar reduction loops of small length used in the evaluation of the thermodynamic fits. More important, however, is the improvement for evaluating D_k, $k = 1, \ldots, K$, which usually is a costly calculation. Note also that exploiting the symmetry of the $\mathcal{D}_{kl}$ is slower than (45) for Hydrogen-Air mixtures.

4.2.2 MID subroutines The performances of the MID routines returning the species enthalpies h_{kj}, the chemical production rates ω_{kj}, the semi-empirical mixture conductivity λ_j, the Wilke mixture conductivity λ_j and the mixture diffusion coefficients D_{kj}, are presented in Table 3 for a Hydrogen-Air flame with $K = 9$, $I = 19$ and $J = 77$, and in Table 4 for a Propane-Air flame with $K = 33$, $I = 126$ and $J = 63$, and in the columns referenced by h_k, ω_k, λ_e, λ_w and D_k^{sym}, respectively. Again, these subroutines represent quite well all the different types of algorithms considered in Section 3. In Tables 3 and 4, the first lines are the vector/scalar performances P of the MID routines, while the other two are the performance ratios $\mathcal{P}_s$ and $\mathcal{P}_v$ of the MID routines over the corresponding SID routines used repetitively. In the column D_k^{sym} the performance ratios are measured in comparison with the SID routine using the algorithm (45) ,which does not exploit the symmetry of the binary diffusion coefficients, and mixture diffusion coefficients are estimated with the modified formula (48).

At variance with the SID routines, the performances P of the MID routines are independent of the reaction mechanism, and essentially depend on the number of input data J. Fairly uniform values are obtained for P, ranging from 6.91 up to 7.88, except with h_k for which P=3.72–3.79, due to the complex formulation (16) (17) and to the fact that there is only one path to/from memory on a Cray-1. These very good results of the MID subroutines are due to their vectorization by 'replication' for which all statements fully vectorize with vectors of length J and which minimizes the number of arithmetic operations. The only disavantage could be the amount of memory needed, but the largest temporary array, the one used to store the $\mathcal{D}_{klj}$ of length $K*K*J$, can be omitted if there are memory problems. A consequence is that in vector mode the MID routines are significantly faster than the corresponding SID modules used repetitively, especially for small reaction schemes like Hydrogen-Air.

	h_k	ω_k	λ_e	λ_w	D_k^{sym}
P	3.72	7.11	7.73	7.11	7.03
$\mathcal{P}_s$	2.45	1.75	1.13	3.36	1.63
$\mathcal{P}_v$	3.95	6.59	5.14	5.82	2.88

TABLE 3
Performances of the MID routines for a Hydrogen-Air flame.

	h_k	ω_k	λ_e	λ_w	D_k^{sym}
P	3.79	7.36	7.88	6.91	7.33
$\mathcal{P}_s$	2.11	1.53	0.97	3.58	1.59
$\mathcal{P}_v$	1.82	4.69	2.08	3.01	1.91

TABLE 4
Performances of the MID routines for a Propane-Air flame.

Now in scalar mode, the MID routines are slightly faster except for scalar reduction loops in the Propane case. The relatively large values $\mathcal{P}_s$=3.36–3.58 obtained for λ_w are due to the smaller number of operations in the MID case, since for instance the quantities like $(1/\sqrt{8})(1+(W_k/W_l))^{-1/2}$ and $(W_l/W_k)^{1/4}$ involved in ϕ_{kl} (31) have to be evaluated only once for $j=1,\ldots,J$.

4.2.3 A laminar flame solver To test the MID routines in a practical situation we have computed several laminar premixed stagnation point flow flames, including extinction limits. For the purpose of the present paper it is enough to know that in this flow configuration Eqs. (1)-(5) reduce to a two-point boundary value problem. More details on the modeling can be found in [6]. The laminar flame solver is based on Newton's method, adaptive gridding and continuation techniques. We refer to [5] [6] [7] [21] for more details on the solution method.

For a wide variety of Hydrogen-Air, Methane-Air and Propane-Air flame structure calculations, we have found that typically 70–80% of the CPU time is spent in the MID routines in scalar mode, so that this part of the CPU time is approximately divided by a factor of 7. in vector mode. The remaining 20–30% of the CPU time in scalar mode is spent for linear algebra, adaptive gridding and the form of the governing equations once thermodynamic properties, chemical production rates and transport coefficients are known. These parts of the solver are only partially vectorized by the compiler. Nevertheless, average vector/scalar performance P around 4. have been obtained for the laminar flame solver, allowing considerable savings in CPU time in our different applications [4] [6] [7].

5. CONCLUSION

The evaluation of thermodynamic properties, chemical production rates, and multicomponent transport coefficients, has been investigated. Vectorizing algorithms have been obtained for single and multiple input data subroutines. These subroutines have allowed important savings in CPU time.

ACKNOWLEDGMENT

We would like to thank Pr. S. Candel, Dr. R. J. Kee, Pr. M. D. Smooke and Pr. J. Warnatz for interesting discussions concerning this material. We would also like to thank the Scientific Council of the 'Centre de Calcul Vectoriel pour la Recherche' for providing the Cray-1 resources and the 'Centre Inter Régional de Calcul Electronique' for providing computational resources. The research was supported in part by the 'Direction des Recherche et Etudes Techniques'.

REFERENCES

1 Boris J. P. and Winsor N. K., *Vectorized Computation of Reactive Flow*, in Parallel Computations, G. Rodrigue Ed., Academic press, (1982), p. 173–215.
2 Chapman S. and Cowling T. G., *The Mathematical Theory of Non-Uniform Gases*, Cambridge University Press, Cambridge, (1970).
3 Coffee T. P. and Heimerl J. M., *Transport Algorithms for Premixed, Laminar Steady-state Flames*, Comb. and Flame, **43**, (1981), p. 273–289.
4 Darabiha N., Giovangigli V., Candel S. and Smooke M., *Extinction of Strained Premixed Propane-Air Flames with Complex Chemistry*, submitted to Comb. Sci. and Tech..
5 Giovangigli V. and Smooke M., *Extinction Limits for Premixed Laminar Flames in a Stagnation Point Flow*, J. Comp. Phys. **68**, (1987), p. 327–345.
6 Giovangigli V. and Smooke M., *Extinction Limits of Strained Premixed Laminar Flames with Complex Chemistry*, Comb. Science and Tech., in press , (1987).
7 Giovangigli V. and Smooke M., *Adaptive Continuation Algorithms with Application to Combustion Problems*, submitted to Applied Numerical Methods.
8 Glowinsky R., Larrouturou B. and Temam R., Eds, *Numerical Simulation of Combustion Phenomena*, Springer Verlag, Heidelberg, (1985).
9 Gordon S. and Mcbride B. J., *Computer Program for Calculation of Complex Chemical Equilibrium Compositions, Rocket Performance, Incident and Reflected Shocks and Chapman-Jauguet Detonations*, NASA SP-273, (1971).
10 Hirschfelder J. O. and Curtiss C. F., *Flame Propagation in Explosive Gas Mixtures*, Third Symposium (International) on Combustion, Reinhold, New York, (1949), p. 121–127.
11 Hirschfelder J. O., Curtiss C. F. and Bird R. B., *Molecular Theory of Gases and Liquids*, John Wiley and Sons, Inc., New York, (1954).
12 Hockney R. W. and Jesshope C. R., *Parallel Computers*, Adam Hilger Ltd, (1981).
13 Kee R. J., and Dwyer H. A., *Review of Stiffness and Implicit Finite-Difference Methods in Combustion Modeling*, Proceedings of the 7th ICODER, (1981), p. 485–500.
14 Kee R. J., Miller J. A. and Jefferson T. H., *CHEMKIN: A General-Purpose, Problem-Independent, Transportable, Fortran Chemical Kinetics Code Package*, SANDIA National Laboratories Report, SAND80-8003, (1980).
15 Kee R. J., Warnatz J. and Miller J. A., *A Fortran Computer Code Package for the Evaluation of Gas-Phase Viscosities, Conductivities, and Diffusion Coefficients*, SANDIA National Laboratories Report, SAND83-8209, (1983).
16 Neves K. N., *Vectorization of Scientific Software*, in High Speed Computation, J. S. Kowalik Ed., NATO ASI Series Vol. F7, Springer-Verlag, (1984), p. 277–291.
17 Oran E. S. and Boris J. P., *Detailed Modeling of Combustion Systems*, Prog. Energy Combust. Sci., **7**, (1981), p. 1–72.
18 Pacific-Sierra Research Corp., *Fortran Programming on the Cray-1*, Pacific-Sierra Research Corp., Los Angeles, (1983).
19 Peters N. and Warnatz J., Eds., *Numerical Methods in Laminar Flame Propagation*, Vieweg Verlag, (1982).
20 Rogg B., *Numerical Analysis of Confined Laminar Jet Diffusio Flames.*, this volume, (1987).
21 Smooke M. D., *Solution of Burner-Stabilized Premixed Laminar Flames by Boundary Value Methods*, J. Comp. Phys., **48**, (1982), p. 72–105.
22 Smooke M. D., *Solution of Two-Dimensional Axisymmetric Laminar Diffusion Flames by Adaptive Boundary Value Methods*, this volume, (1987).
23 Smooke M. D., Miller J. A. and Kee R. J., *Determination of Adiabatic Flame Speeds by Boundary Value Methods*, Comb. Sci. and Tech., **34**, (1983), p. 79–89.
24 Warnatz J., *The Mechanism of High Temperature Combustion of Propane and Butane*, Comb. Sci. and Tech., **34**, (1983), p. 144–200.
25 Warnatz J., *Calculation of the Structure of Laminar Flat Flames II : Flame Velocity and structure of Freely Propagating Hydrogen-Oxygen and Hydrogen-Air Flames*, Ber. Bunsenges. Phys. Chem., **32**, (1978), p. 643–649.

NUMERICAL MODEL FOR PROPELLANT GRAIN BURNING SURFACE RECESSION

A.Y. LE ROUX , G.S.NAMAH , D.RIBEREAU

UER Mathématiques et Informatique
Université de Bordeaux 1
33405 TALENCE - FRANCE

We build a two dimension and a three dimension mathematical model solving the recession of a burning surface. It corresponds to a non strictly hyperbolic system. A numerical scheme is proposed as well as some numerical experiments.

- SECTION 1: - INTRODUCTION - Inside a solid fuel rocket motor, the combustion surface of a mass of a propellant grain is governed by a very simple rule, during its evolution.

By denoting the time by t and the space variable in $\mathbb{R}^2$ or $\mathbb{R}^3$ by x , M(x,t) will represent the position of a point of the surface and U(x,t) the unit vector perpendicular to the surface at that point and directed towards the fresh propellant grain.

We introduce also a local parameter C(x) which will represent the rate of combustion. It will depend on the type of propellant grain and on the pressure in the motor; it will be supposed non constant when several propellants are loaded together in the motor.

We can therefore characterize the displacement of the combustion surface by

$$(1) \qquad \frac{d\,M}{dt}(x,t) = C(x)\,U(x,t) \quad .$$

C.-M. Brauner and C. Schmidt-Lainé (eds.), Mathematical Modeling in Combustion and Related Topics, 505–514.

The approximation of the surface is usually done by geometrical methods where C(x) must be taken as a constant i.e. independant of x. Such methods give good results in 2-dimension but the generalization to the three dimensional case is very bad.

The description of a numerical method is proposed here. This method is based on the approximation of the surface by finite elements and it ensures a second order accuracy.

The two dimensional case is described in details in Section 2 and the three dimensional case is studied in Section 3. The last section contains some numerical results on some geometrical figures representative of the difficulty of the subject. However these results are not related to the usual applications of this study, which was done in collaboration with the S.N.P.E. in St Medard en Jalles.

- SECTION 2 : THE TWO DIMENSIONAL PROBLEM - Let s denote the curvilinear abscissa of a generic point of the burning surface, which is a curve here. This variable s will be very sensitive to any change in time for a given point. From a numerical point of view, this will require some delicate reorganization of the mesh after each time step.

The following diagrams help to visualise this difficulty.

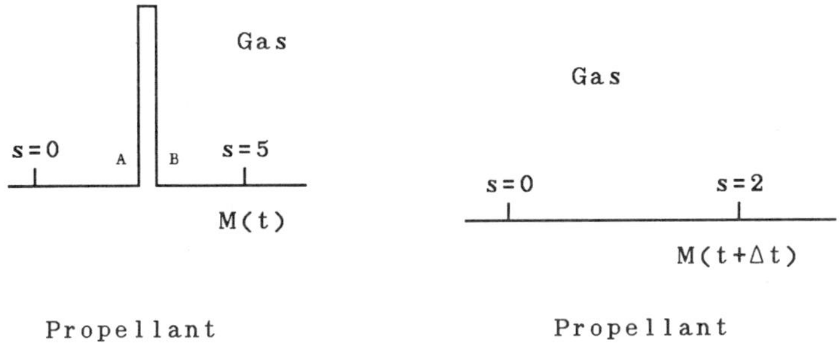

The point M(t+Δt) is situated in a position vertical to that of M(t), which can itself be considered as the former position. However the curvilinear abscissae of M(t) and M(t+Δt) are very different due to the disappearance of the bump between the points A and B.

Another difficulty, which is a numerical one this time, arises when we want to define the normal vector at the nodes of a polygonal line corresponding to the approximate surface. A straightfoward computation of this normal vector resulting for e.g. from an interpolation of the neighbouring vertices can introduce too big an error.

We have preferred to consider this normal vector as an unknown of the problem and written the equations which give this vector as solution.

A discretization scheme based on classical arguments of finite element methods is then proposed. This allows us to deduce the direction of the normal vector at a vertex $M(t+\Delta t)$, from the position of the surface at time t ,even before knowing the exact position of the point $M(t+\Delta t)$ ..

Thus we can introduce a discretization of the displacement equation (1) , which can have a second order accuracy, since we know the two values of the normal vector at the times t and $t+\Delta t$.

Let $s > 0$ be the curvilinear abscissa at a point $M(s,t)$ of the surface at time $t > 0$, and

$$\vec{U}(s,t) = \begin{pmatrix} u \\ v \end{pmatrix}$$

the vector perpendicular to the surface, directed from the burnt gas towards the fresh fuel. We denote by $x(s,t)$ and $y(s,t)$ the coordinates of the point $M(s,t)$ and we introduce

$$a = a(s,t) = \sqrt{u^2 + v^2}$$

in order to ensure the unitary character of the normal vector.

The equation (1) becomes

$$\text{(2)} \qquad \frac{\partial x}{\partial t} = C \frac{u}{a} ,$$

$$\text{(3)} \qquad \frac{\partial y}{\partial t} = C \frac{v}{a} .$$

The components of the normal vector are related to x and y by

$$(4) \qquad u = -\frac{\partial y}{\partial s},$$

$$(5) \qquad v = \frac{\partial x}{\partial s}.$$

By differentiating the equation with respect to t, we get

$$(6) \qquad \frac{\partial u}{\partial t} = -\frac{\partial}{\partial s}\left(\frac{C v}{a}\right),$$

$$(7) \qquad \frac{\partial v}{\partial t} = \frac{\partial}{\partial s}\left(\frac{C u}{a}\right).$$

Thereby we obtain a non strictly hyperbolic system. We can easily verify that it admits zero as a double eigenvalue (at least when C is a constant) and then it cannot be diagonalized.

- SECTION 3 - APPROXIMATION IN TWO DIMENSIONS - Let Δt be the time increment; we set $t_n = n\,\Delta t$, with $n \in \mathbb{N}$. At each time step, the surface is approximated by a polygonal line Γ_n of vertices S_i^n, $i \in \mathbb{N}$. Next consider a set of test functions $\{\varphi_i\}$ defined on Γ_n such that the value of φ_i is one at the vertex S_i^n and zero at the other vertices.

The problem is to find an approximate value of the vector perpendicular to the surface at time t_{n+1} assuming that we know the values of u_i^n and v_i^n at each vertex S_i^n. We multiply the equations (6) and (7) by φ_i and then we integrate by parts with respect to s. Finally by using a quadrature formula to evaluate the integrals, we get the following first order numerical scheme.

$$(8) \qquad u_i^{n+1} = u_i^n - \frac{\Delta t}{L_i^n}\left(\frac{C_{i+1}^n\, v_{i+1}^n}{a_{i+1}^n} - \frac{C_{i-1}^n\, v_{i-1}^n}{a_{i-1}^n}\right),$$

$$(9) \qquad v_i^{n+1} = v_i^n + \frac{\Delta t}{L_i^n}\left(\frac{C_{i+1}^n\, u_{i+1}^n}{a_{i+1}^n} - \frac{C_{i-1}^n\, u_{i-1}^n}{a_{i-1}^n}\right).$$

where C_i^n is the value of the combustion speed at the point corresponding to the position of the vertex S_i^n at time t_n, and

$$(10) \qquad L_i^n = |S_{i+1}^n - S_i^n| + |S_i^n - S_{i-1}^n| \quad .$$

This scheme is to be associated with that of Lax Friedriechs which operates under a stability condition, known as the Courant Friedriechs Lewy condition,

$$\frac{\Delta t}{|S_{i+1}^n - S_i^n|} |\lambda| \leqslant 1 \quad ,$$

where λ is an eigenvalue corresponding to a characteristic speed of the system (6), (7). But in our case, we should not come across the stability problem since all the eigenvalues are equal to zero and this is what is in fact observed in practice.

Once all the values (u_i^{n+1}, v_i^{n+1}) are computed by this scheme, we can reconstitute a new position for the polygonal line Γ_{n+1} of vertices S_i^{n+1} such that

$$(11) \qquad \overrightarrow{S_i^n \; S_i^{n+1}} = \Delta t \; C_i^n \; \frac{u_i^{n+1} + u_i^n}{\| u_i^{n+1} + u_i^n \|} \quad .$$

This scheme can be assimilated to that of Crank Nicholson and is of second order of accuracy when the combustion speed has a constant value. In practice, this speed is piecewise constant and takes only a limited number of values (2 or 3).

We now have to ensure a mesh reorganisation which consists on the one hand of the elimination of vertices found on the already burnt parts of the surface (protruding angles)and on the other hand, of the creation of new vertices in the case where two vertices become too far apart.

This introduction of new vertices utilises approximation techniques (cubic splines) to get a curvature in a better way. In fact a simple linear interpolation does not always give satisfactory results in the introduction of new vertices. This control of the mesh also takes into account the meeting of the combustion surface with an inhibited one which corresponds to the limit of the propellant grain block.

- SECTION 4 - THE THREE DIMENSIONAL MODEL - We consider again the equation (1) but now M and U will be vectors in $\mathbb{R}^3$. We have then

$$(12) \qquad \frac{\partial x}{\partial t} = \frac{C u}{a} \quad ; \quad \frac{\partial y}{\partial t} = \frac{C v}{a} \quad ; \quad \frac{\partial z}{\partial t} = \frac{C w}{a} \quad ;$$

where

$$a = \sqrt{u^2 + v^2 + w^2} \; .$$

We introduce the parameters ξ and η to represent the surface i.e. the surface will be of the form

$$X = X(\xi, \eta, t)$$

where X is the vector of components x, y and z, and ξ, η are two real numbers. In practice ξ and η will parametrise locally two parts of orthogonal curves, drawn on the surface and measured as curvilinear abscissae.

$$(13) \qquad Y(\xi, \eta, t) = \frac{\partial X}{\partial \xi}(\xi, \eta, t) \; ,$$

$$(14) \qquad Z(\xi, \eta, t) = \frac{\partial X}{\partial \eta}(\xi, \eta, t) \; .$$

are two vectors tangent to the surface at the point X considered and the normal vector U can be defined as the vector product to within about the sign

$$U = Y \wedge Z \; .$$

This sign will depend on an orientation directed towards the propellant grain.

By differentiating (12) we get

$$\frac{\partial Y}{\partial t} = \frac{\partial}{\partial \xi}\left[\frac{C\,U}{a}\right], \tag{15}$$

$$\frac{\partial Z}{\partial t} = \frac{\partial}{\partial \eta}\left[\frac{C\,U}{a}\right]. \tag{16}$$

In fact, there is no simple equation verified by U but this can be achieved at each point for Y and Z . But in practice, it is U which is stored in the memory and not Y and Z. We must therefore reconstruct the two tangent vectors Y and Z at each cycle and at each point.

The approximate surface is constituted of a triangulation whose nodes (vertices of triangles) are denoted by S_i^n and the triangles by K_j^n, for i, j $\in \mathbb{N}$ and at time t_n. We introduce a set of test functions φ_i affine on each triangle and such that φ_i is equal to 1 at the vertex S_i^n and zero at the other vertices.

When we multiply (15) or (16) by a test function φ_i, we get for each vertex S_i^n,

$$Y_i^{n+1} = Y_i^n - \frac{\Delta t}{\Gamma_i} \sum_{j \in \sigma_i} \frac{\operatorname{area}(K_j^n)}{3} \sum_{l=1}^{3} \overrightarrow{U(A_l^j)} \left[\overrightarrow{\nabla\varphi(A_l^j)} \cdot \overrightarrow{Y(S_i^n)}\right]$$

where σ_i is the set of indices of triangles having S_i^n as vertex, and A_1^j, A_2^j, A_3^j are the three vertices of K_j^n . We now have to determine an expression for $\nabla\varphi_i$. For that, let us denote by A and B the two vertices of the triangle K_j^n where the value of φ_i is zero, and by Q a point outside the plane determined by the three points A, B and S_i^n for e.g. such that

$$\overrightarrow{Q\,S_i^n} = -\vec{U}\,.$$

We easily obtain

$$\overrightarrow{\nabla\varphi_i} \cdot \overrightarrow{S_i^n A} = 1\ ; \quad \overrightarrow{\nabla\varphi_i} \cdot \overrightarrow{S_i^n B} = 1\ ; \quad \overrightarrow{\nabla\varphi_i} \cdot \overrightarrow{AB} = 0\ .$$

then

$$\overrightarrow{\nabla\varphi_i} = \mu\, \overrightarrow{QA} \wedge \overrightarrow{QB}\ , \quad \text{with} \quad \mu = - \frac{1}{(\overrightarrow{QS_i^n}, \overrightarrow{QA}, \overrightarrow{QB})}$$

where (.,.,.) denotes the mixed product, is a possible choice.

It suffices to introduce this value, trianglewise, to evaluate the second member of (17) and hence to determine Y_i^{n+1}.

We proceed in the same way in the determination of Z_i^{n+1}, where $\nabla\varphi_i$ is the same as that described earlier.

Once that Y_i^{n+1} and Z_i^{n+1} are determined at each point, we reconstruct the approximate normal vector by

$$U_i^{n+1} = Y_i^{n+1} \wedge Z_i^{n+1}$$

to be evaluated at each node. We now have to compute the new position of S_i^{n+1} by using (11) in the same way as for the two dimensional case.

After each cycle, we must introduce a phase for the elimination of triangles which have become too small or which are gone through the burnt area, or for the creation of new triangles when the vertices become too far away from each other. Both cases are very technical. Let us simply remark that the elimination or the creation of a triangle concern two triangles each time. In fact we eliminate a side which has become too short and thereby causing the disappearance of the two adjacent triangles or else we cut a side which has become too long (or rather a cubic spline which approximates this side). In the latter case, we introduce a new vertex at the middle of the side and hence dividing each of the two adjacent triangles into two.

-SECTION 5: A NUMERICAL TEST - The test which is represented here has been done on a well known geometrical figure which shows the real difficulties encountered in practice. This example is quite close to a geometrical shape met in practice for it contains some parts at whose surfaces the normal vectors are perpendicular to an axis. Here it is the axis of symmetry, in order to make the input easier. However this symmetry will not be taken into account in the triangulation.

In this case we can introduce different methods in order to remesh the triangulation : the triangles which have become too small are eliminated, and those becoming too big are refined. This reduces efficiently the occupancy of the memory. Note that the triangulation has been made in such a way that it is independant of the symmetry hypothesis.

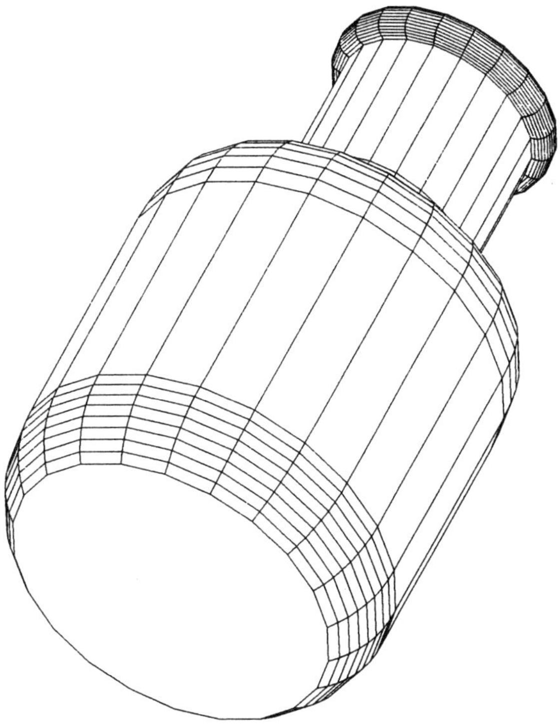

- The original geometry shape -

Finally let us point out that a technique using octrees has been developed in order to locate the type of propellant grain, that is, to evaluate C . This is done to find out if the considered point enters the burnt zone or not, or if it meets the external inhibited surface.

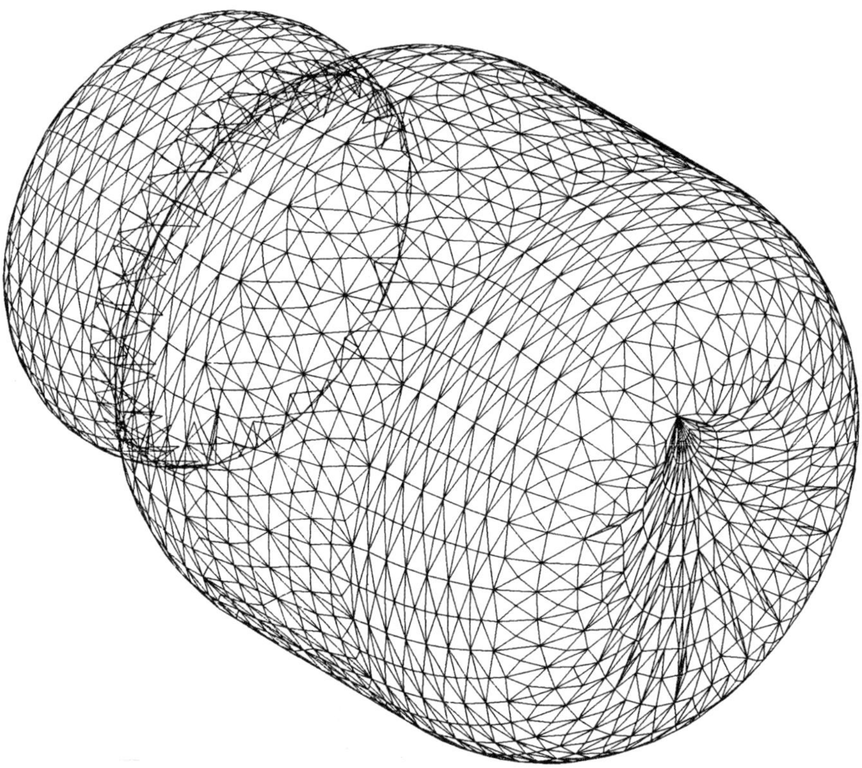

- The shape after a few timesteps of computation -

This work and the production of the corresponding software has been realised in collaboration with G 5 G Society and the S.N.P.E. Note that many programs from G.K.S.-Bx have also been used for this work.

GLOBAL EXISTENCE OF SOLUTIONS FOR A PROBLEM IN DYNAMICS OF THERMAL EXPLOSIONS.

Paola Mannucci
Istituto Matematico Ulisse Dini - Università di Firenze

We consider the behavior of a viscous, ideal, polytropic, compressible, chemically reactive gas, in cylindrical symmetry with radial velocity.

An analogous problem has been considered for the same type of gas, bounded by two infinite parallel plates, by A. Bressan (B. U.M.I. 1986)

The four equations of the system which governs the time evolution of the gas, correspond to the conservation law for mass, momentum and energy , and species diffusion .

In the Eulerian coordinates (r,t) , the system takes the form :

$$\left\{\begin{aligned} &\frac{d\rho}{dt} = -\frac{\rho}{r}\frac{\partial}{\partial r}(rv) \\ &\rho\frac{dv}{dt} = -R\frac{\partial}{\partial r}(\rho T) + \lambda_1\Delta v - (\varsigma + \frac{\eta}{3})\frac{v}{r^2} \\ &\rho c_v\frac{dT}{dt} = -R\rho T\frac{1}{r}\frac{\partial}{\partial r}(rv) + K\Delta T + \lambda_1(\frac{\partial v}{\partial r})^2 + (\frac{2\eta}{3} + \varsigma)(\frac{v}{r})^2 + 2\varsigma\frac{v}{r}\frac{\partial v}{\partial r} + \mu\rho f \\ &\rho\frac{dc}{dt} = \frac{D}{r}\frac{\partial}{\partial r}(r\rho\frac{\partial c}{\partial r}) - \rho f \end{aligned}\right. \tag{S}$$

with $r \in [\, r_0 \,, 1 \,]$ and $r_0 \in (\, 0,1 \,)$.

Here ρ is the fluid density, v the radial velocity, T the temperature, c the concentration, f the intensity of the chemical reaction.

ς , η , λ_1 are the coefficients of viscosity. K and D are the coefficients of heat conduction and of thermical diffusion.

To study the problem, it is convenient to introduce the Lagrangian coordinates (y,τ) , by replacing the spatial variable with a mass-coordinate.

Then the system can be written in the following form :

C.-M. Brauner and C. Schmidt-Lainé (eds.), Mathematical Modeling in Combustion and Related Topics, 515–518.

$$\left\{\begin{aligned}
&\frac{\partial\rho}{\partial\tau} = -2\pi\,\rho^2\,\frac{\partial}{\partial y}(rv)\\
&\rho\frac{\partial v}{\partial\tau} = -\,2\pi r\rho\,\frac{\partial}{\partial y}(R\rho T) + 4\pi^2\lambda_1\rho\,\frac{\partial}{\partial y}(\rho r^2\frac{\partial v}{\partial y}) \;-\; (\varsigma+\frac{\eta}{3})\,\frac{v}{r^2}\\
&\rho c_v\frac{\partial T}{\partial\tau} = -2\pi R\rho^2 T\,\frac{\partial}{\partial y}(rv) + 4\pi^2 K\,\rho\,\frac{\partial}{\partial y}(\rho r^2\frac{\partial T}{\partial y}) + \lambda_1(2\pi r\rho\frac{\partial v}{\partial y})^2\\
&\qquad\qquad + (\frac{2\eta}{3}+\varsigma)(\frac{v}{r})^2 + 4\pi\varsigma\rho v\,\frac{\partial v}{\partial y} + \mu\rho f\\
&\frac{\partial c}{\partial\tau} = 4\pi^2 D\,\frac{\partial}{\partial y}(r^2\rho^2\,\frac{\partial c}{\partial y}) - f
\end{aligned}\right.$$

with the initial conditions :

$\rho(y,0) = \rho_0(y)$; $v(y,0) = v_0(y)$; $T(y,0) = T_0(y)$; $c(y,0) = c_0(y)$

and the boundary conditions (for the annulus) :

$$\left\{\begin{aligned}
&\frac{\partial v}{\partial y}(0,\tau) = v(1,\tau) = 0\\
&\frac{\partial c}{\partial y}(0,\tau) = \frac{\partial c}{\partial y}(1,\tau) = 0\\
&\frac{\partial T}{\partial y}(0,\tau) = 0\\
&a\,c_v\,[\,T(1,\tau) - \tilde{T}\,] + b\,\rho(1,\tau)\,\frac{\partial T}{\partial y}(1,\tau) = 0
\end{aligned}\right.$$

($\tilde{T}$ being the external temperature)

We look for $v(y,\tau)$ in the class :

$$L = \left\{ v(y,\tau) \;:\; v(0,\tau) = 0 \;\text{ and }\; \left|\frac{\partial^2 v}{\partial y^2}(0,\tau)\right| \le C \right\}$$

<u>Theorem 1</u> (LOCAL EXISTENCE)

Let $0 < \alpha < \frac{1}{2}$ and $\Omega = [\,0,1\,]$.

If $\rho_0(y) \in C^{1+\alpha}(\Omega)$, $v_0(y)$, (y), $c_0(y) \in C^{2+\alpha}(\Omega)$,

then there exists a $\delta > 0$ such that there is the classical solution of the problem :

$\rho\ (y,\tau) \in C^{1+\alpha,\ 1+\frac{\alpha}{2}}\ (\ \Omega \times [\ 0,\delta\ ]\ )$ and

$v\ (y,\tau)\ ;\ T\ (y,\tau)\ ;\ c\ (y,\tau)\ \in\ C^{2+\alpha,\ 1+\frac{\alpha}{2}}\ (\ \Omega \times [\ 0,\delta\ ]\ )$ and

$v\ (y,\tau) \in\ L$.

* *

Notice that this theorem implies the existence of the solution of (S) for the annulus $r \in [\ r_0\ ,\ 1\ ]\ ,\ r_0 > 0$.

We can eliminate r and ρ from the last three equations of the system.Then, setting

$$u\ (y,\tau)\ = \Big\{\ v\ (y,\tau)\ ;\ T\ (y,\tau)\ ;\ c\ (y,\tau)\ \Big\} \text{ and } W\ (y,\tau) = u\ (y,\tau) - u_0\ (y)\ ,$$

we can rewrite the system in the compact form :

$$W_\tau = A\ W_{yy} + B \int_0^\tau W_{yy}\ (y,\tau)\ dy + C$$

Our proof of the existence of the solution is based on a iterative process, suggested by Nash (Bull. Soc. Math. France 1962) , that gives the solution of a non linear problem as the limit of solution of linear uniformly parabolic systems :

$$W_\tau^{(n+1)} = A^{(n)}\ W_{yy}^{(n+1)} + B^{(n)} \int_0^\tau W_{yy}^{(n+1)}\ (y,\tau)\ dy + C^{(n)}$$

where $A^{(n)}$, $B^{(n)}$, $C^{(n)}$ are defined by means of $W^{(n)}$.

By means a *priori* estimates we can prove the exixtence in $\Omega \times [\ 0,\delta]$ of $W^{(n+1)}$.

Then, using again a *priori* estimates for $\Big[\ W^{(n+1)} - W^{(n)}\ \Big]$, we prove the convergence of the sequence $\Big\{\ W^{(n)}\Big\}$ to the local solution $W\ (y,\tau)$, in $\Omega \times [0,\delta]$.

We can observe that the time " δ " depends only on the initial data u_0 , u_{oy}, ρ_o, ρ_{oy} .

Theorem 2 (GLOBAL EXISTENCE)

Let $0 < \alpha < \frac{1}{2}$ and $\Omega_0 = [\, r_0 , 1 \,]$ with $r_0 \in (0,1)$.

If $\rho_0 (r) \in C^{1+\alpha} (\Omega_0)$, $v_0 (r)$, $T_0 (r)$, $c_0 (r) \in C^{2+\alpha}(\Omega_0)$,

then for each $T > 0$, there exists the classical solution of the problem , in Eulerian coordinates , in $\Omega_0 \times [\, 0 , T \,]$:

$\rho (r,t) \in C^{1+\alpha,\, 1+\frac{\alpha}{2}} (\Omega_0 \times [\, 0,T \,])$ and

$v (r,t)$; $T (r,t)$; $c (r,t) \in C^{2+\alpha,\, 1+\frac{\alpha}{2}} (\Omega_0 \times [\, 0,T \,])$ and

$v (r,t) \in L$.

* *

The proof of this theorem is obtained by the method of extension of the local solution. The possibility of continuing the solution above the whole interval $[\, 0,T \,]$, is guaranteed by *a priori* global estimates on $\Omega_0 \times [\, 0,T \,]$:

$0 < C_1 \leq \rho (r,t) \leq C_2$

$\rho (r,t)$ is Hölder continuous with exponent α

$\rho (r,t) \in W^1_p (\Omega) \qquad \forall \ t \in [\, 0,T \,]$

$v (r,t) \leq C$

$0 < C_3 \leq T (r,t) \leq C_4$

$0 < C_5 \leq c (r,t) \leq C_6$

$v (r,t)$; $T (r,t)$; $c (r,t) \in W^1_p (\Omega) \qquad \forall \ t \in [\, 0,T \,]$

By means of these estimates we deduce that ρ , ρ_r , v , v_r , T , T_r , c , c_r are uniformly bounded in $\Omega_0 \times [\, 0,T \,]$.

This makes it possible to extend the local solution. The existence of the global classical solution for the cylindrical symmetry is then guaranteed.

We have also considered the case of spherical symmetry and we have obtained a system of equations, analogous to (S) . Then by the same methods used for the cylindrical symmetry, one can prove the global existence as well as in the spherical symmetry case .

MULTIMESHED MOVING FINITE ELEMENTS FOR THE CALCULATION OF FLAME SPEEDS.

J.NOEL and D.JEANDEL

Laboratoire de Mécanique des Fluides et d'Acoustique - Ecole Centrale de Lyon - BP 163 - 69131 ECULLY Cedex - France

Summary:

The numerical simulation of unsteady phenomena in fluid mechanics is often hindered by the difficulty in pinpointing, in the computation zone, the areas where in the system of equations the variables ' gradients are steep.

If the position of these areas is known beforehand, applying the finite element technique will usually reduce the cost of computation by limiting the fine meshing of the zone to those areas only. But if the meshing is fixed in time, a fairly large zone must be meshed to no purpose, which is wasteful when an unsteady front is being studied.

Gelinas, Doss, and Miller /1/ introduced the moving finite element (MFE) technique, thus removing this difficulty. The value of the variables at each meshpoint can be determined by this method, and the position of computation points in relation to a numerical error can be optimised. It must be pointed out that, with this method the positions of the mesh nodes are unknown as well as the variables of the system of equations. Furthermore, remeshing is not periodic, for the mesh adapts automatically at each step in time.

This technique was improved by Wathen and Baines /2/ who used meshing by equation in the case of a simple, one-dimensional wave propagating system. This approach has been adopted here and systematically tested in the case of a one-dimensional Dwyer-Sanders flame model. We tested the speed of the flame displacement, which was compared to values obtained by others using different methods and also to those obtained by asymptotic analysis.

C.-M. Brauner and C. Schmidt-Lainé (eds.), Mathematical Modeling in Combustion and Related Topics, 519–527.

ELEMENTS FINIS AUTOADAPTATIFS A MULTIMAILLAGES POUR LE CALCUL DE VITESSES DE FLAMME

Jean NOEL et Denis JEANDEL

Laboratoire de Mécanique des Fluides et d'Acoustique - Ecole Centrale de Lyon
B.P. 163 - 69131 ECULLY Cédex

1. INTRODUCTION

La simulation numérique de phénomènes instationnaires en Mécanique des Fluides se heurte souvent au problème de la localisation dans le domaine de calcul des zones où les variables du système d'équations présentent de forts gradients.

Si l'emplacement de ces zones est connu a priori, la technique des éléments finis permet le plus souvent de réduire le coût de calcul en limitant le maillage fin du domaine à ces zones. Mais, si le maîllage est fixe au cours du temps, on est conduit à mailler inutilement une zone assez large, ce qui constitue un handicap dans le cas de l'étude de front instationnaire.

Gélinas, Doss, et Miller |1| ont introduit la technique des éléments finis auto-adaptatifs (Moving Finite Element, ou MFE), qui lève cette difficulté. Cette méthode permet de déterminer la valeur des variables en chaque point du maillage et d'optimiser la position des points de calcul en fonction d'une erreur numérique. Il faut bien noter que pour cette méthode, les positions des nœuds du maillage sont des inconnues du problème au même titre que les variables du système d'équation. D'autre-part, il ne s'agit pas d'un remaillage périodique, mais bien d'une auto-adaptation du maillage à chaque pas en temps.

Une amélioration de cette technique a été apportée par Wathen et Baines |2| qui ont utilisé un maillage par équation dans le cas d'un système simple de propagation d'onde unidimensionnelle.

Cette approche a été suivie ici et systématiquement testée sur le cas du modèle de flamme unidimensionnelle de Dwyer-Sanders. Les tests ont porté sur la vitesse de déplacement du front, qui a été comparée aux valeurs obtenues par différents auteurs à l'aide d'autres méthodes, ainsi qu'à celles que permet d'obtenir l'analyse asymptotique.

2. ELEMENTS FINIS AUTO-ADAPTATIFS MULTIMAILLAGE (ELEFAM).

2. 1. Principe

Soit le système d'équations portant sur la variable dépendante $\vec{u}$, dans le cas unidimensionnel instationnaire, le problème s'écrit sous la forme :

$$\frac{\partial \vec{u}}{\partial t} = \vec{\mathcal{L}}(\vec{u}) \qquad , \qquad \vec{u} = (u^k)_{k=1,P}$$

$$\text{où} \quad \begin{cases} u^k \; : \; \Omega \times \mathbb{R}^+ \to \mathbb{R} \\ \qquad (x,t) \to u^k(x,t) \\ u^k(x,0) = u^k{}_0(x) \\ \mathcal{C}^k(u^r) = \mathcal{C}^k \end{cases}$$

Où $\mathcal{C}^k$ définit les conditions aux limites.

En s'inspirant des travaux de MILLER et al., l'approximation $\tilde{\vec{u}}$ de $\vec{u}$ est recherchée sous la forme générale :

$$\tilde{\vec{u}} = \tilde{\vec{u}} \left(u^k{}_i(t) \ , \ s^k{}_i(t) \ , \ x \right) \quad \text{sur } \Omega$$

i : 1, N_k où N_k représente le nombre de nœuds du maillage associé à la $K^{ième}$ équation.

La valeur u_i^k (t) est l'approximation $\tilde{u}^k$ au nœud i de coordonnée $s_i^k(t)$. La dérivée temporelle de $\tilde{u}$ fait alors apparaître les familles de fonctions suivantes :

$$\alpha^k{}_i \ (x, s^k{}_j) = \frac{\partial \tilde{u}^k}{\partial u_i^k} \qquad i = 1, N_k$$

$$\beta^k{}_i \ (x, s^k{}_j, u^k{}_j) = \frac{\partial \tilde{u}^k}{\partial s_i^k} \qquad k = 1, p$$

$$\text{et on a : } \dot{\tilde{u}}^k = \sum_{i=1}^{N_k} (\alpha^k{}_i . \dot{u}^k{}_i + \beta^k{}_i . \dot{s}^k{}_i)$$

La solution est obtenue en minimisant la fonctionnelle $\mathfrak{R}$

$$\mathfrak{R} \ (\dot{u}^k{}_i, u^k{}_i, \dot{s}^k{}_i, s^k{}_i) = \| \dot{\tilde{\vec{u}}} - \vec{\mathcal{L}}(\tilde{\vec{u}}) \|^2_{\mathfrak{P}^2} + \mathfrak{P} \ (\dot{s}^k{}_m, s^k{}_m)$$

relativement aux inconnues indépendantes $\dot{u}_i^k$ et $\dot{s}_i^k$. La pénalisation $\mathfrak{P}$ maintient libre les familles

$$(\ \{\alpha^k{}_i, \beta^k{}_i\} \ , \ K = 1, p.)$$

La minimisation signifie :

$$\frac{\partial \mathfrak{R}}{\partial \dot{u}_i^k} = 0 \quad , \quad \frac{\partial \mathfrak{R}}{\partial \dot{s}_i^k} = 0$$

et conduit aux systèmes

$$\vec{\vec{A}}^k . \dot{\vec{Y}}^k = \vec{B}^k \ , \ \vec{Y}^k = (u^k{}_i, s^k{}_i) \ , \ k = 1, p$$

Les approximations $\tilde{u}^k$ seront prises dans l'espace des fonctions linéaires par morceaux.

Les systèmes différentiels sont résolus par une méthode Runge-Kutta explicite d'ordre 3, controlée par une méthode d'ordre 4 (RK 34, voir |3|).

Les systèmes linéaires qui en découlent sont résolus par la méthode de Gauss.

Dans le cas de l'utilisation d'un seul maillage, on obtient un seul gros système linéaire à résoudre. Si on utilise un maillage par équation, on est amené à résoudre p systèmes indépendants plus petits. Le coût de résolution est moindre, mais l'utilisation du multimaillage nécessite le calcul d'intégrales faisant intervenir des fonctions de formes n'ayant pas les mêmes supports, ce qui augmente le temps-calcul.

Le principal avantage du multimaillage réside dans la possibilité de traiter chaque équation du système de façon indépendante, en ce qui concerne le nombre de nœuds, la position des zones de maillage fin, la pénalisation et la technique de résolution des systèmes.

2. 2. Pénalisation - surdiffusion

La pénalisation, introduite sous la forme d'un terme supplémentaire dans la fonctionnelle $\mathcal{R}$, joue un rôle essentiel dans le cas où une des matrices A^k devient singulière.

Miller et al. ont choisi pour $\mathcal{P}$:

$$\mathcal{P}(\dot{s}^k_i , s^k_i) = \sum_{k=1}^{p} \sum_{j=2}^{N_k} (\varepsilon^k_j . \Delta\dot{s}^k_j + S^k_j)^2$$

$$\text{avec} \quad \begin{cases} \varepsilon^k_j = \dfrac{K^k_2}{(\Delta s^k_j - K^k_1)^n} \\ \\ S^k_j = \dfrac{K^k}{(\Delta s^k_j - K^k_1)^n} \end{cases} \quad \text{et} \quad \Delta s^k_j = s^k_j - s^k_{j-1}$$

En fait, la pénalisation sert à fixer la position d'un nœud en cas d'indétermination. La méthode la plus naturelle pour y parvenir consiste à maintenir au cours d'un pas de temps, la position relative des nœuds concernés.

Pour cela, il faut prendre :

$$K_1 = 0 \quad , \quad K_3 = 0 \quad , \quad K_4 = 0 \quad , \quad n = \tfrac{1}{2}$$

Il faut noter qu'une trop grande pénalisation fige le maillage et réintroduit les problèmes liés aux méthodes à nœuds fixes. On prendra donc K_2 le plus petit possible.

D'autre-part, dans les problèmes comportant un opérateur de type diffusion et une viscosité u, DJOMEHRI |4| a montré qu'une augmentation legère de la viscosité dans les équations régissant le mouvement des nœuds permet d'éviter que le front absorbe tous les nœuds et que le système dégénère. On prendra ici :

$$u^k_{i\ \text{noeud}} = u^k_{i\ \text{composante}} . (1 + \min(E^k_{max}, K^k_E . |m^k_{i+1} + m^k_i|))$$

m_i^k représente la pente de la courbe u^k sur $|s^k_{i-1}, s^k_i|$, E^k_{max} et $K_E{}^k$ sont des constantes à choisir au mieux.

3. Problème traité

3. 1. Système

Le système suivant d'équations a été proposé par DWYER et SANDERS pour simuler de façon simple la propagation d'une flamme entretenue par une réaction exothermique :

$$\left\{\begin{aligned} \frac{\partial T}{\partial t} &= \frac{\partial^2 T}{\partial x^2} + A.\rho.e^{-\beta/T} \\ \\ \frac{\partial \rho}{\partial t} &= \frac{\partial^2 \rho}{\partial x^2} - A.\rho.e^{-\beta/T} \end{aligned}\right. \qquad (S)$$

où ρ représente la fraction massique du réactif et T la température.

On prendra des conditions aux limites de Neumann homogènes sur ρ et T, et les conditions initiales indiquées sur la figure 1.

$T_i = 0.2 \quad T_f = 1.2$

$\rho_i = 1 \quad \rho_f = 0$

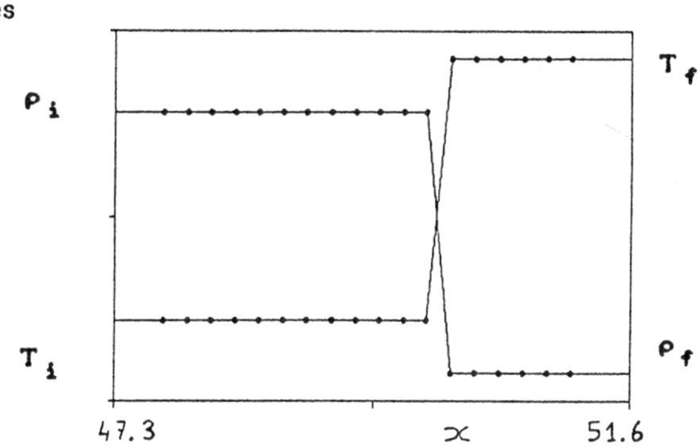

Fig. 1

3. 2. Analyse asymptotique

La vitesse de déplacement V_{OD} du front du système (S) est prédite pour de grandes valeurs du nombre de Zeldovitch Ze (voir |4|, |5|).

On trouve :

$$V^2_{OD}(A,Ze) = \frac{A.e^{-\frac{Ze}{\gamma}}}{\frac{Ze^2}{2}} . \left(\frac{1}{1 + \frac{2}{Ze}.(3\gamma+1-I)} \right)$$

avec $I = 1.344$, $\gamma = \dfrac{T_f - T_i}{T_f}$, $Z_e = \beta.\gamma^2$

Ze représente l'énergie d'activation réduite.

Si dans (S) on prend A = A (Ze) tel que

$$A(Ze) = V^2{}_A.e^{\frac{Ze}{\gamma}}.\left(\frac{Ze^2}{2}\right).\left(1 + \frac{2}{Ze}.(3\gamma+1-I)\right)$$

$$\text{et} \quad V^2{}_A = V^2{}_{OD}\left(\beta = 4 \ , \ \gamma = \frac{5}{6} \ , \ A = 8.8 \ 10^{+5} \right)$$

Alors la vitesse de flamme déterminée numériquement doit converger vers V_A pour de grandes valeurs du nombre de Zeldovitch.

4. RESULTATS

Les paramètres numériques sont respectivement :

20 noeuds par maillage , $K_2 = 10^{-7}$, $E_{max} = 2\ 10^{-2}$, $K_e = 10^{-3}$

Le calcul numérique de la vitesse du front V (t) est donné par la vitesse du point de température moyenne $Tm = \frac{1}{2}(T_i + T_\ell)$.

La force générale des fronts au cours du temps est donnée sur la figure 2 pour

A = 35.2 10+5

β = 4

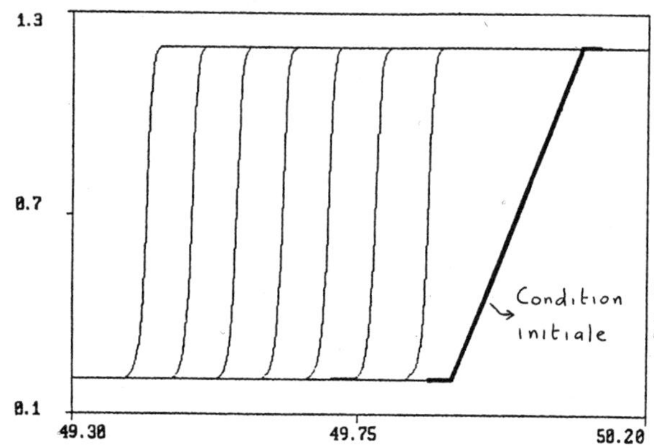

Fig. 2

Il faut noter que les noeuds se regroupent très rapidement dans la zone de fort gradient. Ensuite, le maillage se déplace en bloc.

La figure 3 donne le profil d'un front stabilisé pour :

A = 35.2 10+5

β = 4.

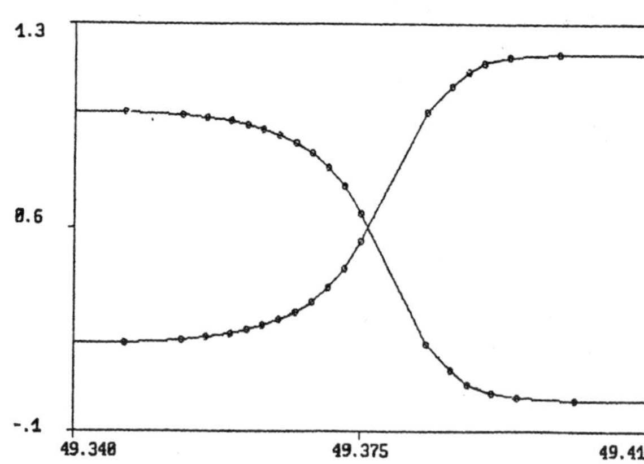

Fig. 3

On remarque que les noeuds respectifs des deux maillages n'adoptent pas des positions identiques.

Les figures 4 et 5 représentent les évolutions de la vitesse au cours du temps, pour deux valeurs de A.

A = 8.8 10+5

B = 4

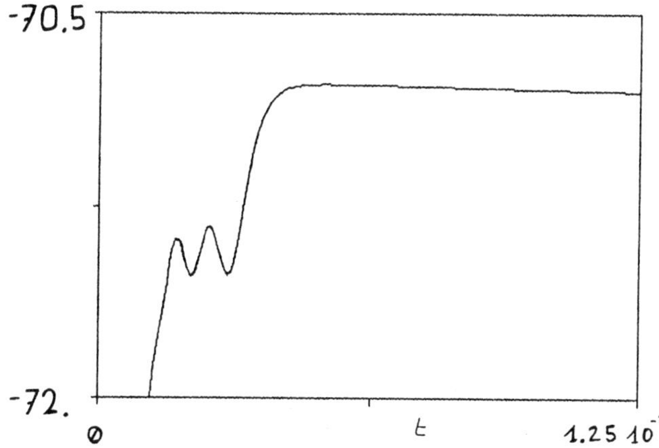

Fig. 4

A = 35.2 10+5

B = 4

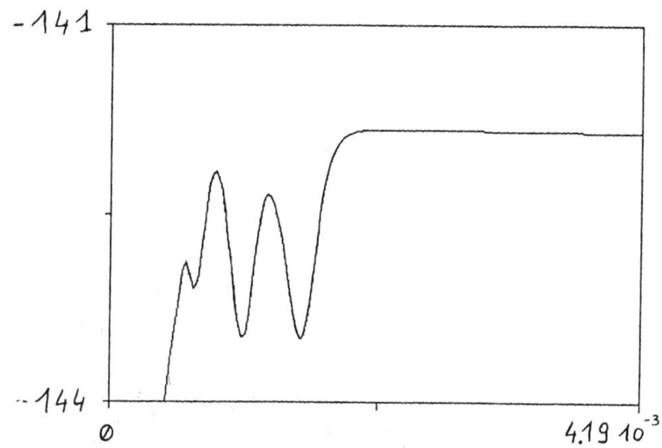

Fig. 5

Il faut noter que les oscillations de vitesse sont causées par un réarrangement des nœuds dans le front.

On représente sur la figure 6 le paramètre $r(t)$

$Log_{10}(\ |r(t)|\ )$

$$r(t) = \frac{V(t)-V_N}{V_N}$$

A = 35.2 10+5

V_N = 141.85

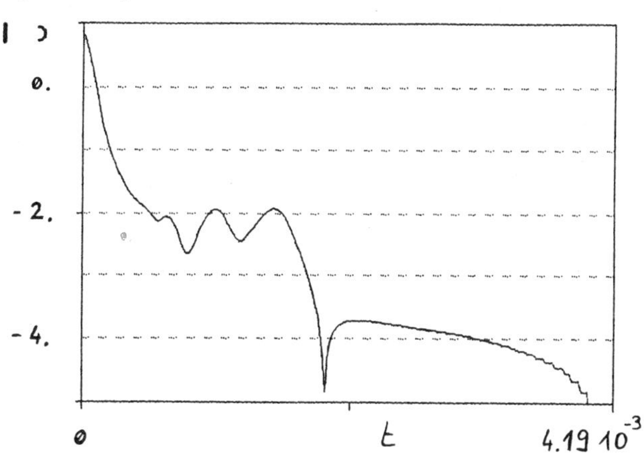

Fig. 6

La figure 7 donne diverses valeurs de Vn obtenues dans la littérature et par la présente méthode :

		OTEY et DWYER	MFE	ELEFAM
A_1	V_N	70.5 / 70.8	/	70.8
	N	61 / 101	/	20
A_1	V_N	140.6 / 141.4	143.7	141.8
	N	81 / 181	11 / 51	20
A_1	V_N	474 / 476	484	483
	N	500 / 600	11 / 51	26

fig. 7 : $A_1 = 8.8\ 10^5$, $A_2 = 35.2\ 10^5$, $A_3 = 400\ 10^5$

N représente le nombre de points de calcul. Dans le cas de la méthode ELEFAM, il s'agit du nombre de nœuds par maillage.

Les vitesses stabilisées V_N données par OTEY et DWYER (Voir |7|) ont été déterminées par différentes méthodes : on donne ici la plage de leurs résultats.

Il faut noter que les essais réalisés par MILLER et al. ont été obtenus par un solveur en temps implicite et à l'aide d'une pénalisation plus forte ($K_2 = 10^{-3}$).

La figure 8 présente les résultats obtenus pour différentes valeurs du nombre de Zeldovitch et pour différents coefficients A (Ze).

On compare les différentes vitesses stabilisées V_N à V_A, valeur prédite par l'analyse asymptotique.

r(Ze)

$$r(Ze) = \frac{V_N(Ze) - V_A}{V_A}$$

$V_A = 56.46$

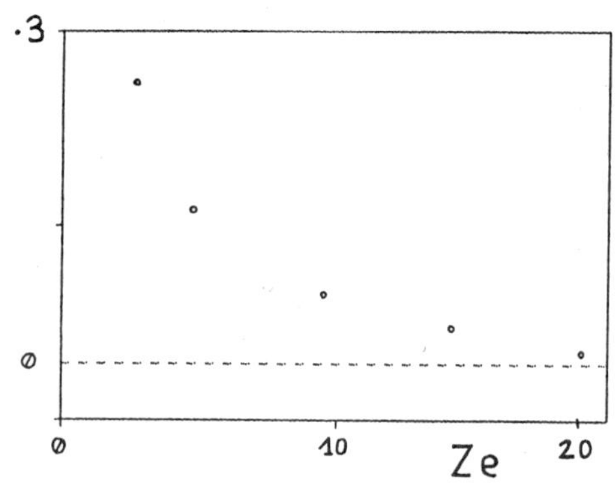

Fig. 8

5. CONCLUSIONS

On peut conclure que la méthode proposée a atteint ses objectifs.

Il faut noter que les résultats présentés ont été réalisés avec des paramètres de calcul (E_{max}, k_E, k_2) qui ont été réglés pour obtenir une excellente précision. Néanmoins, on obtient également de très bons résultats (à 1 % ou 2 %) en prenant ces paramètres dans une plage de valeurs assez large.

BIBLIOGRAPHIE

1. R.J. GELINAS, S.K. DOSS, K. MILLER : "The moving finite element method : application to general PDE's with multiple large gradients." Journal of Computer Physics, vol. 40. 1981.

2. A.J. WATHEN, M.J. BAINES :"On the structure of the moving finite element equations". Journal of Numerical Analysis, vol 5 - 1985.

3. M. CROUZEIX - A.L. MIGNOT :"Analyse numérique des équations différentielles", Masson 1984.

4. J. DJOMEHRI :"Moving finite element solution of systems of PDE' s in one dimension". Ph D Thesis, Berkeley, 1983.

5. W.B. BUSH, F.E. FENDELL :"Asymptotic analysis of laminar flame propagation for general Lewis numbers". Combustion Science and Technology, vol. 1, 1970.

6. B. LARROUTUROU : "Utilisation de maillage adaptatifs pour la simulation de flammes monodimensionnelles instationnaires". Lecture notes in physics 241, Springer Verlag, 1985.

7. G.R. OTEY, H.A. DWYER : "Numerical study of the interaction of fast chemistry and diffusion", AIAA Journal, vol. 17, 1975.

THIN FLAMES AND INTERFACES

R. PRUD'HOMME , F. BAILLOT

LABORATOIRE D AEROTHERMIQUE DU CNRS

92190 - MEUDON . FRANCE.

KEYWORDS, SUMMARY :

Combustion / Premixed flames / Interface equations / Asymptotic / Expansions / Wrinkeled flames / Balance equations /

For several years, the complex behaviour of flames has been inciting to the use of the asymptotic expansions method, which has led to obtain many interesting results, in spite of some rather limitating, simplifying assumptions. The role of flame front strech and curvature has been clearly shown and several problems have been solved.

The present work is based on the idea that, thanks to the important results obtained these last years, it should be possible to reconsider the theory of flames treating the flames as interfaces.

These interfaces cannot be regarded as simple discontinuities only, but have to obey some more complicated balance-equations, which depend on surface parameters, line flux and production rate per unit area. But, contrarily to the usual treating of interfaces, it is not enough here to impose that the Gibbs relations be satisfied by the thermodynamic surface parameters.

The method used has been successfully applied by the authors to the thin premixed flames with an important activation energy. Balance-equations have been established first for the case of a constant density and more recently in the case of a variable density inside the flame thickness.

The form of constitutive relations can be deduced from dimensional considerations, but their exact determination implies the use either of the results of the asymptotic analysis or of the experimental measurements.

In the following, we refer in particular to Clavin's and Joulin's results, concerning the wrinkled flames.

C.-M. Brauner and C. Schmidt-Lainé (eds.), Mathematical Modeling in Combustion and Related Topics, 529–538.

FLAMMES MINCES ET INTERFACES

R. PRUD'HOMME et F. BAILLOT

Laboratoire d'Aérothermique du C.N.R.S., 92190 - Meudon

De nombreuses recherches ont été consacrées à l'étude des interfaces, c'est-à-dire des milieux matériels qui, à l'échelle macroscopique peuvent être assimilés à des surfaces [1,2,3,4,5].

Ces surfaces matérielles sont généralement des surfaces de séparation entre phases ou des membranes pour lesquelles on établit les équations du bilan des espèces, de la masse, de la quantité de mouvement et de l'énergie. Les paramètres interfaciaux vérifient une loi de Gibbs [6,7,8] et les lois de comportement sont déduites de la thermodynamique irréversible et de la cinétique chimique de surface.

L'écriture des équations aux discontinuités dans les fluides procède d'une démarche analogue. Cependant on se borne la plupart du temps à écrire que les flux normaux relatifs de certaines grandeurs sont conservatifs et on ne cherche pas à établir de lois de comportement particulières. Cette méthode conduit à des résultats satisfaisants pour les ondes de choc et pour les détonations de Chapman-Jouguet. Elle ne donne que des résultats partiels, mais insuffisants dans le cas des ondes de déflagration [9].

En effet, dans le cas des flammes de prémélange, la théorie de Rankine-Hugoniot ne fournit pas le débit et une étude de ce qui se passe dans l'épaisseur de la flamme est alors nécessaire [10]. De plus, lorsqu'il ne s'agit plus de flammes droites, adiabatiques et stationnaires, les équations de saut ne fournissent qu'une approximation grossière de la réalité car les effets du taux d'étirement et de la courbure de la flamme sur son comportement sont négligés [11].

La complexité du comportement des flammes a motivé l'usage, depuis plusieurs années, de la méthode des développements asymptotiques [12,13] qui a permis d'obtenir de nombreux résultats intéressants au prix cependant d'hypothèses simplificatrices assez contraignantes. Le rôle du taux d'étirement et de la courbure du front de flamme ont été mis en évidence et plusieurs questions ont pu être élucidées [14].

La démarche du présent travail part de l'idée qu'il doit être possible, compte-tenu de l'importance des résultats acquis ces dernières années, de reconsidérer la théorie des flammes en concevant celles-ci comme des interfaces.

Ces interfaces ne peuvent être de simples discontinuités, mais doivent obéir à des équations de bilan plus complexes, faisant intervenir des grandeurs surfaciques, des flux linéiques et des taux de production par unité d'aire. Mais contrairement aux démarches développées pour les interfaces habituelles, il n'est pas opportun cette fois de s'astreindre au respect de la relation de Gibbs entre les grandeurs thermodynamiques surfaciques.

Cette démarche a été apppliquée avec succès par les auteurs aux flammes minces à grande énergie d'activation. Les équations du bilan ont été établies d'abord dans l'hypothèse de la densité constante [15] et plus récemment en admettant une masse volumique variable à l'intérieur de la flamme [16]. La forme de lois de comportement peut être déduite de considérations dimensionnelles, mais leur détermination exacte implique, soit l'utilisation des résultats de l'analyse asymptotique, soit des mesures expérimentales. Dans ce qui suit, nous nous référons en particulier aux résultats de Clavin et Joulin [17] portant sur les flammes plissées.

1. BILANS VOLUMIQUES ET BILANS INTERFACIAUX

L'équation du bilan d'une propriété extensive F d'un mélange s'écrit habituellement :

$$\rho \, df/dt + \vec{\nabla}.\vec{\mathcal{J}}_F = \dot{W}_F \tag{1}$$

où ρ est la masse volumique du mélange, f la grandeur F par unité de masse, $\vec{\mathcal{J}}_F$ la densité de flux surfacique de F et $\dot{W}_F$ le taux de production de cette grandeur par unité de volume.

Si l'on introduit un mouvement fictif de vitesse $\vec{V}$, la fonction $\vec{V}(\vec{x},t)$ ayant les propriétés requises de continuité et de dérivabilité, on peut écrire l'équation (1) sous la forme :

$$dv(\rho f)/dt + \rho f \, \vec{\nabla}.\vec{V} + \; .[\vec{\mathcal{J}}_F + \rho f(\vec{v}-\vec{V})] = \dot{W}_F \tag{2}$$

où d_V/dt est la dérivée particulaire $(\partial/\partial t + \vec{V}.\vec{\nabla})$ associée à la vitesse $\vec{V}$ et où $\vec{v}$ est la vitesse barycentrique du fluide. Si $\vec{V}$ caractérise le mouvement de la flamme, l'équation (2) représente alors le bilan de F dans un référentiel lié à celle-ci. La relation de bilan interfacial s'obtient par intégration des deux membres de cette équation transversalement à la flamme. On introduit un système de coordonnées curvilignes orthogonales lié au front de flamme et se déformant avec lui au cours de son mouvement :

$$\begin{cases} x = x(x_1, x_2, x_3, t) \\ y = y(x_1, x_2, x_3, t) \\ z = z(x_1, x_2, x_3, t) \end{cases} \tag{3}$$

La surface du front de flamme correspond à x_3=O, chaque surface (S) d'équation x_3=Cte étant parallèle au front de flamme et les distances entre les surfaces (S) restant invariables au cours du temps. Il en résulte en particulier qu'une normale $\vec{N}$ quelconque au front de flamme est normale à toutes les surfaces (S) de la famille (figure1).

On définit alors les opérateurs de dérivation $\vec{\nabla}_{/\!/}$ et $\vec{\nabla}_{\perp}$ tels que :

$$\begin{cases} \vec{\nabla}_{/\!/} = (\vec{1} - \vec{N} \otimes \vec{N}).\vec{\nabla} \\ \vec{\nabla}_{\perp} = \vec{N} \, \vec{N}.\vec{\nabla} = \vec{N} \; \partial/\partial N, \end{cases} \tag{4}$$

La courbure normale moyenne est en tout point de la flamme :

$$\frac{1}{R_1} + \frac{1}{R_2} = \vec{\nabla}_{/\!/}.\vec{N} = \vec{\nabla}.\vec{N}, \tag{5}$$

son signe dépend de l'orientation donnée arbitrairement à la normale unitaire $\vec{N}$. Le taux d'étirement du front de flamme s'écrit :

$$\frac{1}{d\Sigma} \frac{d_V(d\Sigma)}{dt} = \vec{\nabla}_{/\!/}.\vec{V} \tag{6}$$

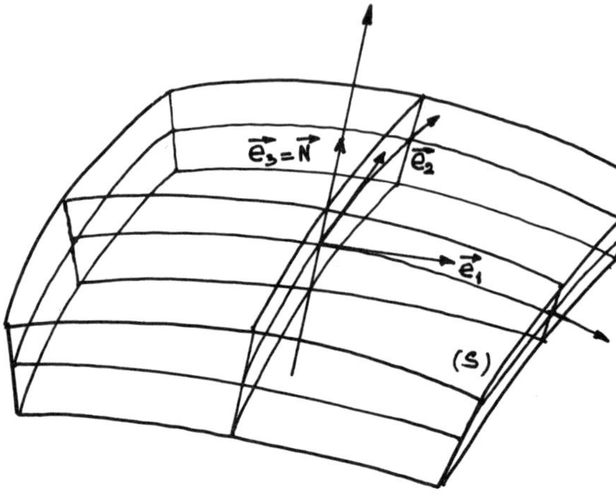

FIGURE 1. Coordonnées curvilignes orthogonales = représentation des surfaces S parallèles et des vecteurs de base.

Soient Λ, d l'échelle de longueur hydrodynamique et une échelle caractérisant l'épaisseur de la flamme, on pose :

(7) $\varepsilon = d/\Lambda \ll 1, \qquad N = \varepsilon\, n$

Généralement les intégrales des termes de l'équation (2) ne convergent pas entre n=-∞ et n=O (Figure 2).

Pour résoudre cette difficulté, on introduit la fonction $f_0(\vec{x},t)$ correspondant aux valeurs de la grandeur massique $f(\vec{x},t)$ dans les gaz frais. La fonction f_0 varie peu à l'échelle de l'épaisseur de la flamme et on considère alors les quantités :

$$(8) \quad \begin{cases} \rho_F = \rho(f - fo) \\ \vec{J}_{VF} = \vec{\mathcal{J}}_F - \vec{\mathcal{J}}_{F0} + \rho_F\,(\vec{v} - \vec{V}) \\ \dot{W}'_F = \dot{W}_F - \dot{W}_{F0} - (\rho - \rho_0)\,\partial f_0/\partial t - (\rho\vec{v} - \rho_0\vec{v}_0).\ f_0 \end{cases}$$

L'équation du bilan (2) prend alors la forme suivante :

$$(9) \quad \frac{d_v \rho_F}{dt} + \rho_F\, \vec{\nabla}_{/\!/}.\vec{V} + \vec{\nabla}_{/\!/}.\vec{J}_{VF} + \frac{\partial \vec{J}_{VF\perp}}{\partial N} = \dot{W}'_F$$

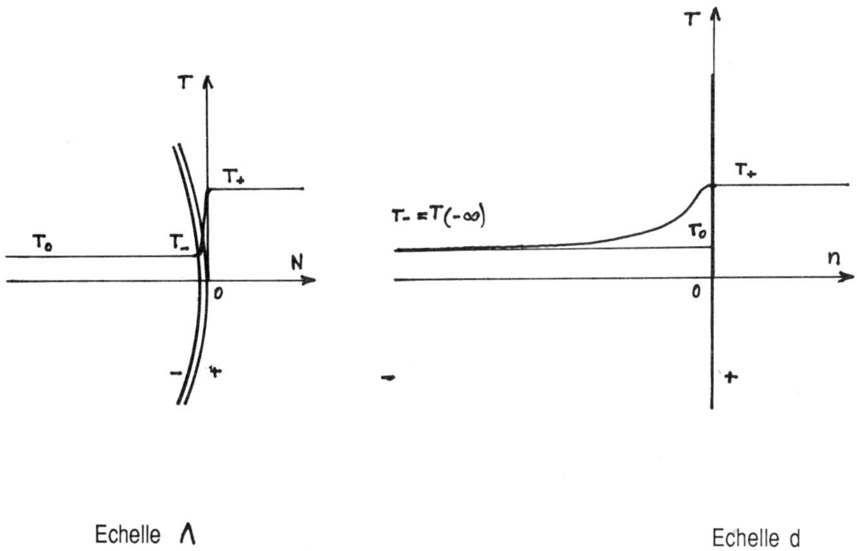

FIGURE 2. Représentation de l'interface aux échelles Λ et d respectivement et profils de température.

La vitesse $\vec{V}$ étant conservative suivant N ainsi que l'opérateur de dérivation $\vec{\nabla}_{//}$ et la courbure, on obtient après intégration suivant N :

$$\frac{d_v \hat{\rho}_F}{dt} + \hat{\rho}_F \vec{\nabla}_{//} \cdot \vec{V} + \vec{\nabla}_{//} \cdot \hat{\vec{J}}_{VF} + [\![J_{VF}]\!]_{\perp} = \hat{W}_F \tag{10}$$

avec

$$[\![J_{VF\perp}]\!] = (J_{VF\perp})_+ - (J_{VF\perp})_-$$

$$\hat{\rho}_F = \int_{N_-}^{0} \rho_F \, dN, \qquad \hat{\vec{J}}_{VF} = \int_{N_-}^{0} \vec{J}_{VF} \, dN, \qquad \hat{W}_F = \int_{N_-}^{0} \dot{W}'_F \, dN \tag{11}$$

Ces grandeurs de surface constituent, pour les bilans de la forme (10) autant d'inconnues. Pour fermer le système ainsi obtenu, il est nécessaire d'établir des lois de comportement. Celles-ci peuvent, bien entendu, être obtenues en calculant directement les intégrales (11) à partir des profils obtenus par la théorie des développements asymptotiques.

Cependant, des considérations dimensionnelles conduisent directement à des relations intéressantes. Enfin, les résultats d'expérience doivent permettre de fournir les valeurs des coefficients introduits par l'analyse dimensionnelle.

Notons que l'équation (10) reste valable dans d'autres cas de couches fluides que les flammes.

2. FLAMMES DE PREMELANGE A HAUTE ENERGIE D'ACTIVATION

Dans le cas de flammes minces étudiées par différents auteurs, on suppose très grand devant l'unité le rapport ß de la température d'activation Ta de la réaction : A $\longrightarrow$ B (supposée unique) à la température Tb des gaz brûlés de la flamme plane adiabatique. L'analyse asymptotique à échelles multiples porte alors sur un ou deux petits paramètres suivant que l'on s'impose ε égal à, ou différent de 1/ß. (Notons que l'on désigne habituellement par ß la quantité $Ta(Tb-Ta)/Tb^2$ au lieu de Ta/Tb ; cela ne change en rien les considérations qui suivent). Rapportons alors les paramètres du problème aux grandeurs de référence suivantes : longueur Λ , vitesse ub de la flamme plane adiabatique, température Tb, temps Λ /ub, masse volumique des gaz frais ρ_0.

Les conditions dans les gaz frais que l'on suppose non perturbées par la flamme sont alors :

$$T=To \; ; \qquad Y=Y_A=1 \; ; \qquad \vec{V}=\vec{V}_- \; ; \qquad \rho=1$$

Dans les gaz brûlés on suppose que Y=O (combustion complète) et $T=T_+$ (inconnue du problème).

En appliquant alors la méthode indiquée au § 1, on trouve, dans tous les cas où $\vec{\nabla}.\vec{N}$ et $\vec{\nabla}_{/\!/}.\vec{V}$ sont O(1), les équations suivantes dans le référentiel lié à la flamme (temps τ) :

$$\frac{\partial \hat{\rho}_v}{\partial \tau} + \hat{\rho}_v \vec{\nabla}_{/\!/}.\vec{V} + [\![\rho_v u_\perp]\!] = [\![u_\perp]\!] \tag{12}$$

$$\frac{\partial \hat{\rho}_Y}{\partial \tau} + \hat{\rho}_Y \vec{\nabla}_{/\!/}.\vec{V} + [\![\rho_Y u_\perp]\!] = -e^{\frac{\beta}{2}(T_+-1)} \tag{13}$$

$$\frac{\partial \hat{\rho}_T}{\partial \tau} + \hat{\rho}_T \vec{\nabla}_{/\!/}.\vec{V} + [\![\rho_T u_\perp]\!] = (1-T_0)\, e^{\frac{\beta}{2}(T_+-1)} \tag{14}$$

$$\frac{\partial \vec{\hat{\rho}}_v}{\partial \tau} + \vec{\hat{\rho}}_v \vec{\nabla}_{/\!/}.\vec{V} + [\![p\vec{N} + \vec{\rho}_v u_\perp]\!] = \vec{\hat{W}}_v \tag{15}$$

avec $\vec{u} = \vec{v} - \vec{V}$ et $\vec{v} = \vec{v}_{/\!/-} + \vec{w}$ où $\vec{w}$ est la vitesse d'avancement normale du front de flamme. On supposera ici que $\vec{\hat{W}}_v$ est $O(\varepsilon)$.

Les seconds membres de (13) et (14) résultent de l'intégration du taux de production chimique dans la zone de réaction (voir [12] ou [13] par exemple).

Les quantités $\hat{\rho}_F$ sont toutes nulles à l'ordre zéro en ε, de sorte que l'équation (15) nous indique la conservation de la vitesse $\vec{v}_{/\!/}$ à la traversée normale de l'onde, à l'ordre zéro.

Il reste à déterminer $\hat{\rho}_v$, $\hat{\rho}_Y$ et $\hat{\rho}_T$ en fonction des autres paramètres pour fermer le système.

Revenons aux grandeurs non adimensionnées. On peut présumer que la quantité d'excès surfacique $\hat{\rho}_F$ sera proportionnelle respectivement à ρ_0, à l'épaisseur caractéristique d_F de la flamme associée à la grandeur dissipative F (par exemple d_Y pour la diffusion et $d_T=d$ pour la conduction thermique) :

(16) $$d_F = D_F/\rho u \ ,$$

et au saut $[\![f]\!]$. On obtient ainsi :

(17) $$\hat{\rho}_F = L_F \ \rho_0 [\![f]\!] \ D_F/\rho u$$

où L_F est un coefficient "phénoménologique" sans dimensions.

Avec les grandeurs sans dimensions on obtient donc :

(18) $$\rho_Y = -L_Y \ \frac{\varepsilon}{\mathcal{L}_e \ m}$$

(19) $$\rho_T = L_T \ \frac{(1-T_0)\ \varepsilon}{m}$$

où m est le débit ρu de combustion.

La valeur de $\hat{\rho}_{\mathcal{V}}$ sera déduite de $\hat{\rho}_Y$ et de $\hat{\rho}_T$ suivant la loi d'état du fluide. En effet, pour un fluide de densité constante, on a simplement $\hat{\rho}_{\mathcal{V}}=0$ alors que pour un fluide vérifiant la loi d'état [17] :

(20) $$\rho \ T = To,$$

on trouve :

(21) $$\hat{\rho}_{\mathcal{V}} = (1/To) \ \hat{\rho}_T.$$

La détermination des coefficients L_Y et L_T fait appel à l'analyse asymptotique.

Les profils de concentration et de température sont toujours les suivants à l'ordre zéro en ε, ce qui est suffisant pour le calcul de $\hat{\rho}_Y$ et $\hat{\rho}_T$:

(22) $$Y^{\circ} = 1 - e^{\mathcal{L}_e m^{\circ} n}$$

(23) $$T^{\circ} = To + (1-To) \ e^{m^{\circ} n}$$

On trouve alors, pour les flammes à densité constante :

(24) $$L_Y = L_T = 1$$

et pour les flammes vérifiant la loi d'état (20) :

(25) $$L_T = \frac{T_0}{1-T_0} \operatorname{Log} \frac{1}{T_0}$$

(26) $$L_Y = \frac{\mathcal{L}_e T_0}{1-T_0} \left[\operatorname{Log} \frac{1}{T_0} - (\mathcal{L}_e - 1)\left(\frac{T_0}{1-T_0}\right)^{\mathcal{L}_e - 1} \int_0^{\frac{1-T_0}{T_0}} \frac{\operatorname{Log}(1+\xi)}{\xi^{2-\mathcal{L}_e}} \, d\xi \right]$$

3. RESOLUTION

La résolution des équations (12) à (15) compte-tenu des lois de comportement (18) (19) avec les valeurs de L_Y et L_T données par (24) ou (25) et (26) suivant les cas, est exposée en détail dans [18].

Dans l'analyse à un seul petit paramètre $\varepsilon = 1/\beta$ avec l'hypothèse de la densité constante, on écrit :

(27) $$e^{\frac{\beta}{2}(T_+ - 1)} = e^{\frac{T_+^1}{2}} + \varepsilon \, \hat{w}^1 + O(\varepsilon^2)$$

on obtient alors aisément :

(28) $$u_\perp^o = e^{\frac{T_+^1}{2}}$$

$$\left(1 - \frac{1}{\mathcal{L}_e}\right) \left(\frac{\partial \, 1/u_\perp^o}{\partial \tau} + \frac{1}{u_\perp^o} \vec{\nabla}_{//} \cdot \vec{V}^o \right) + u_\perp^o \, \frac{T_+^1}{1-T_0} = 0$$

Dans l'analyse à deux petits paramètres, on pose :

(29) $$e^{\beta/2(T_+ - 1)} = 1 + \varepsilon \, \beta/2 \, T_+^1 + O(\varepsilon^2)$$

ce qui conduit à :

(30) $$m^\circ = \rho^\circ u^\circ = 1,$$

et à un nombre de Lewis obligatoirement voisin de l'unité :

(31) $$\mathcal{L}_e = 1 + \ell/\beta, \quad \text{avec} \quad \ell = \text{cte} = O(1).$$

La résolution à deux petits paramètres est très rapide et conduit, comme celle à un seul paramètre, à des résultats identiques à ceux de la théorie classique.

On obtient, avec la densité variable :

(32) $m^\circ_+ = m^\circ_- = 1$ (éqs. (12) (13) (14))

(33) $$\frac{\mathrm{Log}\,(1/T_o)}{m^\circ}\ \vec{\nabla}_{/\!/}.\vec{V}^\circ = [\![m^1]\!]$$ (éq. (12))

(34) $$T^1_+ = -\frac{\ell}{\beta} T_o \left(\int_0^{\frac{1-T_o}{T_o}} \frac{\mathrm{Log}\,(1+\xi)}{\xi} d\xi \right) \vec{\nabla}_{/\!/}.\vec{V}^\circ$$ (éqs. (13) (14))

(35) $m^1_+ - To\ m^1_- = (1\text{-}To)(\beta/2)\ T^1{}_+$ (éqs. (12) (33))

ce qui correspond aux résultats de Clavin et Joulin [17].

4. CONCLUSION

Les résultats obtenus prouvent que dans l'hypothèse où le taux d'étirement et la courbure sont O(1) à l'échelle hydrodynamique, les flammes de prémélange vérifient bien des équations de bilan d'interface. Les lois de comportement ont été établies. Les équations ont été testées dans quelques cas classiques. On peut penser qu'elles resteront valables dans d'autres situations.

Lorsque le taux d'étirement $\vec{\nabla}_{/\!/}.\vec{V}$ n'est plus O(1), nous avons montré [15] que la même démarche restait valable à propos de la flamme plane dans un écoulement à point d'arrêt, à densité constante.

Le cas où $\vec{\nabla}.\vec{N}$ n'est plus O(1) n'a pas été examiné. Mais il ne s'agit plus alors d'interfaces lorsque le rayon de courbure est de l'ordre de l'épaisseur d de la flamme. Néanmoins, le cas se présente fréquemment à la lisière ou en certains points particuliers de flammes et il serait intéressant de pouvoir en donner une modélisation simple.

On pourrait penser que les lois de comportement sont difficiles à déterminer et qu'elles sont différentes dans chaque situation, qu'elles risquent même d'évoluer pour une flamme donnée au cours de son mouvement. C'est probable, mais la détermination expérimentale de la courbure, du taux d'étirement et de la vitesse de combustion ainsi que d'autres paramètres doit certainement permettre d'évaluer les coefficients L_F [19].

On peut considérer que les principaux avantages de cette modélisation théorique sont, d'une part, une analyse simplifiée de manière appréciable dans les cas classiques et d'autre part, l'obtention de conditions aux limites bien définies à introduire dans les calculs d'écoulements réels présentant des flammes plissées.

REFERENCES

1. Slattery JC: General balance equation for a phase interface. I&EC Fundamentals, **6,** 1 1967.
2. Landau L, Lifschitz E: Mécanique des fluides. Ed. MIR, Moscou, 1971.
3. Ghez R: A generalised Gibbsian surface. Surface Science, **4**, pp. 125-14O, 1960.
4. Fer F: Thermodynamique macroscopique, 2. Gordon & Breach, 1971.
5. Barrère M, Prud'homme R: Equations fondamentales de l'Aérothermochimie. Masson, 1973.
6. Defay R, Prigogine I: Tension superficielle et adsorption. Ed. Desoer, Liège.
7. Bertrand G, Prud'homme R: Possibilities of surface reaction coupling with transport phenomena. Int. J. Quantum Chemistry, **12**, Suppl. 2, pp. 159-168, 1977.
8. Gatignol R, Seppecher P: Modelisation of fluid-fluid interfaces with material properties. J. Mécanique Théorique Appliquée, N° Spécial, pp. 225-247, 1986.
9. Jaumotte AL: Chocs et ondes de choc. Masson, 1971.
10. Williams FA: Combustion theory, 2nd ed. The Benjamin/Cumming Publ. Cy, Inc., 1985.
11. Klimov AM, Zhur. Prikl. Mekh. Tekhn. Fiz., n° 3, **49**, 1963.
12. Sivashinsky GI: Structure of Bunsen flames. J. Chemical Physics, 62, 2, 1975.
13. Ludford GSS, Scheurer S: Problèmes mathématiques en combustion. Ecoles CEA/INRIA/EDF (7-13 déc. 1982). Edité par l'INRIA.
14. Clavin P: Dynamic behavior of premixed flame fronts in laminar and turbulent flows. Prog. Energy Combustion Sc., **11**, pp. 1-59, 1985.
15. Prud'homme R, Baillot F, Durox D: Modélisation interfaciale des flammes minces". J. Mécanique Théorique Appliquée, N° spécial, pp. 193-233, 1986.
16. Prud'homme R, Baillot F: Equations de bilan d'interfaces pour les flammes de prémélange. C.R.A.S., Paris, avril-mai 1987.
17. Clavin P, Joulin G: Premixed flames in large and high intensity turbulent flow. Le Journal de Physique: Lettres, **44**, 1, janv. 1983.
18. Prud'homme R: Phénomènes d'interface. Rapport R 87-1 du laboratoire d'Aérothermique du CNRS, Meudon, janvier 1987.
19. Baillot F, Durox D, Prud'homme R: Influence d'une perturbation sinusoïdale sur une flamme prémélangée de type Bunsen. Rapport R 87-2 du laboratoire d'Aérothermique du CNRS, Meudon, janvier 1987.

SECOND ORDER REMESHING METHOD IN 2D-LAGRANGIAN FLUID DYNAMICS

F. RABIE

INSTITUT FRANCAIS DU PETROLE

A fluid dynamics computer code has been conceived at the Institut Français du Pétrole. Its function is to model certain transient phenomena governed by the equations of dynamics of compressible, inviscid fluids occurring in a bounded domain, with two-dimensional or three-dimensional axisymmetric geometry, the boundary conditions being of a Neumann type: u.n = 0.

The conception of this code, named AERO, is situated in the frame of the research conducted at I.F.P. on the knock phenomenon occurring in combustion chambers of engines. The origin of the knock is a combustion phenomenon: auto-ignition, which results in the appearance in the chamber of elevated pressures, and then the propagation of pressure waves with very steep gradients and whose fronts present a nearly discontinuous profile. It is this latter -fluid dynamical- aspect of the knock that the code AERO deals with, but with restrictions, in particular on the geometry of domains of evolution of the phenomena and their dimension.

Considering the fineness of certain transition zones between different pressure levels, mathematical models of this phenomenon are weak solutions, possibly presenting discontinuities, of the equations mentioned above.

Efficient algorithms in the numerical treatment of these equations, in particular for that which concerns discontinuities, are those based on their Lagrangian formulation (namely, that in which intervenes the function which, to every time t, associates to every point in the domain the position at time t of the particle which was located at this point at the initial time). Indeed, this dispenses us from a numerical treatment of transport terms present in an Eulerian formulation. The algorithm applied in the code AERO belongs to these. It is based on a finite element discretization in space and is explicit in time. This code is of a Goad-Wilkins type.

For that which concerns spatial discretization:

- elements are quadrangular
- thermodynamical variables are functions constant in each element (L^2 functions)
- speed-vector components are functions bilinear in each element, defined by an interpolation starting from their values at mesh nodes (H^1 functions).

For that which concerns temporal discretization, the scheme is of a "staggered" type:

C.-M. Brauner and C. Schmidt-Lainé (eds.), Mathematical Modeling in Combustion and Related Topics, 539–549.

- thermodynamical variables are computed "at times t^n"
- the speed-vector is computed "at times $t^{n+\frac{1}{2}}$".

But the Lagrangian character of the basic formulation gives the mesh a mobile structure; at each time step, the run of nodes, that represent mobile fluid particles, is computed as a function of their speed and of the time-step. In general, this leads it, after a more or less short while, to present an "inadmissible" configuration, when the relative movement of the four corners of at least one element has been such that it is no more a quadrangle whose angles are all strictly smaller than 180 degrees. In the case of the algorithm presented above, causes of these distortions are of two much distinct types:

- the first one is linked to the existence of non-physical solutions to the system of discretized equations, made possible by the type of spatial discretization envisaged to the algorithm. Some part of these unrealistic solutions have been eliminated *a priori* thanks to the implementation of a "subgrid" defined by division of each element of the grid in four "subelements" and a computation parallel to the principal computation of thermodynamical variables constant in each subelement. This procedure can delay the instant of an inadmissible distortion
- the second one is of a purely physical type: most phenomena occur with displacements of particles liable to make a quadrangular admissible configuration disappear.

It is therefore necessary, in general, to apply a "remeshing" procedure when such an inadmissible configuration is impending. This remeshing consists of the definition of a more regular grid from which the computations of the Lagrangian algorithm will be continued. Of course, this necessitates a determination of the values to assign to the variables of the problem at the nodes and in the meshes of this new grid, according to their distribution on the Lagrangian grid.

The conservation of a high precision in the treatment of discontinuities, thanks to the choice of a Lagrangian formulation, here meets its counterpart: indeed, the main difficulties of remeshing are linked to the diffusion, in particular in the neibourhood of discontinuities, that its implementation necessarily implies. The quality of the control of this type of diffusion then depends on that of the remeshing method.

The method applied in the remeshing phase we implemented in the code AERO is of a conservative type: essentially, it consists in computing the quantity of certain extensive variables, defined on the Lagrangian grid, which is contained in each element of the "remeshing grid".

In a "first-order" remeshing, this computation corresponds to the integration on the remeshing grid of volumetric functions defined on the Lagrangian grid: density, volumetric internal energy and volumetric momentum. We applied a "second-order" method, which includes a first degree interpolating phase of the volumetric functions, before integration on the remeshing grid of the functions obtained in the sequence of this interpolation. This method was introduced by J.C. Dukowicz. With respect to first-order methods, this one provides a considerable gain in the precision of results, in particular in the neibourhood of discontinuities of solutions. The increase in programming complexity and computing time lies, not only in the computation of the coefficients of interpolation on the Lagrangian grid, but also in the necessity to compute the intersections of the remeshing and Lagrangian grids at each remeshing step (in first-order methods, one can avoid this by using flux-methods).

DETAIL OF THE METHOD

Starting from the equations written in Lagrangian form:

$$\frac{D\rho}{Dt}(x,y) + \rho(x,y,t)\,\nabla.u\,(x,y,t) = 0$$

$$\rho(x,y,t)\,\frac{Du}{Dt}(x,y) + \nabla(p+q)\,(x,y,t) = 0$$

$$\frac{DE_s}{Dt}(x,y) + (p+q)\,(x,y,t)\,\frac{D(\frac{1}{\rho})}{Dt}(x,y) = 0$$

$$p\,(x,y,t) = f\,(\,\rho(x,y,t),\,E_s(x,y,t))\ \text{(equation of state)},$$

where ρ, p, E_s, q and u denote respectively density, pressure, specific internal energy, pseudoviscosity and speed-vector, one obtains an algorithm with the following structure:

1. Lagrangian step.

Conditions at time t^{n-1}:

$$u^{n-\frac{3}{2}} \quad X^{n-1} \quad \rho^{n-1} \quad q^{n-\frac{3}{2}} \quad (E_s)^{n-1} \quad T^{n-1} \quad p^{n-1}$$

(X: position of the nodes,
T: temperature, constant in each cell).

Lagrangian computation:

$$u^{n-\frac{1}{2}} \quad X^{n} \quad \rho^{n} \quad q^{n-\frac{1}{2}} \quad (E_s)^{n} \quad T^{n} \quad p^{n}$$

3x3 system
(resolved by an
iterative loop)

2. Remeshing step.

In general, remeshing is not applied at each step of the Lagrangian algorithm, but only when the Lagrangian grid G^n satisfies a criterion thus established: around each node of the initial grid G^0, one defines a "remeshing zone"; the criterion is satisfied when at least one node of G^n has run beyond the limits of its remeshing zone.

The remeshing grid G' is then defined by:

- the nodes of G^n that have not run out of their remeshing zone
- the nodes of G^0 for which the corresponding nodes of G^n have run out of their remeshing zone (see figure 1).

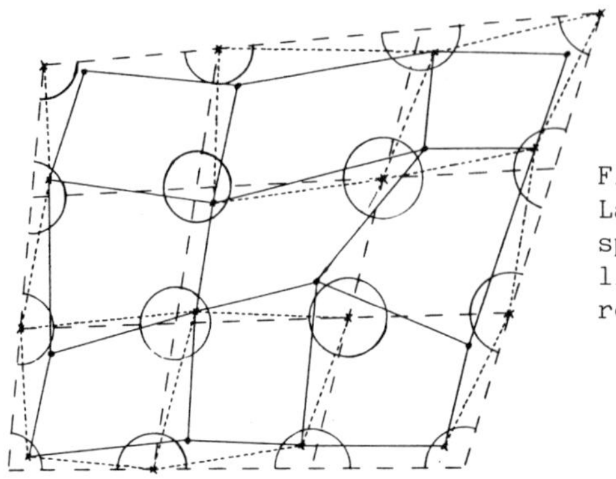

Figure 1: superposition of initial, Lagrangian and remeshing grids (respectively interrupted lines, full lines and dots). Circles represent remeshing zones around initial nodes.

Then, the structure of a remeshing step is as follows:

- Computation of conservative quantities in the elements Q' of G':
  a) Coefficients of interpolation of the volumetric functions defined on G^n
  b) Intersections $G'-G^n$ and elements met by the segments of G'
  c) Integration of the interpolated functions in each element Q'
- Computation of the variables on G':

$$u' \quad X' \quad \rho' \quad q' \quad (E_s)' \quad p'$$

- Link with the Lagrangian algorithm:

$$u^{n-\frac{1}{2}} = u'$$

$$X^n = X'$$

$$\rho^n = \;'$$

$$q^{n-\frac{1}{2}} = q'$$

$$(E_s)^n = (E_s)'$$

$$T^n = T'$$

$$p^n = p'$$

Starting from given initial conditions, the difference between solutions obtained when using first- and second- order remeshing methods can be quite considerable for some components.

We can observe such a difference by drawing representative surfaces of each component of both solutions at fixed time t, as they have been computed by the algorithm respectively with first- and second- order remeshing methods (from now on, let us talk of 1- and 2- methods, 1- and 2- algorithms, etc.).

We can then quantify the difference between the 1- and 2- solutions of a problem by computing the time-evolution of the difference or relative difference between each of their components in the spaces L^p for p=1 and p=∞; namely, for a component w, drawing the curves:

$$t \longmapsto ||\, w_1(.,t) - w_2(.,t)\,||_p$$

(absolute difference)

$$t \longmapsto \frac{||w_1(.,t) - w_2(.,t)||_1}{||\inf\,(|w_1(.,t)|,|w_2(.,t)|)||_1}$$

$$t \longmapsto \left\| \frac{w_1(.,t)-w_2(.,t)}{\inf\,(|w_1(.,t)|,|w_2(.,t)|)} \right\|_\infty$$

(relative difference)

where the subscript i is affected to the result of the i-algorithm.

RESULTS

Starting from initial conditions depicted in figures 2, one obtains solutions whose components w have representative surfaces $z=w(x,y,t_0)$ at some fixed times t_0, drawn in following figures *-a or *-b, whether computed respectively by the 1- or 2- algorithm.

As to the evolution of the difference between the 1- and 2- solutions, taking p=1, we obtained curves starting with a very steep slope, then quickly presenting a horizontal configuration. The relative levels reached are about 75% for specific internal energy, 35% for density and 10% for pressure, in case 1, a bit lower in case 2 where the 2-model is more affected by the multiplication of discontinuities, and still lower in case 3 where the amplitude of contact discontinuities and shocks is lower than in the first two cases.

CONCLUSION

Interpolation before rezoning allows conservation of a large amount of information on the structure of solutions, while it is nearly utterly lost with a first-order remeshing method, and this at the ratio observed in the steep slope of the time-evolution curves of the difference between the 1- and 2- solutions; that is, almost immediatly. Then, the horizontal configuration of these curves proves a good quality of conservativity of the 2-models.

Another improvement given by this method concerns phase errors, not cancelled but significantly diminished compared to those of the 1-method, as can be observed on one-dimensional models, when one knows an exact solution (or can use a Lagrangian algorithm without remeshing). And the phase difference is amplified by passage to superior dimensions.

These considerations lead to the conclusion that a high-order remeshing method is absolutely necessary to any long-time Lagrangian simulation of highly structured phenomena.

Figures 2: four types of initial conditions

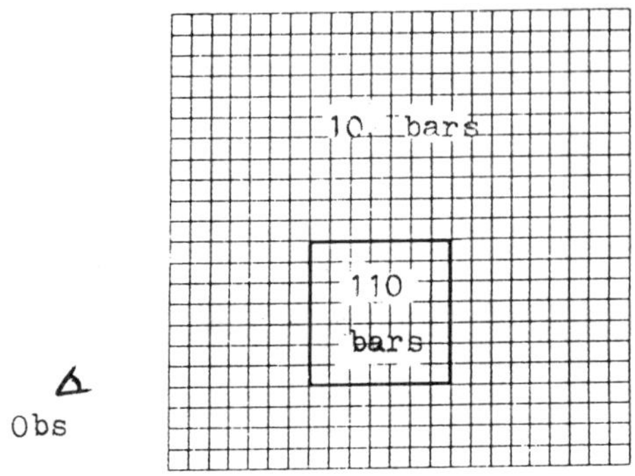

case n°1:
uniforme density: $\rho = 10\ kg/m^3$

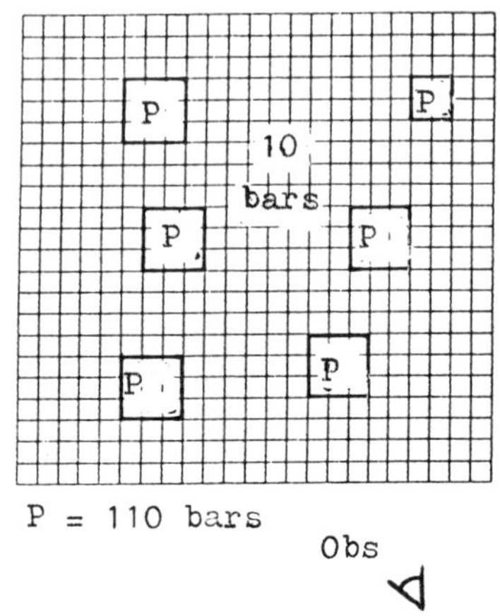

case n°2:
uniform density: $\rho = 10\ kg/m^3$

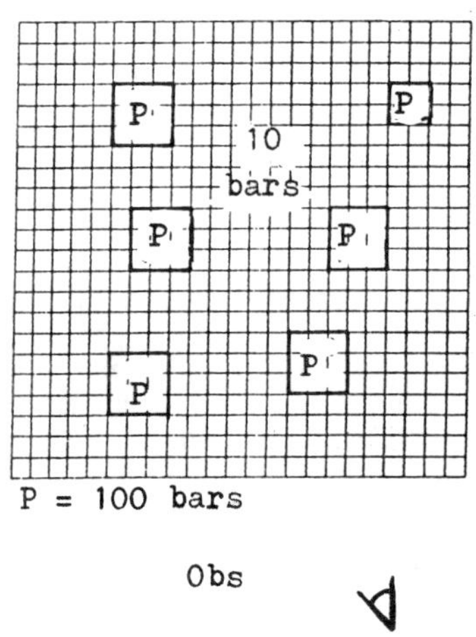

case n°3:
uniform temperature: T = 1 125 K

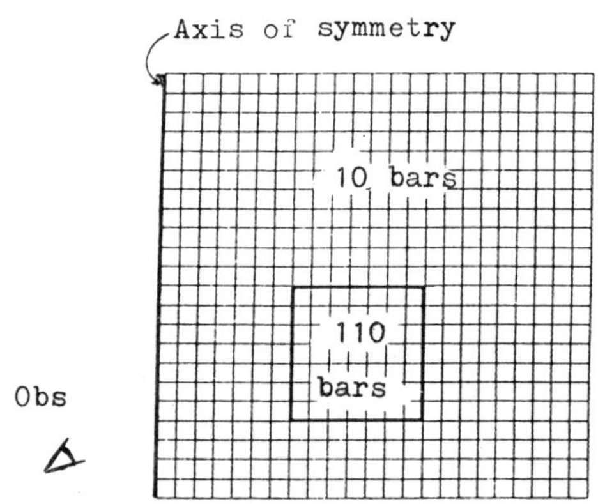

case n°4:
same as case n°1
in 3-D axisymetric geometry

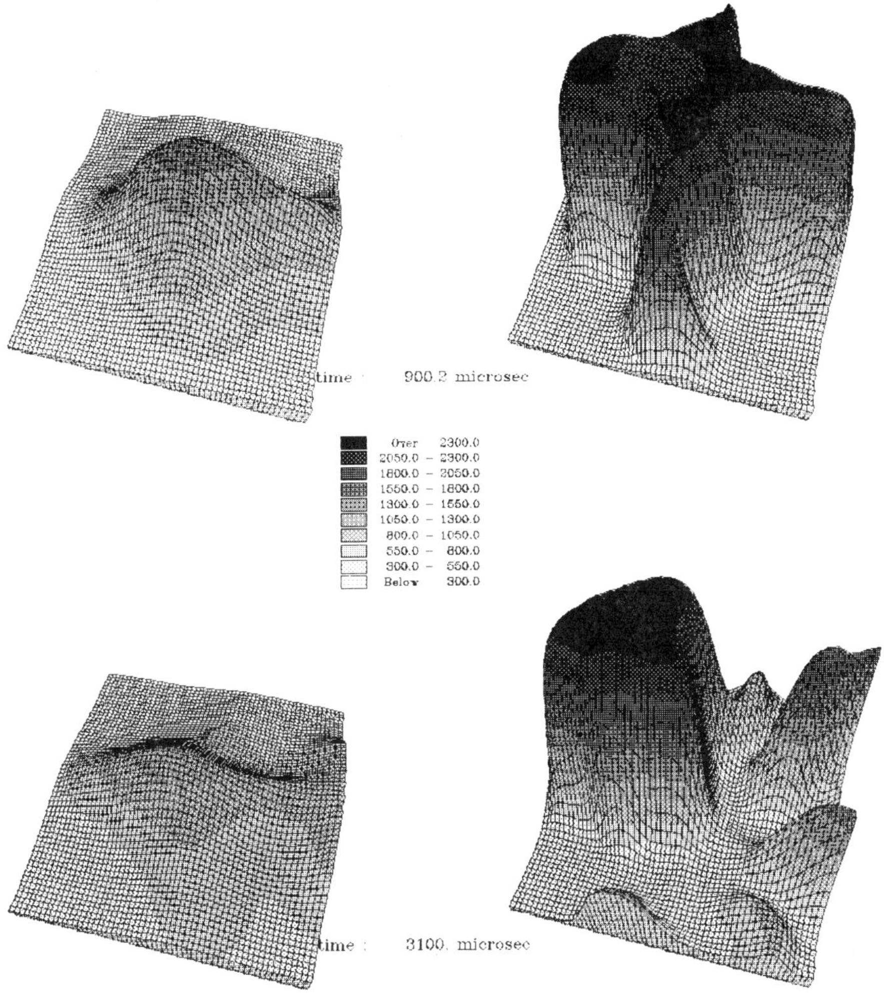

figures 3.a:
1-model of the evolution of
specific internal energy in case n°1

figures 3.b:
2-model of the evolution of
specific internal energy in case n°1

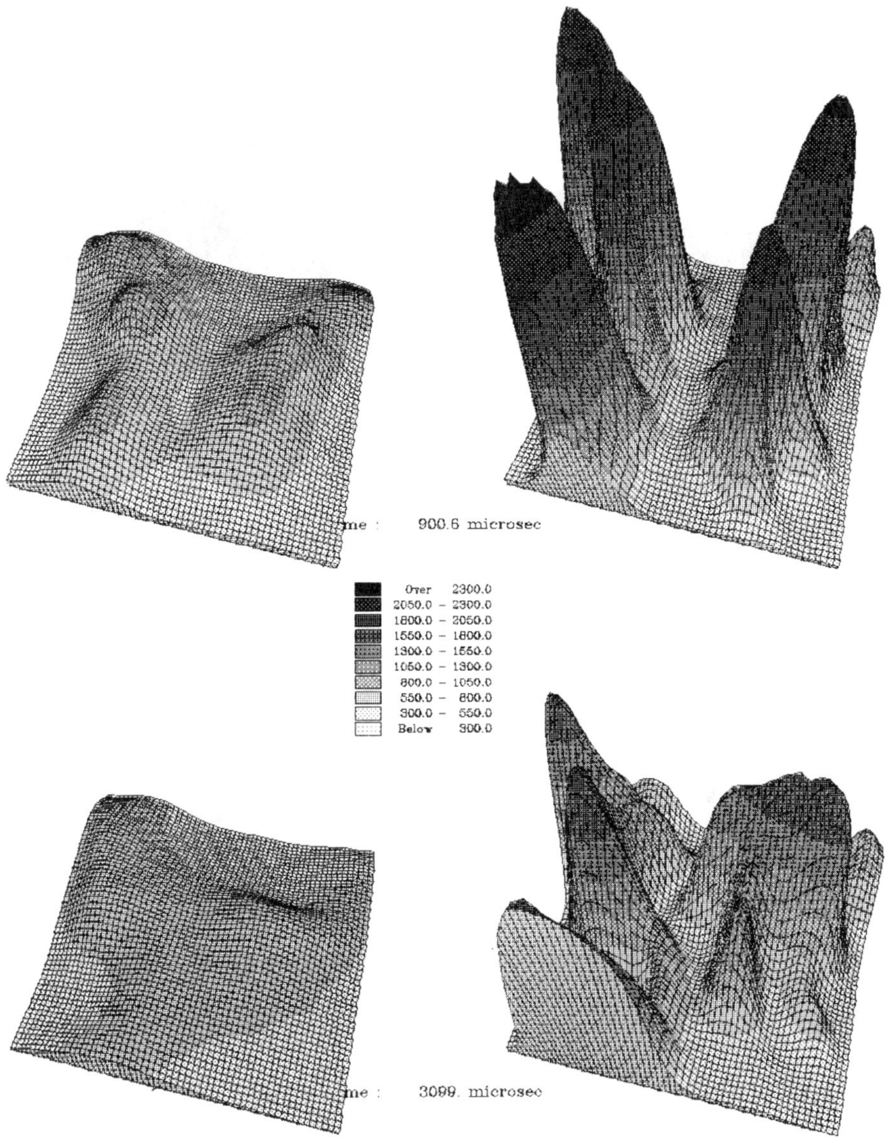

figures 4.a:
1-model of the evolution of
specific internal energy in case n°2

figures 4.b:
2-model of the evolution of
specific internal energy in case n°2

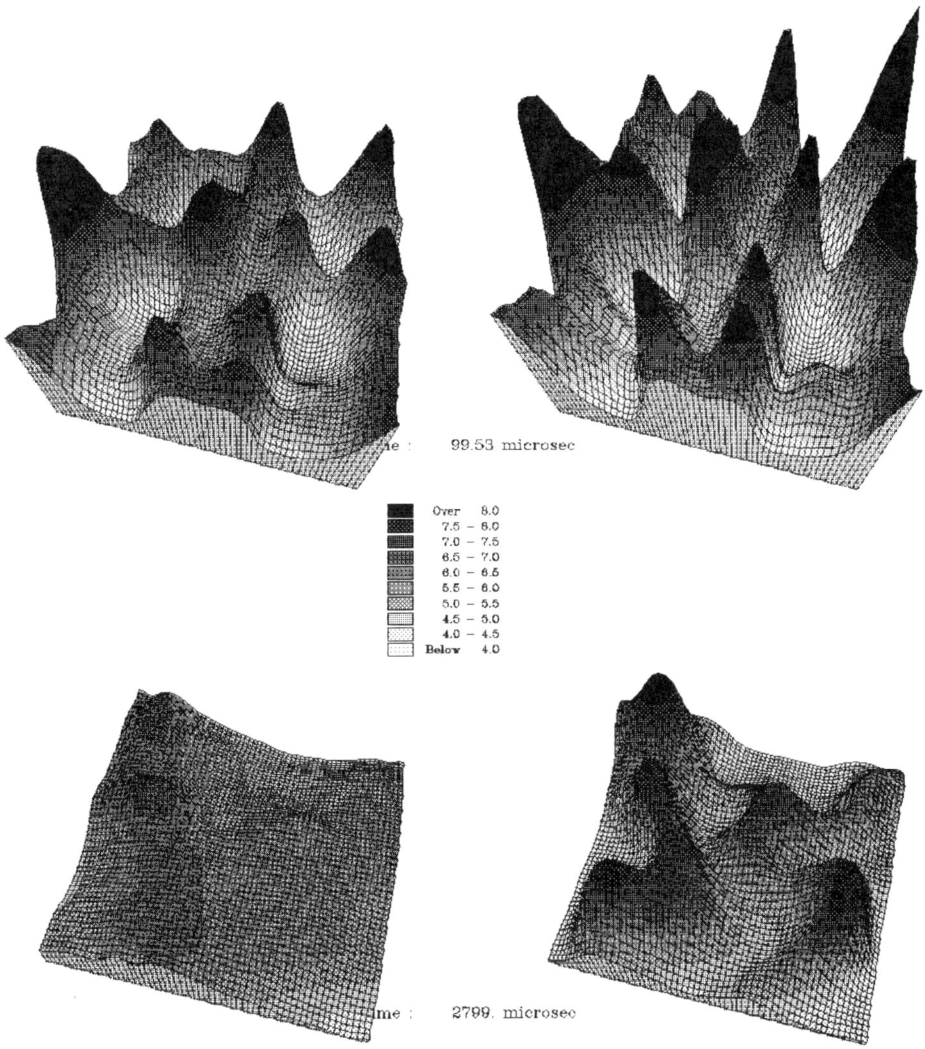

figures 5.a:
1-model of the evolution of
density in case n°3

figures 5.b:
2-model of the evolution of
density in case n°3

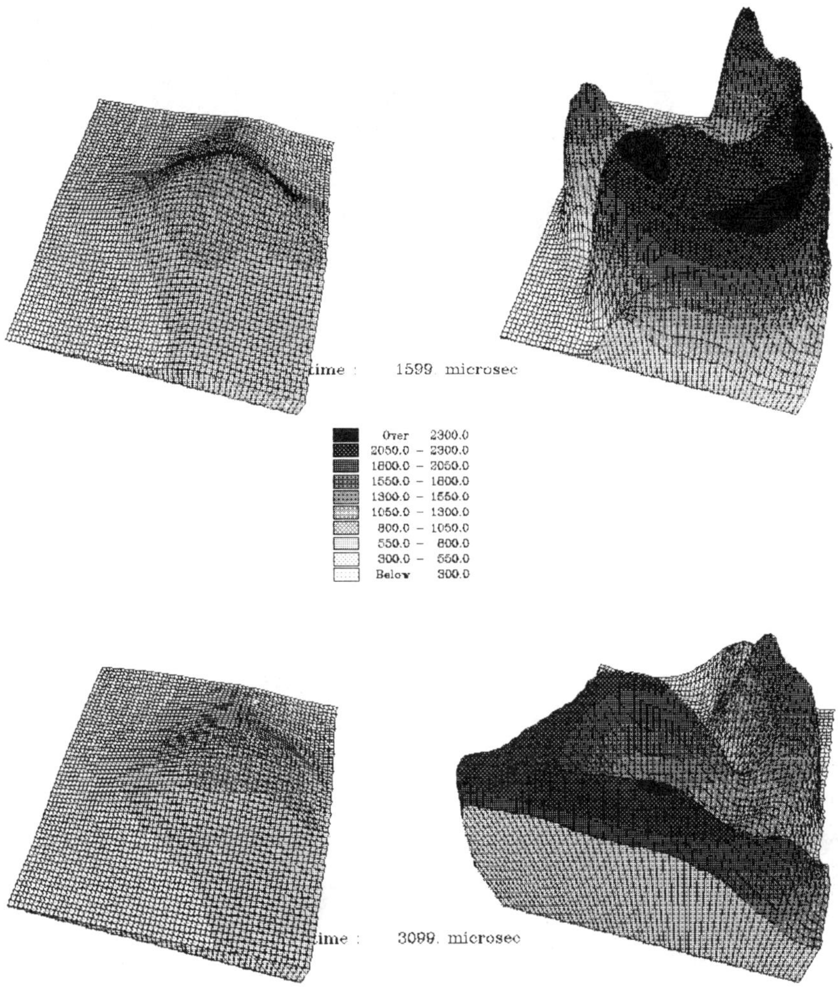

figures 6.a:
1-model of the evolution of
specific internal energy in case n°4

figures 6.b:
2-model of the evolution of
specific internal energy in case n°4

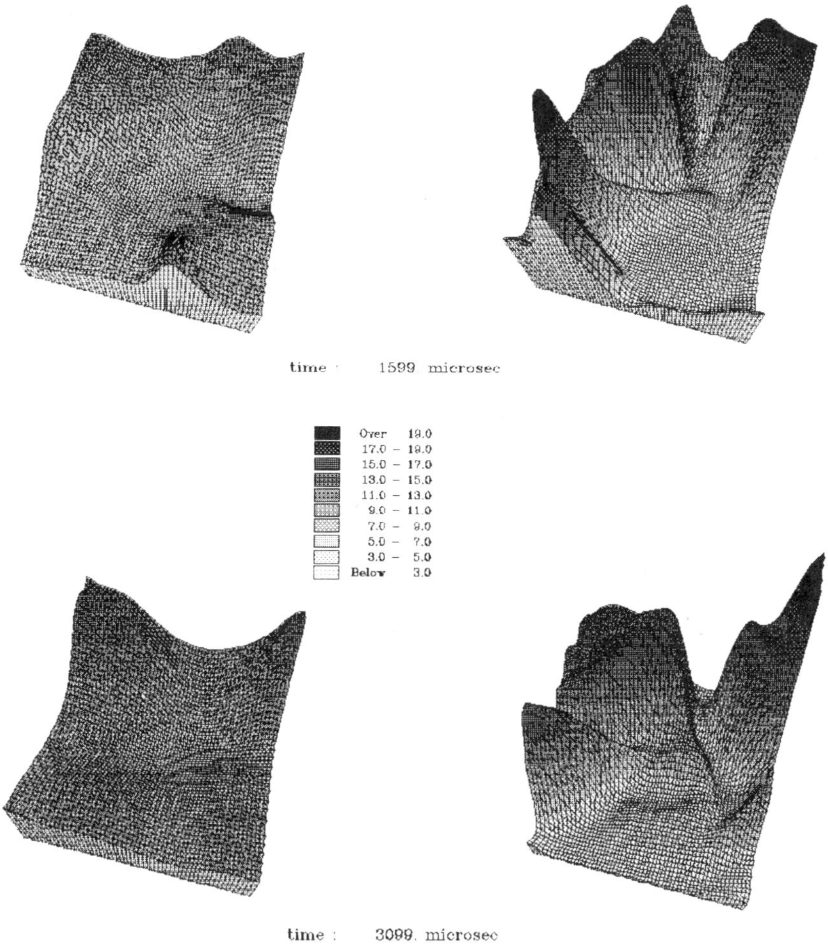

figures 7.a:
1-model of the evolution of
density in case n°4

figures 7.b:
2-model of the evolution of
density in case n°4

ON NUMERICAL ANALYSIS OF TWO-DIMENSIONAL, AXISYMMETRIC, LAMINAR JET DIFFUSION FLAMES

B. ROGG

Angewandte Physikalische Chemie und SFB 123, Universität Heidelberg,
Im Neuenheimer Feld 253, 69 Heidelberg, West Germany

1. INTRODUCTION

The investigation of nonequilibrium effects in laminar diffusion flames is important for both studying pollutant formation during combustion and gaining enhanced insight into the structure of turbulent diffusion flames that can be described by laminar-flamelet models. Laminar diffusion flames that can be treated one-dimensionally in space by theoretical methods, for instance, stretched flames in the context of nonpremixed turbulent combustion, have been subjected to a number of theoretical and numerical studies, and successful numerical simulations exist, e.g. /1-6/, that take into account detailed models for both chemistry and molecular transport. However, mainly because of their two-dimensional structure, in numerical studies little attention has been given to jet diffusion flames which predominantly are being used in experiments. A schematic illustration of such a flame is shown in Fig. 1. Furthermore, in the few numerical analyses available simplifications like the Burke-Schumann flame-sheet approximation, or global one-step kinetics, and constant transport properties have been employed /7,8/, restrictions imposed by computing facilities that only recently can be relaxed /9,10/.

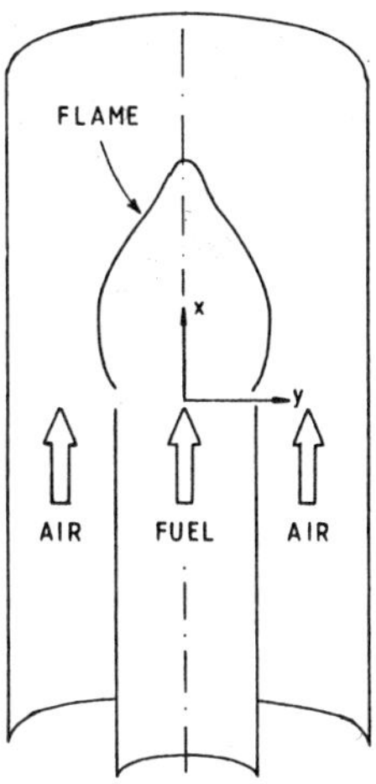

Figure 1: Schematic illustration of a confined, overventilated, laminar jet diffusion flame.

C.-M. Brauner and C. Schmidt-Lainé (eds.), Mathematical Modeling in Combustion and Related Topics, 551–560.

The intention of this paper is to review the present state of the art in modeling and numerically simulating two-dimensional, axisymmetric, laminar jet diffusion flames, and to provide the framework for detailed numerical studies of such flames under various conditions. The governing equations are given in cylindrical coordinates and, with special regard to their hybrid elliptic-hyperbolic (or parabolic-hyperbolic) nature, the question of appropriate boundary conditions is addressed. Primitive-variable formulations and the stream function-vorticity formulation are discussed as well as currently existing numerical approaches to solutions of the governing equations. Multistep-chemistry models and models of molecular transport are discussed that are realistic but still tractable in numerical simulations of truly two-dimensional flames.

2. GOVERNING EQUATIONS

The governing equations for swirl-free, two-dimensional, axisymmetric, laminar flames are derived from the general conservation equations for multicomponent, reacting, ideal-gas mixtures /11/.

The flow field is described by the overall-continuity equation,

$$\frac{D\rho}{Dt} = -\rho(\nabla \cdot \mathbf{v}), \tag{1}$$

and the Navier-Stokes equations in two dimensions,

$$\rho\frac{Du}{Dt} = -\frac{\partial p}{\partial x} - (\nabla \cdot \mathbf{T})_x + \rho g_x, \tag{2}$$

$$\rho\frac{Dv}{Dt} = -\frac{\partial p}{\partial y} - (\nabla \cdot \mathbf{T})_y + \rho g_y. \tag{3}$$

In Eqs. (1) - (3), and below, t is the time, and x and y denote the spatial coordinates in longitudinal and radial direction, respectively; $\mathbf{v}$ is the velocity vector with components u and v, ρ is the density, and p the pressure. Only body forces due to gravity are taken into account, g_x and g_y denote the components of gravitational acceleration. The material derivative $D\,/Dt$ is

$$\frac{D\phi}{Dt} = \frac{\partial\phi}{\partial t} + (\mathbf{v}\cdot\nabla\phi) = \frac{\partial\phi}{\partial t} + u\frac{\partial\phi}{\partial x} + v\frac{\partial\phi}{\partial y} \tag{4}$$

for any scalar variable ϕ, and the divergence of the velocity field is given by

$$(\nabla \cdot \mathbf{v}) = \frac{\partial u}{\partial x} + \frac{\partial v}{\partial y} + \frac{v}{y}. \tag{5}$$

For the ideal-gas mixtures to be considered Newton's law holds, i.e., the x and y components of the product $(\nabla \cdot \mathbf{T})$, where $\mathbf{T}$ denotes the viscous part of the stress tensor, can be written as

$$(\nabla \cdot \mathbf{T})_x = -2\frac{\partial}{\partial x}\Big(\mu\frac{\partial u}{\partial x}\Big) - \frac{1}{y}\frac{\partial}{\partial y}\Big[\mu y\Big(\frac{\partial v}{\partial x} + \frac{\partial u}{\partial y}\Big)\Big] + \frac{2}{3}\frac{\partial}{\partial x}\Big[\mu(\nabla\cdot\mathbf{v})\Big], \tag{6}$$

$$(\nabla \cdot \mathbf{T})_y = -\frac{\partial}{\partial x}\Big[\mu\Big(\frac{\partial v}{\partial x} + \frac{\partial u}{\partial y}\Big)\Big] - \frac{2}{y}\frac{\partial}{\partial y}\Big(\mu y\frac{\partial v}{\partial y}\Big) + \frac{2}{3}\frac{\partial}{\partial y}\Big[\mu(\nabla\cdot\mathbf{v})\Big] + \frac{2\mu v}{y^2}, \tag{7}$$

where μ is the dynamic viscosity of the mixture; effects of bulk viscosity have been neglected.

In the above equations the effects of coexistence of various chemical species and of combustion, i.e., variations of species concentrations and temperature in the flame, enter through μ, ρ and p. The last two quantities are related through the ideal-gas equation of state,

$$p/\rho = R^0 T\sum_{i=1}^{N} Y_i/M_i, \tag{8}$$

where R^0 is the universal gas constant, T the temperature, and Y_i and M_i, $i = 1,...,N$, denote the mass fraction and the molecular weight of species i, respectively.

Temperature and mass fractions obey the energy and species conservation equations, viz.,

$$\rho\frac{DT}{Dt} = \frac{1}{c_p}\frac{\partial p}{\partial t} + \frac{1}{c_p}\Big[\frac{\partial}{\partial x}\Big(\lambda\frac{\partial T}{\partial y}\Big) + \frac{1}{y}\frac{\partial}{\partial y}\Big(y\lambda\frac{\partial T}{\partial y}\Big)\Big]$$
$$- \frac{\partial T}{\partial x}\sum_{i=1}^{N}\frac{c_{pi}}{c_p}\rho Y_i V_{i,x} - \frac{\partial T}{\partial y}\sum_{i=1}^{N}\frac{c_{pi}}{c_p}\rho Y_i V_{i,y} - \frac{1}{c_p}\sum_{i=1}^{N} h_i w_i, \qquad (9)$$

$$\rho\frac{DY_i}{Dt} = -\frac{\partial}{\partial x}\Big(\rho Y_i V_{i,x}\Big) - \frac{1}{y}\frac{\partial}{\partial y}\Big(y\rho Y_i V_{i,y}\Big) + w_i, \quad i = 1,...,N. \qquad (10)$$

In addition to the quantities already defined above, in Eqs. (9) and (10) $V_{i,x}$ and $V_{i,y}$ are the components of the diffusion velocity of species i in the respective coordinate directions, and w_i denotes its rate of production. Expressions for w_i will be discussed in section 5. The transport coefficients, μ in the Navier-Stokes equations and λ in Eq. (9), as well as the thermodynamic properties, c_p, c_{pi} and h_i, are defined as usual. Relationships to be used for the evaluation of transport coefficients, thermodynamic properties and diffusion velocities will be given in section 6. Note that in deriving Eq. (9) low-Mach-number flow has been assumed, and viscous dissipation, Dufour heat flux, radiation heat flux, and work of body forces have been neglected. In open diffusion flames the term $\partial p/\partial t$ may be ommitted as well.

The $N+4$ basic dependent variables in the above set of partial partial differential equations, Eqs. (1), (2), (3), (9) and (10), may be taken to be ρ, u, v, T and Y_i, $i = 1,...,N$, in which case the pressure is obtained from Eq. (8). However, as will be discussed in section 4, due to difficulties inherent with the present state of the art in numerically solving the set of governing equations, a different selection of dependent variables and, eventually, derivation of a different but, of course, equivalent set of governing equations may be suitable.

3. BOUNDARY CONDITIONS

The unsteady governing equations, Eqs. (1), (2), (3), (9) and (10), form a hybrid parabolic-hyperbolic system of partial differential equations. The parabolic nature is related to the presence of second-order derivatives in the momentum, energy and species conservation equations. The absence of such derivatives in the continuity equation makes the system hybrid since by itself it would be a hyperbolic equation. In the steady case, the governing equations form a hybrid elliptic-hyperbolic system.

Problems associated with the numerical simulation of two-dimensional flames are boundary-value problems, either time dependent with given initial values of the basic dependent variables, or time independent. Only a few special results are known about the mathematical statement of boundary conditions to be imposed to ensure existence or uniqueness of the solution. As a consequence, the question of appropriate boundary conditions must be settled in a heuristic way by taking into account the physical meaning of the combustion problem under consideration and the mathematical nature of the equations. Thus, it can be conjectured that the momentum, energy and species conservation equations each require one condition at each boundary. The continuity equation, considered as an equation for ρ assuming a known velocity field, explicitely gives the change of density of a fluid particle along its trajectory. Therefore, if the fluid partical enters the domain of integration through the boundary, its density should be given as a boundary condition. On the contrary, if the particle leaves the computation domain, its density must be calculated as a part of the solution and it cannot be imposed arbitrarily.

As one example of a truly two-dimensional flame, we now discuss the appropriate boundary conditions for a steady, confined, overventilated, laminar jet diffusion flame. The flame is established at the mouth of an inner, cylindrical tube from which fuel issues into an outer,

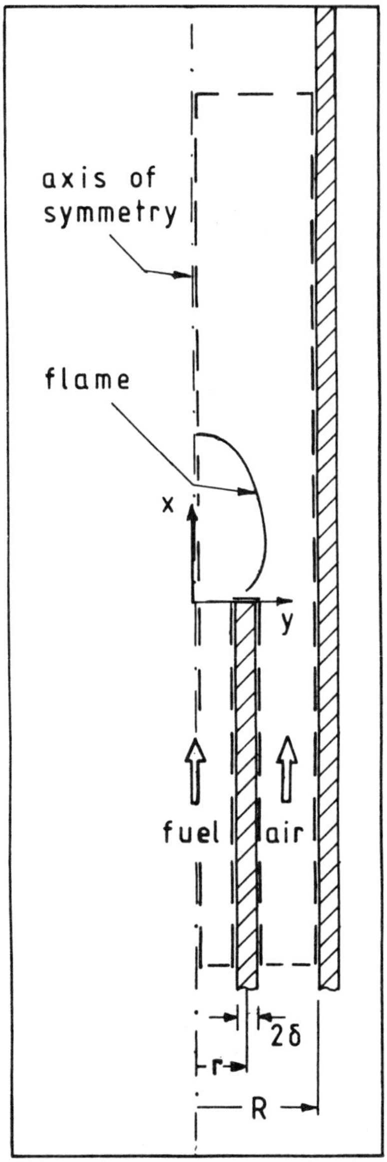

Figure 2a

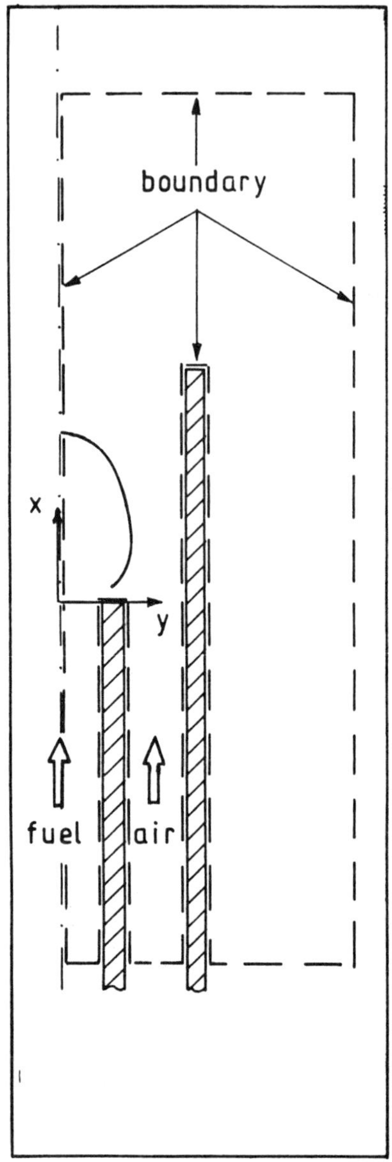

Figure 2b

Figure 2a: Schematic illustration of the ideal domain of integration for a confined, laminar diffusion flame. The outer duct extents to $x = +\infty$.

Figure 2b: As Fig. 2a, however, with an outer duct of finite length.

concentric cylindrical duct. Through the outer duct oxidizer is flowing. In Fig. 2a the dashed lines indicate the ideal domain of integration that should underlie numerical simulations of this flame. The boundary conditions along the axis of symmetry, $y=0$, and at the wall of the outer duct, $y = R$, are straightforward. Along $y = 0$ the conditions are that v itself and the first derivatives with respect to y of u, T and Y_i, $i = 1,...,N$, be zero. However, ρ there should be calculated from the continuity equation which is to be expanded about $y = 0$. At $y = R$ both velocity components are zero; for temperature and mass fractions Dirichlet, Neumann or Robbins conditions are specified depending on whether or not the wall is impermeable for heat and matter, respectively. At the inner and outer wall of the fuel tube, $y = r - \delta$ and $y = r + \delta$, where 2δ is the thickness of the tube, we have $u = v = 0$. However, conditions for the remaining dependent variables may be more complex and have to be derived for the particular geometry under consideration. For instance, depending on the material and the thickness of fuel tube and from a variety of other parameters, it might be necessary to take into account axial and radial heat losses from the gases to the tube in order to obtain numerical predictions that have reasonable confidence. Another problem yet to overcome is the formulation of boundary conditions at the mouth of the inner tube. Finally, asymptotic solutions of the governing equations for large negative and large positive values of x represent the appropriate upstream and dowstream boundary conditions.

For the burner geometry illustrated in Fig. 2a, it may be anticipated that, under certain conditions, far upstream to leading order the fuel velocity distribution is that of a Hagen–Poiseuille flow, the oxidizer distribution that of a flow through a cylindrical annulus. For both types of flow analytical solutions are availabel, see e.g. /12/. For temperature and species mass fractions to leading order convective–diffusive balances are anticipated to hold. In most cases appropriate boundary conditions far downstream are more difficult to derive than upstream conditions because generally little is known about the values of the dependent variables there. As a consequence, at the downstream boundary often zero gradients in axial direction are imposed for all dependent variables, conditions that are justifiable only if the outer duct extends to $x = +\infty$. However, if the length of the outer duct is finite, care must be taken and more sophisticated downstream conditions are to be imposed. An example of a confined laminar diffusion flame that burns inside an outer duct of finite length is the flame investigated by *Mitchell et al.* /7/. For this flame it has been observed that the gas–mixture along the axis is accelerated by the prevailing buoyant forces which give rise to a large recirculation cell. As a consequence, away from the axis at the exit of the outer duct, gas of unknown density, temperature and composition is drawn at an unknown velocity from outside into the duct. Thus, to obtain confident numerical predictions of the latter flame the domain of integration should be enlarged even further as is indicated by the dashed lines in Fig. 2b.

4. FORMULATIONS AND SOME NUMERICAL–SOLUTION METHODS

In this section we briefly outline the essential features of primitive–variable formulations and of the stream–function vorticity formulation. However, due to space limitations in the present volume, only a few of the many numerical–solution approaches available can shortly be discussed. In particular, neither we enter into finite–difference methods, finite–element methods or finite–volume methods nor into relationships between these numerical approaches. Finally we note that even though in this section we consider the unsteady set of governing equations, practically all arguments carry over to the time independent case.

4.1. Primitive – Variable Formulations

If the numerical–solution method is to be general, e.g. equally applicable to both two and three dimensions, the primitive variables u, v, ρ, p, T, and Y_i, $i = 1,...,N$, should be retained in the governing equations as working variables. Of these, at least in non–reactive flows, ρ is often chosen to stand as a basic dependent variable, whereas p is evaluated form Eq. (8). This approach is natural because the density appears explicitely in the continuity equation, Eq. (1), which may be regarded as an equation whose main dependent variable is density. However, numerical experience has shown that when the flow velocity experiences strong variations, e.g. spatial variations across the reaction zone of a flame or temporal variations in an IC-engine, adopting pressure instead of density as the basic dependent variable is advantageous. In order to determine p an equation is usually derived by joint manipulations of Eqs. (1) to (3). The resulting "pressure equation" replaces the continuity equation while the Navier-Stokes equations retain their role for determining the velocity field, the individual equations being still strongly coupled, of course.

Due to the presence of chemical source terms the energy and species conservation equations are stiff, and, as a consequence, the entire set of governing equations should be treated in a fully implicit manner. Some of the relevant numerical-solution methods employ sequential iteration, such as the SIMPLE /13/ and SIMPLER /14/ method, in which at each time step the equations for each dependent variable are solved repeatedly in succession. Others utilize block iteration, such as the SIVA scheme /15/, in which the variable block is solved simultaneously for a point or a line at a time. A new method that avoids iteration at each time step, thereby reducing computing efforts and costs substantially, is PISO (*P*ressure-*I*mplicit with *S*plitting of *O*perators) /16-18/ which, due to its stability for fairly large time steps, might be useful particularly for steady-state calculations of two-dimensional combustion problems.

4.2. The Stream – Function Vorticity Formulation

The stream-function vorticity formulation of the governing equations makes use of the vorticity vector $\mathbf{\Omega} = \nabla \times \mathbf{v}$ which, for two-dimensional flow and cylindrical coordinates, reduces to a scalar quantity, viz., $\omega = \partial v/\partial x - \partial u/\partial y$. Note that ω is a measure of the amount of counter clockwise rotation in an x-y plane possessed by a fluid element. The vorticity-transport equation, i.e. an equation for ω, is obtained by applying the curl operator to the vector version of Eqs. (2) and (3) so that the pressure terms disappear. The vorticity transport equation is usually associated with an equation for a stream-function vector $\mathbf{\Psi}$ which, as $\mathbf{\Omega}$, for the geometry considered herein reduces to a scalar quantity ψ; ψ is defined such that the continuity equation, Eq. (1), is identically satisfied.

An important feature of the stream-function vorticity formulation is that the pressure is no longer explicit in the resulting partial differential equations. In flames where pressure variations are negligibly small p may be taken as constant in Eq. (8). In general, however, for the pressure one must solve a "pressure equation" of the type discussed in section 4.1.

Since in another paper of the present volume the stream-function vorticity formulation itself as well as a suitable numerical-solution method are discussed, for further details of this approach reference is given to /9/. Also /19/ should be consulted.

5. CHEMISTRY MODELS

In numerical simulations of flames that can be treated one-dimensionally, e.g. with a physical distance variable or a similarity variable as the only independent space variable, reaction mechanisms comprising up to several hundred elementary reactions and up to 40 chemical species are being employed. Clearly, in simulations of truly two-dimensional flames the use of such mechanisms currently is prohibitive. Here a promising alternative is reduced mechanisms /20-23/ that are systematically derived from detailed mechanisms of elementary reactions. Currently such mechanisms are available for both premixed /20,21/ and nonpremixed systems /22,23/, and for both modes of combustion they have shown to produce reasonable results /22,24/.

As one example, herein we consider the methane/air 4-step scheme derived by *Peters* /20,22/. This mechanism consists of the four global reactions

$$CH_4 + 2H + H_2O = CO + 4H_2, \quad \text{(I)}$$
$$CO + H_2O = CO_2 + H_2, \quad \text{(II)}$$
$$2H + M = H_2 + M, \quad \text{(III)}$$
$$O_2 + 3H_2 = 2H + 2H_2O, \quad \text{(IV)}$$

and apart from inert N_2, it comprises CH_4, O_2, CO_2, H_2O, CO, H_2 and H. The rates of production of these species may be written as

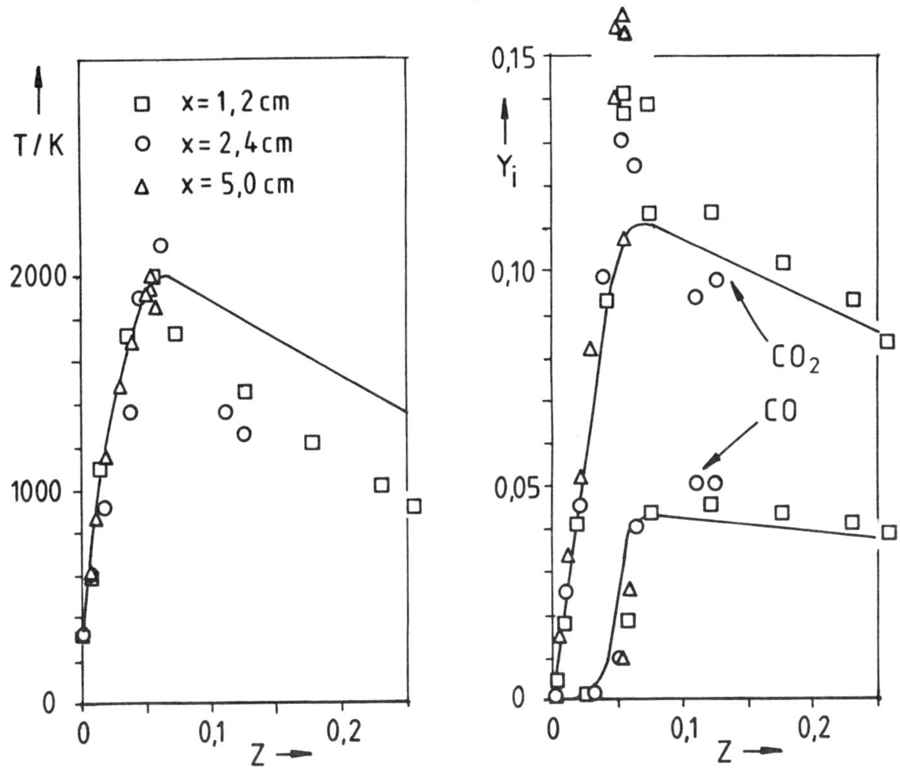

Figure 3: Profiles of temperature and mass fractions of CO and CO_2 as a function of the mixture fraction Z. Solid lines: numerical results by *Peters* and *Kee* /22/ based on a methane/air 4-step mechanism; Symbols: experimental results calculated from the measured data reported by *Mitchell et al.* /7/.

$$
\begin{aligned}
w_{CH_4} &= -M_{CH_4} r_I, \\
w_{O_2} &= -M_{O_2} r_{IV}, \\
w_{CO_2} &= M_{CO_2} r_{II}, \\
w_{H_2O} &= -M_{H_2O}(r_I + r_{II} - 2r_{IV}), \\
w_{CO} &= M_{CO}(r_I - r_{II}), \\
w_{H_2} &= M_{H_2}(4r_I + r_{II} + r_{III} - 3r_{IV}), \\
w_H &= -M_H(2r_I + 2r_{III} - 2r_{IV}), \\
w_{N_2} &= 0,
\end{aligned}
\tag{11}
$$

where r_I, r_{II}, r_{III} and r_{IV} denote the global-reaction rates that are given by

$$
\begin{aligned}
r_I &= r_1 + r_2, \\
r_{II} &= r_9, \\
r_{III} &= r_6 + r_8 + r_{14} + r_{15}, \\
r_{IV} &= r_{10} + r_{16}.
\end{aligned}
\tag{12}
$$

The various rates appearing on the r.h.s. of Eq. (12) are algebraicly complicated expressions containing kinetic data from a number of elementary reactions. They are evaluated according to Eq. (9) in /20/. For a further discussion of this reduced mechanism /22.24/ should be consulted.

Shown in Fig. 3 are profiles of the temperature and the mass fractions of CO and CO_2, respectively, as a function of the mixture fraction Z. The solid lines represent numerical results reported in /22/ for the methane/air counterflow diffusion flame by *Tsuji* and *Yamaoka* /25/ using the reduced 4-step mechanism discussed above. The stagnation-point velocity gradient employed in these computations had a numerical value of $100s^{-1}$ which, for the flame under consideration, corresponds to a low-to-medium rate of strain. The various symbols represent experimental data for the confined, laminar jet diffusion flame investigated in /7/, taken at different heights above the burner port. In evaluating the experimental data, herein the mixture fraction has been calculated according to /22/. It is seen from Fig. 3 that the 4-step mechanism indeed is capable of reflecting the essential features of the flame structure of methane/air diffusion flames. The scatter of the experimental data and and the differences between the measured and computed temperature profiles on the fuel side was observed earlier /26/ and, therefore, reference is given to /26/ for a careful and detailed discussion of these points.

6. TRANSPORT MODELS

As it is the case for chemistry models, current computing facilities impose severe restrictions on transport models to be employed in numerical simulations of two-dimensional flames. Therefore, employing Stefan-Maxwell equations or even more detailed models to determine the diffusion velocities is prohibitive. However, simple but still satisfactory transport models that have previously been employed in simulations of one-dimensional flames, are easily generalized for the two-dimensional case. Subsequently, such a generalized model is presented.

Following /27/ it is assumed that the diffusion-velocity vector $\mathbf{V}_i$, $i = 1, ..., N$, is composed of three parts, i.e.,

$$\mathbf{V}_i = \mathbf{V}_i^D + \mathbf{V}_i^T + \mathbf{V}_c. \tag{13}$$

Here $\mathbf{V}_i^D$ is the ordinary-diffusion velocity for which an approximation recommended by *Hirschfelder* and *Curtiss* /28,29/,

$$\mathbf{V}_i^D = -\frac{D_i}{X_i}\nabla X_i, \quad i = 1, ..., N, \tag{14}$$

is adopted. Here X_i denotes the mole fraction of species i and D_i is its mixture-averaged diffusion coefficient which is given explicitely in terms of the binary diffusion coefficients D_{ij} by

$$D_i = \frac{1 - Y_i}{\sum\limits_{\substack{j=1 \\ j\neq i}}^{N} X_j/D_{ij}}. \tag{15}$$

In Eq. (13) $\mathbf{V}_i^T$ is the thermal-diffusion velocity vector,

$$\mathbf{V}_i^T = -\frac{D_i^T}{\rho Y_i T}\nabla T, \quad i = 1, ..., N, \tag{16}$$

a nonzero value of which is included only for light species such as He, H and H_2. In Eq. (16) D_i^T is the thermal-diffusion coefficient of species i which is evaluated following /29,31/. The correction velocity vector $\mathbf{V}_c$ appearing in Eq. (13) is determined according to /27,30/.

For the evaluation of transport coefficients and thermodynamic properties a number of formulas exist. For instance, the dynamic viscosity and the thermal conductivity of the mixture may be calculated from the respective properties of the pure species according to

$$\mu = \frac{1}{2}\left(\sum_{i=1}^{N} X_i\mu_i + \frac{1}{\sum\limits_{i=1}^{N} X_i/\mu_i}\right) \tag{17}$$

and

$$\lambda = \frac{1}{2}\left(\sum_{i=1}^{N} X_i\lambda_i + \frac{1}{\sum\limits_{i=1}^{N} X_i/\lambda_i}\right). \tag{18}$$

For details of the evaluation of the species transport coefficients and of the species thermodynamic properties /29,30/ may be consulted.

ACKNOWLEDGEMENT

This study was performed within the Research Association Technical Flames (TECFLAM) of the Universities of Heidelberg, Karlsruhe, and Stuttgart and the DFVLR-German Aerospace Research Establishment, and within Sonderforschungsbereich 123 of Heidelberg University.

REFERENCES

1. *Dixon-Lewis, G., David, T., Gaskell, P. H., Fukutani, S., Jinno, H., Miller, J. A., Kee, R. J., Smooke, M. D., Peters, N., Effelsberg, E., Warnatz, J., Behrendt, F.*: 20th Symp. (Int.) on Comb., pp. 1893, The Combustion Institute, Pittsburgh, 1984.
2. *Dixon-Lewis, G.*: Numerical Modelling of Strained Flames with Detailed Chemistry, paper presented at the "Workshop of Gas Flame Structure", Novosibirsk, U.S.S.R., July 1986.
3. *Smooke, M. D., Puri, I. K., Seshadri, K.*: A Comparison Between Numerical Calculations and Experimental Measurements of the Structure of a Counterflow Diffusion Flame Burning Diluted Methane in Diluted Air, Proc. 21st Symp. (Int.) on Comb., The Combustion Institute, Pittsburgh (1986), to appear.
4. *Puri, I. K., Seshadri, K., Smooke, M. D., Keyes, D. E.*: Report ME-102-87, Yale University (1987).
5. *Rogg, B., Behrendt, F., Warnatz, J.*: Turbulent Non-Premixed Combustion in Partially Premixed Diffusion Flamelets with Detailed Chemistry, Proc. 21st Symp. (Int.) on Comb., The Combustion Institute, Pittsburgh (1986), to appear.

6. *Rogg, B., Behrendt, F., Warnatz, J.*: Ber. Bunsenges. Phys. Chem. **90**, pp. 1005, 1986.
7. *Mitchell, R. E., Sarofim, A. F., Clomburg, L. A.*: Combust. Flame **37**, pp. 227, 1980.
8. *Smooke, M. D., Mitchell, R. E., Grcar, J. F.*: in G. Birkhoff, A. Schönstadt (Eds.), Elliptic Problem Solvers, pp. 557, Academic Press, London, 1984.
9. *Smooke, M. D., Turnbull, A. A., Mitchell, R. E., Keyes, D. E.*: Solution of Two-Dimensional Axisymmetric Laminar Diffusion Flames by Adaptive Boundary Value Methods, this volume.
10. *Rogg, B.*: in preparation.
11. *Williams, F. A.*: Combustion Theory, 2nd Ed., Benjamin/Cummings, Menlo Park, 1985.
12. *Bird, R. B., Stewart, W. E., Lightfoot, E. N.*: Transport Phenomena, John Wiley, New York, 1960.
13. *Patankar, S. V., Spalding, D. B.*: Int. J. Heat Mass Transfer **15**, pp. 1787, 1972.
14. *Patankar, S. V.*: Numerical Heat Transfer and Fluid Flow, McGraw-Hill, New York, 1980.
15. *Caretto, L. S., Gosman, A. D. Patankar, S. V., Spalding, D. B.*: Proc. 3rd Int. Conf. Numer. Methods in Fluid Dynamics, pp. 60, 1972.
16. *Issa, R. I.*: J. Comp. Phys. **62**, pp. 40, 1986.
17. *Issa, R. I., Gosman, A. D., Watkins, A.P.*: J. Comp. Phys. **62**, pp. 66, 1986.
18. *Issa, R. I., Beshay, K. R., Ahmadi-Befrui, B., Gosman, A. D.*: Solution of the Implicitly Discretized Reacting Flow Equations by Operator-Splitting, J. Comp. Phys., to appear.
19. *Gosman, A. D., Pun, W. M., Runchal, A. K., Spalding, D. B., Wolfshtein, M.*: Heat and Mass Transfer in Recirculating Flows, Academic Press, London, 1969.
20. *Peters, N.*: in Lecture Notes in Physics **241**, pp. 90, Springer, Berlin-Heidelberg, 1985.
21. *Paczko, G., Lefdal, P. M., Peters, N.*: Reduced Reaction Schemes for Methane, Methanol and Propane Flames, Proc. 21st Symp. (Int.) on Comb., The Combustion Institute, Pittsburgh (1986), to appear.
22. *Peters, N., Kee, R. J.*: Sandia Report SAND86-8842, 1986.
23. *Bilger, R. W.*: Simplified Kinetics of Diffusion Flames of Methane in Air, Combust. Flame, to appear.
24. *Rogg, B.*: Response and Flamelet Structure of Stretched Premixed Methane-Air Flames, Combust. Flame, to appear.
25. *Tsuji, H., Yamaoka, I.*: 13th Symp. (Int.) on Comb., pp. 723, The Combustion Institute, Pittsburgh, 1970.
26. *Saito, K., Williams, F. A., Gordon, A. S.*: Transactions of the ASME, J. Heat Transfer **108**, pp. 640, 1986.
27. *Oran, E. S., Boris, J. P.*: Prog. Energy Combust. Sci. **7**, pp. 1, 1981.
28. *Hirschfelder, J. O., Curtiss, C. F.*: 3rd Symp. on Combustion and Flame and Explosion Phenomena, pp. 121, Williams and Wilkins Co., Baltimore, 1949.
29. *Kee, R. J., Warnatz, J., Miller, J. A.*: Sandia Report SAND-8209, 1983.
30. *Coffee, T. P., Heimerl, J. M.*: Combust. Flame **43**, pp. 273, 1981.
31. *Warnatz, J.*: in N. Peters, J. Warnatz (Eds.), Numerical Methods in Flame Propagation, pp. 87, Vieweg, Braunschweig/Wiesbaden, 1982.

AN ALGORITHM FOR ALLOCATION AND TEMPERATURE, AND ITS CONSEQUENCES FOR THE CHEMISTRY OF H_2-O_2 COMBUSTION

C. D. ROTEN

MATHEMATICAL SCIENCES INSTITUTE, CORNELL UNIVERSITY, U.S.A.

1. OBJECTIVES

We will outline a non numerical algorithm for calculation of chemical species in the burned gas, temperature increase, and overall chemistries for plane premixed laminar flames with arbitrarily complex chemistry. We will then apply this algorithm to H_2-O_2 combustion between 1100° and 1600°, using reaction data from Westbrook and Chase [1]. We will study the system

$$\begin{array}{ll} 1. & H + O_2 = O + OH \,, \\ 2. & O + H_2 = H + OH \,, \\ 3. & H_2O + H = H_2 + OH \,, \\ 4. & O + H_2O = OH + OH \,, \\ 5. & H_2 + M = H + H + M \,, \\ 6. & O_2 + M = O + O + M \,, \\ 7. & OH + M = O + H + M \,, \\ 8. & H_2O + M = H + OH + M \,, \end{array} \qquad (1.1)$$

where M denotes a chaperone species. For convenience, we will refer to reaction i in the forward direction as Ri+ and define Ri- similarly. R1+ competes directly with the reaction $H + O_2 + M = HO_2 + M$, and is faster when $T>1100°$, so we can assume (Westbrook and Dryer, [2]) that H_2O_2 and HO_2 will not be present in significant amounts. We assume that the only input species that are not inert are H_2 and O_2.

2. BASIC DEFINITIONS AND FORMULAS

Suppose there are N species and R reactions in our problem. For (1.1), let species 1 through 6 be H, O, OH, H_2, O_2, and H_2O, in that order. Call Y the vector of the species' mass-fractions divided by their atomic weights, Y- the value of Y in the unburned gas, Y+ the value of Y after the gas passes through the reaction zone, and set Y* equal to Y divided by some quantity such that all components of Y* are less than or equal to O(1). For (1.1), we choose hy≡unburned concentration of H_2 as the divisor. Then define ν_i from Y^*_i by setting

$$Y^*_i \equiv e^{-\nu_i} \cdot W_i \;,\; W_i = O(1) \;, \qquad (2.1)$$

so that

$$\nu_i > 0 \quad \text{for all } i \,. \qquad (2.2)$$

Call K≡matrix of stoichiometric coefficients, and Kj- (Kj+) to be the column of negative (positive) entries from the j'th column of K which corresponds to Rj. Then we can write the log of a typical reaction rate $\omega_{j\pm}$ as

$$h_{j\pm} \equiv D_{j\pm}(\rho,\varepsilon) + \ln(A_{j\pm}) + \nu \cdot Kj\pm + n_{j\pm}\ln(T) - \frac{E_{j\pm}}{R\,T} \,, \qquad (2.3)$$

where ν≡the row vector of ν_i's and $D_{j\pm}$ is a correction made for the the possible presence of a chaperone species, which depends on the

C.-M. Brauner and C. Schmidt-Lainé (eds.), Mathematical Modeling in Combustion and Related Topics, 561–567.

chaperone efficiency ε and the gas density ρ, and evaluates to the form constant/T. For our problem, we will take $\varepsilon=1$ and assume 1 atmosphere of pressure. The fact that the flame is plane, premixed and laminar yields the formula

$$Y+^t = Y-^t + K \cdot \alpha^t \ , \qquad (2.4.a)$$

$$\alpha_j = \frac{1}{M} \int_{-\infty}^{+\infty} (\omega_{j+}(Y,\rho,T) - \omega_{j-}(Y,\rho,T)) dz \ , \qquad (2.4.b)$$

where z is the traveling wave variable across the reaction zone and α is the vector of the quantities defined in (2.4.b), called the "allocation". For the derivation of (2.4), see Fife and Nicolaenko [3]. Other details may be found in Fife and Nicolaenko [4], Fife [5], and Roten and Fife [6].

3. OUTLINE OF THE ALGORITHM

Begin by examining sets of reactions which directly compete for the same species, to find transition temperatures where the rates are equal. For our problem, the only transition temperature is below 1100°. Fix the temperature regime to be studied on one side of the transition temperature or the other. Assume that the reaction dominating dissociation in that regime (i.e., R1+ when T > 1100°) goes to completion in the reaction zone, and use its Zel'dovich number to estimate the zone thickness. Assume the weaker competing reaction does not run at all. Use the zone thickness estimate to make physically reasonable estimates for some of the components of ν in the reaction zone, simplifying the problem.

Call $E \equiv$ { indices of reactions in partial equilibrium }. At the low end of the temperature regime, assume that only the dominant reaction may be in partial equi;ibrium, so for our problem, $E \in \{ \emptyset, \{1\} \}$. Introduce a new variable s to account for this (see [5],[6]). If Rk± is the dominant reaction (k=1 for our problem), then

$$k \notin E \Rightarrow s = 0 \ , \ k \in E \Rightarrow s > 0 \ . \qquad (3.1)$$

Then (if we suppose Rk+ dominates) define

$$\beta_{j\pm} \equiv h_{k+} - h_{j\pm} - s \qquad (3.2)$$

and require

$$j \notin E \Rightarrow \beta_{j\pm} \geq 0 \ , \ j \in E \Rightarrow \beta_{j+} = \beta_{j-} < 0 \ . \qquad (3.3)$$

Note that (2.2), (3.1), and (3.3) form a set of linear constraints on ν and s for fixed T.

Fix T near the low end of the temperature regime, so that $E = \emptyset$ or $E = \{k\}$. For each of these cases, the constraints (2.2), (3.1), and (3.3) define polytopes in ν-s space. Find ν and s at all the vertices of both of these polytopes, since a maximum number of the betas will be zero at a vertex and thus (by assumption) a maximum number of (corresponding) reactions may go to completion in the reaction zone. We need to interpret ν and s at each vertex to find how α is constrained there. To do this, take (2.4.a), simplify it by assuming that distinct atomic species are conserved, and use row-echelon reduction to get a system of the form

$$Y+_i = K'_i \cdot \alpha^t + B'_i \cdot Y-^t \ , \qquad (3.4.a)$$

$$0 = K'_j \cdot \alpha^t + B'_j \cdot Y-^t \ , \qquad (3.4.b)$$

$$K'_i \ , \ K'_j \ , \ B'_i \ , \ B'_j \text{ are known} \ , \qquad (3.4.c)$$

$$K'_i \text{ and } K'_j \text{ are rows of K after row-echelon reduction} \ . \qquad (3.4.d)$$

Choose a value b (we use b=1), then assume that at each vertex

$$\nu_i \geq b \Rightarrow Y+_i = 0 \ , \ \nu_i < b \Rightarrow Y+_i > 0 \ . \qquad (3.5)$$

Choose a nonnegative figure of merit F, then compute the betas at each vertex and assume that

$$\beta_{j+} > F \Rightarrow \alpha_j \leq 0 \ , \ \beta_{j-} < F \Rightarrow \alpha_j \geq 0 \ . \qquad (3.6)$$

For our problem, we will use F=0 or F=1.5. Choose a nonnegative constant c such that c<F (we use c=.8), and then assume that for each vertex,

$$\beta_{j+} - \beta_{j-} > c \Rightarrow \alpha_j \geq 0 \ , \qquad (3.7.a)$$

$$\beta_{j-} - \beta_{j+} > c \Rightarrow \alpha_j \leq 0 \ . \qquad (3.7.b)$$

Use (3.6) and (3.7) to append inequality constraints to (3.4), and to set alphas to zero in certain cases. Then use (3.5) to replace the left-hand sides of equations (3.4.a) with zeroes and the equalities (when indicated) with inequalities. We now have transformed (3.4.a) and (3.4.b) into a set of equalities and a set of inequalities which only depend on α and Y-, and are linear in these variables. If, for a specific vertex, these sets of linear (in)equalities have no solution or the unique solution Y- = α = 0, then we say that the vertex is "incompatible" with the original chemical network, otherwise it is "compatible" and represents some nontrivial chemistry in the reaction zone. When a vertex is compatible, we need to compute the temperature change ΔT = T - T- across the reaction zone, where T- is the temperature of the unburned gas. If we assume that the vector of specific heats Cp(Y,T) is independant of temperature and that the reference enthalpies may be ignored when calculating ΔT, then we find that

$$\Delta T \cong \frac{Q\cdot\alpha^t + Cp\cdot(Y-^t - Y+^t)T_+}{Cp\cdot Y-^t} \ , \qquad (3.8)$$

where Q is the vector of (positive) heat releases.

We must now consider what happens when we increase T, since all of the preceeding calculations are performed at fixed T (to take advantage of the linearity of the betas in all their variables then). The betas generated by (2.3) and (3.2) are linear in the variables { ν , s , ln(T) , 1/T }. We can describe each vertex of our polytope by a system of equations formed by formally setting the determining betas and/or components of ν and/or s equal to zero. This system is linear in the same set of variables as the betas, and we may find ν and s as linear functions of ln(T) and 1/T by Gauss elimination and back substitution. Now simply substitute these expressions for ν and s into the non-constraining betas, components of ν, and s (if applicable) and solve the resulting nonlinear equations for the values of T at which the temperature-dependant vertex passes through the various constraint planes. Suppose a specific vertex passes through a new constraint plane at some temperature T' within our range of study and above the last temperature at which allocations were computed. For T > T', the vertex changes character, with an old constraint being replaced with a new one. The vertex is usually faceted, splitting into several new vertices as T increases past T'. Whenever the constraint set for a vertex changes, we repeat the fixed-T calculation of α and ΔT for T = T' + 0.5, to determine how the chemistry has changed. If

$$T < T' \Rightarrow \beta_{j-} \neq \beta_{j+} \ , \ T = T' \Rightarrow \beta_{j-} = \beta_{j+} = 0 \qquad (3.9)$$

then we will assume that reaction j passes into partial equilibrium as T passes through T'.

4. RESULTS FOR HYDROGEN LEAN H_2-O_2 COMBUSTION

Assume Y- = (0 , 0 , 0 , hy , ox , 0). We set ν_4=2 using the zone thickness estimate, ν_5=0 since we have a surplus of O_2, and ν_6=0 since H_2O is the principal combustion product. The only compatible vertex at

T = 1100° is given by

$$\beta_{1-} = \beta_{2+} = \beta_{3-} = \beta_{8-} = 0 \ , \qquad (4.1.a)$$

$$(\nu_1 , \nu_2 , \nu_3) = (1.5 , 2.3 , 4.1) , \ s = 1.5 , \qquad (4.1.b)$$

which yields a chemistry where the recombination of the free radical H goes to completion in the reaction zone, giving a large ΔT. We call this Chemistry 1, whose detials are as follows:

$$\alpha = (1 , 1 , -1 , 0 , 0 , 0 , 0 , -1) \frac{hy}{2} + N_1 , \qquad (4.2.a)$$

$$Y+ = (0 , 0 , 0 , 0 , ox - \frac{hy}{2} , hy) , \qquad (4.2.b)$$

$$\Delta T = \frac{52400hy}{7.0hy + .27} \quad \text{for } T \in [1300°,1340°] , \qquad (4.2.c)$$

where N_1 is a nullvector of K with zero first component, whose norm is small and which does not appear when F=0. The overall chemistry is

$$2 H_2 + O_2 \rightarrow 2 H_2O . \qquad (4.2.d)$$

As T is increased through 1311°, (4.1) facets into two new vertices:

$$\beta_{1-} = \beta_{3-} = \beta_{8-} = 0 , \ \beta_{2-} = \beta_{2+} < 0 , \qquad (4.3.a)$$

$$\beta_{1-} = \beta_{2+} = \beta_{2-} = \beta_{3-} = 0 , \qquad (4.3.b)$$

where

$$(\nu_1 , \nu_2 , \nu_3) = (0.6 , 1.4 , 2.8) , \ s = 2.0 . \qquad (4.4)$$

The chemistry of (4.3.a) is Chemistry 1, given by (4.2). We will call the chemistry of (4.3.b) Chemistry 2:

$$\begin{aligned} \alpha = {} & (1 , 1 , -2 , 0 , 0 , 0 , 0 , 0) \frac{hy}{3} \\ & + (-1, -1, 2, 0, 3, 0, 0, 0) \ \alpha_5/3) \\ & + (-1, 2, 2, 0, 0, 0, 3, 0) \ \alpha_9/3 \\ & + (-1, -1, -1, 0, 0, 0, 0, 3) \ \alpha_8/3 \\ & + (0 , -1 , -1 , 1 , 0 , 0 , 0 , 0) \ \alpha_4 , \end{aligned} \qquad (4.5.a)$$

$$\begin{aligned} Y+ = {} & (\frac{2hy}{3} , 0 , 0 , 0 , ox - \frac{hy}{3} , \frac{2hy}{3}) \\ & + (\frac{2}{3} , 0 , 0 , 0 , \frac{1}{6} , \frac{-1}{3}) \ 2 \ (\alpha_5 + \alpha_7 + \alpha_8) , \end{aligned} \qquad (4.5.b)$$

$$\Delta T = \frac{3950hy - 96900\alpha_8 - 112700\alpha_7 - 103100\alpha_5}{7.0hy + .27} \qquad (4.5.c)$$

for $T \in [1300° , 1400°]$,

$$\alpha_5 \le 0 , \ \alpha_7 \le 0 , \ \alpha_8 \le 0 , \ -2(\alpha_5 + \alpha_7 + \alpha_8) < hy . \qquad (4.5.d)$$

Chemistry 2 has the overall chemistries

$$O_2 + 3 H_2 \rightarrow 2 H_2O + 2 H \quad (\text{parameterized by } hy/3) , \qquad (4.5.e)$$

$$O_2 + 4 H \rightarrow 2 H_2O \quad (\text{parameterized by } -(\alpha_5 + \alpha_7 + \alpha_8)/3) .$$

Chemistry 2 is "weakly recombinant", in other words, recombination of surplus H via R5-, R7-, and R8- does not go to completion in the reaction zone. This implies the existance of a thick recombination layer where surplus H is recombined and explains the small ΔT when $|\alpha_5 + \alpha_7 + \alpha_8| \ll hy$. Further temperature increases through 1600° only affect (4.3.b) by further weakening R5-, R7-, and R8-. However, vertex (4.3.a) facets again as T passes through 1338°, into

$$\beta_{1-} = \beta_{8-} = 0 , \ \beta_{2-} = \beta_{2+} < 0 , \ \beta_{3-} = \beta_{3+} < 0 , \qquad (4.6.a)$$

$$\beta_{1-} = \beta_{3-} = \beta_{3+} = 0 , \ \beta_{2-} = \beta_{2+} < 0 , \qquad (4.6.b)$$

where

$$(\nu_1 , \nu_2 , \nu_3) = (0.5 , 1.3 , 2.7) , \ s = 2.2 . \qquad (4.6.c)$$

Vertex (4.6.a) has the strongly recombinant Chemistry 1, but (4.6.b) has the weakly recombinant Chemistry 2. Increasing T to 1600° only affects (4.6.b) by further weakening R5-, R7-, and R8-, just as for (4.3.b).

However, as T is increased through 1396°, (4.6.a) changes character without faceting, to become

$$\beta_{1-} = \beta_{4-} = 0 \ , \ \beta_{2-} = \beta_{2+} < 0 \ , \ \beta_{3-} = \beta_{3+} < 0 \ , \qquad (4.7.a)$$

where (at 1396°)

$$(\nu_1 , \nu_2 , \nu_3) = (0.6 , 1.3 , 2.6) \ , \ s = 2.4 \ . \qquad (4.7.b)$$

This vertex has Chemistry 2, and only undergoes further weakening of R5-, R7-, and R8- as T is increased further.

The deficient input species H_2 has hy < 1/18. When hy = O(1/18), the ΔT for Chemistry 1 is so large and the ΔT for Chemistry 2 when $|\alpha_5 \ \alpha_7 \ \alpha_8| \ll hy$ is so small that a plateau temperature T* must exist somewhere in [1311°,1396°] toward which the actual flame temperature must tend. When hy=O(1/18), T < 1311° implies so large a ΔT that T > 1311°, which is a contradiction. A similar contradiction exists for T >> 1396°. Thus we deduce the existence of the plateau temperature T*. However, the chemistry between 1311° and 1396° exhibits two bifurcations, and which branch of (4.3) or (4.6) is stable is still an open problem. Hence, we are not yet able to say which of our most reasonable candidates, { 1311° , 1338° , 1396° }, is the plateau temperature, but only that it must lie in the interval [1311°,1396°].

5. RESULTS FOR OXYGEN LEAN H_2-O_2 COMBUSTION

As before, assume Y- = (0 , 0 , 0 , hy , ox , 0). We use the zone thickness to set ν_5=2, ν_4=0 since we have an O_2 surplus, and ν_6=0 since H_2O is the principal combustion product. At T = 1100°, the only compatible vertex is

$$\beta_{1+} = \beta_{2+} = \beta_{3-} = \beta_{5-} = 0 \ , \qquad (5.1.a)$$

$$(\nu_1 , \nu_2 , \nu_3) = (0.9 , 4.1 , 5.9) \ , \ s = 0 \ . \qquad (5.1.b)$$

The chemistry is given by

$$\alpha = (1 , 1 , -2 , 0 , -1 , 0 , 0 , 0) \ ox \ , \qquad (5.2.a)$$

$$Y+ = (0 , 0 , 0 , hy - 2ox , 0 , 2ox) \ , \qquad (5.2.b)$$

$$\Delta T \cong \frac{111000ox}{3.7 - 110ox} \ . \qquad (5.2.c)$$

This chemistry is strongly recombinant, and the combustion is completed in the reaction zone. The overall chemistry is

$$2 \ H_2 + O_2 \rightarrow 2 \ H_2O \ , \qquad (5.2.d)$$

which is the same as (4.2.d). The behavior of this vertex as T is increased is much more clear cut then the behavior of the vertices in the hydrogen lean problem. As T is increased, ν_1 decreases and at 1224°, (5.1) intersects ν_1 = 0 and changes into

$$\beta_{1+} = \beta_{2+} = \beta_{3-} = \nu_1 = 0 \ , \qquad (5.3.a)$$

$$(\nu_1 , \nu_2 , \nu_3) = (0 , 2.9 , 4.5) \ , \ s = 0 \ . \qquad (5.3.b)$$

No faceting takes place here, so the chemistry cannot bifurcate. The new chemistry is weakly recombinant, since R5- performs the recombination of surplus H, and now no longer goes to completion in the reaction zone. The new chemistry is given (when F=1.5) by

$$\begin{aligned} \alpha = & (1 , 1 , -2 , 0 , 0 , 0 , 0 , 0) \ ox \\ & + (0 , 0 , 0 , 0 , 1 , 0 , 0 , 0) \ \alpha_5 \\ & + (0 , 0 , -1 , 0 , 0 , 0 , 0 , 1) \ \alpha_8 \ , \end{aligned} \qquad (5.4.a)$$

$$\begin{aligned} Y+ = & (2ox , 0 , 0 , hy - 3ox , 0 , 2ox) \\ & + (2 , 0 , 0 , -1 , 0 , 0) (\alpha_5 + \alpha_8) \ , \end{aligned} \qquad (5.4.b)$$

$$\Delta T \cong \frac{12100ox - 93000\alpha_8 - 99000\alpha_5}{3.7 - 110ox} \ , \qquad (5.4.c)$$

$\alpha_5 \le 0$, $\alpha_8 \le 0$, $-(\alpha_5+\alpha_8) < ox$, (5.4.d)

and the overall stoichiometries are

$O_2 + 3 H_2 \rightarrow 2 H_2O + 2 H$ (parameterized by ox) , (5.4.e)

$2 H \rightarrow H_2$ (parameterized by $-(\alpha_5+\alpha_8)$) .

Further increase of T simply increases the betas for R5- and R8-, thus weakening R5- and R8-, decreasing $|\alpha_5|$ and $|\alpha_8|$, and thus slowing the recombination still further. We can easily see that the result of this is a decrease in ΔT as T is increased. As T increases past 1523°, reaction 3 enters partial equilibrium, but this does not change the chemistry at the vertex, which remains that of (5.4).

The deficient input species O_2 has ox < 1/36. As before, ox = O(1/36) implies the existence of a thermal plateau temperature T*, since ΔT for (5.2) is unreasonably large when T < 1224°, and ΔT for (5.4) is unreasonably small when T >> 1224°. However, since no bifurcation of the chemistry takes place, we can easily see that the plateau temperature is T* = 1224°. We expect to find the actual flame temperature slightly above T*.

6. DISCUSSION AND OPEN PROBLEMS

In this work, we have treated the reaction zone as a "black box", and have not attempted to construct flame profiles across it by asymptotic analysis. How this should be done is still not clear. Overall reactions such as (4.5.e) or (5.4.e) will have rates which cannot be modeled by the application of mass-action/Arrhenius kinetics to the overall stoichiometric equations. For the recombinant portions, parameterized by alphas corresponding to chain-terminating reactions which do not go to completion, such as R5- and R8-, a case might be made that the rates of the original reactions (such as R5- if α_5 is the dominant parameter) might be used. This does not go through with the chain branching and chain propagating network which goes to completion in the reaction zone. Hence, the use of the overall chemistries we can deduce, in a simplified asymptotic analysis of the flame profile, is not possible at present.

Some means must be found to deal with the question of nonuniqueness of chemical pathways. Nullvectors of K, such as N_1 in (4.2.a) or the α_4 component of(4.5.a), only represent nonuniqueness of α, and are not reflected in Y+ or ΔT. Geometric bifurcation of a vertex (faceting) can cause greater difficulties when it lead to bifurcation of the stoichiometry as we saw in section 4. We have the equivalent of a bifurcation diagram, without the accompanying stability analysis. Since we do not know which of the alternate chemistries is unstable when such branching occurrs, we cannot arrive at a specific plateau temperature. If we can construct the flame profile asymptotically, we might be able to tackle the stability question.

Different sets of reaction data may lead to different results, especially for the plateau temperatures. The data set of Warnatz [7] is presently being studied using the algorithm of section 3, to begin an attempt to determine the importance of such sensitivity.

There is nothing about the algorithm of section 3 which restricts it to H_2-O_2 combustion, or which restricts the complexity of the reaction network which we may study. As we have seen, we can also obtain results about reactions with low activation energies, such as R5-, R7-, and R8- for (1.1). We can treat systems where many of the reactions have low activation energies. The algorithm of section 3 has been implemented as a set of routines in the MACSYMA language, and the only upper limit on the complexity of the chemical networks we can study appears to be computer time.

REFERENCES

1. C. K. Westbrook and L. L. Chase, Chemical kinetics and thermochemical data for combustion applications, Lawrence Livermore Lab Report, 1980.
2. C. K. Westbrook and F. L. Dryer, Chemical kinetics modelling of hydrocarbon combustion, Prog. in Energy and Ccombustion Sci. 10, 1-57, 1984.
3 P. C. Fife and B. Nicolaenko, Flame structure for reaction networks with chain branching and recombination, Lectures in Applied Mathematics Volume 24, 273-292, Amer. Math. Soc., Providence, 1986.
4. P. C. Fife and B. Nicolaenko, How chemical structure determines spatial structure in flame profiles, Modelling of Patterns in Space and Time, W. Jager and J. D. Murray (ed), Lecture Notes in Biomathematics 55, 73-86, Springer Verlag, Berlin, 1984.
5. P. C. Fife, Modelling and analysis of hydrogen-oxygen flames, preprint.
6. C. D. Roten and P. C. Fife, An analysis of H_2 - O_2 combustion between 1100° and 1600°, preprint.
7. J. Warnatz, Calculation of the structure of laminar flat flames II: Flame velocity and structure of freely propagating hydrogen-oxygen and hydrogen-air flames, Ber. Bunsenges Phys. Chem. 82, 643-649, 1978.

Existence and stability in a plane premixed flame problem

Cl. Schmidt-Lainé

CNRS, Ecole Centrale de Lyon, France.

The purpose of this presentation is to sum up several results on a non-linear system modelling a two-step reaction in a plane premixed flame, introduced by G. Joulin, A. Linan, G.S.S. Ludford, N. Peters, and the author. This work was initiated during my first visit to Geoffrey Ludford at Cornell University (September-December 1983).

I. Physical framework

In [4], a two-step irreversible reaction is considered for a steady plane flame, with chain-branching/chain-breaking kinetics :

(1) $A + X \to 2X \ , \ 2X + M \to 2P + M$

Radical X is obtained in the production step, which has a very large activation energy θ, and provides product P in the recombination step for which the activation energy is taken to be zero ; A is the reactant and M a third body.

This two-step sheme is presented as an alternative to classical single-step kinetics and allows the description of a wider range of phenomena.

The equations are derived in the stretched flame zone, described by the one-dimensional space variable η, $-\infty < \eta < +\infty$, by assuming a fast recombination, i.e. that both production and recombination of radicals take place in the same thin zone. The system reads (see [4 p. 423]).

$$(2)\quad \begin{cases} u'' = q_1 \ r \ \delta(u-v)v \ e^{-u} + \delta(u-v)^2 \\ v'' = r \ \delta(u-v)v \ e^{-u} \end{cases}$$

and the associated boundary conditions are :

$$(3)\quad \begin{cases} u = -\eta + o(1) \ , \ v = -\eta + o(1) \ \text{as} \ \eta \to -\infty \\ u = o(1) \ , \ v = o(1) \ \text{as} \ \eta \to +\infty \end{cases}$$

The unknowns are positive functions u and v, representing temperature and mass fraction of the reactant and a positive constant δ , representing the burning rate. The parameters q_1 and q_2 are the proportions of the total heat released in the first and second steps of the reaction, so that $q_1 + q_2 = 1$. Physical considerations require the recombination step to be exothermic, so that $q_2 > 0$.

C.-M. Brauner and C. Schmidt-Lainé (eds.), Mathematical Modeling in Combustion and Related Topics, 569–577.

Finally r is a positive parameter, corresponding to the ratio of the two reaction rates. The boundary conditions (3) are obtained by matching with expansions on either side of the flame sheet.

The mathematical problem is the following : q_1 , q_2 and $r > 0$ being given, find two functions $u > 0$, $v > 0$ and the constant $\delta > 0$ satisfying (2) (3). We refer to [4] for a numerical treatment of this problem, leading to curves in the (r,δ) - plane.

It is particularly convenient for our study to deal with an equivalent formulation of the system involving the radical mass fraction $w = u - v > 0$.

It consists of both systems :

$$(4)\quad \begin{cases} v'' = r\,\delta\, v\, w e^{-v} e^{-w} \\ w'' = -q_2\, r\,\delta\, v\, w\, e^{-v} e^{-w} + \delta\, w^2 \end{cases}$$

and

$$(5)\quad \begin{cases} u'' = q_1\, r\,\delta(u-w) w\, e^{-u} + \delta w^2 \\ w'' = -q_2\, r\,\delta(u-w) w\, e^{-u} + \delta\, w^2 \end{cases}$$

together with the boundary conditions :

$$(6)\quad \begin{cases} v = -\eta + o(1)\ ,\ w = o(1) \text{ as } \eta \to -\infty \\ v = o(1)\ ,\ w = o(1) \text{ as } \eta \to +\infty \end{cases}$$

and

$$(7)\quad \begin{cases} u = -\eta + o(1)\ ,\ w = o(1) \text{ as } \eta \to -\infty \\ u = o(1)\ ,\ w = o(1) \text{ as } \eta \to +\infty \end{cases}$$

II. The existence question [1], [2], [5]

Let us introduce the following problem in the x-variable, which is obtained from (4) by taking δ equal to 1 :

$$(8)\quad \begin{cases} \dfrac{d^2v}{dx^2} = rv\, w\, e^{-v} e^{-w} \\[2ex] \dfrac{d^2w}{dx^2} = -q_2\, r\, v\, w\, e^{-v} e^{-w} + w^2 \end{cases}$$

We consider this nonlinear system subject to the following boundary conditions :

$$(9)\quad \begin{cases} v' \to 0 \text{ as } x \to +\infty \\ w \to 0 \text{ as } x \to \pm\infty \end{cases}$$

together with the normalization condition

$$(10)\quad v(0) = 1.$$

We prove the following theorem :

Theorem 1 : For any $r > 0$ fixed, there exists a solution $(v,w) \in (C^\infty(\mathbb{R}))^2$ to problem (8), (9), (10) such that

(11) $\quad 0 < w < M_o = q_2\, r/e$

(12) $\quad v > 0\ ;\ v' < 0\ ;\ v \to +\infty$ as $x \to -\infty$, $v \to 0$ as $x \to +\infty$.

Moreover, $w \in H^2(\mathbb{R})$ and, as $x \to -\infty$, $v(x) = -\ell(x - x_o) + $ E.S.T., where $x_o \in \mathbb{R}$ and

(13) $$\ell = \frac{1}{q_2}\int_{-\infty}^{+\infty} w^2(x)\,dx = \frac{1}{q_2}\,||w||^2_{L^2(\mathbb{R})}\ . \quad \blacksquare$$

To return to the initial problem (4) (6), we make the transformation

(14) $\quad \eta = \ell(x - x_o)$

Then (8) becomes

(15) $$\begin{cases} \dfrac{d^2v}{d\eta^2} = \dfrac{1}{\ell^2}\ r\, v\, w\, e^{-v}\, e^{-w} \\ \dfrac{d^2w}{d\eta^2} = -\dfrac{1}{\ell^2}\, q_2\, r\, v\, w\, e^{-v}\, e^{-w} + \dfrac{1}{\ell^2}\, w^2 \end{cases}$$

and, as $\eta \to -\infty$, $v(\eta) = -\eta + $ E.S.T. Thus it is clear that (4) (6) is solved by the pair $(v(\eta), w(\eta))$ with

(16) $\quad \delta = 1/\ell^2$

Therefore, we have determined

(17) $$\delta = q_2 \Big/ \int_{-\infty}^{+\infty} w^2(\eta)\,d\eta\ .$$

In [2], we give a detailed demonstration of Theorem 1. A sketch of the proof is the following :

First, note that (8)-(10) is a nonlinear eigenvalue problem which admits a trivial solution. In order to prove the existence of a nontrivial solution, a positive lower solution $\underline{w}$ is needed. The main trick is a logarithmic transformation $f = -\ell n\, \underline{w}$, but it turns out that a large part of the analysis lies on elementary properties of the mapping $t \to t\, e^{-t}$.

Next, we consider the system (8), on a bounded interval $[-a,b]$ with suitable boundary conditions and we prove the existence of a solution (v,w) in a closed convex set K, by a fixed point argument (the convex K involves the upper and lower solutions). We derive a priori estimates in the C^1-norm as well as in a usual Sobolev space. Finally, we let a and b tend to $+\infty$, and prove the existence of a limit which solves (8) (9) (10).

The asymptotic behaviour of v as x goes to $-\infty$, as well as properties (12), (13) appear as a "spin off" of the existence proof, namely $v'(x) \to -\ell$ and as $w'' + q_2\, v'' = w^2$, $q_2\, \ell = \int_{-\infty}^{+\infty} w^2(x)\,dx$.

An alternative proof by a topological shooting method has been presented by S.P. Hastings, C. Lu and Y.H. Wan [3].

III. The stability question [5], [6], [7]

For the stability question, the solution of (5) (7) is envisaged in terms of orbit. Of special interest is the behaviour of the system in the neighbourhood of the space variable bounds $\pm\infty$. For this purpose, system (5) can be rewritten in the canonical form :

$$
(18)\quad \begin{cases} u' = p \\ p' = q_1\, r\, \delta(u-w)we^{-u} + \delta w^2 \\ w' = q \\ q' = -\, q_2\, r\delta\, (u-w)we^{-u} + \delta w^2 \end{cases}
$$

for which both boundary conditions (7) appear to be degenerate fixed points. However, as η goes to $-\infty$, an approximated system derived from (18) decouples. Therefore the corresponding behaviour is well known. On the other hand, the fixed point $0 \in \mathbb{R}^4$, derived from the boundary condition as η goes to $+\infty$ requires more sophisticated treatment. By linearizing (18) around this point, we obtain 4 zero-eigenvalues ! let us then consider the second order approximation of (18) near this point :

$$
(19)\quad \begin{cases} u' = p \\ p' = q_1\, r\, \delta(u-w)\, w + \delta\, w^2 \\ w' = q \\ q' = -\, q_2\, r\, \delta(u-w)\, w + \delta\, w^2 \end{cases}
$$

This system only contains quadratic terms and can be considered in the general form of second order homogeneous problem.

So, we consider the problem

$$
(20)\quad \frac{d^2X}{d\eta^2} = X'' = F(X)
$$

where $X \in \mathbb{R}^n$ and F is a second order homogeneous function. More precisely, we study the stability of the degenerate fixed point $0 \in \mathbb{R}^{2n}$ of the autonomous dynamical system :

$$
(21)\quad \begin{cases} X' = Y \\ Y' = F(X) \end{cases}
$$

The homogeneity property allows us to reduce the dimension of problem (21). By means of the change of functions :

$$
(22)\quad x = \frac{X}{\|X\|} \; ; \; y = \frac{Y}{\|X\|^{3/2}}
$$

such that $(X,Y) \in \mathbb{R}^n \times \mathbb{R}^n \rightarrow (x,y) \in S^{n-1} \times \mathbb{R}^n$

where S^{n-1} is the unit sphere of $\mathbb{R}^n$, together with the change of variable defined by

$$\frac{ds}{dn} = ||x||^{1/2} \tag{23}$$

the reduced autonomous dynamical system is the following :

$$\left\{\begin{aligned} \frac{dx}{ds} &= y - x(x.y) \\ \frac{dy}{ds} &= F(x) - \frac{3}{2}\, y(x.y) \end{aligned}\right. \tag{24}$$

where (.) denotes the euclidian scalar product in $\mathbb{R}^n$, and $||.||$ the associated norm.

Then we prove existence of at least a couple of symmetric fixed points of (24) : $P_i = (x_i, y_i) \in S^{n-1} \times \mathbb{R}^n$ and exhibit the mapping between trajectories of (21) and trajectories of (24) ; an essential remark is that a trajectory of (24) provides by lifting a <u>family</u> of trajectories of (21), due to an arbitrary integration parameter ; the stability result is the application of the former property to invariant manifolds :

<u>Theorem 2</u> : Let $P_o = (x_o, y_o) \in S^{n-1} \times \mathbb{R}^n$, $n \geq 1$, be a fixed point of (24), such that $(x_o . y_o) < 0$. Then

(i) The stable manifold $W^s(P_o)$ of problem (24) lifts to the stable manifold $W^s(0)$ of homogeneous problem (21).

(ii) $\dim W^s(0) = \dim W^s(P_o) + 1$. ■

The main interest of this result is that investigation of the stable manifold $W^s(P_o)$ of problem (24) is often accessible by classical tool, namely by linearization, because the fixed points of (24) are usually hyperbolic. A complete stability analysis of fixed points P_o of (24) is presented in [6], leading to results on the dimension of the stable manifold $W^s(0)$ of (21). These results are summarized as follows :

<u>Theorem 3</u> : Let us suppose that $F'(x_o)$ is similar to a diagonal matrix and that its eigenvalues λ_j are such that $\lambda_j = a_j + ib_j$:

$$\left\{\begin{aligned} & 6b_j^2 \neq 25\, \rho\,(\rho - a_j) \\ & \rho = \frac{3}{2}\,(x_o . y_o)^2 \end{aligned}\right. \tag{25}$$

Then by lifting of $W^s(P_o)$ to $W^s(0)$

$$\dim W^s(0) = \sum_{j \in J} m(\lambda_j) \tag{26}$$

where

$$J = \{\, j \in \mathbb{N} \ / \ 6b_j^2 > 25\, \rho\,(\rho - a_j) \}$$

and $m(\lambda_j)$ is the geometrical multiplicity of the eigenvalue λ_j . ■

Applying this general method to (19) leads to a reduced system for which two fixed points P_o and P_1 are obtained, verifying $(x_o . y_o) < 0$, $(x_1 . y_1) < 0$.

As r = 1, the two fixed points are no more distinct and this unique fixed point is degenerate. For $r \neq 1$, we obtain by linearization and lifting to (19):

(27) $\dim W^s(0) = 2$

The transcritical bifurcation case r = 1 is solved by means of the Center Manifold Theorem.

The last point is then to prove the topological equivalence between the nonlinear system (18) and its second order approximated system (19), analysed above.

For this purpose the change of functions and variable :

(28) $x = p^2 u^{-3}$, $y = w u^{-1}$, $z = q^2 u^{-3}$

(29) $ds/d\eta = u^{1/2}$

is introduced in both systems (18) and (19), to obtain two reduced systems. The main trick is that, due to the homogeneity property, the reduced system of (19) is autonomous although the corresponding reduced system of (18) is not autonomous. Let us introduce a new function m defined by

(30) $m = u^{1/2}$

and solution of the differential equation :

(31) $dm/ds = -\frac{1}{2} m x^{1/2}$

Rewriting (18) and (19) in term of functions x, y, z, m with respect to the variable s provides two nonlinear systems of dimension 4, with 2 hyperbolic fixed points. Moreover, their Jacobian matrices near these points are identical.

Returning to the original functions and variable we generalize result (27) to the nonlinear problem (18). We refer to [7] for a detailed proof.

Finally the transcritical bifurcation case r = 1 is solved by means of the Center Manifold Theorem for both systems (18) and (19). See figure 1.

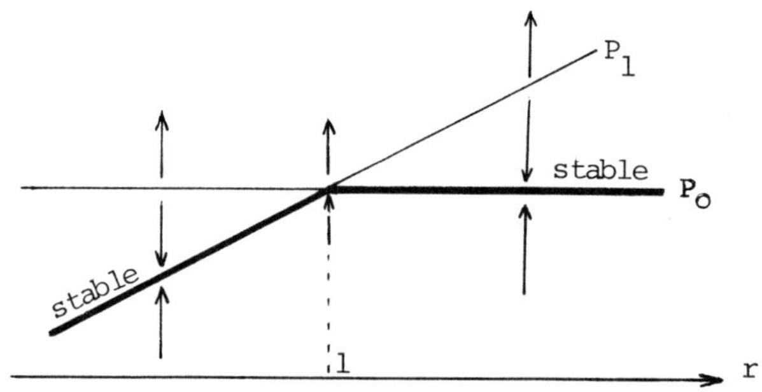

- Figure 1 -

IV. Numerical and asymptotic aspects

In order to end up with characteristic properties obtained on this two-step model, we recall numerical and asymptotic results of [4], [5]. The numerical scheme consists in integrating (18) for a given q_1, δ being fixed, from $\eta = -a$ to $\eta = 0$ as well as from $\eta = +b$ to $\eta = 0$ (with $a,b > 0$ and large). The parameter r is then determined by minimizing a quadratic function whose 0 ensures the C^1 - matching of u and w in 0. Good asymptotic estimates are needed to determine relevant initial conditions at both ends of the domain (see in section III, the analysis of singular points!). Fast convergence was obtained whenever the starting estimates of r and these initial conditions were sufficiently accurate.

Figure 2 shows typical u - and w - profiles. (a and b were set equal to 10 in all the numerical work reported). The smooth agreement at $\eta = 0$ of the left and right integration is evident.

Figure 3 gives the (r,δ) curves for various values of q_1 less than 1, i.e. $q_2 > 0$. In particular, two negative values of q_1 are included, corresponding to endothermic production steps.

Figure 4 shows a plot, for $q_1 = \frac{1}{2}$ of the (r,δ) curve together with its asymptotic approximations as $r \to 0$ and $r \to +\infty$, obtained analytically.

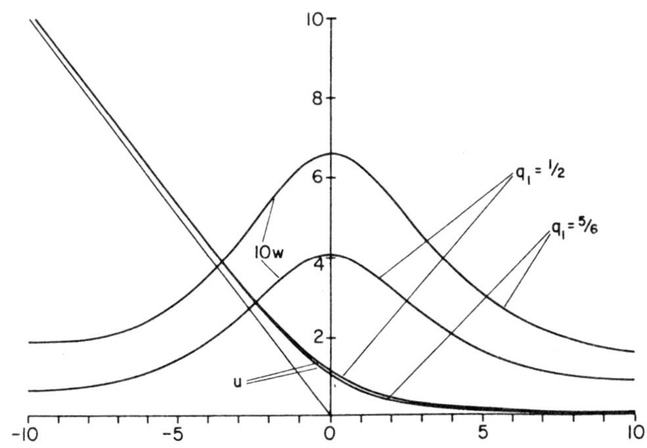

Profiles of u and w for $\delta = 2q_2$.

- Figure 2 -

- Figure 3 -

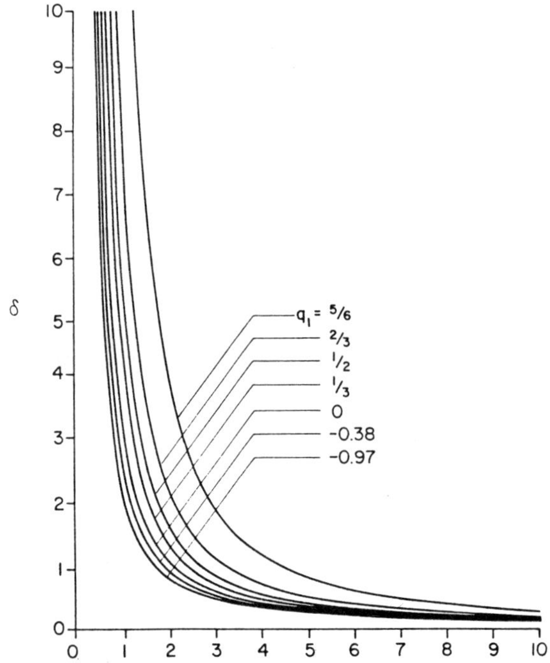

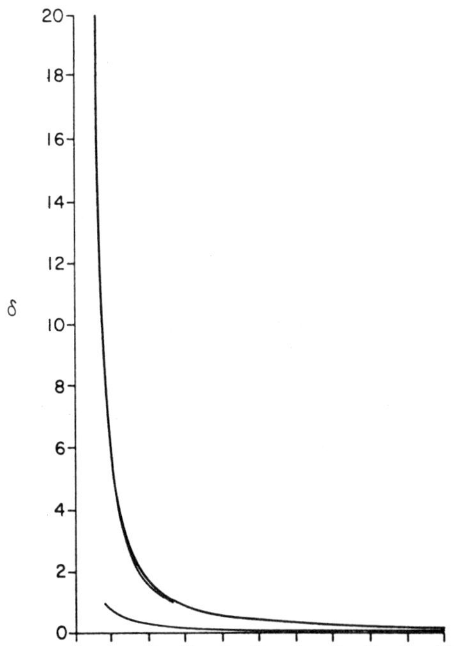

- Figure 4 -

REFERENCES

[1] C.M. Brauner, cl. Schmidt-Lainé,
Existence d'une solution pour le problème de la flamme plane prémélangée avec cinétique à deux pas, C.R. Acad. Sc. Paris, Série I, 301, (1985), 667-670.

[2] C.M. Brauner, Cl. Schmidt-Lainé,
Existence of a solution to a certain plane premixed flame problem with two-step kinetics. SIAM J. Math. Anal., 18 (1987).

[3] S.P. Hastings, C. Lu, Y.H. Wan,
A three dimensional shooting method as applied to a problem in combustion theory. To appear in Physica D.

[4] G. Joulin, A. Liñan, G.S.S. Ludford, N. Peters, Cl. Schmidt-Lainé,
Flames with chain Branching/chain breaking kinetics, SIAM J. Appl. Math, 45 (1985), 420-434.

[5] Cl. Schmidt-Lainé,
Sur quelques problèmes non linéaires en Mécanique des Fluides, Chimie, et Combustion, Thèse d'Etat, Université Lyon I, 1985.

[6] Cl. Schmidt-Lainé, D. Serre,
Etude de stabilité d'un système non linéaire de dimension 4 en combustion et généralisation à une classe de problèmes homogènes de degré 2, Physica, 21 D, (1986), 42-60.

[7] Cl. Schmidt-Lainé, D. Serre,
Etude de stabilité d'un système non linéaire de dimension 4 en combustion au voisinage d'un point fixe dégénéré. C.R. Acad. Sc. Paris, Série I, 303, (1986), 551-554.

Computation of Turbulent Diffusion and Premixed Flames with radiation

P.L. VIOLLET, D. GARRETON, N. MECHITOUA
EDF/DER/Laboratoire National d'Hydraulique
6, quai Watier, 78400 Chatou France

ABSTRACT :
A numerical procedure for the computation of steady turbulent flames with swirl is presented. The combustion process involves a single chemical reaction, with a probability density function fitted using mean and RMS computed mixing rate values for diffusion flames, and an Eddy-Break-Up model for premixed flames. The k-epsilon model is used for turbulence, and the flux model together with a soot production model is used for radiation. The solution procedure uses finite-difference methods in 2D, with a characteristics method for advection, in order to minimize numerical diffusion. A particular treatment of solid boundaries not parallel to coordinate axis makes it possible to map relatively complex geometries.

1. INTRODUCTION

The purpose of the presented numerical model, in two dimensions, is the computation of plane or axisymmetric steady turbulent diffusion and premixed flames in enclosures or in open air.

2. DESCRIPTION OF THE MODEL

2.1 Basic equations

The equations for Favre averaged velocity and enthalpy are written in the form :

Continuity : $$\frac{\partial}{\partial x_i}(\bar{\rho}\,\tilde{u}_i) = 0 \tag{1}$$

Navier-Stokes : $$\bar{\rho}\,\tilde{u}_j \frac{\partial \tilde{u}_i}{\partial x_j} = -\frac{\partial \bar{P}}{\partial x_i} - \frac{\partial}{\partial x_j}\left(\overline{\rho u''_i u''_j} - \mu \frac{\partial \tilde{u}_i}{\partial x_j}\right) + \bar{\rho} g_i \tag{2}$$

Energy balance : $$\bar{\rho}\,\tilde{u}_j \frac{\partial \tilde{H}}{\partial x_j} = -\frac{\partial}{\partial x_j}\left(\overline{\rho u''_j H''} - \frac{\lambda}{C_p}\frac{\partial \tilde{H}}{\partial x_j}\right) + S \tag{3}$$

State equation : $$\bar{P} = \bar{\rho}\,\frac{R\,\tilde{X}_l\,\tilde{T}}{M_l} \tag{4}$$

Here, the horizontal bar denotes conventional (Eulerian) averaged values, while the symbol ~ denotes Favre averaged values. u'', H'' are fluctuating quantities with respect to the Favre averaged $\tilde{u}$ and $\tilde{H}$. H is the total enthalpy per unit of mass of the reacting mixture, defined from the mass fraction X_l and the enthalpy h_l of each species l (with molar mass M_l) by :

$$H = h_l\, X_l \tag{5}$$

$$h_l = h_{lo} + \int_0^T C_{pl}\, dT \tag{6}$$

C.-M. Brauner and C. Schmidt-Lainé (eds.), Mathematical Modeling in Combustion and Related Topics, 579–588.

where h_{lo} the enthalpy of formation of the species l.
S is the source term in the energy balance due to radiation.
In addition to these equations, the mixing phenomena between the fuel and oxydant flows are described using the mixing rate f, defined as :

$$f = \frac{y_1 - y_{12}}{y_{11} - y_{12}} \qquad (7)$$

where y_1 is the mass fraction of any element l(C,H,O etc...), y_{11} the mass fraction of element l in the pure fuel flow and y_{12} the mass fraction of element l in the pure oxydant flow. With these conventions, f=1 in the pure fuel and f=0 in the pure oxydant. It has been assumed here that the flow is turbulent enough for the differences in the diffusion coefficients between the different elements l to be of no importance, all these diffusion coefficients being assumed equal to a single value, denoted as D.
The transport equation for Favre-averaged $\widetilde{f}$ is :

$$\bar{\rho}\,\tilde{u}_j \frac{\partial \tilde{f}}{\partial x_j} = - \frac{\partial}{\partial x_j}\left(\overline{\rho u''_j f''} - \bar{\rho} D \frac{\partial \tilde{f}}{\partial x_j}\right) \qquad (8)$$

An eddy viscosity k-ε turbulence model is used for the closure, assuming :

$$\overline{\rho u''_i u''_j} = - \mu_T \left(\frac{\partial \tilde{u}_i}{\partial x_j} + \frac{\partial \tilde{u}_j}{\partial x_i}\right) + \frac{2}{3}\left(\bar{\rho}\tilde{k} + \mu_T \operatorname{div} \vec{\tilde{u}}\right) \delta_{ij} \qquad (9)$$

$$\overline{\rho u''_i H''} = - \frac{\mu_T}{\sigma_T} \frac{\partial \tilde{H}}{\partial x_i} \qquad (10)$$

$$\overline{\rho u''_i f''} = - \frac{\mu_T}{\sigma_T} \frac{\partial \tilde{f}}{\partial x_i} \qquad (11)$$

The eddy viscosity μ_T is obtained from the turbulent kinetic energy k and the dissipation rate ε :

$$\mu_T = \bar{\rho}\, C_\mu \frac{k^2}{\varepsilon} \qquad (12)$$

The transport equation for $\widetilde{f''^2}$ is written as :

$$\bar{\rho}\,\tilde{u}_j \frac{\partial \widetilde{f''^2}}{\partial x_j} = - \frac{\partial}{\partial x_j}\left(\frac{\mu_T}{\sigma_T}\frac{\partial \widetilde{f''^2}}{\partial x_j} + \bar{\rho} D \frac{\partial \widetilde{f''^2}}{\partial x_j}\right) - 2\,\overline{\rho u''_i f''}\frac{\partial \tilde{f}}{\partial x_i} - \frac{2}{C'_T}\frac{\widetilde{f''^2}}{k}\bar{\rho}\varepsilon \qquad (13)$$

The values of the constants of the turbulence model are taken from standard values used in incompressible flows, after Launder and Spalding (1974) and Launder (1980) :

C_μ	$C_{\varepsilon 1}$	$C_{\varepsilon 2}$	σ_T	σ_k	σ_ε	C'_T
0.09	1.44	1.92	1	1	1,3	1,6

2.2 Combustion models

For the knowledge of the species mass fractions X_1 appearing in (4) and (5), combustion modelling is necessary. The combustion models used for turbulent diffusion and premixed flames are based upon the assumption of a simple chemical reaction with fast chemistry with respect to the turbulent mixing time scale.

2.2.1 Turbulent diffusion flames (after Jones and Whitelaw (1982))

With these assumptions, it can be shown that the instantaneous mass fractions X_1 (fuel), X_2 (oxydant), X_3 (products) are functions only of the mixing rate f defined by (7), in the following way :

$$\begin{cases} f < f_s \\ X_1(f) = 0 \\ X_2(f) = (f_s - f)/f_s \\ X_3(f) = f/f_s \end{cases} \quad (14) \qquad \begin{cases} f > f_s \\ X_1(f) = (f - f_s) / (1 - f_s) \\ X_2(f) = 0 \\ X_3(f) = (1-f)/(1-f_s) \end{cases} \quad (15)$$

Here, f_s is the mixing rate value in the pure products, and is defined from the stoechiometry of the studied reaction.

In order to obtain the averaged $\tilde{X}_1$, the probability density function of the mixing rate, P(f), is introduced, making it possible to write :

$$\tilde{X}_1 = \int_0^1 X_1(f)\, P(f)\, df \qquad (16)$$

The p.d.f. is modelled as described by beta functions, whose parameters are deduced from the mean value $\tilde{f}$ and the variance $\widetilde{f''^2}$.

$$P(f) = \frac{f^{a-1}\,(1-f)^{b-1}}{\int_0^1 f^{a-1}(1-f)^{b-1}\, df} \qquad (17)$$

with $a = \tilde{f}\,\dfrac{\tilde{f}(1-\tilde{f}) - \widetilde{f''^2}}{\widetilde{f''^2}}$ and $b = (1-\tilde{f})\,\dfrac{\tilde{f}(1-\tilde{f}) - \widetilde{f''^2}}{\widetilde{f''^2}}$

2.2.2 Turbulent premixed flames

In this case, the combustion is controled by the turbulent mixing of fresh and burnt gas. If Xo is the mass fraction of methane in the fresh gas, Z = Xl/Xo is equal to 1 in the fresh gas, 0 if the particule has burnt. The mean reaction rate $\tilde{\dot{w}}_o$ of the non-dimensional mass fraction of fuel is assumed to be proportional to a turbulence frequency ($1/\tau = \varepsilon/k$) and the variance of Z, $\widetilde{Z''^2}$.

If C is an empirical constant, $\tilde{\dot{w}}_o = -\,C.\,\varepsilon/k \;\; \widetilde{Z''^2}$

$$Z^2 = Z \;\longrightarrow\; \widetilde{Z''^2} = \widetilde{Z^2} - \tilde{Z}^2 = \tilde{Z}\,(1 - \tilde{Z})$$

The mean rection rate $\tilde{\dot{w}}_F$ for the mass fraction of fuel becomes :

$$\tilde{\dot{w}}_F = -\,C\,\varepsilon/k\;\tilde{X}_1\,(1 - \tilde{X}_1 / X_o)$$

A model based upon a lagrangian approach (Dupoirieux, ONERA*) is also used

assuming : $\tilde{\dot{w}}_F = -\,C\,\dfrac{\varepsilon}{k}\,\tilde{X}_1\,\left(1 - \dfrac{\tilde{X}_1}{\tilde{X}}\right)$

where X is a passive scalar, equal to X_1 at the boundaries.

* Numerical calculations of turbulent reactive flows and comparison with experimental results, Novosibirsk, Juillet 1986.

2.3 Radiation model

The source term S due to radiation in (3) is expressed by :

$$S = - \operatorname{div} \vec{q}/Cp \tag{18}$$

Where the radiative flux $\vec{q}$ is obtained from the intensity $L(\vec{S})$ of radiation (which is a function of the direction $\vec{S}$ of radiation propagation) after integration for all directions :

$$\vec{q} = \int_{4\pi} L(\vec{S})\, \vec{S}\, d\Omega \tag{19}$$

The intensity L is given by the radiation transfer equation (neglecting diffusion of radiation) :

$$\vec{S}\, \overrightarrow{\operatorname{grad}}\, L = - K\, L + K\, \frac{\sigma T^4}{\pi} \tag{20}$$

The absorption coefficient K is obtained, following Modak (1978), as a function of the concentration in CO_2, H_2O, and soot particles. The soot transport and production model, following Tesner (1971) and Galant (1984), involves two equations for :

- precursors formation (where n is the precursors concentration, resulting from fuel cracking :

$$\bar{\rho}\, \tilde{u}_j \frac{\partial n}{\partial x_j} = - \frac{\partial}{\partial x_j} (\overline{\rho u''_j n''}) + A_0 X_1 \bar{\rho}\, e^{-E/RT} + B\, \tilde{n} X_1 - g_0\, \tilde{n}(\tilde{N} + \tilde{n}) \tag{21}$$

- soot formation (where N is the soot concentration, resulting from coagulation around the precursors) :

$$\bar{\rho}\, \tilde{u}_j \frac{\partial \tilde{N}}{\partial x_j} = - \frac{\partial}{\partial x_j} (\overline{\rho u''_j N''}) + A\, \tilde{n}\, (\tilde{N} + \tilde{n}) \tag{22}$$

The values of empirical constants, as used in the present applications, are listed below :

A_0	B	g_o	A
$1.16\ 10^{31}$	100	10^{-15}	10^{-17}

3. NUMERICAL TECHNIQUES

3.1 Fluid dynamics and transport equations

The equations are solved in two dimensions (either formulated in plane or axisymetric geometry), with, in the axisymetric case, three components of the velocity, allowing a swirl to be modelled. A finite difference staggered grid (see figure 1) is used.

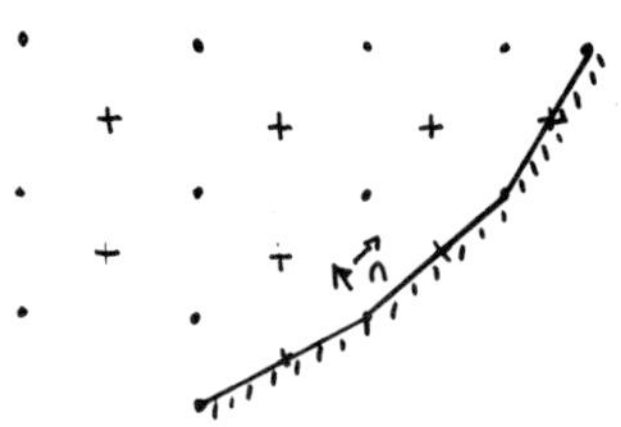

. grid for $\tilde{u}$, $\tilde{H}$, $\tilde{f}$, $\tilde{k}$, ...
+ grid for pressure

Figure 1 - Grid arrangement, and representation of a boundary not parallel to grid axis.

All transport equations (2-3-8-13-21-22), are solved in the same way, using a quasi-unsteady algorithm, written in the general form as :

$$\frac{G^{n+1} - G^n}{\delta t} = - \bar{\rho}\, \tilde{u}_j \frac{\partial G^n}{\partial x_j} + \frac{\partial}{\partial x_j} (K_g \frac{\partial G^{n+1}}{\partial x_j}) + S_g \tag{23}$$

δt is a numerical parameter, G^n and G^{n+1} are the values of G at the nth and $(n+1)^{th}$ iterations.

The solution of (23) is divided in two steps :

a) Advection step, solving :

$$\frac{\hat{G} - G^n}{\delta t} = - \bar{\rho}\, \tilde{u}_j \frac{\partial G^n}{\partial x_j} \qquad (24)$$

The solution $\hat{G}$ is obtained using the method of characteristics, after Esposito (1981), also briefly described in Viollet (1987), with third order interpolation allowing to minimize numerical diffusion.

b) Diffusion step, solving :

$$\frac{G^{n+1} - \hat{G}}{\delta t} = \frac{\partial}{\partial x_j} (K_g \frac{G^{n+1}}{x_j}) + S_g \qquad (25)$$

For the computation of the velocity components, a third step is required in order to prescribe the continuity condition (1), leading to a Poisson equation for the pressure.

The wall boundary conditions are written using wall function techniques, in a rather general way so that, after computation of the normal vector $\vec{n}$ at each boundary point (see figure 1), boundaries not parallel to grid lines can be prescribed.

3.2 <u>Radiation model</u>

In order to approximate the solution of equations (19-20) two different methods have been tested.

In the flux method, following Marco and Lockwood (1975)*, the radiation intensity is developped with respect to the components of the unitary vector S defining the direction of propagation :

$$L(\vec{x},\vec{S}) = A_x(\vec{x})S_x + A_y(\vec{x})S_y + A_z(\vec{x})S_z + B_x(\vec{x})S_x^2 + B_y(\vec{x})S_y^2 + B_z(\vec{x})S_z^2 \qquad (26)$$

The second method consist in discretizing (19) along a finite number of directions $\vec{S}_i$ (i = 1 to I), and solving directly (20) along each of these I directions, as was done before by Shah (1979). In the present model, equation (20) is simply discretized on the finite difference grid in the following way (figure 2) :

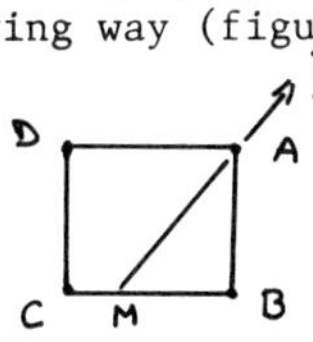

Figure 2 - Definition sketch for discretization of (20)

$$L_i(A) - L_i(C)\frac{MB}{BC} - L_i(B)\frac{MC}{BC} = AM\,(- K\, L_i(A) + K\, \frac{\sigma\, T(A)^4}{\pi}$$

Here, $L(\vec{x},\vec{S}_i)$ is for simplicity denoted as $L_i(\vec{x})$.

In order to increase the accuracy of the method, a higher-order interpolation of $L_i(M)$ could be used.

* A new flux model for the calculation of radiation in furnaces, La Rivista dei combustibli, 20, 184-196.

4. STUDY OF A METHANE DIFFUSION FLAME IN STILL AIR

This case is taken from the experiments of Lockwood and Moneib (1982)* (case SA2) : the natural gas (assumed as pure methane in the computation) is emitted from a nozzle (diameter d = 7.7 mm) with flow rate 4,9 kg/h, in still air.

The numerical computation has been performed both without radiation and with radiation, using the flux method : due to the low sooting of the flame, the differences in the temperature fields are very small. The grid used in the computation covers the area $0 < z < 1.8$ m and $0 < r < 1.16$ m, and is very distorted : for more details, see Viollet (1986). The turbulence model has been slightly modified, following the suggestions from Rodi (1980) for the computation of turbulent jets :

$$C_\mu = 0.09 - 0.04\ f(r)$$
$$C_{\varepsilon 2} = 1.92 - 0.067\ f(r)$$

with $f(r) = \left[\frac{r_{1/2}}{u_c} \left(\frac{\partial u_c}{\partial z} - \left| \frac{\partial u_c}{\partial z} \right| \right) \right]^{0.2}$, where u_c is the centerline jet velocity, and $r_{1/2}$ the half width of the jet.

Figure 5 shows both the computed temperature field and the comparison between measured and computed radial temperature profiles at two levels. As a result of the turbulence of the flow, the mean temperature obtained both from computation and experiment is significantly smaller than the adiabatic flame temperature.

The variance of temperature fluctuations is not directly given by the model. It can be approximated assuming negligible radiation, constant physical properties and same specific heat for all species, hence :

$$X_3 = \frac{T - T_0}{T_a - T_0}$$

, where T_0 and T_a are respectively the initial and adiabatic flame temperatures. Then, $\overline{T'^2}$ may be approximated as :

$$\overline{T'^2} \simeq (T_a - T_0)^2 \left[\int_0^1 X_3(f)^2\ P(f)\ df - \overline{X}_3^{\,2} \right]$$

Figure 6 shows that the variance of temperature fluctuations obtained by this way is significantly higher than from the experiments. This may suggest that the computed values of $\widetilde{f''^2}$ are too high (so that the C_T' constant in equation (13) should be decreased).

From the same temperature and species mass fraction fields, the radiative properties have been calculated using both the flux method and the discrete transfer method as described in section 3.2, with different numbers of directions of propagation (I = 4, 16, 32). Figure 7 shows the comparison of radial profiles of the radial component of radiative flux $\vec{q}$ defined by (19) from these computations. It appears that the flux method is rather inaccurate, with results of the same kind that the discrete transfer method with only I=4 (both strongly privilege the horizontal and vertical directions). The result of discrete transfer method becomes little dependent of the number of directions I for I > 30, approximately.

* Fluctuating temperature measurements in turbulent jet diffusion flame, Combustion and Flame, 47, 291-314.

5. STUDY OF A METHANE PREMIXED FLAME

A methane-air mixture of equivalence ratio of 0,6 is used as a combustible. The mean-velocity is 4,89 m/s at the burner exit. The flame is stabilized by a pilot flame surrounding the burner port (*).
The no perturbation conditions at the boundaries let calm air entry into the domain. Calculations determine the flow-rate of air.
The model based upon a lagrangian approach was used and the following results were performed with a constant equal to 8.

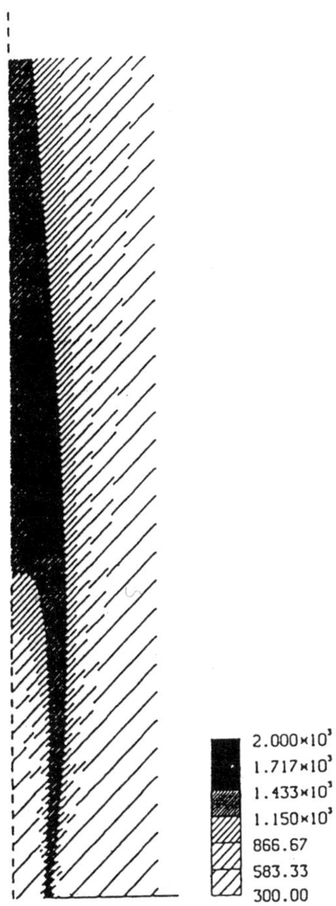

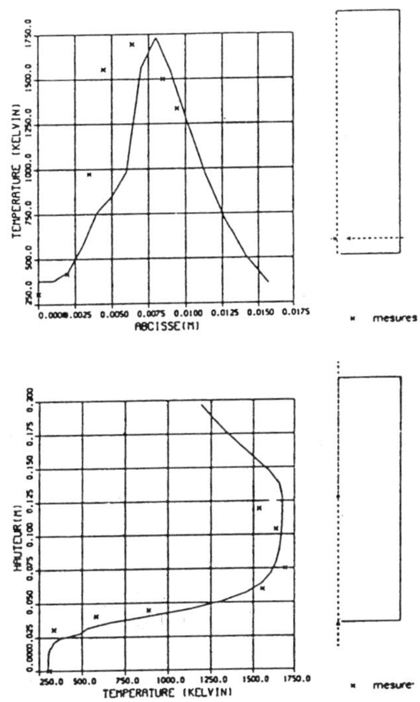

Comparison calculations experiments on radial and axial profiles of mean temperature

Mean temperature field (Kelvin)

Figure 3

* TANAKA, YANAGI, "Cross-correlation of velocity and temperature in a premixed turbulent flame", 1983 Combustion and Flame, 51, 183-191.

6. CALCULATION OF RADIATIVE HEAT FLUX THROUGH THE WALLS OF A ROTATING CYLINDRICAL FURNACE

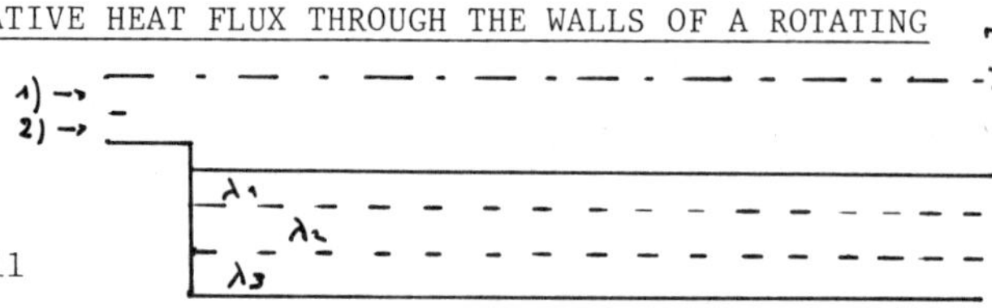

conductibility coefficients of the wall

1) Combustion products 1900°C - 2000°C decomposed by a plasma torch
2) Air at 200°C.

The heat radiative transfer in the fluid is calculated with the "discret transfer". The Fourier equation is solved within the wall.

$$\frac{\partial T}{\partial t} + \frac{1}{r}\frac{\partial}{\partial r}\left(ar\frac{\partial T}{\partial r}\right) + \frac{\partial}{\partial z}\left(a\frac{\partial T}{\partial z}\right) = 0 \qquad a = \frac{\lambda}{\rho C_p}$$

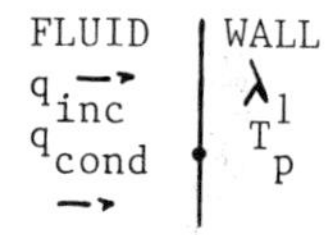

The boundary conditions are defined by :
- for the inner wall

$$-\lambda_1\left(\frac{\partial T}{\partial n}\right)_P + \varepsilon\sigma T_P^4 = q_{cond} + \varepsilon\, q_{inc}$$

conduction through the wall — emitted radiative flux — fluid diffusive flux — absorbed radiative flux

- for the outer wall (the incident radiative flux is néglected)

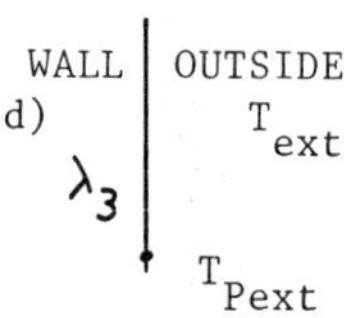

$$\lambda_3\left(\frac{\partial T}{\partial n}\right)_{Pext} + \varepsilon_{Pext}\,\sigma\, T_{Pext}^4 = h_c\,(T_{ext} - T_{Pext})$$

conduction through the wall — emitted radiative flux — conductive exchange outer wall/outside

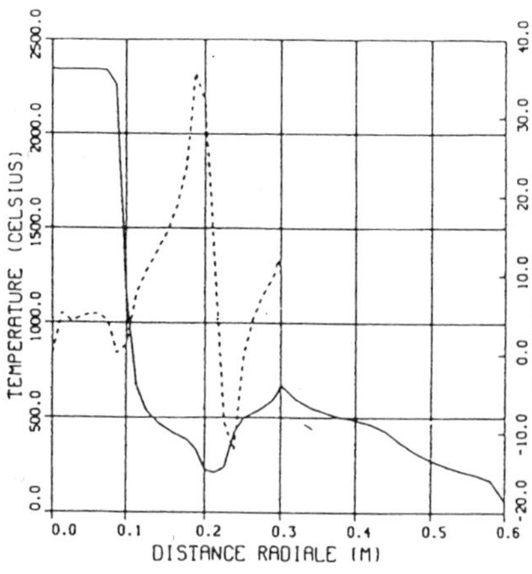

Radial mean temperature (celsius) ——
Radial radiative flux (kW/m²) - - -
at the distance of 0,01 m from the entry

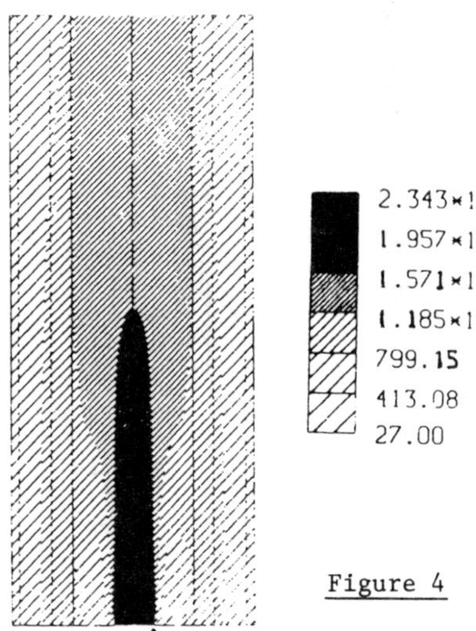

Figure 4

Mean temperature field (celsius)

7. CONCLUSION

The numerical model "Hades" which has been presented here may take into account turbulent combustion phenomena in diffusion flames interacting with radiation transfer. The turbulent combustion model, with presumed p.d.f., seems to perform relatively well, although other validation studies are still to be done (for instance in low Froude number flames such as in fires). The "Eddy-Break-Up" model used for turbulent premixed flames is more empirical but gives quite good results for a good choice of the constant.
A discrete transfer method for radiation has been compared to the flux method and, appearing as much more accurate, will be used in the future applications of the code. Much uncertainties still remain in the soot production model whose empirical constants values are not well known at the present, and may not be of universal use.

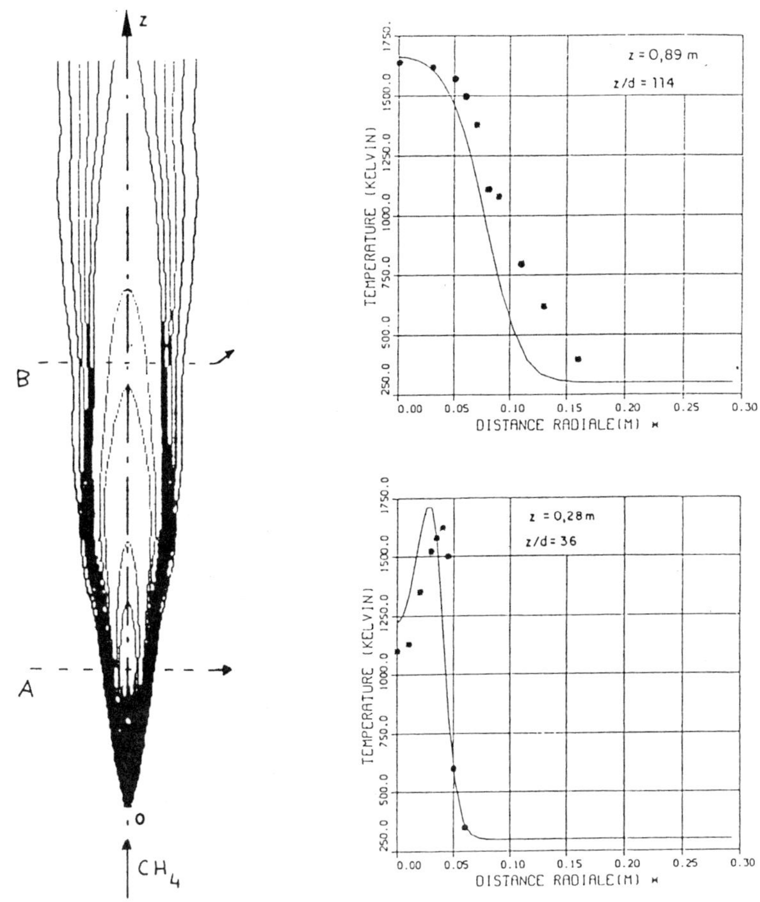

Figure 5 : Computation of a methane-air turbulent diffusion flame

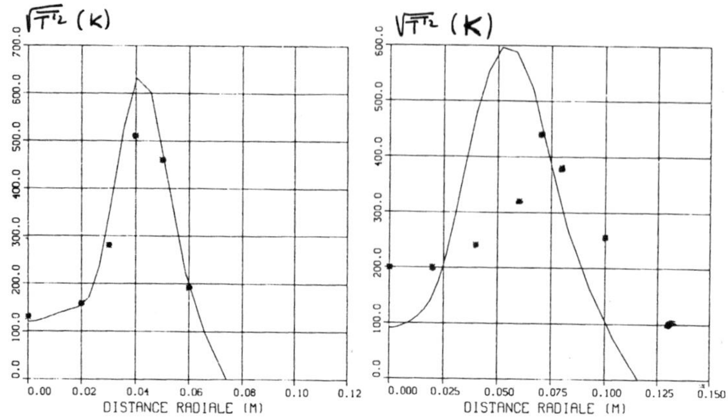

<u>Figure 6</u> - Radial RMS temperature fluctuation at two levels 0,28 m, 0,57 m
. from the experiments (Lockwood & Moneib)
- Computed assuming $\overline{T'^2} = (T_a - T_o)^2 \, \overline{X_3'^2}$, and using (14-15) as well as the p.d.f.(17) for the computation of $\overline{X_3'^2}$

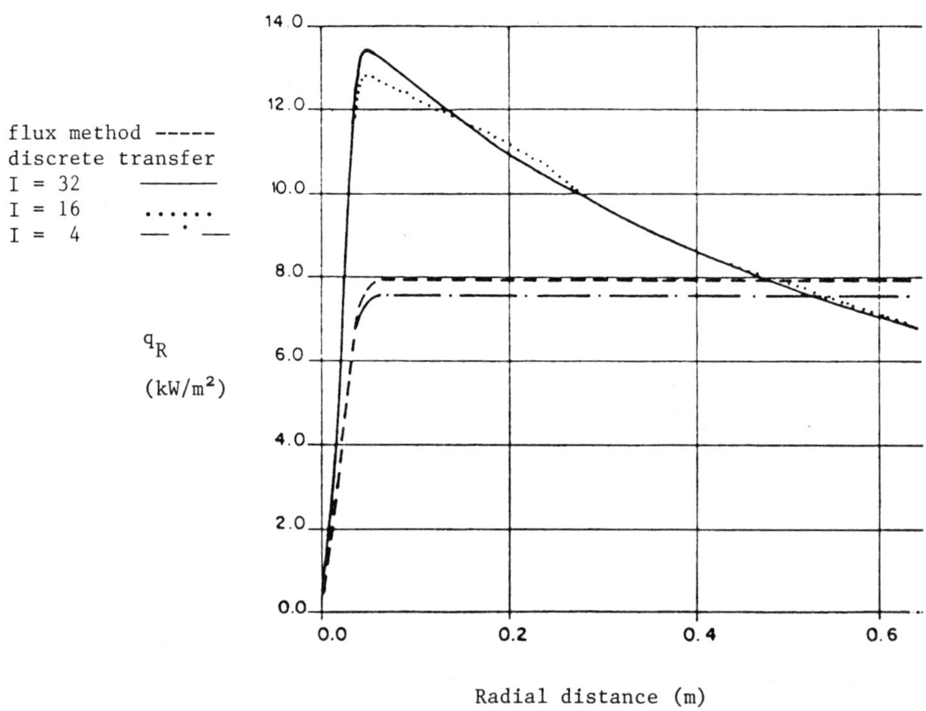

<u>Figure 7</u> : - Radial component of the radiative flux q, at level z = 0.28m, computed using both the flux and discrete transfer (with various numbers of directions) methods.

INDEX